Lecture Notes in Computer Science 4395

Commenced Publication in 1973
Founding and Former Series Editors:
Gerhard Goos, Juris Hartmanis, and Jan van Leeuwen

Editorial Board

David Hutchison
Lancaster University, UK

Takeo Kanade
Carnegie Mellon University, Pittsburgh, PA, USA

Josef Kittler
University of Surrey, Guildford, UK

Jon M. Kleinberg
Cornell University, Ithaca, NY, USA

Friedemann Mattern
ETH Zurich, Switzerland

John C. Mitchell
Stanford University, CA, USA

Moni Naor
Weizmann Institute of Science, Rehovot, Israel

Oscar Nierstrasz
University of Bern, Switzerland

C. Pandu Rangan
Indian Institute of Technology, Madras, India

Bernhard Steffen
University of Dortmund, Germany

Madhu Sudan
Massachusetts Institute of Technology, MA, USA

Demetri Terzopoulos
University of California, Los Angeles, CA, USA

Doug Tygar
University of California, Berkeley, CA, USA

Moshe Y. Vardi
Rice University, Houston, TX, USA

Gerhard Weikum
Max-Planck Institute of Computer Science, Saarbruecken, Germany

T0189203

Michel Daydé   José M. L. M. Palma
Álvaro L. G. A. Coutinho   Esther Pacitti
João Correia Lopes (Eds.)

High Performance Computing for Computational Science - VECPAR 2006

7th International Conference
Rio de Janeiro, Brazil, June 10-13, 2006
Revised Selected and Invited Papers

 Springer

Volume Editors

Michel Daydé
Institut de Recherche en Informatique de Toulouse, France
E-mail: Michel.Dayde@enseeiht.fr

José M. L. M. Palma
Faculdade de Engenharia da Universidade do Porto, Portugal
E-mail: jpalma@fe.up.pt

Álvaro L. G. A. Coutinho
Federal University of Rio de Janeiro
Center for Parallel Computations and Dep. of Civil Engineering, Brazil
E-mail: alvaro@nacad.ufrj.br

Esther Pacitti
LINA-Laboratoire d'Informatique de Nantes-Atlantique, France
E-mail: Esther.Pacitti@univ-nantes.fr

João Correia Lopes
Universidade do Porto/INESC-Porto, Portugal
E-mail: jlopes@fe.up.pt

Library of Congress Control Number: 2007922600

CR Subject Classification (1998): D, F, C.2, G, J.2, J.3

LNCS Sublibrary: SL 1 – Theoretical Computer Science and General Issues

ISSN   0302-9743
ISBN-10   3-540-71350-6 Springer Berlin Heidelberg New York
ISBN-13   978-3-540-71350-0 Springer Berlin Heidelberg New York

This work is subject to copyright. All rights are reserved, whether the whole or part of the material is concerned, specifically the rights of translation, reprinting, re-use of illustrations, recitation, broadcasting, reproduction on microfilms or in any other way, and storage in data banks. Duplication of this publication or parts thereof is permitted only under the provisions of the German Copyright Law of September 9, 1965, in its current version, and permission for use must always be obtained from Springer. Violations are liable to prosecution under the German Copyright Law.

Springer is a part of Springer Science+Business Media

springer.com

© Springer-Verlag Berlin Heidelberg 2007
Printed in Germany

Typesetting: Camera-ready by author, data conversion by Scientific Publishing Services, Chennai, India
Printed on acid-free paper   SPIN: 12033621   06/3142   5 4 3 2 1 0

Preface

Following the practice of all previous editions of the VECPAR series of conferences, the most significant contributions have been organized and made available in a book, edited after the conference, and after a second review of all orally presented papers at VECPAR 2006, the seventh International Meeting on High-Performance Computing for Computational Science, held in Rio de Janeiro (Brazil), June 10–13, 2006.

After the conference is finished this is what is left, a document that, we hope, can be a reference to a wide range of researchers in computational science. For the first time, and reflecting the conference programme of the seventh edition of the VECPAR series, this book includes some of the presentations at the two workshops, namely:

WCGC 2006 — Workshop on Computational Grids and Clusters: Models, Middlewares, Testbeds, Architectures, User Feedback
HPDGrid 2006 — International Workshop on High-Performance Data Management in Grid Environments

Both the workshops and the conference programme evidence the current trends in computer and computational science, with an increasing importance of the Grid technologies.

The book contains 57 papers, organized in seven chapters, with the two last chapters entirely devoted to the workshops. Chapter 1 opens with six papers dealing on current issues such as the scheduling of workflows on grids, their use in structural analysis or peer-to-peer models in large-scale grids. Chapter 2, with a total of 14 papers, is concerned with aspects closer to computer science, which includes computation on volatile nodes, the evaluation of decentralized parallel I/O scheduling strategies for parallel file systems. Chapter 3 is devoted to numerical techniques; it opens with the invited lecture by Bruce Hendrickson on combinatorial scientific computing, followed by 15 papers mostly in the field of linear algebra; application of BlockCGSI algorithm, parallel processing of matrix multiplication in a CPU and GPU heterogeneous environment or construction of a unit triangular matrix with prescribed singular values are examples of subjects that can be found in this chapter that represents a major mainstream in all VECPAR conferences. Chapter 4 contains a set of five papers more concerned with applications in physics, ranging from cosmological simulations to simulations of laser propagations. Chapter 5 is made of three papers on bioinformatics, a topic of greater importance over the last few years that announces the impact of computational methods in life sciences.

Best Student Paper

The best Student Paper Award initiated in the third edition of VECPAR, in 1998, has always been the opportunity to reward high-quality research studies by newcomers and highly promising students. This year, and after a difficult selection among the 11 candidates, the prize was awarded to Jacques Bahi for his work, entitled:

- JaceV: a Programming and Execution Environment for Asynchronous Iterative Computations on Volatile Nodes

Acknowledgements

The seventh edition of VECPAR was the second organized outside Portugal, and the first outside Europe. This time and after the interest shown by our colleagues in Brazil, who volunteered to organize the conference, VECPAR took place in the beautiful city of Rio de Janeiro, at IMPA (Applied Mathematics Institute), located over the Botanic Garden and overlooking the beautiful Rodrigo de Freitas lagoon.

This was a true multi-continent organization, made possible by current computer technologies, where the organizational aspects were dealt with by colleagues in Brazil, a joint collaboration between IMPA and the Institute Alberto Luiz Coimbra of Research and Post-Graduate Studies of the Federal University of Rio de Janeiro (COPPE/UFRJ). Paper submission and selection were managed via the conference database, held and managed by the Faculty of Engineering of the University of Porto. Vítor Carvalho created and maintained the conference Web site.

The success of the VECPAR conferences and its long life are a result of the collaboration of many. This time, given the widespread organization, a larger number of collaborators were involved. We mention only some, and through them we thank many others who offered most of their time and commitment to the success of the conference and workshops: Mara Prata (COPPE/UFRJ, Brazil) Conference Secretary, Marta Mattoso (COPPE/UFRJ, Brazil) and Patrick Valduriez (INRIA and LINA, France).

For their contribution to the present book, we must thank all the authors for meeting the deadlines and all members of the Scientific Committee who helped us so much in selecting the papers.

December 2006

Michel Daydé
José M. L. M. Palma
Álvaro L.G.A. Coutinho
Esther Pacitti
João Correia Lopes

Organization

VECPAR is a series of conferences organized by the Faculty of Engineering of Porto (FEUP) since 1993.

Organizing Committee

A. Coutinho	COPPE/UFRJ, Brazil (Chair)
N. Ebecken	COPPE/UFRJ, Brazil
J. Alves	COPPE/UFRJ, Brazil
R. Silva	LNCC, Brazil
M. Sarkis	IMPA, Brazil
E. Toledo	LNCC, Brazil
M. Mattoso	COPPE, Brazil
J. Correia Lopes	FEUP, Portugal (Web Chair)

Steering Committee

Michel Daydé	ENSEEIHT-IRIT, France (Chair)
Jack Dongarra	University of Tennessee, USA
José Fortes	University of Florida, USA
Vicente Hernandez	Universidad Politécnica Valencia, Spain
André Nachbin	Instituto Matemática Pura e Aplicada, Brazil
Lionel Ni	Hong Kong University of Science and Technology, Hong Kong

Scientific Committee

J. Palma	Portugal (Chair)
A. Bhaya	Brazil
Vincent Breton	France
F. Capello	France
M. Christon	USA
Olivier Coulaud	France
J. C. Cunha	Portugal
Michel J. Daydé	France
F. Desprez	France
P. Devloo	Brazil
Peter Dinda	USA
I. Duff	UK
Renato Figueiredo	USA
Fabrizio Gagliardi	Switzerland
W. Gentzsch	USA

A. George	USA
L. Giraud	France
Abdelkader Hameurlain	France
Michael T. Heath	USA
J. P. Jessel	France
D. Knight	USA
J. Koster	Norway
Dieter Kranzmueller	Austria
V. Kumar	USA
Stéphane Lanteri	France
Kuan-Ching Li	Taiwan
Thomas Ludwig	Germany
O. Marques	USA
A. Padilha	Portugal
D. Parsons	USA
B. Plateau	France
P. Primet	France
Thierry Priol	France
R. Ralha	Portugal
H. Ruskin	Ireland
Tetsuya Sato	Japan
Mitsuhisa Sato	Japan
Satoshi Sekiguchi	Japan
A. Sousa	Portugal
M. A. Stadtherr	USA
D. Talia	Italy
F. Tirado	Spain
Patrick Valduriez	France
M. Valero	Spain
Roland Wismuller	Germany

Invited Speakers

Álvaro Maia da Costa	CENPES/Petrobras, Brazil
Marcos Donato	CENPES/Petrobras, Brazil
Omar Ghattas	University of Texas, USA
Bruce Hendrickson	Sandia National Laboratory, USA
Christopher R. Johnson	University of Utah, USA
Kenichi Miura	National Institute of Informatics, NAREGI, Japan

Additional Reviewers

Carmelo Acosta	Jose Miguel Alonso
Carlos de Alfonso	Carlos Alvarez

Jorge Gomes Barbosa
Ignacio Blanquer
Rob Bisseling
Aurelien Bouteiller
Jacques Briat
Jan Christian Bryne
Miguel Caballer
Eddy Caron
Carmela Comito
Antonio Congiusta
Myrian Costa
Álvaro Coutinho
Holly Dail
Frank Dopatka
Alexandre Evsukoff
Juan Carlos Fabero
Agustín Fernández
Tiago Luis Duarte Forti
Pierre Fortin
Carlos González
James Greco
Eric Grobelny
Ashish Gupta
Rohit Gupta
Bjørn-Ove Heimsund
P. Henon
Andreas Hoffmann
Mathieu Jan
Yvon Jégou
Adrian Kacso
Jean-Yves L'Excellent
Bin Lin

Paulo Lopes
Daniel Lorenz
Hervé Luga
Vania Marangozova
Carlo Mastroianni
Pedro Medeiros
German Moltó
Victor Moya
Aziz Mzoughi
Kyu Park
Gaurav Pandey
Alberto Pascual
Christian Perez
Stéphane Pralet
Andrea Pugliese
Chiara Puglisi
Pål Puntervoll
Yi Qiao
Bolze Raphaël
Jean Roman
Edimar Cesar Rylo
Esther Salamí
Oliverio J. Santana
Damià Segrelles
Michael Steinbach
Cedric Tedeschi
Christian Tenllado
Ian Troxel
Paolo Trunfio
Pierangelo Veltri
Jian Zhang
Ming Zhao

Gold Sponsors

Petrobras   Petróleo Brasileiro S/A, Brazil
IBM Brazil   International Business Machines Corporation

Sponsoring Organizations

Furthermore, the Organizing Committee is very grateful to the following organizations for their support:

ABMEC   Associação Brasileira de Métodos Computacionais em
          Engenharia, Brazil

CNPq	Conselho Nacional de Desenvolvimento Científico e Tecnológico, Brazil
FAPERJ	Fundação de Amparo e Pesquisa do Estado do Rio de Janeiro, Brazil
FEUP	Faculdade de Engenharia da Universidade do Porto, Portugal
INRIA	Institut National de Recherche en Informatique et Automatique, France
SGI	Sillicon Graphics
UP	Universidade do Porto, Portugal

Workshop on Computational Grids and Clusters

The WCGC2006 Workshop followed the VECPAR 2006 Conference and focused on cluster and grid environments and tools for efficient management of computations.

The objectives of the workshop were to bring together researchers, practitioners and people with less experience in grid and clusters, to report on recent advances, and to share user feedback.

The topics of the workshop included (but were not restricted to):

- Hardware issues for clusters and grids
- Middlewares, distributed systems, runtime systems
- Interoperability issues
- Programming environments
- Communication protocols
- User experience in deploying grids and testbeds
- Grid and cluster management
- Performance evaluation
- Scheduling, load balancing, scalability, fault-tolerance issues
- Web applications, peer-to-peer
- Design of high performance clusters

The program of the workshop consisted of two invited talks and nine papers. One invited talk and six papers are included in the present book. A wide range of important topics in grid computing are covered (management of clusters, management of services, and deploying applications on computational grids).

December 2006

Cristina Boeres
Rajkumar Buyya
Walfredo Cirne
Myrian Costa
Michel Daydé
Frédéric Desprez
Bruno Schulze

Organization

Organizing Committee

Michel Daydé   IRIT, ENSEEIHT

Workshop Committee

Cristina Boeres   Instituto de Computação, Universidade
                            Federal Fluminense
Rajkumar Buyya   Melbourne University
Walfredo Cirne   Universidade Federal de Campina Grande
Myrian Costa   NACAD, COPPE/UFRJ
Michel Daydé   IRIT, ENSEEIHT
Frédéric Desprez   LIP, ENS-Lyon INRIA
Bruno Schulze   ComCiDis/LNCC

Additional Reviewers

Alexandre Nóbrega Duarte
Alexandre Sena
Aline de Paula Nascimento
Eliane Araújo
Evandro Barros Costa
Jacques da Silva
Lauro Beltrão Costa
Marcelo Costa Oliveira
Marcos André Duarte Martins
William Voorsluys

Sponsoring Organizations

The Organizing Committee is very grateful to the following organizations for
their support:

IBM Brazil   International Business Machines Corporation

International Workshop on High-Performance Data Management in Grid Environments

Initially developed for the scientific community as a generalization of cluster computing using the Web, grid computing is now gaining much interest in other important areas such as enterprise information systems. This makes data management more critical than ever. Compared with cluster computing which deals with homogeneous parallel systems, grids are characterized by high heterogeneity, high autonomy and large-scale distribution of computing and data resources. Managing and transparently accessing large numbers of autonomous, heterogeneous data resources efficiently is an open problem. Furthermore, different grids may have different requirements with respect to autonomy, query expressiveness, efficiency, quality of service, fault-tolerance, security, etc. Thus, different solutions need be investigated, ranging from extensions of distributed and parallel computing techniques to more decentralized, self-adaptive techniques such as peer-to-peer (P2P).

The objective of this one-day workshop was to bring together researchers and practitioners from the high-performance computing, distributed systems and database communities to discuss the challenges and propose novel solutions in the design and implementation of high-performance data management in grid environments.

The Program Committee received 19 paper submissions. Each paper was reviewed by three PC members. The following program is the result of the paper selection, with nine papers presented in three sessions: (1) data grid applications, (2) replication and consistency in data grids, (3) design and implementation of data grids. In addition, we had one keynote session on "Enterprise Grids: Challenges Ahead" by Ricardo Jiménez-Peris, Marta Patiño-Martinez and Bettina Kemme. The authors of the papers are from five different countries (Brazil, Canada, France, Spain and USA), thus reflecting the true international nature of the workshop.

December 2006

Esther Pacitti
Marta Mattoso
Patrick Valduriez

International Workshop on High-Performance Data Management in Grid Environments

Organization

Organizing Committee

Vanessa Braganholo	COPPE/UFRJ, Brazil (Chair)
Fernanda Baião	Unirio, Brazil
Alexandre Lima	UNIGRANRIO, Brazil
Luiz A V C Meyer	COPPE, Brazil
Gabriela Ruberg	COPPE, Brazil

Workshop Co-chairs

Marta Mattoso	COPPE/UFRJ, Brazil
Patrick Valduriez	INRIA and LINA, France

Program Chair

Esther Pacitti	INRIA and LINA, France

Program Committee

Henrique Andrade	IBM Research, USA
Claudio Luiz de Amorim	UFRJ, Brazil
Gabriel Antoniu	INRIA et IRISA, France
Stefano Ceri	Politecnico di Milano, Italy
Christine Collet	Institut Polytechnique Grenoble, France
Ricardo Jimenez-Peris	Universidad Politecnica de Madrid, Spain
Sergio Lifschitz	PUC Rio, Brazil
Alexandre Lima	UNIGRANRIO, Brazil
Ioana Manolescu	INRIA Futurs, France
Hubert Naacke	University Paris 6, France
Rui Oliveira	University of Minho, Portugal
Vincent Oria	New Jersey Institute of Technology, USA
M. Tamer Özsu	University of Waterloo, Canada
Cesare Pautasso	ETH, Switzerland
Alexandre Plastino	UFF, Brazil
Fabio Porto	EPFL, Switzerland
Guillaume Raschia	INRIA and LINA, France
Marc Shapiro	INRIA and LIP6, France
Mohamed Zait	Oracle Corporation, USA

Additional Reviewers

Andre Ormastroni Victor
Bioern Bioernstad
Cedric Tedeschi
Genoveva Vargas-Solar
Khalid Belhajjame
Lauro Whately
Lei Chen
Luiz Monnerat
Marcelo Lobosco
Mathieu Jan
Simone de Lima Martins

Sponsoring Organizations

The Organizing Committee is very grateful to the following organizations for
their support:

CAPES	Coordenação de Aperfeiçoamento de Pessoal de Nível Superior,Brazil
CNPq	Conselho Nacional de Desenvolvimento Científico e Tecnológico,Brazil
INRIA France	Institut National de Recherche en Informatique et Automatique, France

Table of Contents

Chapter 3: Numerical Methods

Chapter 4: Large Scale Simulations in Physics

Chapter 5: Computing in Biosciences

Workshop 1: Computational Grids and Clusters

Workshop 2: High-Performance Data Management in Grid Environments

An Opportunistic Algorithm
for Scheduling Workflows on Grids

Luiz Meyer[1], Doug Scheftner[2], Jens Vöckler[2],
Marta Mattoso[1], Mike Wilde[3], and Ian Foster[2,3]

[1] Federal University of Rio de Janeiro - COPPE, Department of Computer Science
[2] University of Chicago - Department of Computer Science
[3] Argonne National Laboratory - Mathematics and Computer Science Division

Abstract. The execution of scientific workflows in Grid environments imposes many challenges due to the dynamic nature of such environments and the characteristics of scientific applications. This work presents an algorithm that dynamically schedules tasks of workflows to Grid sites based on the performance of these sites when running previous jobs from the same workflow. The algorithm captures the dynamic characteristics of Grid environments without the need to probe the remote sites. We evaluated the algorithm running a workflow in the Open Science Grid using twelve sites. The results showed improvements up to 150% relative to other four usual scheduling strategies.

1 Introduction

Grids [9] are emerging as virtual platforms for high performance and integration of networked resources. In these environments, distributed and heterogeneous resources owned by independent organizations can be shared and aggregated to form a virtual computer. Scientific applications usually consist of numerous jobs that process and generate large datasets. Frequently, these components are combined generating complex scientific workflows. Therefore, scientific communities like physicists, biologists, astronomers are using Grid computing to solve their complex large-scale problems.

Processing scientific workflows in a Grid imposes many challenges due to the large number of jobs, file transfers and the storage needed to process them. The scheduling of a workflow focuses on mapping and managing the execution of tasks on shared resources that are not directly under the control of the workflow systems [23]. Thus, choosing the best strategy for a workflow execution in a Grid is a challenging research area.

Often, a scientific workflow can be represented as a Directed Acyclic Graph (DAG) where the vertices represent tasks and the edges represent data dependencies. One alternative to process this kind of workflow is to statically pre-assign tasks to resources based on the information of the entire workflow. This strategy can be used by a planner to optimize the execution plan for the DAG [6]. However, since a Grid execution environment can be very dynamic, this alternative may produce poor schedules because by the time the task is ready to run the resource may perform

M. Daydé et al. (Eds.): VECPAR 2006, LNCS 4395, pp. 1–12, 2007.
© Springer-Verlag Berlin Heidelberg 2007

poorly or even be unavailable. Besides, it is not easy to accurately predict the execution time of all tasks. Another scheduling approach is to perform the assignment of tasks to resources dynamically as soon as the task is ready to be executed. In this case, if a resource is not available, it will not be selected to process the task. However, many sites can be available to run the task, and selecting the best one can be done according to many alternatives, like the number of processors in the site, load balance or data availability.

This work presents an algorithm, which we name *Opportunistic*, which dynamically assigns jobs to Grid sites. The algorithm adopts an observational approach and exploits the idea of scheduling a job to a site that will probably run it faster. The opportunistic algorithm takes into account the dynamic characteristics of Grid environments without the need to probe the remote sites. We compared the performance of the *Opportunistic* algorithm with different scheduling algorithms in a context of a workflow execution running in a real Grid environment. We conducted our experiments using the Virtual Data System (VDS) [10], which presents an architecture to integrate data, programs, and the computations performed to produce data. VDS combines a virtual data catalog for representing data derivation procedures and derived data with a virtual data language that enables the definition of workflows. VDS also provides users with two planners that schedule jobs onto the Grid and manage their execution. Scheduling in VDS can be done according to a family of site selectors available for user needs. This work extends the library of site selectors with a new *Opportunistic* site selector algorithm. Our results with experiments in a real Grid environment suggest that the *Opportunistic* algorithm can increase performance up to 150% when compared to the scheduling algorithms currently adopted in most systems, particularly in VDS.

The rest of this paper is organized as follows. Section 2 discusses the related work to Grid scheduling. Section 3 describes the Virtual Data System architecture where the opportunistic strategy was implemented while in section 4, we detail the *Opportunistic* algorithm. In section 5 we describe the experiments performed and in section 6 the experimental results are analyzed. Finally, section 7 concludes this work and points to future directions.

2 Related Work

Finding a single best solution for mapping workflows onto Grid resources for all workflow applications is difficult since applications and Grid environments can have different characteristics [23]. In general, scheduling workflow applications in distributed environments is done by the adoption of heuristics. There are many works in the literature addressing the benefits of scheduling based on data locality in scenarios of data Grids. Casanova et al. [2] propose an adaptive scheduling algorithm for parameter sweep applications where shared files are pre-staged strategically to improve reuse. Ranganathan and Foster [17, 18] evaluate a set of scheduling and replication algorithms and the impact of network bandwidth, user access patterns and data placement in the performance of job executions. The evaluation was done in a simulation environment and the results showed that scheduling jobs to locations where the input data already exists, and asynchronously replicating popular data files across the Grid provides good results. Cameron et al. [4, 5] also measure the effects of various job scheduling and data replication strategies in a simulation environment. Their results show benefits of scheduling taking into account also the workload in the computing resources. Mohamed and Epema [14] propose an algorithm to place jobs on clusters close to the site where the input files reside. The algorithm assumes knowledge about the number of idle processors and the size of the input file for scheduling a job.

The workloads studied in these works consist of a set of independent jobs submitted from different users spread over different sites. Our work differs by focusing on scheduling jobs belonging to a single application, which is a workflow with job dependencies and synchronism, submitted from a single user at a single site.

Many researchers have studied scheduling strategies for mapping workflows onto the Grid. Ammar et al. [1] developed a framework to schedule a DAG in a Grid environment that makes use of advance reservation of resources and also considers previous knowledge about task execution time, transfer rates, and available processors to generate a schedule. Their simulation results show advantages of unified scheduling of tasks rather than scheduling each task separately. Mandal et al. [13] apply in-advance static scheduling to ensure that the key computational steps are executed on the right resources and large scale data movement is minimized. They use performance estimators to schedule workflow applications. Wieczoreket et al. [22] compare full graph scheduling and just-in-time strategies for scheduling a scientific workflow in a Grid environment with high availability rate and good control over the resources by the scheduler. Their results show best performance for full graph scheduling. Deelman et al. [6, 7] can map the entire workflow to resources at once or portions of it. This mapping can be done before or during the workflow execution. Their algorithm aims to schedule computation where data already exist. Additionally, users are able to specify their own scheduling algorithm or to choose between a random and a round robin schedule technique. Dumitrescu et al. [8] studied the performance execution of Blast jobs in Grid3 [11] according to several scheduling algorithms. In their experiments they used a framework that considered resource usage policies for scheduling the jobs. Their results showed that random and round-robin algorithms achieved the best performance for medium and large workloads.

Triana [12] allows scientists to specify their workflows which can be scheduled directly by the user or by the GriLab Resource Management System. In this case, the scheduling is done according to requirements specified for each task. Taverna [15] provides a set of tools to define bioinformatics workflows based on a composition of web services, but not much detail is given about the scheduling of tasks.

In our work, we also deal with the problem of scheduling jobs belonging to a single application, which is a workflow expressed as a DAG. Like in the previous workflow scheduling works, the goal of the scheduling is to minimize the overall job completion. In our algorithm, the planning scheme is completely dynamic and based on an observational approach. We do not consider performance estimation of Grid resources, use of advance reservations or requirements specifications. The performance evaluation was conducted in a real Grid environment without any control or reservation of the resources.

3   VDS Planning Architecture

In VDS, users specify their workflows through the use of VDL [10]. The VDC (Virtual Data Catalog) stores the user's workflow definition and provides the planner with the logical file names of the files and the name of the transformations (executable programs). The Replica Catalog provides the physical name for the input files given their logical file names. The transformation catalog (TC) specifies how to invoke (executable name, location, arguments) each program. Finally, the Pool Configuration catalog is responsible to provide the information about the desirable Grid sites to run the workflow. Figure 1 illustrates the VDS planner architecture.

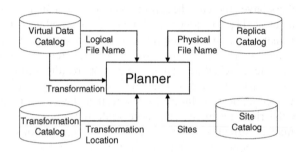

Fig. 1. VDS Planner architecture

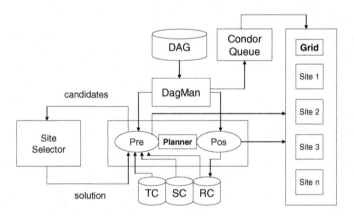

Fig. 2. VDS Planning mechanism

The planner makes use of a site selector mechanism in order to schedule each job of the workflow. The goal of the site selector is to choose a Grid site capable to execute a given job. In the VDS planner, the pre-script dynamically builds a list of the available sites for executing each job based on the information of the Transformation Catalog and Pool Configuration. The Pre-script then calls the site selector mechanism and waits for the solution, that is, the site selected for the job execution. The solution returned by the site selector is passed to DagMan [3], which is responsible for scheduling the job into the Grid.

Figure 2 details the planner's functionality and its interaction with a site selector. After receiving the identification of the site to run the job, the VDS-Planner executes the replica selection by querying the Replica Location Service to locate all replicas for each file. If there is a replica located in the selected site then this replica will be chosen. Otherwise, the planner will perform a third party transfer of the input files from the sites where they are located to the site where the job is supposed to run. Whenever a job ends, all input files dynamically transferred for the job execution site are erased in the post-processing step.

4 The Opportunistic Algorithm

Scheduling workflow tasks in Grid environments is difficult because resource availability often changes during workflow execution. The main idea of the *Opportunistic* algorithm is to take advantage of this environment changes without needing to probe the remote sites. In order to implement our *Opportunistic* algorithm using VDS, a few extensions were promoted in the system: we created a control database for logging the location and the status of the workflow jobs, and coded a new site selector routine responsible for choosing the execution site for a job. Since the control database is updated by the postscript of each job, the VDS postscript code also had to be modified.

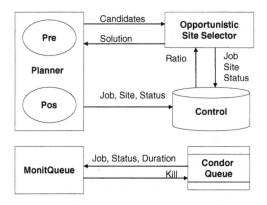

Fig. 3. The opportunistic site selector architecture

The goal of the *Opportunistic* site selector is to select a site to run a job based on the performance of each site when running previously jobs of the same workflow. In other words, the site selector assigns more jobs to sites that are performing better, according to the architecture in Figure 3. The performance is measured by the ratio (number of ended jobs / number of submitted jobs) at each site, as shown in the algorithm from Figure 4. As long as no jobs have completed, the site selector performs a round robin distribution between the sites. In order to keep track of the submissions and completions of the workflow jobs, the site selector makes use of a control database.

Algorithm: *Opportunistic*
Input: Job *J* to be submitted.
Set S_e {s_i} of available sites informed by the planner.
Set S_o {s_i} of sites informed by the *Control Database*
$f_1(S) \rightarrow$ Number of jobs scheduled to site S
$f_2(S) \rightarrow$ Number of jobs ended at site S
$f_3(max) \rightarrow$ Site;
$f_4(min) \rightarrow$ Site;
$f_5 \rightarrow$ {j_i} ; Set of submitted jobs
Output: *Solution* - Identification of the Site selected to run the job.
Initialization:
$flag \leftarrow 0$
$min \leftarrow high\ value$
$max \leftarrow low\ value$
Steps:
   1.   **For each** site $s_i \in S_o$ **do**
         1.1 **if** $s_i \in S_e$ **then**
             1.1.1 $T_{i,s} \leftarrow f_1(s_i)$
             1.1.2 **if** $T_{i,s} < min$ **then**
                  1.1.2.1 $min \leftarrow T_{i,s}$
             1.1.3 $T_{i,c} \leftarrow f_2(s_i)$
         1.2 **if** $T_{i,c} > 0$ **then**
             1.2.1 $R_i \leftarrow (T_{i,c} / T_{i,s})$
             1.2.2 $flag \leftarrow 1$
             1.2.3 **if** $R_i > max$ **then**
                  1.2.3.1 $max \leftarrow R_i$
   2.   **if** $flag = 1$ **then**
   2.1 *Solution* $\leftarrow f_3(max)$
   2.2 **else** *Solution* $\leftarrow f_4(min)$
   3. $T_o \leftarrow f_5$
   4.   **if** $T \in T_o$ **then**
         4.1 **update** *siteid* for job *T*
         4.2 **else insert** tuple (*T, solution*)
   5.   **Return** *Solution*

Fig. 4. The opportunistic algorithm

    Whenever the site selector chooses a site to run a job, one record is inserted in the control database with the identifications of the job, the identification of the selected site and a status set to "submitted." Whenever a job ends, another record with job identifier, site, and status set to "ended" is also added. This last insertion is done by the postscript of every job.

    The second component of the opportunistic approach is a queue monitor for the submitted jobs of the workflow. The main motivation to develop this *MonitQueue* component is that often submitted jobs remain waiting for execution in remote queues. The goal of *MonitQueue* is to keep track of the jobs submitted by

DagMan/Condor in order to remove those jobs that are not presenting a desired performance. In the actual implementation, the user must inform the maximum time a job can wait in a queue in an idle status. When a job reaches this time it is killed and automatically re-planned by Euryale. In this case, the *Opportunistic* site selector will have the chance to choose another site to run the job.

5 Experiments

Many scientific applications can be characterized as having sets of input and derived data that have to be processed in several steps by a set of programs. These batch-pipelined workloads [21] are composed of several independent pipelines and each pipeline contains sequential processes that communicate with the preceding and succeeding processes via data files.

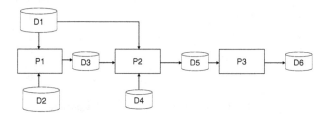

Fig. 5. The pipelined workflow

We defined a pipelined workflow to evaluate a set of scheduling strategies in this experiment. The design of the workflow is shown in figure 5 while figure 6 depicts the corresponding DAG. There is an input dataset D1 with only one file that is input for the first and second programs in the pipeline. The first program also has, as input file, a file belonging to dataset D2. Program P2 processes the output generated by P1 and also has two more files as input for its processing: the file from D1 dataset and a file from dataset D4. The third and last program of the pipeline processes the output file produced by P2 and outputs a file for the dataset D6. The width of the pipeline was set to 100 nodes in each level, totalizing 300 jobs in the workflow.

Currently, the VDS system provides three choices for the planners: *Round-Robin*, *Random* and *Weighted-Linear-Random*. We evaluated the *Opportunistic* algorithm against the *Weighted-Linear-Random*, *Round-Robin*, *Last-Recent-Used* and *Data-Present* algorithms. The last two strategies were coded within VDS for our experiments. The overall ideas of these algorithms are:

1. *Weighted-Linear-Random (WLR)* - The execution site is selected randomly but sites with more processors receive more jobs to process.
2. *Round-Robin* - Jobs are sent to sites in a round-robin way. Thus, the number of jobs assigned to each site tends to be the same.
3. *Last-Recent-Used (LRU)* - The execution site corresponds to the site where the last job ended.

4. *Data-Present* - A job is sent to a site with most of the files that it needs. If more than one site qualifies then a random one is chosen.

5. *Opportunistic* - The execution site is selected according to the performance of each site. This performance is measured by dividing the number of concluded jobs by the number of submitted jobs at each site. While there are no jobs concluded, a round-robin scheduling is performed.

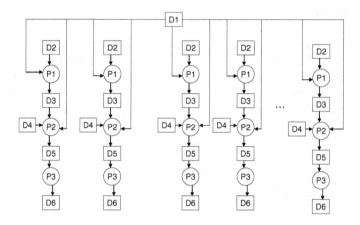

Fig. 6. The DAG shape for the pipelined workflow

Table 1. Resources available in OSG sites

Site	Processors
UIOWA_OSG_PROD	6
HAMPTONU	8
PURDUE_PHYSICS	63
UFLORIDA_IHEPA	70
UWMADISON	83
UC_ATLAS	110
UTA_DPCC	148
UERJ_HEPGRID	160
CIT_T2	224
UWMILWAUKEE	304
OSG_LIGO_PSU	314
USCMS_FNAL	989

We conducted the experiments using twelve sites from the Open Science Grid [16]. Table 1 shows a snapshot of the total resources available at each site. To avoid interfering with the production, we defined all workflow jobs as sleep jobs. We used two different machines at University of Chicago for running DagMan and the replica catalog respectively. A third machine at the same site was used to store all input files of the workflow. Table 2 shows the average execution time and transfer time for each type of job and file of the workflow. The size of all input and output files is one megabyte.

Table 2. Average time in seconds for data transfer and execution according to the type of the workflow level

	P1	P2	P3
Number of jobs	100	100	100
Transfer time	17	27	13
Execution time	300	120	60

6 Results

We executed the workflow twenty times for each scheduling strategy, totalizing 30,000 job executions. Figure 7 presents the performance results for all algorithms. The performance of the five algorithms is almost the same during the execution of the first hundred jobs of the workflow. *Opportunistic, Last-Recent-Used* and *Round-Robin* algorithms adopt the same scheduling strategy when no jobs have finished. The *Data-Present* algorithm uses a strategy similar to *Weighted-Linear-Random* while there is no site with the needed input files for the job. Since there is no dependency among the jobs in the first level of the workflow, Condor/DagMan can submit them as soon as the pre-script of each job is finished. The time to transfer the input files is very low and consequently the pre-processing for each job is very fast causing most jobs in this level to be scheduled before any job has finished. As soon as the jobs in the first and second levels begin to finish, *Opportunistic, Last-Recent-Used* and *Data-Present* start to schedule according to different approaches. *Opportunistic* and *Last-Recent-Used* use their observational characteristics while *Data-Present* takes advantage of data locality. In the first case, the scheduling becomes based on the ratio (*jobs concluded/jobs submitted*) while *Last-Recent-Used* aims to schedule jobs to the site that has finished processing the last job.

The *Opportunistic* algorithm benefits from the dynamic aspects of the Grid environment. If a site happens to perform poorly, then the number of jobs assigned to that site decreases. Similarly, if a site begins to process jobs quickly, then more jobs are scheduled to that site.

The *Last-Recent-Used* algorithm may not present a good performance when a job is scheduled to a site and has to wait a long period of time in the remote queue. When this happens, the next job in the workflow will probably show the same performance problem because it must be scheduled to run in the same site. The Opportunistic algorithm avoids this kind of problem because Moniqueue can cancel a job if it has not started after a determined period of time.

Since the size of the files generated during the execution is small, the time to transfer these files does not impact the performance and does not bring benefits to *Data-Present* algorithm.

Round-Robin provides a good load balance among the sites but since the performance varies among sites, scheduling the same number of jobs to each site is not beneficial. *Weighted-Linear-Random* does not show a good performance because scheduling more jobs to sites with more resources does not guarantee better results since jobs may have to wait in the remote queues. It seems that this kind of strategy is more indicated to Grid environments where resources can be reserved for the entire execution of the workflow.

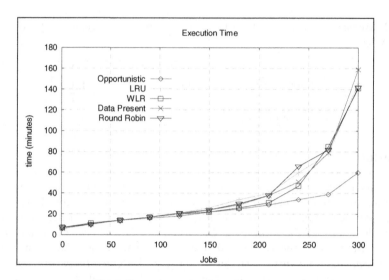

Fig. 7. Execution time of the five algorithm

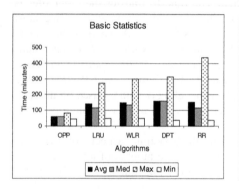

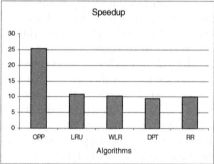

Fig. 8. Basic execution statistics   **Fig. 9.** Speedup of the five algorithms

Figure 8 shows a set of few basic statistics about the workflow execution. The minimum execution time is almost the same for all algorithms. This occurs because occasionally all sites may be presenting a good performance due to having processing resources available by the time of the execution. However, the most expected behavior is to have sites presenting different performances as the workflow is being processed. Consequently, the median, average and maximum execution times differ according to the execution strategy.

Figure 9 shows the speedup of the five algorithms. The execution of the workflow with the opportunistic algorithm was approximately twenty five times faster than running in a single machine. The speedup achieved by the *Opportunistic* algorithm was more than 150% higher than the other strategies.

7 Conclusions and Future Work

We have proposed a new "opportunistic" algorithm for scheduling jobs in Grid environments, and compared its performance with other algorithms. In particular, we analyzed the performance with a very common workflow pattern, a pipeline of programs in a real Grid. The results showed that the *Opportunistic* algorithm provided superior performance when compared to other four known algorithms for scheduling workflow jobs. The performance improvement is achieved as a consequence of the observational approach implemented by the algorithm. This approach exploits the idea of scheduling jobs for sites that are presenting good response times and to cancel jobs that are not being executed after a period of time. The algorithm is not aware of sites capabilities and does not need to collect data from remote sites being easy to implement and can be used by other workflow engines.

We intend to perform more comparative experiments with other scheduling algorithms to confirm the efficiency of the *Opportunistic* algorithm. We also intend to study the performance of the algorithm when dealing with other workflow patterns and sizes, and to promote extensions in order to analyze the impact of dealing with different sizes of historical data to compute a site's value.

Acknowledgements

This work is supported in part by the National Science Foundation GriPhyN project under contract ITR-086044, U.S. Department of Energy under contract W-31-109-ENG-38 and CAPES and CNPq Brazilian funding agencies.

References

1. Ammar H. Alhusaini, Viktor K. Prasanna, C.S. Raghavendra. "A Unified Resource Scheduling Framework for Heterogeneous Computing Environments," *hcw*, p. 156, Eighth Heterogeneous Computing Workshop, 1999.
2. Casanova, H., Obertelli, G., Berman, F., Wolski. R., The AppLeS Parameter Sweep Template: User-Level Middleware for the Grid, in SuperComputing 2000, Denver, USA, 2000.
3. DagMan, http://www.cs.wisc.edu/condor/dagman/.
4. D.G. Cameron, R. Carvajal-Schiaffino, A.P.Millar, Nicholson C., Stockinger K., Zini, F., Evaluating Scheduling and Replica Optimisation Strategies in OptorSim, in Proc. of 4th International Workshop on Grid Computing (Grid2003). Phoenix, USA, November 2003.
5. D.G. Cameron, R. Carvajal-Schiaffino, A.P.Millar, Nicholson C., Stockinger K., Zini, F., Evaluation of an Economic-Based File Replication Strategy for a Data Grid, in Int. Workshop on Agent Based Cluster and Grid Computing at Int. Symposium on Cluster Computing and the Grid (CCGrid2003), Tokyo, Japan, May 2003.
6. Deelman,E., Blythe, J., Gil, Y., Kesselman, C., Mehta, G., Patil, S., Su, M., Vahi, K., Livny, M., Across Grids Conference 2004, Nicosia, Cyprus
7. Deelman, E., Blythe, J., Gil, Y., Kesselman,C., Workflow Management in GriPhyn, The Grid Resource Management, Netherlands 2003.

8. Dumitrescu, C, Foster, I., Experiences in Running Workloads over Grid3, GCC 2005, LNCS 3795, pp.274-286, 2005.
9. Foster, I., Kesselman, C., 1999, Chapter 4 of "The Grid 2: Blueprint for a New Computing Infrastructure", Morgan-Kaufman, 2004.
10. Foster, I., Voeckler,J., Wilde,M., Zhao, Y., Chimera: A Virtual Data System for Representing, Querying, and Automating Data Derivation, in 14th International Conference on Scientific and Statistical Database Management (SSDBM 2002), Edinburgh, July 2002.
11. Foster, I. et al.,The Grid2003 Production Grid: Principles and Practice, in 13th International Symposium on High Performance Distributed Computing, 2004.
12. GOODALE, T., TAYLOR, I., WANG, I., "Integrating Cactus Simulations within Triana Workflows", In: Proceedings of 13th Annual Mardi Gras Conference - Frontiers of Grid Applications and Technologies, Louisiana State University, pp. 47-53, February, 2005.
13. Mandal, A., Kennedy, K., Koelbel, C., Marin, G., Crummey, J., Liu, B., Johnsson, L., Scheduling Strategies for Mapping Application Workflows onto the Grid, The 14th IEEE International Symposium on High-Performance Distributed Computing (HPDC-14), Research Triangle Park, NC, USA, July 2005.
14. Mohamed, H.H., Epema, D.H.J., An Evaluation of the Close-to-Files Processor and Data Co-Allocation Policy in Multiclusters, IEEE International Conference on Cluster Computing, San Diego, USA, September 2004.
15. Oinn, T., ADDIS, M., FERRIS, J. et al, 2004, "*Taverna: a Tool for the Composition and Enactment of Bioinformatis Workflow*", In: *BIOINFORMATICS*, vol. 20, no 17 2004, pp. 3045-3054, Oxford University Press.
16. Open Science Grid, http://www.opensciencegrid.org
17. Ranganathan,K., Foster,I., Simulation Studies of Computation and Data Scheduling Algorithms for Data Grids, in Journal of Grid Computing, V1(1) 2003.
18. Ranganathan,K., Foster,I., Computation Scheduling and Data Replication Algorithms for Data Grids, 'Grid Resource Management: State of the Art and Future Trends', J. Nabrzyski, J. Schopf, and J. Weglarz, eds. Kluwer Academic Publishers, 2003.
19. Shan,H., Oliker, L., Smith, W., Biswas, R., Scheduling in Heterogeneous Grid Environments: The Effects of Data Migration, International Conference on Advanced Computing and Communication, Gujarat, India, 2004.
20. Singh, G., Kesselman, C., Deelman, E., Optimizing Grid-Based Workflow Execution, work submitted to 14th IEEE International Symposium on High Performance Distributing Computing, July 2005.
21. Thain,D., Bent,J., Arpaci-Dusseau, A., Arpaci-Dusseau,R., Livny, M., Pipeline and Batch Sharing in Grid Workloads, 12th Symposium on High Performance Distributing Computing, Seattle, June 2003.
22. Wieczorek, M., Prodan, R.,Fahringer,T., Scheduling of Scientific Workflows in the ASKALON Grid Environment, SIGMOD Record, Vol. 34, No.3, September 2005.
23. Yu,J., Buyya, R., A Taxonomy of Scientific Workflow Systems for Grid Computing, SIGMOD Record, Vol.34, No.3, September 2005.
24. Zhang, X., Schopf, J., Performance Analysis of the Globus Toolkit Monitoring and Discovery Service, Proceedings of the International Workshop on Middleware Performance (MP 2004), part of the 23rd International Performance Computing and Communications Conference (IPCCC), April 2004.

A Service Oriented System for on Demand Dynamic Structural Analysis over Computational Grids*

J.M. Alonso, V. Hernández, R. López, and G. Moltó

Departamento de Sistemas Informáticos y Computación.
Universidad Politécnica de Valencia. Camino de Vera s/n 46022 Valencia, Spain
{jmalonso,vhernand,rolopez,gmolto}@dsic.upv.es
Tel.: +34963877356; Fax: +34963877359

Abstract. In this paper we describe the implementation of a service oriented environment that enables to couple a parallel application, which
performs the 3D linear dynamic structural analysis of high-rise buildings, to a Grid Computing infrastructure. The Grid service, developed
under Globus Toolkit 4, exposes the dynamic simulation as a service to
the structural scientific community. It employs the GMarte middleware,
a metascheduler that enables to perform the computationally intensive
simulations on the distributed resources of a Grid-based infrastructure.

Topics: Parallel and Distributed Computing, Cluster and Grid Computing, Large Scale Simulations in All Areas of Engineering and Science.

1   Introduction

Traditionally, the dynamic analysis of large scale buildings has been limited to
simplifications with the purpose of reducing the computational and memory
requirements of the problem. Although these simplifications have been proved
to be valid for simple and symmetric structures, they have demonstrated to be
inappropriate for more complex buildings.

Nowadays, many buildings are asymmetric and the effects of torsion have been
identified as one of the main reasons that make a building collapse when an
earthquake occurs. Considering the dramatic effects of earthquakes, it is crucial
to investigate their impact before a building gets constructed. However, the
required memory and the computation involved in a 3D realistic analysis of
a large dimension building can be too intensive for a traditional PC.

This way, the authors have developed an MPI-based application that performs
the 3D linear dynamic analysis of structures using three different direct time integration schemes. Typically, a structural designer works with different preliminary

---

* The authors wish to thank the financial support received from the Spanish Ministry of Science and Technology to develop the GRID-IT project (TIC2003-01318)
and the Conselleria de Empresa, Universidad y Ciencia - Generalitat Valenciana for
the GRID4BUILD project (GV04B-424). We wish also to thank Anshul Gupta for
providing us with a trial license of WSMP library for 16 CPUs.

M. Daydé et al. (Eds.): VECPAR 2006, LNCS 4395, pp. 13–26, 2007.
© Springer-Verlag Berlin Heidelberg 2007

alternatives when designing a building, considering distinct layouts or applying multiple sections or dimensions to its members, requiring their simulation under the influence of several dynamic loads. For example, the Spanish Earthquake-Resistant Construction Standards (NCSE-02) demands a building to be analysed with at least five different representative earthquakes. Obviously, this situation largely increases the computational cost of the problem. However, although the parallel application offers quite good parallel performance and carries out a 3D realistic analysis, studios for engineering rarely own parallel platforms to execute this software.

Therefore, we have implemented a service oriented system, based on Grid services, that enables to perform on demand dynamic analysis over computational Grids in a collaborative environment. It implies a two-fold strategy. Firstly, the main objective of Grid technology is to share and use different resources available in the network. Thus, it is possible to create a scientific and technical virtual organisation where most of the members do not need to invest in computational machines and software, and to be worried about licenses and new updates. It would be enough to establish agreements for their usage. Secondly, the service exploits, in a transparent way for the user, the computational capabilities of a distributed Grid infrastructure which delivers enough power to satisfy the computational requirements of the resource-starved dynamic structural simulations of high-rise buildings.

The reminder of this paper is structured as follows: First, section 2 describes the motion equation and the parallel application developed to simulate the behaviour of structures. One building has been also simulated to analyse the performance of this HPC application. Next, section 3 shows the Grid service implemented and the metascheduling approach to enable high-throughput when multiple user requests are concurrently received. Section 4 presents the structural case study that has been executed to test the performance of the Grid service, the computational resources employed and the task allocation performed. Finally, section 5 concludes the paper.

2   Parallel 3D Linear Dynamic Analysis of Buildings

The second order differential equations in time that governs the motion of structural dynamic problems can be written as follows [1]:

$$Ma(t) + Cv(t) + Kd(t) = f(t) \tag{1}$$

where M, C and K are the mass, damping and stiffness matrices respectively, $f(t)$ is the applied dynamic load vector, and $d(t)$, $v(t)$ and $a(t)$ represent the unknown displacement, velocity and acceleration vectors at the joints of the structure. The initial conditions at $t = 0$ are given by $d(0) = d0$ and $v(0) = v0$.

Because of their inherent advantages, direct time integration algorithms have been widely employed for the numerical solution of this computationally demanding equation [2]. In this way, an MPI-based parallel application for the 3D linear dynamic analysis of high-rise buildings has been implemented, where all the nodes of the structure are taking into account and six degrees of freedom per joint are considered. Node condensation techniques have not been assumed. All these resultant computational burden implies the need of using HPC strategies able to tackle large dimension problems and reduce the time spent on the analysis. In the application, the following three well-known time integration methods have been parallelised, providing comprehensive results in very reasonable response times: Newmark [3], Generalized-α [4] and SDIRK [5]. Consistent-mass matrix has been assumed, and Rayleigh damping has been employed, what means that $C = \alpha M + \beta K$. Besides, the standard implementation of MPI-2 I/O by ROMIO has been used to guarantee good performance on secondary storage device accesses. The application is highly portable and it can be easily migrated to a wide variety of parallel platforms.

Regarding the parallelisation of the problem, each processor is assigned firstly a group of N/p consecutive nodes and another one of B/p consecutive structural elements, being N and B the total number of nodes and beams in the building, respectively, and p the number of processors employed. Then, each processor generates and assembles in parallel its local part of the stiffness, mass and damping matrices, according to their nodes assigned. In this way, all the matrices of the problem, together with the different resulting vectors, will be partitioned among the processors following a row-wise block stripped distribution. Consistent-mass matrices have been considered, a more realistic alternative than lumped (diagonal) mass matrix. However, the dynamic analysis of a consistent-mass system requires considerably more computational effort and memory requirements than a lumped-mass system does.

Next, the effective stiffness matrix $\hat{K}$, or coefficient matrix of the problem, is obtained in parallel by means of a linear combination of K, M and C matrices. Different functions for summing sparse matrices in parallel have been implemented in order to generate these C and $\hat{K}$ matrices. Finally, the initial conditions are imposed in the system. Displacements and velocities at $t = 0$ will be usually known, and initial accelerations will be computed by solving the resulting system of linear equations when the Equation (1) is evaluated at $t = 0$, where M matrix constitutes the coefficient matrix.

Then, for each time step ($t = \Delta t, 2\Delta t, 3\Delta t, \ldots, n\Delta t$) different numerical phases must be also carried out. Firstly, the movement, velocity and acceleration vectors at the joints of the structure are computed in parallel by means of the chosen time integration method. More in detail, movements are worked out by solving a system of linear equations where the $\hat{K}$ coefficient matrix is large, sparse, symmetric and positive definite. Fortunately, the K, M and C matrices are constant, along the time, in a linear analysis. Thus, the $\hat{K}$ coefficient matrix does not change during the simulation process, and it just need to be factorised once if a direct method is employed to compute the linear systems. In this way, one forward-backward substitution will be carried out, for each time step, for computing the nodal movements. Parallel direct and iterative methods implemented in WSMP [6], MUMPS [7] and PETSc [8] public domain numerical libraries have been used for solving these linear systems. These three libraries have been chosen due to its availability, good performance and state-of-the-art capabilities. WSMP and MUMPS are MPI-based numerical libraries for solving large sparse symmetric and non-symmetric systems of linear equations. The parallel symmetric numerical factorisation implemented in WSMP is based on Cholesky Multifrontal algorithm. MUMPS uses a Multifrontal technique which is a direct method based on LU or LDL^T factorisation of the matrix. On the other hand, PETSc provides parallel matrix and vector assembly routines, basic linear algebra operations and parallel linear, nonlinear equation solvers and time integrators. The combination of a Krylov subspace method and a preconditioner is the heart of the parallel iterative methods implemented in PETSc. Besides, PETSc provides and efficient access to different external numerical libraries that implement direct methods, such as MUMPS, or preconditioners.

Before solving the linear system, the effective dynamic load vector, i.e. the right hand-size vector, must be evaluated in parallel. Again, each processor just computes and assembles the load vector corresponding to its group of nodes assigned. Sparse matrix-vector products, a constant times a vector and sums of vectors are the basic lineal algebra operations than take place in this phase. Therefore, different functions that carry out these mentioned linear algebra operations in parallel have been programmed and they will be used when employing WSMP, but not when using PETSc or MUMPS, since PETSc already provides routines for these functionalities.

Notice that parallel sparse matrix-vector product has a crucial importance for each time step, where the performance achieved could be severely degraded if an efficient implementation is not developed. Having in mind this consideration, communications have been tried to be minimised. For that, the processor i just sends the processor j those elements of its local vector that the processor j needs to carry out the matrix-vector product. Remember two things: (1) the vector is initially partitioned into the processors by means of a row-wise block-striped distribution and (2) the matrix is sparse and so not all the vector elements belonging to other processors will be needed. Considering the non-zero structure of problem matrices, each processor computes just once, at the beginning of the simulation and in a very fast way, which elements belonging to itself must be sent for each time step to every other processor. As a consequence, each processor just receives from the others the vector elements that it strictly needs during the simulation.

Once joint displacements have been computed, velocity and acceleration values are updated by taking advantage of the implemented routines of sum of vectors. Unlike Newmark and Generalized-α methods, SDIRK procedure requires to solve two linear systems for each time step. The first one, for $\hat{K}$ coefficient matrix, is composed of s right hand-size vectors, being s the number of stages employed in the method. Solution vectors of this system will be employed for updating displacement and velocity vectors. In the second one, M represents the coefficient matrix and the acceleration vector is computed. Obviously, both matrices will be factorised once if a direct method is used, and multiple forward-backwards substitutions will be required for each time step.

Finally, each processor evaluates in parallel, for its structural elements initially assigned, the member end forces and the reactions at the points attached to the rigid foundation. Bending moments and deformations at the predefined division points of the members will be evaluated in parallel, with the same data distribution, to check that they do not exceed the established design limits.

A building composes of 68,800 nodes (412,800 degrees of freedom) and 137,390 structural elements has been chosen to show the performance achieved in the parallel application. The behaviour of the building was dynamically analysed under the influence of an earthquake applied during 6 seconds, with time steps equals to 0.01 seconds.

Tables 1, 2 and 3 show the time (in minutes) and the efficiencies spent on the whole structural analysis, for the different integration methods parallelised, employing up to 16 processors, for WSMP, MUMPS and PETSc numerical libraries. MUMPS was employed thanks to the interface provided by PETSc. This time does not include the initial one corresponding to the generation of the stiffness, mass, damping and effective stiffness matrices, or the imposition of initial conditions or the factorisation of effective stiffness matrices. The simulations have been run on a cluster of 20 dual Pentium Xeon@2GHz, with 1 GByte of RAM and interconnected by a SCI network.

Table 1. Simulation time (in minutes) and efficiencies (%) for Newmark method

Proc.	WSMP		MUMPS (QAMD)		MUMPS (MND)		PETSc	
1	-	-	53.94	100.00%	-	-	2087.06	100.00%
2	32.16	100.00%	27.33	98.28%	26.72	100.00%	1119.20	93.23%
4	17.91	89.80%	16.04	84.07%	14.82	90.15%	581.91	89.66%
8	11.32	71.00%	10.33	65.27%	9.24	72.29%	305.64	85.36%
16	7.97	50.41%	7.52	44.83%	6.45	51.78%	165.84	78.65%

Table 2. Simulation time (in minutes) and efficiencies (%) for Generalized α-method

Proc.	WSMP		MUMPS (QAMD)		MUMPS (MND)		PETSc	
1	-	-	54.99	100.00%	-	-	2205.38	100.00%
2	32.85	100.00%	27.73	99.15%	27.17	100.00%	1182.03	93.29%
4	18.12	90.55%	16.31	84.29%	15.03	90.39%	618.89	89.09%
8	11.42	71.84%	10.76	63.88%	9.56	71.05%	324.98	84.83%
16	8.01	51.21%	7.63	45.04%	6.71	50.61%	173.17	79.60%

The shortest response times were achieved with MUMPS library, together with these ordering algorithms: MND (Multilevel Nested Dissection), implemented in METIS package [9], and QAMD (Approximate Minimum Degree Ordering with Automatic Quasi Dense Row Detection) [10]. WSMP achieved excellent results as well, with similar efficiencies than MUMPS with MND. WSMP ordering is also based on MND. Simulations with 1 processor overcame the RAM memory available in the approaches employing WSMP and MUMPS with MND.

Table 3. Simulation time (in minutes) and efficiencies (%) for SDIRK method

Proc.	WSMP		MUMPS (QAMD)		MUMPS (MND)		PETSc	
1	-	-	-	-	-	-	7201.76	100.00%
2	-	-	49.90	100.00%	-	-	3903.79	92.24%
4	42.29	100.00%	33.56	74.34%	30.01	100.00%	2127.92	84.61%
8	28.3	74.19%	23.08	54.05%	20.53	73.31%	1158.86	77.68%
16	22.11	47.81%	18.43	33.84%	16.25	46.31%	594.80	75.69%

Clearly, the number of non-zero elements of the coefficient matrix, after numerical factorisation, in MND is superior to QAMD. Therefore, the efficiency values appearing in Tables 1 and 2 for WSMP, and MUMPS with MND, are obtained with respect to 2 processors, or with respect to 4 processors at Table 3.

Regarding PETSc libraries, best results have been achieved by means of the combination of Conjugate Gradient as iterative method with block Jacobi preconditioning, where Incomplete Cholesky factorisation is also applied as subblock preconditioner. Structural coefficient matrices are usually ill-conditioning, what explains that iterative methods have been much slower than direct methods. It should be noticed that the main drawback of Block Jacobi preconditioner is that the number of iterations can rise when the number of processors is increased, what obviously has influence on the simulation times and efficiencies obtained. Anyway, direct methods just need to carry out a forward-backward substitution for each time step, what is much more efficient than solving the whole linear system as the iterative methods do.

While Newmark and Generalized-α methods offer second-order accuracy on the results, stage parameter was set to four in the SDIRK method, trying to achieve third-order accuracy. As expected, it increased dramatically the simulation times, since two linear systems (the first one composed of four right-hand size vectors) must be solved. In spite of acceleration values were not calculated in this case, with the aim to avoid the factorisation of the mass matrix, memory requirements of WSMP and MUMPS with MND ordering exceeded the available RAM even with two processors.

3   Service Oriented Dynamic Structural Analysis

Web services have emerged as the standard framework in distributed system development. They provide flexible and extensible mechanisms for describing, discovering, and invoking public network services by means of XML-based protocols. Globus Toolkit 4 (GT4)[11], the latest version of the current standard in Grid middleware systems, has performed a natural evolution to Web services technology, adopting them to define its architecture and interfaces. The result is the so-called Grid services, i.e. enhanced Web services that extend their conventional functionality into the Grid domain.

In this work, we have developed and deployed under GT4 middleware a Structural Dynamic Analysis Grid Service (SDAGS) for the 3D dynamic simulation

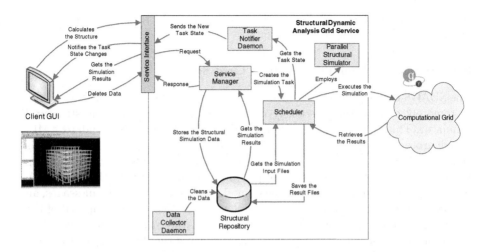

Fig. 1. Diagram of the implemented Grid service architecture

of large-scale buildings. Figure 1 exposes the Grid service architecture proposed. The diagram shows some of the principal parts involved, such as the GUI client, the SDAGS itself and the Globus-based computational infrastructure.

3.1 The GUI Client

The structural engineers can simulate the structures in the SDAGS thanks to an advanced graphical user interface (GUI) program. This software enables the user to perform the pre-processing phase, where the different properties are assigned to the structural members of the building (i.e. initial conditions, sections, external loads, etc.) in a user-friendly way. Using the Java 3D libraries, this highly portable application shows a 3D scene in which the user can interact with the building by means of different functionalities such as rotations, translations, zooming, selections, etc., employing the wired or solid modes of visualisation.

The GUI client interacts with the SDAGS, via its public interface, to analyse the structures. For this, the client sends, via a SOAP (Simple Object Access Protocol) request [12], an XML file with the properties of the building to be simulated, together with different parameters related to the dynamic analysis. Then, the status of the simulation task is periodically received, and once the structure has been remotely analysed, the output data are retrieved in a SOAP message, and then deleted in the machine that runs the Grid service. Finally, the post-processing phase takes place and the results obtained are automatically mapped onto the graphical display and easily interpreted.

This GUI client incorporates a fault-tolerant procedure with the SDAGS for the data retrieval. In case of client failure, thanks to a unique simulation identifier, the results would be retrieved later. Failures in the client do not affect the simulation executions as both the client and the service are decoupled.

In a dynamic analysis, all the result data successfully received by the client will not be sent again by the SDAGS.

3.2   Structural Dynamic Analysis Grid Service

The SDAGS is a flexible and extensible Grid service implementation that enables to remotely employ the previously mentioned HPC-based dynamic structural simulator. This service publishes a set of methods, by means of standard XML-based protocols, that are invoked via SOAP requests. On the one hand, this enables to implement heterogeneous clients, developed in different programming languages and over a wide variety of platforms, to interact with the service. To include all the input and output binary simulation data in the XML messages, an hexadecimal encode schema has been employed. The SDAGS is composed of the following main components: the *Service Manager*, the *Scheduler*, the *Data Collector Daemon*, the *Parallel Structural Simulator* and the *Task Notifier Daemon*.

The *Service Manager* represents the core of the SDAGS and it is in charge of satisfying the requests from the clients. It acts as the front end, receiving client requests as well as interconnecting all the system components. The *Task Notifier Daemon* is responsible of performing the notification process of the state changes of the tasks, thus enabling the users to instantly know the state of their simulations. The structures are analysed on the Grid resources by the *Parallel Structural Simulator*, which is able to perform efficiently and in a realistic way static and dynamic analyses.

The *Scheduler* agent executes the structural simulations in the available Grid infrastructure. Currently, we are employing several cost-effective cluster of PCs located at our research center. Firstly, the *Scheduler* involves the resource discovery to obtain a list of candidate execution machines. After that, a resource selection phase is carried out in order to select the best available computational machine for each structural simulation. Finally, the different phases related to achieve remote task execution, such as data stage and job monitor to detect failures, will be also performed by this component.

Input and output simulation data will be stored in a Structural Repository, implementing a data persistence schema and enabling the use of the system also as a Storage Service. Finally, the *Data Collector Daemon* component inspects periodically the Structural Repository and cleans the old simulation files.

3.3   The Structural Analysis Process

The implementation details of the Grid Service developed are exposed in the next paragraphs, by means of the sequence of steps to be followed to simulate a building. First, the client submits the request, sending the corresponding files that define the structure, such as its structural and geometric properties, the different external load hypotheses to be evaluated and the needed parameters to define the type of analysis.

This request is received in the SDAGS by the *Service Manager*, which processes all the input data, storing them in the Structural Repository, returns to the client a simulation identifier, which will be used in later invocations to identify the simulation, and generates the appropriate binary input file for the parallel simulator. Then, the *Service Manager* creates an execution task that contains all the required properties to execute it in the computational Grid. Next, this task is added to the *Scheduler* module, which, in a transparent way, performs the resource selection and the simulation execution management. A resource selection policy has been defined addressed to optimise the throughput and reduce the execution time of the each analysis. The simulation type, static or dynamic, the dimension of the structure and the user privileges will be values used to decide the number of processors involved in each execution.

For each simulation request, the SDAGS creates and publishes a notification item that is in charge of informing the client about its evolution. After subscribing to this item, the *Task Notifier Daemon* notifies the user any change that takes place in the analysis process. In this way, the client is perfectly aware of the status of the simulations: waiting, in execution, failed, finalised, etc. This approach dramatically reduces the overhead that would appear in the system if the clients periodically queried the service about the status of every simulation.

In a static analysis, and once the task execution has finished, the output results are automatically saved into the Structural Repository by the *Scheduler*, and the user is informed about their availability thanks to the *Task Notifier*. However, a dynamic analysis is performed by means of an iterative process that implies the generation of output data for each simulation time step. With the purpose of reducing the waiting time, the client is informed by the Task Notifier when there are enough results to be sent, thus submitting it different retrieval requests. In this way, the simulation and data retrieval phases are overlapped, what implies a clear benefit for the user who can begin to process the results before the analysis is completed.

The result retrieval procedure is performed by the *Service Manager*, which processes and analyses all the output files in order to generate an XML file that is sent to the client in a SOAP message. One of the main problems related to the use of this type of messages, being based on XML, is their size, thus introducing a communication overhead between the client and the Grid service [13]. In our case, the solution adopted has lied in the use of an hexadecimal codification schema for including binary data, instead of inserting all of them in a text-based format. This approximation enables to reduce substantially the message dimension, which has a direct impact on the data transfer times.

An erase method that deletes all the simulation data is also available. Notwithstanding, the client is not required to invoke it, thus taking advantage of a Data Storage Service that can be employed during a certain period of time. Nevertheless, due to the fact that there are users with different privileges in the system, a component called *Data Collector Daemon* will be in charge of periodically erasing the simulation results of those lowest level clients.

Several fault tolerance levels have been implemented in the system, including the service itself, the task scheduling and execution, and the client, what guarantees that all the simulations submitted will be successfully attended. On the one hand, the SDAGS implements a persistence schema that stores a description of all the tasks in course or waiting for execution, and those finalised simulations that still have results to be recovered by the client. Therefore, in case of service failure, all the non-finished tasks would be launched later, and the identifiers of those having pending results would be registered again. On the other hand, the fault tolerance level included in the *Scheduler* ensures that a failed execution will be transparently migrated to another Grid resource. Failures in the data communication (including network outages) between the service and the client, or the service and the computational resources, are also handled. Several tests have been performed (injecting failures in the client, the network, the service and the remote resources) to ensure the robustness of the fault-tolerant mechanisms.

A robust security system has been integrated in the service, including user authorisation and authentication, and privacy and integrity of results. On the one hand, the user authorisation and authentication capability establishes an access control to the published services, enabling to register all the actions performed by the clients. The authorisation system employs a configuration file that contains all the users authorised to interact with the service. All the requests from users not registered will be directly rejected. The authentication process is implemented by means of a X.509 certificate that identifies the user. This certificate is sent to the service when the communication begins. The data privacy and integrity has been achieved using a private-public key approach. It employs the same certificate X.509 to perform the encryption and signature of all the data exchanged between the service and the client.

3.4   Interacting with the Computational Grid Via GMarte

The SDAGS execute the *Parallel Structural Simulator* over a computational Grid by using the functionality of the GMarte middleware [14]. GMarte is a software abstraction layer, developed on top of the Java CoG Kit 1.2 [15], which provides an object-oriented API for the description of generic batch computational tasks from any scientific area. It provides all the required software infrastructure to perform the fault-tolerant allocation of tasks to machines based on the Globus Toolkit.

In order to achieve remote task execution, GMarte enables the user to focus on *what* should be executed, instead of messing around with all the implementation details of the underlying Grid middleware. For that, GMarte first introduces an abstraction layer over the information provided by the computational resources of a Grid infrastructure. This enables the user to access computational information of the resources, such as the number of available processors or RAM, in the same manner, regardless the underlying differences of the Grid middleware.

Figure 2 describes how GMarte fits in the service proposed. The *Service Manager*, in Figure 1, uses the GMarte API to provide the description of the computational tasks, which are assigned to a daemon *Scheduler* that waits for new tasks to be executed.

The implemented GMarte-based *Scheduler* is in charge of performing a sequence of steps in order to achieve successful execution of the tasks. This

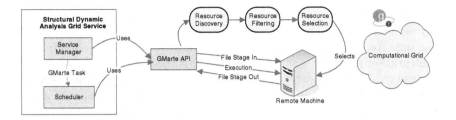

Fig. 2. Usage of GMarte within the Grid service

procedure involves, when the Grid service starts, the *Resource Discovery* and the *Resource Filtering* phases to obtain a list of currently available machines to host executions. Then, for each structural analysis request, the *Resource Selection* phase selects the current best computational resource to execute it. Later, all the needed input files are automatically transferred to the remote machine, before the remote parallel execution is started. When the simulation has finished, all the generated output files are moved to the machine hosting the SDAGS.

GMarte implements a multi-threaded metascheduler that enables to concurrently carry out the resource selection phase for the different simulations that have to be executed. This notably reduces the start-up time of the metascheduling procedure, when compared with other traditional single-threaded metaschedulers, what enables to notably increase the service productivity when it is concurrently used by multiple users. The metascheduling policy implemented in GMarte considers the application requirements specified by SDAGS, as well as the dynamic state of computational resources to select the most appropriate resource.

A multilevel fault-tolerance scheme is enforced to cope with the errors arising both during data transfers and remote execution. This ensures that executions will proceed as long as there are living resources in the Grid Computing infrastructure. The use of this proposed SDAGS, that uses a computational Grid for executions, enables to increase the productivity when the service is concurrently used by multiple users.

4   Multiuser Structural Case Study

In order to test the performance of the SDAGS in a multiuser environment, a structural case study composed of several simulations has been simulated on a Grid infrastructure.

The case study proposed, addressed to reproduce the Grid service availability with different clients, is composed of 30 user simulations, which must be concurrently managed. Each simulation represents the dynamic analysis of a building whose structural features (68,800 nodes and 137,390 beams) were described in section 2.

Different representative earthquakes, according to the geographical location of the building, have been applied. The accelerograms employed had an duration between 5 and 10 seconds and they include an equally-spaced ground acceleration every 0.01 seconds. Due to the accelerograms duration variability and in order to employ an homogeneous case study, the simulation time was fixed to 5 seconds using a time step of 0.01 seconds. The Newmark method was the chosen direct time integration procedure, and the Parallel Structural Simulator was configured to use the WSMP library. The output data contains information about the stresses and deformations at multiple predefined intermediate points of all the structural elements that compose the building. This was configured to be stored every 0.5 seconds. This resulted in an output data of 646 MBytes for simulation resulting in a total of 19 GBytes.

The execution of the case study was performed in a Grid infrastructure composed of computational resources which belong to our research group, since we had not access to a global Grid. It consists of 2 clusters of PCs, whose principal characteristics are detailed in Table 4. Both machines are interconnected via a local area network delivering 100 Mbits/sec. with the service host. The Globus Toolkit version 2.4 was previously installed on each machine of the Grid deployment.

Following the policy of selecting the number of processors according to the features of the structure, the service estimated a number of two processors involved in each parallel execution. This decision enabled to efficiently share the limited available computational resources, as many executions could be proceeded simultaneously.

4.1   Execution Results

Table 4 shows that a similar number of simulations were allocated to each computational resource. In fact, the GMarte resource selection component implements a policy that distributes the workload on the different resources of a Grid, trying to minimise the impact in case of failure in a determined host. Clearly, resource selection is a fundamental key in the whole task allocation procedure. Fine-tuning this phase, by allocating more executions to Odin, could probably have obtained better results.

The execution of the structural case study on the proposed infrastructure required a total of 38 minutes, since the scheduling procedure started until the output data of the last simulation was retrieved to the Grid service machine. On the one hand, executing all the simulations using a sequential platform, one execution after another and employing 1 PC of Odin, the faster cluster, required 566 minutes. On the other hand, using a High Performance Computing approach, assuming a typical cluster of 8 CPUs, and performing 2-processor executions on cluster Odin (4 simultaneous simulations) required a total of 105 minutes.

Therefore, the Grid Computing approach delivered an speedup of 14.89 with respect to the sequential execution and 2.76 compared to the HPC approach. Obviously, this improvement in speed depends on the amount of computational resources employed in the Grid deployment. Anyway, it is important to point out that the Grid approach introduces an overhead, both at the scheduling level (for the resource selection) and the data transfers involved in the stage in and the stage out phases.

Table 4. Detailed machine characteristics of the Grid infrastructure

Machine	Processors	Memory	Tasks Allocated
Kefren	20 dual Intel Pentium Xeon@2.0 Ghz	1 GByte	16
Odin	55 dual Intel Pentium Xeon@2.8 Ghz	2 GBytes	14

5 Conclusions

In this paper, we have developed a Grid service oriented system, based on GT4, that enables to perform high performance and realistic 3D dynamic structural simulations of large dimension buildings on a Grid infrastructure. For that, an MPI-based structural application has been previously implemented, where 3 different direct time integration methods have been parallelised. Underlying linear systems of equations have been solved by means of WSMP, MUMPS and PETSc numerical libraries. The parallelisation strategy of the different stages that compose the parallel structural simulator has been discussed, as well as the parallel performance, in terms of speed-up and efficiency, in the dynamic analysis of a building, considering the time integration algorithms and the distinct numerical libraries employed.

Besides, the architecture of the Grid service has been described, emphasizing its design and implementation. GMarte framework has been presented as an appropriate metascheduler to carry out the remote task simulation in a Grid infrastructure. Finally, the behaviour of the Grid service has been tested when multiple clients try to analyse, at the same time, different structures, with the purpose of evaluating the needed high-throughput of the system. Simulation times corresponding to the analysis of all these buildings have been provided, comparing them with different computational approaches. From our point of view, the system presents an acceptable development level to begin to be tested by end-users.

References

1. Clough, R., Penzien, J.: Dynamics of Structures. Second edn. McGraw-Hill, Inc (1993)
2. Fung, T.: Numerical Dissipation in Time-Step Integration Algorithms for Structural Dynamic Analysis. Progress in Structural Engineering and Materials **5** (2003) 167–180
3. Wilson, E.L.: A Computer Program for the Dynamic Stress Analysis of Underground Structures. Technical Report SESM Report 68-1, Division of Structural Engineering and Structural Mechanics, University of California, Berkeley (1968)
4. Chung, J., Hulbert, G.: A Time Integration Algorithm for Structural Dynamics with Improved Numerical Dissipation: the Generalized α-Method. Journal of Applied Mechanics **60** (1993) 371–376
5. Owren, B., Simonsen, H.: Alternative Integration Methods for Problems in Structural Dynamics. Computer Methods in Applied Mechanics and Engineering **122**(1-2) (1995) 1–10

6. Gupta, A.: WSMP: Watson Sparse Matrix Package Part I - Direct Solution of Symmetric Sparse Systems. Technical Report Technical Report IBM Research Report RC 21886(98462), IBM (2000)
7. Amestoy, P., Duff, I., L'Excellent, J., Koster, J.: MUltifrontal Massively Parallel Solver (MUMPS Version 4.6.1) Users Guide. Technical report, IBM (2006)
8. Balay, S., Buschelman, K., Gropp, W., Kaushik, D., Knepley, M., Curfman-McInnes, L., Smith, B., Zhang, H.: PETSc Users Manual. Technical Report Technical Report ANL-95/11 - Revision 2.3.1, Argonne National Laboratory (2006)
9. Karypis, G., Kumar, V.: METIS: A Software Package for Partitioning Unstructured Graphs, Partitioning Meshes, and Computing Fill-Reducing Orderings of Sparse Matrices. Technical Report Version 4.0, University of Minnesota, Department of Computer Science /Army HPC Research Center (1998)
10. Amestoy, P.: Recent Progress in Parallel Multifrontal Solvers for Unsymmetric Sparse Matrices. In: Proceedings of the 15th World Congress on Scientific Computation, Modelling and Applied Mathematics, IMACS 97. (1997)
11. Foster, I.: Globus Toolkit Version 4: Software for Service-Oriented Systems. In: IFIP International Conference on Network and Parallel Computing, Springer-Verlag LNCS. Volume 3779. (2005) 2–13
12. Gudgin, M., Hadley, M., Mendelsohn, N., Moreau, J., Frystyk, H.: SOAP Version 1.2 Part 1: Messaging Framework. W3C Recommendation (2003)
13. Lu, W., Chiu, K., Gannon, D.: Building a Generic SOAP Framework over Binary XML. HPDC-15: The 15th IEEE International Symposium on High Performance Distributed Computing (Paris, France, June 2006)
14. Alonso, J., Hernández, V., Moltó, G.: An Object-Oriented View of Grid Computing Technologies to Abstract Remote Task Execution. In: Proceedings of the Euromicro 2005 International Conference. (2005) 235–242
15. von Laszewski, G., Foster, I., Gawor, J., Lane, P.: A Java Commodity Grid Kit. Concurrency and Computation-Practice & Experience **13**(8-9) (2001) 645–662

Scalable Desktop Grid System[*]

Péter Kacsuk[1], Norbert Podhorszki[1], and Tamás Kiss[2]

[1] MTA SZTAKI
Computer and Automation Research Institute of the
Hungarian Academy of Sciences
H-1528 Budapest, P.O. Box 63, Hungary
{kacsuk,pnorbert}@sztaki.hu
[2] University of Westminster, Cavendish School of Computer Science
115 New Cavendish Street, London W1W 6UW, UK
T.Kiss@westminster.ac.uk

Abstract. Desktop grids are easy to install on large number of personal computers, which is a prerequisite for the spread of grid technology. Current desktop grids connect all PCs into a flat hierarchy, that is, all computers to a central server. SZTAKI Desktop Grid starts from a standalone desktop grid, as a building block. It is extended to include clusters displaying as single powerful PCs, while using their local resource management system. Such building blocks support overtaking additional tasks from other desktop grids, enabling the set-up of a hierarchy. Desktop grids with different owners thus can share resources, although only in a hierarchical structure. This brings desktop grids closer to other grid technologies where sharing resources by several users is the most important feature.

1 Introduction

Originally, the aim of the researchers in the field of Grid was that anyone could offer resources for a Grid system, and anyone can claim resources dynamically, according to the actual needs, in order to solve a computationally intensive task. This twofold aim has been, however, not fully achieved. Currently, we can observe two different trends in the development of Grid systems, according to these aims.

Researchers and developers in the first trend are creating a Grid service, which can be accessed by lots of users. A resource can become part of the Grid by installing a predefined software set (middleware). The middleware is, however, so complex that it needs a lot of effort to maintain. Therefore it is natural, that single persons do not offer their resources but all resources are maintained by institutions, where professional system administrators take care of the hardware/middleware/software environment and ensure the high-availability of the Grid. Examples of such Grid infrastructures are the EGEE infrastructure (Enabling Grids for E-SciencE) and its Hungarian affiliate virtual organisation, the HunGrid, or the NGS (National Grid Service) in the UK. The original aim of enabling anyone to join the Grid with one's resources has not been fulfilled.

[*] This research work is carried out under the FP6 Network of Excellence CoreGRID funded by the European Commission (Contract IST-2002-004265) and by the Hungarian Jedlik Anyos HAGrid project (Grant No.: NKFP2-00007/2005).

M. Daydé et al. (Eds.): VECPAR 2006, LNCS 4395, pp. 27–38, 2007.
© Springer-Verlag Berlin Heidelberg 2007

Nevertheless, anyone who is registered at the Certificate Authority of such a Grid and has a valid certificate can access the Grid and use the resources.

A complementary trend can also be observed for the other part of the original aim. Here, anyone can bring resources into the Grid system, offering them for the common goal of that Grid. Nonetheless, only some people can use those resources for computation. The most well-know example, or better to say, the original distributed computing facility example of such Grids is the underlying infrastructure of the SETI@home project [1]. In Grids, similar to the concepts of SETI@home, personal computers owned by individuals are connected to some servers to form a large computing infrastructure. Such systems are called with the terms: Internet-based distributed computing, public Internet computing or desktop grid; we use the term desktop grid (DG) from now on. A PC owner should just install one program package, register herself on the web page of the Grid system and configure the program by simply giving the address of the central server. Afterwards, the local software runs in background (e.g. as a screensaver) and the owner does not need to take care of the Grid activity of her computer. In a desktop grid, applications can be performed in the well-known master-worker paradigm. The application is split up into many small subtasks (e.g. splitting input data into smaller, independent data units) that can be processed independently. Subtasks are processed by the individual PCs, running the same executable but processing different input data. The central server of the Grid runs the master program, which creates the subtasks and processes the incoming sub-results. The main advantage of a desktop grid is its simplicity thus, allowing anyone to join. The main disadvantage is that currently only problems computable by the master-worker paradigm can be implemented on such a system. Desktop grids have already been used at world-wide scales to solve very large computational tasks in cancer research [2], in search for the sign of extraterrestrial intelligence [1], climate predictions [3] and so on.

Desktop grids can be used efficiently and conveniently in smaller scales as well. We believe that small scale desktop grids can be the building blocks of a larger Grid. This is a new concept that can bring closer the two directions of Grid developments. It is easy to deploy desktop grids in small scale organisations and to connect individual PCs into it therefore we get a grid system that can spread much faster then heavy-weight grid implementations. On the other hand, if such desktop grids can share the resources and their owners can use others' desktop grid resources, the support of the many users of the other trend is also realised. The realisation steps towards such collaboration of desktop grids consist of the support of clusters (so they are easy to include as a resource), the hierarchy of desktop grids within a large organisation with several levels of hierarchy, and the resource sharing among independent desktop grids in different organisations.

SZTAKI Desktop Grid realizes these steps, starting from an established standalone desktop grid infrastructure as building block. It is extended to include clusters as single powerful PCs into the desktop grid computational resource set, while using their local resource management system. Then, such building blocks support overtaking additional tasks from other desktop grids, enabling the set-up of a hierarchy of DGs. The final goal of creating a large-scale Grid from DGs as building blocks will be investigated in the recently submitted EU project called as COCIDE. In this paper, the SZTAKI Desktop

Grid is described, from the basic single desktop grid to the support of clusters and to the hierarchy of desktop grids.

1.1   Related Work

Condor. Condor's approach is radically different from the DG concept. Condor [14] represents a push model by which jobs can be submitted into a local Grid (Cluster) or global Grid (friendly condor pools or Globus Grid [11]). The DG concept applies the pull model whereby free resources can call for task units. The advantage of the DG concept is that it is highly scalable (even millions of desktops can be handled by a DG server) and extremely easy to install at the desktop level. The scalability of Condor is not proven yet. Largest experiments are at the level of 10000 jobs in EGEE but it requires an extremely complicated Grid middleware infrastructure that is difficult to install and maintain at the desktop level.

BOINC. BOINC (Berkeley Open Infrastructure for Network Computing, see [4], [5]) is developed by the SETI@home group in order to create an open infrastructure that could be the base for all large-scale scientific projects that are attractive for public interest and that can use millions of personal computers for processing their data. This concept enables millions of PC owners to install single software (the BOINC client) and then, each of them can decide what project they support with the empty cycles of their computers. There is no need to delete, reinstall and maintain software packages to change among the projects. As of October 2006, there are over two dozens BOINC-based projects overall in the world using more than a 1.4 million hosts providing 476 TeraFLOPS computational power.

The properties of BOINC can be used for smaller scale, combining the power of the computers at institutional level, or even at department level. The SZTAKI Desktop Grid is based on BOINC since this is a well-established open source project that already proved its feasibility and scalability. The basic infrastructure of SZTAKI Desktop Grid is provided by a BOINC server installation and the connected PCs at a given organisational level. The support of clusters and the organization of such BOINC-based desktop Grids are the new features that SZTAKI Desktop Grid provides over BOINC.

XtremWeb. XtremWeb [6] is a research project, which, similarly to BOINC, aims to serve as a substrate for Global Computing experiments. Basically, it supports the centralised set-up of servers and PCs as workers. In addition, it can also be used to build a peer-to-peer system with centralised control, where any worker node can become a client that submits jobs. It does not allow storing data, it allows only job submission.

Commercial Desktop Grids. There are several companies providing a Desktop Grid solution for enterprises [7] [8] [9] [10]. The most well-known examples are the Entropia Inc, and the United Devices. Those systems support the desktops, clusters and database servers available at an enterprise. However, their cluster connection solution is not known for the research community and it is very likely that their model is based on the push model. Our goal is to develop a pull model solution since it is consistent with the current BOINC concept.

2   SZTAKI Desktop Grid

The basic idea of SZTAKI Desktop Grid is first, to provide a basic DG infrastructure that is easy to install, to maintain and to use at an organisational level. This basic infrastructure enables us to connect PCs within a department and to run (small) distributed projects on it. Second, clusters are supported as they are increasingly available at many departments of institutions and companies as well. Third, the hierarchical structure of an organisation needs the possibility of connecting such departmental desktop grids into an infrastructure where larger projects can use more resources than available within one department. Fourth, more generally, to make possible the resource sharing among desktop grids that are not related in a hierarchical way. In this way, small-scale desktop grids, which are easy to install, can be the building blocks of a large grid infrastructure.

2.1   Single Organisation's Desktop Grid

SZTAKI Desktop Grid is based on the BOINC infrastructure, as we believe that it provides everything that is needed for a basic desktop grid with one (running on a single machine or on multiple machines) server and many workers. The infrastructure for executing computational tasks and for storing data sets is used only. Its support for user credits, teams and the web-based discussion forums are not relevant for an organisation but, of course, all these features are available if needed. Note that while BOINC had been designed for large public projects originally, here it is used within different circumstances. A typical BOINC project is about only one application and individual people (at home) select between such BOINC projects (identified by their (only) application) that they want to support. In an organisation, there are several applications in the only BOINC "project" and the only individual is the administrator with all the machines within the organisation. Thus, many different applications use the shared resources.

The BOINC-based desktop grid within an organisation (institution, or just a department) enables

- to connect PCs in the organisation into the desktop grid,
- to install several distributed computing projects on the desktop grid,
- and to use the connected PCs to compute subtasks of those projects.

As Figure 1 shows, there is a Scheduler Server and a Data Server in the BOINC infrastructure, however, they can be simply installed on one computer but also they can exist in multiple instances as well, depending on the central processing needs of a project. Scheduler Server stores all information about available platforms, application programs, subtasks, connected machines (and users) and results for subtasks. Data Server stores all executables, input and output files. On each PC, a core client is running that downloads application client executables, subtasks (describing actual work) and input files to perform the subtasks. The main application on the top level has to generate the sequential subtasks and to process subresults. BOINC gives tools and support for generic distributed projects to do that, however, SZTAKI provides a much simpler and easier-to-use API, called DC-API. The use of this API enables scientist just concentrate on task generation and processing results without knowing even what grid infrastructure is serving the processing needs. Of course, the use of the API is not obligatory, one can use BOINC's tools as well.

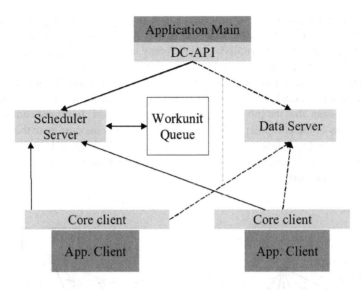

Fig. 1. BOINC-based Desktop Grid infrastructure

2.2   Supporting Clusters Within SZTAKI Desktop Grid

BOINC in itself does not provide any support for clusters. It has a server that generates work and there are clients that do the work (actually several ones on an SMP node, one subtask per CPU). The need for cluster support is clear. No one would like to develop a sophisticated distributed application that uses partly the desktop grid and partly a cluster, all with different concepts, APIs and syntaxes. Cluster's job management concept is more general than the execution of work units (subtasks) within a desktop grid therefore, the latter one can be mapped onto the previous one. There are five possibilities in extending the BOINC infrastructure for cluster support.

1. A desktop grid client is installed on all machines of the cluster and connected to the server of the desktop grid of the given organisation, i.e. all machines of the cluster participate individually, as a normal PC in the desktop grid.
2. A complete desktop grid is installed on the cluster, with the server on the front-end node, and all machines connected to it. This way, the cluster can participate in a larger desktop grid as one leaf element in a hierarchy, see section 2.3.
3. An independent, higher-level broker distributes work among clusters and desktop grids.
4. The server of the desktop grid should be aware of the presence of a cluster and submit jobs instead of work units,
5. An extended version of a single desktop grid client is installed onto the cluster's front-end, which converts desktop grid work units into traditional jobs and submits them to the cluster's job management.

The *first possibility* is easy to achieve, only the desktop grid client should be installed on the machines, see Figure 2. The configuration of BOINC core client consist of defining a registered user's ID and the project server URL. Settings for the user's preferences are defined on the project web server, and settings are propagated to all clients with the same user ID. BOINC provides easy install on multiple machines based on one installation therefore, the whole procedure is very easy.

However, if the cluster is not a brand new one or the owners do not want to use it exclusively for the desktop grid, a job manager is surely installed and used on that cluster. This means, that the job manager and the desktop grid clients are competing for the spare cycles of the computers. The job manager's role is to coordinate the resources within a cluster and to balance the load on it. Desktop grid clients and subtasks coming from the desktop grid server are out of the view for the job manager therefore, it is not able to function properly.

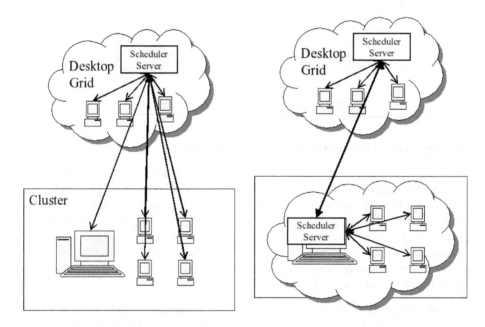

Fig. 2. Clusters 1. All machines are clients   **Fig. 3.** Clusters 2: stand-alone desktop grid

The *second possibility* (see Figure 3) by-passes the job manager as well, having the same drawback and therefore, it is not recommended. However, if the hierarchy of desktop grids are a reality, this option can be considered as a free solution for connecting a cluster into an existing desktop grid.

Usually, we may think at first that if different things are to be connected and to work together, there is a need for a higher-level actor that distributes work among those things and takes care of the good balance, as in the *third possibility*. That is, in our case, an appropriate broker is needed that is able to gather information about the status of the different entities (desktop grids and clusters), to decide where to send the next piece of work and to convert subtasks into work units or jobs according to the target system, see Figure 4. Such an approach is followed in the Lattice project [12], which is developing a community-based Grid system that integrates Grid middleware technologies and widely used life science applications. This system deals with traditional jobs, i.e. executables, input data and definition of requirements and preferences. Jobs are submitted to a modified version of the Condor-G broker [13] that sends a job either to a Globus-based grid or to a BOINC-based desktop grid.

In this case, a desktop grid is just one element among others. Different grid implementations can be connected together this way if appropriate conversion between the different concepts, representations and syntaxes can be managed.

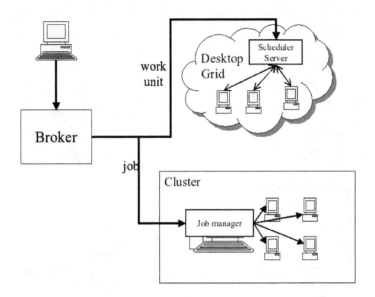

Fig. 4. Cluster 3: High-level brokering of jobs

The *fourth possibility* keeps the heading role of the desktop grid server, see Figure 5. In this scenario, there is a desktop grid as "the grid", in which clusters are connected from "below". The server should be configured in a way that it knows about the cluster, its static status information (size, benchmark information) and its dynamic status information (number of available machines) - the same way, as the broker of the third option should do. As in the basic desktop grid, work is distributed by the server; however, it can decide to send some work to the cluster. In this case, the work unit representation should be converted to the job representation, which can be submitted to the job manager of the cluster.

This solution needs lot of development of the server's implementation. A monitoring system should be used to get status information about the cluster, such information should be stored and handled somehow, decision logic should be altered - all these tasks are also part of the third option. Besides that, the internal work unit should be converted into a traditional job and the server should be able to contact the job manager of the cluster remotely and submit jobs. As we mentioned, work unit representations can be mapped onto job representations therefore, this is quite a simple task.

The *fifth possibility* is the most elegant way of including clusters into the desktop grid, see Figure 6. In a desktop grid, client machines are connecting to the server and ask for work; this is called pull-mode. In contrast, job managers and grids of the first trend mentioned in the introduction submit work (jobs) to selected resources (push-mode). In this option, clusters can participate in the pull-mode execution of the desktop grid. A desktop grid client originally asks for a given amount of work to be processed on the given machine. However, with some modification, it can ask for many work units, transform them into jobs and submit them into a cluster. The desktop grid server can see it as a normal, but somewhat very powerful client. In this solution, only the client should be modified, and since it is running on the front-end node of the cluster, information gathering and job submission are easy to perform.

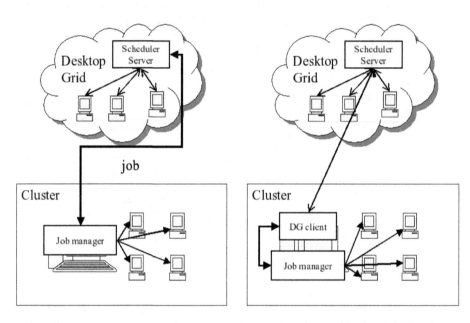

Fig. 5. Clusters 4: Submit jobs from server

Fig. 6. Clusters 5: Special DG client on the front-end

We have chosen the fifth possibility for SZTAKI Desktop Grid, because this way clusters are seamlessly integrated into it, it keeps the role of the job manager of the cluster and it requires less modifications than the others.

2.3 Hierarchical Desktop Grid

Departments can be satisfied by using the basic SZTAKI Desktop Grid with cluster support. All PCs and clusters of a department can be connected into one local (department level) DG system and distributed projects can use all these resources. It is natural to ask, what if there are several departments using their own resources independently but there is an important project at a higher organisational level (e.g. at a school or campus level of a university or at university level). Having the previous set-up in the departments, only one of the departments can be selected to run the project. Of course, the ideal would be to use all departments' resources for that project. Besides again developing something new component (e.g. a broker) to control over the different desktop grids, there is the possibility to build a hierarchy of desktop grids - if the building blocks can enable it as shown in Figure 7. In such a hierarchy, desktop grids on the lower level can ask for work from higher level (pull mode), or vice versa, desktop grids on the higher level can send work to the lower levels (push mode).

SZTAKI Desktop Grid supports the pull mode, as this is the original way how desktop grids work. The control of important work on the higher level can be realised with priority handling on the lower level. A basic SZTAKI Desktop Grid can be configured to participate in a hierarchy, that is, to connect to a higher-level instance of SZTAKI Desktop Grid (parent node in the tree of the hierarchy). When the child node (a stand-alone desktop grid) has less work than resources available, it asks for work from the parent. The parent node can see the child as one powerful client, exactly as in the case of a cluster, which asks for work units.

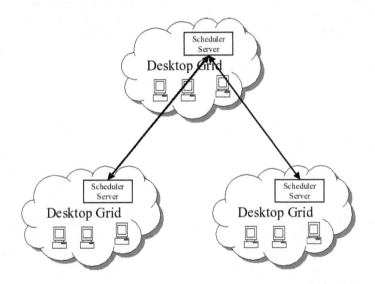

Fig. 7. Hierarchy of desktop grids

Of course, the BOINC-based server has to be extended to ask for work from somewhere else (i.e., behave similarly as a client) when there is not enough work locally. Fortunately, this can be done separately in the case of BOINC. Work units are generated by the running applications and they are put into a database of the BOINC server. Whether a work unit arrives from outside or from a local application, it does not matter. Therefore, it is enough to create a new daemon on the server machine that observes the status of the desktop grid. When client machines' requests for work are rejected - or when the daemon predicts that this will happen soon - the daemon can turn to the parent desktop grid and ask for work units. The daemon behaves towards the parent as a BOINC client, asking for work and reporting results. However, it puts all those work units into the database of the local server thus, client machines will process them and give the results. The daemon should also wait and look for the incoming results and send them back to the parent.

However, there is the issue of applications when we want to connect two BOINC-based desktop grids. In the BOINC infrastructure, application executables should be registered in the server and signed with a private key (of the project). Clients always check if the downloaded executable is registered and valid thus, avoiding the possibility of spreading arbitrary code by hackers. A parent desktop grid is an alien to the child in this sense; executables registered in the parent desktop grid should be registered before work units using that executables can be processed.

In BOINC, for security reasons, the private key of a project should be stored on a machine that is separated from the network. Application client executables should be signed by the administrator of the projects and only the signature should be copied from that separated machine. The signature is checked by using the public key at the client level. If a client machine receives work units from projects belonging to different levels of the DG hierarchy, the client should know the public keys of all the servers placed above it in the DG hierarchy. When a work unit arrives it should contain the source level's identifier based on which the desktop will know which public key to use for checking the signature of the code.

2.4   The SZTAKI Desktop Grid Service

One of our goal is to connect several organisation level Desktop Grids in Hungary using the SZTAKI Desktop Grid (similarly as HunGrid tries to gather institutional clusters into a virtual organization of the EGEE Grid infrastructure). To achiveve this, first we need to establish several such standalone Desktop Grids that will allow then each other share the resources. In order to demonstrate the strength and usage of the DG concept for Hungarian institutes SZTAKI has created a new BOINC-like DG service called as SZDG [15], named simply after SZTAKI Desktop Grid, not after its single application as usually BOINC projects do. The task to be solved by SZDG is a math problem of generating 11-dimension binary number systems. These can contribute to develop new encryption algorithms for safer security systems. SZDG has been running since July 2005 and shortly after that extracted more than 3500 participants and more than 5500 machines from all over the world. The performance of SZDG varies between 100 GigaFLOPS and 1.5 TeraFLOPS depending on how much resources are actively computing SZDG tasks (people share their resources among many BOINC based projects).

Though SZDG works the same way as the other global DG systems and is basically yet another BOINC project with one application, its basic role is to provide an experimental system for Hungarian institutes and companies to learn the technology and its possible usage as local DG system. We have found that institutes are very cautious with the usage of Grid technology and hence in order to convince them about the usefulness and safety of the DG systems they can test the DG technology in three phases:

1. Phase 1: Test the client side. Staff members of institutes can connect their PCs to the demo project thus, participating in one large-scale computing project; similarly, as people all over the world participate in BOINC, XtremWeb and Grid.org based projects. In this way they can be convinced that the client components of SZTAKI Desktop Grid are safe enough and do not cause any harm to their desktop machines.
2. Phase 2: Test the server side with their own application. If an institute has a problem that needs large computing power to solve, SZTAKI helps to create a new project on SZDG and provides the central server for that project. The institute should provide the PCs and clusters for SZDG to work on that project. In this way the desktops of the institute will work on the institute's project separated from other projects running on SZDG.
3. Finally, if the institute is convinced on the usefulness of the SZTAKI Desktop Grid concept SZTAKI can help them to set-up and maintain their own local DG system based on the SZTAKI Desktop Grid concept.

3   Conclusion

In this paper the structure of SZTAKI Desktop Grid is presented, discussing the possibilities of the support of clusters within a desktop grid. SZTAKI Desktop Grid uses the BOINC infrastructure as a basic building block for connecting PCs to solve large scale distributed programs. It is extended by the support of clusters by installing a modified version of the PC client that converts incoming subtasks into traditional jobs and submits them to the cluster's job manager. Such a desktop grid, as a building block, is then used to build a hierarchy of DGs in an institute or company to provide individual desktop grids to the lower level organisational units but also to provide a larger infrastructure to solve problems on the higher level. The ability to propagate work from one desktop grid to the other (but only in a hierarchy) is a step towards a grid infrastructure that is easy to install and has several users that share resources. This means that in the future DG based grid systems these two features will not exclude each others as they currently do in today's grid systems.

References

1. D.P. Anderson, J. Cobb, E. Korpela, M. Lebofsky, D. Werthimer: SETI@home: An Experiment in Public-Resource Computing. Communications of the ACM, Vol. 45 No. 11, November 2002, pp. 56-61
2. United Devices Cancer Research Project: http://www.grid.org/projects/cancer
3. D. A. Stainforth et al.: Uncertainty in the predictions of the climate response to rising levels of greenhouse gases. Nature, 27 January 2005, vol 433.
4. D. P. Anderson: BOINC: A System for Public-Resource Computing and Storage. 5th IEEE/ACM International Workshop on Grid Computing, November 8, 2004, Pittsburgh, USA. http://boinc.berkeley.edu/grid_paper_04.pdf
5. BOINC Home Page: http://boinc.berkeley.edu
6. G. Fedak, C. Germain, V. Néri and F. Cappello: XtremWeb: A Generic Global Computing System. CCGRID2001 Workshop on Global Computing on Personal Devices, May 2001, IEEE Press.

7. Grid MP, United Devices Inc. http://www.ud.com
8. Platform LSF, Platform Computing. http://www.platform.com
9. A. A. Chien: Architecture of a commercial enterprise desktop Grid: the Entropia system. Grid Computing - Making the Global Infrastructure a Reality. Ed. F. Berman, A. Hey and G. Fox. John-Wiley & Sons, Ltd. Chapter 12. 2003
10. DeskGrid, Info Design Inc. http://www.deskgrid.com
11. I. Foster, C. Kesselman: Globus: A Metacomputing Infrastructure Toolkit. Intl J. Supercomputer Applications, 11(2):115-128, 1997.
12. Myers, D. S., and M. P. Cummings: Necessity is the mother of invention: a simple grid computing system using commodity tools. Journal of Parallel and Distributed Computing, Volume 63, Issue 5, May 2003, pp. 578-589.
13. James Frey, Todd Tannenbaum, Ian Foster, Miron Livny, and Steven Tuecke: Condor-G: A Computation Management Agent for Multi-Institutional Grids. Proceedings of the Tenth IEEE Symposium on High Performance Distributed Computing (HPDC10) San Francisco, California, August 7-9, 2001.
14. D. Thain, T. Tannenbaum and M. Livny: Condor and the Grid. Grid Computing - Making the Global Infrastructure a Reality. Ed. F. Berman, A. Hey and G. Fox. John-Wiley & Sons, Ltd. Chapter 11. 2003
15. SZTAKI Desktop Grid: http://szdg.lpds.sztaki.hu/szdg/

Analyzing Overheads and Scalability Characteristics of OpenMP Applications*

Karl Fürlinger and Michael Gerndt

Technische Universität München
Institut für Informatik
Lehrstuhl für Rechnertechnik und Rechnerorganisation
{Karl.Fuerlinger, Michael.Gerndt}@in.tum.de

Abstract. Analyzing the scalability behavior and the overheads of OpenMP applications is an important step in the development process of scientific software. Unfortunately, few tools are available that allow an exact quantification of OpenMP related overheads and scalability characteristics. We present a methodology in which we define four overhead categories that we can quantify exactly and describe a tool that implements this methodology. We evaluate our tool on the OpenMP version of the NAS parallel benchmarks.

1 Introduction

OpenMP has emerged as the standard for shared-memory parallel programming. While OpenMP allows for a relatively simple and straightforward approach to parallelizing an application, it is usually less simple to ensure efficient execution on large processor counts.

With the widespread adoption of multi-core CPU designs, however, scalability is likely to become increasingly important in the future. The availability of 2-core CPUs effectively doubles the number of processor cores found in commodity SMP systems based for example on the AMD Opteron or Intel Xeon processors. This trend is likely to continue, as the road-maps of all major CPU manufacturers already include mulit-core CPU designs. In essence, performance improvement is increasingly going to be based on parallelism instead of improvements in single-core performance in the future [13].

Analyzing and understanding the scalability behavior of applications is therefore an important step in the development process of scientific software. Inefficiencies that are not significant at low processor counts may play a larger role when more processors are used and may limit the application's scalability. While it is straightforward to study how execution times scale with increasing processor numbers, it is more difficult to identify the possible reasons for imperfect scaling.

---

* This work was partially funded by the Deutsche Forschungsgemeinschaft (DFG) under contract GE1635/1-1.

M. Daydé et al. (Eds.): VECPAR 2006, LNCS 4395, pp. 39–51, 2007.
© Springer-Verlag Berlin Heidelberg 2007

Here we present a methodology and a tool to evaluate the runtime characteristics of OpenMP applications and to analyze the overheads that limit scalability
at the level of individual parallel regions and for the whole program. We apply
our methodology to determine the scalability characteristics of several benchmark applications.

2   Methodology

To analyze the scalability of OpenMP applications we have extended our Open-
MP profiler ompP [6] with overhead classification capability. ompP is a profiler for
OpenMP programs based on the POMP interface [9] that relies on source code
instrumentation by Opari [10]. ompP determines execution counts and times for
all OpenMP constructs (parallel regions, work-sharing regions, critical sections,
locks, . . .) in the target application. Depending on the type of the region different
timing and count categories are reported.

ompP consists of a monitoring library that is linked to an OpenMP application.
Upon termination of the target application, ompP writes a profiling report to a
file. An example output of ompP for a critical section region is shown in Fig. 1.
A table that lists the timing categories reported by ompP for the different region
types is shown in Fig. 2, a particular timing is reported if a "•" is present, the
counts reported by ompP are not shown in Fig. 2.

R00002	CRITICAL		cpp_qsomp1.cpp (156–177)		
TID	execT	execC	enterT	bodyT	exitT
0	1.61	251780	0.87	0.43	0.31
1	2.79	404056	1.54	0.71	0.54
2	2.57	388107	1.38	0.68	0.51
3	2.56	362630	1.39	0.68	0.49
*	9.53	1406573	5.17	2.52	1.84

Fig. 1. Example ompP output for an OpenMP CRITICAL region. R00002 is the region
identifier, cpp_qsomp1.cpp is the source code file and 156–177 denotes the extent of
the construct in the file. Execution times and counts are reported for each thread
individually, and summed over all threads in the last line.

The timing categories reported by ompP shown in Fig. 2 have the following
meaning:

- seqT is the sequential execution time for a construct, i.e., the time between
  forking and joining threads for PARALLEL regions and for combined work-
  sharing parallel regions as seen by the master thread. For a MASTER region it
  similarly represents the execution time of the master thread only (the other
  threads do not execute the MASTER construct).
- execT gives the total execution time for constructs that are executed by
  all threads. The time for thread n is available as execT $[n]$. execT always
  contains bodyT, exitBarT, enterT and exitT.

	seqT	execT	bodyT	exitBarT	enterT	exitT
MASTER	•					
ATOMIC		• (S)				
BARRIER		• (S)				
USER_REGION		•				
LOOP		•		• (I)		
CRITICAL		•	•		• (S)	• (M)
LOCK		•	•		• (S)	• (M)
SECTIONS		•	•	• (I/L)		
SINGLE		•	•	• (L)		
PARALLEL	•	•		• (I)	• (M)	• (M)
PARALLEL_LOOP	•	•		• (I)	• (M)	• (M)
PARALLEL_SECTIONS	•	•	•	• (I/L)	• (M)	• (M)

Fig. 2. The timing categories reported by `ompP` for the different OpenMP constructs and their categorization as overheads by `ompP`'s overhead analysis. (S) corresponds to synchronization overhead, (I) represents overhead due to imbalance, (L) denotes limited parallelism overhead, and (M) signals thread management overhead.

- `bodyT` is the time spent in the "body" of the construct. This time is reported as `singleBodyT` for SINGLE regions and as `sectionT` for SECTIONS regions.
- `exitBarT` is the time spent in "implicit exit barriers". I.e., in worksharing and parallel regions OpenMP assumes an implicit barrier at the end of the construct, unless a `nowait` clause is present. Opari adds an explicit barrier to measure the time in the implicit barrier.
- `enterT` and `exitT` are the times for entering and exiting critical sections and locks. For parallel regions `enterT` is reported as `startupT` and corresponds to the time required to spawn threads. Similarly, `exitT` is reported as `shutdownT` and represents thread teardown overhead.

2.1 Overhead Analysis

From the per-region timing data reported by `ompP` we are able to analyze the overhead for each parallel region separately, and for the program as a whole. We have defined four overhead categories that can be exactly quantified with the profiling data provided by `ompP`:

Synchronization: Overheads that arise because threads need to coordinate their activity. An example is the waiting time to enter a critical section or to acquire a lock.

Imbalance: Overhead due to different amounts of work performed by threads and subsequent idle waiting time, for example in work-sharing regions.

Limited Parallelism: This category represents overhead that results from unparallelized or only partly parallelized regions of code. An example is the idle waiting time threads experience while one thread executes a `single` construct.

Thread Management: Time spent by the runtime system for managing the application's threads. That is, time for creation and destruction of threads in parallel regions and overhead incurred in critical sections and locks for signaling the lock or critical section as available (see below for a more detailed discussion).

The table in Fig. 2 details how timings are attributed to synchronization (S), imbalance (I), limited parallelism (L), thread management overhead (M), and work (i.e., no overhead). This attribution is motivated as follows:

- `exitBarT` in work-sharing or parallel regions is considered imbalance overhead, except for `single` regions, where the reason for the time spent in the exit barrier is assumed to be limited parallelism. The time in the exit barrier of a `sections` construct is either imbalance or limited parallelism, depending on the number of `section` constructs inside the `sections` construct, compared to the number of threads. If there are fewer sections than threads available, the waiting time is considered limited parallelism overhead and load imbalance otherwise.
- The time spent in `barrier` and `atomic` constructs is treated as synchronization overhead.
- The time spent waiting to enter a critical section or to acquire a lock is considered synchronization overhead. Opari also adds instrumentation to measure the time spent for leaving a critical section and releasing a lock. These times reflect the overhead of the OpenMP runtime system to signal the lock or critical section being available to waiting threads. Hence, these overheads do not relate to the synchronization requirement of the threads but rather represent an overhead related to the implementation of the runtime system. Consequently, the resulting waiting times are treated as thread management overhead.
- The same considerations as above hold true for `startupT` and `shutdownT` reported for parallel regions. This is the overhead for thread creation and destruction, which is usually insignificant, except in cases where a team of threads is created and destroyed repeatedly (if, for example, a small `parallel` region is placed inside a loop). Again, this overhead is captured by in the thread management category.

The overheads for each category are accumulated for each parallel region in the program separately. That is, if a parallel region P contains a critical section C, C's enter time will appear as synchronization overhead in P's overhead statistics. Note that, while ompP reports inclusive timing data in its profiling reports, the timing categories related to overheads are never nested and never overlap. Hence, a summation of each sub-region's individual overhead time gives the correct total overhead for each parallel region.

An example of ompP's overhead analysis report is shown in Fig. 3 (the columns corresponding to the thread management overhead category are omitted due to space limitations). The first part of the report, denoted by ⓐ, gives general information about the program run. It lists the total number of parallel regions (OpenMP parallel constructs and combined parallel work-sharing constructs), the total wallclock runtime and the parallel coverage (or parallel fraction). The parallel coverage is defined as the fraction of wallclock execution time spent inside parallel regions. This parameter is useful for estimating the optimal execution time according to Amdahl's law on p processors as

$$T_p = \frac{T_1 \alpha_1}{p} + T_1(1 - \alpha_1),$$

where T_i is the execution time on i processors and α_1 is the parallel coverage of an execution with one thread.

Section ⓑ lists all parallel regions of the program with their region identifiers and location in the source code files, sorted by their wallclock execution time.

Part ⓒ shows the parallel regions in the same order as in part ⓑ and details the identified overheads for each category as well as the total overhead (Ovhds column). The total runtime is given here accumulated over all threads (i.e., Total = wallclock runtime × number of threads) and the percentages for the overhead times shown in parenthesis refer to this runtime.

The final part in the overhead analysis report (ⓓ) lists the same overhead times but the percentages are computed according to the total runtime of the program. The regions are also sorted with respect to their overhead in this section. Hence, the first lines in section ⓓ show the regions that cause the most significant overall overhead as well as the type of the overhead. In the example shown in Fig. 3, the most severe inefficiency is imbalance overhead in region R00035 (a parallel loop in y_solve.f, lines 27-292) with a severity of 3.44%.

2.2   Scalability Analysis

The overhead analysis report of ompP gives valuable insight into the behavior of the application. From analyzing overhead reports for increasing processor counts, the scalability behavior of individual parallel regions and the whole program can be inferred. We have implemented the scalability analysis as a set of scripts that take several ompP profiling reports as input and generate data to visualize the program's scalability as presented in Sect. 3.

The graphs show the change of the distribution of overall time for increasing processor counts. That is, the total execution time as well as each overhead category is summed over all threads and the resulting accumulated times are plotted for increasing processor numbers.

3   Evaluation

To evaluate the usability of the scalability analysis as outlined in this paper, we test the approach on the OpenMP version of the NAS parallel benchmarks [8] (version 3.2, class "C"). Most programs in the NAS benchmark suite are derived from CFD applications, it consists of five kernels (EP, MG, CG, FT, IS) and three

```
---------------------------------------------------------------------
-----      ompP Overhead Analysis Report      ----------------------
---------------------------------------------------------------------
```

(a) Total runtime (wallclock)   : 736.82 sec [4 threads]
    Number of parallel regions : 14
    Parallel coverage   : 736.70 sec (99.98%)

(b) Parallel regions sorted by wallclock time:
```
              Type                      Location       Wallclock (%)
    R00018  parall                  rhs.f (16-430)     312.48 (42.41)
    R00037  ploop               z_solve.f (31-326)     140.00 (19.00)
    R00035  ploop               y_solve.f (27-292)      88.68 (12.04)
    R00033  ploop               x_solve.f (27-296)      77.03 (10.45)
    ...
           *        *                          *       736.70 (99.98)
```

(c) Overheads wrt. each individual parallel region:
```
            Total      Ovhds (%) =  Synch  (%)  + Imbal  (%) + Limpar (%)
    R00018 1249.91   0.44 ( 0.04)  0.00 (0.00)   0.35 ( 0.03)  0.00 (0.00)
    R00037  560.00 100.81 (18.00)  0.00 (0.00) 100.72 (17.99)  0.00 (0.00)
    R00035  354.73 101.33 (28.56)  0.00 (0.00) 101.24 (28.54)  0.00 (0.00)
    R00033  308.12  94.62 (30.71)  0.00 (0.00)  94.53 (30.68)  0.00 (0.00)
    ...
```

(d) Overheads wrt. whole program:
```
            Total      Ovhds (%) =  Synch  (%)  + Imbal  (%) + Limpar (%)
    R00035  354.73 101.33 ( 3.44)  0.00 (0.00) 101.24 ( 3.44)  0.00 (0.00)
    R00037  560.00 100.81 ( 3.42)  0.00 (0.00) 100.72 ( 3.42)  0.00 (0.00)
    R00033  308.12  94.62 ( 3.21)  0.00 (0.00)  94.53 ( 3.21)  0.00 (0.00)
    ...
          * 2946.79 308.52 (10.47)  0.00 (0.00) 307.78 (10.44)  0.00 (0.00)
```

Fig. 3. Example overhead analysis report generated by ompP, the columns related to the thread management category (Mgmt) are omitted due to space limitations

	BT	CG	EP	FT	IS	LU	MG	SP
MASTER	4					13	2	4
ATOMIC	2		1			2		1
BARRIER					1	3		
LOOP	25	13	1		1	30	5	25
CRITICAL								1
LOCK								
SECTIONS								
SINGLE						6		
PARALLEL	6	9	1		2	8	2	6
PARALLEL_LOOP	4	5	1	8	2	1	8	8
PARALLEL_SECTIONS								

Fig. 4. The OpenMP constructs found in the NAS parallel benchmarks version 3.2

simulated CFD applications (LU, BT, SP). Fig. 4 shows the characteristics of the benchmark applications with respect to the OpenMP constructs used for parallelization.

Fig. 6 presents the result of the scalability analysis performed on a 32 CPU SGI Altix machine, based on Itanium-2 processors with 1.6 GHz and 6 MB L3 cache, used in batch mode. The number of threads was increased from 2 to 32. The graphs in Fig. 6 show the accumulated runtime over all threads. Hence, a horizontal line corresponds to a perfectly scaling code with ideal speedup. For convenience, a more familiar *speedup graph* (with respect to the 2-processor run) computed from the same data is shown in Fig. 5.

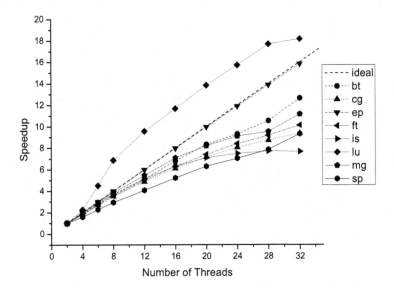

Fig. 5. Speedup achieved by the NAS benchmark programs relative to the 2-processor execution

In Fig. 6, the total runtime is divided into work and the four overhead categories, and the following conclusions can be derived:

- Overall, the most significant overhead visible in the NAS benchmarks is imbalance, only two applications show significant synchronization overhead, namely IS and LU.
- Some applications show a surprisingly large amount of overhead, as much as 20 percent of the total accumulated runtime is wasted due to imbalance overhead in SP.
- Limited parallelism does not play a significant role in the NAS benchmarks. While not visibly discernable in the graphs in Fig. 6 at all, the overhead is actually present for some parallel regions albeit with very low severity.
- Thread management overhead is present in the applications, mainly in CG, IS and MG, although it is generally of little importance.

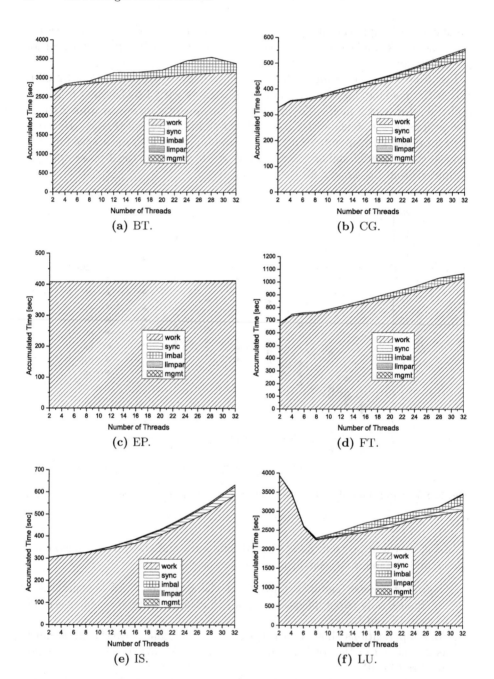

Fig. 6. Scaling of total runtime and the separation into work and overhead categories for the NAS OpenMP parallel benchmarks (BT, CG, EP, FT, IS, LU, MG, and SP)

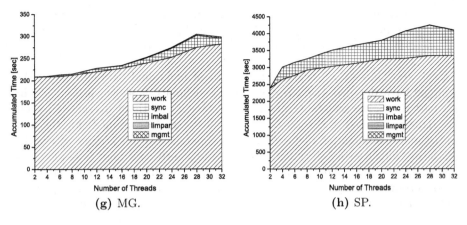

Fig. 6. (*continued*)

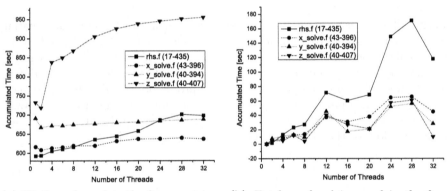

(a) Work performed in the four most important parallel regions of BT.

(b) Total overhead incurred in the four most important parallel regions of BT.

Fig. 7. Detailed scalability analysis at the level of individual parallel regions of the BT application. The four most important parallel regions are analyzed with respect to the work performed and the overheads incurred for each region individually.

– EP scales perfectly, it has almost no noticeable overhead.
– The "work" category increases for most applications and does not stay constant, even though the actual amount of work performed is independent of the number of threads used. This can be explained with overhead categories that are currently not accounted for by ompP, for example increasing memory access times at larger processor configurations. Additional factors that influence the work category are the increasing overall cache size (which can lead to super-linear speedups) and an increased rate of cache conflicts at larger systems. This issue is discussed further in Sect. 5.
– For some applications, the summed runtime increases linearly, for others it increases faster than linearly (e.g., IS scales very poorly).

– For LU, the performance increases super-linearly at first, then at six processors the performance starts to deteriorate. The reason for the super-linear speedup is most likely the increased overall cache size.

ompP also allows for a scalability analysis of individual parallel regions. An example for a detailed analysis of the BT benchmark is shown in Fig. 7. The left part shows the scalability of the work category (without overheads), while the right part shows the total overhead (all categories summed) for the four most time consuming parallel regions. It is apparent that for powers of two, the overhead (which is mainly imbalance) is significantly less than it is for other configurations. A more detailed analysis of ompP's profiling report and the application's source code reveals the cause: most overhead is incurred in loops with an iteration count of 160, which is evenly divisible by 2, 4, 8, 16, and 32.

4   Related Work

Mark Bull describes a hierarchical classification scheme for temporal and spatial overheads in parallel programs in [3]. The scheme is general (not dependant on a particular programming model) and strives to classify overheads in categories that are complete, meaningful, and orthogonal. Overhead is defined as the difference between the observed performance on p processors and the "best possible" performance on p processors. Since the best possible performance is unknown (it can at best be estimated by simulation), $T_p^{ideal} = \frac{T_1}{p}$ is often used as an approximation for the ideal performance on p processors. Thus

$$T_p = T_p^{ideal} + \sum_i O_p^i \qquad (1)$$

where O_p^i represent the overhead in category i. This is similar to our scheme with the difference that ompP does not report overheads with respect to the wallclock execution time but aggregated over all threads.

Bull's hierarchical classification scheme has four categories at the top level:

– Information Movement
– Critical Path
– Control of Parallelism
– Additional Computation

While this scheme allows for a well defined conceptual breakdown of where an application spends its time, ompP's classification scheme is based on what can be actually automatically be measured. For example, it is not possible to account for additional computation automatically.

Bane and Riley developed a tool called Ovaltine [1,2] that performs overhead analysis for OpenMP code. The overhead scheme of Ovaltine is based on the classification scheme by Bull. Ovaltine performs code instrumentation based on the Polaris compiler. Not all overheads in Ovaltine's scheme can be computed automatically. For example the cost of acquiring a lock has to be determined empirically. Practically, only the "load imbalance" and "unparallelized" overheads are computed automatically in Ovaltine as described in [1].

Scal-Tool [11] is a tool for quantifying the scalability bottlenecks of shared memory codes. The covered bottlenecks include insufficient cache, load imbalance and synchronization. Scal-Tool is based on an empirical model using cycles-per-instruction (CPI) breakdown equations. From a number of measurement (fixed data-set, varying number of processors and varying the size of the dataset on a single processor, the parameters in the CPI equations can be estimated. The result of the analysis is a set of graphs that augment the observed scalability graph of the application with estimated scalability, if one or more of the scalability bottlenecks (cache, load imbalance, synchronization) are removed.

Scalea [14] is a tool for performance analysis of Fortran OpenMP, MPI and HPF codes. Scalea computes metrics based on a classification scheme derived from Bull's work. Scalea is able to perform overhead-to-region analysis and region-to-overhead analysis. I.e., show a particular overhead category for all regions or show all overhead categories for a specific region.

Fredrickson et al. [5] have evaluated the performance characteristics of the class B of the NAS OpenMP benchmarks version 3.0 on a 72 processor Sun Fire 15K. The speedup of the NAS benchmarks is determined for up to 70 threads. In their evaluation, CG shows super-linear speedup, LU shows perfect scalability, FT scales very poorly and BT SP and MG show good performance (EP and IS are not evaluated). In contrast, in our study CG shows relatively poor speedup while LU shows super-linear speedup. Our results for FT, BT, SP, and MG are more or less in-line with theirs.

Fredrickson et al. also evaluate "OpenMP overhead" by counting the number of parallel regions and multiplying this number with an empirically determined overhead for creating a parallel region derived from an execution of the EPCC micro-benchmarks [4]. The OpenMP overhead is low for most programs, ranging from less than one percent to five percent of the total execution time, for CG the estimated overhead is 12%. Compared to our approach this methodology of estimating the OpenMP overhead is less flexible and accurate, as for example it does not account for load-imbalance situations and requires an empirical study to determine the "cost of a parallel region". Note that in ompP all OpenMP-related overheads are accounted for, i.e., the work category does not contain any OpenMP related overhead.

Finally, vendor-specific tools such as Intel Thread Profiler [7] and Sun Studio [12] often implement overhead classification schemes similar to ompP. However, these tools are limited to a particular platform, while ompP is compiler and platform-independent and can thus be used for cross-platform overhead comparisons, for example.

5 Summary and Future Work

We presented a methodology for overheads- and scalability analysis of OpenMP applications that we integrated in our OpenMP profiler ompP. We have defined four overhead categories (synchronization, load imbalance, limited parallelism and thread management) that are well defined and can explicitly be measured. The overheads are reported per parallel region and for the whole program. ompP allows for an exact quantification of all OpenMP related overheads.

From the overhead reports for increasing processor counts we can see how programs scale and how the overheads increase in importance. We have tested the approach on the NAS parallel benchmarks and were ably to identify some key scalability characteristics.

Future work remains to be done to cover further overhead categories. What is labeled Work in Fig. 5 actually contains overheads that are currently unaccounted for. Most notably it would be important to account for overheads related to memory access. Issues like accessing a remote processors memory on a ccNUMA architecture like the SGI Altix and coherence cache misses impact performance negatively while increased overall caches size helps performance and can actually lead to negative overhead. For the quantification of these factors we plan to include support for hardware performance counters in ompP.

References

1. Michael K. Bane and Graham Riley. Automatic overheads profiler for OpenMP codes. In *Proceedings of the Second Workshop on OpenMP (EWOMP 2000)*, September 2000.
2. Michael K. Bane and Graham Riley. Extended overhead analysis for OpenMP (research note). In *Proceedings of the 8th International Euro-Par Conference on Parallel Processing (Euro-Par '02)*, pages 162–166. Springer-Verlag, 2002.
3. J. Mark Bull. A hierarchical classification of overheads in parallel programs. In *Proceedings of the First IFIP TC10 International Workshop on Software Engineering for Parallel and Distributed Systems*, pages 208–219, London, UK, 1996. Chapman & Hall, Ltd.
4. J. Mark Bull and Darragh O'Neill. A microbenchmark suite for OpenMP 2.0. In *Proceedings of the Third Workshop on OpenMP (EWOMP'01)*, September 2001.
5. Nathan R. Fredrickson, Ahmad Afsahi, and Ying Qian. Performance characteristics of OpenMP constructs, and application benchmarks on a large symmetric multiprocessor. In *Proceedings of the 17th ACM International Conference on Supercomputing (ICS 2003)*, pages 140–149. ACM Press, 2003.
6. Karl Fürlinger and Michael Gerndt. ompP: A profiling tool for OpenMP. In *Proceedings of the First International Workshop on OpenMP (IWOMP 2005)*, May 2005. Accepted for publication.
7. Intel Thread Profiler http://www.intel.com/software/products/threading/tp/.
8. H. Jin, M. Frumkin, and J. Yan. The OpenMP implementation of NAS parallel benchmarks and its performance. Technical Report NAS-99-011, 1999.

9. Bernd Mohr, Allen D. Malony, Hans-Christian Hoppe, Frank Schlimbach, Grant Haab, Jay Hoeflinger, and Sanjiv Shah. A performance monitoring interface for OpenMP. In *Proceedings of the Fourth Workshop on OpenMP (EWOMP 2002)*, September 2002.

10. Bernd Mohr, Allen D. Malony, Sameer S. Shende, and Felix Wolf. Towards a performance tool interface for OpenMP: An approach based on directive rewriting. In *Proceedings of the Third Workshop on OpenMP (EWOMP'01)*, September 2001.

11. Yan Solihin, Vinh Lam, and Josep Torrellas. Scal-Tool: Pinpointing and quantifying scalability bottlenecks in DSM multiprocessors. In *Proceedings of the 1999 Conference on Supercomputing (SC 1999)*, Portland, Oregon, USA, November 1999.

12. Sun Studio `http://developers.sun.com/prodtech/cc/hptc_index.html`.

13. Herb Sutter. The free lunch is over: A fundamental turn toward concurrency in software. *Dr. Dobb's Journal*, 30(3), March 2005.

14. Hong-Linh Truong and Thomas Fahringer. SCALEA: A performance analysis tool for parallel programs. *Concurrency and Computation: Practice and Experience*, (15):1001–1025, 2003.

Parallel Fuzzy c-Means Cluster Analysis

Marta V. Modenesi, Myrian C.A. Costa, Alexandre G. Evsukoff,
and Nelson F.F. Ebecken

COPPE/Federal University of Rio de Janeiro,
P.O. Box 68506, 21945-970 Rio de Janeiro RJ, Brazil
Tel.: (+55) 21 25627388; Fax: (+55) 21 25627392
modenesi@lamce.ufrj.br, myrian@nacad.ufrj.br,
alexandre.evsukoff@coc.ufrj.br, nelson@ntt.ufrj.br

Abstract. This work presents an implementation of a parallel Fuzzy c-means cluster analysis tool, which implements both aspects of cluster investigation: the calculation of clusters' centers with the degrees of membership of records to clusters, and the determination of the optimal number of clusters for the data, by using the PBM validity index to evaluate the quality of the partition.

The work's main contributions are the implementation of the entire cluster's analysis process, which is a new approach in literature, integrating to clusters calculation the finding of the best natural pattern present in data, and also, the parallel processing implementation of this tool, which enables this approach to be used with vary large volumes of data, a increasing need for data analysis in nowadays industries and business databases, making the cluster analysis a feasible tool to support specialist's decision in all fields of knowledge.

The results presented in the paper show that this approach is scalable and brings processing time reduction as an benefit that parallel processing can bring to the matter of cluster analysis.

Topics of Interest: Unsupervised Classification, Fuzzy c-Means, Cluster and Grid Computing.

1 Introduction

The huge amount of data generated by data intensive industries such as Telecommunications, Insurance, Oil & Gas exploration, among others, has pushed Data Mining algorithms through parallel implementations [1, 2]. One requirement of data mining is efficiency and scalability of mining algorithms. Therefore, parallelism can be used to process long running tasks in a timely manner.

There are several different parallel data mining implementations being used or experimented, both in distributed and shared memory hardware [3], as so as in grid environments [4]. All the main data mining algorithms have been investigated, such as decision tree induction [5], fuzzy rule-based classifiers [6, 7], neural networks [8, 9], association rules' mining [10, 11] and clustering [12, 13].

Data clustering is being used in several data intensive applications, including image classification, document retrieval and customer segmentation (among others). Clustering algorithms generally follows hierarchical or partitional approaches [14]. For the partitional approach the k-means and its variants, such as the fuzzy c-means algorithm [13], are the most popular algorithms.

M. Daydé et al. (Eds.): VECPAR 2006, LNCS 4395, pp. 52–65, 2007.
© Springer-Verlag Berlin Heidelberg 2007

Partitional clustering algorithms require a large number of computations of distance or similarity measures among data records and clusters centers, which can be very time consuming for very large data bases. Moreover, partitional clustering algorithms generally require the number of clusters as an input parameter. However, the number of clusters usually is not known *a priori*, so that the algorithm must be executed many times, each for a different number of clusters and uses a validation index to define the optimal number of clusters. The determination of the clusters' numbers and centers present on the data is generally referred to as cluster analysis.

Many cluster validity criteria have been proposed in the literature in the last years [16, 17 and 18]. Validity indexes aim to answer two important questions in cluster analysis: (*i*) how many clusters are actually present in the data and (*ii*) how good the partition is. The main idea, present in most of the validity indexes, is based on the geometric structure of the partition, so that samples within the same cluster should be compact and different clusters should be separate. When the cluster analysis assigns fuzzy membership functions to the clusters, "fuzziness" must be taken in account in a way that the less fuzzy the partition is the better.

Usually, parallel implementations of clustering algorithms [12, 13] only consider strategies to distribute the iterative process to find the clusters' centers. In this work, the entire cluster analysis is investigated, including the determination of the clusters' centers and the optimal number of clusters.

The paper is organized as follows: next section the fuzzy c-means algorithm is sketched. The cluster validation index, known as the PBM index, is presented in section three. The parallel implementation of the cluster analysis is presented in section four. The results obtained with this approach considering scalability and speed-up are presented in section five. Final conclusions and future works are discussed in section six.

2   The Fuzzy c-Means Algorithm

The Fuzzy c-means (FCM) algorithm proposed by Bezdek [15] is the well known fuzzy version of the classical ISODATA clustering algorithm.

Consider the data set $T = \{(\mathbf{x}(t)),\ t = 1..N\}$, where each sample contains the variable vector $\mathbf{x}(t) \in R^p$. The algorithm aims to find a fuzzy partition of the domain into a set of K clusters $\{C_1 ... C_K\}$, where each cluster C_i is represented by its center's coordinates' vector $\mathbf{w}_i \in R^p$.

In the fuzzy cluster analysis, each sample in the training set can be assigned to more than one cluster, according to a value $u_i(t) = \mu_{C_i}(\mathbf{x}(t))$, that defines the membership of the sample $\mathbf{x}(t)$ to the cluster C_i.

The FCM algorithm computes the centers' coordinates by minimizing the objective function J defined as:

$$J(m, \mathbf{W}) = \sum_{t=1..N} \sum_{i=1..K} u_i(t)^m d(\mathbf{x}(t), \mathbf{w}_i)^2 \tag{1}$$

where $m > 1$. The m parameter, generally referred as the "fuzziness parameter", is a parameter to adjust the effect of membership values and $d(\mathbf{x}(t), \mathbf{w}_i)$ is a distance measure, generally the Euclidean distance, from the sample $\mathbf{x}(t)$ to the cluster's center $\mathbf{w}_i$.

The membership of all samples to all clusters defines a *partition matrix* as:

$$U = \begin{bmatrix} u_1(1) & \cdots & u_K(1) \\ \vdots & \ddots & \vdots \\ u_1(N) & \cdots & u_K(N) \end{bmatrix}. \tag{2}$$

The partition matrix is computed by the algorithm so that:

$$\forall \mathbf{x}(t) \in T, \ \sum_{i=1..K} u_i(t) = 1.. \tag{3}$$

The FCM algorithm computes interactively the clusters centers coordinates from a previous estimate of the partition matrix as:

$$\mathbf{w}_i = \frac{\sum_{t=1..N} u_i(t)^m . \mathbf{x}(t)}{\sum_{t=1..N} u_i(t)^m}. \tag{4}$$

The partition matrix is updated as:

$$u_i(t) = \frac{1}{\sum_{j=1..K} \left(\dfrac{d(\mathbf{x}(t), \mathbf{w}_i)}{d(\mathbf{x}(t), \mathbf{w}_j)} \right)^{\frac{2}{(m-1)}}}. \tag{5}$$

The FCM algorithm is described as follows:

0. Set $m > 1$, $K \geq 2$ and initialize the cluster centers' coordinates randomly, initialize the partition matrix as (5).
1. For all clusters $(2 \leq i \leq K)$, update clusters' centers coordinates as (4).
2. For all samples $(1 \leq t \leq N)$ and all clusters $(2 \leq i \leq K)$, update the partition matrix as (5).
3. Stop when the norm of the overall difference in the partition matrix between the current and the previous iteration is smaller than a given threshold ε; otherwise go to step 1.

In fuzzy cluster analysis the FCM algorithm computes clusters centers' coordinates and the partition matrix from the specification of the number of clusters K that must be given in advance. In practice, the FCM algorithm is executed to various values of K, and the results are evaluated by a cluster validity function, as described next.

3   Cluster Validity Index

In this work, the PBM index [18] is used to evaluate the number of clusters in the data set. The PBM index is defined as a product of three factors, of which the maximization ensures that the partition has a small number of compact clusters with large separation between at least two of them. Mathematically the PBM index is defined as follows:

$$PBM(K) = \left(\frac{1}{K} \cdot \frac{E_1}{E_K} \cdot D_K \right)^2 \qquad (6)$$

where K is the number of clusters.

The factor E_1 is the sum of the distances of each sample to the geometric center of all samples $\mathbf{w}_0$. This factor does not depend on the number of clusters and is computed as:

$$E_1 = \sum_{t=1..N} d(\mathbf{x}(t), \mathbf{w}_0). \qquad (7)$$

The factor E_K is the sum of within cluster distances of K clusters, weighted by the corresponding membership value:

$$E_K = \sum_{t=1..N} \sum_{i=1..K} u_i(t) d(\mathbf{x}(t), \mathbf{w}_i)^2 \qquad (8)$$

and D_K that represents the maximum separation of each pair of clusters:

$$D_K = \max_{i,j=1..K} \left(d(\mathbf{w}_i, \mathbf{w}_j) \right). \qquad (9)$$

The greatest PBM index means the best clustering fuzzy partition. As other indexes, the PBM index is an optimizing index, so that it can be used to search the best number of clusters within a range of number of clusters. The PBM procedure can be described as follows:

0.   Select the maximum number of clusters M ;
1.   Compute the *PBM* factor E_1 (7)
2.   For $K = 2$ to $K = M$, do:
    2.1.   Run the FCM algorithm;
    2.2.   Compute the *PBM* factors E_K (8) and D_K (9);
    2.3.   Compute the *PBM*(K) index (6)

3.   Select the best number of clusters K^* that maximizes the PBM index:

$$K^* = \arg\max(PBM(K)) \qquad (10)$$

The PBM index has achieved a good performance in several data when compared with the Xie-Beni index [16]. This index is thus used as a validity index of the methodology presented in this work.

4 Parallel Cluster Analysis Implementation

The aim of the FCM cluster analysis algorithm is to determine the best partition for the data being analyzed, by investigating different partitions, represented by the partitions' centers. Hence, the cluster analysis must integrate the FCM algorithm and the PBM procedure as described above.

The cluster analysis is an iterative process where the FCM algorithm is computed for a range of number of clusters and the PBM index is computed for every partition generated by the FCM algorithm. When all partitions have been computed, the partition corresponding to the maximum PBM index is chosen as the best partition for the data.

The most complex computation in the FCM algorithm is the distance computation from each sample $\mathbf{x}(t)$ to all clusters' centers $\mathbf{w}_i$, $i = 1..K$. This computation is performed every interaction, for all records in the dataset. Aggregates of the distances are used to compute the new centers' estimate (4), the fuzzy partition's matrix (5) and the PBM factors E_1 (7) and E_K (8). These are the steps of the FCM cluster analysis that should be parallelized.

The parallel FCM cluster analysis procedure is sketched in Fig 1 and described by the following sequence:

Step 1. (Master processor): Splits the data set equally among the available processors so that each one receives N/p records, where N is the number of records and p is the number of processes

Step 2. (All processors): Compute the geometrical center of its local data and communicate this center to all processors, so that every processor can compute the geometrical center of the entire database. Compute the PBM factor E_1 (7) on local data and send it to root.

Step 3. (Master processor): Sets initial centers and broadcasts them, so that all processors have the same clusters' centers values at the beginning of the FCM looping.

Step 4. (All processors): Until convergence is achieved compute the distances from each record in the local dataset to all clusters' centers; update the partition matrix as (5), calculate new clusters' centers as (4).

Step 5. (All processors): Compute the PBM factor E_K (8) on its local data and send it to root.

Step 6. (Master Processor): Integrates the PBM index as (6) and stores it. If the range of number of clusters is covered, stops, otherwise returns to *Step3*.

The procedure described above is computed for each number of clusters in the cluster analysis, so that the procedure is repeated as many times as the desired range of numbers of clusters, so that the PBM index, as a function of the number of centers, is computed. The best partition is the one corresponding to the largest value of the PBM index.

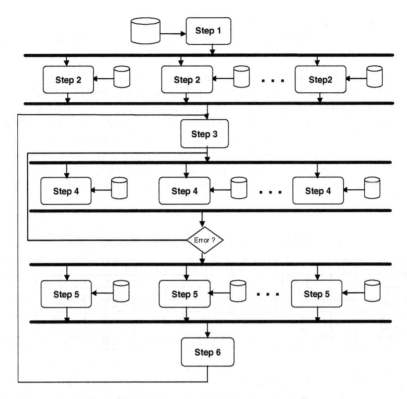

Fig. 1. The parallel FCM cluster analysis

5 Results and Discussion

5.1 Environment

Two machines were used for execution and performance analysis of this work: the PC Cluster Mercury and the SGI Altix 350, both from the High Performance Computing Center (NACAD) of COPPE/UFRJ. The PC cluster has 16 dual Pentium III, 1 GHz, processor nodes interconnected via Fast Ethernet and Gigabit networks and 8GB of total memory. The SGI Altix 350 has 14 Intel Itanium2 cpus with 28 Gbytes of RAM (shared - NUMA) and 360 Gbytes of disk storage.

In both machines the execution is controlled by PBS (Portable Batch System) job scheduler avoiding nodes sharing during execution and Linux Red Hat runs on processing and administrative nodes. The application was developed using C programming language and Message Passing Interface (MPI).

5.2 The Cluster Mercury Results and Speed-Up Analysis

The speed up evaluation of the FCM cluster analysis algorithm was made in two test steps. In the first one, the objective was to observe the algorithm behavior increasing the number of records. In the second test the objective was to observe the behavior of the algorithm when increasing the number of variables and of the range of partitions.

Test 1. Datasets of different line sizes were used. The datasets had 1.000, 12.500, 50.000, 65.000 and 100.000 records (lines) and size of 38Kb, 500kb, 1.88MB, 2.5MB and 3.76MB. A fixed number of variables and a fixed range of clusters were used. The datasets had 18 variables (columns). The evaluation was performed considering 9 partitions calculated from $K = 2$ to $K = 10$ clusters' centers, used for the PBM index calculation. The number of iterations of the FCM algorithm is limited to 500, such that the complexity of the dataset does not affect the result. Moreover, the same initial cluster centers were used in all evaluations. The speed-up results are shown in Table 1.

Table 1. Speed-up results for Cluster Mercury

Datasets	Number of Processors							
	1	2	3	4	5	6	7	8
38Kb	1	1.84	1.97	2.24	2.01	1.94	1.95	1.91
500Kb	1	1.96	2.80	3.67	4.33	4.97	5.67	6.12
1.88MB	1	1.96	2.89	3.80	4.62	5.48	6.32	7.11
2.50MB	1	1.96	2.90	3.82	4.68	5.55	6.41	7.21
3.76MB	1	1.96	2.91	3.83	4.72	5.60	6.47	7.30

The algorithm's speed up when processing the smaller dataset was clearly worse than the others. Its highest speed up value was 2.24 when using 4 processors. The algorithm showed higher speed up values for a larger number of processors when processing datasets with larger number of records. When processing a small number of records, communications are too costly compared to the advantage of parallelizing the calculations of distances to clusters, and the benefits of parallelism do not happen.

As showed in Table 1, using 8 processors, speed up values of more than 7.0 were achieved for databases with more than 1.88MB, which is a very good gain for the overall parallel process. The speed up of the parallel program against the correspondent sequential one has an efficiency factor that gets as much closer to 1 as the database records increase, when considering a growing number of processors.

Test 2. To investigate the effect of the number of variables in the parallel process, two datasets were used: one of 50.000 lines and 10 variables of 1.07MB and the other with 50.000 lines and 40 variables (columns) of 4.12MB. The two datasets were processed using different ranges of clusters. The first computation of the algorithm was made with one partition of 2 clusters. Ranges of 2 to 4 clusters, of 2 to 8 clusters, 2 to 16 clusters and of 2 to 32 clusters had been used in each other algorithm computation. All processing were executed for 1, 2, 4, 6, and 8 processors. Results are presented in Table 2 and Table 3.

The second test showed that the number of variables have also an impact on the overall performance of the algorithm. The processing of datasets with small number of variables gets smaller speed up values. The processing of datasets with larger number of variables results in better speed up values when using a larger number of processors.

The same happens when referring to the range of clusters to form the partitions. The bigger is the range of number of clusters' centers that have to be investigated, the greater is the number of partitions that will have to be calculated. Also, as larger are the clusters' numbers, more computation is involved in distance calculations from records to clusters. As the calculations increase in the processing, the parallel algorithm benefits show up clearly. To use a higher number of processors is as much interesting, in time savings and processing acceleration, as the range of number of clusters increases.

Table 2. Speed up for dataset of 50.000 lines x 10 variables

Clusters'	Number of Processors				
Ranges	1	2	4	6	8
2 clusters	1	1.79	2.82	3.44	3.79
from 2 to 4	1	1.91	3.47	4.78	5.83
from 2 to 8	1	1.94	3.71	5.26	6.66
from 2 to 16	1	1.95	3.78	5.46	7.02
from 2 to 32	1	1.96	3.81	5.51	7.19

Table 3. Speed up for dataset of 50.000 lines x 40 variables

Clusters'	Number of Processors				
Ranges	1	2	4	6	8
2 clusters	1	1.78	2.80	3.45	3.93
from 2 to 4	1	1.92	3.55	4.94	6.22
from 2 to 8	1	1.96	3.81	5.54	7.20
from 2 to 16	1	1.98	3.89	5.73	7.53
from 2 to 32	1	1.98	3.92	5.81	7.67

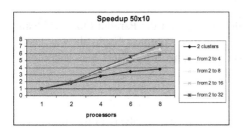 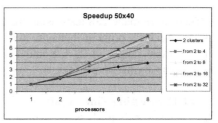

Fig. 2. Speed-up graphs for different files sizes

The tests of the parallel Fuzzy c-Means cluster analysis tool on the Mercury Cluster hardware has shown that it is scalable for parallel cluster analysis processing on these machines for databases of bigger sizes.

5.3 The Altix Machine Tests

The Tests Description

There were used twelve different files to proceed with the programs test in the Altix machine. The files' dimensions schemas are presented in Table 4 and the files sizes are presented in Table 5.

Table 4. Files dimensions

Records	Variables			
50.000	50	100	150	200
100.000	50	100	150	200
200.000	50	100	150	200

Table 5. Files sizes

File Id	Records (in thousands) x Variables	Size(in MB)
1	50 x 50	5.138
2	50 x 100	10.227
3	50 x 150	15.317
4	50 x 200	20.406
5	100 x 50	10.347
6	100 x 100	20.552
7	100 x 150	30.730
8	100 x 200	40.909
9	200 x 50	20.747
10	200 x 100	41.104
11	200 x 150	61.469
12	200 x 200	81.817

There were made 288 tests using this files base to test the Parallel Fuzzy c-Means Cluster Analysis program behavior with greater data volumes. Each file was tested with clusters' ranges of 2, 4, 8 and 16, and each range of clusters was tested with processors varying from 1 to 6.

Parallel Processing Time Analysis

One of this work's goals is to create a useful tool for data analysis in the oil and gas field, where huge volumes of data need to be investigated in order to find out oil reservoir information. So, it is necessary to investigate the programs behavior when number of records and variables increase.

It was observed that when the number of records in the database grows, the processing time grows as well, proportionally to the increase of records (Fig.3).

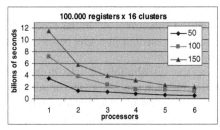
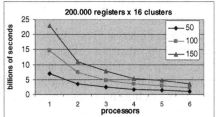

Fig. 3. Increasing of time when increasing number of variables

The same behavior occurred when increasing the number of variables: processing time grows at the same rate of the increasing of the variables as can be seen in Fig.4. Nevertheless, the parallel Fuzzy c-Means Cluster Analysis program has the same behavior for all problem sizes being investigated. The parallel approach decreases the time processing for all files' sizes, but the time savings is more meaningful for larges databases because it is proportional to the problem size.

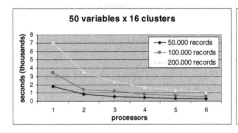
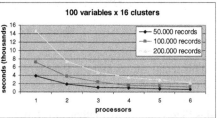

Fig. 4. Increasing of time when increasing number of records

Speed Up and Efficiency Analysis
In the Altix machine tests the processing for the smaller files size for only 2 clusters did not presented a good speedup curve. This can be understood considering that the computational effort for processing only one partition of two clusters is significantly smaller than the communications' cost involved in the overall process. Measurements show it clearly as can be visualized in Figure 5 bellow.

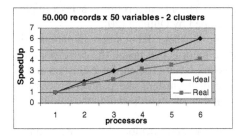
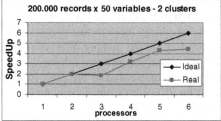

Fig. 5. Speedup of processing one partition of two clusters using files of 50.000 records and 50 variables and 200.000 records and 50 variables

Processing with bigger files improved the speed-up curve because the computational time tends to be greater than communications time involved in the process for bigger clusters interval. Tests in the Altix machine presented in a few cases a super linear speed-up value as an example showed in Figure 6.

This behavior could be explained because the distribution of one bigger file through several processors produces smaller files for each processor. The architecture of the Altix machine provides a performance improvement of an application with this feature. If the file is so small that can be stored in the memory cache, the computational time decreases and the speed-up becomes super linear.

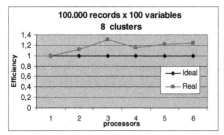 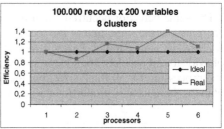

Fig. 6. Super linear efficiency values

This result shows that the Parallel Fuzzy c-Means Cluster Analysis tool scales well and can be used successfully processing bigger datasets.

Files Size versus Parallelization Benefits

In order to have an indication of whether it is efficient to use parallel processing or not, all the 288 tests results in Altix hardware were used as records in a dataset for a decision tree induction classification algorithm.

The input variables for the decision tree induction algorithm were the number of records, the number of variables and the number of clusters. The class output was computed by the efficiency of the parallel processing, defined as the ratio between the speed-up and the number of processors. For each test record, if the efficiency was greater than 0.8 the test was set to YES, meaning that the parallel processing is efficient, otherwise the class was set to NO.

For the analysis, the decision tree induction algorithm J48, was used within the Weka open source data mining workbench [19]. The J48 algorithm is a Java implementation of the classical C4.5 algorithm, developed by Quinlan [20].

The resulting decision tree is shown in Figure 7, where the numbers in parenthesis at each leaf are the number of positive/negative records that match the corresponding rule. It is clearly shown that the parallel processing is not efficient for 2 clusters and is efficient for more than 8 clusters. In the case of 4 clusters, the parallel processing could be efficient or not depending on the number of records and/or the number of variables.

Although it is not a definitive answer for the load optimization of the cluster analysis, the decision tree shows that the number of clusters is the parameter that affects mostly the efficiency of the parallel processing. As a current cluster analysis must

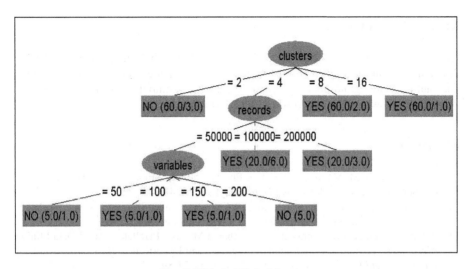

Fig. 7. Weka's J48 decision tree

compute the FCM algorithm for a range of number of clusters, in order to determine the optimal number of clusters, it is preferable to not parallelize the processing for small number of clusters and use more processors as the number of clusters increases.

6 Conclusions

This work has presented a parallel implementation of FCM cluster analysis where both the determination of clusters' centers and the number of clusters are optimized by the algorithm. The main contribution of this work is the integration of the cluster validation index in the optimization process, allowing the optimization of the overall parallel process.

The implementation and performance tests were made in two different hardware architectures: the first on low cost distributed memory hardware, a PC Cluster, and the second on a machine of bigger computational power, the Altix 350.

The parallel Fuzzy c-Means Cluster Analysis tool behaviors in a scalable manner presenting good speedup and efficiency values in both hardware, showing that it can be used as a valuable and efficient tool for improve processing time in knowledge discovery in very bigger databases.

Most of existing approaches of parallel FCM implementations, such as [13], do not include the determination of the number of clusters. It is thus difficult to discuss the result with related work. In this work, the number of iterations is also used as stopping criteria, such that the complexity of the dataset does not influence the performance.

Future work for this project is to apply this approach to seismic data from oil and gas exploration in order to test the parallel FCM cluster analysis with real data and to introduce a new approach based on the Knapsack algorithm for load balance study.

Acknowledgements

This work has been supported by the Brazilian Research Council (CNPq), by the Brazilian Innovation Agency (FINEP) and by the National Petroleum Agency (ANP). The authors are grateful to High Performance Computing Center (NACAD-COPPE/UFRJ) where the experiments were performed.

References

1. M. S. R. Sousa, M. Mattoso and N. F.F. Ebecken (1999). Mining a large database with a parallel database server. *Intelligent Data Analysis* 3, pp. 437-451.
2. M. Coppola, and M. Vanneschi. (2002). High-performance data mining with skeleton-based structured parallel programming. Parallel Computing 28, pp. 783-813.
3. R. Jin, G. Yang, and G. Agrawal (2005). Shared Memory Parallelization of Data Mining Algorithms: Techniques, Programming Interface, and Performance. *IEEE Transaction on Knowledge and Data Engineering*, vol. 17, no. 1, pp. 71-89.
4. M. Cannataro, A. Congiusta, A. Pugliese, D. Talia, P. Trunfio (2004). Distributed data mining on grids: services, tools, and applications. *IEEE Transactions on Systems, Man and Cybernetics, Part B*, vol. 34, no. 6, pp. 2451 – 2465.
5. K. Kubota, A. Nakase, H. Sakai and S. Oyanagi (2000). Parallelization of decision tree algorithm and its performance evaluation. *Proceedings of the Fourth International Conference on High Performance Computing in the Asia-Pacific Region*, vol. 2, pp. 574 – 579.
6. M. W. Kim, J. G. Lee and C. Min (1999). Efficient fuzzy rule generation based on fuzzy decision tree for data mining. *Proceedings of the IEEE International Fuzzy Systems Conference FUZZ-IEEE '99.* pp1223 – 1228.
7. A. Evsukoff, M. C. A. Costa and N. F. F. Ebecken (2004). Parallel Implementation of Fuzzy Rule Based Classifier. *Proceedings of the VECPAR'2004*, vol. 2, pp. 443-452.
8. P. K. H. Phua and D. Ming. (2003). Parallel nonlinear optimization techniques for training neural networks. *IEEE Transactions on Neural Networks*, vol. 14, no. 6, pp. 1460 - 1468.
9. M. C. A. Costa and N. F. F. Ebecken (2001). A Neural Network Implementation for Data Mining High Performance Computing. *Proceedings of the V Brazilian Conference on Neural Networks*, pp. 139-142.
10. R. Agrawal and J. C. Shafer (1996).Parallel mining of association rules. *IEEE Transactions on Knowledge and Data Engineering*, vol. 8, no. 6, pp. 962 - 969.
11. L. Shen, H. Shen and L. Cheng (1999). New algorithms for effcient mining of association rules. *Information Sciences* 118, pp. 251 – 268.
12. B. Boutsinas and T. Gnardellis (2002). On distributing the clustering process. *Pattern Recognition Letters* 23, pp. 999–1008.
13. S. Rahimi, M. Zargham, A. Thakre and D. Chhillar (2004) A parallel Fuzzy C-Mean algorithm for image segmentation. *Proceedings of the IEEE Annual Meeting of the Fuzzy Information NAFIPS '04*, vol. 1, pp. 234 – 237.
14. A. K. Jain, M. N. Murty and P. J. Flynn (1999). Data clustering: a review. *ACM Computing Surveys*, vol. 31, no. 3. pp. 264-323.
15. J. C. Bezdek (1981). *Pattern Recognition with Fuzzy Objective Function Algorithms.* New York, Plenum.
16. X. L. Xie and G. A. Beni (1991). Validity measure for fuzzy clustering. *IEEE Transactions on Pattern Analysis and Machine Intelligence*, vol. 3 no. 8, pp. 841–846.

17. J. Bezdek and N.R. Pal (1998). Some new indexes of cluster validity. *IEEE Trans. Systems Man and Cybernetics B*, vol. 28, pp. 301–315.
18. M. K. Pakhira, S. Bandyopadhyay and U. Maulik (2004). Validity index for crisp and fuzzy clusters. *Pattern Recognition*, vol. 37, pp. 487-501.
19. I. H. Witten and E. Frank (2005). *Data Mining: Practical Machine Learning Tools and Techniques*, 2nd Edition, Morgan Kaufmann, San Francisco.
20. R. Quinlan (1993). *C4.5 – Programs for Machine Learning*. Morgan Kaufmann, San Francisco.

Peer-to-Peer Models for Resource Discovery in Large-Scale Grids: A Scalable Architecture

Domenico Talia[1,2], Paolo Trunfio[1,2], and Jingdi Zeng[1,2]

[1] DEIS, University of Calabria
Via P. Bucci 41c, 87036 Rende (CS), Italy
[2] CoreGRID NoE
{talia, trunfio}deis.unical.it,
zeng@si.deis.unical.it

Abstract. As Grids enlarge their boundaries and users, some of their functions should be decentralized to avoid bottlenecks and guarantee scalability. A way to provide Grid scalability is to adopt *Peer-to-Peer* (*P2P*) models to implement non hierarchical decentralized Grid services and systems. A core Grid functionality that can be effectively redesigned using the P2P approach is *resource discovery*. This paper proposes a P2P resource discovery architecture aiming to manage various Grid resources and complex queries. Its goal is two-fold: to address discovery of multiple resources, and to support discovery of dynamic resources and arbitrary queries in Grids. The architecture includes a scalable technique for locating dynamic resources in large-scale Grids. Simulation results are provided to demonstrate the efficiency of the proposed technique.

1 Introduction

In Grid environments, applications are composed of dispersed hardware and software resources that need to be located and remotely accessed. Efficient and effective resource discovery is then critical. Peer-to-Peer (P2P) techniques have been recently exploited to achieve this goal.

A large amount of work on P2P resource discovery has been done, including both unstructured and structured systems. Early unstructured P2P systems, such as Gnutella [1], use the *flooding technique* to broadcast the resource requests in the network. The flooding technique does not rely on a specific network topology and supports queries in arbitrary forms. Several approaches [2,3,4], moreover, have been proposed to solve two intrinsic drawbacks of the flooding technique, i.e., the potentially massive amount of messages, and the possibility that an existing resource may not be located. In structured P2P networks, Distributed Hash Tables (DHTs) are widely used. *DHT-based systems* [5,6,7] arrange $< key, value >$ pairs in multiple locations across the network. A query message is forwarded towards the node that is responsible for the key in a limited number of hops. The result is guaranteed, if such a key exists in the system. As compared to unstructured systems, however, DHT-based approaches need intensive maintenance on hash table updates.

M. Daydé et al. (Eds.): VECPAR 2006, LNCS 4395, pp. 66–78, 2007.
© Springer-Verlag Berlin Heidelberg 2007

Taking into account the characteristics of Grids, several P2P resource discovery techniques have been adapted to such environments. For instance, DHT-based P2P resource discovery systems have been extended to support range value and multi-attribute queries [8,9,10,11]. Two major differences between P2P systems and Grids, however, determine their different approaches towards resource discovery. First, P2P systems are typically designed to share files among peers. Differently, Grids deal with a set of different resources, ranging from files to computing resources. Second, the dynamism of P2P systems comes from both nodes and resources. Peers join and leave at any time, and thus do the resources shared among them. In Grid environments, nodes connect to the network in a relatively more stable manner. The dynamism of Grids mainly comes from the fast-changing statuses of resources. For example, the storage space and CPU load can change continuously over time.

Highlighting the variety and dynamism of Grid resources, this paper proposes a DHT-based resource discovery architecture for Grids. The rest of the paper is organized as follows. Section 2 introduces existing Grid resource discovery systems that relate to our work. Section 3 discusses characteristics of Grid resources and related query requirements. Section 4 unfolds the picture of the proposed architecture, while Section 5 studies the performance of its dynamic resource discovery strategy through simulations. Section 6 concludes the paper.

2   Related Work

Several systems exploiting DHT-based P2P approaches for resource discovery in Grids have recently been proposed [8,9,10]. Two important issues investigated by these systems are range queries and multi-attribute resource discovery.

Range queries look for resources specified by a range of attribute values (e.g., a CPU with speed from $1.2GHz$ to $3.2GHz$). These queries are not supported by standard DHT-based systems such as Chord [5], CAN [6], and Pastry [7]. To support range queries, a typical approach is to use locality preserving hashing functions, which retain the order of numerical values in DHTs [8,9].

Multi-attribute resource discovery refers to the problem of locating resources that are described by a set of attributes or characteristics (e.g., OS version, CPU speed, etc.). Several approaches have been proposed to organize resources in order to efficiently support multi-attribute queries. Some systems focus on weaving all attributes into one DHT [10] or one tree [12]. Some others adopt one DHT for each attribute [9,11].

Aside from single value queries, range queries, and multi-attribute queries for single resources, the proposed architecture aims to support queries for multiple resources. We use multiple DHTs to manage attributes of multiple resources. This provides a straightforward architecture, and leaves space for potential extensions.

Gnutella-based *dynamic query* [13] strategy is used to reduce the number of messages generated by flooding. Instead of all directions, this strategy forwards the query only to a selected peer. If a response is not returned from a direction, another round of search is initiated in the next direction, after an estimated time. For relatively popular contents this strategy significantly reduces the number of messages without increasing the response time.

Broadcast in DHT-based P2P networks [14] adds broadcast service to a class of DHT systems that have logarithmic performance bounds. In a network of N nodes, the node that starts the broadcast reaches all other nodes with exactly $N-1$ messages (i. e., no redundant messages are generated).

The approach proposed for dynamic resource discovery in this paper is inspired by both the dynamic query strategy and the broadcast approach mentioned above. It uses a DHT for broadcasting queries to all nodes without redundant messages, and adopts a similar "incremental" approach of dynamic query. This approach reduces the number of exchanged messages and response time, which ensures scalability in large-scale Grids.

3   Resources and Query Types

In Grids, *resources* belong to different *resource classes*. A *resource class* is a "model" for representing resources of the same type. Each resource class is defined by a set of attributes which specify its characteristics. A *resource* is an "instance" of a resource class. Each resource has a specific value for each attribute defined by the corresponding resource class. Resources are univocally identified by URLs.

An example of resource class is "computing resource" that defines the common characteristics of computing resources. These characteristics are described by attributes such as "OS name", "CPU speed", and "Free memory". An instance of the "computing resource" class has a specific value for each attribute, for example, "OS name = Linux", "CPU speed = 1000MHz", and "Free memory = 1024MB". Table 1 lists some examples of Grid resource classes. A more complete list of resource classes can be found in [15].

Resource classes can be broadly classified into *intra-node* and *inter-node* resources. "Computing resource" is an example of intra-node resource class. An example of inter-node resource class is "network connection" (see Table 1), which

Table 1. Examples of Grid resource classes

Resource class	Description
Computing resource	Computing capabilities provided by computers, clusters of computers, etc.
Storage resource	Storage space such as disks, external memory, etc.
Network resource	Network connections that ensures collaboration between Grid resources.
Device resource	Specific devices such as instruments, sensors, etc.
Software resource	Operating systems, software packages, Web services, etc.
Data resource	Various kinds of data stored in file systems or databases.

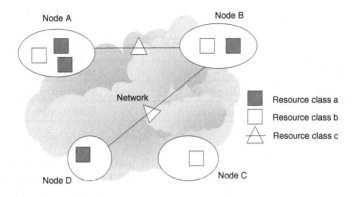

Fig. 1. Inter-node and intra-nodes resources

defines network resource characteristics. Fig. 1 shows a simple Grid includ-ing four nodes and three resource classes. As examples of intra-node resources, *NodeA* includes two instances of resource class *a* and one instance of resource class *b*. The figure also shows two inter-node resources: one between *NodeA* and *NodeD*, and the other between *NodeB* and *NodeD*.

The attributes of each resource class are either *static* or *dynamic*. *Static* at-tributes refer to resource characteristics that do not change frequently, such as "OS name" and "CPU speed" of a computing resource. *Dynamic* attributes are associated to fast changing characteristics, such as "CPU load" and "Free memory".

The goal of resource discovery in Grids is to locate resources that satisfy a given set of requirements on their attribute values. Three types of queries apply to each attribute involved in resource discovery:

- *Exact match query*, where attribute values of numeric, boolean, or string types are searched.
- *Range query*, where a range of numeric or string values is searched.
- *Arbitrary query*, where for instance partial phrase match or semantic search is carried out.

A *multi-attribute* query is composed of a set of sub-queries on single attributes. Each sub-query fits in one of the three types as listed above, and the involved attributes are either static or dynamic.

Complex Grid applications involve multiple resources. Thus, *multi-resource* queries are often needed. For instance, one can be interested in discovering two computing resources and one storage resource; these resources may not be ge-ographically close to each other. A multi-resource query, in fact, involves a set of sub-queries on individual resources, where each sub-query can be a multi-attribute query.

Taking into consideration both characteristics and query requirements of Grid resources, the P2P search techniques exploited in our framework are listed in Table 2.

Table 2. P2P search techniques for different types of resources and queries

	Static Grid resources	**Dynamic Grid resources**
Exact queries	Structured	Unstructured
Range queries	Structured	Unstructured
Arbitrary queries	Unstructured	Unstructured

As shown in the table, structured search is used only for exact and range queries on static Grid resources. This is because DHT-based structured systems are not effective for dynamic resources and arbitrary queries. In fact, DHT-based P2P systems were not originally designed for queries of arbitrary expression forms. Moreover, fast-changing resources, such as CPU load, require frequent updates on DHTs, and thus cause prohibitive maintenance costs. On the other hand, unstructured approaches are used for both dynamic Grid resources and arbitrary queries on static resources. This is because unstructured systems generally do not require table updates and maintenance. However, the huge amount of messages generated by flooding-based unstructured systems requires the use of appropriate strategies to ensure scalability in large networks.

4   System Architecture

The framework aims to provide a generic architecture that leverages existing techniques to fulfill various resource discovery needs in Grid environments. In order to exploit diverse resource discovery techniques, the DHT-based architecture described in Fig. 2 is proposed.

The system is composed of a set of *virtual planes*, one for each resource class. Within the virtual plane of resource class R_a, for example, static attributes $R_a.A_{s1}$, ..., $R_a.A_{sn}$ are associated to their DHTs, respectively. Exact or range queries on *static* attributes are carried out using the DHTs corresponding to these attributes.

An additional "general purpose" DHT is dedicated to queries on dynamic attributes and to arbitrary queries on static attributes. This DHT is different from the DHTs in the virtual planes. The DHTs in the virtual plane are standard DHTs, in which both nodes and resource identifiers are mapped to the same ring. In general purpose DHT, only node identifiers are mapped to the ring, while resources are not mapped to it. In other words, there are not pointers to resources in the general purpose ring.

The general purpose DHT is used to broadcast queries to all Grid nodes whose identifiers are mapped to the ring. All Grid nodes reached by a query are in charge of processing it against the local resources, and sending the response to the node that initiated the query. The mechanisms used for broadcasting a query on this ring are described in Section 4.3.

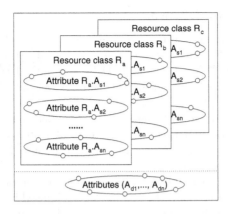

Fig. 2. System architecture

4.1 Local Component

Fig. 3 shows the software modules inside each Grid node. With multiple virtual planes defined in the system, each node participates in all DHTs of these virtual planes. Therefore, multiple finger tables corresponding to each DHT co-exist in each node, as illustrated in Fig. 3. For example, finger tables $FT(R_a.A_1)$, $FT(R_a.A_2)$,..., and $FT(R_a.A_n)$ correspond to DHTs of attributes $R_a.A_{s1}...R_a.A_{sn}$ in Fig. 2.

The finger table of the general purpose DHT, that is, $FT(General\ purpose\ DHT)$, is used to reach individual nodes and locate dynamic attributes $A_{d1},...,A_{dn}$. A query engine processes resource discovery requests and associates them to different query instances and thus DHTs. The results are then generated at the node where related queries are initiated.

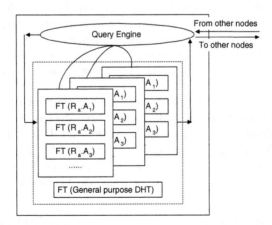

Fig. 3. Software modules inside each Grid node. FTs are finger tables associated to the used DHTs.

4.2   Static Attribute Discovery

A number of multi-attribute, range query approaches have emerged. They either use one DHT [10] or one tree [11] for all attributes, or arrange attribute values on multiple DHTs [9]. While both single-DHT and multi-DHT approaches have proved effective, we adopt the multi-DHT strategy because of its simplicity and extension potentials.

Assume there are p classes of resources, each of which has q types of attributes. Although one node does not necessarily have all attributes, it is included in all DHTs, and the values of its blank entries are left as null. The number of finger tables that a node maintains is $p \times q$.

While existing approaches support resource discovery on single or multiple attributes of one resource class, the architecture proposed in this paper manages multiple resources. One way to do this is to hash the string of "resource class + attribute" into a *DHT ID*; this ID is used to identify the corresponding finger table inside a node.

4.3   Dynamic Attribute Discovery

As mentioned in Section 2, our approach for dynamic resource discovery exploits both the dynamic query [13] and the broadcast over DHT [14] strategies. The general purpose DHT and associated finger tables, as illustrated in Figs. 2 and 3, are used only to index Grid nodes, without keeping pointers to Grid resource attributes. Queries are then processed by the local query engine of each node.

To Reach all Nodes. To reach all nodes without redundant messages, the broadcast strategy is based on a DHT [14]. Taking a fully populated Chord ring with $N = 2^M$ nodes and a M-bit identifier space as an example. Each Chord node k has a finger table, with fingers pointing to nodes $k + 2^{i-1}$, where $i = 1, ..., M$. Each of these M nodes, in turn, has its fingers pointing to another M nodes. Each node forwards the query to all nodes in its finger table, and in turn, these nodes do the same with nodes in their finger tables. In this way, all nodes are reached in M steps. Since multiple fingers may point to the same node, a strategy is used to avoid redundant messages. Each message contains a "limit" argument, which is used to restrict the forwarding space of a receiving node. The "limit" argument of a message for the node pointed by finger i is finger $i + 1$.

Fig. 4 gives an example of an eight-node three-bit identifier Chord ring. The limit of broadcast is marked with a black dot. Three steps of communication between nodes are demonstrated with solid, dotted, and dashed lines. Obviously, node 0 reaches all other nodes via $N - 1$ messages within M steps. The same procedure applies to Chord ring with $N < 2^M$ (i.e., not fully populated networks). In this case, the number of distinct fingers of each node is $logN$ on the average.

Incremental Resource Discovery. The broadcast over DHT presented above adopts a "parallel" approach. That is, the node that initiates the discovery tasks sends the query message to all its fingers in parallel.

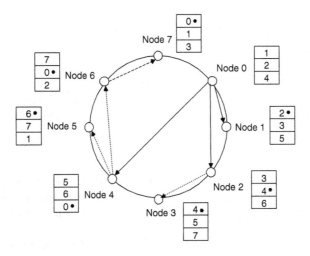

Fig. 4. An example of broadcast

Although no redundant messages are generated in the network, its $N - 1$ messages can be prohibitive in large-scale Grids. Referred to as "incremental", our approach uses a mixed parallel and sequential query message forwarding.

A "parallel degree" D is introduced to adjust the range of parallel message forwarding, and thus curb the number of exchanged messages. Given a node that initiates the query, it forwards the query message in parallel to nodes pointed by its first distinct D fingers. If there is a positive response, the search terminates; otherwise, this node forwards the query message to the node pointed by its $D + 1$ finger. This procedure applies to nodes pointed by the rest of fingers, sequentially, until a positive response returns.

When $D = M$, our incremental approach turns into the parallel one; when $D = 1$, the incremental approach becomes a sequential one, where nodes pointed by all fingers are visited one after another.

The number of generated messages by the incremental approach is obviously less than or equal to that of the parallel one. The response time of incremental approach, however, may be prolonged owing to its sequential query message forwarding. We argue that this does not necessarily hold true. In large-scale Grids, multiple query requests at one node can be prominent, which adds extra delay to response time. Under this circumstance, the incremental approach shall benefit from its reduced number of messages that shortens this extra delay.

5   Performance Evaluation

A discrete-event simulator has been implemented to evaluate the performance of the incremental approach, used in our framework for the discovery of dynamic resources and arbitrary queries, in comparison with the original parallel approach.

Table 3. System parameters

Parameter	Description
M	Number of bits of node identifiers.
N	Number of Grid nodes in the network.
R	Number of nodes that concurrently submit query requests.
P	Fraction of nodes in the network that possesses the desired resource.
D	Number of first distinct fingers the search is conducted on in parallel.

Two performance parameters have been evaluated: the *number of messages* Q and the *response time* T. Q is the total number of exchanged messages in the network, and T is the time a node waited to receive the first response (i.e., the first query hit).

The system parameters are explained in Table 3. In all simulations we used $M = 32$ and $D = 7$. The number of nodes N ranges from 2000 to 10000, R ranges from 10 to 1000, and P ranges from 0.005 to 0.25. The values of performance parameters are obtained by averaging the results of 10 to 30 independent simulation runs.

The system parameter values have been chosen to fit as much as possible with real Grid scenarios. In particular, the wide range of values chosen for P reflects the fact that, in real Grids, discovery tasks can search both for rare resources (e.g., the IRIX operating system) and more popular ones (e.g., Linux).

The time to pass a message from $NodeA$ to $NodeB$ is calculated as the sum of a processing time and a delivery time. The processing time is proportional to the number of queued messages in $NodeA$, while the delivery time is proportional to the number of incoming messages at $NodeB$. In this way, the response time depends on both message traffic and processing load of nodes.

Table 4 shows the number of exchanged messages in both parallel and incremental strategies, with $R = 1$. The parallel strategy always generates $N - 1$ messages for each submitted query, which could be prohibitive for large-scale Grids. In the incremental approach, the number of messages is dramatically reduced. Moreover, when the value of P is over a certain limit, the number of messages fluctuates around the value of 2^D and it does not depend from the number of nodes (i.e., network size). This limit is determined by the number of Grid nodes N, the fraction of nodes with matching resources P, and the number of first distinct fingers D.

For example, in a network with $N = 10000$, when $P = 0.1$ the number of matching resources is $N \times P = 1000$. The number of nodes included in the first $D = 7$ fingers is $2^D = 128$, on the average. Obviously, this density is high enough for the incremental strategy to locate the desired resource within the first D fingers. With a lower value of P, nevertheless, the search needs to go beyond the first D fingers; this introduces a fluctuation in the number of exchanged messages, as in the case of $P = 0.005$.

Table 4. Comparison on the number of exchanged messages (Q) in parallel and incremental approaches

N	P	Q (Parallel)	Q (Incremental)
	0.005	1999	279
2000	0.10	1999	127
	0.25	1999	126
	0.005	3999	326
4000	0.10	3999	129
	0.25	3999	124
	0.005	5999	291
6000	0.10	5999	126
	0.25	5999	126
	0.005	7999	282
8000	0.10	7999	128
	0.25	7999	128
	0.005	9999	389
10000	0.10	9999	127
	0.25	9999	125

Figs. 5, 6 and 7 show the response time in networks composed by 2000, 6000 and 10000 nodes, respectively, with $P = 0.10$ and values of R ranging from 10 to 1000. The response time is expressed in time units.

The main result shown in Figs. 5, 6 and 7 is that, for any value of N, when the values of R are at the lower end of its range, the parallel approach has a shorter response time. When the value of R increases, the incremental approach outperforms the parallel one. This is because in the parallel approach the overall number of generated messages is much higher than the one in the incremental approach, resulting in increased message traffic and processing load that cause a higher response time.

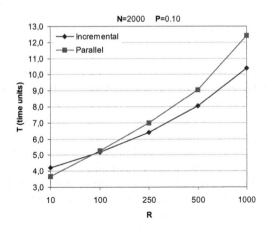

Fig. 5. Response time of parallel and incremental approaches ($N = 2000$)

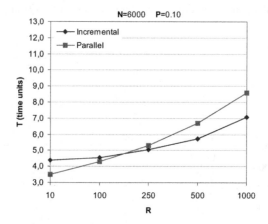

Fig. 6. Response time of parallel and incremental approaches ($N = 6000$)

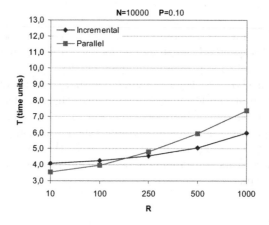

Fig. 7. Response time of parallel and incremental approaches ($N = 10000$)

It can also be noted that, for any value of R, the response time decreases as N increases. This trend is similar in both parallel and incremental approaches. This is because the probability of finding the desired resource in a given time interval is proportional to the number of nodes that possess it, and this in turn is proportional to the network size.

The simulation results demonstrate that with higher values of R the incremental approach scales better than the parallel one. It is important to recall that in our simulator the processing time is proportional to the number of messages to be processed, and the delivery time is proportional to the number of messages to be delivered. Therefore, the response time increases linearly with message traffic and load of nodes. In a more realistic scenario the processing time and the delivery time may increase exponentially with the load of the network.

In this case, the response time in the incremental approach should result significantly better that the parallel one. To better evaluate the effect of high loads in large-scale Grids, we are currently studying the use of more complex processing and delivery time functions in our simulator.

6 Conclusions

This paper discussed the characteristics of Grid resources and identified critical problems of resource discovery in Grids. A DHT-based P2P framework has been introduced to address the variety and dynamism of Grid resources. It exploits multiple DHT and existing P2P techniques for multiple static resources and implements an "incremental" resource discovery approach for dynamic resources. As compared to the original strategy, the incremental approach generates reduced number of messages and experiences lower response time in large-scale Grids.

With the emergence of service-oriented Grids [16], *service discovery* has become an important topic. Grid services are today implemented complying with the *Web Services Resource Framework* (*WSRF*) family of specifications, which define standard mechanisms for accessing and managing Grid resources using *Web services*. Web services are defined using XML-based languages, and XML queries are used to query their features. We are currently studying how to extend the architecture in this paper to address Grid service discovery, in particular, dynamic service indexing and XML-based queries support.

Acknowledgements

This research work is carried out under the FP6 Network of Excellence Core-GRID funded by the European Commission (Contract IST-2002-004265). This work has been also supported by the Italian MIUR FIRB Grid.it project RBNE01KNFP on High Performance Grid Platforms and Tools.

References

1. Gnutella Protocol Development. http://rfc-gnutella.sourceforge.net/src/rfc-0_6-draft.html.
2. Gkantsidis, C., Mihail, M., Saberi, A.: Hybrid Search Schemes for Unstructured Peer-to-peer Networks. Proc. of IEEE INFOCOM'05, Miami, USA (2005).
3. Lv, Q., Cao, P., Cohen, E., Li, K., Shenker, S.: Search and Replicating in Unstructured Peer-to-peer Networks. Proc. of 16th Annual ACM Int. Conf. on Supercomputing (ISC'02), New York, USA (2002).
4. Crespo, A., Garcia-Molina, H.: Routing Indices for Peer-to-peer Systems. Proc. of Int. Conf. on Distributed Computing Systems (ICDCS'02), Vienna, Austria (2002).
5. Stoica, I., Morris, R., Karger, D., Kaashoek, M. F., Balakrishnan, H.: Chord: A Scalable Peer-to-peer Lookup Service for Internet Applications. Proc. of ACM SIG-COMM'01, San Diego, USA (2001).

6. Ratnasany, S., Francis, P., Handley, M., Karp, R. M., Shenker, S.: A Scalable Content-Addressable Network. Proc. of ACM SIGCOMM'01, San Diego, USA (2001).
7. Rowstron, A., Druschel, P.: Pastry: Scalable, distributed object location and routing for large-scale peer-to-peer systems. Proc. of IFIP/ACM International Conference on Distributed Systems Platforms (Middleware), Heidelberg, Germany (2001).
8. Andrzejak, A., Xu, Z.: Scalable, Efficient Range Queries for Grid Information Services. Proc. of 2nd IEEE Int. Conf. on Peer-to-peer Computing (P2P'02), Linköping, Sweden (2002).
9. Cai, M., Frank, M., Chen, J., Szekely, P.: MAAN: A Multi-Attribute Addressable Network for Grid Information Services. Journal of Grid Computing, vol. 2 n. 1 (2004) 3-14.
10. Oppenheimer, D., Albrecht, J., Patterson, D., Vahdat, A.: Scalable Wide-Area Resource Discovery. UC Berkeley Technical Report, UCB/CSD-04-1334 (2004).
11. Spence, D., Harris, T.: XenoSearch: Distributed Resource Discovery in the XenoServer Open Platform. Proc. of HPDC'03, Washington, USA (2003).
12. Basu, S., Banerjee, S., Sharma, P., Lee, S.: NodeWiz: Peer-to-peer Resource Discovery for Grids. Proc. of IEEE/ACM GP2PC'05, Cardiff, UK (2005).
13. Fisk, A. A.: Gnutella Dynamic Query Protocol v0.1. http://www.the-gdf.org/wiki/index.php?title=Dynamic_Querying.
14. El-Ansary, S., Alima, L., Brand, P., Haridi, S.: Efficient Broadcast in Structured P2P Networks. Proc. of IEEE/ACM Int. Symp. on Cluster Computing and the Grid (CCGRID'05), Cardiff, UK (2005).
15. Andreozzi, S., Burke, S., Field, L., Fisher, S., Konya, B., Mambelli, M., Schopf, J., Viljoen, M., Wilson, A.: GLUE Schema Specification Version 1.2: Final Specification - 3 Dec 05. http://infnforge.cnaf.infn.it/glueinfomodel/index.php/Spec/V12.
16. Comito, C., Talia, D., Trunfio, P.: Grid Services: Principles, Implementations and Use. International Journal of Web and Grid Services, vol. 1 n. 1 (2005) 48-68.

JaceV: A Programming and Execution Environment for Asynchronous Iterative Computations on Volatile Nodes

Jacques M. Bahi, Raphaël Couturier, and Philippe Vuillemin

LIFC, University of Franche-Comté, France[*,**]
{jacques.bahi,raphael.couturier,philippe.vuillemin}@iut-bm.univ-fcomte.fr
http://info.iut-bm.univ-fcomte.fr/and/

Abstract. In this paper we present JaceV, a multi-threaded Java based library designed to build asynchronous parallel iterative applications (with direct communications between computation nodes) and execute them in a volatile environment. We describe the components of the system and evaluate the performance of JaceV with the implementation and execution of an iterative application with volatile nodes.

Keywords: Asynchronous iterative algorithms, computational science problems, desktop grid computing, volatile nodes.

1 Introduction

Nowadays, PCs and workstations are becoming increasingly powerful and communication networks are more and more stable and efficient. This leads scientists to compute large scientific problems on virtual parallel machines (a set of networked computers to simulate a supercomputer) rather than on expensive supercomputers. However, as the node count increases, the reliability of the parallel system decreases. As a consequence, failures in the computing framework make it more difficult to complete long-running jobs. Thus, several environments have been proposed to compute scientific applications on volatile nodes using cycle stealing concepts. In this paper, we consider as volatile node any volunteer personal computer connected to a network (WAN or LAN[1]) that can be used as a computational resource during its idle times. The aim of this work is to run scientific computations in such a volatile framework.

In this paper, we are interested in iterative algorithms. Those algorithms are usually employed for sparse systems (like some linear systems) or when direct methods cannot be applied to solve scientific problems (e.g. for polynomial root finders). In the parallel execution of iterative algorithms, communications must be performed between computation nodes after each iteration in order to satisfy

---

[*] Candidate to the Best Student Paper Award.
[**] This work was supported by the "Conseil Régional de Franche-Comté".
[1] World Area Network or Local Area Network.

M. Daydé et al. (Eds.): VECPAR 2006, LNCS 4395, pp. 79–92, 2007.
© Springer-Verlag Berlin Heidelberg 2007

all the computing dependencies. For that reason, the reliability of the system is a very important feature in such a context and can become a limiting factor for scalability. Hence, it is necessary to study this reliability according to the different classes of architecture. We consider three concepts (or classes) of architecture with different characteristics.

1. Grid computing environments enable the sharing and aggregation of a wide variety of geographically distributed computational resources (such as supercomputers, computing clusters...). In such architectures, communications are very fast and efficient and the topology of the system is considered stable.
2. Desktop grid computing environments (also called Global Computing) exploit unused resources in the Intranet environments and across the Internet (e.g. the SETI@home project [2]). In this class of parallelism, the architecture is fully centralized (client-server-based communications), tasks are independent and the topology of the system is completely dynamic (nodes appear and disappear during the computation).
3. Peer-To-Peer (P2P) environments are networks in which each workstation has equivalent capabilities and responsibilities. The architecture is completely decentralized (peers directly communicate between each other) and the topology of the system is completely dynamic.

As reliability is generally ensured in a Grid computing context, we do not consider this class; furthermore, several frameworks are already available to implement and run parallel iterative applications in such environments. Concerning Desktop grid, although this class can provide much more resources than the first one, it is generally not directly suitable for parallel iterative computations as long as communication is restricted to the master-slave model of parallelism.

For that reason we would like to gather functionalities and characteristics of the latter two cases: 1) a centralized architecture to manage all the nodes of the system akin to a Desktop grid environment with volatile nodes and 2) direct communications between computation nodes like in P2P environments. The purpose of this paper is to describe a programming environment allowing users to implement and run parallel asynchronous iterative algorithms on volatile nodes. Asynchronous algorithms can be used in a significant set of applications. Indeed, scientific applications are often described by systems of differential equations which lead, after discretization, to linear systems $Ax = b$ where A is a M-matrix (i.e. $A_{ii} > 0$ and $A_{ij} \leq 0$ and A is nonsingular with $A^{-1} \geq 0$). A convergent weak regular splitting can be derived from any M-matrix and any iterative algorithm based on this multiplitting converges asynchronousely (see [1,4,10] and the references therein).

As idle times and synchronizations are suppressed in the asynchronous iteration model (i.e. a computing node can continue to execute its task without waiting for its neighbor results), we do believe this solution is the most suitable in an environment with volatile nodes. Furthermore, computations formulated in parallel asynchronous iterative algorithms are much less sensitive to heterogeneity of communication and computational power than conventional synchronous parallel iterative algorithms.

We do not consider the synchronous iteration model because it is neither convenient for this volatile framework, nor for the heterogeneity and scalability.

In this paper, we describe JaceV, a multi-threaded Java based library designed to build asynchronous parallel iterative applications (with direct communications between computation nodes) and execute them in a desktop grid environment with volatile nodes. To the best of our knowledge, this work is the first one presenting a volatile execution environment with direct communications between computing nodes and allowing the development of actual scientific applications with interdependent tasks.

The following section presents a survey of desktop grid and volatility tolerant environments. Section 3 presents the architecture of JaceV and an overview of all its components. Section 4 describes the scientific application implemented with JaceV (the Poisson problem) in order to perform experiments. Section 5 evaluates the performance of JaceV by executing the application in different contexts with volatile nodes. In section 6, we conclude and some perspectives are given.

2   Related Work

Cycle stealing in a LAN environment has already been studied in the Condor [9] and Atlas [3] projects. However, the context of LAN and the Internet are drastically different. In particular, scheduling techniques [7,8] need to be adapted for a Global Computing environment due to: 1) the very different communication and computing performance of the targeted hosts, 2) the sporadic Internet connection and 3) the high frequency of faulty machines.

MPICH-V and MPICH-V2 [11,13] (message passing APIs[2] for automatic Volatility tolerant MPI environment) have been proposed for volatile nodes. However, MPI is not a multi-threaded environment. As a consequence, it is not suitable for asynchronous iterations in which it is convenient to separate communications and computation.

XtremWeb [16] is a Desktop Grid and Global Computing middleware which allows users to build their own parallel applications and uses cycle stealing. However, this environment does not provide direct communications between the different computing nodes of the system. As a consequence, it is not suitable for implementing and running parallel iterative applications.

Ninflet [6] is a Java-based global computing system. It is designed to overcome the limitations of Ninf [5] that currently lacks security features as well as task migration. The goal of Ninflet is to become a new generation of concurrent object-oriented system which harnesses abundant idle computing powers, and also integrates global as well as local network parallel computing. Unfortunately, as with the XtremWeb environment, Ninflet only applies Master-Worker pattern and does not provide direct communications between computation nodes.

In [15], no environment is proposed but the authors define the requirements for an effective execution of iterative computations requiring communication on

---

[2] Application Programming Interfaces.

a desktop grid context. They propose a combination of a P2P communication model, an algorithmic approach (asynchronous iterations) and a programming model. Finally, they give some very preliminary results from application of the extended desktop grid for computation of Google pagerank and solution of a small linear system.

Jace [14] is a multi-threaded Java based library designed to build asynchronous iterative algorithms and execute them in a Grid environment. In Jace, communications are directly performed between computation nodes (in a synchronous or an asynchronous way) using the message passing paradigm implemented with Java RMI[3]. However, this environment is not designed to run applications on volatile nodes.

3   The JaceV System

3.1   The Goal of JaceV

As described in the previous section, Jace is fully suitable for running parallel iterative applications (in a synchronous or asynchronous mode) in a Grid computing context where nodes do not disappear during computations. Then, it was essential to completely redesign the Jace environment in order to make it tolerant to volatility. To do this, it is necessary to develop a strategy to periodically save the results computed by each node during the execution in order to restart computations from a consistent global state [12] when faults occur (but the messages are not logged and are lost if the destination node has failed).

We propose JaceV, the volatility tolerant implementation of Jace (JaceV for Jace Volatile). JaceV allows users to implement iterative applications and run them over several volatile nodes using the asynchronous iteration model and direct communications between processors.

Hence, when a computer is not used during a defined finite time, it should automatically contribute to compute data of a parallel iterative application already running (or to be started) on the system. *A contrario*, when a user needs to work on this workstation, the resource must instantaneously be freed and this node must automatically be removed from the system. In this way, a volatility tolerant system must both tolerate appearance and disappearance of computation nodes without disturbing the final results of the applications running on it. In fact, JaceV tolerates N simultaneous faults (N being the number of computational resources involved in an application) without disturbing the results at all.

3.2   Architecture of the System

A JaceV application is a set of *Task* objects running on several computation nodes. Like in Jace, the different Task objects of an application cooperate by exchanging messages and data to solve a single problem. The JaceV architecture

---

[3] Remote Method Invocation.

consists of three entities which are JVMs [4] communicating with each others:
1) the *Daemons*, 2) the *Spawner* and 3) the *Dispatcher*. Since JaceV is based
on Jace, all the communications performed between the different entities of the
system are based on Java RMI and threads are used to overlap communications
by computations during each iteration.

The user of the JaceV system could play two types of roles, one being the
resource provider (during idle times of his computer) and the other being *application programmer* (the user who wants to run his own specific parallel iterative
application on several volatile nodes). The resource provider will have a Daemon running on his host. The Daemon is the entity responsible for executing a
Task and we consider it is busy and not available when a Task is executed on it
(thereafter, we use the term Daemon and node indifferently). On the other side,
the application programmer implements an application (using the Java language
and the JaceV API) and actually runs it using the Spawner: this entity actually
starts the application on several available Daemons.

Finally, the Dispatcher is the component in charge of 1) registering all the
Daemons connected to the system and managing them (i.e. detect the eventual
disconnections and replace the nodes) 2) distributing the Task objects of an application over the different available nodes, 3) detecting the global convergence of
a running application, and 4) storing the backups of all the Tasks being executed.

Three-tier architectures are commonly used in fault tolerant platforms, like
in Ninflet, XtremWeb, etc. However, JaceV has the advantage to enable both
direct communications between computing nodes and multi-threaded programming, which is impossible with other existing environments. Furthermore, JaceV
is the only one to implicitly provide an asynchronous iteration model by using primitives of its API in a volatile context. Therefore, JaceV is an original
architecture.

3.3   The Dispatcher

The Dispatcher is the first entity to be launched for the environment. We consider
it is running on a powerful and stable server. Therefore, all the data stored in
this entity are considered as persistent. The Dispatcher is composed of three
main components, 1) the *JaceVDispatchServer*, 2) the *GlobalRegister* and 3) the
ApplicationManager.

The JaceVDispatcher is the RMI server that contains all the methods remotely
invoked by the Daemons and the Spawner. It is launched when the Dispatcher
starts and is continuously waiting for remote invocations.

The GlobalRegister registers all the Daemons connected to the JaceV system
and also stores their current state (the 'alive' and the 'busy' states, this will be
described in section 3.4).

Finally the ApplicationManager indexes all the *RunningApplication* objects of
the system. A RunningApplication is a JaceV object that models an application
being currently executed on the system: it contains for example attributes such

---

[4] Java Virtual Machines.

as the URL where are available the corresponding class files of the application, the number of Tasks, the optional arguments, etc.

Each RunningApplication contains a single *Register* object, which is a subset of the GlobalRegister. During the execution, the Register is automatically updated in case of fault (due to a crash or a user disconnection) of one of the computation node (it models the current configuration of the nodes running a given application and the mapping of the Tasks over the Daemons).

The Dispatcher is also in charge of storing the Task objects saved (called *Backups*) during the computation in order to restart the application from a consistent global state in case of fault. A list stored in the RunningApplication object (the *BackupList*) indexes each Task composing an application. Hence, when a faulty node is replaced, the last Backup of the Task it was computing is sent to the new Daemon in order to restart computations. As iterations are desynchronized in the asynchronous model, the other nodes keep computing without stopping.

Finally, the RunningApplication object is responsible for detecting the global convergence and halting the application when convergence is reached. To do this, each RunningApplication object manages an array containing the local states of the nodes involved in the computation. This array is affected each time a local convergence message is received from the Daemons. When a node is in a local stable state (i.e. the relative error between the last two iterations on this node is greater than a given threshold) after a given number of iterations, it sends 1 to the Dispatcher, or else, it sends 0. The global state is computed on the Dispatcher by testing all the cells of the array, if they are all in stable state then the convergence is detected and the Daemons can stop computing.

To summarize the architecture of the Dispatcher, Figure 1 describes the main objects with the GlobalRegister on the left, the JaceVDispatcher (the RMI server) on the right and the ApplicationManager in the center.

In this example, nine Daemons are currently registered to the Dispatcher (nodes *N1* up to *N9* in the GlobalRegister). Only seven nodes are actually busy (i.e. computing an application). They appear in grey in the GlobalRegister (in the figure, we represent the nodes in different grey levels in order to differentiate

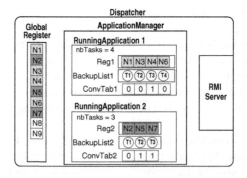

Fig. 1. Description of the Dispatcher elements

the application being executed on the corresponding Daemon). In the ApplicationManager, we can see that two applications are currently running, the first one (*RunningApplication1*) is distributed over four nodes (which are the nodes *N1*, *N3*, *N4* and *N6* in the corresponding Register called *Reg1*) and the second one (*RunningApplication2*) over three nodes (which are the nodes *N2*, *N5* and *N7* in the corresponding Register called *Reg2*). This figure also shows the Backups stored on the Dispatcher for each application being executed (*BackupList1* for the first application and *BackupList2* for the second one). Every BackupList contains a single Backup object for each Task running on a Daemon. The last elements appearing in the figure are the convergence arrays (*ConvTab1* for the first application and *ConvTab2* for the second one): with the values of *ConvTab1*, we can deduce that only Task *T3* (executed on node *N4*) is in a local convergence state for the first application. Concerning the second application, we can see that Tasks *T2* and *T3* (respectively running on nodes *N5* and *N7*) have locally converged to the solution.

3.4   The Daemon

When the Daemon is started, an RMI server is launched on it and is continuously waiting for remote invocations. Then, the Daemon 1) contacts the Dispatcher in order to obtain its remote RMI reference 2) remotely registers itself on the GlobalRegister of the Dispatcher (where this Daemon is then labeled as *available* because it has not been attributed an application yet), and 3) starts locally the *heartbeatThread*: this thread periodically invokes the *beating* remote method on the Dispatcher RMI server to signal its activity. The Dispatcher continuously monitors these calls to implement a timeout protocol: when a Daemon has not called for a sufficient long time, it is considered down in the GlobalRegister (i.e. it is labeled as *notAlive*). In case this node was executing an application, the Task initially running on it should be rescheduled to a new available Daemon by reloading the last Backup stored on the Dispatcher for the faulty node.

Once all those features are performed, the Daemon is initialized and ready to be invoked by the Spawner in order to actually run computation Tasks. The main objects composing the Daemon are mostly the same as in the Jace environment (interested readers can see [14] to have more details about the components of the Jace Daemon and their interaction). However, several objects have been deeply modified or added to the JaceV environment in order to ensure volatility tolerance. Those components are described in the following.

The Daemon contains the Register of the application it is running and this Register is automatically updated by the Dispatcher when faults occur during the execution. As the Register also contains the complete list of the nodes running a given application and the mapping of the Tasks over them, the Daemon is always aware of the topology of the system. This ensures direct communications are carried out between nodes because the Register contains the remote reference RMI for each Daemon. As a consequence, a given node can invoke remote methods on every Daemon running the same application. Furthermore, when a node receives a new Register, the recipient of all the *Message* objects to be sent is automatically updated (if it has changed).

Concerning the Messages to send to other Daemons, as the asynchronism model is message loss tolerant, the Message is simply lost if the destination node is not reachable.

3.5   The Spawner

The Spawner is the entity that actually starts a user application. For this reason, when launching the Spawner, it is necessary to give some parameters to define this application: 1) the number of nodes required for the parallel execution, 2) the URL where the class files are available and finally 3) the optional arguments of the specific application.

Then, the Spawner sends this information to the Dispatcher that creates a new RunningApplication with the given parameters and a new Register composed of the required number of available nodes appearing in the GlobalRegister (which are then labeled as *notAvailable*). This Register is then attributed to the RunningApplication and sent to the Spawner.

Finally, when the Spawner receives the Register object, it broadcasts it to the whole nodes of the topology and then actually starts the computation on each of the Daemons.

The whole interaction between the JaceV entities is described in Figure 2. In this example, we can see the Daemon *N1* (fig.2(a)) and then a set of Daemons (*N2*, *N3* and *N4*, fig.2(b)) registering themselves to the Dispatcher. Those Daemons are then added to the GlobalRegister (*Reg*) and are labeled as *available* because no application has been spawned on the system yet.

In fig.2(c), the Spawner *S1* launches application *appli1* which requires two nodes. The Dispatcher creates then a RunningApplication object for this application and attributes it a Register object (*Reg1*) containing two available nodes of the GlobalRegister (*N1* and *N2* which are then labeled as *notAvailable* and appear in grey level in the GlobalRegister). The Register is sent to *N1* and *N2* (in order to permit direct communications between the two nodes) and the application is actually run by the Spawner on those two Daemons.

In fig.2(d), the Daemon *N2* crashes (or is disconnected by its user). However, as the asynchronous iteration model is used in JaceV, *N1* keeps computing and does not stop its job (the eventual messages to send to the Task running on *N2* will be lost until the node is replaced). The Dispatcher detects this disconnection and labels *N2* as *notAlive* in the GlobalRegister. *Reg1* is then updated in the RunningApplication object (*N2* is replaced by *N3* which is available) and this new Register is sent to the corresponding Daemons (*N1* and *N3*, the new one). Since then, *N1* is aware of the new topology of the system and updates the list of its neighbors (i.e it will no longer try to send messages to *N2* but will directly send them to *N3*). Finally, the Dispatcher sends the appropriate Backup to the new node of the topology and computations can restart on this Daemon.

After several minutes, the Daemon is launched again on node *N2* (fig.2(e)). It is then labeled as *alive* and *available* in the GlobalRegister.

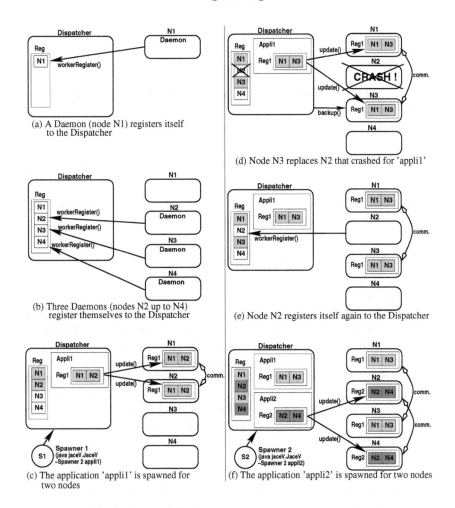

(a) A Daemon (node N1) registers itself
to the Dispatcher

(b) Three Daemons (nodes N2 up to N4)
register themselves to the Dispatcher

(c) The application 'appli1' is spawned for
two nodes

(d) Node N3 replaces N2 that crashed for 'appli1'

(e) Node N2 registers itself again to the Dispatcher

(f) The application 'appli2' is spawned for two nodes

Fig. 2. The registering and spawning processes in JaceV

Finally, in fig.2(f), the Spawner *S2* launches application *appli2* that requires
two nodes. The Dispatcher creates the RunningApplication object for this application, attributes it a new Register (*Reg2*) which contains the last two available
nodes of the GlobalRegister (*N2* and *N4*) and sends them *Reg2* in order to
enable direct communication between these Daemons. At the end, *S2* actually
starts computations on *N2* and *N4*.

4   Problem Description

In this section, we describe the problem used for the experiments with JaceV. It
consists of the Poisson equation discretized in two dimensions. This is a common
problem in physics that models for instance heat problems. This linear elliptic
partial differential equations system is defined as

$$-\Delta u = f. \tag{1}$$

This equation is discretized using a finite difference scheme on a square domain using a uniform Cartesian grid consisting of grid points (x_i, y_i) where $x_i = i\Delta x$ and $y_j = j\Delta y$. Let $u_{i,j}$ represent an approximation to $u(x_i, y_i)$. In order to discretize (1) we replace the $x-$ and $y-$derivatives with centered finite differences, which gives

$$\frac{u_{i-1,j} - 2u_{i,j} + u_{i+1,j}}{(\Delta x)^2} + \frac{u_{i,j-1} - 2u_{i,j} + u_{i,j+1}}{(\Delta y)^2} = -f_{i,j} \tag{2}$$

Assuming that $\Delta x = \Delta y = h$ are discretized using the same discretization step h, (2) can be rewritten in

$$\frac{-4 * u_{i,j} + u_{i-1,j} + u_{i+1,j} + u_{i,j-1} + u_{i,j+1}}{h^2} = -f_{i,j}. \tag{3}$$

For this problem we have used Dirichlet boundary conditions.

So, (1) is solved by finding the solution of the following linear system of the type $A \times x = b$ where A is a 5-diagonal matrix and b represents the function f.

To solve this linear system we use a block-Jacobi method that allows us to decompose the matrix into block matrices and solve each block using an iterative method. In our experiments, we have chosen the sparse Conjugate Gradient algorithm. Besides, this method allows to use overlapping techniques that may dramatically reduce the number of iterations required to reach the convergence by letting some components to be computed by two processors.

From a practical point of view, if we consider a discretization grid of size $n \times n$, A is a matrix of size (n^2, n^2).

It should be noticed that, in the following, the number of components by processor is important and is a multiple of n, the number of components of a discretized line, and that the overlapped components is less important than this number of components. The solution of this problem using parallelism involves that each processor exchanges, at each Jacobi iteration, its first n components with its predecessor neighbor node and its last n ones with its successor neighbor node. The number of components exchanged with each neighbor is equal to n. In fact, we have only studied the case where the totality of overlapped components are not used by a neighbor processor, only the first or last n components are used because the other case entails more data exchanged without decreasing the number of iterations. So, whatever the size of the overlapped components, the exchanged data are constant.

Moreover we recall that the block-Jacobi method has the advantage to be solvable using the asynchronous iteration model if the spectral radius of the absolute value of the iteration matrix is less than 1, which is the case for this problem.

Finally, the Poisson problem implemented using the JaceV API has the skeleton described in Algorithm 1:

Algorithm 1. The Poisson problem skeleton using the JaceV API

Build the local Poisson submatrix
Initialize dependencies
repeat
  Solve local Block-Jacobi subsystem
  Asynchronous exchange of nonlocal data //with jaceSend() and jaceReceive()
  jaceLobalConvergence() //Local convergence detection
  jaceSave() //Primitive used to save the Task object on the Dispatcher
  jaceIteration++ //Increment the iteration number of the Backup to store
until jaceGlobalConvergence()

---

5   Experiments

For our experiments, we study the execution times of the application over 16 nodes according to n (with n varying from 500 up to 1800, which respectively corresponds to matrices of size $250,000 \times 250,000$ up to $3,240,000 \times 3,240,000$ because the problem size is n^2). An optimal overlapping value is used for each n. These experiments are performed with different configurations of processors and networks. For each configuration, we first run the application over 16 stable nodes, and then, for the execution in a volatile context, we launch 19 Daemons and run the application over 16 of them. In the last case, our strategy for volatility is to randomly disconnect each Daemon on average slightly less than two times during the whole execution of the application and reconnect it a few seconds later (i.e. there are approximatively about 30 disconnections/reconnections for each execution).

We choose to perform those series of tests with different configurations of processors and networks. According to processors, we use both homogeneous and heterogeneous processors. The first context consists of a 19-workstation cluster of Intel(R) Pentium(R) 4 CPU 3.00GHz processors with 1024MB of RAM. For the heterogeneous case, we use 19 workstations from Intel(R) Pentium(R) III CPU 1266MHz processors with 256MB of RAM up to Intel(R) Pentium(R) 4 CPU 3.00GHz with 1024MB of RAM. Then, we perform our tests with different network bandwidths.

Finally our series of tests are performed using four configurations of processors and network, which are described as follows.

1. A configuration with homogeneous processors and an Ethernet 1Gbps network,
2. a configuration with homogeneous processors and a 10,000Kbps upload and download bandwidth,
3. a configuration with homogeneous processors and a 1,000Kbps upload and download bandwidth,
4. a configuration with heterogeneous processors and an Ethernet 100Mbps network.

For the second and the third configurations, each workstation of the cluster runs a Qos[5] script in order to limit the network bandwidth to 10,000Kbps (for configuration 2) and 1,000Kbps (for configuration 3).

Whatever the configuration used, the Dispatcher is running on an Intel(R) Pentium(R) 4 CPU 3.00GHz processor with 1024MB of RAM.

The results of the experiments are represented in figure 3 and each execution time is the average of a series of ten executions.

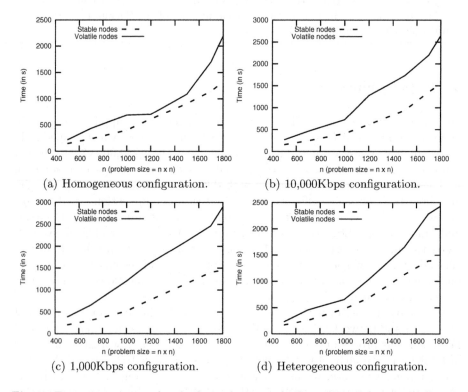

(a) Homogeneous configuration.

(b) 10,000Kbps configuration.

(c) 1,000Kbps configuration.

(d) Heterogeneous configuration.

Fig. 3. Execution times of volatile and non volatile contexts for the different configurations

Analyzing the four figures, we deduce that JaceV supports rather well the volatile context. Indeed, although there are approximatively 30 disconnections during the whole execution, the ratio *volatile context execution time/stable context execution time* is always less than 2.5. Furthermore, at some point during the execution, less than 16 nodes are actually computing because more than 3 nodes are currently disconnected (they have not reconnected to the system yet). In this case, the alive nodes keep computing and are not waiting for the other Daemons to reconnect as it would occur in a synchronous execution.

---

[5] Quality of Service.

We can also deduce that the lower the network bandwidth is, the greater the ratio according to the problem size is (this is particularly obvious in fig.3(c)). This is due to the fault detection and the restarting of the application. Indeed, when the Dispatcher detects the disconnection of a node (and eventually replaces it), it broadcasts the new Register object to all the alive nodes involved in the execution of the application. If the bandwidth is low, this action takes a certain time to be performed (because the size of the Register is not negligible). Hence, some Daemons would continue to send messages to the disappeared node during this period until the Register is actually updated on the Daemons. Furthermore, when the new Daemon replaces a faulty node, it must completely reload the Backup object from the Dispatcher. This object is rather important in terms of size, and it can take some time to deliver it on a low bandwidth network and to actually update it on the new Daemon. All those actions make the application much slower to converge to the solution.

Finally, comparing the execution times on homogeneous and heterogeneous workstations (respectively fig.3(a) and fig.3(d)) we can see that the curves are rather similar. As a consequence, we can deduce that JaceV does not seem to be that sensitive to the heterogeneity of processors for this typical application and perhaps for other similar coarse grained applications. This is undoubtedly due to the asynchronism which allows the fastest processors to perform more iterations.

6   Conclusion and Future Works

In this paper, we describe JaceV, a multi-threaded Java based library designed to build asynchronous parallel iterative applications and run them over volatile nodes. A goal of JaceV is to provide an environment with communications between computation nodes after each iteration, as it is necessary to run parallel iterative applications. JaceV uses the asynchronous iteration model in order to avoid synchronizations. Indeed, synchronous iterations would dramatically slow down the execution in a volatile context where nodes appear and disappear during computation.

The performance of the Poisson problem resolution show that JaceV is fully suitable for running asynchronous iterative applications with volatile nodes. We also remark that performances of JaceV are degraded if the network bandwidth gets very low. Experiments have been conducted with matrices of size 250,000×250,000 up to 3,240,000×3,240,000.

In future works, we plan to decentralize the architecture of JaceV in order to avoid bottlenecks on the Dispatcher. Some solutions to carry out those modifications lie in using for example a decentralized convergence detection algorithm, or storing Backups on computation nodes, and so, to reach a really P2P like environment. It could also be interesting to enable the environment to modify the number N of computing nodes (and as a consequence to redistribute the data among the processors) during the execution. In addition we plan to test JaceV with a large scale platform with more nodes.

References

1. Bertsekas, D., Tsitsiklis, J.: Parallel and Distributed Computation: Numerical Methods. Prentice Hall, Englewood Cliffs NJ (1989)
2. SETI@home: http://setiathome.ssl.berkeley.edu
3. Baldeschwieler, J., Blumofe, R., Brewer, E.: Atlas: An infrastructure for global computing. 7th ACM SIGOPS European Workshop on System Support for Worldwide Application (1996)
4. Bahi, J., Miellou, J. -C., Rhofir, K.: Asynchronous multisplitting methods for nonlinear fixed point problems Numerical Algorithms, 15(3, 4) (1997) 315-345
5. Sato, M., Nakada, H., Sekiguchi, S., Matsuoka, S., Nagashima, U., Takagi, H.: Ninflet: A Network based information Library for a global world-wide computing infrastructure. HPCN'97 (LNCS-1225) (1997) 491-502
6. Takagi, H., Matsuoka, S., Nakada, H., Sekiguchi, S., Sato, M., Nagashima, U.: a Migratable Parallel Object Framework using Java. In Proceedings of the ACM 1998 Workshop on Java for High-Performance Network Computing (1998)
7. Aida, K., Nagashima, U., Nakada, H., Matsuoka, S., Takefusa, A.: Performance evaluation model for job scheduling in a global computing system. 7th IEEE International Symp on High Performance Distributed Computing. (1998) 352-353
8. Rosenberg A. L.: Guidelines for data-parallel cycle-stealing in networks of workstation. Journal of Parallel and Distributed Computing. 59 (1999) 31-53
9. Basney, J., Levy, M.: Deploying a High Throughput Computing Cluster. Volume 1, Chapter 5, Prentice Hall (1999)
10. Frommer, A. and Szyld, D.: On asynchronous iterations Journal of computational and applied mathematics. 23 (2000) 201-216
11. Bosilca, G., Bouteiller, A., Capello, F., Djilali, S., Fedak, G., Germain, C., Herault, T., Lemarinier, P., Lodygensky, O., Magniette, F., Neri, V., Selikhov, A.: MPICH-V: Toward a Scalable Fault Tolerant MPI for Volatile Nodes. ACM/IEEE International Conference on SuperComputing, SC 2002, Baltimore, USA (2002)
12. Elnozahy, E.N., Alvisi, L., Wang, Y.M., and Johnson, D.B.: A survey of rollback-recovery protocols in message-passing systems. ACM Comput. Surv., 34(3) (2002) 375-408
13. Bouteiller, A., Capello, Herault, T., Lemarinier, P., Magniette, F.: MPICH-V2: a Fault Tolerant MPI for Volatile Nodes based on Pessimistic Sender Based Message Logging. ACM/IEEE International Conference on SuperComputing, SC 2003, Phoenix, USA (2003)
14. Bahi, J., Domas, S. and Mazouzi, K.: Combination of java and asynchronism for the grid: a comparative study based on a parallel power method. 6th International Workshop on Java for Parallel and Distributed Computing, JAVAPDC workshop of IPDPS 2004, IEEE computer society press (2004) 158a, 8 pages
15. Browne, J. C., Yalamanchi, M., Kane, K., Sankaralingam, K.: General Parallel Computations on Desktop Grid and P2P Systems. 7th Workshop on Languages, Compilers and Runtime Support for Scalable Systems. LCR 2004, Houston,Texas (2004)
16. Cappello, F., Djilali, S., Fedak, G., Hérault, T., Magniette, F., Néri, V. and Lodygensky, O.: Computing on large-scale distributed systems: Xtremweb architecture, programming models, security, tests and convergence with grid. Future Generation Comp. Syst., 21(3) (2005) 417-437

Aspect Oriented Pluggable Support for Parallel Computing*

João L. Sobral[1], Carlos A. Cunha[2], and Miguel P. Monteiro[3]

[1] Departamento de Informática, Universidade do Minho, Braga, Portugal
[2] Escola Superior de Tecnologia, Instituto Politécnico de Viseu, Portugal
[3] Faculdade de Ciências e Tecnologia, Universidade Nova de Lisboa, Portugal

Abstract. In this paper, we present an approach to develop parallel applications based on aspect oriented programming. We propose a collection of aspects to implement group communication mechanisms on parallel applications. In our approach, parallelisation code is developed by composing the collection into the application core functionality. The approach requires fewer changes to sequential applications to parallelise the core functionality than current alternatives and yields more modular code. The paper presents the collection and shows how the aspects can be used to develop efficient parallel applications.

1 Introduction

The widespread use of multithreaded and multi-core architectures requires adequate tools to refactor current applications to take advantage of this kind of platforms. Unfortunately, parallelising compilers do not yet produce acceptable results, forcing programmers to rewrite their applications to take advantage of this kind of systems. When they do this, parallelisation concerns become intertwined with application core functionality, increasing complexity and decreasing maintainability and evolvability.

Tangling concurrency and parallelisation concerns with core functionality was identified as one of the main problems in parallel applications, increasing development complexity and decreasing code reuse [1, 2]. Similar negative phenomena of *code scattering* and *tangling* were identified as symptoms of the presence of *crosscutting concerns* in traditional object oriented applications [3]. Aspect Oriented Programming (AOP) was proposed to deal with such concerns, enabling programmers to localise within a single module code related to a crosscutting concern.

The use of AOP to implement parallelisation concerns provides the same benefits of modularisation as in other fields, namely improved code readability and an increased potential for reusability and (un)pluggability, for both parallelisation concerns and sequential code. AOP techniques were successful in modularising distribution code [4, 5, 6], middleware features [7], and, to a lesser extent, in isolating parallel code in loop based parallel applications [2].

---

* This work is supported by PPC-VM (Portable Parallel Computing based on Virtual Machines) project POSI/CHS/47158/2002, funded by Portuguese FCT (POSI) and by European funds (FEDER). Miguel P. Monteiro is partially supported by FCT under project SOFTAS (POSI/EIA/60189/2004).

M. Daydé et al. (Eds.): VECPAR 2006, LNCS 4395, pp. 93–106, 2007.
© Springer-Verlag Berlin Heidelberg 2007

This paper presents a collection of aspect oriented abstractions for parallel computing that replace traditional parallel computing constructs and presents several case studies that illustrate how this collection supports the develop parallel applications. Section 2 presents related work. Section 3 presents a brief overview of AspectJ, an AOP extension to Java that was used to implement the collection. Section 4 presents the collection. Section 5 presents several case studies and section 6 presents a performance evaluation. Section 7 concludes the paper.

2 Related Work

We classify related work in two main areas: concurrent object oriented languages (COOL) and approaches to separate parallel code from core functionality.

COOLs received a lot of attention in the beginning of the 1990s. ABCL [8] provides active objects to model concurrent activities. Each active object is implemented by a process and inter-object communication can be performed by asynchronous or synchronous method invocation. Concurrent Aggregates [9] is a similar approach but supports groups of active objects than can work in a coordinated way and includes mechanisms to identify an object within a group. Recent COOLs are based on extensions to sequential object oriented languages [10, 11, 12]. These extensions introduce new language constructs to specify active objects and/or asynchronous method calls. ProActive [13] is an exception, as it relies on an implicit wait by necessity mechanism, however, when a more fine grain control is required, an object body should be provided (to replace the default active object body). Object groups, similar to concurrent aggregates, were recently introduced [14, 15]. With these approaches, the introduction of concurrency primitives and/ or object groups entails major modifications to source code. Parallelisation concerns are intertwined with core functionality, yielding the aforementioned negative phenomena of code *scattering* and *tangling*.

One approach to separate core functionality from parallel code is based on *skeletons* where the parallelism structure is expressed through the implementation of off-the-shelf designs [16, 17, 18, 19]. In generative patterns [20], the skeletons are generated and the programmer must fill the provided *hooks* with core functionality.

AspectJ was used in [4, 5, 6] to compose distribution concerns into sequential applications. In [2], an attempt is made to move all parallelism related issues into a single aspect and [21] proposes a more fine-grained decomposition. In [22], a collection of reusable implementations of concurrency concerns is presented.

OpenMP [23] introduces concurrency concerns by means of programming annotations that can be ignored by the compiler in a sequential execution.

Our approach differs from the aforementioned efforts in that we propose a collection of reusable aspects that achieve the same goals, by supporting object group relationships. We use concurrency constructs equivalent to traditional COOLs but we deploy all code related to parallelism within (un)pluggable aspects. Our approach differs from skeleton approaches as it uses a different way to compose core functionality and parallel code. Our approach requires less intrusive modifications to the core functionality to achieve a parallel application, yields code with greater potential for reuse and supports (un)plugability of parallelisation concerns.

3 Overview of AspectJ

AspectJ [24, 25] is a backwards compatible extension to Java that includes mechanisms for AOP. It supports two kinds of crosscutting composition: static and dynamic. Static crosscutting allows type-safe modifications to the application static structure that include member introduction and type-hierarchy modification. AspectJ's mechanism of *inter-type declarations* enables the introduction of additional members (i.e. fields, methods and constructors). AspectJ's type-hierarchy modifications add super-types and interfaces to target classes. Fig. 1 presents a point class and Fig. 2 presents an Aspect that changes class *Point*, to implement interface *Serializable*, and to include an additional method, called *migrate*.

```
public class Point {
        private int x=0;
        private int y=0;

        public void moveX(int delta) {  x+=delta; }
        public void moveY(int delta) {  y+=delta; }

        public static void main(String[] args) {
                Point p = new Point();
                p.moveX(10);
                p.moveY(5);
        }
}
```

Fig. 1. Sample point class

```
public aspect StaticIntroduction {
        declare parents: Point implements Serializable;
        public void Point.migrate(String node) {  System.out.println("Migrate to node" + node); }
}
```

Fig. 2. Example of a static crosscutting aspect

Dynamic crosscutting enables the *capture* of various kinds of execution events, dubbed *join points*, including object creation, method calls or accesses to instance fields. The construct specifying a set of interesting join points is a *pointcut*. A pointcut specifies a set of join points and collects context information from the captured join points. The general form of a named pointcut is:

<visibility-modifier> pointcut <name>(ParameterList): <pointcut_expression>;

The *pointcut_expression* is composed by pointcut designators (PCDs), through operators &&, ||, and !. AspectJ PCDs identify sets of join points, by filtering a subset of all join points in the program. Join point matching can be based on the kind of join point, on scope and on join point context. For more information on PCDs, see [24].

Dynamic crosscutting also enables composing behaviour before, after or instead of each of the captured join points using the *advice* construct. Advices have the following syntax:

```
[before I after I <Type> around] (<ParameterList>): <pointcut_expression>
    {... // added behaviour }
```

The *before* advice adds the specified behaviour before the execution point associated to the join points quantified by the *pointcut_expression*. *around* advices replace the original join point with new behaviour and is also capable of executing the original join point through the *proceed* construct. *after* advice adds new behaviour immediately after the original execution point. The *pointcut_expression* is an expression comprising one or several PCDs that can also reuse previous pointcut definitions. Objects and primitive values specific to the context of the captured join point are obtained through PCDs *this, target* and *args*. Fig. 3 shows the example of a logging aspect, applied to class Point. In this example, a message is printed on the screen on every call to methods moveX or moveY. The wildcard in the poincut expression is used to specify a pattern for the call's signature to intercept.

```
public aspect Logging {
    void around(Point obj, int disp) : call(void Point.move*(int)) && target(obj) && args(disp) {
        System.out.println("Move called: target object = " + obj + " Displacement " + disp);
        proceed(obj,disp);                   // proceed the original call
    }
}
```

Fig. 3. Example of a dynamic crosscutting aspect

Modularisation of crosscutting concerns is an achievement that contributes to code reusability. Though it is a necessary condition, it is not a sufficient one, as only the non case-specific code is reusable. Essential parts of the aspect's behaviour are the same in different join points, whereby other parts vary from join point to join point. Reuse of crosscutting concerns requires the localisation of reusable code within *abstract base aspects* that can be reused by concrete sub-aspects. Concrete aspects contain the variable parts tailored to a specific code base, specifying the case-specific join points to be captured in the logic declared by the abstract aspect. Abstract aspects rely on abstract pointcuts and/or marker interfaces. In both cases, the abstract aspect only refers to abstract pointcut(s) or to the interface(s) and is therefore potentially reusable. Each concrete implementation entails the creation of one or several concrete sub-aspects that concretise inherited pointcuts by specifying the set of join points specific to the system at hand, and by making case-specific types implement the marker interfaces. In addition, aspects can hold their own state and behaviour.

An aspect is supposed to localise code related to a concern that otherwise would be crosscutting. A composition phase called *weaving* enables the placement of aspect code in multiple non-contiguous points in the system. As an example, the behaviour specified by the around advice in Fig. 3 is composed in all base classes that call moveX or moveY methods.

4   Aspect Oriented Collection for Parallel Computing

The aspect oriented collection (Table 1) presented in this paper is based on three programming abstractions: *separable/migrable objects, asynchronous method calls* and *object aggregates*. By implementing the abstractions through aspects, it becomes possible to turn a given sequential application (i.e., sequential, domain-specific, object oriented code) into a parallel application. However, the base code should be amenable for parallelisation, i.e., the amount of parallelism that can be introduced by the aspect collection is subject to dependencies in application tasks and data. The composition of the collection with core functionality requires a set of suitable join points. If these are not available, the source code must be refactored to expose the necessary join points.

Table 1. Aspect oriented collection of abstractions for parallel computing

Abstraction	Scope	Description
Separate	Class	Separate object - can be placed in any node
Migrable	Class	Migrable object - can migrate among nodes
Grid1D, Grid2D	Class	Object aggregate in a 1 or 2d GRID
OneWay	Method	Spawns a new thread to execute the method
Future	Method	Spawns a new thread and returns a future
Synchronised	Method	Implements object-based mutual exclusion
Broadcast/scatter	Aggregate	Broadcast/scatter method among members
Reduction/gather	Aggregate	Reduce/gather method among members
Redirection	Aggregate	Redirect method call to one member (round-robin)
DRedirection	Aggregate	Redirect call to one member (demand-driven)
Barrier	Aggregate	Barrier among aggregate members

Separable objects are objects that can be placed in remote nodes, selected by the run-time system. *Migrable objects* are similar but they can migrate to a different node after their creation. These two abstractions are specified through the separable and migrable interfaces using the declare parents AspectJ construct (see section 2).

Asynchronous method calls introduce parallel processing between a client and a server. The client can proceed while the server executes the requested method. Asynchronous calls can be *OneWay* and *Future*. One-way calls are used when no return value is required. Fig. 4 shows the synopsis for the use of one-way calls.

```
public aspect aspectName extends OnewayProtocol {
   protected pointcut onewayMethodExecution(Object servant) : <pointcut definition>;
   protected pointcut join() : <pointcut definition>;
}
```

Fig. 4. One-way introduction

Pointcut *onewayMethodExecution* specifies the join points associated to invocation of methods that run into a new parallel task. Pointcut *join* can optionally be used to specify join points where the main thread blocks, waiting for the termination of the spawned tasks.

Future calls are used for asynchronous calls that require a return value. In typical situations, a variable stores the result of a given method call, which is used in a later phase. Instead of blocking in the method call, the client blocks when the variable that stores the result (i.e., the value returned by the method) is actually accessed. Fig. 5 shows the synopsis for the implementation of futures.

```
public aspect aspectName extends FutureProtocol {
    protected pointcut futureMethodExecution(Object servant): <pointcut definition>;
    protected pointcut useOfFuture(Object servant): <pointcut definition>;
}
```

Fig. 5. Future introduction

Pointcut *futureMethodExecution* indicates the asynchronous method calls and pointcut *useOfFuture* defines the join points where the result of the call is needed. The client blocks on join points captured by *useOfFuture*, in case the methods defined in *futureMethodExecution* have not completed execution.

A richer set of primitives for synchronisation is also available [22], namely Java's synchronised methods, barriers and waiting guards, but their description is out of scope of this paper.

Object aggregates are used to transparently represent a set of object instances in the core functionality. An object aggregate deploys one or several object instances in each node (usually one per physical processor/core) and provides additional constructs to access the members of the aggregate. There are two main interfaces to support aggregates: *Grid1D* and *Grid2D*; they differ only in the way the internal members of the aggregate are referenced. For instance, a *Grid1D* aggregate provides two calls: *getAggregateElems()* and *getAggregateElemId()*. *Grid1D* and *Grid2D* aggregates are specified in a way similar to separate objects (i.e., using *declare parents*).

Calls to the original object instance (i.e., calls in the core functionality) are replaced by calls to the first object in the aggregate (called the aggregate representative). These calls can also be *broadcasted, scattered* and *reduced* among members of the aggregate. Broadcasted calls are executed in parallel by all aggregate members, using the same parameters of the core functionality call. Such call returns when all broadcasted calls complete. Fig. 6 shows the synopsis for the use of broadcasted calls. Pointcut *broadcastMethodExecution* specifies method calls broadcasted to all aggregate members.

```
protected pointcut broadcastMethodExecution(Object servant) : <pointcut definition>;
```

Fig. 6. Broadcasted calls introduction

Scattered calls (Fig. 7) are similar to broadcasted calls but they provide a mechanism to specify a different parameter for each call into aggregates member. This is specified by implementing the abstract method *scatter* which returns a vector whose elements correspond to the parameters sent to aggregate members.

```
protected Vector scatter(Object callParameter) {
    ...
}
protected pointcut scatterMethodExecution(Object serv, Object arg) : <pointcut definition>;
```

Fig. 7. Scattered calls introduction

Reduced calls are also similar to broadcasted calls, but they provide a mechanism to combine return values of each aggregate member call. This type of calls should be used instead of a broadcasted call, when the call returns a value. In this case a reduction function specifies how to combine the returned values of each aggregate member call (Fig. 8).

```
protected Object reduce(Vector returnValues) {
    ...
}
protected pointcut reduceMethodExecution(Object serv, Object arg) : <pointcut definition>;
```

Fig. 8. Reduced calls introduction

An additional function (scatter/reduce) performs a combination of scatter and reduce calls. Other aggregate functions can redirect a call to one aggregate member in a round-robin fashion (*redirectCall*) or in a demand driven scheme (*dredirectCall*).

Broadcasted, scattered and reduced calls are valid just for object aggregates (e.g., method calls on objects that implement interfaces *Grid1D* or *Grid2D*).

Fig. 9 shows a simple application that illustrates the use of this collection of aspects. The object *Filter* in the core functionality (left column of Fig. 9) is replaced by an aggregate in the parallelisation code (right column, *declare parents* statement) and calls to method *filter* are broadcasted, in parallel, to all aggregate members (pointcuts *broadcastMethodExecution* and *onewayMethodExecution*). Before *filter* method execution (advice *before() execution(* Filter.filter)*), each aggregate member displays its identification within the aggregate.

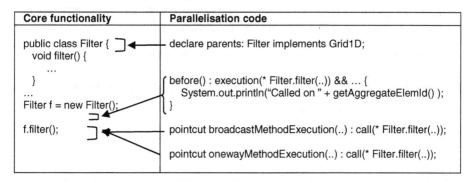

Core functionality	Parallelisation code

public class Filter {
   void filter() {
     ...
   }
 ...
 Filter f = new Filter();
 	
 f.filter();
	declare parents: Filter implements Grid1D;
	
	 before() : execution(* Filter.filter(..)) && ... {
	   System.out.println("Called on " + getAggregateElemId());
	 }
	 pointcut broadcastMethodExecution(..) : call(* Filter.filter(..));
	
	 pointcut onewayMethodExecution(..) : call(* Filter.filter(..));

Fig. 9. Simple application example

5 Case Studies

This section presents two case studies that illustrate the use of the aspect collection to develop modular parallel applications. The case studies are taken from the parallel Java Grande Forum Benchmark (JGF) [26]. This benchmark includes several sequential scientific codes and parallel versions of the same applications, using mpiJava (a bind of MPI to Java). Their parallel implementations introduce modifications to the sequential code, intermingling domain specific code with MPI primitives to achieve a parallel execution. Tangling makes it difficult to understand the parallelisation strategy as well as the domain specific code. Our approach entails introducing as fewer modifications as possible to the domain scientific code by introducing the parallelisation logic through non-invasive composition of the aspects from the collection. We believe that this approach makes the implementation of the parallelisation strategy more modular and explicit.

The first case study is a Successive Over-Relation method (SOR), an iterative algorithm to solve Partial Differential Equations (PDEs). This application is parallelised using a heartbeat scheme, where each parallel task processes part of the original matrix. After each iteration, neighbour parallel tasks must exchange information required for the next iteration.

The second application is a ray-tracer that renders a scene with 64 spheres. It is parallelised using a farming strategy, where each worker renders a set of image lines.

5.1 Successive Over-Relation

The SOR method is used to iteratively solve a system of PDE equations. The method successively calculates each new matrix element using its neighbour points. The sequential Java program of the JGF method is outlined in Fig. 10. This code iterates a number of pre-defined iterations, given by *num_iterations*, over matrix *G*.

In this particular case, the sequential version could limit parallelism due to dependencies among calculations. To overcome this limitation the *SORrun* implementation was changed to use the Red-Black parallel version, becoming more amenable for parallel execution. This strategy was also followed in the JGF parallel benchmark to derive the parallel version of the application.

```
public class SOR {
    ...
    public static final void SORrun(double omega, double G[][], int num_iterations) {
        ...
        for (int p=0; p<num_iterations; p++) {
            ... // performs one iteration
        }
        ...
    }
}
```

Fig. 10. JGF SOR sequential code

The sequential code from the JGF does not provide adequate join points to compose with our collection. Our first step is to use the static crosscutting of AspectJ to make this code suitable for composition with parallelisation code (Fig. 11). This code introduces two new methods into the SOR class: the *init* method (lines 04-05) initialises the SOR matrix and the *iterate* method (lines 07-08) performs one iteration. In lines 10-17 the original SORrun call is redefined to call these methods. An

```
01    double SOR.MyG[][],
02    static int SOR.omega;
03
04    // initialise matrix
05    public void SOR.init(double G[][]) { MyG = G; }
06
07    // performs one iteration
08    public void SOR.iterate() { SORrun(omega, MyG, 1); }
09
10    // redirects SORrun calls to use SOR instances, init call and iterate calls
11    void around(double omega, double G[][], int iterations) call(* SOR.SORrun(..)) && ... {
12        SOR.omega = omega;
13        SOR so = new SOR();
14        so.init(G);
15        for(int i=0; i<iterations; i++)
16            so.iterate();
17    }
```

Fig. 11. SOR method core functionality

alternative would be to refactor all the JGF SOR sequential code to use SOR instances, *init* and *iterate* calls.

SOR core functionality can be parallelised through a typical heartbeat strategy. According to this strategy, each parallel task iterates over a subset of the matrix, periodically exchanging boundary information with its neighbours. The parallelisation aspect has four parts: 1) creates multiple SOR objects; 2) assigns a subset of the matrix to each SOR object; 3) performs a call to the *iterate* method on all the objects in the set and 4) exchanges matrix lines among objects after each iteration.

The first step creates an aggregate of SOR objects in place of a single object (Fig. 12), by specifying that the SOR class implements the Grid1D interface (line 01 in Fig. 13). Our system intercepts the creation of SOR instances in the core functionality and creates one SOR object on each node/CPU.

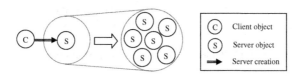

Fig. 12. Transparent creation of several SOR objects

The second step distributes the G matrix among the elements of the aggregate (Fig. 14). The code for this step intercepts the *init* method, splits the received matrix into blocks, using method *scatter* (line 02 in Fig. 13) and calls the *init* method on each object in the set, passing a different block to each element using the scatter method (line 03 in Fig. 13). Code for the matrix partition (*scatter* method in line 02 in Fig. 13) is a bit tricky to implement since there are lines from the matrix that are replicated in several objects and the first and the last objects receive one line less than other

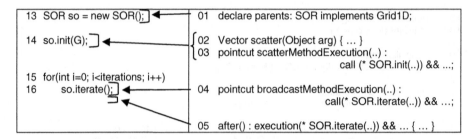

```
13  SOR so = new SOR();        01   declare parents: SOR implements Grid1D;

14  so.init(G);                02   Vector scatter(Object arg) { ... }
                               03   pointcut scatterMethodExecution(..) :
                                                     call (* SOR.init(..)) && ...;
15  for(int i=0; i<iterations; i++)
16      so.iterate();          04   pointcut broadcastMethodExecution(..) :
                                                     call(* SOR.iterate(..)) && ...;

                               05   after() : execution(* SOR.iterate(..)) && ... { ... }
```

Fig. 13. Parallelisation of the SOR application using our AOP collection

objects. However, this code is also required in a traditional parallel application and it
is usually tangled with the algorithm core functionality.

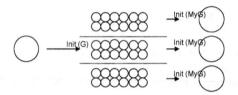

Fig. 14. Matrix distribution among SOR objects

Third, *iterate* method calls are executed by all SOR aggregate objects (Fig. 15).
Code for this operation implements the broadcast pointcut (line 04 in Fig. 13).

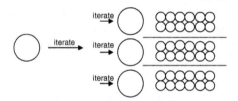

Fig. 15. Iteration distribution among SOR objects

The last step exchanges matrix boundary lines among SOR objects, after an iterate
method execution (Fig. 16 and line 05 in Fig. 13).

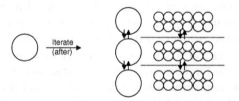

Fig. 16. Boundary exchange among SOR objects

5.2 RayTracer

The JGF RayTrace renders an image of sixty-four spheres. A simplified version of the JGF sequential code is provided in Fig. 17. Method *JGFinitialise* initialises the scene to be rendered and method *JGFapplication* renders the scene. The class *Interval* allows the specification of a subset of the lines to be rendered.

```
public class JGFRayTracerBench extends RayTracer ... {
    ...
    public void JGFinitialise(){
        ...
        scene = createScene();      // create the objects to be rendered
        setScene(scene);            // get lights, objects etc. from scene.
        ...
    }

    public void JGFapplication() {
        ...
        // Set interval to be rendered to the whole picture
        Interval interval = new Interval(0,width,height,0,height,1);

        render(interval);           // Do the business!
        ...
    }
}
```

Fig. 17. JGF RayTracer sequential code

The parallelisation aspect for this benchmark (Fig. 18) declares the class *JGFRayTracerBench* to implement the *Grid1D* interface (line 01). Calls to *JGFinitialise* are broadcasted to all aggregate members (line 03) and a call to the *render* method is scattered throughout aggregate elements. The *scatter* function builds a vector with the arguments for each call to one aggregate member. This is the same strategy followed in the JGF parallel version of this application.

```
01   declare parents: RayTracerBench implements Grid1D;
02
03   pointcut broadcastMethodExecution(Object servant) : call(* *. JGFinitialise(..)) && ... ;
04
05   Vector scatter(Object arg) {      // calculates the parameters of each call
06       Vector v = new Vector();
07       Interval in = (Interval) arg;
08       ...
09       for(int i=0; i<workers; i++) {
10           Interval inp = new Interval(/* sub-interval range */);
11           v.add(inp);    // saves the range of each worker
12       }
13       return(v);
14   }
15
16   pointcut scatterMethodExecution(Object serv, Object arg) : call (* *.render(..)) && ... ;
17
```

Fig. 18. JGF RayTracer parallelisation aspect

6 Performance Results

This section presents a performance evaluation of the proposed aspect collection. The results presented in this section were measured on an unloaded cluster of 8 dual-Xeon 3.2 GHz machines, with hyper-threading enabled, connected through a 1 Gbit Ethernet. This cluster runs Rocks 4.0.0 and Sun Java JDK 1.5.0_3 in client mode. Presented execution times are the median of five executions. Sequential execution times were measured on JGF versions where our parallelisation aspects were unplugged. Speed-up values are relative to these sequential execution times.

Fig. 19 presents the execution time for a SOR (4000x4000 matrix) and a RayTracer (500x500 image) on a single machine. With two aggregate members the ray tracer presents better speed-ups, due to less communication required among tasks. Both applications can benefit from hyper-threading (i.e., using more than two aggregate members per node). In this case, higher gains in the SOR can be due to stronger dependencies among matrix elements calculations; leading to higher parallelism when the user performs an explicit parallelisation (e.g., provides more independent tasks, by means of a higher number of aggregate members).

Fig. 20 presents execution times on 8 cluster nodes. Also in this case the ray tracer presents better speed-ups, due to less communication among tasks. Note that using more than 16 aggregate members leads to a smaller performance improvement, since this additional gain is achieved by using multi-threading capabilities of these processors.

Execution times compared to equivalent Java versions (not shown), using MPP (message passing library built on top of Java nio) and Java Threads are within 5% execution time. This low overhead is due to static nature of AspectJ weaving, which can inline most aspect code into the core classes. The aspect overhead results from additional data structures and from some code that can not be in-lined in the original and is placed in new classes. Scatter and reduce functions can also be an additional source of overhead, since they may require additional data copies.

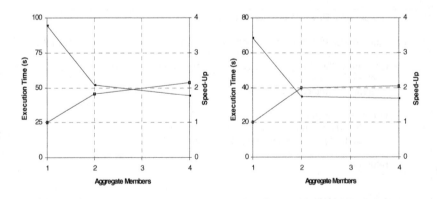

Fig. 19. Execution time and speed-ups for a SOR (at left) a RayTracer (at right)

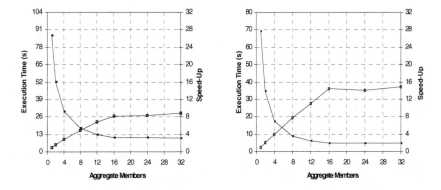

Fig. 20. Execution times and speed-ups for a SOR (at left) and a RayTracer (at right)

7 Conclusion

This paper presents a collection of aspects for parallel computing that requires fewer and smaller changes to parallelise sequential applications than current alternatives. It yields parallel object-oriented scientific applications that are more modular and easier to reuse. The collection was successfully applied to several JGF applications.

One of the main drawbacks of the approach stems from the non object-oriented nature of current scientific applications, as these do not provide adequate join point leverage to compose the sequential code with our collection. However, this limitation is expected to have less impact in the future, as scientific code becomes more object oriented. We can partially overcome this limitation by using the static crosscutting mechanisms of AspectJ to introduce the appropriate join points (as in the SOR application).

A second limitation is when the sequential code is not amenable for parallelisation. One solution is to refactor the core functionality in order to obtain a more fine grained decomposition. As an example, in the RayTracer example we could have a method *renderLine* which would provide more flexibility to derive the parallel version of RayTracer.

Current work includes the extension of this collection to support more orthogonal compositions of broadcast, scatter and reduce pointcuts; and a more efficient implementation of these pointcuts on distributed memory machines (e.g., using MPI collective primitives).

References

1. S. Matsuoka, K. Taura, A. Yonezawa: Highly Efficient and Encapsulated Re-use of Synchronisation Code in Concurrent Object-Oriented Languages, OOPSLA '93, Oct. 1993.
2. B. Harbulot, J. Gurd.. Using AspectJ to Separate Concerns in Parallel Scientific Java Code, ACM AOSD'04, Lancaster, UK, March 2004.
3. G. Kiczales, J. Lamping, A. Mendhekar, C. Maeda, C. V. Lopes, J.-M. Loingtier, J. Irwin. Aspect Oriented Programming, ECOOP '97, June 1997.

4. S. Soares, E. Loureiro, P. Borba. Implementing Distribution and Persistence Aspects With AspectJ, OOPSLA '02, November 2002.
5. M. Ceccato, P. Tonella. Adding Distribution to Existing Applications by means of Aspect Oriented Programming, 4th IEEE SCAM, September 2004.
6. E. Tilevich, S. Urbanski, Y. Smaragdakis, M. Fleury. Aspectizing Server-Side Distribution, IEEE ASE 2003, Montreal, Canada, October 2003.
7. C. Zhang, H. Jacobsen. Resolving Feature Convolution in Middleware Systems, OOPSLA'04, Vancouver, Canada, October 2004.
8. A. Yonezawa, M. Tokoro, ed, Object-Oriented Concurrent Programming, MIT Press, 1987.
9. A. Chien, V. Karamcheti, J. Plevyak, X. Zhang, Concurrent Aggregates (CA) Language Report - Version 2.0, TR, Dep. Computer Science, University of Illinois, UC, Nov., 1993
10. G. Wilson (Ed). Parallel Programming Using C++, MIT Press, 1996.
11. M. Philippsen. A Survey of Concurrent Object-Oriented Languages, Concurrency: Practice and Experience, 10(12), August 2000.
12. M. Factor, A. Schuster, K. Shagin. A Distributed Runtime for Java: Yesterday and Today, IEEE IPDPS'04, New Mexico, April 2004.
13. F. Baude , L. Baduel, D. Caromel, A. Contes, F. Huet, M. Morel, R. Quilici, Programming, Composing Deploying for the Grid, in GRID COMPUTING: Software Environments and Tools, Jose C. Cunha and Omer F. Rana (Eds), Springer Verlag, January 2006.
14. J. Maassen, T. Kielmann and H. Bal, GMI: Flexible and Efficient Group Method Invocation for Parallel Programming, Sixth Workshop on Languages, Compilers, and Run-time Systems for Scalable Computers (LCR-02), Washington DC, March 2002.
15. L. Baduel, F. Baude, D. Caromel, Object-Oriented SPMD, International Symposium on Cluster Computing and the Grid (CCGrid2005), Cardiff, May, 2005.
16. J. Darlington, Y. Guo, H. To, J. Yang. Parallel Skeletons for Structured Composition, PPoPP'95, Santa Clara, USA, 1995.
17. P. Trinder, K. Hammond, H. Loidl, S. Jones. Algorithm + Strategy = Parallelism, Journal of Functional Programming, 8(1), January 1998.
18. F. Rabhi, S. Gorlatch (ed): Patterns and Skeletons for Parallel and Distributed Computing, Springer, 2003.
19. J. Fernando, J. Sobral, A. Proenca. JaSkel: A Java Skeleton-Based Framework for Structured Cluster and Grid Computing, CCGrid'2006, Singapore, May 2006
20. K. Tan, D. Szafron, J. Schaeffer, J. Anvik, S. MacDonald. Using Generative Design Patterns to Generate Parallel Code for a Distributed Memory Environment, PPoPP'03, San Diego, California, USA, June, 2003.
21. J. Sobral, Incrementally Developing Parallel Applications with AspectJ, IEEE IPDPS'06, Rhodes, Greece, April 2006
22. C. Cunha, J. Sobral, M. Monteiro, M., Reusable Aspect-Oriented Implementations of Concurrency Patterns and Mechanisms, AOSD'06, Bonn, Germany, March 2006.
23. OpenMP architecture review board, OpenMP Application Program Interface, Version 2.5, May 2005, www.openmp.org.
24. G. Kiczales, E. Hilsdale, J. Hugunin, M. Kersten, J. Palm, W. Griswold, An Overview of AspectJ. ECOOP 2001, Budapest, Hungary, June 2001.
25. G. Kiczales, E. Hilsdale, J. Hugunin, M. Kersten, J. Palm, W. Griswold, Getting Started with AspectJ. Communications of the ACM, 44(10), October 2001.
26. A. Smith, J. Bull, J. Obdrzálek: A Parallel Java Grande Benchmark Suite, Supercomputing (SC'01), November 2001.

Model for Simulation of Heterogeneous High-Performance Computing Environments

Rodrigo Fernandes de Mello[1] and Luciano José Senger[2],[*]

[1] Universidade de São Paulo – Departamento de Computação
Instituto de Ciências Matematicas e de Computação
Av. Trabalhador Saocarlense, 400 Caixa Postal 668
CEP 13560-970 São Carlos, SP, Brazil
mello@icmc.usp.br
[2] Universidade Estadual de Ponta Grossa – Departamento de Informatica
Av. Carlos Cavalcanti, 4748
CEP 84030-900 Ponta Grossa, PR, Brazil
ljsenger@icmc.usp.br

Abstract. This paper proposes a new model to predict the process execution behavior on heterogeneous multicomputing environments. This model considers the process execution costs such as processing, hard disk acessing, message transmitting and memory allocation. A simulator of this model was developed which help to predict the execution behavior of processes on distributed environments under different scheduling techniques. Besides the simulator, it was developed a suite of benchmark tools in order to parameterize the proposed model with data collected from real environments. Experiments were conduced to evaluate the proposed model which used a parallel application executing on a heterogeneous system. The obtained results show the model ability to predict the actual system performance, providing an useful model for developing and evaluating techniques for scheduling and resource allocation over heterogeneous and distributed systems.

1 Introduction

The evaluation of a computing system allows the analysis of its technical and economic feasibility, safety, performance and correct execution of processes. In order to evaluate a system, techniques that estimate its behavior on different situations are used. Such techniques provide numerical results which allow the comparison among different solutions for the same problem [1]. The evaluation of a computing system may use elementary or indirect techniques. The elementary ones are directly applied over the system, so it is necessary to have it previously implemented. The indirect ones allow the system evaluation before its implementation, what is relevant at the project phase [2,3,4,5,6].

The indirect techniques use mathematic models to represent the behavior of the main system components. Such models should be as similar as possible to the real problems, generating results for a good evaluation without being necessary to implement them [6].

---

[*] The authors thank to William Voorsluys for improving the source code of the benchmark *memo* and the fundings from Capes and Fapesp Brazilian Foundations (under the process number 04/02411-9).

M. Daydé et al. (Eds.): VECPAR 2006, LNCS 4395, pp. 107–119, 2007.
© Springer-Verlag Berlin Heidelberg 2007

Several models have been proposed for the evaluation of the execution time and the process delay. They consider the CPU consumption, the performance slowdown due to the use of the virtual memory [7] and the time spent with messages transmitted through the communication network [8].

Amir et al. [7] have proposed a method for job assignment and reassignment on cluster computing. This method uses a queuing network model to represent the slowdown caused by virtual memory usage. In such model the static memory $m(j)$ used by the process is known. This model defines the load of each computer in accordance with the equation 1, where: $L(t, i)$ is the load of computer i at the instant t; $l_c(t, i)$ is the CPU occupation; $l_w(t, i)$ is the amount of main memory used; $r_w(i)$ is maximum capacity of the main memory; β is the slowdown factor due to the use of virtual memory. Such factor increases the process response time, what consequently reflects in a lower final performance. This work attempts to minimize the slowdown by means of scheduling operations.

$$L(t, i) = \begin{cases} l_c(t, i) & \text{if } l_w(t, i) \leq r_w(i) \\ l_c(t, i) * \beta & \text{otherwise} \end{cases} \tag{1}$$

Mello et al. [9] have proposed improvements to the slowdown model by Amir et al. [7]. This work includes new parameters which allow a better modelling of process slowdown. Such parameters are the capacity of CPU and memory, throughput for reading and writing on hard disk and delays generated by the use of the communication network. However, this model presents similar limitations to the work by Amir et al. [7], as it does not offer any resource to model, through equations, the delay caused by the use of virtual memory (represented in equation 1 by the parameter β), nor consider other delays of the process execution time generated by: message transmission, hard disk access and other input/output operations. The modeling of message transmission delays is covered by other works [8, 10].

Culler et al. [8] have proposed the LogP model to quantify the overhead and the network communication latency among processes. The overhead and latency cause delays among processes which communicate. This model is composed of the following parameters: L which represents the high latency limit or delay incurred in transmitting a message containing a word (or a small number of words) from the source computer to a destination; o represents the overhead which is the time spent by processor to prepare a message for sending or receiving; g is the minimum time interval between consecutive message transmittion (sending or receiving); P is the number of processors. The LogP model assumes a finite capacity network with the maximum transmission defined by L/g messages.

Sivasubramaniam [10] used the LogP model to propose a framework to quantify the overhead of parallel applications. In such framework are considered aspects such as the processing capacity and the communication system usage. This framework joins efforts of actual experiments and simulations to refine and define analytic models. The major limitation of this work is that it does not present a complete case study.

The LogP model can be aggregated to the model by Amir et al. [7] and Mello et al. [9], permitting to evaluate the process execution time and slowdowns considering the resources of CPU, memory and transmitted messages on the network. Although unifying the models, they are still incomplete because do not consider the spatial and message generation probability distributions. Motivated by such limitations, some studies have been proposed [11, 12].

Chodnekar *et al.* [11] have presented a study to characterize the probability distribution of messages on communication systems. In such work, the 1D-FFT and IS [13], Cholesky and Nbody [14], Maxflow [15], 3D-FFT and MG [16] parallel applications are evaluated executing on real environments. In the experiments, some informations have been captured such as the message sending and receiving moments, size of messages and destination. These informations were analyzed through statistic tools, and the spatial and message generation probability distributions obtained. The spatial distribution defines the frequency each process communicates with others. The message generation distribution defines the probability that each process sends messages to others.

They have concluded that the most usual message generation probability distribution for parallel applications are the exponential, hyperexponential and Weibull. It has also been concluded that the spatial distribution is not uniform and there are different traffic patterns during the applications' execution. In the most part of applications there is a process which receives and sends a large number of messages to the remainder processes (like a master for PVM – Parallel Virtual Machine – and MPI – Message Passing Interface – applications). The work also presents some features about message volume distribution, but there is not a precise analysis about the message size, overhead and latency.

Vetter and Mueller [12] have studied the communication behavior of scientific applications using MPI (Message Passing Interface). This study quantifies the average volume of transmitted messages and their size. It has been concluded that in peer-to-peer systems 99% of the transmitted messages vary from 4 to 16384 bytes. In collective calls this number varies from 2 to 256 bytes. This was combined with the studies on spatial and message generation distributions by Chodnekar *et al.* in [11] and to the LogP model [8] which allow the identification of overhead and communication latency in computing systems. By unifying these studies to the previously described slowdown models it is possible to evaluate the process behavior considering CPU, virtual memory and message transmittion. However, it is not possible to model voluntary delays in the execution of processes (generated by *sleep* calls) and accesses to hard disks.

Motivated by the unification of the previously presented models, the aggregation of the applications' voluntary delays and hard disk access, this paper presents the UniMPP (*Unified Modeling for Predicting Performance*) model. This model unifies the CPU consumption considered in the models by Amir *et al.* [7] and Mello *et al.* [9], the time spent to transmit messages modeled by Culler *et al.* [8] and Sivasubramaniam [13], the message volume and the spatial and message generation probability distributions by Chodnekar *et al.* [11], and Vetter and Mueller [12]. Experiments confirmed that this model can be used to predict the behavior of process execution on heterogeneous environments, once it generates the process response times very similar to the observed on real executions.

This model was implemented in a simulator which is parameterized with system configurations (CPUs, main and virtual memories, hard disk thoughput and network capacity) and receives processes for execution. Distribution functions are used to characterize the process CPU, memory, hard disk and network occupations. The simulator also generates new processes according to a probability distribution function, allowing to evaluate different scheduling and load balancing policies without needing the real execution.

As presented before, the simulator needs to be parameterized with the actual system configurations. For this purpose, a suite of benchmark tools was developed to collect informations such as the capacity of CPUs in MIPS (millions of instructions per second), the main and virtual memory behavior under a progressive occupation (this generates delay functions), the hard disk throughput in reading and writing operations (in MBytes per second) and the network delay (considering the overhead and latency in seconds).

The main contribution of this work is the UniMPP model which can be used with the simulator allied to the benchmark tools to predict the process execution time on heterogeneous environments. The simulator is prepared to receive new scheduling and load balancing policies and evaluate them using different workload models [17].

This paper is divided into the following sections: 2 The model; 3 Parameterization; 4 Model Validation; 5 Conclusions and References.

2   The Model

Motivated by the unification of the virtual memory slowdown models [7,9], by the models of delays in process execution caused by messages transmission [8, 10], by studies about spatial and message generation probability distributions [11], by the slowdown caused in main and virtual memory ccupation, by the definition of voluntary delay and access to hard disks, the UniMPP (*Unified Modeling for Predicting Performance*) model has been designed. These models are presented in the previous section. Unifying the ideas of each model and adding voluntary delays and hard disk access, we have defined a new model to predict the execution behavior of processes running on heterogeneous computers. By using this model, researchers can evalutate different techniques such as scheduling and load balancing without being necessary to run an application on an real environment.

In this model, a process p_j arrives at the system, following a probability distribution function, at the instant a_j. Such process is started by the computer c_i. Each computer maintains its queue $q_{i,t}$ of processes at the instant t. In this model, every computer c_i is composed of the sextuple $\{pc_i, mm_i, vm_i, dr_i, dw_i, lo_i\}$, where: pc_i is the total computing capacity of each computer measured in instructions per unit of time; mm_i is the total main memory; vm_i is the total virtual memory capacity; dr_i is the hard disk reading throughput; dw_i is the hard disk writing throughput; lo_i is the time spend in sending and receiving messages.

In the UniMPP, each process is represented by the sextuple $\{mp_j, sm_j, pdfdm_j, pdfdr_j, pdfdw_j, pdfnet_j\}$, where: mp_j represents the processing consumption; sm_j is the amount of static memory allocated by the process; $pdfdm_j$ is the probability distribution function used to represent the dynamic memory occupation; $pdf\,dr_j$ is the probability distribution function used to represent the hard disk reading; $pdf\,dw_j$ is the probability distribution function used to represent the hard disk writing; $pdf\,net_j$ is the probability distribution function used to represent the sending and receiving operations on communication system.

Having formally defined computers and processes, equations were defined to obtain the process response time and delay. The first equation (equation 2) presents the response time (TE_{p_j,c_i}) of a process p_j being executed in a computer c_i, where the total computing capacity pc_i of c_i and the processing consumption of p_j should be represented by the same metric, such as MI (millions of instructions when the capacity of processors was obtained in Mips – Millions of instructions per second) or MF (millions of float-point instructions when the capacity of processors was obtained in Mflops – Millions of float-point instructions per second).

$$TE_{p_j,c_i} = \frac{mp_j}{pc_i} \qquad (2)$$

The equation 2 presents a calculation method for the execution time of a process under ideal conditions, in which there is no competition nor delays caused by the memory and input/output usage. The work by Amir *et al.* [7] presents a more adequate equation in which, from the moment that the virtual memory starts to be used, there is a delay in the process execution. These authors use a constant delay in their equations. However, by using the benchmark tools described in section 3, it was observed that there are limitations in their model, since the performance slowdown is linear during the main memory usage and exponential from the moment the virtual memory starts to be used.

$$TEM_{p_j,c_i} = TE_{p_j,c_i} * (1 + \alpha) \qquad (3)$$

The Amir's performance model does not consider this linear performance slowdown caused by the use of the main memory and considers a constant factor for the performance slowdown caused by the use of the virtual memory when, in fact, this slowdown is exponential. The UniMPP models the process performance slowdown generated by the use of main and virtual memories, by the equation 3, where α represents a percentage obtained from a delay function and TE is presented in equation 2. This delay function is generated by a benchmark tool (section 3) where in the $x - axis$ is the memory occupation up to use all the virtual memory and in the $y - axis$ is the α value (the slowdown imposed in the process execution by the memory occupation).

A model which considers the process execution slowdown caused by the use of main and virtual memories become more adequate, however, it does not allow the precise quantification of the total execution time of processes which perform input and output operations to the hard disk. For this reason, experiments have been conduced and equations developed to measure the delays generated by accesses to hard disk. The equation 4 models the process delay generated by reading operations from hard disk, where: nr represents the number of reading accesses; $bsize$ represents the data buffer size; dr_i represents the throughput capacity for reading accesses from hard disk; and $wtdr_k$ represents the waiting time for using the resource.

$$SLDR_{p_j,c_i} = \sum_{k=1}^{nr} \frac{bsize_k}{dr_i} + wtdr_k \qquad (4)$$

The hard disk writing delay is defined by equation 5, where: nw represents the number of writing accesses; $bsize$ represents the data buffer size to be written; dw_i is the throughput capacity for writing accesses in hard disk; $wtdw$ is the waiting time for using the resource.

$$SLDW_{p_j,c_i} = \sum_{k=1}^{nw} \frac{bsize_k}{dw_i} + wtdw_k \qquad (5)$$

In addition to the delays caused by memory usage and input/output to hard disks, there are delays generated by sending and receiving messages on communication systems. Such delays vary according to the network bandwidth, latency and overhead of communication protocols [18, 19, 20]. The protocol latency involves the transmission time on communication system, which vary in accordance with the message size and control messages generated by the protocol [18, 19, 20]. The protocol overhead is the time involved for packing and unpacking messages for transmission. This time also varies according to the messages size [18, 19, 20]. The delay for sending and receiving messages is defined by equation 6, where: nm represents the number of sent and received messages; $\theta_{s,k}$, described in equation 7 is the time used for sending and receiving messages on communication system, not considering the wait for resources; and wtn_k represents the wait time, the queue time, to send or receive a message, when the resource is busy. The components of equation 7 are: $o_{s,k}$ overhead, which when multiplied by two allows the quantification of packing time (by the sender) and the unpacking time (by the receiver) of a message; and $l_{s,k}$ is the latency to transmit a message.

$$SLN_{p_j,c_i} = \sum_{k=1}^{nm} \theta_{s,k} + wtn_k \qquad (6)$$

$$\theta_{s,k} = 2 * o_{s,k} + l_{s,k} \qquad (7)$$

Aiming the unification of all previously described delay models, it is proposed the equation 8, which allows the definition of the response time (the prediction of this time in a real enviroment) of a process p_j in a computer c_i, where: lz is the process voluntary delay generated by the system calls *sleep*. In the case of load transference (that is, process migration) the communication channels may modify their behaviors and perform a higher or lower number of input/output operations (a process migrating to a computer where there are others which it communicates, reduces the latency and overhead because does not use the communication system, although it can overload the CPU). The equation 9 is the response time of a process p_j transferred among n computers.

$$\begin{aligned} SL_{p_j} = SL_{p_j,c_i} = TEM_{p_j,c_i} + SLDR_{p_j,c_i} + \\ SLDW_{p_j,c_i} + SLN_{p_j,c_i} + \\ lz \end{aligned} \qquad (8)$$

$$SL_{pj} = \sum_{k=1}^{n} SL_{pj,c_k} \qquad (9)$$

The UniMPP model unifies the concepts from models by Amir *et al.* [7], Mello *et al.* [9] and Culler *et al.* [8] and extends them by adding voluntary delay equations and the time for reading and writing accesses to hard disks. In addition, based on experiments, this work proposes new equations to define the main and virtual memory slowdown. By these equations, it was observed that the slowdown is linear when using the main memory, and exponential using the virtual. Such experiments were carried though using the benchmark tools from section 3. This model allows studies of scheduling, load balancing algorithms and prediction of process response times on heterogeneous environments.

The proposed model has been implemented in a simulator, named SchedSim[1], which allows other researchers to conduct related studies. Such simulator is implemented in Java language and uses the object oriented concepts that simplify its extension and functionality additions. The simulator is parameterized with system configurations (CPUs, main and virtual memories, hard disk thoughput and network capacity) and receives processes for execution. It generates new processes according to a probability distribution function, allowing to evaluate different scheduling and load balancing policies without needing the real execution.

3   Parameterization

In order to parameterize the SchedSim simulator using real environment characteristics, a suite of benchmark tools[2] was developed. These tools measure the capacity of CPU, reading and writing hard disk throughput and the message transmission delays. Such tools evaluate these characteristics until they reach a minimum sample size based on the central limit theorem, allowing to apply statistical summary measures such as confidence interval, standard deviation and average [21]. This suite is composed by the following tools:

1. `mips`: it measures the capacity of a processor, in millions of instructions per second. This tool uses a `bench()` function implemented by Kerrigan [22];
2. `memo`: it creates child processes until all main and virtual memories are filled up, measuring the delays of the context switches among processes. The child processes only allocate the memory and then sleep for some seconds, thus it does not consider the processor usage;
3. `discio`: it measures the average writing throughput (buffered and unbuffered) and the average reading throughput in local storage devices (hard disks) or remote storage devices (via network file systems);
4. `net` - it is composed of two applications, a customer and a server, which allow the evaluation of the time spent to send and receive messages over communication networks (based on the equation 7).

---

[1] Source code available at http://www.icmc.usp.br/~mello/outr.html
[2] Benchmark – source code available at http://www.icmc.usp.br/~mello/outr.html

4 Validation

In order to validate the proposed model, executions of a parallel application developed in PVM (Parallel Virtual Machine) [23] in a scenario composed of two homogeneous computers have been considered. This adopted application is composed of a master and worker processes. The master process launches one worker on each computer and defines three parameters: the problem size, that is, the number of mathematic operations executed to solve an integral (eq. 10) defined between two points a and b using the trapezium rule [24, 25], the number of bytes that will be transferred over the network and recorded in the hard disk. These workers are composed of four stages: message receiving, processing, writing into the hard disk and message sending. The message exchange happens between master and worker at the beginning and at the end of the workers' execution. The workers are instrumented to account the time consumed in operations.

$$\int_a^b 2 * \sin x + e^x \tag{10}$$

Scenario details are presented on the table 1 and they have been obtained with the benchmark suite. A message size of 32 bytes has been considered for the benchmark *net*. The table 2 presents the slowdown equations generated by using main and virtual memories, respectively, on the computers c_1 and c_2. Such equations have been obtained through the experiments with the benchmark *memo*. The linear format of the equations is used when the main memory is not completely filled up, for instance, in the case of computers c_1 and c_2 not exceed 1 *Gbyte* of its memory capacity. After exceeding such limit, the virtual memory is used and the delay is represented by the exponential funtion.

Table 1. System details

Resource	c_1	c_2
CPU (Mips)	1145.86	1148.65
Main memory (Mbytes)	1*Gbyte*	1*Gbyte*
Virtual memory (Mbytes)	1*Gbyte*	1*Gbyte*
Disk writing throughput (MBytes/seg)	65.55	66.56
Disk reading throughput (MBytes/seg)	76.28	75.21
Overhead + Latency (seconds)	0.000040	

The experiment results are presented in the table 3. It may be observed that the error among the curves is low, close to zero. Ten experiments have been conduced for different numbers of applications, each one composed of two workers executing on two computers. Such experiment was used to saturate the capacity of all computing resources of the environment. The figure 1 shows the experiment and simulation results.

The simulation obtained results show the model ability to reproduce the real system behavior. It is important to notice the increasing of the prediction errors when the system runs a number of processes between 90 and 100.

Table 2. Memory slowdown functions for computers c_1 and c_2

Memory	Regression	Equation	R^2
Main memory	Linear	$y = 0.0012x - 0.0065$	0.991
Main and Virtual memory	Exponential	$y = 0.0938 * e^{0.0039x}$	0.8898

Table 3. Simulation results for computers c_1 and c_2

Processes	Actual Average	Predicted	Error (%)
10	151.40	149.51	0.012
20	301.05	293.47	0.025
30	447.70	437.46	0.022
40	578.29	573.58	0.008
50	730.84	714.92	0.021
60	856.76	862.52	0.006
70	1002.10	1012.17	0.009
80	1147.44	1165.24	0.015
90	1245.40	1318.37	0.055
100	1396.80	1471.88	0.051

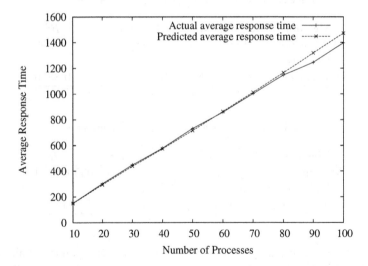

Fig. 1. Actual and predicted average response times for computers c_1 and c_2

The real executions, using 90 and 100 processes, overloaded the computers and some processes were killed by the PVM system. The premature stopping of processes (at about 5 processes where killed) decreases the computer's load, justifiyng the model prediction error. The simulator was used aiming to predict the system behavior considering a number of processes greater than the number of processes executed by PVM.

After experiments in an homogeneous system, a new environment composed of heterogeneous computers were parameterized using the benchmark tools. In this environment, it was executed the same application, which computes an integral function between two points using the trapezium rule. The features of the heterogeneous computers are presented in the table 4.

Table 4. System details

Resource	c_3	c_4
CPU (Mips)	927.55	1600.40
Main memory (Mbytes)	256	512
Virtual memory (Mbytes)	400	512
Disk write throughput (MBytes/seg)	47.64	15.99
Disk read throughput (MBytes/seg)	41.34	32.55
Overhead + Latency (seconds)	0.000056924	

The tables 5 and 6 present the slowdown equations, obtained by the *memo* benchmarking, considering the main and virtual memory usage.

Table 5. Memory slowdown functions for computer c_3

Memory	Regression	Equation	R^2
Main memory	Linear	$y = 0.0018x - 0.0007$	0.9998
Main and Virtual memory	Exponential	$y = 0.7335 * e^{0.0097x}$	0.8856

Table 6. Memory slowdown functions for computer c_4

Memory	Regression	Equation	R^2
Main memory	Linear	$y = 0.0018x - 0.0035$	0.9821
Main and Virtual memory	Exponential	$y = 0.0924 * e^{0.0095x}$	0.8912

The experiment results are presented in the table 7. The error values obtained comparing the simulated and the actual execution time values are close to 0, allowing to confirm the model ability in predicting real executions. The figure 2 shows the experiment and simulation results.

When a number at about 60 processes are running, some problems were observed, due to PVM process management. It was observed that using some computers with less processing power, PVM started to kill processes earlier, when running more than 60 processes. These problems explain the difference between the actual and the simulated time values and the increasing in predicting errors.

Table 7. Simulation results for computers c_3 and c_4

Processes	Actual Average	Predicted	Error (%)
10	153.29	152.38	0.0059
20	306.63	304.66	0.0064
30	457.93	457.46	0.0010
40	593.66	610.78	0.0280
50	760.02	764.65	0.0060
60	892.29	918.97	0.0290
70	1040.21	1074.18	0.0316
80	1188.14	1230.75	0.0346
90	1333.70	1388.14	0.0392
100	1488.97	1572.22	0.0529

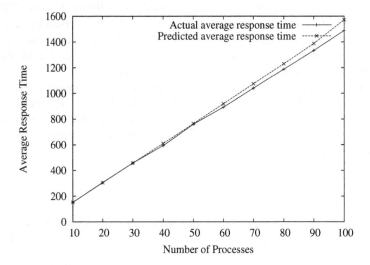

Fig. 2. Actual and predicted average response times for computers c_3 and c_4

The experiments presented in this section validate the model used by the simulator. The model and the simulator is able to predict the behavior of a real and dynamic system, modelling distinct parallel applications which solve problems from different areas, such as: aeronautics, fluid dynamics and geoprocessing. Thus, the system behavior can be predicted earlier, in project phase, minimizing the development costs.

5 Conclusions

Several models have been proposed to measure the response time of processes in computing systems [7,9]. Such models have presented some contributions, considering that the virtual memory occupation causes delays in process executions [7, 9], as well as delays generated by the message transmissions on communication systems [8,10]. Nevertheless, such models do not unify all possible delays of a process execution.

Motivated by such limitations, this work has presented a new unified model to predict the applications' execution running on heterogeneous distributed envionments. This model considers the process execution time in accordance with the processing, accesses to hard disk, message transmissions on communication networks, main and virtual memory slowdowns.

This work has contributed by modeling the delays in reading and writing accesses to hard disks and presenting a new technique which uses equations to represent the delays generated by the main and virtual memory usage. This has complemented studies by Amir *et al.* [7] and Mello *et al.* [9], which consider a constant delay.

In addition it was developed a simulator of the proposed model which can be used to predict the execution of applications on heterogeneous multicomputing environments. Such simulator has been developed considering extensions such as the design of new scheduling and load balancing policies. This simulator is licensed under GNU/GPL which allows its broad use by the researchers interested in developing and evaluating resource allocation techniques. In order to complement this simulator and allow its parameterization using real environment information, a suite of benchmark tools was developed and is also available under the GNU/GPL license.

In order to validate the simulator, a parallel application was implemented, simulated and executed on a real environment. It was observed that the percentage error obtained between the actual and the predicted execution times was lower than 1%, what confirms the accuracy of the proposed model to predict the application execution on heterogeneous multicomputing environments.

References

1. de Mello, R.F.: Proposta e Avaliacão de Desempenho de um Algoritmo de Balanceamento de Carga para Ambientes Distribuídos Heterogêneos Escaláveis. PhD thesis, SEL-EESC-USP (2003)
2. et. al, E.L.: Quantitative System Performance: Computer System Analysis Using Queueing Networks Models. Prentice Hall (1984)
3. et. al, P.B.: A Guide to Simulation. Spring-Verlag (1987)
4. Kleinrock, L.: Queueing Systems - Volume II: Computer Applications. John Wiley & Sons (1976)
5. Lavenberg, S.S.: Computer Performance Modeling Handbook. Academic Press (1983)
6. Jain, R.: The Art of Computer Systems Performance Analysis: Techniques for Experimental Design, Measurements, Simulation and Modeling. John Wiley & Sons (1991)
7. Amir, Y.: An opportunity cost approach for job assignment in a scalable computing cluster. IEEE Transactions on Parallel and Distributed Systems **11**(7) (2000) 760–768
8. Culler, D.E., Karp, R.M., Patterson, D.A., Sahay, A., Schauser, K.E., Santos, E., Subramonian, R., von Eicken, T.: LogP: Towards a realistic model of parallel computation. In: Principles Practice of Parallel Programming. (1993) 1–12
9. et. al, R.F.M.: Analysis on the significant information to update the tables on occupation of resources by using a peer-to-peer protocol. In: 16th Annual International Symposium on High Performance Computing Systems and Applications, Moncton, New-Brunswick, Canada (2002)

10. Sivasubramaniam, A.: Execution-driven simulators for parallel systems design. In: Winter Simulation Conference. (1997) 1021–1028
11. et. al, S.C.: Towards a communication characterization methodology for parallel applications. In: Proceedings of the 3rd IEEE Symposium on High-Performance Computer Architecture (HPCA '97), IEEE Computer Society (1997) 310
12. Vetter, J.S., Mueller, F.: Communication characteristics of large-scale scientific applications for contemporary cluster architectures. J. Parallel Distrib. Comput. **63**(9) (2003) 853–865
13. Sivasubramaniam, A., Singla, A., Ramachandran, U., Venkateswaran, H.: An approach to scalability study of shared memory parallel systems. In: Measurement and Modeling of Computer Systems. (1994) 171–180
14. Singh, J.P., Weber, W., Gupta, A.: Splash: Stanford parallel applications for shared-memory. Technical report (1991)
15. Anderson, R.J., Setubal, J.C.: On the parallel implementation of goldberg's maximum flow algorithm. In: Proceedings of the fourth annual ACM symposium on Parallel algorithms and architectures, San Diego, California, United States, ACM Press (1992) 168–177
16. Bailey, D.H., Barszcz, E., Barton, J.T., Browning, D.S., Carter, R.L., Dagum, D., Fatoohi, R.A., Frederickson, P.O., Lasinski, T.A., Schreiber, R.S., Simon, H.D., Venkatakrishnan, V., Weeratunga, S.K.: The NAS Parallel Benchmarks. The International Journal of Supercomputer Applications **5**(3) (1991) 63–73
17. Feitelson, D.G., Rudolph, L., Schwiegelshohn, U., Sevcik, K.C., Wong, P.: Theory and Practice in Parallel Job Scheduling. In: Job Scheduling Strategies for Parallel Processing. Volume 1291. Springer (1997) 1–34 Lect. Notes Comput. Sci. vol. 1291.
18. Chiola, G., Ciaccio, G.: A performance-oriented operating system approach to fast communications in a cluster of personal computers. In: In Proc. 1998 International Conference on Parallel and Distributed Processing, Techniques and Applications (PDPTA'98). Volume 1., Las Vegas, Nevada (1998) 259–266
19. Chiola, G., Ciaccio, G.: (Gamma: Architecture, programming interface and preliminary benchmarking)
20. Chiola, G., Ciaccio, G.: Gamma: a low cost network of workstations based on active messages. In: Proc. Euromicro PDP'97, London, UK, January 1997, IEEE Computer Society (1997)
21. W.C.Shefler: Statistics: Concepts and Applications. The Benjamin/Cummings (1988)
22. Kerrigan, T.: Tscp benchmark (2004)
23. Beguelin, A., Gueist, A., Dongarra, J., Jiang, W., Manchek, R., Sunderam, V.: PVM: Parallel Virtual Machine: User's Guide and tutorial for Networked Parallel Computing. MIT Press (1994)
24. Pacheco, P.S.: Parallel Programming with MPI. Morgan Kaufmann Publichers (1997)
25. Burden, R.L., Faires, J.D.: Análise Numérica. Thomson (2001)

On Evaluating Decentralized Parallel I/O Scheduling Strategies for Parallel File Systems

Florin Isailă, David Singh, Jesús Carretero, and Félix Garcia

Department of Compute Science,
University Carlos III de Madrid, Spain
{florin,desingh,jcarrete,fgarcia}@arcos.inf.uc3m.es

Abstract. This paper evaluates the impact of the parallel I/O scheduling strategy on the performance of the file access in a parallel file system for clusters of commodity computers (Clusterfile). We argue that the parallel I/O scheduling strategy should be seen as a complement to other file access optimizations like striping over several I/O servers, non-contiguous I/O and collective I/O. Our study is based on three simple decentralized parallel I/O heuristics implemented inside Clusterfile. The measurements in a real environment show that the performance of parallel file access may vary with as much as 86% for writing and 804% for reading with the employed heuristic and with the schedule block granularity.

1 Introduction

The performance of applications accessing large data sets is often limited by the speed of I/O subsystems. On one hand, this limitation comes from the ever increasing discrepancy between processor, memory speed and magnetic disks. On the other hand, the potential for parallelism existent in clusters of commodity computers and supercomputers is not always fully exploited by the I/O system software, like the parallel file systems [1,2,3,4,5,6,7,8,9,10,11] and libraries [12,13]. These systems employ mechanisms such as striping a file over several independent disks managed by I/O nodes and allowing parallel file access from several compute nodes.

For a better utilization of network and storage resources several point-to-point non-contiguous I/O methods have been proposed: data sieving [14], list I/O [15], view I/O [16]. These methods greedily optimize the communication between exactly one pair compute node - I/O node without regard at the global system performance. The collective I/O methods two-phase I/O [17] and disk-directed I/O [18] use collective buffers in order to gather the requests from compute nodes before sending them to disks. For disk-directed I/O the collective buffers reside at I/O nodes, whereas for two-phase I/O at intermediary compute nodes. Both of these methods describe how the data flows through the system between compute nodes and I/O nodes, but do not say anything about the order in which requests are sent between parallel running compute nodes and I/O nodes. However, an improper request ordering may cause idleness, load imbalance or resource contention, which may have a tremendous impact on performance.

The parallel I/O scheduling strategy may be seen as a complement to the above mentioned I/O optimizations. File striping describes a parallel spatial data placement,

M. Daydé et al. (Eds.): VECPAR 2006, LNCS 4395, pp. 120–130, 2007.
© Springer-Verlag Berlin Heidelberg 2007

whereas the parallel I/O strategy decides the temporal order of parallel requests. Non-contiguous I/O methods gather small messages into larger ones, while the parallel I/O strategy targets to schedule requests with sizes and in an order that optimize the resource usage. For collective I/O methods, the scheduling strategy may intervene both in the process of gathering the data into collective buffers, as well as in the sending the collective buffers to the I/O servers.

In previous work [19] we have developed a decentralized parallel I/O scheduling strategy for collective I/O operations. However, this strategy is specialized for collective I/O operations, and the impact of this strategy on the global system performance was not evaluated.

In this paper we evaluate three simple decentralized parallel I/O scheduling strategies for well balanced I/O loads implemented in the Clusterfile [20] parallel file system. We show that optimizations like non-contiguous I/O and collective I/O can not achieve a high resource utilization without a proper parallel I/O scheduling strategy. Additionally, the choice of the proper strategy and proper schedule block size may have an important influence on the overall file system performance.

2   Parallel I/O Scheduling Problem

The parallel I/O scheduling problem is not new. It was formulated by Jain and et al. [21] as follows. Given n_p compute nodes, n_{IOS} I/O servers and a set of requests for transfers of the same length among compute nodes and I/O servers and assuming that a compute node and an I/O server can perform exactly one transfer at any given time, find a service order that minimizes the schedule length [21].

Figure 1 shows an example, in which $n_p = 2$ compute nodes simultaneously issue in order four requests T1, T2, T3, T4 for $n_{IOS} = 2$ I/O servers. For this set of requests, several schedules are possible under the assumption that for each pair compute node - I/O node, only one request can be serviced at a time. Two of them are shown in the figure. In "Schedule 1", T1 and T2 are serviced at time 0; subsequently, T3 and T4 can not be scheduled simultaneously, because they have the same destination. The resulting schedule has the length 3. If T4 and T1 are scheduled in the first phase, T2 and T3 can be executed in parallel in the second phase and the schedule length is 2 ("Schedule 2").

The general scheduling problem is shown to be NP-complete, which makes it impractical for the real medium size parallel systems. Consequently, all solutions presented in the related work section are based on heuristics that try to minimize the schedule length, but without guaranteeing that the optimal value is used.

3   Related Work

The proposed solutions to the parallel I/O scheduling problem can be divided at least by five criteria. First, the proposed algorithms may be centralized or distributed. The centralized algorithms assume that there is a place in the system where all information about requests are gathered, before the schedule is computed. In the distributed algorithms the global schedule is computed in parallel by different nodes having only partial information about the requests. Second, some algorithms solve the parallel I/O schedule

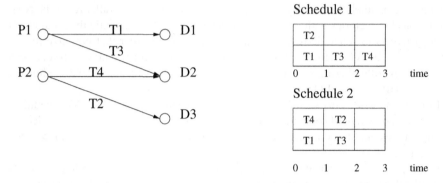

Fig. 1. Parallel I/O scheduling problem

problem in the presence of replication and others consider that there is only one copy of the data in the system. Third, the algorithms may be off-line or on-line. In the off-line algorithms the schedule is computed based on the fact that all the request information is available and is executed as such. In the on-line algorithms requests generated during the execution trigger a re-computation of the schedule. Fourth, the algorithms may differentiate between data with and with-out real-time constraints. Fifth, the evaluation can be based on simulations or on real implementations and systems.

The strategies incorporated in Clusterfile and discussed in this paper are decentralized, without replication, off-line, without real-time constraints, implemented and evaluated in a real environment.

Jain et al. [21] were among the first that formalized the parallel I/O scheduling problem in absence of replication and proposed three centralized off-line heuristics based on graph coloring algorithms. In First-Come First-Serve (FCFS), in each phase, as many as possible requests are served in parallel (colored with the same color) in the order of their arrival. For Figure 1, if the order of request arrival is T1, T2, T3, T4, "Schedule 1" is produced. Highest Degree First (HDF) considers first the graph nodes with the higher degrees in order to schedule parallel transfers. Both schedules from Figure 1 may be produced. Highest Common Degree First (HCDF) processes first the graph edges with the higher sum of the node degrees in order to schedule parallel transfers. Only the optimal "Schedule 2" can be produced in the example from Figure 1. The evaluation is based on a simulation and shows as expected a superior performance for the "more-informed" HCDF heuristic.

Chen and Majumdar [22] evaluate five centralized parallel I/O scheduling strategies for clusters in the presence of replication. On one hand they add replication support to FCFS and HCDF. On the other hand they propose Lowest Destination Degree First (LDDF), Shortest Job First (SJF) and Shortest Outstanding I/O Demand Job First (SOJF). The strategies are evaluated on a real system for single job and multiprogrammed workloads. An other real evaluation of five replication-based centralized parallel I/O scheduling strategies including those from [22] is presented in [23].

Durand et al. [24] propose distributed randomized parallel I/O scheduling algorithms based on edge colorings. In Uniformly at Random (UAR) each client selects randomly a request and sends it as a bid to an I/O server. Then each I/O server selects one received request at random and colors it with the current color. The algorithm repeats until all the graph is colored. Our implemented randomized strategy is a simplified version of this algorithm. MPASSES gives several (M) opportunities to color an edge to the clients whose proposal were rejected in the first pass of UAR. HDF is a distributed variant of the centralized HDF from [21] in which the clients send their degree together with the bid and the I/O servers picks up the client with the highest number of pending requests. The evaluation is based on a simulation.

In [25], the authors propose a decentralized update-based parallel I/O algorithm (D-SPTF) targeting load balance, efficient global cache exploitation and reducing disk positioning times for writing. The data may be replicated over several disks, which allows for an efficient read from the client which can serve the request fastest and a slow write due to the update of all replicas. Locality Aware Request Distribution (LARD) [26] requires a front-end which distributes the requests among the I/O servers according to the locality. Simulation results show that D-SPTF outperforms LARD and hash-based request distribution in terms of throughput and response time.

Lebre et al. [27] present the implementation and real system evaluation of two centralized parallel I/O strategies targeting global performance optimization and fairness in a multi-application environment. Their solution is a file system independent application-level library, whereas ours is done at file system level.

4   Parallel File System Overview

Clusterfile(CLF) [20] is a parallel file system for clusters of commodity computers. The architecture is based on the classical parallel file system model, in which the files are declustered over several I/O nodes managed by I/O servers. The applications run on compute nodes and access the file system through a POSIX-like proprietary interface or a classical UNIX interface after mounting the file system. Each individual process may declare a file *view*, i.e. a *logical contiguous* window mapped onto a non-contiguous file region. After declaration, each view can be accessed like a regular file. Clusterfile performs efficient non-contiguous I/O through a method called view I/O, described in detail in [16].

Clusterfile integrates two well-known collective I/O techniques, disk-directed [18] and two-phase I/O [17], into a common design [19]. The collective buffers are stored into a global cache, managed in cooperation by several cache managers running a version of the decentralized hash-distributed cooperative caching algorithm presented in [28].

5   Goals

The parallel I/O scheduling presented in this paper are implemented inside a real parallel file system. The parallel file system consists of four types of components: several parallel acting clients, several I/O servers, several cache managers, one metadata manager. The interaction between these components even in a relatively small cluster is highly complex. Experience with the xFS file system [29] has shown that complex protocols may make the development of a parallel system very difficult. In fact, the initial proposal of xFS was never fully implemented in part due to the exponential explosion of protocol complexity, which made bug detection very challenging even with formal verification tools. Consequently, when adding adding a parallel I/O scheduling strategy to an existing complex system we have in mind *simplicity*.

The scheduling strategy should have a small overhead. On one hand, this overhead is proportional with the number of messages exchanged for taking a scheduling decision. Even though the latency of the network is low, the communication may cause side-effects like context switches or evictions affecting data locality. On the other hand, data replication would perform poor for file writing. We have chosen not to replicate the data inside Clusterfile. Eventual replication schemes could be implemented on top of the file system.

Some scheduling strategies presented in the related work section are centralized. However, gathering the scheduling information at a central point may be difficult. First, this involves communication that adds additional complexity to the existing distributed protocols. Second, the additional communication for gathering the requests from all nodes and distributing the decision makes the solution costly and non-scalable. Therefore, the scheduling I/O strategies we chose are decentralized.

6   Parallel Scheduling I/O Heuristics

For all parallel scheduling heuristics, we assume that, at a certain point in time, n_p compute nodes simultaneously issue large data requests for n_{IOS} I/O servers. The decision of the order of data service is taken by the compute node for writing and by the I/O for reading in a similar way. For this reason we describe here only the write scheduling strategy. For writing, large requests are split by each compute node into smaller requests of size b.

In the first scheduling strategy, *first-IOS* (I/O server), each compute node sends the data to the I/O nodes in the order of file offsets. This is a natural approach, but may pose the potential risk that all the compute nodes send the data to the same I/O node at the same instant. However, the load balance problem may be compensated by high data locality in the case of non-contiguous interleaved access, as will be shown in the evaluation section.

In the second write scheduling strategy, *random-IOS*, each compute node first builds a list of requests targeted to each I/O node. Then the compute node chooses randomly the I/O server to which the data will be send until all the data is sent.

The third scheduling strategy, *hash-IOS*, is the one employed for the collective I/O operations of Clusterfile [19]. Conforming to the theoretical problem definition, for which each compute node can perform exactly one transfer at any given time, at time step $t_j, j = 0, 1, ...$, the compute node i sends a block to the I/O server $(i + j)$ modulo n_{IOS}.

Figure 1 shows an example, in which $n_p = 2$ compute nodes simultaneously issue 4 requests for $n_{IOS} = 2$ I/O servers. For the first-IOS method, CN0 decides to send the request to the IOS0 first, and then to IOS1, and a schedule of length 3 is produced ("Schedule 1"). On the other hand, hash I/O produces a schedule of length 2 ("Schedule 2"), as the I/O servers may run in parallel. Random IOS may produce any possible schedule, depending on the generated random numbers.

Notice that, for all strategies, there is no central point of decision, each process acts independently.

7 Evaluation

We performed our experiments on a cluster of 16 dual processor Pentium III 800MHz, having 256 KBytes L2 cache and 1024 MB RAM, interconnected by Myrinet LANai 9 cards at 133 MHz, capable of sustaining a throughput of 2 GB/s in each direction. The machines are equipped with IDE disks and were running LINUX kernels version 2.6.13 with the *ext2* local file system. We used TCP/IP on top of the 2.0.24 version of the GM [30] communication library. The ttcp benchmark delivered a TCP/IP node-to-node throughput of 120 MB/sec.

The I/O scheduling heuristics are all implemented inside the Clusterfile parallel file system.

In the following two subsections, we present the results for two different workloads: a synthetic parallel benchmark accessing contiguously a file and BTIO [31], a NASA parallel benchmark, in which several processes write non-contiguously to a file and then read back the result.

7.1 Synthetic Benchmark

We have written a synthetic benchmark in Message Passing Interface [32], in which all processes write and read in parallel different regions of the same file. The writes and reads are performed contiguously, as we first want to investigate the effect of parallel I/O heuristics on the performance, unaffected from the gather-scatter operations that are necessary for non-contiguous I/O.

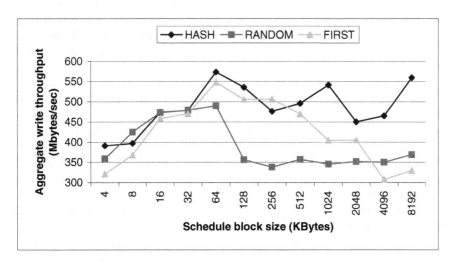

Fig. 2. Synthetic benchmark file write throughput

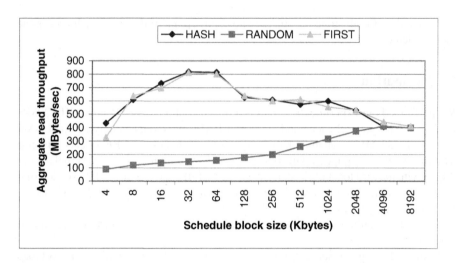

Fig. 3. Synthetic benchmark file read throughput

Clusterfile uses 8 I/O server running on 8 I/O nodes. The file block size is 64 KBytes. In the benchmark each of the 8 compute nodes writes 32 MBytes, resulting in a total of 256 MBytes. Each measurement was repeated 5 times and the mean value is reported.

Figures 2 and 3 show the aggregate throughput in MBytes/second obtained employing the three parallel I/O scheduling heuristics for writing and reading, respectively. The x-axis values represent the length of schedule block b (as introduced in the previous section).

First of all, note that for diverse parameters the performance of the same application may vary by as much as 86% for writing and 804% for reading. For writing, the highest value is obtained for hash-IOS for $b = 64KBytes$ (573 MBytes/second) and the lowest for first-IOS for $b = 4096KBytes$ (308 MBytes/second). For reading, the highest value is obtained for hash-IOS for $b = 32KBytes$ (817 MBytes/second) and the lowest for first-IOS for $b = 4KBytes$ (90 MBytes/second).

As expected, for first-IOS strategy, the aggregate write throughput decreased with schedule block granularity. The reason is that all the I/O servers try to send the data in the same order to all the I/O servers, which creates contention at I/O servers. The contention prevents the compute nodes from advancing and employing the other available I/O servers.

We have expected that the random-IOS write performance results lie somewhat in the middle between the results of hash-IOS and first-IOS. Surprisingly, the random IOS heuristic outperformed first-IOS only for the smallest four and largest two values. We believe that the reason lies in the fact that first-IOS generates a critical bottleneck only when accessing the first I/O server. The first compute node that "escapes" this bottleneck continues sending the data to the second I/O server and so on, generating a pipeline behavior. On the other hand, it appears that the randomly generated bottlenecks cause a higher overhead, as they can appear non-deterministically throughout the whole run of the application.

For large schedule block sizes, hash-IOS clearly outperforms the other two methods. For this heuristic each compute node starts by contacting a different I/O server which provides a good initial load balance, which is then preserved throughout the whole run by a cyclic access to the I/O servers.

The aggregate read throughput was similar for hash-IOS and first-IOS. This is due to the fact that in the present Clusterfile implementation an I/O server that receives the first request starts serving it. It appears that the initial potential bottleneck can not be overcome by the hash-IOS. This is unlike the write case, where the performance does not degrade with the the size of the schedule block.

The first-IOS and hash-IOS managed to exploit 85% of the theoretical point-to-point bandwidth of 8x120MBytes/second (as measured by the ttcp benchmark) for reading with $b = 32KBytes$ and $b = 64KBytes$. A further performance analysis is necessary in order to try to improve the write aggregate throughput.

7.2 BTIO Benchmark

NASA's BTIO benchmark [31] solves the Block-Tridiagonal (BT) problem, which employs a complex domain decomposition across a square number of compute nodes. Each compute node is responsible for multiple Cartesian subsets of the entire data set. The execution alternates computation and I/O phases. Initially, all compute nodes collectively open a file and declare views on the relevant file regions. After each five computing steps the compute nodes write the solution to a file through a collective operation. There are three sizes of the data sets: A (419.43 MBytes), B (1697.93 MBytes) and C (6802.44 MBytes). For these classes the benchmark performs 200 compute steps and 40 I/O steps. We are interested only in the results for a single I/O phase writing 10.5 MBytes (A), 42.2 MBytes (B) and 170 MBytes (C). The parallel I/O scheduling policies are relevant for large amounts of data, therefore, we report in this paper the I/O access times of the C class data set. The access pattern of C class is nested-strided with

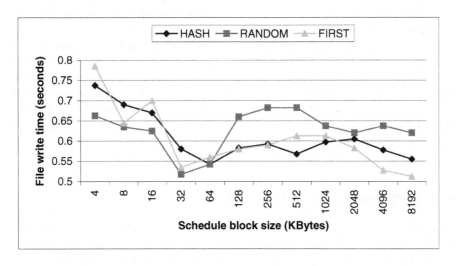

Fig. 4. BTIO file write times

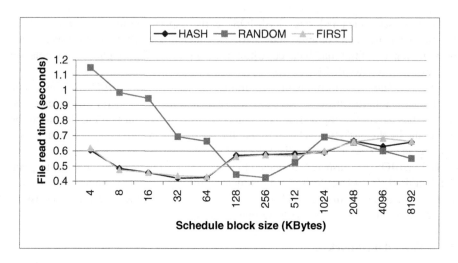

Fig. 5. BTIO file read times

a nesting depth of 2 with an access granularity of 3240 bytes. We report the results for 9 compute nodes and 9 I/O nodes in Figures 4 and 5 for writing and reading, respectively.

At the beginning of the BTIO benchmark, each process opens a file and declares a view on the file regions of interests. The individual file regions of the processes corresponding to the views do not overlap. Later, during each I/O phase, each process writes to the file through the view I/O method of Clusterfile [16]. Each process uses the view in order to contiguously send the data from each compute node to the I/O nodes. At I/O node, the data is scattered into the file blocks, kept in collective buffers. The reverse process takes place for reading. In a previous paper we have [19], we have shown that the combined view I/O and collective I/O method of Clusterfile significantly outperform two-phase I/O method of ROMIO [14], the most popular MPI-IO implementation. In the paper cited above the parallel scheduling strategy was fixed.

However, Figures 4 and 5 show that, depending on the parallel I/O scheduling strategy employed, the time to write a file in BTIO may vary with as much as 53% for writing (the ratio of 0.78 seconds for first-IOS with $b = 4KBytes$ to 0.51 seconds for first-IOS with $b = 8MBytes$) and 173% for reading (the ratio of 1.15 seconds for random-IOS with $b = 4KBytes$ to 0.42 seconds of hash-IOS with $b = 8MBytes$). As it can be noticed the performance span is not as large as in the case of the contiguous access. This is mainly due to the fact that the non-contiguous access generates a relatively constant overhead for scattering or gathering the data at the compute and I/O nodes.

8   Conclusions and Current Work

This paper presents and contrasts three parallel I/O scheduling heuristics implemented in Clusterfile parallel file system. The performance results show that the performance of parallel file access strongly depends on the choice of the parallel I/O scheduling strategy, as a combination of the employed heuristic and the granularity of the schedule.

An improper scheduling strategy may result in inefficient utilization of the parallel network paths, poor load balance and high contention at I/O nodes.

The classical parallel I/O optimizations like data striping, non-contiguous I/O, collective I/O should be seen as a complement to a parallel I/O scheduling strategy. Our experiments have demonstrated that various simple parallel I/O scheduling strategies may produce a performance difference of as much as 53% for file writing and 173% for file reading over the above mentioned optimizations.

The decentralized strategies presented in this paper address I/O workloads of well-balanced parallel applications. For irregular applications, some form of centralization or communication between the application library and I/O servers would be needed. Our current work includes the design and analysis of strategies for this type of applications.

Acknowledgments

The authors want to thank the anonymous reviewers for the very useful comments and suggestions.

This work has been funded in part by the project Técnicas de optimización y fiabilidad para sistems de entrada/salida escalables de altas prestaciones (Comunidad de Madrid-UC3M) and by the Spanish Ministry of Education and Science under the TIN2004-02156 contract.

References

1. DeBenedictis, E., Rosario, J.D.: nCUBE Parallel I/O Software. In: Proceedings of 11th International Phoenix Conference on Computers and Communication. (1992)
2. LoVerso, S., Isman, M., Nanopoulos, A., Nesheim, W., Milne, E., Wheeler, R.: sfs: A Parallel File System for the CM-5. In: Proceedings of the Summer 1993 USENIX Conference. (1993) 291–305
3. Huber, J., Elford, C., Reed, D., Chien, A., Blumenthal, D.: PPFS: A High Performance Portable File System. In: Proceedings of the 9th ACM International Conference on Supercomputing. (1995)
4. Corbett, P., Feitelson, D.: The Vesta Parallel File System. ACM Transactions on Computer Systems (1996)
5. Carretero, J., Serez, F., Miguel, P., Garca, F., Alonso, L.: ParFiSys: A Parallel File System for MPP. ACM SIGOPS 30(2) (1996)
6. Freedman, C., Burger, J., DeWitt, D.: SPIFFI-A Scalable Parallel File System for the Intel Paragon. IEEE Transactions on Parallel and Distributed Systems (October 1996)
7. Nieuwejaar, N., Kotz, D.: The Galley Parallel File System. Parallel Computing (1997)
8. O'Keefe, M.: Shared file systems and fibre channel. In: In the Proceedings of the Sixth NASA Goddard Space Flight Center Conference on Mass Storage Systems and Technologies. (1998)
9. Ligon, W., Ross, R.: An Overview of the Parallel Virtual File System. In: Proceedings of the Extreme Linux Workshop. (June 1999)
10. Schmuck, F., Haskin, R.: GPFS: A Shared-Disk File System for Large Computing Clusters. In: Proceedings of FAST. (2002)
11. Garcia-Carballeira, F., Calderon, A., Carretero, J., Fernandez, J., Perez, J.M.: The Design of the Expand Parallel File System. The International Journal of High Performance Computing Applications 17(1) (2003) 21–38

12. Winslett, M., Seamons, K., Chen, Y., Cho, Y., Kuo, S., Subramaniam, M.: The Panda library for parallel I/O of large multidimensional arrays. In: Proceedings of Scalable Parallel Libraries Conference III. (October 1996)
13. Message Passing Interface Forum: MPI2: Extensions to the Message Passing Interface. (1997)
14. Thakur, R., Gropp, W., Lusk, E.: Data Sieving and Collective I/O in ROMIO. In: Proc. of the 7th Symposium on the Frontiers of Massively Parallel Computation. (February 1999) 182–189
15. Thakur, R., Gropp, W., Lusk, E.: On Implementing MPI-IO Portably and with High Performance. In: Proc. of the Sixth Workshop on I/O in Parallel and Distributed Systems. (May 1999) 23–32
16. Isaila, F., Tichy, W.: View I/O:improving the performance of non-contiguous I/O. In: Third IEEE International Conference on Cluster Computing. (December 2003) 336–343
17. del Rosario, J., Bordawekar, R., Choudhary, A.: Improved parallel I/O via a two-phase runtime access strategy. In: Proc. of IPPS Workshop on Input/Output in Parallel Computer Systems. (1993)
18. Kotz, D.: Disk-directed I/O for MIMD Multiprocessors. In: Proc. of the First USENIX Symp. on Operating Systems Design and Implementation. (1994)
19. Isaila, F., Malpohl, G., Olaru, V., Szeder, G., Tichy, W.: Integrating Collective I/O and Cooperative Caching into the "Clusterfile" Parallel File System. In: Proceedings of ACM International Conference on Supercomputing (ICS), ACM Press (2004) 315–324
20. Isaila, F., Tichy, W.: Clusterfile: A flexible physical layout parallel file system. Concurrency and Computation: Practice and Experience 15(7–8) (2003) 653–679
21. Jain, R., Somalwar, K., Werth, J., Browne, J.C.: Heuristics for scheduling I/O operations. IEEE Transactions on Parallel and Distributed Systems 8(3) (March 1997) 310–320
22. Chen, F., Majumdar, S.: Performance of parallel I/O scheduling strategies on a network of workstations. In: Proceedings of ICPADS 2001. (April 2001) 157–164
23. Abawajy, J.H.: Performance Analysis of Parallel I/O Scheduling Approaches on Cluster Computing Systems. In: CCGRID '03: Proceedings of the 3st International Symposium on Cluster Computing and the Grid, Washington, DC, USA, IEEE Computer Society (2003) 724
24. Durand, D., Jain, R., Tseytlin, D.: Parallel I/O scheduling using randomized, distributed edge coloring algorithms. J. Parallel Distrib. Comput. 63(6) (2003) 611–618
25. Lumb, C.R., Golding, R.A., Ganger, G.R.: D-SPTF: decentralized request distribution in brick-based storage systems. In: ASPLOS. (2004) 37–47
26. Pai, V., Aron, M., Banga, G., Svendsen, M., Druschel, P., Zwaenepoel, W., Nahum, E.: Locality-Aware Request Distribution in Cluster-based Network Servers. In: Proceedings of the ACM Eighth International Conference on Architectural Support for Programming Languages and Operating Systems (ASPLOS-VIII) . (October 1998)
27. Lebre, A., Denneulin, Y., Van, T.T.: Controlling and Scheduling Parallel I/O in Multi-application Environments. Technical report, INRIA (2005)
28. Dahlin, M., Yang, R., Anderson, T., Patterson, D.: Cooperative Caching: Using Remote Client Memory to Improve File System Performance. In: The First Symp. on Operating Systems Design and Implementation. (November 1994)
29. Wang, R.Y., Anderson, T.E., Dahlin, M.D.: Experience with a distributed file system implementation with adaptive. Technical report (1998)
30. Myricom. GM: the low-level message-passing system for Myrinet networks: http://www.myri.com/. (2000)
31. Wong, P., der Wijngaart, R.: NAS Parallel Benchmarks I/O Version 2.4. Technical Report NAS-03-002, NASA Ames Research Center, Moffet Field, CA (2003)
32. Message Passing Interface Forum: MPI: A Message-Passing Interface Standard. (1995)

Distributed Security Constrained Optimal Power Flow Integrated to a DSM Based Energy Management System for Real Time Power Systems Security Control

Juliana M.T. Alves[1], Carmen L.T. Borges[2], and Ayru L. Oliveira Filho[1]

[1] CEPEL – Centro de Pesquisas de Energia Elétrica, Caixa Postal 68007,
CEP 21944-970 – Rio de Janeiro RJ
Tel.: +55-21-25986419
juliana.timbo@cepel.br, ayru@cepel.br
[2] COPPE/UFRJ – Programa de Engenharia Elétrica, Caixa Postal 68504,
CEP 21941-972 – Rio de Janeiro RJ
Tel.: +55-21-25628027; Fax: +55-21-25628080
carmen@dee.ufrj.br

Abstract. This paper presents the development of the distributed processing based Security Constrained Optimal Power Flow (SCOPF) and its integration to a Distributed Shared Memory Energy Management System (EMS) in order to enable real time power systems security control. The optimization problem is solved by the Interior Points Method and the security constraints are considered by the use of Benders Decomposition techniques. The SCOPF is initially parallelized using MPI and then integrated to the actual DSM based SCADA/EMS system SAGE, thoroughly used in the Brazilian power system including the National System Operation Center (CNOS). Results obtained on both the MPI and DSM platforms are presented for actual large size Brazilian power systems analyzed over a list of about a thousand contingencies. The results obtained demonstrate the high efficiency and applicability of the developed tool at Control Centers for real time security control.

Topics of Interest: Cluster Computing, Large Scale Simulations in Engineering, Parallel and Distributed Computing.

1 Introduction

Modern Control Centers of electrical power systems are equipped with computational tools to help the operators to provide high quality service with a minimum number of supply interruptions and at a minimum cost. The operation is done in a way to maintain the system in a secure mode, i.e., ensuring that the system will be operating continually even when components of the network fail, what are called contingencies [1]. The electric system is monitored by the Supervisory Control and Data Acquisition (SCADA) System, which periodically acquires analog measurements and status of switching devices from the network. The monitoring system also allows the operator to act in the system through remote controls, changing switches status and position of transformers tap, etc. The inherent complexity of the electric system operation makes it necessary to have sophisticated functions of diagnosis, analysis and advising

M. Daydé et al. (Eds.): VECPAR 2006, LNCS 4395, pp. 131–144, 2007.
© Springer-Verlag Berlin Heidelberg 2007

available at the Energy Management System (EMS), such as Network Topology Configurator, State Estimator, Contingency Analysis, Emergency Control, etc.

This paper deals with the integration of the Static Security Control to the functions of an EMS, through the solution of the Security Constrained Optimal Power Flow (SCOPF) problem. This function will give as result a set of control actions that should be taken by the operator to maintain the system in a secure mode even if any contingency of a predefined list occurs. However, one of the problems of SCOPF is that for large systems, the processing time is elevated. In that sense, this paper proposes the application of distributed processing in order to make feasible the use of SCOPF in a real time EMS environment. The SCOPF is initially parallelized using MPI – Message Passing Interface [2] and then integrated to an actual SCADA/EMS system, SAGE [3], thoroughly used in the Brazilian power system including the National System Operation Center (CNOS). The basic program used for development of the distributed tool is the software FLUPOT [4], whose solution of SCOPF is based on the Non-Linear Interior Points Method and in the Benders Decomposition technique for the security constraints consideration.

Recently, some papers have been published reporting implementations of OPF for real time application. In [5], the authors use the Unlimited Point Algorithm for the solution of the OPF. The parallelization is made at the matrix solution level. They use MPI for the distributed implementation. In [6], the class of genetic algorithms is used for the solution of the OPF, also using MPI for the parallel implementation. In [7] and [8], the concept of decentralized solution of the OPF problem is used, where the electric network is divided into areas and each area is optimized in a separate computer. The OPF is solved by the Non Linear Interior Points Method and PVM – Parallel Virtual Machine is used for the distributed implementation. In [9], the concept of decentralized solution is also used. In this case, however, the solution of the OPF is given by linear programming techniques. The distributed implementation is made using PVM. An implementation of SCOPF using distributed processing is found in [10], using linear programming techniques for the solution of SCOPF and considering only active generation rescheduling to obtain a secure solution. However, none of these papers deals with real time application of SCOPF with the same complexity as the present paper nor even considers the integration to an actual, commercial and operational SCADA/EMS system. The results obtained and reported here for an actual power system demonstrate the applicability of the developed distributed tool at Control Centers for real time Security Control.

2 Security Constrained Optimal Power Flow

The Security Constrained Optimal Power Flow has the objective to determine a feasible point of operation that minimizes an objective function, guaranteeing that even if any of the contingencies obtained from a list occurs, the post-contingency state will also be feasible, i.e., without limits violations [11]. From a given list of N possible contingencies, the SCOPF problem can be represented mathematically as:

$$\min f(z_o)$$

$$s.t.$$

$$a_0(z_0) \le b_0 \tag{1}$$
$$a_i(z_i) \le b_i$$

$$for \quad i = 1,2,...,N$$

Where:

$f(.)$ is the objective function;

$a(.)$ represents the non-linear balance equations of the electric network together with the operative constraints;

z represents the variables that will be optimized in the solution of the problem (state and control variables).

Each set of constraints $a_i(z_i) \le b_i$, for i = 1, 2, ..., N, is related with the configuration of the network under contingency and must respect the operations constrains in this condition.

The objective function to be minimized in the problem depends on the purpose of the utilization of the tool. For the use in control centers as a security control tool, the common objective functions are minimum loss, minimum deviation of the programmed operation point and minimum deviation of the scheduled area interchange. Other objective functions are also used in SCOPF problems, such as minimum reactive allocation, minimum load shed, minimum generation cost, etc. The state variables are, usually, the busbars voltages and angles. The control variables, which are modified in order to obtain the optimal operation point, are the generators active injection, terminal voltages and reactive injection, transformers tap position, areas interchange, etc.

The SCOPF can be interpreted as a two-stage decision process [12]:

- In the first stage, find an operation point z_o for the base-case problem, $a_o(z_o) \le b_o$;
- In the second stage, given the operating point z_o, find new operating points z_i that meet the constraints $a_i(z_i) \le b_i$, for each contingency configuration.

The solution method used in this work is based on Benders Decomposition, which allows handling separately the base-case problem and each of the N contingency sub-problems. To represent the possible unfeasibility of each contingency sub-problem, penalty variables are added to the problem in order to represent the amount of violation associated with the contingency operation point. Therefore, the minimization of the constraints violations can be defined as a new objective function and the contingency sub-problem can be formulated as:

$$w(z_o) = \min d^r.r$$

$$s.t. \tag{2}$$

$$a(z_0) \le b$$

Where $r \ge 0$ is the vector of penalty variables for the group of operative constraints and d^r is the cost vector. From this formulation, it can be seen that if $w(z_0) = 0$, the sub-problem is feasible and if $w(z_0) > 0$, the sub-problem is unfeasible.

The SCOPF can, then, be re-written in terms of z_o as follows, where the scalar functions $w_i(z_o)$ are the solutions of the contingency sub-problems (2) for the given operation point z_o.

$$\min f(z_o)$$

$s.t.$

$$a_0(z_0) \le b_0$$
$$w_i(z_0) \le 0$$

$$\text{(3)}$$

$$for \quad i = 1,2,...,N$$

The Benders Decomposition method consists in obtaining an approximation of $w_i(z_o)$ based on an iterative solution of the base-case and the N contingencies sub-problems. The Lagrange multipliers associated with the solution of each contingency sub-problem are used to form a linear constraint, known as Benders Cut, which are added to the base-case problem solution. Figure 1 shows the flowchart of the SCOPF solution algorithm based on Benders Decomposition.

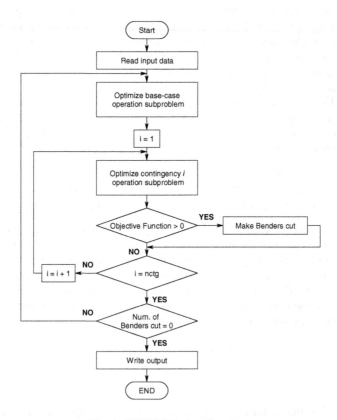

Fig. 1. SCOPF Solution Flowchart

The SCOPF solution algorithm consists, then, in solving the base-case optimization problem and then, from the operation point obtained in the base-case solution, to solve each of the contingency sub-problems. For each unfeasible contingency, a Benders Cut is generated. At the end of all contingencies solution, the generated Benders Cuts are introduced in the new solution of the base-case. The convergence is achieved when no contingency sub-problem generates Benders Cut. The contingencies sub-problems correspond to conventional OPF problems representing the configurations of the network under contingency. The base-case sub-problem is formulated as an OPF augmented by the constraints relative to the unfeasible contingencies (Benders Cuts). Each OPF problem is solved by the Interior Points Method and the network equations are formulated by the non linear model.

3   Distributed SCOPF Based on Message Passing

It can be noticed that the N contingencies sub-problems can be solved independently, once they only depend on the incoming operation point of the base-case, z_o. In that sense, the solution of the SCOPF based on Benders Decomposition can be directly benefited from the use of distributed processing, due to the natural parallelism that exists in the problem.

The parallelization strategy developed in this work is based on the master-slaves computation topology. When the parallel processing begins, each processor receives an identification number, the master being processor number zero and the slaves, processors number 1 to (*nprocs*-1), where *nprocs* is the total number of processors available. Figure 2 shows an example of contingencies allocation for a list of 10 contingencies distributed among 3 processors (1 master and 2 slaves).

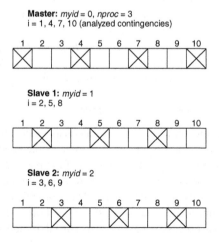

Fig. 2. Example of Contingencies Distribution among the Processors

The allocation of the contingencies sub-problems to the processors is performed by an asynchronous algorithm, based on the identification number of the processors (*myid*). In this way, the list of contingencies to be analyzed is distributed evenly among the participating processors in accordance with the value of *myid*, each one being responsible for the analysis of the contingencies of numbers (*myid* + 1 + *i*. *nprocs*), $i = 0, 1, 2, 3,...$ until the end of the list. It is important to emphasize that the master processor also participates in the contingencies analysis task, guaranteeing a better efficiency for the whole process, once the processor is not idle while the slaves work.

Depending on the number of contingencies in the list and the number of processors, it can happen that some processors receive one contingency more than the others. The processing time for each contingency can also vary, since the number of iterations required for the OPF of the contingencies to converge varies from case to case. However, in the contingencies distribution strategy adopted, each processor, after finishing the analysis of a contingency, immediately begins the analysis of another without needing the intervention of a control process. In that way, the computational load of each processor is, on average, approximately the same for large contingencies lists, ensuring an almost optimal load balancing.

3.1 Distributed Algorithm Based on Message Passing

The algorithm of the developed distributed application based on message passing can be summarized in the following steps:

Step 1: All processors read the input data.
Step 2: All processors optimize the base-case sub-problem.
Step 3: Parallel contingency sub-problems optimization by all processors.
Step 4: Master processor collects from all processors the partial Benders Cuts data structure (Synchronization Point).
Step 5: Master processor groups and reorganizes the complete Benders Cuts data structure.
Step 6: Master processor sends the complete Benders Cuts data structure to all processors (Synchronization Point).
Step 7: Verify if any of the N contingencies sub-problems is unfeasible. In the positive case, return to step 2.
Step 8: Master processor generates output reports.

All processors read the input data simultaneously, since the input files can be accessed by all via a shared file system, what eliminates the need to send the data read by just one processor to the others. The solution of the base-case sub-problem is also done simultaneously by all processors in order to avoid the need to send the results calculated by just one processor to the others. The reorganization of the Benders Cut data structure is a task introduced due to the distributed processing. After the analysis of their lists, each slave processor has its own partial Benders Cut data structures, which are sent to the master processor to be grouped and reorganized and later sent back again to all slave processors.

The flowchart of the developed parallel algorithm is shown in Figure 3. The two synchronization points of the algorithm, associated with the collection of the partial Benders Cuts structures and the distribution of the updated complete structure to all processors, are the only communication points of the algorithm, and for that reason, a high efficiency is expected from the distributed implementation. However, the efficiency will also depend on other factors, such as the communication technology used and the number of base-contingencies interactions (base-case plus contingencies sub-problems) necessary for convergence, since more interactions cause more communication among processors.

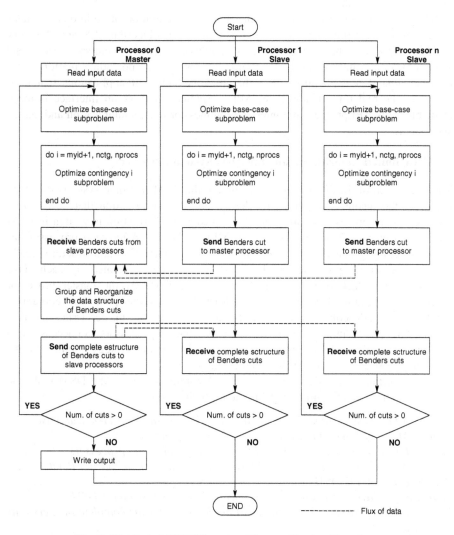

Fig. 3. Distributed SCOPF based on Message Passing Flowchart

4 Distributed SCOPF Integrated to the DSM Real Time System

SAGE [3] is a SCADA – Supervisory Control and Data Acquisition and EMS – Energy Management System, designed and developed by CEPEL, the Brazilian National Utility Energy Research Center. It includes modern energy management functions, such as State Monitoring (Network Configuration and State Estimation), Emergency Control (OPF Solution) and Security Monitoring (Contingency Analysis). The system is based on a distributed and expandable architecture. The use of redundant configurations and sophisticated control software ensures high reliability and availability for the system.

SAGE was designed to make possible the easy integration of additional modules directly to the real time database, which is build over a Distributed Shared Memory (DSM) support. The availability of common shared memory space and synchronization and control functions makes it a potential platform for distributed/parallel applications. For the implementation of applications that need access to the real time database, an API – Application Program Interface is made available. This API provides the interface routines for the communication and alarms subsystems.

4.1 Distributed Algorithm Using the DSM System Resources

The integration of the distributed application to the DSM real time systems was done exploring the resources provided by the system. The information that needs to be accessed by all processors during the distributed SCOPF solution is written in DSM modules instead of being exchanged via MPI, as before. After each processor has solved its contingencies list, each one has its own Benders Cut data structure locally stored. The master processor reads the data structures generated by each slave processor, after each one has copied them to the DSM by the master processor request. After reading these Benders Cut structures, the master processor reorganizes them into a single complete structure. Soon afterwards, the master writes the new complete Benders Cut structure in the DSM, so that all processors can access it and continue with the solution process, in a parallelization strategy similar to the previously described for the MPI implementation.

The algorithm of the developed distributed application using the DSM real time system resources can be summarized in the following steps:

Step 1: All processors read the input data directly from the Real Time Database.
Step 2: All processors optimize the base-case sub-problem.
Step 3: Parallel contingency sub-problems optimization by all processors.
Step 4: Master processor asks each Slave processor to write its partial Benders Cuts data structure on the DSM.
Step 5: Each Slave processor writes its partial data structure on the DSM.
Step 6: Master processor reads the DSM and reorganizes the complete Benders Cuts data structure.
Step 7: Master processor writes the complete data structure on the DSM.
Step 8: Slave processors read the complete data structure from the DSM .

Step 9: Verify if any of the N contingencies sub-problems is unfeasible. In the positive case, return to step 2.
Step 10: Master processor generates output reports.

From steps 4 to 8 there is a synchronization process for the information exchange among the processors, via the access to the DSM, in order for the master to read all the partial Benders Cuts data structure generated by the slaves and to write the complete reorganized Benders Cuts data structure on the DSM.

Figure 4 shows the EMS functions organization at the real time environment of SAGE, already including the developed distributed tool for the Security Control module.

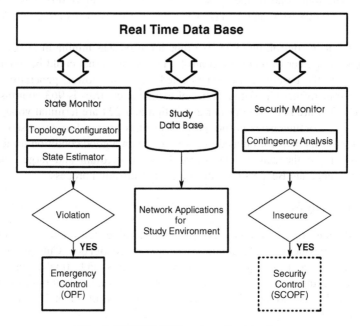

Fig. 4. SAGE EMS Functions Organization

The integration of the distributed solution of the SCOPF problem into SAGE makes it possible to use all available features of this SCADA/EMS: process control, high availability, cluster management, graphic interface, access to real time data, event triggering, alarms and logs. With the distributed tool integrated to the database of the EMS system, the SCOPF activation can be made by a request from the operator using the graphical interface, periodically or triggered by an event, such as the result of the Security Monitoring function. In the case it is detected that the system is insecure, the Security Control function (distributed SCOPF) is executed automatically afterwards. The use of SCOPF functionality in an EMS can potentially improve the operator online decision making process. The tool can advise the operator on which controls he should actuate to maintain the electrical system in a secure state.

5 Message Passing Implementation Results

5.1 Computational Platform and Test System

The computational platform used for the validation tests of the distributed implementation based on MPI was the Cluster Infoserver-Itautec [13], composed of 16 dual-processed 1.0GHz Intel Pentium III with 512MB of RAM and 256 KB of cache per node, Linux RedHat 7.3 operating system and dedicated Fast Ethernet network. The test system used is an equivalent of the actual Brazilian Interconnected System for the December 2003 heavy load configuration. The studied system is composed of 3073 busbars, 4547 branches, 314 generators and 595 shunt reactors or capacitors. The total load of the system is 57,947 MW and 16,007 Mvar, the total generation is 60,698 MW and 19,112 Mvar and the total losses are 2,751 MW and 3,105 Mvar.

For the security control, the objective function used was minimum losses together with minimum number of modified controls. The controls that could be modified for optimization of the base-case were: generated reactivate power, generator terminal voltage and transformer tap. The number of available controls is 963 controls, where 515 are taps, 224 are reactivate power generation and 224 are terminal voltages. The list of analyzed contingencies is formed by 700 lines or transformers disconnection. The contingency list was formulated in a way to obtain a good condition for tests, that is, the size of the list can be considered between medium and large, and some contingencies generate Benders Cuts during the optimization process.

5.2 Results Analysis

The SCOPF solution process converged in 3 base-contingencies iterations. In the first base-contingencies iteration, 12 contingencies generated Benders Cuts, in the second iteration, only 1 contingency generated Benders Cuts and, finally, in the third iteration, no contingency generated cut. The total losses after the optimization were 2,706 MW and 2,935 Mvar, what represents a reduction of about 2% in the active losses of the system.

The number of controls modified to lead the system to an optimal secure operation point was small, only 7 modifications of generator terminal voltages. That is due to the use of the objective function of minimum number of modified controls together with the minimum losses. This is an important issue for the use of SCOPF in the real time system operation. If the list of control actions is very long, it becomes unfeasible for the operator to perform them in time to turn the system secure.

The performance of the distributed implementation has been evaluated using from 1 to 12 nodes of the Cluster, obtaining exactly the same results as the sequential program. Table 1 shows the execution time, while Table 2 shows the Speedup and the Efficiency, for different numbers of processors.

It can be observed that the distributed implementation presents an excellent efficiency, superior to 92% using 12 processors of the distributed platform. The processing time is significantly reduced, changing from 8 minutes 25 seconds of the sequential processing to 45.43 seconds in parallel using 12 processors.

Table 1. Execution Time

No. Processors	Time
1	8 min 25 s
2	4 min 16 s
4	2 min 10 s
6	1 min 27 s
8	1 min 7 s
10	54.26 s
12	45.43 s

Table 2. Speedup and Efficiency

No. Processors	Speedup	Efficiency (%)
1	1	-
2	1.97	98.56
4	3.88	97.10
6	5.79	96.49
8	7.50	93.70
10	9.31	93.06
12	11.11	92.62

Figures 5 and 6 show the Speedup evolution and processing time reduction with the number of nodes used for the distributed solution, respectively.

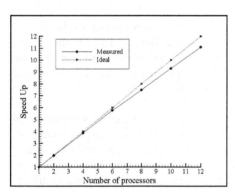

Fig. 5. Speedup Curve (MPI)   **Fig. 6.** Processing Time (MPI)

It can be also observed that the Speedup curve is almost ideal (linear). The Efficiency is only slightly reduced as the number of processors increases, what indicates that the parallel algorithm is scalable. From the good performance obtained, it can be expected that, if it is necessary to obtain a smaller response time for the real time application, this objective can be reached using a larger number of processors.

6 DSM Real Time System Implementation Results

6.1 Computational Platform and Test System

The computational platform used for the validation tests of the DSM integrated implementation was a Fast Ethernet network of computers at the Supervision and Control Laboratory at CEPEL, composed of eight 3.0GHz Intel Pentium IV microcomputers with 1GB RAM each and Enterprise Linux operational system. The test system used was generated based on real time operation data for the CNOS managed network of January 2006 on medium load level. This equivalent system is composed of 1419 busbars, 2094 branches, 388 generators and 92 shunts reactors and capacitors. The total load of the system is 43,654 MW and 14,582 Mvar, the total losses are 1,901 MW and 2,293 Mvar. The tests of the DSM platform were not based on the same system used in the MPI implementation in order to explore the real time database available on SAGE SCADA/EMS system.

For the security control, the objective function used was minimum losses together with minimum number of modified controls. The controls that could be modified for optimization of the base-case were: generated active and reactivate power, generator terminal voltage and transformer tap. All the variable tap transformers and generators were considered as control equipments in the optimization process. The list of analyzed contingencies is formed by 1012 simple contingencies involving lines, transformers, reactors, capacitors, load, generators and compensators disconnection.

6.2   Results Analysis

The SCOPF solution process converged in 3 base-contingencies iterations. In the first base-contingencies iteration, 16 contingencies generated Benders Cuts, in the second iteration, 11 contingencies generated Benders Cuts and in the third iteration, no contingency generated cut. The total losses after the optimization were 1,704 MW and 1,603 Mvar, what represents a reduction of about 8,5% in the active losses of the system.

A group of 122 control variables (about 10% of the total available) were modified to lead the system to an optimal secure operation point, 28 being active generations, 88 generators/synchronous terminal voltages and 6 transformers taps modifications.

Table 3 shows the execution time while Table 4 shows the Speedup and the Efficiency, for different numbers of processors, obtaining exactly the same results in parallel as in the sequential program.

Table 3. Execution Time

No. Processors	Time
1	11 min 48s
2	5 min 58s
4	3 min 2 s
6	2 min 2 s
8	1 min 34s

Table 4. Speedup and Efficiency

No. Processors	Speedup	Efficiency (%)
1	1.00	-
2	1.98	98.89
4	3.89	97.25
6	5.80	96.72
8	7.53	94.14

The execution time reduces from 11 minutes 48 seconds of the sequential simulation to about 1.5 minute on 8 processors of the distributed platform. The parallel implementation presents an efficiency superior to 94% using 8 processors, what can be considered a very good performance.

Figures 7 and 8 show the Speed up evolution and the processing time reduction with the number of processors used for the distributed solution, respectively.

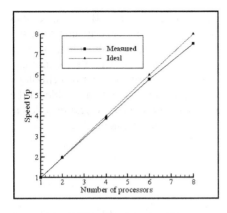

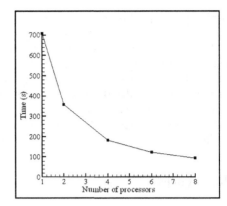

Fig. 7. Speedup Curve (DSM)   **Fig. 8.** Processing Time (DSM)

It can be observed that the Speedup curve is almost ideal. The Efficiency is only slightly reduced as the number of processors increases, what indicates that the parallel algorithm is scalable and a good performance can be expected when using a larger number of processors in the DSM based distributed platform.

Although the test systems are not the same in the MPI and DSM implementations, the results obtained for the two environments show similar performance. This can be verified comparing the characteristic of the speedup curves and the efficiency obtained for 8 processors, which are about 94% on both platforms.

7 Conclusions

This paper presented a proposal for enabling the system static Security Control in real time operation, based on the solution of SCOPF and in the use of distributed processing techniques. The use of this type of tool in the real time operation increases the security level of the system, since the operator, based on a list of control actions supplied by the tool, can preventively act on the system, avoiding that it evolves to a severe operative condition in the case some contingency happens.

The distributed application developed based on the MPI system is an autonomous tool, ready to be used in control centers if it is supplied, periodically, with the real time data. The integration of the distributed SCOPF directly to an DSM based EMS system added to the developed tool all the computational support offered by this type of environment, besides allowing the use of the computers already available at the control center for the SCADA/EMS functions.

In that sense, the Security Control based on SCOPF is made possible in real time application, using low cost and easily scalable platforms as a cluster of PCs or workstations. Although there is no consensus about which execution time is acceptable for the use of this type of tool in the real time operation, certainly the execution time obtained in the two test cases of the Brazilian System are acceptable. The smallest time obtained was about 45 seconds with 12 processors on the MPI environment and 1.5 minute with 8 processors on the DSM environment, which is perfectly compatible with the real time requirements of the Brazilian CNOS nowadays, where a time slice of two minutes is available for the security control task.

However, if a more time constrained response time is required, this objective than the reached not only upgrading the cluster platform but also de network technology, increasing networking speed and the overall system performance. Grid technology may also be considered for use between the several control centers available in the power system, in order to share the computational load with other clusters installed in other controls centers and also to low the overall cost in upgrading the computing platform.

References

1. Wood, A. J., Wollenberg, B.F., Power Generation, Operation, and Control. 2 ed. New York, John Wiley & Sons, 1996.
2. Gropp, W., Lusk, E. e Skjellum, A. (1996). Using MPI – Portable Parallel Programming with the Message Passing Interface, The MIT Press, Cambridge, UK.
3. http://www.sage.cepel.br/ – SAGE – Sistema Aberto de Gerenciamento de Energia, CEPEL (In Portuguese).
4. CEPEL, Optimal Power Flow Program – FLUPOT: User Manual (In Portuguese), Rio de Janeiro, RJ, Brazil, 2000.
5. Huang, Y., Kashiwagi, T., Morozumi, S., "A Parallel OPF Approach for Large Scale Power Systems". Fifth International Conference on Power System Management and Control, pp. 161-166, April 2002.
6. Lo, C.H.; Chung, C.Y.; Nguyen, D.H.M.; Wong, K.P., "A Parallel Evolutionary Programming Based Optimal Power Flow Algorithm and its Implementation", In: Proceedings of 2004 International Conference on Machine Learning and Cybernetics, Vol. 4, pp. 26-29, August 2004.
7. Baldick, R., Kim, B.H., Chase, C., Luo, Y., "A Fast Distributed Implementation of Optimal Power Flow", IEEE Transactions on Power Systems, Vol. 14, pp. 858-864 , August 1999.
8. Hur, D. Park, J., Balho Kim, H., "On the Convergence Rate Improvement of Mathematical Decomposition Technique on Distributed Optimal Power Flow", Electric Power and Energy Systems, No 25, pp. 31-39, 2003.
9. Biskas, P. N., Bakirtzis, A. G., Macheras, N. I., Pasialis, N. K., "A Decentralized Implementation of DC Optimal Power Flow on a Network of Computers", IEEE Transactions on Power Systems, Vol. 20, pp. 25-33, February 2005.
10. Rodrigues, M., Saavedra, O. R., Monticelli, A. "Asynchronous Programming Model for the Concurrent Solution of the Security Constrained Optimal Power Flow", IEEE Transactions on Power System, Vol. 9, No 4, pp. 2021-2027, November 1994.
11. Monticelli, A., Pereira, M.V. F., Granville, S. "Security-Constrained Optimal Power Flow With Post-Contingency Corrective Rescheduling", IEEE Transactions on Power System, Vol. PWRS-2, No 1, pp. 175-182, February 1987.
12. Granville, S., Lima, M.C. A., "Application of Decomposition Techniques to VAr Planning: Methodological & Computational Aspects", IEEE Transactions on Power System, Vol. 9, No 4, pp. 1780-1787, November 1994.
13. http://www.nacad.ufrj.br/ – NACAD – Núcleo de Atendimento de Computação de Alto Desempenho, COPPE/UFRJ, (In Portuguese).

Metaserver Locality and Scalability in a Distributed NFS*

Everton Hermann[1,**], Rafael Ávila[1,***], Philippe Navaux[1],
and Yves Denneulin[2]

[1] Instituto de Informática/UFRGS
Caixa Postal 15064
91501-970 Porto Alegre – Brazil
Phone.: +55 (51) 3316-6165; Fax: +55 (51) 3316-7308
{ehermann,avila,navaux}@inf.ufrgs.br
[2] Laboratoire ID/IMAG
51, avenue Jean Kuntzmann
38330 Montbonnot-Saint Martin – France
Phone.: +33 (4) 76 61 20 13; Fax: +33 (4) 76 61 20 99
Yves.Denneulin@imag.fr

Abstract. The leveraging of existing storage space in a cluster is a desirable characteristic of a parallel file system. While undoubtedly an advantage from the point of view of resource management, this possibility may face the administrator with a wide variety of alternatives for configuring the file server, whose optimal layout is not always easy to devise. Given the diversity of parameters such as the number of processors on each node and the capacity and topology of the network, decisions regarding the locality of server components like metadata servers and I/O servers have a direct impact on performance and scalability. In this paper, we explore the capabilities of the dNFSp file system on a large cluster installation, observing how scalable the system behaves in different scenarios and comparing it to a dedicated parallel file system. Our obtained results show that the design of dNFSp allows for a scalable and resource-saving configuration for clusters with a large number of nodes.

Topics: Cluster and grid computing, parallel I/O, parallel and distributed computing.

1 Introduction

Solutions for efficient management of I/O in large clusters have long been the focus of several research groups and industrials working on parallel computing [1]. Ranging from RAID arrays and fibre optics to virtual distributed disks, many approaches have been proposed in the last decade that vary considerably in terms of performance, scalability and cost.

---

* Candidate to the best student paper award.
** Work partially supported by CAPES and CNPq.
*** Work supported by HP Brazil grant.

M. Daydé et al. (Eds.): VECPAR 2006, LNCS 4395, pp. 145–157, 2007.
© Springer-Verlag Berlin Heidelberg 2007

In previous works [2,3], we have presented the *dNFSp* file system, an extension of NFSv2 that aims at improving both performance and scalability of a regular NFS server while keeping its standard administration procedures and, mainly, compatibility with the regular NFS clients available on every Unix system. Similarly to other parallel file systems such as PVFS [4] and Lustre [5], dNFSp is based on a distributed approach where the gain in performance is obtained by executing tasks in parallel over several machines of the cluster.

One important aspect of a parallel file system is its capability of leveraging existing resources. In the case of commodity clusters, the hard disks that are installed on the compute nodes are frequently under-used: a typical GNU/Linux node installation takes only a few gigabytes, and today's PCs are hardly ever configured with less than 40 gigabytes of storage. This leaves us with at least 75% of the total hard disk capacity available for the storage of data, and consequently it is important that a cluster file system have the ability to use it.

dNFSp provides such a feature, so that the storage on the compute nodes can be used to form a single cluster file system. It is then up to the cluster administrator to decide how to configure the system, finding a good balance between resource utilization, performance and scalability, which might not be an obvious task.

For this reason, we have conducted a series of experiments varying the configuration of dNFSp on a large cluster. This allowed us to watch how scalable the system os. Also it was possible to identify layouts best suited for one or another situation, and whose results and conclusions are presented in this work.

In the remainder of the paper, Section 2 presents the dNFSp file system and its main characteristics, with the purpose of providing some background knowledge; Section 3 describes in more details the experiments we have conducted and introduces the evaluation criteria; in Section 4 we present the results obtained in the experiments and provide the discussion which is the focus of this work; Section 5 brings a comparison of our work to systems with related objectives, and finally Section 6 draws some conclusions on the obtained results and analysis and reveals future directions.

2   dNFSp – A Distributed NFS Server

The NFSp project [6] has been established in 2000 at the *Laboratoire Informatique et Distribution* of Grenoble, France, with the goal of improving performance and scalability in a regular NFS installation. The main idea of the project is to provide a cluster file system that benefits from the standard administration procedures and behavior of a well-known protocol such as NFS. As a result, NFSp — for *parallel* NFS — presents some simple extensions to the NFS server implementation that distributes its funcionalities over a set of nodes in the cluster, thus gaining performance. On the other hand, the client machines do not have to be modified at all, favoring portability.

As a subproject within the NFSp group, dNFSp has been proposed as a further extension to the model, aiming at an improvement on concurrent write operations by client machines.

Figure 1 depicts the distribution model proposed by dNFSp. The top of the figure shows the two main server components: the *I/O daemons*, or IODs, and the *metaservers*. The metaservers are daemons that play the role of the NFS server, serving clients' requests and cooperating with each other to form the notion of a single file server. The IODs work as backends for the metaservers, being responsible solely for data storage and retrieval. On the lower part of the figure, client machines connect to the metaservers in the same way that clients connect to a regular NFS server.

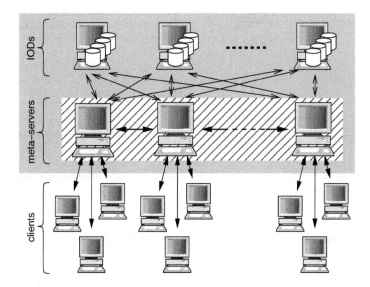

Fig. 1. Distributed metaserver architecture of dNFSp

Each client is connected to one metaserver, which is responsible for handling its requests. Operations involving only metadata are replied directly by the metaserver, which in some cases can contact other metaservers to obtain the needed metadata. I/O operations are forwarded by the metaserver to the IODs, which will perform the operation and reply directly to the client. In the case of read operations, the file contents are transferred directly from the IODs to the clients, allowing parallel reads up to the number of available IODs. In the case of write operations, the data are transferred from the client to the metaserver, and then forwarded to the IODs. Therefore, global write performance depends both on the number of IODs and the number of metaservers.

The metaservers exchange information to keep metadata coherence across all metaservers. The information is retrieved only when needed by a client, avoiding unnecessary network traffic. However, there are situations when all metaservers must be contacted (e.g. *lookup()* upon file creation), and this communication becomes more visible as we increase the number of metaservers.

The extensions introduced by dNFSp have been implemented on an existing NFSp prototype, and a performance evaluation has been carried out previously [3] with the execution of some micro benchmarks. As an illustration of dNFSp raw performance, Figure 2 shows the results obtained for concurrent read and write operations (each client reads/writes one independent 1 GB file) in comparison to the performance of a regular NFS server using Fast Ethernet network ($\sim$11 MB/s). As expected, one obtains an increased throughput when more than one client read/write at the same time.

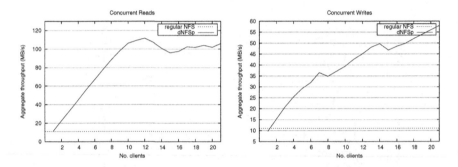

Fig. 2. dNFSp performance for concurrent read and write operations using 12 IODs and 7 metaservers

3   Benchmark and Cluster Environment

The measurements we carried out have the objective of evaluating the scalability of dNFSp in a large number of nodes using a real application-based benchmark. The benchmark we used is the *NAS/BTIO*, and the machine used to run the applications is the *INRIA i-cluster2*[1]. Both the application and the cluster are detailed in the following sections.

3.1   The NAS/BTIO Benchmark

The BTIO Benchmark is a part of the NAS Parallel Benchmarks (NPB) [7]. It is commonly used for evaluating the storage performance of parallel and distributed computer systems. The application used by BTIO is an extension of the BT benchmark [8]. The BT benchmark is based on a Computational Fluid Dynamics (CFD) code that uses an implicit algorithm to solve the 3D compressible Navier-Stokes equations. BTIO uses the same computational method employed by BT. The I/O operations have been added by forcing the writing of results to disk. In BTIO, the results must be written to disk at every fifth step of BT.

The number of process running one execution of BTIO must be a perfect square (1, 4, 9, 16, etc.). The problem size is chosen by specifying a class. Each

---

[1] http://ita.imag.fr

class corresponds to the dimensions of a cubic matrix to be solved by the application: class A (64^3), class B (102^3), class C (162^3). We have chosen class A since it was enough to have a good mixing of computation and file system operations. Moreover, using a larger class has not changed the profile of the results.

Another customization of BTIO is the way the I/O operations are requested to the file system. There are four flavors that can be chosen at compilation time:

- BTIO-full-mpiio: This version uses MPI-IO file operations with *collective buffering*, which means that data blocks are potentially re-ordered previously to being written to disk, resulting in coarser write granularity
- BTIO-simple-mpiio: Also uses MPI-IO operations, but no data re-ordering is performed, resulting in a high number of seeks when storing information on the file system
- BTIO-fortran-direct: This version is similar to simple-mpiio, but uses the Fortran direct access method instead of MPI-IO
- BTIO-epio: In this version, each node writes in a separate file. This test gives the optimal write performance that can be obtained, because the file is not shared by all the processes, so there is no lock restriction. In order to compare with other versions, the time to merge the files must be computed, as required by the Application I/O benchmark specification.

In order to perform MPI-IO operations using an NFS-based file system, it would be necessary to have an implementation of the NFSv3 protocol, due to the need of controlling file access by means of locks. Since dNFSp was implemented based on the NFSv2 protocol, it has no lock manager; hence, we decided to use the BTIO-epio version of the benchmark. Furthermore, using epio allows one to achieve optimal performance results, since the data are written to individual files by each node.

With the goal of having a more write-intensive benchmark, we have made a small change in the BTIO benchmark code, modifying the frequency of file writes. The original code performes writes on every five iterations, resulting in a total amount of writes of 400 megabytes; with the modification, BTIO performs writes on every iteration, resulting in 2 gigabytes of written data. The results of each computing step are appended to the end of the file used by the process. The granularity of writes changes with the number of nodes used on the computation. Changing the frequency of writings allowed us to make the differences between the file systems more visible.

3.2   The i-Cluster2

The *i-cluster2* [9] is installed in Montbonnot Saint Martin, France, in the IN-RIA Rhône-Alpes facility. The cluster is composed by 100 nodes. Each node is equipped with a dual Itanium2 900 MHz with 3 gigabytes of memory and a disk storage with 72 gigabytes, 10000 rpm, SCSI. All the nodes are interconnected using a 1 Gigabit Ethernet network, Fast Ethernet network and Myrinet Network. The experiments were performed using the 1 Gigabit Ethernet network.

The software installed on i-cluster2 is based on Red Hat Enterprise Linux AS release 3 distribution, with a Linux kernel version 2.4.21. The MPI implementation used with the BTIO benchmark is mpich version 1.2.6.

4 Performance, Scalability and Locality Evaluation

In this section we present the results obtained with our experiments. The reported execution times are those informed by BTIO at the end of the execution, together with a confirmation of correct computation. Each value reported is the arithmetic mean of at least 5 runs of BTIO with the same configuration, so as to obtain a stable value. Standard deviations lie within a maximum value of 3 seconds.

4.1 Performance Analysis

The first step in our analysis of dNFSp has been an evaluation of the performance of the system on the i-cluster2. We have run BTIO on a large subset of the available nodes, and compared the performance of dNFSp with that of a dedicated parallel file system. We have chosen PVFS [4] for this task, as it is a representative parallel file system in the Beowulf cluster context in which our work is inserted. It wasn't possible to perform tests using Lustre because it needs a kernel patch to the system to run, and we didn't have the permission needed to do this task.

For both systems, we have varied the number of IODs in the file server from 4 up to 12 IODs, in steps of 2. In the case of dNFSp, we always use a number of metaservers equal to that of IODs. PVFS uses only one extra node in all cases, for the *manager*. The experiments have been executed from 4 up to 49 clients, respecting the feature of BTIO that the number of clients must be a perfect square.

We show, in Figure 3, the results obtained using the minimum and the maximum number of IODs, respectively 4 and 12. Intermediate configurations have shown proportional variation.

When the file system is accessed by a small number of clients, a shorter number of metaservers has shown better performance results, because a higher number of metaservers results in more management communication. In this case, the application does not have enough nodes to benefit from all the parallelism offered by the file system.

As expected, execution times drop as the number of clients increase. For dNFSp, the reduction in execution time is progressive on the whole range of clients, except for the case of 49 clients using 4 IODs, where it slightly starts to rise again. For PVFS, one observes a lower limit at around 100 seconds. We conclude that the main cause for the limitation in both systems is that, as the number of clients increase, the amount of data written by each one decreases, and reaches a point where parallelism does not pay off anymore due to the management cost of the striping mechanism. dNFSp seems to handle the situation better than PVFS, reaching around 47 seconds in the case of 12 IODs.

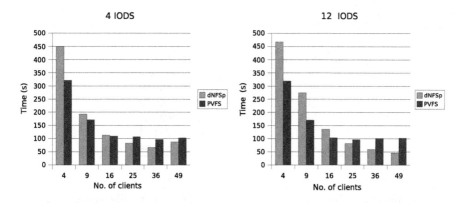

Fig. 3. Performance comparison for dNFSp vs. PVFS using 4 and 12 IODs on the file server

It is important to remark that such experiments were run using PVFS v1 (more precisely, version 1.6.3), while PVFS2 is already available and should presumably yield more performance than its predecessor. PVFS2 was effectively our initial choice for comparison, not only because of its performance, but also because it features a distributed metadata model which is closer to that of dNFSp. However, early experiments with that version on the i-cluster2 resulted in strangely poor performance (results are shown in Table 1 for information). One reason to this poor result is the writing profile of BTIO, which appends contents to the end of the files, changing metadata information on each write. The relaxed model used by dNFSp does not require updating metadata on all the metaservers so there is no extra communication when an append is performed. While we are still investigating further causes for that behavior, we chose to report results for PVFS v1, which nevertheless represents no loss in significance since it is still fully supported by the developers as a production system.

4.2   Scalability Evaluation

As a subsequent analysis of our experiments we have compared how both file systems react to the addition of more nodes to the file system. The number of IODs, metaservers and clients is the same as described in the previous section.

In Figure 4 we have three samples from our experiment. In the first chart we have kept the number of BTIO clients on 4 and varied the number of IODs and metaservers. We can see that both file systems sustain an almost constant performance due to the fact that the application doesn't have enough clients to stress the capability of storage offered by the file systems. In this case, dNFSp even shows a small decrease in performance as we add more nodes. This loss of performance comes from the metaserver communication, which is more significant when we have more of them. The better performance of PVFS lies on the size of messages. As we have only four BTIO clients, the amount of computation designated to each node is large resulting in larger writes on the file system.

Table 1. Sample execution times (in seconds) obtained for dNFSp and PVFS v1 in comparison to PVFS2

No. of clients	dNFSp	PVFS v1	PVFS2
4	464.416	319.273	363.395
9	272.728	170.663	263.655
16	135.164	103.247	253.88
25	76.712	95.35	252.505
36	53.745	96.313	293.44
49	43.776	105.44	353.64

The second chart shows the transition case where both file systems have a similar behavior. Using 25 clients BTIO seems to have a block size that results in similar performance to both file systems. They have an improvement of performance as we add more nodes to the file system, reaching a limit where adding more nodes increases the execution time instead of reducing.

In the third chart we show a more stressing case, where we have 49 clients accessing the file system. In this situation we can see that from 4 to 10 nodes dNFSp has an improvement of performance as we increase the file system size. When we reach 12 IODs and 12 metaservers, the overhead added by the insertion of more nodes is not compensated by the performance gain. The clients don't have enough writes to stress the file system, and the communication between metaservers is more expressive, falling in the same situation shown by the four clients sample. PVFS has shown a performance limitation when we have small writes to the file system. As the number of clients is larger than the previous case, we have smalls chunks of data being written.

4.3   Metaservers and IODs Locality Impact

As a last experiment, we have investigated the capabilities of dNFSp in saving cluster nodes for the deployment of the file server. As usual in distributed file systems (and distributed systems in general) like dNFSp and PVFS, each task of the system is usually performed on a distinct machine or compute node. For example, in the previous experiments we have always used distinct nodes for running the IODs and the metaservers/manager.

It would be desirable, however, that one could make use of as few nodes for the file server as possible, in order to maximize the number of nodes available for the real computing tasks. While the decision depends mostly on the amount of storage desired for the server, some considerations can be made regarding performance and scalability that might allow for a shorter number of nodes than that initially accounted. This is specially true if the compute nodes are dual-processed.

The results in Figure 5 correspond to the execution of BTIO with dNFSp with IODs and metaservers running on the same nodes, compared to the original execution where the two entities run on separate nodes. Again, we show results for a small and a large number of client nodes.

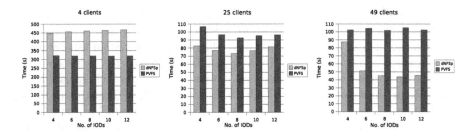

Fig. 4. Scalability comparison for dNFSp vs. PVFS using 4, 25 and 49 BTIO clients processes

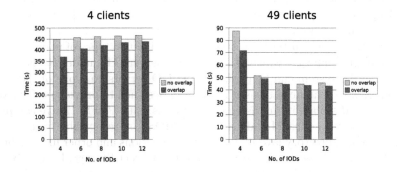

Fig. 5. Results for dNFSp with and without overlapping IODs and metaservers on the same nodes

Two different situations are presented, both favoring the overlapping of IODs and metaservers. In the first case, with few client nodes, more information is written by each single client, and thus there is a visible difference in performance in favor of the overlapping configuration, since communication between the IOD and the metaserver on the same node is done faster (by means of memory copy). Approximately $1/N$ requests, where N is the number of IODs, can be processed locally, without the need of contacting an IOD through the network. On the second case, there are much more, smaller client writes, and consequently the differences are not much evident. Increasing the number of IODs also contributes to minimize the differences, as the probability of performing a request locally decreases as the number of striping slices grows. As a conclusion, we can see that such an overlapping configuration can be employed without loss of performance, contributing to the amount of nodes dedicated to computation.

5   Related Work

The increasing performance gap between I/O and processor has placed the file system performance as the most severe bottleneck to applications that massively use the storage subsystem [1,10]. Several approaches have been proposed since the deployment of the first large-scale parallel machines. Many are based on the use of specialized technologies (e.g. RAID, fiber optics) as a means to increase performance, such as GPFS [11] and GFS [12]. This kind of system usually relies on the concept of a Storage Area Network (SAN), which basically defines a common storage "device" composed of several physical devices. As such, scalability is a direct consequence of this concept. Other projects like Petal/Frangipani [13,14] and the Shared Logical Disk [15] make use of the same concept, but the SAN is implemented in software over a network. Good performance and scalability thus depend heavily on the communication technology.

Research projects like PVFS [4] and Lustre [16] follow another trend. To achieve high performance on I/O operations, these file systems distribute the functions of a file system across a set of nodes in a cluster. To perform parallel I/O operations, they stripe the data across the nodes, keeping the striping transparent to the application.

PVFS is a parallel cluster file system composed of two types of nodes: the *I/O server* and the *manager*, which is a metadata server. The nodes in the cluster used by the file system can be configured as I/O servers, and one of them as a manager. Lustre is an object-based file system designed to provide performance, availability and scalability in distributed systems. Like PVFS, Lustre comprises two types of server nodes: Metadata Servers (MDS) are responsible for managing the file system's directory layout, as well as permissions and other file attributes. The data is stored and transferred through the Object Storage Targets (OST). Figure 6 shows the architecture of PVFS and Lustre in comparison to that of dNFSp.

High performance in PVFS is achieved by distributing the contents of a file across the I/O server nodes (striping). Clients are allowed to access the contents of the file in parallel. The way the files are striped is handled by the metadata manager, which is also responsible for managing file properties and a name space to applications, but has no participation in I/O operations. The I/O servers are accessed directly by the clients to deal with the data transfers. The user can access the file system through the PVFS library or using an OS-specific kernel module. The latter allows the user to mount the file system using a POSIX interface, and access files as any other file system.

In Lustre, similarly to PVFS, the client contacts the Metadata Servers to know which OST contains a given part of a file. After obtaining this information, the client establishes direct connections to the OST performing reads and writes. The MDS has no intervention in the I/O process, being contacted by the OST only to change file attributes like file size. Both types of nodes can have replicas working in pairs and taking the place of each other when a failure occurs. Figure 6 shows an *active* MDS, and its replica is represented by the *failover* node. Information concerning the overall system configuration is stored in a Lightweight Directory Access Protocol (LDAP) server. In the event of a MDS failure, the client can access the LDAP server to ask for an available MDS.

As in dNFSp, PVFS version 2, or PVFS2 [17], has the option of running more than one manager. The main difference lies on the way PVFS controls the

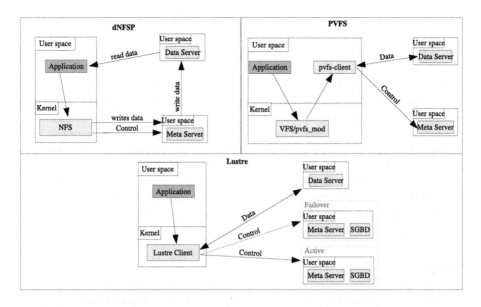

Fig. 6. Architecture of the related cluster file systems

distribution of metadata. Each manager stores the metadata information about a range of files, while in dNFSp each server has the metadata information about all the files. The PVFS approach can result in a surcharged manager when all the clients access files in the same range.

6   Conclusions and Final Considerations

The execution of the BTIO benchmark with dNFSp and PVFS on the i-cluster2 has confirmed the objectives of our system in providing good performance and scalability while keeping compatibility with NFS. The results show very good, scalable performance in comparison to a dedicated parallel file system. dNFSp is able to reach the same level of performance of PVFS, and many times even reach beyond it. We understand that this advantage comes from the fact that dNFSp can tolerate a smaller size of writes than PVFS before reaching the point where parallelism in no longer favorable. When configured with a large number of clients, dNFSp has outperformed PVFS in up to 50% of its execution time.

Another positive aspect of our benchmarking is that dNFSp performs efficiently when IODs and metaservers have been run together. The performance results obtained using IODs and metaservers on the same node were up to 17% faster than in the case where metaservers and IODs were run on distinct machines. This allows for a resource-saving configuration which maximizes the availability of compute nodes without sacrificing performance. Although the nodes of the i-cluster2 are dual-processed, which favors this configuration, we believe that a similar approach, at least partial, should be possible on single-processor clusters, since the IODs present a typical I/O-bound profile, while the metaservers do little disk activity. This evaluation was not possible on the i-cluster2, as it would require booting with a non SMP-enabled kernel. We intend to carry it out in the next stage of the project.

One of the future activities in dNFSp is the implementation of a dedicated communication mechanism between metaservers, given some loss of performance in a few of the experiments due to the heavy lookup mechanism that the metaservers perform. The dedicated protocol should minimize the impact of lookup operations by implementing some kind of prefetching and message aggregation for the exchange of metadata. Also, we are planning a porting of dNFSp to NFS version 3. This will allow us to profit from the changes on the protocol, like asynchronous I/O operations and the larger block size limit. Another aspect to be worked upon is fault tolerance and replication, which are being studied and shall be included in the upcoming versions.

References

1. Schikuta, E., Stockinger, H.: Parallel I/O for clusters: Methodologies and systems. In Buyya, R., ed.: High Performance Cluster Computing: Architectures and Systems. Prentice Hall PTR, Upper Saddle River (1999) 439–462
2. Kassick, R., Machado, C., Hermann, E., Ávila, R., Navaux, P., Denneulin, Y.: Evaluating the performance of the dNFSP file system. In: Proc. of the 5th IEEE International Symposium on Cluster Computing and the Grid, CCGrid, Cardiff, UK, Los Alamitos, IEEE Computer Society Press (2005)
3. Ávila, R.B., Navaux, P.O.A., Lombard, P., Lebre, A., Denneulin, Y.: Performance evaluation of a prototype distributed NFS server. In Gaudiot, J.L., Pilla, M.L., Navaux, P.O.A., Song, S.W., eds.: Proceedings of the 16th Symposium on Computer Architecture and High-Performance Computing, Foz do Iguaçu, Brazil, Washington, IEEE (2004) 100–105
4. Carns, P.H., Ligon III, W.B., Ross, R.B., Thakur, R.: PVFS: a parallel file system for Linux clusters. In: Proc. of the 4th Annual Linux Showcase and Conference, Atlanta, GA (2000) 317–327 Best Paper Award.
5. Cluster File Systems, Inc.: Lustre: A scalable, high-performance file system (2002) Available at http://www.lustre.org/docs/whitepaper.pdf (July 2004)
6. Nfsp: homepage (2000) Available at <http://www-id.imag.fr/Logiciels/NFSP> Access in May 2005.
7. Wong, P., der Wijngaart, R.F.V.: NAS Parallel Benchmarks I/O Version 2.4. RNR 03-002, NASA Ames Research Center (2003)
8. Bailey, D.H., et al.: The NAS parallel benchmarks. International Journal of Supercomputer Applications 5(3) (1991) 63–73
9. i-Cluster 2 (2005) Available at <http://i-cluster2.inrialpes.fr> Access in May 2005.
10. Baker, M., ed.: Cluster Computing White Paper. IEEE Task Force in Cluster Computing (2000) Available at http://www.dcs.port.ac.uk/~mab/tfcc/WhitePaper/final-paper.pdf Final Release, Version 2.0.
11. Schmuck, F., Haskin, R.: GPFS: A shared-disk file system for large computing clusters. In: Proc. of the Conference on File and Storage Technologies, Monterey, CA (2002) 231–244
12. The openGFS project (2003) http://opengfs.sourceforge.net

13. Lee, E.K., Thekkath, C.A.: Petal: Distributed virtual disks. In: Proc. of the 17th International Conference on Architectural Support for Programming Languages and Operating Systems, Cambridge, MA (1996) 84–92
14. Thekkath, C.A., Mann, T., Lee, E.K.: Frangipani: A scalable distributed file system. In: Proceedings of the 16th ACM Symposium on Operating Systems Principles, Saint Malo, France, New York, ACM Press (1997) 224–237
15. Shillner, R.A., Felten, E.W.: Simplifying distributed file systems using a shared logical disk. Technical Report TR-524-96, Dept. of Computer Science, Princeton University, Princeton, NJ (1996)
16. Schwan, P.: Lustre: Building a file system for 1000-node clusters. In: Proceedings of the 2003 Linux Symposium. (2003)
17. Latham, R., Miller, N., Ross, R., Carns, P.: A next-generation parallel file system for Linux clusters. LinuxWorld Magazine (2004)

Top-k Query Processing in the APPA P2P System*

Reza Akbarinia[1,3], Vidal Martins[1,2], Esther Pacitti[1], and Patrick Valduriez[1]

[1] ATLAS group, INRIA and LINA, University of Nantes, France
[2] PPGIA/PUCPR – Pontifical Catholic University of Paraná, Brazil
[3] Shahid Bahonar University of Kerman, Iran
FirstName.LastName@univ-nantes.fr, Patrick.Valduriez@inria.fr

Abstract. Top-k queries are attractive for users in P2P systems with very large numbers of peers but difficult to support efficiently. In this paper, we propose a fully distributed algorithm for executing Top-k queries in the context of the APPA (Atlas Peer-to-Peer Architecture) data management system. APPA has a network-independent architecture that can be implemented over various P2P networks. Our algorithm requires no global information, does not depend on the existence of certain peers and its bandwidth cost is low. We validated our algorithm through implementation over a 64-node cluster and simulation using the BRITE topology generator and SimJava. Our performance evaluation shows that our algorithm has logarithmic scale up and improves Top-k query response time very well using P2P parallelism in comparison with baseline algorithms.

1 Introduction

Peer-to-peer (P2P) systems adopt a completely decentralized approach to data sharing and thus can scale to very large amounts of data and users. Popular examples of P2P systems such as Gnutella [10] and KaZaA [13] have millions of users sharing petabytes of data over the Internet. Initial research on P2P systems has focused on improving the performance of query routing in unstructured systems, such as Gnutella and KaaZa, which rely on flooding. This work led to structured solutions based on distributed hash tables (DHT), *e.g.* CAN [16], or hybrid solutions with super-peers that index subsets of peers [23]. Although these designs can give better performance guarantees than unstructured systems, more research is needed to understand their trade-offs between autonomy, fault-tolerance, scalability, self-organization, etc. Meanwhile, the unstructured model which imposes no constraint on data placement and topology remains the most used today on the Internet.

Recently, other work in P2P systems has concentrated on supporting advanced applications which must deal with semantically rich data (*e.g.* XML documents, relational tables, etc.) using a high-level SQL-like query language, *e.g.* ActiveXML [2], Piazza [20], PIER [12]. High-level queries over a large-scale P2P system may produce very large numbers of results that may overwhelm the users. To avoid such overwhelming, a solution is to use Top-k queries whereby the user can specify a limited number (k) of the most relevant answers. Initial work on Top-k queries has concentrated on SQL-like language extensions [7][6] . In [6] for instance, there is a

---

* Work partially funded by ARA Massive Data of the French ministry of research.

M. Daydé et al. (Eds.): VECPAR 2006, LNCS 4395, pp. 158–171, 2007.
© Springer-Verlag Berlin Heidelberg 2007

STOP AFTER k clause to express the k most relevant tuples together with a scoring function to determine their ranking.

Efficient execution of Top-k queries in a large-scale distributed system is difficult. To process a Top-k query, a naïve solution is that the query originator sends the query to all nodes and merges all the results, which it gets back. This solution hurts response time as the central node is a bottleneck and does not scale up. Efficient techniques have been proposed for Top-k query execution in distributed systems [25][24]. They typically use histograms, maintained at a central site, to estimate the score of databases with respect to the query and send the query to the databases that are more likely to involve top results. These techniques can somehow be used in super-peer systems where super-peers maintain the histograms and perform query sending and result merging. However, keeping histograms up-to-date with autonomous peers that may join or leave the system at any time is difficult. Furthermore, super-peers can also be performance bottlenecks. In unstructured or DHT systems, these techniques which rely on central information no longer apply.

In this paper, we propose a fully distributed algorithm for executing Top-k queries processing in the context of APPA (Atlas Peer-to-Peer Architecture), a P2P data management system which we are building [3][4]. The main objectives of APPA are scalability, availability and performance for advanced applications. APPA has a network-independent architecture in terms of advanced services that can be implemented over different P2P networks (unstructured, DHT, super-peer, etc.). This allows us to exploit continuing progress in such systems. Our Top-k query processing algorithm has several distinguishing features. For instance, it requires no central or global information. Furthermore, its execution is completely distributed and does not depend on the existence of certain peers. We validated our algorithm through a combination of implementation and simulation and the performance evaluation shows very good performance. We have also implemented baseline algorithms for comparing with our algorithm. Our performance evaluation shows that our algorithm improves Top-k query response time very well using P2P parallelism in comparison with baseline algorithms.

The rest of this paper is organized as follows. Section 2 describes the APPA architecture. In Section 3, we present our algorithm, then we analyzes the bandwidth cost of our algorithm and propose techniques in order to reduce this cost. Section 4 describes a performance evaluation of the algorithm through implementation over a 64-node cluster and simulation (up to 10,000 peers) using the BRITE topology generator [5] and SimJava [11]. Section 5 discusses related work. Section 6 concludes.

2 APPA Architecture

APPA has a layered service-based architecture. Besides the traditional advantages of using services (encapsulation, reuse, portability, etc.), APPA is a network-independent architecture so it can be implemented over different P2P networks (unstructured, DHT, super-peer, etc.). The main reason for this choice is to be able to exploit rapid and continuing progress in P2P networks. Another reason is that it is unlikely that a single P2P network design will be able to address the specific requirements of many different applications. Obviously, different implementations will yield different trade-offs between performance, fault-tolerance, scalability, quality of service, etc. For instance, fault-tolerance can be higher in unstructured P2P systems because no peer is a single point of failure. On the other hand, through index servers, super-peer systems enable more efficient query processing. Furthermore, different P2P networks could be combined in order to exploit their relative advantages, *e.g.* DHT for key-based search and super-peer for more complex searching.

There are three layers of services in APPA: P2P network, basic services and advanced services.

P2P network. This layer provides network independence with services that are common to all P2P networks, for instance:

- **Peer id assignment:** assigns a unique id to a peer using a specific method, *e.g.* a combination of super-peer id and counter in a super-peer network.
- **Peer linking:** links a peer to some other peers, *e.g.* by setting neighbors in an unstructured network, by locating a zone in CAN [16], etc. It also maintains the address and id of the peer's neighbors.
- **Peer communication:** enables peers to exchange messages (*i.e.* service calls).

Basic services. This layer provides elementary services for the advanced services using the P2P network layer, for instance:

- **P2P data management:** stores and retrieves P2P data (*e.g.* meta-data, index data) in the P2P network.
- **Peer management:** provides support for peer joining, rejoining, and for updating peer address (the peer ID is permanent but its address may be changed).
- **Group membership management:** allows peers to join an abstract *group*, become *members* of the group and send and receive membership notifications.

Advanced services. This layer provides advanced services for semantically rich data sharing including schema management, replication, query processing, security, etc. using the basic services.

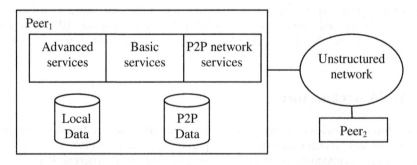

Fig. 1. APPA architecture over an unstructured network

For the cases where APPA is based on a DHT or an unstructured network, the three service layers are completely distributed over all peers, but in a super-peer network the super-peers provide P2P network services and basic services while other peers provide only the advanced services. Figure 1 shows an APPA architecture based on an unstructured network.

3 Top-k Query Processing

In this section, we first make precise our assumptions and define the problem. Then, we present a basic algorithm for Top-k query processing in APPA when it is based on an unstructured P2P system. Finally, we analyze the bandwidth cost of our algorithm and propose some techniques for reducing it.

3.1 Problem Definition

We first give our assumptions regarding schema management and Top-k queries. Then we can precisely state the problem we address in this paper.

In a P2P system, peers should be able to express queries over their own schema without relying on a centralized global schema as in data integration systems [20]. Several solutions have been proposed to support decentralized schema mapping. However, this issue is out of the scope of this paper and we assume it is provided using one of the existing techniques, *e.g.* [15] and [20]. Furthermore, also for simplicity, we assume relational data.

Now we can define the problem as follows. Let Q be a Top-k query, *i.e.* the user is interested to receive k top answers to Q. Let TTL (Time-To-Live) determine the maximum hop distance which the user wants her query be sent. Let D be the set of all data items (*i.e.* tuples) that can be accessed through *ttl* hops in the P2P system during the execution of Q. Let $Sc(d, Q)$ be a scoring function that denotes the score of relevance of a data item $d \in D$ to Q. Our goal is to find the set $T \subseteq D$, such that:

$$/T/= k \text{ and } \forall d_1 \in T, \ \forall d_2 \in (D - T) \text{ then } Sc(d_1, Q) \geq Sc(d_2, Q)$$

while minimizing the response time of Q and the bandwidth cost.

3.2 Algorithm

The algorithm starts at the *query originator*, the peer at which a user issues a Top-k query Q. The query originator performs some initialization. First, it sets *TTL* with a value which is either specified by the user or default. Second, it gives Q a unique identifier, denoted by *QID*, which is made of a unique peer-ID and a query counter managed by the query originator. Peers use *QID* to distinguish between new queries and those received before. After initialization, the query originator triggers the sequence of the following four phases: query forward, local query execution, merge-and-backward, and data retrieval. In all of these four phases, the communication between peers is done via APPA's Peer Communication service.

Query Forward

Q is included in a message that is broadcast to all reachable peers. Thus, like other flooding algorithms, each peer that receives Q tries to send it to its neighbors. Each peer p that receives the message including Q performs the following steps.

1. Check *QID*: if Q has been already received, then discard the message else save the address of the sender as the *parent* of p.
2. Decrement *TTL* by one: if $TTL > 0$, make a new message including Q, *QID*, new *TTL* and the query originator's address and send the message to all neighbors (except parent).

In order to know their neighbors, the peers use the Peer Linking service of APPA.

Local Query Execution

After the query-forward phase, each peer p executes Q locally, *i.e.* accesses the local data items that match the query predicate, scores them using a scoring function, selects the k top data items and saves them as well as their scores locally. For scoring the data items, we can use one of the scoring functions proposed for relational data, *e.g.* Euclidean function [7][6]. These functions require no global information and can score peer's data items only using local information. The scoring function can also be specified explicitly by the user.

After selecting the k local top data items, p must wait to receive its neighbors' score-lists before starting the next phase. However, since some of the neighbors may leave the P2P system and never send a score-list to p, we must set a limit for the wait time. We compute p's wait time using a cost function based on TTL, network dependent parameters and p's local processing parameters. However, because of space limitations, we do not give the details of the cost function here.

Merge-and-Backward

After the wait time has expired, each peer merges its local top scores with those received from its neighbors and sends the result to its *parent* (the peer from which it received Q) in the form of a *score-list*. In order to minimize network traffic, we do not "bubble up" the top data items (which could be large), only their addresses. A score-list is simply a list of k couples (p, s), such that p is the address of the peer owning the data item and s its score. Thus, each peer performs the following steps:

1. Merge the score-lists received from the neighbors with its local top scores and extracting the k top scores (along with the peer addresses).
2. Send the merged score-list to its parent.

Data Retrieval

After the query originator has produced the final score-list (gained by merging its local top scores with those received from its neighbors), it directly retrieves the k top data items from the peers in the list as follows. For each peer address p in the final score-list:

1. Determine the number of times p appears in the final score-list, *e.g.* m times.
2. Ask the peer at p to return its m top scored items.

Formally, consider the final score-list L_f which is a set of at most k couples (p, s), in this phase for each $p \in Domain(L_f)$, the query originator determines $T_p = \{s \,/\, (p, s) \in L_f\}$ and asks peer p to return $/T_p/$ of its top scored items.

3.3 Analysis of Bandwidth Cost

One main concern with flooding algorithms is their bandwidth cost. In this section, we analyze our algorithm's bandwidth cost. As we will see, it is not very high. We also propose strategies to reduce it more. We measure the bandwidth cost in terms of number of messages and number of bytes which should be transferred over the network in order to execute a query by our algorithm. The messages transferred can be classified as: 1) *forward messages*, for forwarding the query to peers. 2) *backward messages*, for returning the score-lists from peers to the query originator. 3) *retrieve messages*, to request and retrieve the k top results. We first present a model representing the peers that collaborate on executing our algorithm, and then analyze the bandwidth cost of backward, retrieve and forward messages.

Model

Let P be the set of the peers in the P2P system. Given a query Q, let $P_Q \subseteq P$ be a set containing the query originator and all peers that receive Q. We model the peers in P_Q and the links between them by a graph $G(P_Q, E)$ where P_Q is the set of *vertices* in G and E is the set of the *edges*. There is an edge p-q in E if and only if there is a link between the peers p and q in the P2P system. Two peers are called *neighbor*, if and only if there is an edge between them in G. The number of neighbors of each peer $p \in P_Q$ is called the *degree of p* and is denoted by $d(p)$. The average degree of peers in G is called the *average degree of G* and is denoted by $d(G)$. The average degree of G can be computed as $d(G) = (\sum_{p \in P_Q} d(p)) / |P_Q|$.

During the execution of our algorithm, $p \in P_Q$ may receive Q from some of its neighbors. The first peer, say q, which p receives Q from, is the *parent* of p in G, and thereby p is a *child* of q. A peer may have some neighbors that are neither its parent nor its children.

Backward Messages

In the Merge-and-Backward phase, each peer in P_Q, except the query originator, sends its merged score-list to its parent. Therefore, the number of backward messages, denoted by m_{bw}, is $m_{bw} = /P_Q/ - 1$.

Let L be the size of each element of a score-list in bytes (*i.e.* the size of a score and an address), then the size of the score-list is $k \times L$, where k is the number of top results specified in Q. Since the number of score-lists transferred by backward messages is $/P_Q/ - 1$, then the total size of data transferred by backward messages, denoted by b_{bw}, can be computed as $b_{bw} = k \times L \times (/P_Q/ - 1)$. If we set $L = 10$, *i.e.* 4 bytes for the score and 6 bytes for the address (4 bytes for IP address and 2 bytes for the port number), then $b_{bw} = k \times 10 \times (/P_Q/ - 1)$.

Let us show with an example that b_{bw} is not significant. Consider that *10,000* peers receive Q (including the query originator), thus $/P_Q/ = 10,000$. Since users are interested in a few results and k is usually small, we set $k=20$. As a result, b_{bw} is less than 2 megabytes. Compared with the tens of megabytes of music and video files, which are typically downloaded in P2P systems, this is small.

Retrieve Messages
By retrieve messages, we mean the messages sent by the query originator to request the k top results and the messages sent by the peers owning the top results to return these results. In the Data Retrieval phase, the query originator sends at most k messages to the peers owning the top results (there may be peers owning more than one top result) for requesting their top results and these peers return their top results by at most k messages. Therefore, the number of retrieve messages, denoted by m_{rt}, is $m_{rt} \leq 2 \times k$.

Forward Messages
Forward messages are the messages that we use to forward Q to the peers. According to the basic design of our algorithm, each peer in P_Q sends Q to all its neighbors except its parent. Let p_o denote the query originator. Consider the graph $G(P_Q, E)$ described before, each $p \in (P_Q - \{p_o\})$, sends Q to $d(p)-1$ peers, where $d(p)$ is the degree of p in G. The query originator sends Q to all of its neighbors, in other words to $d(p_o)$ peers. Then, the sum of all forward messages m_{fw} can be computed as

$$m_{fw} = (\sum_{p \in (P_Q - \{p_o\})} (d(p) - 1)) + d(p_o) \cdot$$

We can write m_{fw} as follows:

$$m_{fw} = (\sum_{p \in P_Q} (d(p) - 1)) + 1 = (\sum_{p \in P_Q} (d(p))) - |P_Q| + 1$$

Based on the definition of $d(G)$, m_{fw} can be written as $m_{fw} = (d(G) -1) \times |P_Q| +1$, where $d(G)$ is the average degree of G. According to the measurements in [17], the average degree of Gnutella is 4. If we take this value as the average degree of the P2P system, i.e. $d(G)=4$, we have $m_{fw} = 3 \times |P_Q| +1$. From the above discussion, we can derive the following lemma.

Lemma 1: The number of forward messages in the basic form of our algorithm is $(d(G) -1) \times |P_Q| +1$.

Proof: Implied by the above discussion.   □

To determine the minimum number of messages necessary for forwarding Q, we prove the following lemma.

Lemma 2: The lower bound of the number of forward messages for sending Q to all peers in P_Q is $|P_Q| - 1$.

Proof: For sending Q to each peer $p \in P_Q$, we need at least one forward message. Only one peer in P_Q has Q, i.e. the query originator, thus Q should be sent to $|P_Q| - 1$ peers. Consequently, we need at least $|P_Q| - 1$ forward messages to send Q to all peers in P_Q. □

Thus, the number of forward messages in the basic form of our algorithm is far from the lower bound.

3.4   Reducing the Number of Messages

We can still reduce the number of forward messages using the following strategies. 1) sending Q across each edge only once. 2) Sending with Q a list of peers that have received it.

Sending Q Across each Edge only once
In graph G, there may be many cases that two peers p and q are neighbors and none of them is the parent of the other, *e.g.* two neighbors which are children of the same parent. In these cases, in the basic form of our algorithm, both peers send Q to the other, *i.e.* Q is sent across the edge p-q twice. We develop the following strategy to send Q across an edge only once.

Strategy 1: When a peer p receives Q, say at time t, from its parent (which is the first time that p receives Q from), it waits for a random, small time, say λ, and then sends Q only to the neighbors which p has not received Q from them before $t + \lambda$.

Lemma 3: With a high probability, the number of forward messages with Strategy 1 is reduced to $d(G) \times /P_Q/ / 2$.

Proof: Since λ is a random number and different peers generate independent random values for λ, the probability that two neighbors send Q to each other simultaneously is very low. Ignoring the cases where two neighbors send Q to the other simultaneously, with Strategy 1, Q is sent across an edge only once. Therefore, the number of forward messages can be computed as $m_{fw} = /E/$. Since $/E/ = d(G) \times /P_Q/2$, then $m_{fw} = d(G) \times /P_Q/2$.   □

Considering $d(G)=4$ (similar to [17]), the number of forward messages is $m_{fw} = 2 \times /P_Q/$.

 With Strategy 1, m_{fw} is closer to the lower bound than the basic form of our algorithm. However, we are still far from the lower bound. By combining Strategy 1 and another strategy, we can reduce the number of forward messages much more.

Attaching to Forward Messages the List of Peers that have received Q
Even with Strategy 1, between two neighbors, which are children of the same parent p, one forward message is sent although it is useless (because both of them have received Q from p). If p attaches a list of its neighbors to Q, then its children can avoid sending Q to each other. Thus, we propose a second strategy.

Strategy 2: Before sending Q to its neighbors, a peer p attaches to Q a list containing its Id and the Id of its neighbors and sends this list along with Q. Each peer that receives the Q's message, verifies the list and does not send Q to the peers involved in the list.

Theorem 1: By combining Strategy 1 and Strategy 2, with a high probability, the number of forward messages is less than $d(G) \times /P_Q/2$.

Proof: With Strategy 2, two neighbors, which have the same parent, do not send any forward message to each other. If we use Strategy 1, with a high probability at most one forward message is sent across each edge. Using Strategy 2, there may be some edges such that no forward message is sent across them, *e.g.* edges between two neighbors with the same parent. Therefore, by combining Strategy 1 and Strategy 2, the number of forward messages is $m_{fw} \leq |E|$, and thus $m_{fw} \leq d(G) \times |P_Q|/2$.   □

Considering $d(G)=4$, the number of forward messages is $m_{fw} \leq 2 \times |P_Q|$.

4   Performance Evaluation

We evaluated the performance of our Fully Distributed algorithm (FD for short) through implementation and simulation. The implementation over a 64-node cluster was useful to validate our algorithm and calibrate our simulator. The simulation allows us to study scale up to high numbers of peers (up to 10,000 peers).

The rest of this section is organized as follows. In Section 4.1, we describe our experimental and simulation setup, and the algorithms used for comparison. In Section 4.2, we evaluate the response time of our algorithm. We first present experimental results using the implementation of our algorithm and four other baseline algorithms on a 64-node cluster, and then we present simulation results on the response time by increasing the number of peers up to *10,000*. We also did other experiments on the response time by varying other parameters, *e.g.* data item size, connection bandwidth, latency and k, but due to space limitation we cannot present them.

4.1   Experimental and Simulation Setup

For our implementation and simulation, we used the Java programming language, the SimJava package and the BRITE universal topology generator.

SimJava [11] is a process based discrete event simulation package for Java. Based on a discrete event simulation kernel, SimJava includes facilities for representing simulation objects as animated icons on screen. A SimJava simulation is a collection of entities each running in its own thread. These entities are connected together by ports and can communicate with each other by sending and receiving event objects.

BRITE [5] has recently emerged as one of the most promising universal topology generators. The objective of BRITE is to produce a general and powerful topology generation framework. Using BRITE, we generated topologies similar to those of P2P systems and we used them for determining the linkage between peers in our tests.

We first implemented our algorithm in Java on the largest set of machines that was directly available to us. The cluster has 64 nodes connected by a 1-Gbps network. Each node has an Intel Pentium 2.4 GHz processor, and runs the Linux operating system. We make each node act as a peer in the P2P system. To have a P2P topology close to real P2P overlay topologies, we determined the peer neighbors using the topologies generated by the BRITE universal topology generator [5]. Thus, each node only is allowed to communicate with the nodes that are its neighbors in the topology generated by BRITE.

To study the scalability of our algorithm far beyond 64 peers and to play with various performance parameters, we implemented a simulator using SimJava. To simulate a peer, we use a SimJava entity that performs all tasks that must be done by a peer for executing our algorithm. We assign a delay to communication ports to simulate the delay for sending a message between two peers in a real P2P system. For determining the links between peers, we used the topologies generated by BRITE.

In all our tests, we use the following simple query as workload:

SELECT R.data FROM R ORDER BY R.score
STOP AFTER *k*

Each peer has a table *R(score, data)* in which attribute *score* is a random real number in the interval *[0..1]* with uniform distribution, and attribute *data* is a random variable with normal distribution with a mean of 1 (kilo bytes) and a variance of 64. Attribute *score* represents the score of data items and attribute *data* represents (the description of) the data item that will be returned back to the user as the result of query processing. The number of tuples of *R* at each peer is a random number (uniformly distributed over all peers) greater than 1000 and less than 20,000.

The simulation parameters are shown in Table 1. Unless otherwise specified, the latency between any two peers is a normally distributed random number with a mean of 200 (ms) and a variance of 100. The bandwidth between peers is also a random number with normal distribution with a mean of 56 (kbps) and a variance of 32. Since users are usually interested in a small number of top results, we set $k=20$.

The simulator allows us to perform tests up to *10,000* peers, after which the simulation data no longer fit in RAM and makes our tests difficult. This is quite sufficient for our tests. Therefore, the number of peers of P2P system is set to be *10,000*, unless otherwise specified. In all tests, *TTL* is set as the maximum hop-distance to other peers from the query originator, thus all peers of the P2P system can receive *Q*. We observed that in the topologies with *10,000* nodes, with *TTL=12* all peers could receive *Q*. Our observations correspond to those based on experiments with the Gnutella network [17]; for instance, with *50,000* nodes, the maximum hop-distance between any two nodes is *14*.

Table 1. Simulation parameters

Parameter	Values
Bandwidth	Normally distributed random, Mean = 56 Kbps, Variance = 32
Latency	Normally distributed random, Mean = 200 ms, Variance = 100
Number of peers	10,000 peers
TTL	Large enough such that all of peers can receive the query
K	20
Result items size	Normally distributed random, Mean = 1 KB, Variance = 64

In our simulation, we compare our FD algorithm with four other algorithms. The first algorithm is a *Naïve* algorithm that works as follows. Each peer receiving *Q* sends its *k* top relevant items directly to the query originator. The query originator merges the received results and extracts the *k* overall top scored data items from them.

The second algorithm is an adaptation of Edutella's algorithm [21] which is designed for super-peer. We adapt this algorithm for an unstructured system and call it *Sequential Merging (SM)* as it sequentially merges top data items. The original Edutella algorithm works as follows. The query originator sends *Q* to its super-peer, and it sends Q to all other super-peers. The super-peers send Q to the peers connected to them. Each peer that has data items relevant to Q scores them and sends its maximum scored data item to its super-peer. Each super-peer chooses the overall maximum scored item from all received data items. For determining the second best item, it only asks one peer, the one which returned the first top item, to return its second top scored item. Then, the super-peer selects the overall second top item from the previously received items and the newly received item. Then, it asks the peer which returned the second top item and so on until all k top items will be retrieved. Finally the super-peers send their top items to the super-peer of the query originator, to extract overall k top items, and to send them to the query originator. In Edutella, a very small percentage of nodes are super-peers, *e.g.* in [19] it is 0.64, *i.e.* 64 super-peers for 10,000 peers. In our tests, we consider the same percentage, and we select the super-peers randomly from the peers of P2P system. We consider the same computing capacity for the super-peers as for the other peers.

We also propose the optimized versions of Naïve and SM algorithms that bubble up only the score-lists, as in our algorithm, and we denote them *Naïve** and *SM** respectively. In our tests, in addition to Naïve and SM algorithms, we compare our algorithm with Naïve* and SM*.

4.2 Scale Up

In this section, we investigate the scalability of our algorithm. We use both our implementation and simulator to study response time while varying the number of peers. The response time includes local processing time and data transfers, *i.e.* sending query messages, score-lists and data items.

Using our implementation over the cluster, we ran experiments to study how response time increases with the addition of peers. Figure 2 shows excellent scale up of our algorithm since response time logarithmically increases with the addition of peers until 64. Using simulation, Figure 3 shows the response times of the five algorithms with a number of peers increasing up to 10000 and the other simulation parameters set as in Table 1.

FD always outperforms the four other algorithms and the performance difference increases significantly in favor of FD as the number of peers increases. The main reason for FD's excellent scalability is its fully distributed execution. With the SM, SM*, Naive and Naïve*, a central node is responsible for query execution, and this creates two problems. First, the central node becomes a communication bottleneck since it must receive a large amount of data from other peers that all compete for bandwidth. Second, the central node becomes a processing bottleneck, as it must merge many answers to extract the k top results.

Another advantage of FD is that it does not transfer useless data items over the network. For determining top items, FD only bubbles up the score-lists (which are small) while SM and Naive algorithms transfer many data items of which only a small fraction makes the final top results. SM transfers the first top-scored item of every peer and Naïve transfers k top-scored data items of all peers. With a large number of peers, data transfer is a dominant factor in response time and FD reduces it to minimum.

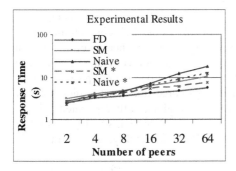

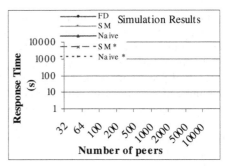

Fig. 2. Response time vs. number of peers   **Fig. 3.** Response time vs. number of peers

Overall, the experimental results correspond with the simulation results. However, the response time gained from our experiments over the cluster is a little better than that of simulation because the cluster has a high-speed network.

We also did experiments on the response time by varying data item size, connection bandwidth, latency and k. The item size has little impact on the response time of FD, SM* and Naïve*, but has strong impact on SM and Naïve. The response time decreases with increasing the connection bandwidth in all five algorithms. However, FD outperforms the other algorithms for all the tested bandwidths. FD also outperforms the other algorithms for the tested values of latency (up to 10,000 ms). However, high latency, *e.g.* more than 2000 ms, has strong impact and increases response time much, but below 2000 ms, latency has not much effect on FD's response time. According to studies reported in [18], more than 80% of links between peers have good latency, less than 280 ms, for which FD has very good performance. k has little impact on the response time of SM, but has some impact on FD, SM*, Naïve and Naïve*. Despite the effect of k on FD, it is by far the superior algorithm for the tested values of k ($k<200$). Since users are usually interested in a small number of top results, *e.g.* less than *20* results, the performance advantage for FD remains high.

5 Related Work

In the context of P2P systems, little research has concentrated on Top-k query processing. In [21] the authors present a Top-k query processing algorithm for Edutella which is a super-peer network. The technique which Edutella uses for processing Top-k queries is explained in Section 4.1. Although very good for super-peer networks, this technique cannot apply efficiently to other networks, in particular, unstructured, since there may be no peer with higher reliability and computing power. In contrast, our algorithm makes no assumptions about the P2P network topology and the existence of certain peers.

A good formal framework for ranking is introduced in [1] based on a ranking algebra. The authors show that not only one global ranking should be taken into account, but also several in different contexts. The ranking algebra allows aggregating the local rankings into global rankings.

PlanetP [8] is a P2P system that constructs a content addressable publish/subscribe service using gossiping to replicate global documents across P2P communities up to ten thousand peers. In PlanetP, a Top-k query processing method is proposed that works as follows. Given a query Q , the query originator computes a relevance ranking of peers with respect to Q, contacts them one by one from top to bottom of ranking and asks them to return a set of their top-scored document names together with their scores. To compute the relevance of peers, a global fully replicated index is used that contains term-to-peer mappings. In a large P2P system, keeping up-to-date the replicated index is a major problem that hurts scalability. In contrast, our algorithm does not use any replicated data.

For the cases where a data item can have multiple scores at different sites, *e.g.* the amount of a customer's purchase in several stores, the TA family of algorithms for monotonic score aggregation [9] stands out as an efficient and highly versatile method. There have been many algorithms in order to optimize the TA algorithm in terms of bandwidth cost and response time, *e.g.* [22] and [14].

6 Conclusion

In this paper, we proposed a fully distributed algorithm for Top-k query processing in the context of the APPA data management system. APPA has a network-independent design that can be implemented over different P2P networks (unstructured, DHT, super-peer, etc.), thus allowing us to exploit continuing progress in such systems. We presented our algorithm for the case of unstructured systems, thus with minimal assumptions. Our algorithm requires no global information, does not depend on the existence of certain peers and its bandwidth cost is low.

For determining the k top results, we use the concept of score-list which reduces the bandwidth consumption and also reduces the response time. We analyzed the bandwidth cost of our algorithm and we proposed efficient techniques in order to reduce it.

We validated the performance of our algorithm through implementation over a 64-node cluster and simulation using the BRITE topology generator and SimJava. The experimental and simulation results showed that our algorithm can have logarithmic scale up. The simulation also showed the excellent performance of our algorithm compared with a naïve algorithm and an adaptation of an existing algorithm.

As future work, we plan to deal with replicated data in P2P Top-k query processing. In this paper, we assumed that data items are not replicated. In the case of data replication, with our algorithm, there may be replicated data items in the final score-list. This may be fine for the user as it is an indication of the items' usefulness (in a P2P system, the most useful data get most replicated). But we could also identify replicated items.

References

[1] Aberer, K., AND Wu., J. Framework for Decentralized Ranking in Web Information Retrieval. *Proc. of the 5th Asia Pacific Web Conference (APWeb)*, 2003.
[2] Abiteboul, S., et al. Dynamic XML documents with distribution and replication. *SIGMOD Conf.*, 2003.

[3]   Akbarinia, R., Martins, V., Pacitti, E., and Valduriez, P. Design and Implementation of Atlas P2P Architecture. *Global Data Management* (Eds. R. Baldoni, G. Cortese, F. Davide), IOS Press, 2006.

[4]   Akbarinia, R., Martins, V., Pacitti, E., AND Valduriez, P. Replication and Query Processing in the APPA Data Management System. *6th Workshop on Distributed Data & Structures (WDAS)*, 2004.

[5]   BRITE, http://www.cs.bu.edu/brite/.

[6]   Carey, M.J., AND Kossmann, D. On saying 'Enough Already!'. *SIGMOD Conf.*, 1997.

[7]   Chaudhuri, S., et al. Evaluating Top-k Selection queries. *VLDB Conf.*, 1999.

[8]   Cuenca-Acuna, F.M., Peery, C., Martin, R.P., AND Nguyen, T.D. PlanetP: Using Gossiping to Build Content Addressable Peer-to-Peer Information Sharing Communities. *IEEE Int. Symp. on High Performance Distributed Computing (HPDC)*, 2003.

[9]   Fagin, R., Lotem, J., AND Naor, M. Optimal aggregation algorithms for middleware. *J. Comput. Syst. Sci. 66(4)*, 2003.

[10] Gnutella. http://www.gnutelliums.com/.

[11] Howell, F., AND McNab, R. SimJava: a discrete event simulation package for Java with applications in computer systems modeling. *Int. Conf. on Web-based Modelling and Simulation, San Diego CA, Society for Computer Simulation*, 1998.

[12] Huebsch, R., et al. Querying the Internet with PIER. *VLDB Conf.*, 2003.

[13] Kazaa. http://www.kazaa.com/.

[14] Michel, S., Triantafillou, P., AND Weikum, G. KLEE: A Framework for Distributed Top-k Query Algorithms. *VLDB Conf.*, 2005.

[15] Ooi, B., Shu, Y., AND Tan, K-L. Relational data sharing in peer-based data management systems. *SIGMOD Record, 32(3)*, 2003.

[16] Ratnasamy, S., Francis, P., Handley, M., Karp, R.M., AND Shenker, S. A scalable content-addressable network. *Proc. of SIGCOMM*, 2001.

[17] Ripeanu, M., AND Foster, I. Mapping the gnutella network: Macroscopic properties of large-scale peer-to-peer systems. *IPTPS*, 2002.

[18] Saroiu, S., Gummadi, P., AND Gribble, S. A Measurement Study of Peer-to-Peer File Sharing Systems. *Proc. of Multimedia Computing and Networking (MMCN)*, 2002.

[19] Siberski, W., AND Thaden, U. A Simulation Framework for Schema-Based Query Routing in P2P-Networks. *EDBT Workshops*, 2004.

[20] Tatarinov, I., et al. The Piazza peer data management project. *SIGMOD Record 32(3)*, 2003.

[21] Thaden, U., Siberski, W., Balke, W.T., AND Nejdl, W. Top-k query Evaluation for Schema-Based Peer-To-Peer Networks, *Int. Semantic Web Conf. (ISWC)*, 2004.

[22] Theobald, M., Weikum, G., AND Schenkel, R. Top-k Query Evaluation with Probabilistic Guarantees. *VLDB Conf.*, 2004.

[23] Yang, B., AND Garcia-Molina, H. Designing a super-peer network. *Int. Conf. on Data Engineering*, 2003.

[24] Yu, C., et al. Databases Selection for Processing k Nearest Neighbors Queries in Distributed Environments. *ACM/IEEE-CS joint Conf. on DL*, 2001.

[25] Yu, C., Philip, G., AND Meng, W. Distributed Top-n Query Processing with Possibly Uncooperative Local Systems, *VLDB Conf.*, 2003.

Posterior Task Scheduling Algorithms for Heterogeneous Computing Systems

Linshan Shen and Tae-Young Choe

School of Computer Engineering, Kumoh National Institute of Technology, Yangho-dong,
730-701 Kumi city, KyungPook, Korea
smart781005@hotmail.com, choety@kumoh.ac.kr

Abstract. The task scheduling problem in heterogeneous system is known as NP-hard. Recently, Bajaj and Agrawal proposed an algorithm TANH (Task duplication-based scheduling Algorithm for Network of Heterogeneous systems) with optimality conditions, which are wider than previous optimality conditions. TANH algorithm combines the clustering technique with task duplication. We propose two postprocessing algorithms, HPSA1 (Heterogeneous Posterior Scheduling Algorithm) and HPSA2, to reduce the schedule length for DAGs which don't satisfy the optimality conditions of TANH algorithm. Our algorithms reduce the schedule length by exchanging task clusters in which its parent tasks reside. We compare with HCNF (Heterogeneous Critical Node First) algorithm by illustrating an example to show how our algorithms operate.

Keywords: Heterogeneous computing system, DAG, task scheduling algorithm, postprocessing, clustering.

1 Introduction

For high-speed computation purposes, parallel processing has been extensively explored. Some applications like fluid flow, image processing, weather modeling, and distributed database systems get a great deal of parallelism. A general methodology adopted in parallel processing is to partition an application into a set of cohesive tasks and to run them separately on different processors. The partitioned application can be modeled as a directed acyclic graph (DAG). In DAGs, a forward edge means that the predecessor task transmits the data to the successor task.

The task scheduling problem is to allocate tasks to processors in order to minimize the completion time of given application which can be expressed as a DAG. The task scheduling problem is known as NP-hard [1,2]. Therefore, heuristic task scheduling algorithms are used to tackle the problem. The scheduling algorithms have been extensively studied [3-19]. These heuristics are classified into a variety of categories such as, list scheduling algorithms, clustering algorithms, duplication-based algorithms, and guided random search methods. Most of them are designed for homogeneous computing systems.

In the classical list scheduling algorithms [3,4,5,19], tasks are assigned priorities statically or dynamically. The priorities are assigned based on computation and

M. Daydé et al. (Eds.): VECPAR 2006, LNCS 4395, pp. 172–183, 2007.
© Springer-Verlag Berlin Heidelberg 2007

communication costs in the task graph. The next chosen is the highest priority task among the ready tasks whose precedence have been met and are ready for scheduling. The following step is to select most suitable processor to accommodate the chosen task. Performances of list scheduling algorithms tend to suffer due to min-max problem and deteriorate substantially for fine grain task graphs having high communication to computation cost ratio (CCR) [18].

Clustering-based algorithms [7,8,9] try to schedule heavily communicating tasks onto the same processor. It is also known as three phase scheduling. In the first phase, heavily communicating tasks are grouped into a set of clusters (unbounded) using linear or nonlinear clustering heuristics. In the second phase, clusters are mapped onto the set of available processors using communication sensitive or insensitive heuristics. In the third phase cluster merging is done based on the available number of processors.

Duplication-based algorithms allow tasks to be duplicated on one or more than processors, in order to reduce the start time of its successor tasks. Now, the duplication-based algorithms have been blended with both list and clustering-based techniques by other researchers. List or clustering-based algorithms [5,6,10-13] with task duplication tend to perform better than no duplication algorithms. Genetic algorithms [14,15] are of the most widely studied guided random search techniques for the task scheduling problem. Although they provide good quality of schedules, their execution times are significantly higher than other alternatives.

The processors in heterogeneous systems have different processing powers, so scheduling them is more complex. In recent years, many scheduling algorithms for the heterogeneous system are proposed, such as Levelized Duplication Based Scheduling (LDBS) [10], Dynamic Level Scheduling (DLS) [4], Task Duplication-based Scheduling Algorithm for Network of Heterogeneous System (TANH) [13], Fast Critical Path (FCP), Fast Load Balancing (FLB) [5], Task Duplication Scheduling (TDS-1) [16], Heterogeneous Earliest Finish Time First (HEFT) [17].

In this paper, we propose two posterior algorithms, HPSA1 (Heterogeneous Posterior Scheduling Algorithm) and HPSA2 for heterogeneous computing system. The main motivation is to improve the quality of the scheduling length, while DAGs don't satisfy TANH algorithm's optimality conditions. The algorithms run at the hind of TANH algorithm.

In the next section, we define the parameters and data structures served for our algorithms. In section 3, we briefly introduce TANH and HCNF algorithms. In section 4, we propose our two algorithms. In section 5, we illustrate an example to show that our algorithms how to work. In section 6, we discuss the experimental results. In the final part, we present conclusions.

2 Problem Definition

We consider any application that is represented as a directed acyclic graph (DAG). In DAG, each node represents a task and each directed edge represents communication cost between tasks. A task is assumed to be nonpreemptive. A tuple $G = (V, E, P, w, c)$

is used to define a DAG, where V is a set of nodes, |V| is the number of nodes in V; E is a set of edges; P is a set of processors, |P| is the number of available processors; w = w(n_i, p_i) indicates the computation cost of task n_i on processor p_i, where $n_i \in V$ and $p_i \in P$; c = c(n_i, n_j) indicates the communication cost between task n_i and n_j, where n_i, $n_j \in V$. If both n_i and n_j are scheduled onto the same processor, c(n_i, n_j) is assumed to be zero. In the other hand, the network bandwidth is assumed to be wide enough to provide contention-free transmission. Given a task n_j, pred(n_j) is a set of predecessor tasks which have the outgoing edge into n_j; succ(n_j) consists of the tasks which receives the data from n_j. And n_i = fpred(n_j) denotes the favorite predecessor, which means among all the predecessor tasks of n_j, n_i has the highest value of the earliest finish time. The earliest start/finish time indicates when a task could be started/finished at the earliest possible time. Arrival time of task n_j equals the sum of the earliest finish time of the parent tasks which are not in the same processor with n_j and the communication time between the parent task and n_j.

A set of tasks assigned to a processor is called a cluster. Each task in a cluster has its start time and finish time in the corresponding processor. A task n_j at processor p_k has its start time st(n_j, p_k) and finish time ft(n_j, p_k). In order to compute st(n_j, p_k) and ft(n_j, p_k) for task n_j at processor p_k, we need to introduce several equations:

$$st(n_j, p_k) = \max[ft(p_k), rdy(p_k, n_j)] \tag{1}$$

$$ft(n_j, p_k) = st(n_j, p_k) + w(n_j, p_k) \tag{2}$$

$$rdy(p_k, n_j) = \max_{n_i \in pred(n_j)} [\min_{n_i \in p_i, p_i \neq p_j} [ft(n_i, p_i)] + c(n_i, n_j)] \tag{3}$$

where ft(p_k) represents the current finish time of processor k, and rdy(p_k,n_j) represents the largest ready time from parent tasks allocated in other processors.

3 Related Works

In recent past, some algorithms combined clustering technique with duplication have been proposed. The TANH algorithm [13] just belongs to them, its time complexity is $|V|^2$. The algorithm firstly generates the initial clusters. If the number of required processors (RP) is larger than the number of available processors (AP), then using compaction procedure selects two processors and merges them to one processor until RP equals AP. Bajaj and Agrawal proposed TANH algorithm and presented a set of optimal conditions for join nodes. Unfortunately, when an application DAG doesn't satisfy the optimal conditions, the expected result is difficult to be gained.

HCNF [19] is list scheduling algorithm. It proceeds by identifying the critical path in the DAG. The critical path is a path which has the largest sum of average task computation costs and inter-task communication costs among all the paths from the entry task to the exit task. Task is free when its predecessor tasks have been assigned to processors. A list of free tasks is constructed. Within the list, the highest priorities are assigned to tasks which fall in the critical path, followed by those with the highest computation cost. A task n_i is scheduled onto processor P_i, which gives the lowest eft(n_i, P_i). Its earliest finish time eft(n_i, P_i) = w(n_i, P_i) + est(n_i, P_i), where the earliest start time est(n_i, P_i) is the maximum of time at which processor P_i becomes available and the time at which the last message arrives from any predecessor of n_i, and w(n_i, P_i) is the execution time of n_i on P_i. In order to reduce the communication time, the favorite predecessor is considered for duplication.

4 Proposed Algorithms

The main motivation of our algorithms is to reduce the finish time of tasks which determine the scheduling length by exchanging its parent clusters. If the finish time of the task can be reduced, the scheduling length may be reduced. Thus, two posterior processes, HPSA1 and HPSA2, are proposed to improve the scheduling length.

The proposed two algorithms execute at the hind of the general scheduling algorithms, as shown in Fig. 1.

```
Main( )
    {
        . . .
        Scheduling algorithm ( );
        Posterior scheduling algorithm;
    }
```

Fig. 1. The position of the posterior algorithm

4.1 Selection of Scheduling Algorithm

In this paper, we select TANH algorithm as the scheduling algorithm instead of HCNF. The researchers of TANH algorithm proposed a set of optimal conditions. If DAGs satisfied the optimal conditions, TANH algorithm got the optimal schedule. When our algorithms executed at the hind of TANH algorithm, the optimal schedule was also got. But it was not sure to get optimal schedule when HCNF algorithm was used. If DAGs didn't satisfy the optimal conditions, it was not sure to get optimal schedule when TANH was used to schedule. But combining our algorithms to TANH can make the schedule length less than or equal to that of TANH, while NCNF could not guarantee to generate optimal schedule. So the combination of our algorithms and TANH is used in this paper. The comparisons are shown in Table 1.

Table 1. The comparisons of TANH, TANH+HPSA, and HCNF

Task scheduling algorithm	DAG (directed a-cyclic graph)	
	It satisfies the optimal conditions of TANH	It doesn't satisfy the optimal conditions of TANH
TANH	Optimal schedule	Not sure
TANH+HPSA	Optimal schedule	The schedule length<=the schedule length by TANH
HCNF	Not sure	Not sure

Our algorithms execute at the hind of TANH algorithm, only when exchanging operation can make the schedule length reduce. When the exchange operation is finished, if no improvement is gained, the operation will be canceled. So the combination of our algorithms and TANH algorithm can reduce the schedule length given by TANH algorithm. Otherwise the schedule length is preserved.

4.2 HPSA1 Algorithm

The main idea of this algorithm is to reduce the start time of target task at its processor, and to reduce the scheduling length. Firstly, the exit node is target task, and its predecessors are checked, which predecessor determines the start time of the target task, and exchange the cluster including the decided task and the other parent tasks' clusters of the target task, respectively. Selected is an exchange that mostly reduces the

```
Void HPSA1( )
{
  n_t =the exit task;
  do{
      do{
            n_p = parent task that determines the start time
                  of n_t;
            n_q = ParentTask1(n_t ,n_p);
            exchange clusters of n_p and n_q;
      } while (start time of n_t reduces);
        n_t = ParentTask2(n_t);
      }while(there is an improvement);
}
```

Fig. 2. The pseudo code of HPSA1: Function *ParentTask1*(n_t, n_p) is responsible for finding another parent task n_q of n_t that minimizes the start time of n_t by exchanging clusters of n_p and n_q each other. Function *ParentTask2*(n_t,) is responsible for finding the parent task of n_t which determines the start time of n_t.

start time of the target. For the current target, next passes are done, until no improvement is gained. Checking the current schedule, the predecessor task that determines the start time of the target task is selected as the next target task, repeat the above steps, until no improvement is gained. The pseudo code for HPSA1 algorithm is described in the Fig. 2. The time complexity of HPSA1 is $|V|^2|P|^2 \max_{n_i \in V} |succ(n_i)|$.

4.3 HPSA2 Algorithm

For n_i and n_j in some processor p_i, an empty slot exits if the finish time of n_i in p_i is smaller than the start time of n_j in p_i. After a schedule, it is highly probable that the empty slots exist. If an empty slot not only has the largest size among all the empty slots, but also affects the schedule length, then that is more improvement on the scheduling length when we try to exchange the clusters that the predecessors accommodate and the cluster above the slot. The biggest reduction cluster that its exchange makes the scheduling length reduce most is selected as the right exchange. After the operation to the slot which has biggest size, then consider the slot with the second biggest size and affecting the scheduling length.

```
Void  HPSA2 ( )
{
   Assign every task as -1;
   do {
          Find all empty slots;
          Generate the path;
          nt= Findtail();
           for each predecessor np of nt
               np =  FindTask(nt);
          exchange clusters of nt and np ;
         Assign nt as 0;
        } while (there is an improvement)
}
```

Fig. 3. The pseudo code of HPSA2: Function *FindTask*(n_t) is responsible for finding a task which is neither at the same cluster with n_t, nor with the exit task, and maximum reduction for the slot is obtained by exchanging this cluster and the cluster that n_t resides. Function *Findtail*() is responsible for finding the tail task of an empty slot that has the biggest size and affects the scheduling length.

Each slot has its information: head task, tail task, and slot size. In order to know which slot affects the scheduling length, we need to check the tail task whether affects the scheduling length. We use a queue path to save tasks that determine the scheduling length from the exit task, and write down the slot size related to the tasks. If a task in path has only one predecessor, then also adds this predecessor task to the path, and the slot size is zero.

The pseudo code of HPSA2 is illustrated in the Fig. 3. It marks all the tasks. If the slot related with this task was operated, this task is unmarked. Next, it memories the current scheduling length, which can be used to compare the scheduling length after exchanging the clusters that make the slot reduce most. The time complexity of HPSA2 is $|V|^2|P|^2$.

5 Illustration of an Example

In this part, we illustrate an example to show combining our two posterior processes to TANH is efficient for improving the schedule of TANH algorithm, and their performances are better than HCNF algorithm. The example DAG is illustrated in the Fig. 4a, and the computation cost of tasks at every processor is illustrated in the Fig. 4b. We get the initial schedule as the Fig. 5 after TANH algorithm. We get the schedule as the Fig. 6 after the HCNF algorithm.

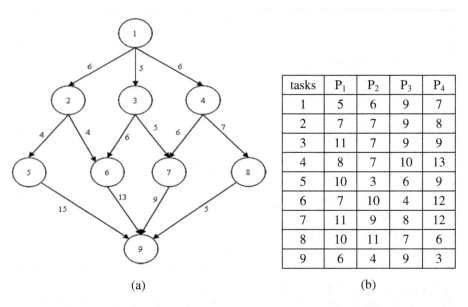

tasks	P_1	P_2	P_3	P_4
1	5	6	9	7
2	7	7	9	8
3	11	7	9	9
4	8	7	10	13
5	10	3	6	9
6	7	10	4	12
7	11	9	8	12
8	10	11	7	6
9	6	4	9	3

(a)   (b)

Fig. 4. (a) An example DAG G_1. (b) Runtime of tasks for G_1.

5.1 Using HPSA1 to Post-process

In the Fig. 5, firstly, the target is task 9, task 7 determines the start time of task 9. By exchanging cluster $\{1, 4, 7\}$ and cluster $\{1, 3, 6\}$, $\{1, 4, 7\}$ and $\{1, 4, 8\}$, $\{1, 4, 7\}$ and $\{1, 2, 5\}$, we know that the exchange of cluster $\{1, 4, 7\}$ and cluster $\{1, 3, 6\}$ makes the

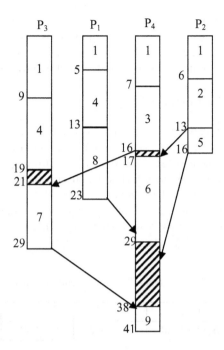

Fig. 5. The schedule S_1 of DAG G_1 by TANH algorithm

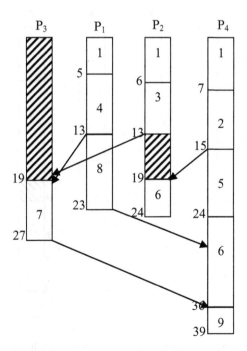

Fig. 6. The schedule S_2 of DAG G_1 by HCNF algorithm

start time of task 9 reduce most. We get a schedule of Fig. 7. When the start time of task 9 was reduced, next passes will be done, but no improvement is gained. For the current schedule, task 6 and task 7 are considered for exchanging, because they determine the start time of the task 9. However, neither of them as the target improves the schedule length.

5.2 Using HPSA2 to Post-process

Firstly, a path is created, in which contain the tasks determine the scheduling length of the schedule in Fig. 5. We illustrate tasks in the path 9-7-3-1 in accordance with the size of slots which related to them in Table 2. From the table, the empty slot related to task 9 has the biggest size and affects the schedule length. Exchanges are executed, cluster $\{1, 3, 6\}$ and cluster $\{1, 4, 7\}$, $\{1, 3, 6\}$ and $\{1, 4, 8\}$, $\{1, 3, 6\}$ and $\{1, 2, 5\}$, exchanging $\{1, 3, 6\}$ and $\{1, 4, 7\}$ has most reduction for the slot above task 9, the schedule as the Fig. 7. Next, only one slot for task 7, we operate an exchange: $\{1, 4\} \Leftrightarrow \{1, 3\}$, but no improvement is gained, HPSA2 ends.

Table 2. The Path that affects the scheduling length of S_1

task	9	7	3	1
Size of slot	9	2	0	0

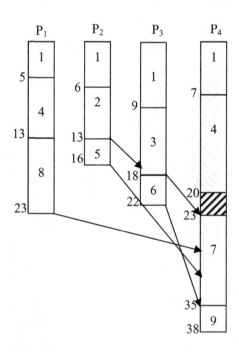

Fig. 7. First exchange of S_1 when HPSA1 is applied and first exchange of S_1 when HPSA2 is applied

6 Experimental Results

To study the performances of our algorithms in execution time, a random DAG generator is designed by us. The generator requires the following input parameters: i) graph level GL, ii) the number of processors NoP, iii) the maximal outdegree of task is 3. Without loss of generality, all DAGs are required a single entry node and an exit node. The computation time is selected randomly from 3 to 7, and the communication time is selected randomly from 2 to 6.

We did two sets of experiments: i) fork tasks in DAG have the same computation cost, ii) fork tasks in DAG have -1~1 different computation cost. Every set of experiments use 10 random generated DAGs when GL=5, 6, 7 in accordance with NoP=6, 8, 9, respectively. We compare the average execution times (msec) of TANH, TANH+HPSA, and HCNF. The experimental results are shown in Fig. 8 and Fig. 9, respectively.

For the first set of experiments, the sum of schedule length of TANH+HPSA1 has 3.82%, 1.79%, and 2.98% decrements than that of TANH when GL=5, 6, 7 in accordance with NoP=6, 8, 9. The sum of schedule length of TANH+HPSA2 has 4.36%, 1.79%, and 2.98% decrements than that of TANH when GL=5, 6, 7 in accordance with NoP=6, 8, 9. To the same DAGs, in the case of HCNF, the sum of schedule length of TANH+HPSA1 has 1.67%, 4.14%, and -1.68% decrements than that of HCNF. The sum of schedule length of TANH+HPSA2 has 2.23%, 4.14%, and -1.68% decrements than that of HCNF.

For the second set of experiments, the sum of schedule length of TANH+HPSA1 has 2.96%, 0.83%, and 1.73% decrements than that of TANH when GL=5, 6, 7 in accordance with NoP=6, 8, 9. The sum of schedule length of TANH+HPSA2 has 3.5%, 0.66%, and 2.25% decrements than that of TANH when GL=5, 6, 7 in accordance with NoP=6, 8, 9. To the same DAGs, in the case of HCNF, the sum of schedule length of TANH+HPSA1 has -5.56%, 2.18%, and -6.98% decrements than that of HCNF.

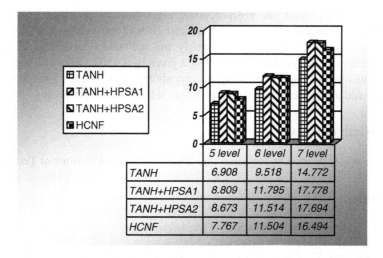

Fig. 8. The execution time comparison for fork tasks with the same computation time in DAG

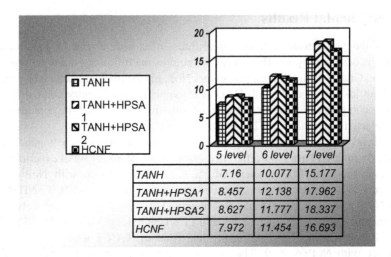

	5 level	6 level	7 level
TANH	7.16	10.077	15.177
TANH+HPSA1	8.457	12.138	17.962
TANH+HPSA2	8.627	11.777	18.337
HCNF	7.972	11.454	16.693

Fig. 9. The execution time comparison for fork tasks with -1~1 different computation time in DAG

The sum of schedule length of TANH+HPSA2 has -4.97%, 1.96%, and -6.42% decrements than that of HCNF.

The execution times of TANH+HPSA1 are 11%~74% increments than those of TANH, and 4%~40% increments than those of TANH in case of TANH+HPSA2. So our algorithms used short execution times to improve the schedule length of the previous algorithms.

7 Conclusions

In this paper, we present two postprocessing algorithms for previous algorithms in heterogeneous computing systems. HPSA1 and HPSA2 are executed at the hind of TANH algorithm. If DAG does not satisfy the optimality conditions of TANH, TANH+HPSA can reduce the schedule length given by TANH. Otherwise it preserves the schedule length. Our algorithms spend very short execution time. Thus they can be efficiently added to other scheduling algorithms in heterogeneous systems.

Acknowledgements

This paper was supported by Research Fund, Kumoh National Institute of Technology.

References

[1] J.D. Ullman. NP-complete scheduling problems. *Journal of Computing System Science*, 10:384-393, 1975.
[2] J. Liou and M. Palis, "A Comparison of General Approaches to Multiprocessor Scheduling," *Proc. Int'l Parallel Processing Symp.*, pp. 152-156, 1997.

[3] Y. Kwok and I. Ahmed, "Dynamic Critical-Path Scheduling: An Effective Technique for Allocating Task Graphs to Multiprocessors," *IEEE Trans. Parallel and Distributed Systems*, Vol. 7, no. 5, pp. 506-521, May 1996.

[4] G.C. Sih and E.A. Lee, "A Compile-Time Scheduling Heuristic for Interconnection-Constrained Heterogeneous Processor Architectures," *IEEE Trans. Parallel and Distributed Systems*, Vol. 4, no. 2, pp. 175-186, Feb. 1993.

[5] A. Radulescu and A.J.C. Van Gemund, "Fast and Effective Task Scheduling in Heterogeneous Systems," *Proc. of HCW*, pp. 229-238, May 2000.

[6] I. Ahmad and Y. Kwok, "On exploiting Task Duplication in Parallel Program Scheduling," *IEEE Trans. Parallel and Distributed Systems*, 9(9):872-892, Sept. 1998.

[7] T. Yang and A. Gerasoulis, "DSC: Scheduling Parallel Tasks on an Unbounded Number of processors," *IEEE Trans. Parallel and Distributed Systems*, vol. 5, no. 9, pp. 951-967, Sept. 1994.

[8] S.J. Kim and J.C. Browne, "A General Approach to Mapping of Parallel Computation upon Multiprocessor Architecture.," *Proc. Int'l Conf. Parallel Proc.*, vol. 2, pp. 23-32, Jan. 1988.

[9] J. Liou and M.A. Palis, "An Efficient Clustering Heuristics for Scheduling DAGs on Multiprocessors," *Proc. of Parallel and Distributed processing symposium*, 1996.

[10] Atakan Dogan and Fusun Ozguner, "LDBS: A Duplication Based Scheduling Algorithm for Heterogeneous Computing Systems," *Proc. of Int'l Parallel Processing (ICPP'02)*, 2002.

[11] S. Darba and D. P. Agrawal, "Optimal Scheduling Algorithm for Distributed-Memory Machines," *IEEE Trans. Parallel and Distributed Systems*, (1):87-94, Jan. 1998.

[12] G.-L. Park, B. Shirazi, and J. Marguis, "DFRN: A New Approach for Duplication Based Scheduling for distributed memory multiprocessor systems," *Proc. Of Int'l Parallel Processing Symposium*, Geneva, Switzerland, Apr. 1997.

[13] Rashmi Bajaj and Dharma P. Agrawal, "Improving Scheduling of Tasks in a Heterogeneous Environment," *IEEE Trans. Parallel and Distributed Systems*, vol. 15, No. 2, February 2004.

[14] E.S.H. Hou, N. Ansari, and H. Ren, "A Genetic Algorithm for Multiprocessor Scheduling," *IEEE Trans. Parallel and Distributed Systems*, Vol. 5, no. 2, pp. 113-120, Feb. 1994.

[15] H. Singh and A. Youssef, "Mapping and Scheduling Heterogeneous TaskGraphs using Genetic Algorithms," *Proc. of Heterogeneous Computing Workshop*, pp. 86-97, 1996.

[16] A. Ranaweera and D.P. Agrawal, "A Task Duplication based Algorithm for Heterogeneous Systems," *Proc. of IPDPS*, pp. 445-450, May 1-5, 2000.

[17] H. Topcuoglu, S. Hariri, M-Y. Wu, "Performance-Effective and Low-complexity Task Scheduling for heterogeneous computing," *IEEE Trans. on Parallel and Distributed Systems*, vol. 13, no. 3, March 2002.

[18] E. Ilavarasan and P. Thambidurai, "Levelized Scheduling of Directed A-cyclic Precedence Constrained Task Graphs onto Heterogeneous Computing System," *Proceedings of the First International Conference on Distributed Frameworks for Multimedia Applications (DFMA'05)*, 2005.

[19] Sanjeev Baskiyar and Prashanth C. SaiRanga, "Scheduling Directed A-cyclic Task Graphs on Heterogeneous Network of Workstations to Minimize Schedule length," *Proceeding of the 2003 International Conference on Parallel Processing Workshops (ICPPW'03)*, 2003.

Design and Implementation of an Environment for Component-Based Parallel Programming

Francisco Heron de Carvalho Junior[1], Rafael Dueire Lins[2],
Ricardo Cordeiro Corrêa[1], Gisele Araújo[1], and Chanderlie Freire de Santiago[1]

[1] Departamento de Computação, Universidade Federal do Ceará
Campus do Pici, Bloco 910, Fortaleza, Brazil
{heron,correa,gisele,cfreire}@lia.ufc.br
[2] Depart. de Eletrônica e Sistemas, Universidade Federal de Pernambuco
Av. Acadêmico Hélio Ramos s/n, Recife, Brazil
rdl@ufpe.br

Abstract. Motivated by the inadequacy of current parallel programming artifacts, the # component model was proposed to meet the new complexity of high performance computing (HPC). It has solid formal foundations, layed on category theory and Petri nets. This paper presents some important design and implementation issues on the implementation of programming frameworks based on the # component model.

1 Introduction

Clusters and grids have brought the processing power of high performance computing (HPC) architectures to a wide number of academic and industrial users, bringing new challenges to computer scientists. Contemporary parallel programming techniques that can exploit the potential performance of distributed architectures, such as the message passing libraries MPI and PVM, provide poor *abstraction*, requiring a fair amount of knowledge on architectural details and parallelism strategies that go far beyond the reach of users in general. On the other hand, higher level approaches, such as parallel functional programming languages and scientific computing parallel libraries do not merge efficiency with generality. Skeletal programming has been considered a promising alternative, but several reasons have made difficult its dissemination [12]. In fact, the scientific community still looks for parallel programming paradigms that reconciles portability and efficiency with generality and high-level of abstraction [4].

In recent years, the HPC community has tried to adapt component technology, now successfully applied in business applications in dealing with software complexity and extensibility, to meet the needs of HPC applications. These efforts yielded CCA and its frameworks [2], P-COM [21], Fractal [3], et cetera [25]. Besides being a potential alternative to reconcile abstraction, portability, generality, and efficiency in parallel programming, components leverage *multi-disciplinary*, *multi-physics*, and *multi-scale* software for HPC [5], possibly targeting heterogenous execution environments that are enabled for grid, cluster, and capability computing [14].

M. Daydé et al. (Eds.): VECPAR 2006, LNCS 4395, pp. 184–197, 2007.
© Springer-Verlag Berlin Heidelberg 2007

The most important challenge to make components suitable for HPC relies on their support for parallel programming [1,11]. Surprisingly, parallel programming based on the current approaches for supporting peer-to-peer components inter-action is not suitable for performance demands of HPC software that are not embarrassingly parallel [1,11]. Unfortunately, the presence of complex process interactions are common in modern HPC software. For this reason, HPC com-ponents models and architectures have been extended for supporting non-trivial forms of parallelism [18,13,23,1,3]. However, such approaches for parallelism do not reach generality of message-passing based parallel programming. In addition, they are influenced by the common trend of lower level parallel programming approaches to treat processes, and not only concerns, as basic units of software decomposition. We consider that this is one of the main reasons of the difficulty in adapting current software engineering techniques for the development of par-allel programs. Software engineering approaches have appeared in the context of sequential software, where processes do not exist. We advocate orthogonality between processes and concerns [8]. Thus, they cannot be appropriately viewed under the same software decomposition dimension.

The # component model was primarily developed for general purpose paral-lel programming, taking the orthogonality between processes and concerns as a design premise. Unlike most of the recently proposed approaches for HPC com-ponents, it is not founded on existing component models design patterns. It has origins in the coordination model of Haskell$_\#$ [7], a parallel extension to the func-tional language Haskell. Most possibly, any component model may be interpreted in terms of the # component model abstractions. Besides to deal with parallel programming in a natural way, the # component model is influenced by modern ideas regarding the notion of separation of concerns [8], one of the main driv-ing forces for recent advances in software engineering technologies [22]. Indeed, cross-cutting composition of concerns is supported. The # component model tries to achieve a higher level of abstraction by employing skeletal programming through *existentially bounded quantified component types*. This paper intends to present the design of a framework based on the # component model for parallel programming targeting HPC software on top of IBM Eclipse Platform.

In what follows, Section 2 introduces the basic principles behind the # com-ponent model, comparing it to other HPC component approaches. Section 3 depicts the general design of # frameworks. Section 4 presents the design of a general purpose parallel programming framework. Section 5 concludes this pa-per, describing ongoing and lines for further works regarding the implementation of programming environments based on the # component model.

2   The # Component Model: Principles and Intuitions

Motivated by the success of the component technology in software industry, scientific computing community has proposed component models, architectures and frameworks for leveraging *multi-disciplinary*, *multi-physics*, and *multi-scale* HPC software, possibly targeted at HPC architectures enabled for grid, cluster, and capability computing [25]. Unfortunately, their requirements for the support of parallelism and high processing efficiency make usual forms of peer-to-peer component interaction unsuitable [11,5]. For this reason, specific parallel programming extensions have been proposed to current component technology. For example, CCA specification includes SCMD[1] extensions for supporting SPMD style of parallel programming [1]. PARDIS[18], PADICO[13], and GridCCM[23] have also adopted a similar concept for supporting parallel objects inside components. Fractal proposes collective ports that may dispatch method calls to a set of inner parallel components [3]. In general, such extensions cover requirements of a wide range of parallel programs in certain domains of interest, but they do not provide full generality of message-passing parallel programming. It is usual to find papers on HPC components that include "support for richer forms of parallelism" in the list of lines for further works. For example, CCA attempts to move from SCMD to MCMD, a simple conceptual extension, but difficult to reach in practice. In fact, to support general purpose parallel programming is still a challenge for HPC component technology.

The **inductive** approach to augment component technology with new parallel programming extensions breaks down conceptual homogeneity of component models, making them more complex to be grasped by informal means and mathematically formalized. The # component model comes from the "opposite direction", taking a **deductive** generalization of channel-based parallel programming for supporting a suitable notion of component. The # component model has its origins in Haskell$_\#$ [7], a parallel extension to the functional language Haskell, inheriting their design premises, including Petri nets translation [9].

2.1   From Processes to Concerns

The basic principles behind the # component model come from message passing programming, notably represented by PVM and MPI. They have successfully exploited peak performance of parallel architectures, reconciling generality and portability, but with hard convergence with software engineering disciplines for supporting productive software development. The following paragraphs introduce fundamental principles behind the # component model: *the separation of concerns through process slicing*; and *orthogonality between processes and concerns as units of software decomposition*. Familiarity of readers with parallel programming is needed to understand the # component model from the intuition behind their underlying basic principles. Induction from examples must be avoided. Readers may concentrate on fundamental ideas and try to build their own examples from their experience and interests. Figures 1 and 2 complementarily present a simple parallel program that is used to exemplify the idea of slicing processes by concerns. Let $\mathbf{A}$ and $\mathbf{B}$ be $n \times n$ matrixes and X and Y be vectors. The parallel program computes $(\mathbf{A} \times X^T) \bullet (\mathbf{B} \times Y^T)$.

We have searched for the fundamental reasons that make software engineering disciplines too hard to be applied for parallel programming, concluding that

---

[1] Single Component Multiple Data.

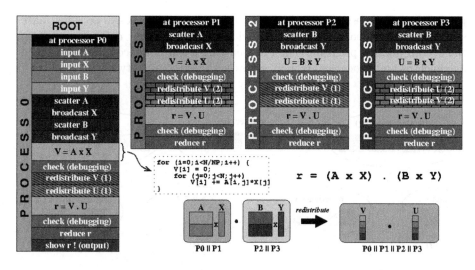

Fig. 1. Slicing a Simple Parallel Program by Concerns

they reside on the tendency to mix *processes* and *concerns* in the same dimension of software decomposition, due to the traditional process-centric perspective of parallel programming practice. Software engineering disciplines assume concerns as basic units of software decomposition [22]. We advocate that processes and concerns are orthogonal concepts. Without loss of generality, aiming at to clarify intuitions behind the enunciated orthogonality hypothesis, let $\mathcal{P}$ be an arbitrary parallel program formed by a set $\{p_1, p_2, \ldots, p_n\}$ of processes that synchronize through message-passing. By looking at each process individually, it may be split in a set of slices, each one addressing a concern. Figure 1 shows an example of process slicing in a simple parallel program. Examples of typical concerns are: (a) a piece of code that represents some meaningful calculation, for example, a local matrix-vector multiplication; (b) a collective synchronization operation, which may be represented by a sequence of send/recv operations; (c) a set of non-contiguous pieces of code including debugging code of the process; (d) the identity of the processing unit where the process executes; (e) the location of a process in a given process topology. The reader may be convinced that there is a hierarchical dependency between process slices. For instance: (a) the slice representing collective synchronization operation is formed by a set of slices representing send/recv point-to-point operations; (b) a local matrix-vector multiplication slice may include a slice that represent the local calculation performed by the process and another one representing the collective synchronization operation that follows it. If we take all the processes into consideration, it is easy to see existence of concerns that cross-cuts processes. For example: (a) the concern of parallel matrix-vector multiplication includes all slices, from individual processes, related to local matrix-vector multiplication; (2) the concern of process-to-processor allocation is formed by the set of slices that define the identities of processors where each process executes. It is easy to see that, from the overall perspective of processes, most of slices inside individual processes does not make sense when observed in isolation. Individually, they do not define concerns in the overall parallel program. The cross-cutting nature of decompositions based on concerns and processes strongly enforces the orthogonality hypothesis.

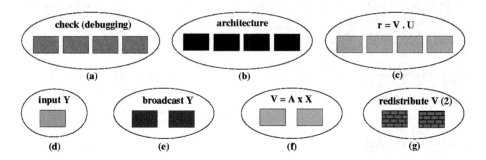

Fig. 2. #-Components From The Example in Figure 1

Above, some examples of #-components extracted from slicing of the parallel program in Figure 1. Some of them are non-functional concerns: (a) debugging code and (b) process-to-processor mapping. The #-components $V = A \times X$ and $r = V \bullet U$ addresses functional concerns: parallel matrix-vector multiplication and parallel dot product, respectively. The #-components "broadcast Y" and "redistribute V (2)" address data distribution concerns, acting as synchronization protocols. The #-component "input Y" is a local concern of a process (root) that is responsible to read vector Y.

The # component model moves parallel programming from the process-based perspective to a concern-oriented one. In fact, through a *Front-End*, # programmers may build applications through composition of concerns. Then, a *Back-End* may synthesize the process topology of the intended parallel program. A **#-component** is a software entity that encapsulates a concern. Such definition covers usual notions of components, because concerns are elementary units of software decomposition. The *units* of a #-component correspond to the slices (of processes) that constitutes its addressed concern. A #-component may be inductively built from other #-components through unification of their units, forming units of the resultant #-component. Thus, units also form a hierarchical structure, attempting to resemble hierarchical structure of process slices, where units may be formed by other units (unit slices). Sharing between components is supported through *fusion of unit slices* in unification. Sharing of data structures is a fundamental feature for ensuring high performance in scientific software. Another component model that supports sharing between components is Fractal [6]. The protocol of a unit is specified by a labelled Petri net whose labels are identifiers of their slices. It determines a Petri net formal language which dictates the possible activation traces for slices. Intuitively, it defines the order in which processes execute their functional slices. Petri nets allows for analysis of formal properties and performance evaluation of parallel programs.

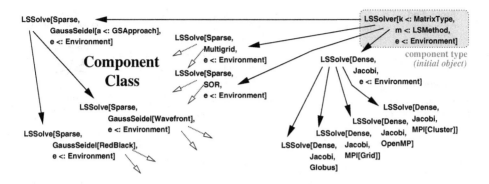

Fig. 3. Component Class for LSSOLVER

A component class for LSSOLVER components, represented by the *component type* inside the gray box. Lowercase identifiers are formal parameters. The notation $x <: C$ says that x may be replaced by any sub-type of component type C. Arrows indicate instantiations, which replace a formal parameter by an actual component. For example, there are four components that implement solutions for *dense linear systems* using the Jacobi iterative method. They target different architectures. The component type MPI is parameterized by the intended architecture (MPI[$a <:$ ARCHITECTURE]).

2.2   Skeletal Programming and Parameterized Component Types

The simpler form of abstraction in # programming is to hide lower level operations in higher level ones encapsulated in components. For example, a programmer that makes use of a component LSSOLVER for solving a linear system $A.x = B$ does not need to be aware about synchronization operations inside the component, resembling linear algebra parallelized libraries. Partitioning of parallel programs by concerns suggests richer abstraction mechanisms, such as skeletal programming through existential polymorphic component types, representing classes of components that address the same concern through distinct implementations, each one appropriate to a specific execution environment. For example, there may exist several possible implementations for LSSOLVER, adapted to specific parallel architectures, process topologies, and density properties of matrix A. Such parameters are known by programmers before execution. Figure 3 exemplifies a component class for dealing with implementations of LSSOLVER. The idea of a polymorphic type system for #-components comes from the formalization of the # component model using Theory of Institutions [10], firstly intended to study its formal properties and to formalize the notions of component types and their *recursive composition*. Theory of Institutions [16] have been widely used to capture logical independence in algebraic specification languages. Some ideas from this context have been applied to # programming, including parameterized programming [15], giving rise to polymorphic component type systems with bounded quantification.

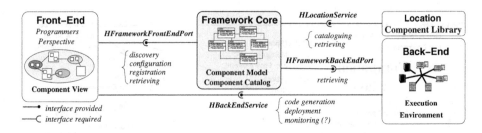

Fig. 4. The # Framework and #-Components Life Cycle

3   An Architecture for # Programming Frameworks

Figure 4 depicts an architecture proposal for # programming frameworks. Like
CCA and Fractal, # compliant frameworks be built from instantiation of a set
of interfaces that define the # *component model architecture*, whose interfaces
are depicted in the UML diagram of Figure 5. Frameworks control life cycle
of components by means of the interfaces that components must provide. CCA
targets simplicity, by adopting a lightweight interface for components to interact
with the framework, including only the method *setServices*, where programmers
register their *uses ports* and *provides ports* for dynamic binding. Fractal compliant components also support dynamic bindings, also targeting hierarchical
composition from primitive components. As already shown, the # component
model supports recursive composition, but their "bindings" are static, which, at
a first glance, appears to restrict the application domain of #-components. For
this reason, the # component model is not yet proposed for general distributed
applications, but only for applications in high performance computing domain.
In fact, most of parallel programs are static, avoiding performance overheads
of run-time control. However, #-frameworks can still deal with dynamic execution scenarios needed by HPC applications. In fact, static configuration does not
imply static execution. A #-framework could encapsulate predictable dynamic
adaptations of programs, supported by some underlying programming artifact,
as concerns of #-components.

Unlike CCA and Fractal, # programmers use an architecture description language (ADL) for component composition. This is motivated by the requirement
to place coordination and computation concerns at separate programming layers,
and to support overlapping composition. Current usual programming artifacts
does not support to overlap implementation of modules. As depicted in Figure
4, the *Front-End* of a #-framework deals with component views of a component model, managed by the *Framework-Core* and accessed by the Front-End
through the interface *HFrameworkFrontEndPort*. The *Framework-Core* is also
responsible to manage a library of #-components placed at registered locations.
In Fractal and CCA, programmers directly manipulates the *component model*.
The # component model delegates to #-frameworks to define one or more appropriate ADL's, managed by distinct *Front-End*'s. ADL's may be *graphical*,using
visual metaphors, or *textual*. A textual ADL may be XML-based, making possible

interoperability at the level of component views. Indeed, general framework interoperability can be achieved at the level of component models. In visual programming, the MVC (Model-View-Controller) pattern is a good design pattern for *interpreting component views* as *component models*.

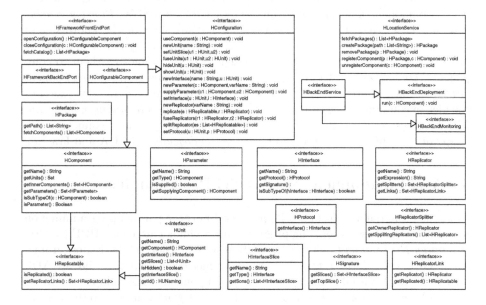

Fig. 5. The Interfaces of the # Framework Architecture

The *Back-End* of #-frameworks synthesizes a parallel program from the *component model*, targeting their supported execution environments. This is needed due to orthogonality between #-components and processes, a fundamental distinction from CCA and Fractal, where component models are the units of programming and deployment. For this reason, CCA and Fractal do not need a *Front-End* and a *Back-End* (Figure 4). In fact, #-frameworks act as bridges between *component views* and parallel programming artifacts. The # component model does not intend to be "yet another parallel programming technology", but to be a components-based layer on top of which existing ones can take advantage of software engineering disciplines. It is conjectured that any programming technology may be defined in terms of the # component model, including CCA and Fractal frameworks. The interoperability hypothesis has been verified by experimental evaluation with #-frameworks. The *Back-End* of #-frameworks may perform optimization steps for reducing synchronization costs in the resultant parallel program. For example, if all slices of a #-process are programmed in the same language, they can be fused (inlined) in a single procedure, avoiding unnecessary costs of procedure calls and improving cache behavior. It is intended that the synthesized parallel program be similar or better than programmed by hand, since programmers have explicit control over all parallelism concerns.

A #-framework defines a set of specialized **components kinds**, each one containing a set of component types with an intended meaning, which may imply in different visual metaphors and model interpretations at the *Front-End* and *Back-End* sides. In fact, component kinds are supported by the specialization of the interfaces of the *# component model architecture*, presented in Figure 5. The use of component kinds for designing of # compliant PSE's (*Problem Solving Environments*) that uses visual metaphors that are near to the knowledge of specialists has been investigated. The # component model goes far beyond the idea to raise connectors to first-class citizens [24], by promoting them to components. For example, a CCA binding could be implemented as a #-component BINDING. Such approach leads to uniformity of concepts. Fractal also exercises the idea of components as connectors, by means of *composite bindings*, but *primitive bindings* are not components, breaking homogeneity. The # connectors are *exogenous*[20], like in P-COM, while they are *endogenous* in CCA and Fractal.

4   A # Environment for Parallel Programming

Now, the design of a #-framework for general purpose parallel programming on top of common message-passing programming technologies, called HPE (*# Programming Environment*), is presented. It is an extension to the Eclipse framework. GEF (*Graphical Editing Framework*) has been used to build an ADL for dealing with visual configuration of #-components. GEF adopts the MVC (Model-View-Controller) design pattern. The framework complies to the # component model architecture, specializing it to support the *component kinds* of HPE, classifying component types as *qualifiers*, *architectures*, *environments*, *data structures*, *computations*, *synchronizers*, and *applications*. Some built-in component types are supported by the framework, whose subtyping hierarchy is depicted in Figure 6. Each component kind is represented by a component type (*top*). Programmers configure new component types by subtyping them and the natively supported component types of the framework.

Figure 7(a) presents a screenshot of the HPE's *Front-End*, showing a # program (application component) that solves a linear system $A \times x = B$. Inputs (matrix A and vector B), and output (vector x), are retrieved/saved in a remote data center defined by the component DATACENTER. It will illustrate the idea of *component kinds*. Input data is distributed across processes using the collective communication component SCATTER, while output is joined in the *root* process using GATHER. In Figure 7(b), the protocol of the unit *peer* is depicted, defining that scattering operations must be performed in parallel, followed by the solution computation and gathering.

4.1   Component Kinds

Qualifier components specify non-functional, stateless, concerns of components. In practice, they may act selectively among #-components in a component class, allowing programmers to control choice of component instances

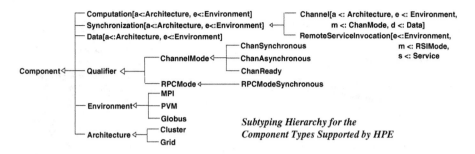

Fig. 6. The # Component Kinds of the Framework

from its representant component type. For example, suppose a component class for point-to-point communication channels, represented by the component type $\exists mode.\text{CHANNEL}[\cdots, mode <: \text{CHANNELMODE}]$. The syntax says that, among other parameters (reticences), their component instances may vary according to the intended communication semantics. For that, the parameter *mode* must be supplied with a subtype of the qualifier component CHANNELMODE, which comprises two units, respectively intended to be slices of sender and receiver units of a channel component. The natively supported subtypes of CHANNELMODE in the framework are CHANSYNCHRONOUS, CHANASYNCHRONOUS, and CHANREADY. Programmers may define other channel modes. The component type CHANNEL $[\cdots, \text{CHANSYNCHRONOUS}]$ represents the class of channel components with synchronous semantics. In Figure 7, qualifier components are also used to describe the solution method in a LSSOLVER component.

Architecture components intend to describe parallel architectures where #-components intends to run. Their units represent processing nodes. Using subtyping capabilities, supported architectures must be organized in hierarchies, in order to be classified according to their common characteristics. For example, a CLUSTER architecture component type may be specialized in component types referring to common cluster designs, possibly distinguished by the processor type, communication network type, homogeneity/heterogeneity, and so on. At the leaves of the hierarchy, there are component types for describing specific clusters. Thus, a programmer may target a class of architectures by using architecture types at non-leave nodes. A specific architecture could be in more than one intersecting classes. Similarly, grid-based architectures could be classified. The example in Figure 7 runs in a specific cluster, named PARGO'S CLUSTER.

Environment components define the parallelism enabling software technology intended for a component. Typical examples are message-passing libraries, such as MPI and PVM, for cluster and capability computing, and Globus and OurGrid, for grid computing. Some MPI implementations also target grids. Notice that a pair architecture/environment defines a complete parallel run-time execution context for a component. Hierarchies of component types may also be used to define classes of environments of special interest, for example, software technologies for enabling message-passing or bag-of-tasks parallelism.

194   F.H. de Carvalho Junior et al.

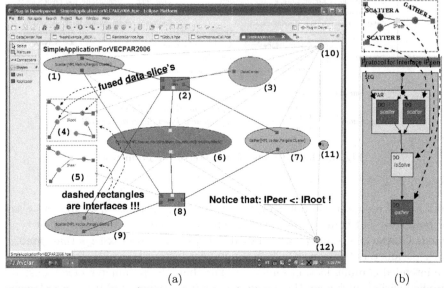

(a)   (b)

LEGEND: **(1)** Component SCATTER[MPI,MATRIX,PARGO'S CLUSTER]; **(2)** Unit *root*; **(3)** Component DATACENTER; **(4)** Interface *IRoot* (of unit *root*); **(5)** Interface *IPeer* (of unit *peer*); **(6)** Component LSSOLVER[MPI, SPARSE, GAUSSSEIDEL[PARREDBLACKMETHOD]]; **(7)** Component GATHER[MPI, VECTOR, PARGO'S CLUSTER]; **(8)** Unit *peer*; **(9)** Component SCATTER[MPI, VECTOR, PARGO'S CLUSTER]; **(10)** Replicator (×1); **(11)** Split replicator; **(12)** Replicator (×n).

Fig. 7. The Visual Configuration of a Simple #-Component (Screen Shot)

Data components are formed by one unit, whose interface is attached to a SIDL (Scientific Interface Description Language) interface. SIDL has been supported by the Babel toolkit [19] to be a neutral language for specification of CCA components interfaces. Subtyping is supported for data components, resembling multiple inheritance in object-oriented programming. For that, a data component type D must be composed from a set of data components supertypes, whose units becomes slices of the units of D. In Figure 7, there are data component types VECTOR and MATRIX, sub-types of the component type DATA.

Synchronization components allows inter-process communication. There are synchronization components for dealing with *point-to-point message-passing* (the usual *send* and *receive* primitives), collective communication (structure parallel programming [17]), and remote service invocation (such as RPC, RMI, and so on). The class of all channel components is represented by the component type CHANNEL[a<:ARCHITECTURE, s<:ENVIRONMENT, d<:DATA, m<:CHANMODE]. A highly tuned member of the component class of CHANNEL[···] may be specific to a given architecture, environment, data type, and channel semantics. The class of *remote service invocation* components is represented by the component type REMOTESERVICEINVOCATION[e<:ENVIRONMENT, m<:RSIMODE, s<:SERVICE]. They comprise two externally visible units: *client* and *server*. The activation of a *client slice* is a *null* operation. Client slices only carries *stub objects* for each interface provided by the service. A *server slice* only implements methods of the service interfaces. Collective communication components correspond to that supported by MPI. All of them comprise only one replicated unit. Some of them, such as BROADCAST, SCATTER, and GATHER distinguish a root unit, the first one in the replication enumeration. Qualifier components are used to define channel communication semantics, as described above.

Computation components specify parallel computations over distributed data structures encapsulated in data components that makes part of their constitution. Their units define *state transformer procedures* over a set of local *data slices*, units of the inner data components. Data slices may be *private* or *public*. Private data slices become local variables in the unit procedure, where public ones become their parameters and return values, which are visible to procedure callers. There are three kinds of public data slices: *in*, for input data; *out*, for output data; *in/out*, for input and output data. Such modifiers are supported by SIDL, covering possible parameter passing semantics. Slices that comes from other inner computation components are called *computation slices*. Computation slices also define procedures whose parameters are their public data slices. Data slices from different unified computation slices may be fused to refer to the same data item (data sharing mentioned in Section 2.1). The protocol of the unit dictates a control flow for calling the *procedures* of computation slices (denotation of slice activation for computation slices). In HPE, *behavior expressions* are used for specifying protocols, with combinators from synchronized regular expressions, a formalism that reaches expressiveness of terminal labelled Petri nets. The *Front-End* may partially generate the code for procedures, using the signature and protocol of the computation slice to define parameters, local variables, and control flow. In Figure 7, the *root* process comprises three *data slices* (one input vector B, one input matrix A, and one output vector X), two *service slices*, for accessing a data center (DATACENTER component) where input data is retrieved and where output data is stored for further analysis, three *synchronization slices*, for data distribution across processes (GATHER and SCATTER operations for collective synchronization), and one *computation slice*, whose procedure computes a solution to the linear system $A \times x = B$. The *peer* process does not have a service slice for accessing the remote data center, because only *root* needs to access it. Fusion of *data slices*, represented by circles attached to the synchronization and computation slices involved, is used to set input data (A and B) and output data (x) for the LSSOLVE component.

Application components are similar to computation components, but requires complete execution context information (architecture and environment). For this reason, they cannot be parameterized, all units are hidden, and all data slices are private. In fact, they are not components in the intuitive sense. Application components are input to the HPE's *Back-End*, containing all the necessary information to generate MPI code to run in clusters.

5   Conclusions and Lines for Further Work

This paper sketched the architecture of frameworks that complies the # component model. The design of a #-framework for general purpose parallel programming was presented. The # component model intends to reconcile software engineering disciplines with efficient parallel programming, meeting the needs of the HPC community. Besides that, it is another attempt to adapt component technology to the demands of HPC software development. Compared to other HPC component models, the # component model is parallel by nature, targeting expressiveness of message passing parallel programming. Its main principles comes from the study of reasons that make difficult software engineering disciplines and parallel software development to be compatible with each other. The implementation of # compliant frameworks intends to make possible experimental evaluation of the hypothesis underlying the # component model principles.

The authors are currently working in the design and implementation of # compliant frameworks. Its formal semantics and analysis of the properties of the # component model are been studied under Category Theory and the Theory of Institutions. The study the use of #-frameworks as a platform for implementation of interoperable PSE's (Problem Solving Environments) is already planned. For that, the use of visual metaphors for component kinds is proposed to bring closer together programming abstractions and the needs of users of HPC.

References

1. B. A. Allan, R. C. Armstrong, A. P. Wolfe, J. Ray, D. E. Bernholdt, and J. A. Kohl. The CCA Core Specification in a Distributed Memory SPMD Framework. *Concurrency and Computation: Practice and Experience*, 14(5):323–345, 2002.
2. R. Armstrong, D. Gannon, A. Geist, K. Keahey, S. Kohn, L. McInnes, S. Parker, and B. Smolinski. Towards a Common Component Architecture for High-Performance Scientific Computing. In *The Eighth IEEE International Symposium on High Performance Distributed Computing*. IEEE Computer Society, 1999.
3. F. Baude, D. Caromel, and M. Morel. From Distributed Objects to Hierarchical Grid Components. In *International Symposium on Distributed Objects and Applications*. Springer-Verlag, 2003.
4. Bernholdt D. E., J. Nieplocha, and P. Sadayappan. Raising Level of Programming Abstraction in Scalable Programming Models. In *Workshop on Productivity and Performance in High-End Computing (in HPCA'2004)*, pages 76–84. Madrid, 2004.
5. R. Bramley, R. Armstrong, L. McInnes, and M. Sottile. High-Performance Component Software Systems. *SIAM*, 49:, 2005.
6. E. Bruneton, T. Coupaye, and J. B. Stefani. Recursive and Dynamic Software Composition with Sharing. In *European Conference on Object Oriented Programming (ECOOP'2002)*. Springer, 2002.
7. F. H. Carvalho Junior and R. D. Lins. Haskell#: Parallel Programming Made Simple and Efficient. *J. of Univ. Computer Science*, 9(8):776–794, August 2003.
8. F. H. Carvalho Junior and R. D. Lins. Separation of Concerns for Improving Practice of Parallel Programming. *INFORMATION, An International Journal*, 8(5), September 2005.

9. F. H. Carvalho Junior, R. D. Lins, and R. M. F. Lima. Translating Haskell$_\#$ Programs into Petri Nets. *Lecture Notes in Computer Science (VECPAR'2002)*, 2565:635–649, 2002.

10. F. H. Carvalho Junior, R. D. Lins, and A. T. C. Martins An Institutional Theory for #-Components. In *Proceedings of the Brazilian Symposium on Formal Methods (SBMF'2006)*, pages 137–152. September 2006.

11. K. Chiu. *An Architecture for Concurrent, Peer-to-Peer Components*. PhD thesis, Department of Computer Science, Indiana University, 2001.

12. M. Cole. Bringing Skeletons out of the Closet: A Pragmatic Manifesto for Skeletal Parallel Programming. *Parallel Computing*, 30:389–406, 2004.

13. A. Denis, C. Pérez, and T. Priol. PadicoTM: An Open Integration Framework for Communication Midleware and Runtimes. *Future Generation Computing Systems*, 19:575–585, 2004.

14. J. Dongarra. Trends in High Performance Computing. *The Computer Journal*, 47(4):399–403, 2004.

15. J. Goguen. Higher-Order Functions Considered Unnecessary for Higher-Order Programming. In D. A. Turner, editor, *Research Topics in Functional Programming*, pages 309–351. Addison-Welsey, Reading, MA, 1990.

16. J. Goguen and R. Burnstal. Institutions: Abstract Model Theory for Specification and Programming. *Journal of ACM*, 39(1):95–146, 1992.

17. S. Gorlatch. Send-Recv Considered Harmful? Myths and Truths about Parallel Programming. *ACM Trans. in Programming Languages and Systems*, (1):47–56, January 2004.

18. K. Koahey and D. Gannon. PARDIS: A Parallel Approach to CORBA. In *Proc. of the 6th IEEE Intl. Symposium on High Performance Distributed Computing (HPDC'97)*, pages 31–39. Springer, August 1997.

19. S. Kohn, G. Kumfert, J. Painter, and C. Ribbens. Divorcing Language Dependencies from a Scientific Software Library. In *10th SIAM Conference on Parallel Processing*. Springer-Verlag, March 2001.

20. K. Lau, P. V. Elizondo, and Z. Wang. Exogenous Connectors for Software Components. *Lecture Notes in Computer Science (CBSE'2005)*, 3489:90–108, 2005.

21. N. Mahmood, G. Deng, and J. C. Browne. Compositional Development of Parallel Programs. In *16th International Workshop on Languages and Compilers for Parallel Computing*, October 2003.

22. H. Milli, A. Elkharraz, and H. Mcheick. Understanding Separation of Concerns. In *Workshop on Early Aspects (in AOSD'04)*, pages 411–428, March 2004.

23. C. Pérez, T. Priol, and A. Ribes. A Parallel Corba Component Model for Numerical Code Coupling. In *Proc. of the 3rd Intl. Workshop on Grid Computing (published in LNCS 2536)*, pages 88–99. Springer, November 2002.

24. M. Shaw. Procedure Calls are the Assembly Language of Software Interconnection: Connectors Deserve First-Class Status. In *International Workshop on Studies of Software Design*, Lecture Notes in Computer Science. Springer-Verlag, 1994.

25. A. J. van der Steen. Issues in Computational Frameworks. *Concurrency and Computation: Practice and Experience*, 18(2):141–150, 2005.

Anahy: A Programming Environment for Cluster Computing*

Gerson Geraldo H. Cavalheiro[1], Luciano Paschoal Gaspary[2],
Marcelo Augusto Cardozo[3], and Otávio Corrêa Cordeiro[3]

[1] Universidade Federal de Pelotas (UFPel)
Pelotas – Rio Grande do Sul – Brazil
gerson.cavalheiro@ufpel.edu.br
[2] Universidade Federal do Rio Grande do Sul (UFRGS)
Porto Alegre – Rio Grande do Sul – Brazil
paschoal@inf.ufrgs.br
[3] Universidade do Vale do Rio dos Sinos (UNISINOS)
São Leopoldo - Rio Grande do Sul – Brazil
{mcardozo,otaviocc}@anahy.org

Abstract. This paper presents Anahy, a programming environment for
cluster computing. Anahy is presented in terms of its programming interface (API) and its scheduling mechanism. The main features of this
environment are the specification of a POSIX thread-based API and the
use of dynamic scheduling techniques based on Directed Acyclic Task
Graphs (DAG). The main advantage obtained with these features is the
dissociation between the description of the concurrency of an application and its parallel execution. The paper examines how Anahy builds a
DAG describing the dependencies among tasks at execution time from
a multithreaded program and how this DAG is handled by the runtime
to apply dynamic scheduling techniques. The paper concludes discussing
three case studies of applications developed in the context of Anahy
environment.

1 Introduction

New runtime environments have been proposed for cluster computing to assist
the development of applications. Some of them are composed by a layered architecture, wherein at the top they propose a high level application programming
interface (API) to describe the concurrency of an application as a concurrent
program and, at the bottom, a runtime to execute this program. Therefore, an
efficient execution depends on a good strategy for scheduling the computational
cost generated by the program in execution (computation, data, and communication) over the computational resources available on the hardware (processors,

---

* The Anahy project is supported by CNPq/PDPG-TI (55 2196/2002-9), FAPERGS
(02/0571.4), and was developed in collaboration with UNISINOS and HP Brazil
R&D.

M. Daydé et al. (Eds.): VECPAR 2006, LNCS 4395, pp. 198–211, 2007.
© Springer-Verlag Berlin Heidelberg 2007

memories, and network). Such scheduling cannot be undertaken in a straightforward manner, since it must consider information related to program behavior.

Overcoming the above difficulty is a challenge that involves both programming model [1] and scheduling. A common approach taken in the development of programming tools and runtime environments (e.g. Athapascan-1 [2], Cilk [3], Clik [4], Pyrros [5], and GrADS [6]) is building an intermediate level between the program in execution and the scheduler describing the program structure in terms of a Directed Acyclic task Graph (DAG). Since the literature on scheduling strategies taking graphs as input is vast (e.g. [5,7,8,9]), the interaction between graph and scheduling is well known (e.g. [10]). However, traditional programming tools for cluster computing (such as those based on multithreading and/or message passing) don't offer high level programming resources for creating such graph representation.

This paper addresses the aforementioned problem by proposing Anahy, a programming environment for cluster computing. We present the programming interface (API) proposed for this environment as well as some aspects related to its scheduling mechanism. The main features of this environment are the specification of a POSIX thread-based API and the use of dynamic scheduling techniques based on DAGs. The main advantage obtained with these features is the dissociation between the description of the concurrency of an application and its parallel execution [11]. The paper examines how Anahy builds a DAG describing the dependencies among tasks at execution time from a multithreaded program and how this DAG is handled by the runtime to apply dynamic scheduling techniques.

The remaining of the paper is organized as follows. In the next section, related work is briefly presented. Section 3 presents the Anahy programming interface. Section 4 covers the algorithm employed to schedule concurrent programs. Section 5 presents three case studies, and Section 6 concludes the paper with final remarks and perspectives for future work.

2 Related Work

A DAG is a typical abstraction to model the structure of programs in terms of concurrent activities and data communications [12]. In this abstraction, each concurrent activity defined by the program, named task, is represented by a vertex and a communication between two tasks is represented by an arc connecting two vertices. The use of DAG is very common in static schedulers. Dynamic techniques have been proposed ([13]) in order to avoid inefficiency on blind dynamic scheduling techniques [14] (that is, schedule techniques that don't considering the program structure).

Scheduling DAG is a NP-hard problem [15]. Most of the DAG schedulers are based on list scheduling techniques (e.g. [8] and [7]). Those schedulers handle the tasks generated by the program in priority list. This technique is based on a two step algorithm: in the first step a priority list is built by assigning each task generated by the program a priority; in the second step tasks are mapped to processors respecting their execution priorities.

In 1995 Feitelson has observed in [16] that although lots of researches were being made on DAG scheduling strategies, few efforts were observed in exploiting their use on runtime systems. Nowadays, the research on the area is still popular (e.g. [17,18,19,6]). Nevertheless, the number of programming and execution environments employing DAG based scheduling is limited, particularly if we consider those that support applications whose DAGs are created at execution time. Cilk [3] and Athapascan-1 [2] are examples of them for SMP and cluster architectures. Both Cilk and Athapascan-1 propose APIs that allow building graph structures at execution time and a runtime able to apply dynamic scheduling techniques based on list strategies.

The Cilk API provides resources for the explicit creation and synchronization of concurrent activities, called threads, and to access a shared memory space. These features allow the programmer to introduce synchronizations among threads in order to control data exchange. The Athapascan-1 API offers special data types in a shared memory space and a primitive to create concurrent activities, called tasks. Tasks are created explicitly but, differently from Cilk, the programmer must identify the input and the output data of each task.

The approach considered by both Cilk and Athapascan-1 takes into account that the scheduler can exploit the structure of the graph *during its construction*. In such way, they can apply a heuristic to explore the program structure in order to achieve an index of performance and avoid inefficiency of blind scheduling techniques. Nevertheless, the graph built in Cilk represents only the precedence among threads, not representing concurrency in a smaller unit such as a task. As a consequence, the Cilk scheduler is able to exploit only serial parallel graphs (nested fork and join operations). On the other hand, the graph built in Athapascan-1 is more complete since it includes the data dependencies among tasks. In this case the scheduler has more information about the program in execution but the cost to build and manage the graph is higher. We propose to mix these two approaches by offering a programming environment able to obtain data dependencies among tasks from a graph describing execution precedence among threads.

3   The Anahy Programming Interface

The Anahy API offers high level programming resources to handle a large number of concurrent activities and communications in a multithreading style. This API offers a *fork/join*-based model to describe the concurrency in terms of threads. An intermediate level between this API and the runtime is responsible for identifying the concurrency in smaller units, called *tasks*, and creating a DAG representing the data dependencies between tasks.

3.1   Handling Tasks with Anahy

The Anahy API provides services to explore a shared memory multiprocessor architecture. These services allow the creation and the synchronization of threads

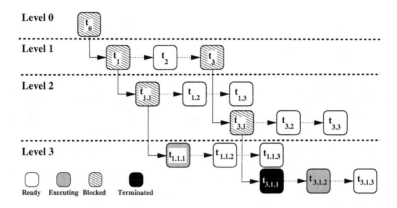

Fig. 1. Graph representing threads creation

and can be represented by the operations *fork/join/exit*. A *fork* consists in the creation of a new execution flow responsible for executing a function $\mathcal{F}$ defined in the body of the program having a set of data $\mathcal{X}$ as input. The *fork* operator returns an identifier for the newly created thread. Although the thread is ready to be executed, the programmer cannot predict when this thread will be triggered. The *exit* corresponds to the last operation performed by the thread extinguishing the execution flow. The synchronization upon termination of a thread is performed through a *join*, identifying the flow to be synchronized. This operation allows a thread to be blocked until the termination of another thread, so that it can gather the results $\mathcal{Y}$ produced by $\mathcal{F}(\mathcal{X})$.

Figure 1 shows a snapshot of threads state taken during the execution of an Anahy program. The graph in this figure contains all threads created until the snapshot: notice that they are in different states reflecting their life cycle (executing, ready, blocked, and terminated). The threads are grouped in levels, based on their depth in the program: a thread from level i creates threads on level $i + 1$. For example, thread t_0 creates threads t_1, t_2 and t_3 (in this order). Arrows in this representation identify the relation of creation; dotted arrows were employed to identify threads on the same level created by the same thread.

3.2 A POSIX-Like Thread Interface

Considering the programming model, both *fork* and *join* operations create new tasks. We have implemented this model in Anahy as a library for C/C++ programs offering a programming interface closer to the POSIX threads standard in order to provide a *multithreading* programming style. Therefore, although *fork* and *join* handle tasks, the API of Anahy offers primitives to create and synchronize (join) Anahy threads.

The body of a thread. The body of a thread is defined as a conventional C function, as follows:

```
void * func( void * in ) {
  /* code */
  return out;
}
```

In this example, `func` corresponds to the function to be executed in a new thread and `in` corresponds to the memory address (in the shared memory) where the input data for the function is located. The return instruction (`return out`) corresponds to the *exit* operation. Notice that when a thread finishes its output is stored in the shared memory at the address specified by `out`.

Synchronization of threads. The `pthread_create` and `pthread_join` services correspond to the creation and join-synchronization of threads in POSIX threads standard. The corresponding syntaxes in Anahy are:

```
int athread_create( athread_t *th, athread_attr_t *attr,
                    void *(*func)(void *), void *in);
int athread_join( athread_t th, void **res);
```

`athread_create` creates a new thread to execute the function defined by `func`; the input data of `func` is stored in the address specified by `in`. The parameter `th` will be updated to get a value to identify the new thread created. The `attr` argument specifies thread attributes to be applied to the new thread (as memory requirements or computational costs). In the operation of `athread_join` the thread on which the synchronization is to be performed is identified by `th` and `res` will be updated to point to a position in the shared memory where the output of the function executed by the thread `th` can be found.

Migration of threads. Although Anahy interface provides a multithreaded programming style, executions can be achieved on distributed memory architectures. Thus, threads can be migrated between nodes. The scheduling mechanism was developed to migrate threads transparently. Nonetheless, the programmer must provide the execution support with information about the data required (parameters) and produced (results) by the threads allowing the data transfers. The mechanism adopted introduces the use of *pack/unpack* functions. The prototypes of pack/unpack functions for a given thread are the following:

```
int packInFunc( void *in, char **buff );
int unpackInFunc( void *in, char **buff );
int packOutFunc( void *res, char **buff );
int unpackOutFunc( void *res, char **buff );
```

The first parameter of each pack/unpack function represents the data to be sent (`in`) or produced (`res`) to/by a thread. The second parameter (`buff`) represents the buffer where the input data for a thread must be *packed* – in thread creation – or from where data must be read to be *unpacked* – in thread launching –. Each function must return the size (in bytes) of data packed/unpacked. The programmer associates specific pack/unpack functions to threads in the thread attributes (`athread_attr_t`):

```
int athread_attr_setpackinput(      athread_attr_t *attr,
                                    int (*func)(void *in, char **buff) );
int athread_attr_setunpackinput(    athread_attr_t *attr,
                                    int (*func)(void *in, char **buff) );
int athread_attr_setpackoutput(     athread_attr_t *attr,
                                    int (*func)(void *res, char **buff) );
int athread_attr_setunpackoutput(   athread_attr_t *attr,
                                    int (*func)(void *res, char **buff) );
```

The default value (NULL) allows the thread to execute only in the node where it was created.

To illustrate the use of Anahy, the program presented in Figure 2 implements the code able to generate the graph in Figure 1. Due to space limitations the code not related to Anahy and the one describing pack/unpack operations are not presented.

```
void *foo( void *depth ) {
    athread_t child[3];
    int mydepth, *childdepth, *ret, *res = new int(0);
    mydepth = (int *) i*depth;
    if( mydepth > 3 )
        *res = computeSomething( mydepth );
    else {
        *childdepth = new int( mydepth+1 );
        for( int i = 0 ; i < 3 ; i++ )
            athread_create(&child[i], NULL, foo, childdepth);
        for( int i = 0 ; i < 3 ; i++ ) {
            athread_join( child[i], (void **)&ret );
            *res += computeSomething( *ret );
            delete( *ret );
        }
        delete( childdepth );
    }
    return res;
}
int main() {
    int complexity, *result;
    result = foo( (void *) &complexity )
    free(result);
    return 0;
}
```

Fig. 2. An example of Anahy program

4   The Anahy Scheduler

While the API of Anahy provides a multithreaded abstraction to describe the concurrency of applications, the scheduling strategy deals with tasks. The interface between the API and the scheduling builds a DAG considering accesses to the shared memory. These tasks are implicitly defined when the program executes calls to athread_create and athread_join.

A call to a athread_create implies the creation of two tasks: the first one is defined in the context of the new thread spawned. This task has as input data the arguments of the thread itself. The second task is created in the original

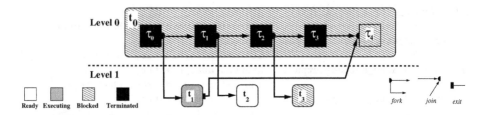

Fig. 3. Zoomed section of the DAG representing the tasks of thread t_0 in Figure 1

thread, having as input the data present in the local memory of this thread and the identifier of the new thread created. A call to a `athread_join` implies the creation of one new task: the thread terminates the execution of the current task and creates a new one starting in the instruction that follows (in lexicographical order) the *join*. This new task has as input data the local memory of the current thread and the output produced by the last task executed in the synchronized thread.

The set of tasks created in the context of thread t_0 (Figure 1) is represented in Figure 3. In this figure arrows represent data dependencies among tasks. Down arrows represent *fork* operations while the up arrow represents a *join*.

4.1 Scheduling Algorithm

The scheduling algorithm assumes shared memory architecture. A task is defined as the unit of scheduling manipulation as well as is assumed to be executable in a finite time. Tasks finish by executing *fork*, *join* or *exit* operations. The basic algorithm generates three lists: the **ready**, containing the tasks with no restrictions to be launched; the **terminated** list, containing the tasks that have finished; and the **blocked** list containing the tasks waiting for synchronization.

First we introduce the scheduling algorithm considering a mono-processor architecture. The processor is initially idle, then it takes the first task, τ_0, from the list of ready tasks and starts its execution. The instructions of τ_0 are processed sequentially until τ_0 finishes by executing one operation involving the scheduling process (*fork/join/exit*). The execution of a *fork* produces the creation of two tasks τ_1 and τ_2. Task τ_2 is the one explicitly created by the *fork* and task τ_1 is the one created to be the continuation of τ_0 just after the *fork*. τ_2 is stored in the list of ready tasks while τ_1 is launched. When a *join* is executed, for example $\tau_1.join(\tau_2)$, τ_1 terminates and a new task τ_3 is created: the code of task τ_3 starts at the instruction that follows the *join*; the initial state of τ_3 is blocked. The next scheduling action executes τ_2: the task τ_2 is taken from the ready list of and launched. At the termination of τ_2, τ_3 is unblocked (becomes ready) and started. Notice that τ_3 has as input both the data produced by τ_2 and τ_1. This process can be recursively applied.

In the case of a parallel architecture, there are two or more processors executing the algorithm described above. So, when a task τ_i requests a *join* with τ_j, two new situations may arise: either τ_j has already terminated or τ_j is being executed at that moment. In both cases τ_i finishes and a new task τ_{i+1} is created. In the former case (τ_j has terminated), the procedure consists of recovering the data produced by τ_j, allowing the processor to continue with the execution of τ_{i+1} (τ_j is removed from the list of terminated tasks). In the latter case (τ_j is being executed), τ_{i+1} remains blocked and the processor looks for a new activity on the list of ready tasks. τ_{i+1} will become ready when τ_j terminates.

The Anahy scheduler was conceived to exploit list scheduling strategies. Thus, its implementation was guided by the existence of a critical path defining the largest sequence of tasks in the program. Considering this critical path, the best performance can be achieved if the scheduler guarantees that during the execution of a program, at least one of the processors is executing a task from this path. Since Anahy focuses dynamic execution of programs, the real critical path is unknown during the execution of the program. Therefore, considering that the concurrent execution of a program must give the same result that a sequential one, the scheduling assumes that the first and the last tasks of the critical path are, respectively, the first and the last task created in the context of t_0 (the first thread launched). The algorithm was implemented in order to guarantee that a processor will be dedicated to execute the tasks of t_0 or the tasks defined in the context of the threads synchronized by t_0. The optimization obtained by the Anahy implementation exploits the recursive nature of the scheduling: while a processor is dedicated to execute the path starting on t_0, a second processor is dedicated to execute the path starting at t_1, another to the path starting at t_2 and so on.

4.2   Multilevel Scheduling

To execute an Anahy program, the user must inform a description of the real architecture that will be explored. Like in MPI or PVM, it is necessary to inform the number of nodes of a cluster involved in the execution as well as, for each node, the number of virtual processors (VPs) desired ([20]). The Anahy *virtual machine* is loaded as a runtime kernel when the program is launched on each node. This runtime executes cooperatively and supports the implementation of the scheduling algorithm. This implementation was conceived in three layers. The lowest is handled by the operating system. In this level the VPs are scheduled as system threads over the real processors on each node of the cluster. There is no migration of VPs between nodes.

The second level refers to the allocation of tasks to VPs considering task status (ready, terminated etc.). This level was implemented to consider the locality of tasks. The list of tasks is implemented as a tree where each node represents an Anahy thread (Figure 1) an Anahy thread is a sequence of tasks. This tree has as root the first thread executed by the program, whereas the threads created by the root thread compose the second level. The threads in the second level are the roots of new *sub-trees* of threads and so on – as shown in figures 1 and 3. So, a VP handles only a section of the tree where there are tasks involved in the execution of the current thread. When a VP has no more threads to be executed in its local section, it tries to steal one from a different VP. If so, the

VP will choose one thread ready to execute from the highest level of the tree. Such thread is expected to have a larger amount of work than those in lower levels. Another key aspect of this strategy is related to the locality of tasks inside threads. Since each thread defines a sequence of tasks, the data transfers between them are accomplished without accesses to the shared memory.

Finally, the third level of scheduling is demanded by the distribution of computational load among the nodes of the cluster. The algorithm is an extension of the second level, taking into account the (communication) costs involved in thread migration between nodes. The implemented load distribution strategy considers the depth of threads in the graph and the size of the data to be sent between nodes. Other factors, including the computational and the physical location of the data, can be added to this basically strategy. Notice that this scheme doesn't consider the migration of running threads.

5   Case Study

In this paper, we discuss the use of Anahy to support the description of applications describing DAGs. A general performance assessment of Anahy can be found in [21] and the performance of a specific application developed in the context of the Anahy project is presented in [22].

To illustrate the use of Anahy we present a synthetic program in Figure 4. This program implements a recursive algorithm able to construct a binary tree structure with a great number of concurrent activities. More details can be found in [21]. The main input of the program is the one defining the number of recursive interactions to be accomplished. Figure 5 presents the graph generated by running this program. In this figure, we also highlight the dependencies between tasks (continuation dependency), between a task and a thread (creation dependency), and between a thread and a task (join dependency). The final structure of the graph (a binary tree) reflects the locality of references of data (inputs and outputs of threads). Those dependencies are exploited at execution time by the Anahy scheduler in order to optimize the execution of tasks in the critical path. Notice that all threads execute the same amount of work.

Figure 5 presents the graph generated by the program in Figure 4 and Figure 6 presents the execution trace of the same program. For the trace, the Anahy runtime was configured with 4 VPs. In the figure, each line segment represents a thread executed by a VP. Each different line style represents a different VP responsible for executing the corresponding thread. Thus, it is possible to observe that the scheduling strategy confer different priorities to threads according to VPs – each VP is give a high priority to execute the tasks in the path . In this figure we have highlighted the threads executed by the VP 1 to exemplify the scheduling behavior.

In the context of the Anahy project we are working on the development of real applications, among them we name a dynamic programming based sequence alignment algorithm and a fluid dynamics simulation. The DAGs for these applications are represented in Figure 7 – circles represent tasks and boxes the data

```
#include <athread.h>
int main(int argc, char **argv){
    athread_t thr;
    void *dta, *res;
    dta = malloc(...); *dta = foo(...);
    athread_create(&thr, NULL, tree, &dta);
    athread_join(thr, &res);
    free(dta);
    doSomething(*res);
    free(res);
    return 0;
}

void *tree(void *argVoid){
    void *arg0, *arg1, *res, *aux0, *aux1;
    athread_t thr0, thr1;

    if( notFinish(*argVoid) ) {
        arg0 = malloc(...); *arg0 = foo(*argVoid);
        arg1 = malloc(...); *arg1 = bar(*argVoid);
        athread_create(&thr0, NULL, tree, &arg0);
        athread_create(&thr1, NULL, tree, &arg1);
        *res = doSomething(arg0, arg1);
        athread_join(thr0, &aux0);
        athread_join(thr1, &aux1);
        *res += doSomething(*aux0, *aux1);
        free(aux0); free(aux1);
    }
    else res = NULL;

    return res;
}
```

Fig. 4. Synthetic program executing a recursive algorithm

exchanged between them. In [22] it is presented an evaluation of the performance obtained with the dynamic programming application.

The dynamic programming application describes a regular DAG (Figure 7.a). In this application, a recursive algorithm fills in a matrix representing the comparison of two sequences. The value of each cell of the matrix corresponds to the similarity between the elements of these sequences. The matrix is filled in from top to bottom and from left to right, with element $M_{i,j}$ requiring three values that were previously calculated according to the concurrency relation: $M(i,j) = \mathcal{F}(M_{i-1,j-1}, M_{i-1,j}, M_{i,j-1})$. As shown in Figure 7.a, data locality can be predicted by the scheduler considering the regular structure of communications.

On other hand, fluid dynamics simulation is an irregular application, since it presents an unpredictable program structure (Figure 7.b). The proposed

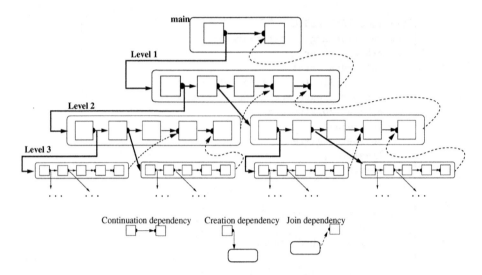

Fig. 5. DAG generated by the Anahy runtime for the recursive program presented in Figure 4

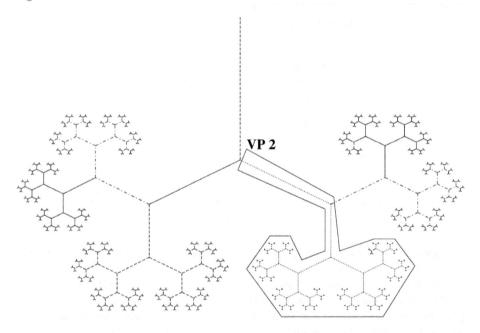

Fig. 6. Trace representing the execution of the DAG presented in Figure 5

implementation divides the physical space into triangles. A thread is generated for each triangle to compute the fluid velocity using Euler equation. Once a thread finishes computing the equation, new threads can be generated to give sequence to the simulation. Although the DAG is irregular, the scheduler can

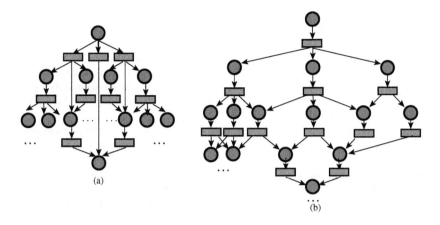

Fig. 7. Sections of DAGs generated by a regular (a) and an irregular (b) application

apply a load balancing strategy considering the depth of threads in the graph: the closer to the top a thread is, the higher is the probability of this thread accumulating a large amount of work.

6   Conclusion

This paper presented Anahy, an environment for exploring high performance processing in cluster architectures. Anahy was presented in terms of its API and the principles adopted for introducing an intermediate level responsible for building a DAG at execution time. This DAG is exploited by the Anahy runtime to avoid inefficacity of blind dynamic scheduling strategies in the execution of tasks. Another key contribution of this work is the adoption of an API based on the POSIX threads standard allowing the development of programs to distributed memory architectures without dealing with issues related to message exchange mechanisms.

The next steps of this work include the development of load balancing strategies and the extension of the API to include all POSIX-defined synchronization mechanisms for thread execution control (as critical sections and condition variables). Even though the use of such mechanisms is not recommended in the Anahy programming model, due to potential performance loss, they will be included to increase compatibility with legacy code.

References

1. Alverson, G.A., Griswold, W., Lin, C., Snyder, L.: Abstractions for portable, scalable parallel programming. IEEE Trans. on Parallel and Distributed Systems **9**(1) (1998) 71–86
2. Galilée, F., Cavalheiro, G.G.H., Roch, J.L., Doreille, M.: Athapascan-1: on-line building data flow graph in a parallel language. In: Proc. of the 7th International Conference on Parallel Architectures and Compilation Techniques (PACT), Paris (1998)

3. Blumofe, R., Joerg, C.F., Kuszmaul, B.C., Leiserson, C.E., K. H. Randall, Y.Z.: Cilk: an efficient multithreaded runtime system. Journal of Parallel and Distributed Computing **37**(1) (1996) 55–69
4. Mendes, R., Whately, L., de Castro, M.C., Bentes, C., Amorim, C.L.: Runtime system support for running applications with dynamic and asynchronous task parallelism in software DSM systems. In: Proc. of the 18th International Symposium on Computer Architecture and High Performance Computing (SBAC-PAD'06), Ouro Preto (2006)
5. Yang, T., Gerasoulis, A.: DSC: Scheduling parallel tasks on an unbounded number of processors. IEEE Transactions on Parallel and Distributed Systems **5**(9) (1994) 283–297
6. Berman, F., Casanova, H., Chien, A., Cooper, K., Dail, H., Dasgupta, A., Deng, W., Dongarra, J., Johnsson, L., Kennedy, K., Koelbel, C., Liu, B., Liu, X., Mandal, A., Marin, G., Mazina, M., Mellor-Crummey, J., Mendes, C., Olugbile, A., Patel, M., Reed, D., Shi, Z., Sievert, O., Xia, H., YarKhan, A.: New grid scheduling and rescheduling methods in the grads project. International Journal of Parallel Programming **33**(2–3) (2005) 209–229
7. Coffman, E., Graham, R.: Optimal scheduling for two-processor systems. Acta Informatica **1** (1972) 200–213
8. Hu, T.: Parallel sequencing and assembly line problems. Operations Research **19**(6) (1961) 841–848
9. Kwok, Y.K., Ahmad, I.: Benchmarking and comparison of the task graph scheduling algorithms. Parallel and Distributed Computing **59**(3) (1999) 381–422
10. Cavalheiro, G.: A general scheduling framework for parallel execution environments. In: Proc. of the SLAB'01, Brisbane (2001)
11. Black, D.L.: Scheduling support for concurrency and parallelism in the mach operating system. IEEE Computer **23**(5) (1990) 35–43
12. Xiao, Z., Li, W., Jenq, J.: On unit task linear-nonlinear two-cluster scheduling problem. In: Proc. of the ACM Symposium on Applied Computing, Santa Fe (2005)
13. Kwok, Y.K., Ahmad, I.: Static scheduling algorithms for allocating directed task graphs to multiprocessors. ACM Comput. Surv. **31**(4) (1999) 406–471
14. Culler, D., Arvind: Resource requirements of dataflow programs, Honolulu (1988)
15. Garey, M., Johnson, D.: Computers and intractability: a guide to the theory of NP-Completeness. (1979)
16. Feitelson, D., Rudolph, L.: Parallel job scheduling: issues and approaches. In Feitelson, D., Rudolph, L., eds.: Proc. of the IPPS'95. Volume 949., Springer (1995) 1–18
17. Iverson, M.A., Özgüner, F.: Dynamic, competitive scheduling of multiple DAGs in a distributed heterogeneous environment. In: Heterogeneous Computing Workshop. (1998)
18. Sinnen, O., Sousa, L.: List scheduling: extension for contention awareness and evaluation of node priorities for heterogeneous cluster architectures. Parallel Computing (V. 30:1. 2004)
19. Sakellariou, R., Zhao, H.: A hybrid heuristic for DAG scheduling on heterogeneous systems. Proc. of the Heterogeneous Computing Workshop (2004)
20. Valiant, L.G.: A bridging model for parallel computation. Commun. ACM (V. 33:8. 1990)

21. Cordeiro, O., Peranconi, D., Villa Real, L., Dall'Agnol, E., Cavalheiro, G.: Exploiting multithreaded programming on cluster architectures. In: Proc. of the 19th Annual International Symposium on High Performance Computing Systems and Applications (HPCS), Guelph (2005)
22. Peranconi, D.S., Cavalheiro, G.G.H.: Using Active Messages to explore high performance in cluster of computers. In: Proc. of the 15th International Conference of the Chilean Computer Science Society (SCCC). (2005)

DWMiner: A Tool for Mining Frequent Item Sets Efficiently in Data Warehouses

Bruno Kinder Almentero, Alexandre Gonçalves Evsukoff, and Marta Mattoso

COPPE/Federal University of Rio de Janeiro,
P.O. Box 68511, 21941-972 Rio de Janeiro RJ, Brazil
Tel.: (+55) 21 25627388; Fax: (+55) 21 22906626
kinder@cos.ufrj.br, evsukoff@coc.ufrj.br, marta@cos.ufrj.br

Abstract. This work presents DWMiner, an association rules efficient mining tool to process data directly over a relational DBMS data warehouse. DWMiner executes the Apriori algorithm as SQL queries in parallel, using a database PC Cluster middleware developed for SQL query optimization in OLAP applications. DWMiner combines intra- and inter-query parallelism in order to reduce the total time needed to find frequent item sets directly from a data warehouse. DWMiner was tested using the BMS-Web-View1 database from KDD-Cup 2000 and obtained linear and super-linear speedups.

1 Introduction

The application of data mining tasks on huge databases requires an increasingly large processor and memory capacity. Currently most data to be mined resides in Data Base Management Systems (DBMS). An increasing number of organizations are installing large data warehouses using relational database technology. There is a huge demand for nuggets of knowledge from these data warehouses [16]. Nevertheless, most of the mining algorithms do not operate directly over a data warehouse. The integration of Data Mining (DM) tools with DBMS is now more than a trend, it is a reality. The major DBMS vendors have already integrated DM solutions within their products. In addition, the main DM suites have also provided the integration of DM models into DBMS through modeling languages such as the Predictive Model Markup Language (PMML). It is thus a fact that solutions on new DM tools and methods must consider their integration with DBMS.

In this paper, we present DWMiner, an efficient mining tool to process data directly over a relational DBMS data warehouse. Our solution takes advantage of a cluster of PCs running a Database Cluster middleware.

DBMS query processing techniques have been optimized to take advantage of PC Clusters without having to do a new physical database design through Database Cluster solutions [11] [6] [5]. They preserve the application and DBMS autonomy while providing high performance query processing in PC clusters. The database cluster combines a low cost solution with an excellent performance. Briefly, a database cluster is a middleware between the application and the DBMS that runs on a set of PC servers interconnected by a dedicated high-speed network, each one having its own processors and hard disks, and running an off-the-shelf DBMS [5].

M. Daydé et al. (Eds.): VECPAR 2006, LNCS 4395, pp. 212–224, 2007.
© Springer-Verlag Berlin Heidelberg 2007

This work addresses the mining of association rules task, more specifically, the search for frequent item sets. The procedure was based on the Apriori algorithm, developed by Agrawal and Srikant [4]. The Apriori algorithm for finding frequent item sets makes multiple passes over the data. Each pass consist of two phases. The first is the candidate generation phase where all candidate item sets are generated. Then, data is scanned to count, for each transaction, the occurrences of a candidate itemset in a transaction. Our implementation simply transforms every database search into an SQL query.

Many parallel algorithms have been proposed based on Apriori. Count Distribution, Data Distribution and Candidate Distribution [3] are some examples. However, these algorithms do not work with a DBMS.

The Apriori algorithm was modified in DWMiner to deal with SQL queries and a DBMS instead. DWMiner executes SQL queries in parallel using database cluster middleware techniques proposed by Lima et al. [9] and [10]. Such middleware is based on parallel query processing techniques developed for SQL query optimization in OLAP (On-Line Analytical Processing) applications. This database cluster has become an open source solution named ParGRES [11], [13] and is publicly available at http:// forge.objectweb.org/projects/pargres/. Each cluster node can run any non parallel relational DBMS. In this work we use PostgreSQL [14] which is open source. DWMiner combines intra- and inter-query parallelism in order to reduce the total time needed to find frequent item sets directly from a data warehouse. We ran DWMiner using the BMS-Web-View1 database from KDD-Cup 2000 [8] and obtained linear and super-linear speedups in cases where the support threshold is small like, for instance 0.01.

This paper is organized as follows. Section 2 describes the Apriori algorithm used as a basis for our implementation. Section 3 describes how we changed the Apriori algorithm to access a data warehouse and the parallel techniques used in DWMiner. Section 4 describes our prototype implementation and experimental results and Section 5 concludes.

2 The Apriori Algorithm

The problem of mining association rules was initially presented by Agrawal [1] and today is one of the most popular data mining algorithms. Association rule mining, also known as market basket analysis, finds interesting association relationships among a large set of data items. Typically, the data is a set of records where each record represents a transaction containing a set of items. The main goal of the algorithm is to find associations on items that are often present in the same transaction.

Association rules are considered interesting if they satisfy both a minimum support threshold and a minimum confidence threshold. But before describing the procedures that generate association rules we first need to formally define the terms item set, confidence, support and an association rule. An association rule is an implication of the form $X => Y$ where X and Y are sets of items and $X \cap Y = \varnothing$ [2]. The intuitive meaning of this rule is that transactions of the database which contain items in the set

X tend to contain also the items in Y. The set of items X and Y are generally referred as itemsets. A k-itemset is an itemset that contains k items in a lexicographic order.

Association rules are generally defined based on two measures: Support and Confidence. The Support of a rule $X => Y$ is the percentage of the transactions that contains $X \cup Y$, i.e. both X and Y. The Support is computed as the probability $P(X \cup Y)$. A frequent itemset is an itemset with a support value higher than a minimum specified threshold. The Confidence of a rule $X => Y$ is the percentage of transactions that contains X and also contains Y. This is taken to be the conditional probability, $P(Y \mid X)$. These measures can be summarized as:

Support $(X => Y) = P(X \cup Y)$

Confidence $(X => Y) = P(Y \mid X)$

The algorithm of mining association rules can be divided in two sub problems: (i) find all the combinations of items having support higher than the minimum support, called frequent item sets; and (ii) find the association rules with confidence greater than or equal to the minimum confidence, based on frequent item sets generated previously. We are particularly interested in the first sub problem: finding the frequent item sets. There are many algorithms to generate frequent item sets such as the AIS [1], the SETM [7] and the AprioriTid [4]. Among these algorithms, the Apriori is considered one of the most important and widely used. Thus, we chose Apriori to be the basis of our implementation. The pseudo-code for the Apriori algorithm is as follows.

```
Input: Database,D, of transactions;
       minimum support threshold min_sup

Output: frequent item sets in D

C_k: Candidate itemset of size k;

L_k: frequent itemset of size k;


1.   L_1 = {frequent 1-itemsets};
2.   for (k = 2; L_k-1 !=Ø; k++) {
3.       C_k = candidates generated from L_k-1;
4.       for each transaction t in database do{
5.           increment the count of all candidates in C_k
6.           that are contained in t
7.       }
8.       L_k = candidates in C_k with min_sup
9.   }
10. return ∪_k L_k;
```

Step 1 of Apriori finds L_1, the frequent 1-itemsets (line 1). In the next step the frequent itemset L_{k-1} is used to generate the candidate k-itemsets C_k (line 3). Then, the dataset is scanned to find the support values for candidates (lines 4 to 7). Finally, the frequent k-itemsets are determined (line 8). The final solution is the union of the frequent k-itemsets (line 10).

3 Apriori Implementation in DWMiner

Discovery of association rules is an important Data Mining problem. Parallel algorithms are required [3] to cope with the databases to be mined which are often very large (measured in gigabytes or even in petabytes). However, most of the parallel solutions do not deal with a DBMS. In DWMiner we combine parallel techniques with DBMS advantages to efficiently mine frequent item sets from large databases. In this section we present how we adapted Apriori to issue queries to run in a database cluster.

3.1 Database Clusters

Database Cluster is a middleware that provides parallel query processing in applications that use a sequential DBMS [15]. In a database cluster, each node of the cluster runs its own sequential DBMS as a *black-box* component. Clients submit transactions to the middleware which is responsible to distribute queries through the cluster nodes.

Parallel query processing of database clusters is based on two techniques known as intra-query and inter-query parallelism. In intra-query parallelism, a query is decomposed in sub-queries that scan different subsets of the data. The sub-queries are executed in parallel in the cluster nodes. Fig. 1 (a) shows an example of intra-query parallelism where the query Q_1 is decomposed in n sub-queries. The database is replicated at all nodes involved with the intra-query processing. Each sub-query is responsible to process a different range of data at each node in parallel. Finally the sub-results are combined to produce the final query result. This technique aims to reduce the execution time of heavy-weight queries, i.e., queries that access large amounts of data and may perform complex operations, thus taking a long time to be processed. In the inter-query technique, queries are executed as they are really, which means no decomposition. Distinct queries are distributed and executed concurrently in the cluster nodes to enhance database system throughput. Fig. 1 (b) shows an example of inter-query parallelism where queries Q_1 to Q_n are distinct and distributed over the cluster nodes to be executed in parallel.

These two techniques are not exclusive, but most database clusters provide either inter-query [6], [15] or intra-query [5] parallelism. However, they have been successfully combined in [10] and [11], so in DWMiner both inter and intra-query parallelism are explored. When receiving a heavy weight query we can use intra-query parallelism, and, in the case of simple queries, the inter-query parallelism should be more appropriate. In addition, a query being processed by intra-query parallelism can run concurrently with other queries through inter or intra-query parallel processing.

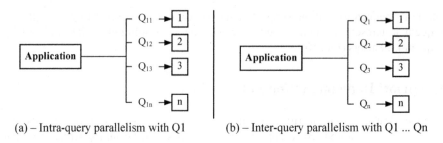

(a) – Intra-query parallelism with Q1   (b) – Inter-query parallelism with Q1 ... Qn

Fig. 1. Parallel query processing techniques

3.2 Adapting Apriori to DBMS Access

Now we describe how we changed the Apriori algorithm to generate SQL queries, instead of reading data from a flat file to main memory. First of all we name a set of items which contains k items as a k-item set. Hence, the first step of the algorithm ($k=1$) is to generate the 1-itemset and find the support for each element. Thus, we simply have to scan all the transactions in order to count the number of occurrences of each item. In our case, the table was called *bmswebview1* with two attributes: *a_item* and *a_tid*, where *a_item* is the item identification and *a_tid* the transaction identification. The SQL query generated by our implementation is Q_1, described as follows. This query corresponds to the line 1 of the pseudo-code described earlier.

```
Q1: Select    a_item, count(*) as total
    from      bmswebview1
    group by a_item
    having count(*) >= minimum_support
```

In Figure 2, we show the architecture of the typical Apriori algorithm and DWMiner with Apriori accessing data to be mined directly from the data warehouse through a DBMS driver interface.

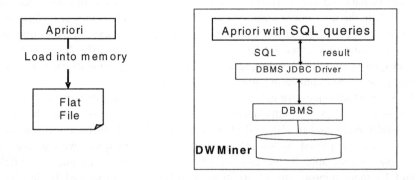

Fig. 2. Apriori data access and DWMiner database cluster access

The next step is to generate the candidate 2-item sets from which we will find the frequent 2-itemset. In this case the SQL query Q2 generated by DWMiner is described as follows.

```
Q2: Select count(a_tid)
    from    bmswebview1
    where   a_item = item1
    and exists (select a_tid
                from    bmswebview1
                where   a_item = item2)
```

This query Q2 corresponds to the steps 4 to 6 in the Apriori pseudo-code. So, we are counting the transactions that contain both *item1* and *item2*. The number of queries generated is equal to the number of candidate 2-item sets. This process continues until there is no more candidate item sets left. Every query generated in this loop corresponds to the steps 4 to 6 in the Apriori pseudo-code. They will be different depending on the value of k, from the current k-itemset being analyzed. Thus, to generate the query to find the 3-itemset support we just need to add one more level of nested select in Q2 generating the following query.

```
Q3: Select count(a_tid)
    from    bmswebview1
    where   a_item = item1
    and exists (select a_tid
                from    bmswebview1
                where   a_item = item2
                and exists (select a_tid
                            from bmswebview1
                            where a_item = item3)
```

Then, the number of nested selects is directly related to the itemset being analyzed. If we are analyzing the k-itemset then we will have k levels of nested selects. Once a candidate itemset is created we can build SQL queries for each element and process them in parallel because they are independent.

3.3   Adapting Apriori to Database Clusters

The main goal of DWMiner is to reduce the total time of database searching. In order to do that, DWMiner adopts inter and intra-query parallelism available in ParGRES database cluster. Intra-query parallelism is obtained by using a virtual partition technique (VP) [10]. This technique breaks one heavy weight query into sub-queries by adding selection predicates as proposed in [5]. Each DBMS receives a sub-query and is forced to process a different subset of data items. Each subset is called a "virtual partition".

The SQL Q_1 query generated in the first step of the Apriori algorithm to find the frequent 1-itemset involves a *group by* and a *having* operation. Such operations are time consuming since a full scan on a large relation is needed. To overcome this problem at this point DWMiner takes advantage of intra-query parallelism involving all of the cluster nodes. Thus, the Q_1 generated for the first step would be rewritten by the database cluster as the following Q_{1i} sub-queries, where i varies from 1 to n being the number of nodes involved on the intra-query processing.

```
Q1i: Select a_item, count(*) as total
     from bmswebview1
     and  bmswebview1_key > :v1 and bmswebview1_key <= :v2
     group by a_item
```

The basic difference between Q_1 and Q_{1i} is the range predicate "bmswebview1_key > :v1 and bmswebview1_key <= :v2". However, in this case we also suppress the minimum support clause, since it can only be checked after all intervals are done counting. We call *virtual partitioning attribute* (VPA), the attribute chosen to virtually partition the data. Usually the VPA is the primary key of the table being selected, in this case bmswebview1_key. The values used for parameters v1 and v2 vary from node to node and are computed according to the total range of the VPA and the number of nodes. Let us assume that the interval of values of bmswebview1_key is [1; 6,000,000] and we have 4 nodes, then, 4 sub-queries must be generated. The intervals covered by each sub-query are the following: Q_{11}: v1=0 and v2=1,500,000; Q_{12}: v1=1,500,000 and v2=3,000,000; and so on. In spite of each node having the same replica of bmswebview1 table, virtual partitioning forces each one to process a different data subset of bmswebview1. Besides, full replication makes it possible to allocate any node to process any sub-query. After sub-query execution, it is necessary to compose the partial counting produced by each one in order to have the final result. Consider that all Q_{1i} partial results are stored in table Temp. The final result can be obtained by executing the following query:

```
Q1result: select a_item, sum(total)
          from Temp
          group by a_item
          having sum(total) >= minimum_support
```

In Fig. 3, we show the architecture of the Apriori algorithm and DWMiner with respect to accessing data to be mined through a database cluster middleware. In this case DWMiner is issuing query Q_1 to the database cluster, which decides to process it through intra-query parallelism. Thus Q_1 is decomposed as Q_{1i} sub-queries to access n different virtual partitions of table bmswebview1. Such middleware can be C-JDBC or ParGRES or any other database cluster. However, if C-JDBC is used, Q1 cannot be processed through intra-query parallelism.

For the queries of the following steps of Apriori, DWMiner tries to find a balance between inter and intra-query parallelism. For example, once the k-itemset is analyzed, a candidate itemset is created. Each element query can be processed independently in parallel through inter-query. Therefore, DWMiner sends each query to a cluster node. However, the time needed to process one such query may be relatively large. In this case, the query is decomposed and its sub queries are processed in parallel.

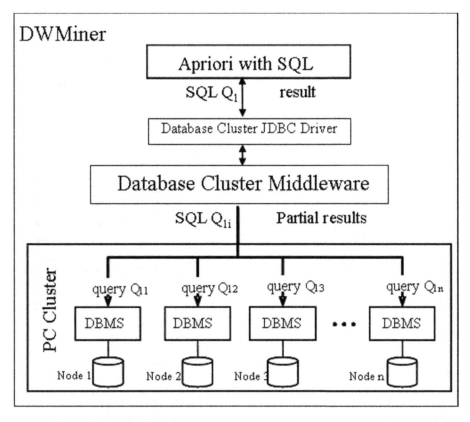

Fig. 3. DWMiner using Apriori accessing a database cluster

4 Experimental Results

To evaluate DWMiner techniques we have used a Linux based PC cluster and PostgreSQL 8.0 DBMS [14]. The dataset used in our experiment is the BMS-Web-View1 which contains several months' worth of click stream data from an e-commerce web site. A portion of their data was used in KDD-Cup 2000 competition [8]. This dataset has a total of 56,902 transactions and 497 distinct items, its maximum transaction size is 267 and the average transaction size is 2.5. Our experiments run on top of the cluster system of the Paris team at INRIA [12]. Our tests have used up to 32 nodes of this cluster system, each node configured with dual 2.2 GHz Opteron processors with 2 GB of main memory. The cluster is interconnected by a standard Ethernet network.

The results from our experiments are shown in Fig. 4. We plot times taken by our implementation for values of support ranging from 0.1% to 2% using only inter-query parallelism. We ran DWMiner varying the number of nodes from 1 to 32 for each support value. In order to improve reading and analysis, we use logarithmic scale.

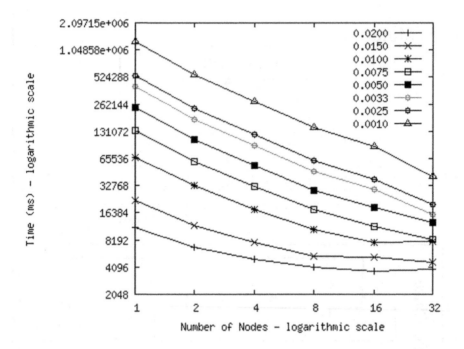

Fig. 4. Execution times for DWMiner

Although, DWMiner implements its own inter-query mechanism we also used C-JDBC[6] to perform inter-query parallelism as an alternative successful open source database cluster solution.

Most of the results in Fig. 4 present linear speedup as we increase the number of nodes, since queries sent to the database cluster are independent from each other. But, analyzing the higher support curves like 0.02 (2%) and 0.015 (1.5%) we note that the results are worse than linear. This happens because the number of candidates generated and, consequently, the number of queries is not enough to compensate the time spent to distribute these queries over the cluster nodes and receive the results. Still, DWMiner does not experience slow down factors. Table 1 gives a more accurate view of the graphic shown in Fig. 4. In the worst case, using 32 nodes is 4 times faster than using 1 node.

As we can see in Table 1, by using two nodes the execution time of DWMiner is reduced by almost 50% for the support 0.02 (2%). However, when 4 nodes are used the time reduction is linear and the execution time remains almost the same until 32 nodes. This happens because when we use 4 nodes we get too close to the situation where the time spent to distribute the queries and wait for the results is the main factor in the total time of execution. However, as the support threshold decreases, the time reduction continues to improve the performance and it is often super-linear.

Table 1. Query Execution times for DWMiner

Support	Number of Nodes					
	1	2	4	8	16	32
0.0200	11,354	6,842	4,959	4,034	3,672	3,867
0.0150	22,581	11,991	7,618	5,328	5,274	4,648
0.0100	67,707	32,651	17,878	10,821	7,653	7,952
0.0075	132,153	60,715	31,801	17,989	11,678	8,355
0.0050	238,717	105,636	55,043	29,326	18,722	12,891
0.0033	411,360	177,668	90,894	47,575	29,724	15,564
0.0025	540,933	234,830	119,517	61,941	38,767	20,246
0.0010	1,291,949	555,865	282,049	144,623	88,928	41,662

Table 2 shows the performance improvement we obtained in each case. We can see in Table 2 that most results are *quasi*-linear or super-linear. When we use 2 nodes and the supports going from 0.01 (1.0 %) to 0.001 (0.1%) the support is lower enough to generate a relatively large number of candidate item sets. For these support thresholds, a large number of queries are generated and sent to the nodes. When many queries are sent to a node the database cluster makes a wise use of the system cache instead of reading data from disk. Thus, many queries process data from memory reducing considerably the query execution time achieving, this way, super-linear speedups.

Table 2. Perfomance evaluation of DWMiner

Support	Number of Nodes (Linear Speedup)				
	2 (50.00%)	4 (25.00%)	8 (12.50%)	16 (6.25%)	32 (3.13%)
0.0200	60.26%	43.68%	35.53%	32.34%	34.06%
0.0150	53.10%	33.74%	23.60%	23.36%	20.58%
0.0100	48.22%	26.40%	15.98%	11.30%	11.74%
0.0075	45.94%	24.06%	13.61%	8.84%	6.32%
0.0050	44.25%	23.06%	12.28%	7.84%	5.40%
0.0033	43.19%	22.10%	11.57%	7.23%	3.78%
0.0025	43.41%	22.09%	11.45%	7.17%	3.74%
0.0010	43.03%	21.83%	11.19%	6.88%	3.22%

The graphic in Fig. 5 compares the results of inter-query only by using C-JDBC with the results of intra-query combined with inter-query through ParGRES. We also compared inter-query only using C-JDBC and inter-query only using ParGRES. In both implementations queries are distributed to cluster nodes in a round robin fashion.

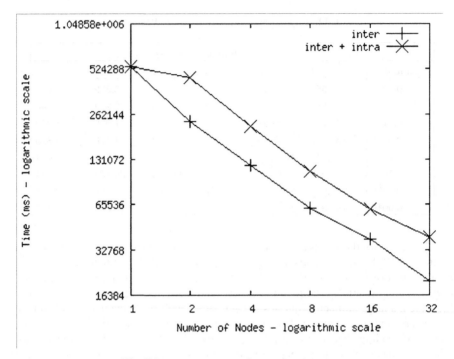

Fig. 5. Inter-query versus (inter + intra) query

We obtained very similar results in both database clusters. Therefore, in Fig. 5 we kept the legend as inter *versus* inter/intra rather than C-JDBC *versus* ParGRES. In the combination case, intra-query was implemented using only two nodes. Queries were decomposed in two sub-queries and executed in parallel in the cluster concurrently with other queries. So, when running with 32 nodes, it means that 16 different queries can be executed in parallel. However, intra-query demands an aggregation phase for each query to compose the two partial results.

As shown in Fig. 5, the inter-query parallelism alone is better than the combination between inter and intra-query parallelism. Since we are using a relatively small database, individual queries could not be considered to be heavy weight queries. So, the time needed to aggregate the partial results of the sub-queries was relevant with respect to overall query reduction. Nevertheless, the combination of inter with intra-query achieved linear and super-linear speedups.

5   Conclusions and Future Work

One of the best advantages in using a DBMS is that it already provides efficient techniques to deal with large datasets. These techniques need to be re-implemented in part if we want to work with flat files that do not fit in the available memory.

Most of the mining algorithms demand a flat file to be in a special format. These algorithms need an extra step to extract the information they need to a flat file. Since we can have data warehouses with dozens of gigabytes or even petabytes, to generate a file from these data may require a lot of extra storage. DWMiner solution is DBMS vendor independent, thus it can be applied directly over a data warehouse system using techniques that take advantage of a low cost high performance scenario such as database clusters.

In this work we showed that by using such techniques we acquire significant improvement in the process of mining data directly from a DBMS. We can efficiently mine frequent item sets from a data warehouse by sending queries to be processed in parallel by the database cluster. In our experiments, we have used one representative dataset – BMS-Web-View1 – and as future work we intend to test DWMiner against some larger databases where we expect to explore the combination of inter and intra-query parallelism and take more advantage of the intra-query parallelism. Nevertheless, we achieved linear and super-linear results working with a relatively small dataset comparing to a real data warehouse.

The techniques adopted in DWMiner are not difficult to implement and maintain since they are based on SQL and take advantage of simple parallel techniques found in database clusters. In addition, DWMiner solution is all based on open-source software and commodity hardware. Such techniques can also be applied in tasks different from mining frequent item sets inside the data mining context.

Acknowledgements

The authors are grateful to the Brazilian research agencies CNPq, CAPES and FINEP for the financial support of the work. We are also grateful to the Paris team at INRIA for providing the PC cluster environment.

References

1. Agrawal, R., Imielinsk, T., Swami, A. N., 1993, "Mining association rules between sets of items in large databases". In: *1993 ACM SIGMOD International Conference on Management of Data*, pp.207-216.
2. Agrawal, R., Mannila, H., Srikant, R., et al, 1996, "Fast discovery of association rules". In U.M.Fayyad, G.Piatetsky-Shapiro, P.Smyth, and R.Uthurusamy, *Advances in Knowledge Discovery and Data Mining*, chapter 12, AAAI/MIT Press.
3. Agrawal, R.,Shafer, J., 1996, "Parallel Mining of Association Rules", *IEEE Trans.Knowledge and Data Engineering*,v.8, pp.962-969.
4. Agrawal, R.,Srikant, R., 1994, "Fast algorithms for mining association rules". In: *20th International Conference on Very Large Databases (VLDB)*, pp.487-499.
5. Akal F., Böhm, K., Schek, H. J., 2002, "OLAP Query Evaluation in a Database Cluster: a Performance Study on Intra-Query Parallelism". In: *East-European Conference on Advances in Databases and Information Systems (ADBIS)*, Bratislava, Slovakia.
6. C-JDBC. In: http://c-jdbc.objectweb.org/, Accessed in 2005.
7. Houtsma, M.,Swami, A., 1995, "Set-oriented mining of association rules". In: *11th Conference on Data Engineering*, Taipei, Taiwan.
8. Kohavi, R., Brodley, C. E., Frasca, B., et al., 2000, "KDD Cup 2000 Organizers' Report: Peeling the Onion", In: *SIGKDD Exploration 2 (2)*, pp.86-98.

9. Lima, A. A. B., Mattoso, M., Valduriez, P., 2004, "OLAP Query Processing in a Database Cluster". In: *10th Euro-Par Conference*, pp. 355-362.
10. Lima, A. A. B., Mattoso, M., Valduriez, P., 2005, "Adaptive Virtual Partitioning for OLAP Query Processing in a Database Cluster". In: *19th SBBD*, pp.92-105.
11. Mattoso, M., Zimbrão, G., Lima, A. A. B., Almentero, B.K. et al., 2005, "ParGRES: a middleware for executing OLAP queries in parallel". In: COPPE/UFRJ Technical Report ES-690, *http://pargres.nacad.ufrj.br/Documentos/ES-690.pdf*.
12. Paris Project. In: http://www.irisa.fr/paris/General/cluster.htm.
13. ParGRES In: http://pargres.nacad.ufrj.br/, Accessed in 2005.
14. PostgreSQL. In: http://www.postgresql.org, Accessed in 2005.
15. Röhm, U., Böhm, K., Schek, H. J., 2002, "FAS - A Freshness-Sensitive Coordination Middleware for a Cluster of OLAP Components". In: *28th International Conference on Very Large Data Bases (VLDB2002)*, pp.754-765.
16. Sarawagi, S., Thomas, S., Agrawal, R., 1998, "Integrating Association Rule Mining with Relational Database Systems: Alternatives and Implications". In: *1998 ACM SIGMOD International Conference on Management of Data*, pp.343-355.

A Parallel Implementation of the K Nearest Neighbours Classifier in Three Levels: Threads, MPI Processes and the Grid*

G. Aparício, I. Blanquer, and V. Hernández

Instituto de las Aplicaciones de las Tecnologías de la Información y Comunicaciones
Avanzadas - ITACA
Universidad Politécnica de Valencia. Camino de Vera s/n 46022 Valencia, Spain
{gaparicio, iblanque, vhernand}@itaca.upv.es
Tel.: +34963877007; Fax: +34963877274

Abstract. The work described in this paper tackles the problem of data mining and classification of large amounts of data using the K nearest neighbours classifier (KNN) [1]. The large computing demand of this process is solved with a parallel computing implementation specially designed to work in Grid environments of multiprocessor computer farms. The different parallel computing approaches (intra-node, inter-node and inter-organisations) are not sufficient by themselves to face the computing demand of such a big problem. Instead of using parallel techniques separately, we propose to combine the three of them considering the parallelism grain of the different parts of the problem. The main purpose is to complete a 1 month-CPU job in a few hours. The technologies that are being used are the EGEE Grid Computing Infrastructure running the Large Hadron Collider Computing Grid (LCG 2.6) middleware [3], MPI [4] [5] and POSIX [6] threads. Finally, we compare the results obtained with the most popular and used tools to understand the importance of this strategy.

Topics: Grid, Parallel Computing, Threads and Data Mining.

1 Introduction

Data Mining is a recently created concept that groups different techniques of data analysis and model extraction. The main purpose of Data Mining is the extraction of hidden predictive information from large databases. In this way, Data Mining is a helpful technology to get profit of the great amount of poorly exploited data. The interest of this work is focused on the classification of new registers (automatic selection of the category in which a piece of information will more likely fall into). The target of this work is the development of a set of tools to assist large-scale epidemiology studies. Thus the initial hypothesis is a

* The authors wish to thank the financial support received from the Spanish Ministry of Science and Technology to develop the GRID-IT project (TIC2003-01318).

M. Daydé et al. (Eds.): VECPAR 2006, LNCS 4395, pp. 225–235, 2007.
© Springer-Verlag Berlin Heidelberg 2007

large database with registers in which most of them are labelled and in which the process will predict the label for a few of them. Considering this situation, we select the K nearest neighbours method [1] as the most suitable classification method to obtain the needed results, i.e., the predicted labels. The work will concentrate on speeding up the performance. The analysis of the accuracy and goodness of the predictions are not in the scope of this work.

Along this paper we will first make a review, in Section 2, of the specific problem to work with (the KNN method). After that, in Section 3, we will propose an architecture to the application (the Three-layer Parallelism Architecture) using the three technologies proposed. Once we have presented the problem and the architecture, then will be review the implementation in Section 4, starting with brief comments on the three technologies used and the selected interface. Finally we will see the results obtained with the solution proposed and a set of conclusions about the work made will be presented in the last two sections.

2   K Nearest Neighbours

One of the most popular instance-based classification techniques is the K Nearest Neighbours method [1]. This method is a generalisation of the one nearest neighbour rule. It is appropriate when dealing with a large set of labelled registers or instances and a small group of non-labelled registers to be classified with the most probable label. Our hypothesis is that we are dealing with data with an strong relationship among the multidimensional distance of each entry with the rest. In this way, the 1-NN (one nearest neighbour [1]) consists on assigning to the non-labelled registers, the label of the nearest labelled register. Based on that, Cover & Hart [1] implemented a variant known as K-NN (K Nearest Neighbours) that applies the same philosophy that 1-NN but considering the most frequent label in the K Nearest Neighbours instead. The K parameter is very important and the optimal value will depend on many factors, such as the data nature, and it has to be chosen experimentally.

The cross validation test is the most popular test technique used for the analysis of the error of the classification. Starting from a set of labelled registers, those are divided into B blocks. One block will be used as the test set and the rest as the training set. This operation is performed with the B blocks. The K-NN classifier is applied to any register in the test block and compares the label assigned with the one previously defined. If the labels differ, an error is recorded, and the consolidation of all the errors obtained for the whole test block is the validation error. Repeating this process by changing the test set to the rest of blocks, the sum of all the errors will be an approximation to the real error of the K-NN classifier with the chosen K parameter

$$err = \sum_{b=1}^{b=B} (err_b) \text{ where b is the chosen block.}$$

3   Three-Layer Parallelism Architecture

The decision of using the three layer parallelism lies on the difficulty of managing a large amount of data (especially when dealing epidemiology databases containing several millions of registers) and the poor results of traditional sequential approaches. Our purpose is based on performing a set of experiments with different values of K in the K-NN method to determine the optimal value for this K according to the lowest error in a cross validation test. Once the optimal value for K is determined, we will be able to classify the non-labelled registers with the K-NN method. The main aim of the work will be on reducing computing time especially for epidemiology analysis and support.

The evaluation of the error for each value of K takes around 18 CPU hours using a state-of-the-art computer and a database of 1 million records and 20 fields per record. Thus, a complete optimisation process of 10 different values of K would take more than 7 CPU days, which would be unmanageable in a production environment. However, the process is intrinsically parallel, presenting two clear levels of parallelism. In a first level, each evaluation of the error using different values of K is totally independent, since it consists on computing the whole classification and cross-validation process for each value of K, using the same input database. This process consists on computing the K minimal distances to all registers and selecting a block to act as the test set. This block will be re-labelled using the rest of the database as the training set. Labels are assigned considering the labels of the K Nearest Neighbours. This process will be repeated selecting different blocks of the database as training sets in order to cover the whole database. Each validation using a different block is independent. However, the computational cost of this process is on the order of cN^2, being c the number of flops for computing a single distance between two registers and N the number of registers in the database (directly affecting communication cost). Thus, a trade-off solution must be applied to obtain the maximum performance considering the best granularity.

In the frame of this scenario, three parallelism approaches can be considered. The conditions of each one concerning the problem of this article are the following:

- Grid Computing [2]. This technique implies the coarsest granularity. Grid Computing deals with the concurrent usage of different computing resources in different administrative domains in a similar approach as a large-scale batch queue. Inter-resource communication is not usually available due to the long latencies, the internal configuration of nodes in resource providers and the overhead of the security policies. Data access thus is mainly performed through the job submission process and shared repositories, using ftp-like protocols. In this paradigm, the minimal running entity is the job, interacting with the rest of the jobs through input and output files.
- Message-Passing Parallel Computing. This technique is proven to be very efficient in most medium-grain problems in which communication costs are on an order of magnitude lower than computing cost. Typically, jobs are fairly symmetrical and run on homogeneous nodes connected through a fast

network. Security policies are not applied in communication and data exchange is performed through message passing.
- Shared-Memory Parallel Computing. This constitutes the finest-grain parallelism. Applicable in very coupled and homogeneous environments, different threads concurrently execute a common program on different fragments of data. Data is exchanged through shared regions and contention mechanisms. Generally the scaling factor of those systems is low, due to hardware constraints and speed-ups are good.

Our proposal is to combine the three levels to achieve the maximum performance. Using shared-memory approaches only would lead to small speed-ups (limited by the number of available processors) and the need of shared-memory supercomputers. The combination of distributed-memory and shared-memory approaches would increase notably the speed-up, since computing farms can reach without performance losses many tens of bi-processor nodes. However, and considering that our problem is massively parallel, more powerful configurations could be efficiently used. Thus, the coordinated use of several computing farms is a reasonable choice considering the availability of those systems. In this case, grid computing constitutes an efficient way to organise and manage different computing resources in different administrative domains. So, in order to achieve the maximum performance, we have decided to combine three different techniques of parallel computing: Grid technology, MPI programming and POSIX threads. Considering the different characteristics of each approach, the problem of classification must be structured to obtain the maximum efficiency from each one. According to this, we have chosen the EGEE infrastructure currently running the LCG 2.7 Grid Middleware [3]. Command Line Interface (CLI) will be used to implement the scripts and the programmes for submitting several experiments with different values for K each. Each experiment is performed concurrently by several MPI processes to divide the cross validation into a

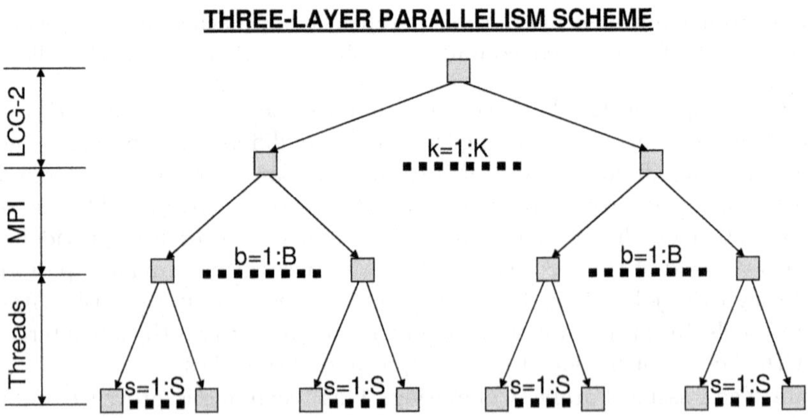

Fig. 1. Three-layer parallelism scheme

simple test-training partition validation. Finally different sub-blocks of each MPI process testing partition are computed on the different POSIX threads created in a MPI process and executed within the processors of a node.

An example of a tree diagram of our approach is printed in figure 1. In this figure we can see the evolution of the work. It begins at the root of the tree with the submission of K different LCG Grid jobs (where the value of K will determine for each LCG job the number of neighbours to which the distances are computed). In a second phase, MPI parallel process are executed. We choose the same number of MPI processes as cross validation blocks, although other factors of parallelism grain could be analysed in the future. In the third phase, MPI processes are split into threads, according to the features of the target hardware resources and the experimental results.

4   Implementation

The implementation of the three-layer model mainly considers three components:

- Parallel KNN module. This component implements the KNN classification and cross-validation algorithm using MPI and POSIX Threads. It is an autonomous executable that takes as input the reference to the labelled registers file name, the reference to the file that contains the registers we want to label and the value of K to be used. It produces a different output file depending on our demands, being possible to show the labels assigned to the target registers or to obtain a statistical summarized file.
- Grid scripts. They implement the selection of the rightmost computing resources, the job description file, the start-up script for the parallel executable, the job submission and monitoring and the job output retrieval. All these tasks are implemented through scripts that make use of the CLI.
- Java Interface. It implements a user-friendly interface to select the data and the parameters for the Grid jobs and to retrieve their output.

The parallel KNN module comprises the MPI and POSIX Threads Computing levels. The synchronised execution of different instances of this processing module is performed through the Grid scripts.

The figure 2 shows us the global process, according to a chronological view. In the upper part of the diagram we can see subprocesses classified by functionality and in the lower part they are classified by technology.

4.1   Grid Computing Level

One typical application of Grid technology is multi-parametrical runs. The difficulty in establishing efficient communications among independent submitted jobs in a Grid environment has been traditionally an important barrier. In our case, different experiments with different values of K in the K-NN method, constitute clearly a multi-parametrical task which can be achieved by different LCG Grid jobs.

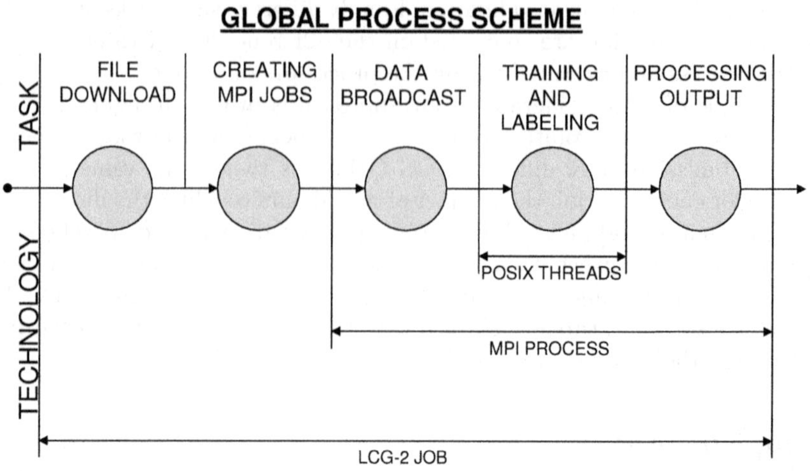

GLOBAL PROCESS SCHEME

Fig. 2. Global process scheme

The Grid infrastructure selected is EGEE (Enabling Grids in E-sciencE). This is the largest production infrastructure available for research world-wide, integrating, in October 2006, more than 29000 computers and more than 40 Petabytes of storage in 177 sites. This infrastructure runs currently the gLite 3.0. gLite 3.0 and LCG 2.7 share a major part of components, so migration is feasible. Both LCG 2.7 and gLite 3.0 are batch-oriented Grid middlewares and consider the same computing structure, which comprises the following components:

- Computing Resources. The computing resources are organised in the form of Computing Elements (CEs), and Working Nodes (WNs). CEs are the front-ends and visible entry-points to computational farms of WNs. CEs implement the necessary batch queues to manage the jobs in the WNs and keep track of the status of both jobs and resources.
- Storage Resources. The files in the EGEE infrastructure are stored in a distributed way in many Storage Elements (SEs).
- Workload Management. The destination of a job (a queue in a CE) can be directly selected by the user, although the more effective way is to rely on the Workload Management System.
- Storage Catalogue. Data stored on the SEs is organised through Storage Catalogues. Storage Catalogues keep track of the files stored.
- User Management. Users are organised in Virtual Organisations (VOs). Typically, users in a VO have the same authorisation rights to access the resources. This reduces the burden of managing individual security policies.
- System Information. The information of the system (status of the jobs and resources mainly) is published in a hierarchical model by the sites and the monitoring system.

In order to execute a job in the LCG environment, a Job Description File must be written according to the Job Description Language (JDL). This file defines the input and output files, the executable file and the running parameters and the program requirements. The executable is typically a shell script that copies all necessary data locally on the resource and performs other preliminary steps (such as re-compiling, executable permissions, etc.). Job is submitted through the specific commands or API calls and enters in a cycle of states (submitted - waiting - ready - scheduled - running - done - outputready - cleared).

Once the LCG job is assigned to a Computing Element, a shell-script is executed. The shell-script initial instructions will fetch the databases stored in the Grid, including both training registers and labelling registers, since not only training but also classification is performed in the last phase of the job. This approach will reduce the Grid waiting time. When all the necessary data are downloaded on the computing nodes, the classification process can begin. This is achieved using an MPI process that will be the responsible of making a test with an assigned block. The sum of all the block test errors will be the cross validation error, i.e., the information we are requesting to decide the optimal K value and then use it to classify the non-labelled registers.

The tasks that must be implemented for delivering the above functionality are:

- Selection of the rightmost Computing Resource. Resources in the EGEE Grid are accessed through Workload Management Systems. Those resources are selected according to the job features and the VO policies. The resources were ranked considering their proximity to Storage Resources where replicas of the database are stored and other performance criteria (mainly the number of free CPUs, historical average length of the submission queue).
- Submission and Resubmission. Once the resources are identified, input data for each job is packed and jobs are submitted along with all the needed information (including references to the stored databases). The status of the job is periodically monitored and jobs are resubmitted to a new computing resource if scheduling time exceeds a predefined threshold.
- Monitoring and Output Retrieval. Jobs being executed are monitored through the corresponding scripts. Once finished, output data are retrieved and user is notified via e-mail (obtained automatically from the Distinguished Name of the certificate or given as a parameter).

Final result of all the process is the cross-validation error for the execution of KNN for a specific value of K. Results are presented as available and sorted by the magnitude of the error.

4.2 MPI and POSIX Threads Computing Level

The MPI executable is an autonomous programme that computes the distance evaluation, cross-validation and labelling of the registers of a database.

MPI process 0 will be the responsible to load and broadcast databases to the rest of MPI processes that run in other nodes of the cluster selected in the Grid Infrastructure for each job. Databases are replicated among all the processors.

The computation of the distance requires considering all the registers each time. Other distributions could be considered if memory is insufficient, although they will require additional communication cost, since they must involve intermediate data exchange among processes. The distribution of large amounts of data does not imply an important penalisation since the communication is performed inside the cluster farm and not through the Grid.

MPI process 0 also normalises the database to ensure that all fields of each register are considered with the same weight. Registers are evenly distributed among the processors and within each processor evenly among the threads. A list of the K nearest registers is updated during the process. This process is run for the block of registers selected as the test set in each computer. The errors are computed as the number of wrongly assigned labels. Finally, labels are assigned at the end of the process to reduce the overhead of redistributing the data again.

Each block of test-set registers considered in a processor is processed by different threads. POSIX threads are created in each MPI process dealing with the distance computation and labelling of a portion of the testing set. The use of POSIX threads permits exploiting efficiently the multiprocessor capability that modern clusters have. Moreover, the process do not imply conflicts neither on write access to common variables nor on synchronisation. Experiments also prove that the consideration of the hyper-threading capabilities of current processors provide an additional gain factor in the speed-up, being able to run more threads than physical processors are available.

The moderated cost of this process (in the quadratic order) and the large amount of data to be exchanged makes this problem suitable for parallel computing rather than Grid computing. Complexity can be increased by considering more costly computations of the distances (currently a homogeneous Euclidean distance) considering different weights or distance metrics for different fields, or considering more complex error metrics, such as distances to the right label.

The development of MPI applications across different Grid nodes is an issue currently tackled by Grid-MPI projects. Grid-MPI is a Grid approach of the Message Passing Interface. Although GRID-MPI enables the development of large-scale MPI parallel applications, current infrastructures are not supporting it. The main reason is that WNs typically can communicate within the cluster and although inbound connections are allowed, direct communication within WNs of different resources is not possible. From the users' point of view, the resource is the CE. Moreover, geographically distributed resources on the Grid make communications fairly inefficient for a fine-grain application, so coarse-grain parallelism is clearly the current aim.

4.3   User Interface

In order to ease the process of creating experiments, submitting the jobs and monitoring the results, a java-based interface has been implemented. This interface co-ordinately execute the necessary scripts to deal with the Grid job submission, monitoring and output retrieval.

The interface enables an authenticated and authorised user to log in the system and to upload the test and training sets on a specific SE. Target CEs will be selected according to the availability of computing resources, the support of MPI and the proximity to a SE publishing a replica of the databases.

Then, jobs are automatically constructed considering the range of values of K that will be sweeped. Jobs are submitted through the interface and their status is monitored either at global level (percentage of jobs in each state) or individually by jobs. Jobs are automatically submitted if an error is produced (although a maximum retry count is reached) or if they keep on waiting on a queue for an excessive time, choosing in both cases a different computing resource.

Finally, the result of the jobs can be dynamically consulted. The user is notified when a job has finished through e-mail.

Figure 3 shows a couple of snapshots of the application interface.

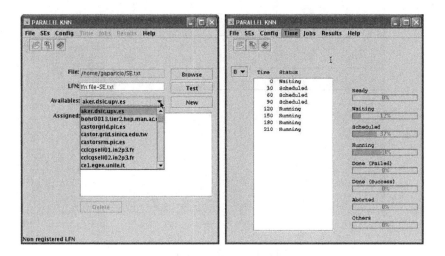

Fig. 3. Snapshots of the user interface application

5   Results

We have done different experiments in three scenarios. First, using the command-line WEKA [7] java K-NN class (3.4.5 release, a very well-known and widely used classification tool); second, using a sequential tool we have implemented in "C"; and, third, by using the three-layer parallel K-NN tool. We have used two different training databases, one with one hundred thousand labelled registers and another one with one million labelled registers. The number of fields of the two databases was 20, plus the label field.

The performance of the version implemented in "C" language was very efficient comparing to the results with WEKA, mainly due to the fact that WEKA is implemented in Java. A linear speed-up is obtained using grid technology since the experiments are independent. The gain in the MPI parallelisation approach has been above a factor of 9.5 with 10 biprocessors nodes (PIII Xeon 3GHz. on an SCI 3D torus network) and the gain using four threads in each node is above an additional 1.5 factor. We selected four threads considering that each node has two hyper-threading processors. This gain is not as linear as in the other cases since not all the process has been implemented using threads. Thus, the gain of using the MPI and threads parallel technologies is above 15, and the total advantage from WEKA multiplies by 6. Moreover, as it was mentioned before, Grid scales linearly. Figure 4 summarizes the results obtained.

This figure reflects the Speed Up comparing to WEKA to the sequential process, the MPI parallel process and the three-layer parallelism process.

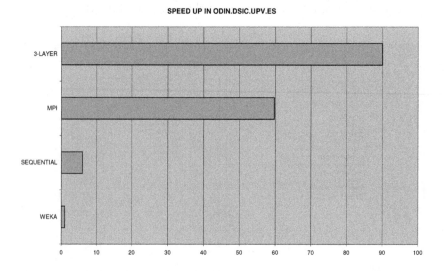

Fig. 4. Speed Up results

6   Conclusion

The results of our work have been very encouraging. The approach based on a three-layer parallelism is a very effective way to get the best performances and give us a hopeful vision of data mining in the Grid. The classification of the registers, including the identification of the optimal K value in the K-NN method on a database of one million registers took less than 6 hours, i.e., a single nightly run. For a comparison, this results implies an speed-up larger than 90 compared to the equivalent WEKA K-NN sequential execution, i.e., more than a month waiting time. Considering that our efforts are routed to biomedical prediction, this advantage would enable, for example, classifying the information recorded in Primary Care each day (in the order of millions of records) and its automatic classification.

References

1. T.M.Cover, P.E.Hart: Nearest neighbour pattern recognition. IEEE Trans. on Information Theory **13(1)** (1967) 2127
2. I. Foster, C. Kesselman, S. Tuecke: The Anatomy of the Grid: Enabling Scalable Virtual Organizations. International J. Supercomputer Applications, 15(3) (2001) http://www.globus.org/research/papers/anatomy.pdf
3. LCG: World Wide Web Computing Grid. Distributed Production Environment of Physics Data Processing. http://lcg.web.cern.ch/LCG
4. Message Passing Interface Forum: MPI: A message-passing interface standard. (2003) http://www.mpi-forum.org/
5. Gropp, W., Huss-Lederman, S., Lumsdaine, A., Lusk, E., Nitzberg, B., Saphir, W., Snir, M.: MPI: The Complete Reference, MIT Press, Cambridge, MA (1998)
6. Drepper, U and I. Molnar: The Native POSIX Thread Library for Linux. (2003) http://people.redhat.com/drepper/nptl-design.pdf
7. E. Frank, M. Hall, L.T.: Weka 3: Data Mining Software in Java. (2005) http://www.cs.waikato.ac.nz/ml/weka.

On the Use of the MMC Language to Utilize SIMD Instruction Set

Patricio Bulić and Veselko Guštin

University of Ljubljana, Faculty of Computer and Information Science, Slovenia
patricio.bulic@fri.uni-lj.si
http://lra-1.fri.uni-lj.si

Abstract. This paper presents the use of the Multimedia C (MMC) language to develop multimedia applications. The MMC language was designed to support operations with multimedia extensions included in all modern microprocessors. Although the idea to extend high programming languages to support vector operations is not novel, we show that integration of multimedia extensions into C is valuable. This is specially true for idiomatic expressions which are difficult for a compiler to identify. The MMC language has been used to develop some of the most frequently used multimedia kernels. The presented experiments on these scientific and multimedia applications have yielded good performance improvements. Although this paper discuses the use of MMC, the key features of the MMC language and implementation of its compiler are also presented.

1 Introduction

Today's computer architectures are very different from those of a few years ago in terms of complexity and the computational availabilities of the execution units within a processor. Practically all modern processors have facilities that improve performance without placing an additional burden on the software developers, as well as those facilities which require support from external entities (i.e. assembler language and compilers) such as multimedia (also called short vector) processing ability (i.e. Intel MMX, Intel SSE, Intel SSE2, Motorola Altivec, SUN VIS, ...). This was reflected in an extension of the assembly languages (extended instruction set).

But a powerful SIMD (*Single Instruction Multiple Data*) multimedia instruction set is worthless without the mean to utilize it. Today, we can utilize SIMD multimedia instruction set in three ways:

1. assembly language - this is the most effective method but it is also more tedious and error prone than any other methods,
2. shared libraries - these libraries are often available from microprocessor manufacturers, but they tend to only cover particular functions and for some particular class of microprocessors,

M. Daydé et al. (Eds.): VECPAR 2006, LNCS 4395, pp. 236–248, 2007.
© Springer-Verlag Berlin Heidelberg 2007

3. vectorizing compilers - ideally, high level language compiler would be able to automatically identify parallelizable sections of code and generate appropriate SIMD instructions. There have been many proposed methods of automatic SIMD vectorization ([1], [4], [8]) but they have only limited success ([6]).

Programming in high level languages and relying on the compiler to produce the SIMD code is a much easier way to utilize multimedia extensions. But if we want to use them in high-level programming languages such as C, then we have to add these new facilities in some way to the high-level programming languages.

As a consequence of the above we decided to extend the syntax of C and to redefine the existing semantics in such a way that we could use multimedia processing facilities in C. The goal was to provide programmers with the most natural way of using the multimedia processing facilities in the C language. We named this extended C as MMC (MultiMedia C). The MMC was first introduced in the paper [2]. Readers are suggested to refer to this paper for the more extensive description of the language syntax.

This paper is organized as follows: in Section 2 we describe the MMC programming language, in Section 3 we describe the implementation of the MMC compiler, in Section 4 we give real examples from multimedia applications and the performance results.

2   The MMC Language

MMC language is an upward extension of the ANSI C language with multimedia processing facilities. It keeps all the ANSI C syntax plus the syntax rules for vector processing.

2.1   Access to the Array Elements

To access the elements of an array or a vector we can use one of the following expressions:

1. `expression[expr1]` - with this expression we can access the `expr1`-th element of an array object `expression`. Here, the `expr1` is an integral expression and `expression` has a type "array of type".
2. `expression[expr1:expr2, expr3:expr4]` - with this expression we can access the bits `expr4` through `expr3` of the elements `expr2` to `expr1` of an array object `expression`. Here, the `expr1`, `expr2`, `expr3`, `expr4` are integral expressions and `expression` has a type "array of type". The `expr1` denotes the last accessed element, `expr2` denotes the first accessed element, `expr3` denotes the last accessed bit and `expr4` denotes the first accessed bit.
3. `expression[,expr1:expr2]` - with this expression we can access the bits `expr1` through `expr2` of all the elements of an array object `expression`. Here, the `expr1` and `expr2` are integral expressions and `expression` has a type "array of type". The `expr1` denotes the last accessed bit and `expr2` denotes the first accessed bit.
4. `expression[]` - with this expression we can access the whole array object `expression`. Here, the `expression` has a type "array of type".

2.2 Operators

Unary Operators. We extended the semantics of the existing ANSI C unary operators &, *, +, -, ~, ! in the sense that they may now have both scalar- and vector-type operands.

We have also added new reduction unary operators [+], [-], [*], [&], [|], [^]. These operators are overloaded existing binary operators +, -, *, &, |, ^ and are only applicable to the vector operands. These operators perform the given arithmetic/logic operation between the components of the given vector. The result is always a scalar value.

We have also added one new vector operator |/, which calculates the square root of each component in the vector.

Binary Operators. We have extended the semantics of the existing ANSI C binary operators and the assign operators in such a way that they can now have vector operands. Thus, one or both operands can have an array type.

We have overloaded the existing binary operators with 4 new operators:

?	this operator overloads the binary operators in such a way that the given binary operator performs the operation with saturation,
@	this operator overloads the binary add operator in such a way that the given binary operator first performs addition over adjacent vector elements and then averages (shift right one bit) the result,
~	this operator overloads the multiply operator in such a way that the result is the high part of the product,
_	this operator overloads the multiply operator in such a way that the result is the low part of the product.

Besides the existing binary operators we have added one new, binary operator, which we found to be important in multimedia applications. This operator is applicable only on vector operands (if any operand has a scalar type then it is expanded into an appropriate vector strip) and is as follows:

| − |	absolute difference (in the grammar denoted as VEC_SUB_ABS).

Example 1. As saturated arithmetic is widely used in multimedia programs (especially in image processing) and as there should be a mechanism to efficiently deal with multiple possible overflows in packed values, the operations that support saturated arithmetic have been added to microprocessors' instruction set. Since C semantics does not support saturated arithmetic as native operators, programmers are forced to express saturated operations in native C operations.

Figure 1 gives such an example (taken from Berkeley Multimedia workload [7], [6]). The code presented in the Figure 1 could not be efficiently vectorized by an automatic vectorizer. Thus, this portion of parallelism could not be efficiently utilized on multimedia extended processors.

```
/* short a, b; int ltmp; */
#define GSM_ADDS(a, b) \
        ((unsigned)((ltmp=(int)(a)+(int)(b)) - MINWORD) > \
        MAXWORD - MINWORD ? (ltmp>0 ? MAXWORD : MINWORD) : ltmp)

#define MAXWORD 127
#define MINWORD -128
#define N 16

//implementation:

  for(i = 0; i<N; i++ ){
    rez[i] = GSM_ADDS(A[i], B[i]);
  }
```

Fig. 1. C Implementation of saturated add operation from Berkeley Multimedia Workload/GSM

Figure 2 gives the MMC code for the same saturated add operation. The saturation is now easily expressed in native MMC operations.

```
char A[16];
  char B[16];
  char rez[16];

rez[] = A[] ?+ B[];
```

Fig. 2. Implementation of saturated arithmetic in MMC

Conditional Expression. The conditional operator from ANSI C '?:' which is used in the conditional expression can now have array-type operands.

3   Implementation of the MMC Compiler

The laboratory version of the MMC compiler was implemented for Intel Pentium III and Intel Pentium IV processors. It was implemented as a translator to ordinary C code that is then compiled by an ordinary C compiler (in our example with Intel C++ Compiler for Linux [9]).

The MMC compiler parses input MMC code, performs syntax and semantics analysis, builds its internal representation, and finally translates the internal representation into ANSI C with macros written in a particular assembly language instead of the MMC vector statements. The compilation process is presented in Figure 3. After syntax and semantic analysis of the MMC source code the list

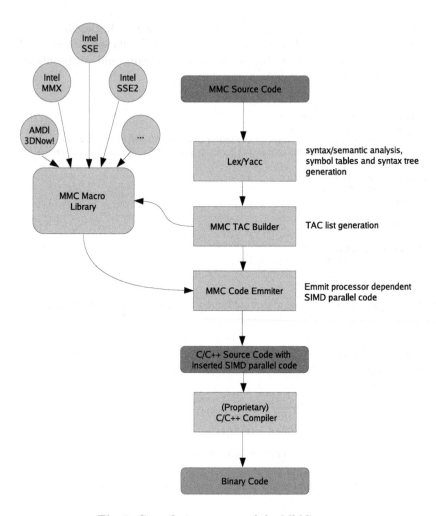

Fig. 3. Compilation process of the MMC source

of tree-address codes (TAC) for SIMD statements is generated. Then, the MMC Code Emitter inserts SIMD macros for each TAC. The appropriate macro is taken from the MMC macro library. Here we will only show the implementation of one macro for conditional assignment:

```
MMC_QUEST_INT;
*( ($T *)($1) ) = _
MMC_QUEST_SSE2_INT ( *( ($T *)($2) ) ,
                     *( ($T *)($3) ) ,
                     *( ($T *)($4) ));
```

```
__m128i _MMC_QUEST_SSE2_INT (__m128i ab,
                             __m128i c,
                             __m128i d )
{
__m128i rez1;
__m128i rez2;
__m128i tmp=_mm_set1_epi32(0);
tmp=_mm_cmpeq_epi32(tmp,ab);

rez1=_mm_and_si128(ab,c);
rez2=_mm_and_si128(tmp,d);

return _mm_or_si128(rez1,rez2);
}
```

The whole macro library, source code of the MMC compiler with Doxygen documentation can be freely downloaded from the MMC web site.

4   Developing Multimedia Kernels

In this section we present the use of MMC language to code some commonly used multimedia kernels. At the end of this section the performance results for the given examples are presented.

Example 2. Finite impulse response (FIR) filters are used in many aspects of present-day technology because filtering is one of the basic tools of information acquisition and manipulation. FIR filters can be expressed by the equation:

$$y(n) = \sum_{k=0}^{N-1} h(k) \cdot x(n-k) \tag{1}$$

where N represents the number of filter coefficients $h(k)$ (or the number of delay elements in the filter cascade), $x(k)$ is the input sample and $y(k)$ is the output sample.

Structurally, FIR filters consist of just two things: a sample delay line and a set of coefficients. To implement the filter one has to:

1. Put the input sample into the delay line.
2. Multiply each sample in the delay line by the corresponding coefficient and accumulate the result.
3. Shift the delay line by one sample to make room for the next input sample.

The MMC implementation of the above algorithm for the FIR filter is as follows:

```
#define FILTER_LENGTH 1024
#define SIGNAL_LENGTH 8192
int j;
double h[FILTER_LENGTH];
double delay_line[FILTER_LENGTH];
double x[SIGNAL_LENGTH];
double y[SIGNAL_LENGTH];

for (j=0; j<SIGNAL_LENGTH; j++) {
  delay_line[0] = x[j];

  //calculate FIR:
  y[j] = [+] ( h[] * delay_line[] );

  //shift delay line:
  delay_line[] = delay_line[] << 1;
}
```

This MMC code is translated by the MMC compiler into C code with inserted macros. So, after strip-mining and macro insertion, which is done by the MMC compiler, we have C code like in the Figure 4. The compiled code can now be further compiled into binary code by the use of C/C++ compiler for desired processor family.

Example 3. An Infinite Impulse Response (IIR) filter produces an output, $y(n)$, that is the weighted sum of the current and the past inputs, $x(n)$, and past outputs. IIR filters can be expressed by the equation:

$$y(n) = \sum_{k=0}^{N-1} h(k) \cdot x(n-k) + \sum_{p=1}^{M-1} h'(p) \cdot y(n-p) \qquad (2)$$

where N represents the number of forward-filter coefficients $h(k)$ (or the number of delay elements in the forward-filter cascade) and M represents number of backward-filter coefficients $h'(k)$ (or the number of delay elements in the backward-filter cascade), $x(k)$ is the input sample and $y(k)$ is the output sample.

To implement the IIR filter one has to:

1. Put the input sample into the input delay line, and the output sample into the output delay line.
2. Multiply each sample in the delay line(s) by the corresponding coefficient and accumulate the result.
3. Shift the delay lines by one sample to make room for the next input or output sample.

The MMC implementation of the above algorithm for the IIR filter is as follows (note that for simplicity in implementation we use the $h'(0)$ coefficient, which is always zero):

```
#include <mmintrin.h>
#include <xmmintrin.h>
#include <emmintrin.h>
#include <stdlib.h>
#include <stdio.h>
#include <time.h>

void main()
{
  int j;
  __declspec(align(16))  float h[1024];
  __declspec(align(16))  float delay_line[1024];
  __declspec(align(16))  float x[8192];
  __declspec(align(16))  float y[8192];

  for(j=0; j<8192; j++)
  {
    delay_line[0] = x[j];
    float __mmc_internal_symbol_2 = 0;
    __declspec(align(16)) float __mmc_internal_symbol_1[1024] = {0};

    int __mmc_internal_symbol_3;
    for(__mmc_internal_symbol_3=0; __mmc_internal_symbol_3<1024; __mmc_internal_symbol_3+=4)
    {
      *((__m128 *)(__mmc_internal_symbol_1 + __mmc_internal_symbol_3 + 0)) = _mm_mul_ps( *((__m128 *)(h +
__mmc_internal_symbol_3 + 0)), *((__m128 *)(delay_line + __mmc_internal_symbol_3 + 0)) );
      __mmc_internal_symbol_2 += __mmc_internal_symbol_1[__mmc_internal_symbol_3 + 0];
      __mmc_internal_symbol_2 += __mmc_internal_symbol_1[__mmc_internal_symbol_3 + 1];
      __mmc_internal_symbol_2 += __mmc_internal_symbol_1[__mmc_internal_symbol_3 + 2];
      __mmc_internal_symbol_2 += __mmc_internal_symbol_1[__mmc_internal_symbol_3 + 3];
      y[j] = __mmc_internal_symbol_2;
    }

    int __mmc_internal_symbol_6;
    for(__mmc_internal_symbol_6=22; __mmc_internal_symbol_6>=0; __mmc_internal_symbol_6--)
    {
      delay_line[__mmc_internal_symbol_6 + 1] = delay_line[__mmc_internal_symbol_6 + 0];
    }

  }
}
```

Fig. 4. Compiled MMC source of the FIR filter

```
int j;
float hf[FILTER_LENGTH_F];
float hb[FILTER_LENGTH_B];
float in_delay[FILTER_LENGTH];
float out_delay[FILTER_LENGTH];
float x[SIGNAL_LENGTH];
float y[SIGNAL_LENGTH];

for (j=0; j<SIGNAL_LENGTH; j++) {
  in_delay[0] = x[j];

  //calculate FIR:
  y[j] = [+] ( hf[] * in_delay[] );

  out_delay[0] = y[j];

  //calculate IIR:
  y[j] += ([+]( hb[] * out_delay[]))
```

```
//shift delay lines:
in_delay[] = in_delay[] << 1;
out_delay[] = out_delay[] << 1;
}
```

Example 4. The MPEG audio standard uses Discrete Cosine Transformation (DCT) to transform samples from one domain into another. DCT is defined as a linear transformation of N input samples, $s[k]$, and N DCT samples , $x[i]$ where $k = 0 \ldots K - 1$ and $i = 0 \ldots K - 1$ (see Equation 3).

$$x(i) = \sum_{k=0}^{N-1} s(k) \cdot \cos \frac{(2k+1) \cdot i \cdot \pi}{2N} \tag{3}$$

The DCT formula can also be expressed in matrix form as:

$$\boldsymbol{x} = \mathbf{D} \cdot \boldsymbol{s} \tag{4}$$

where $\boldsymbol{x}$ is the vector of N DCT samples and $\boldsymbol{s}$ is the vector of N input samples. $\mathbf{D}$ is an N by N matrix with the elements presented in Equation 5.

$$D_{i,j} = \cos \frac{(2j+1) \cdot i \cdot \pi}{2N} \tag{5}$$

The matrix representation is used for practical implementation. The matrix representation of the DCT algorithm is well suited for MMC code implementation since the regular structure of matrix multiplication fits the SIMD nature. The MMC implementation of the DCT algorithm is as follows:

```
int j;
float D[N*N];
float v[N];
float s[N];
float D_row[N];

for (j=0; j<N; j++) {
  D_row[] = D[j*N : j*N+(N-1)];

  //calculate j-th DCT sample:
  v[j] = [+] ( D_row[] * s[] );
}
```

This MMC code is translated by the MMC compiler into C code with inserted macros. So, after strip-mining and macro insertion, which is done by the MMC compiler, we have C code like in the Figure 5. The compiled code can now be further compiled into binary code by the use of C/C++ compiler for desired processor family.

Example 5. This example demonstrates how to implement saturated operations in MMC. Saturated addition of two vectors (i.e. bitmaps) can be expressed in MMC as: :

```
#include <mmintrin.h>
#include <xmmintrin.h>
#include <emmintrin.h>
#include "MMC_SSE2_MACROS.H"

#include <stdlib.h>
#include <stdio.h>
#include <time.h>

void main()
{
  int j;
  __declspec(align(16)) float D[16900];     /* D matrix */
  __declspec(align(16)) float v[130];       /* DCT samples vector */
  __declspec(align(16)) float s[130];       /* output samples vector */
  __declspec(align(16)) float D_row[130];   /* D matrix row */

  for (j=0; j<130; j++) {
    /* D_row[] = D[j* : j*+(N-1)];   */ /* store matrix row into a vector */


    /* calculate j-th DCT sample: */
    float __mmc_internal_symbol_2 = 0;
    __declspec(align(16)) float __mmc_internal_symbol_1[130] = {0};

    int __mmc_internal_symbol_3;
    for(__mmc_internal_symbol_3=0; __mmc_internal_symbol_3<130; __mmc_internal_symbol_3+=4)
    {
      *( (__m128 *)(__mmc_internal_symbol_1 + __mmc_internal_symbol_3 + 0) ) =
             __mm_mul_ps ( *( (__m128 *)(D_row + __mmc_internal_symbol_3 + 0) ) ,
                           *( (__m128 *)(s + __mmc_internal_symbol_3 + 0) ) );
      __mmc_internal_symbol_2 += __mmc_internal_symbol_1[__mmc_internal_symbol_3 + 0];
      __mmc_internal_symbol_2 += __mmc_internal_symbol_1[__mmc_internal_symbol_3 + 1];
      __mmc_internal_symbol_2 += __mmc_internal_symbol_1[__mmc_internal_symbol_3 + 2];
      __mmc_internal_symbol_2 += __mmc_internal_symbol_1[__mmc_internal_symbol_3 + 3];
      v[j] = __mmc_internal_symbol_2;
    }

    for(__mmc_internal_symbol_3=128; __mmc_internal_symbol_3<130; __mmc_internal_symbol_3++)
    {
      __mmc_internal_symbol_1[__mmc_internal_symbol_3] =
                   D_row[__mmc_internal_symbol_3] * s[__mmc_internal_symbol_3];
      __mmc_internal_symbol_2 += __mmc_internal_symbol_1[__mmc_internal_symbol_3];
      v[j] = __mmc_internal_symbol_2;
    }
  }
}
```

Fig. 5. Compiled MMC source of DCT

```
char bits1[SIZE];
char bits2[SIZE];
char bitsDest[SIZE];


...


bitsDest[] = bits1[] ?+ bits2[];


...
```

This MMC code is translated by the MMC compiler into C code with inserted macros. The compiled code is:

```
int __mmc_internal_symbol_2;
for(__mmc_internal_symbol_2=0;
```

```
    __mmc_internal_symbol_2<SIZE;
    __mmc_internal_symbol_2+=8)
{
    *((__m64 *)(destBits + __mmc_internal_symbol_2 + 0)) =
        _mm_adds_pi8( *( (__m64 *)(bits1 + __mmc_internal_symbol_2 + 0) ),
                      *( (__m64 *)(bits2 + __mmc_internal_symbol_2 + 0) ) );
}
```

Example 6. This example demonstrates how to implement averaging operations in MMC. We can describe an average operation as:

```
A[]@+[] = ((A[0]+[0])>>1)|((A[0]+[0])&1),
```

$$\ldots$$

$$((A[N-1]+[N-1])>>1)|((A[N-1]+[N-1])\&1)$$

This is an idiomatic expression that is ussually hard to detect by compilers. The MMC implementation of an average operation is straightforward:

```
#include <stdlib.h>
#include <stdio.h>
#include <time.h>

void main()
{
  char x[1024];
  char y[1024];
  char z[1024];

  z[] = x[] @+ y[];
}
```

This MMC code is translated by the MMC compiler into C code with inserted macros. So, after strip-mining and macro insertion, which is done by the MMC compiler, we have C code like in the Figure 6. The compiled code can now be further compiled into binary code by the use of C/C++ compiler for desired processor family.

In Figure 7 we can see the performance improvement of some typical multimedia cores when using the MMC language instead of C. We have implemented these cores in MMC and C. Then, the MMC code was compiled into C code with the MMC compiler and into binary code with Intel C/C++ compiler. Sequential C sources are compiled into binaries with Intel C/C++ compiler with vectorization switched on.

In cases where vectorization was successful we reached slightly better performance with the Intel vectorizing compiler. This is because the Intel vectorizing compiler performs some additional optimizatins on the vectorized loop [1], [9]. But in cases where vectorization failed (SATURATION, AVERAGE, SUM OF ABS. DIFF.), better performance is reached with MMC. This is because in

```
#include <mmintrin.h>
#include <xmmintrin.h>
#include <emmintrin.h>
#include "MMC_SSE2_MACROS.H"

#include <stdlib.h>
#include <stdio.h>
#include <time.h>

void main()
{
    __declspec(align(16))  char x[1024];
    __declspec(align(16))  char y[1024];
    __declspec(align(16))  char z[1024];
    __declspec(align(16)) char __mmc_internal_symbol_1[1024] = {0};

    int __mmc_internal_symbol_2;
    for(__mmc_internal_symbol_2=0; __mmc_internal_symbol_2<1024; __mmc_internal_symbol_2+=16)
    {
        *( (__m128i *)(__mmc_internal_symbol_1 + __mmc_internal_symbol_2 + 0) ) =
            _mm_avg_epu8 ( *( (__m128i *)(x + __mmc_internal_symbol_2 + 0) ) ,
                           *( (__m128i *)(y + __mmc_internal_symbol_2 + 0) ) );
        *((__m128i *)(z + __mmc_internal_symbol_2 + 0)) =
            *((__m128i *)(__mmc_internal_symbol_1 + __mmc_internal_symbol_2 + 0));
    }
}
```

Fig. 6. Compiled MMC source of an averaging operation

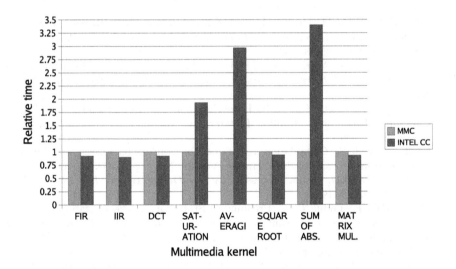

Fig. 7. Speedup on an Intel Pentium IV using MMC

these three cases idiomatic expressions where used, which were very difficult for a compiler to identify.

5   Conclusion

We have developed a MMC programming language which is able to use hardware-level multimedia execution capabilities. The MMC language is an upward extension of ANSI C and it saves all the ANSI C syntax. In this way it is suitable for use by programmers who want to extract SIMD parallelism in a high-level programming language and also by programmers who do not know anything about multimedia processing facilities and who are using the C language.

We have shown the ease with which it is possible to express some common multimedia kernels with MMC. With MMC we can express these kernels in a more straightforward or 'natural' way. The presented extension to C also preserves the interchangeability of arrays and pointers and adds as few as possible new operators. All added operators have an analogue in ordinary C. The declarations of arrays are left unchanged and also no new types have been added.

We obtained good performance for several application domains. Experiments on representative scientific and multimedia applications have significant performance improvements. We are currently rewriting Berkely Multimedia Workload with the MMC language. In such a way we will be able to fully evaluate the performance improvement of widely used multimedia applications. We will also be able to evaluate how difficult is for people to use the MMC language.

References

1. Bik A.J.C., Girkar M., Grey P.M., Tian X.M. Automatic Intra-Register Vectorization for the Intel (R) Architecture. *International Journal of Parallel Programming.* Vol 30., No. 2, pp. 65-98. 2002.
2. Bulić P., Guštin V. An Extended ANSI C for Processors with a Multimedia Extension. *International Journal of Parallel Programming.* Vol 31., No. 2, pp. 107-136. 2003.
3. Ferretti M., Rizzo D. Multimedia Extensions and Sub-Word Parallelism in Image Processing: Preliminary Results. *Lecture Notes in Computer Science*, No. 1685, pp. 977-986, 1999.
4. Krall A., Lelait S. Compilation Techniques for Multimedia Processors. *International Journal of Parallel Programming.* Vol. 28, No. 4, pp. 347-361, 2000.
5. Lee R., Smith M.D. Media Processing: A New Design Target. *IEEE Micro*, Vol. 16, No. 4, pp. 6-9, 1996.
6. Ren G., Wu P., Padua D. A Preliminary Study On the Vectorization of Multimedia Applications for Multimedia Systems. *Proceedings of the 16th International Workshop on Languages and Compilers for Parallel Computers, October 2-4, 2003, College Station, Texas.* pp. 2-16, 1987.
7. Slingerland N.T., Smith A.J.,. Multimedia extensions for General Purpose Microprocessors: a Survey *Microprocessors and Microsystems*, Vol. 29, pp. 225-246, 2005.
8. Sreraman N., Govindarajan R. A Vectorizing Compiler for Multimedia Extensions. *International Journal of Parallel Programming*, Vol. 28, No. 4, pp. 363-400, 2000.
9. Intel C++ Compiler for Linux 9.0. *http://www.intel.com/software/products/compilers.*
10. MMX Technology Application Notes: Using MMX Instructions to Convert RGB To YUV Color Conversion. *http://cedar.intel.com.*
11. DSP Guru: Finite Impulse Response FAQ *http://www.dspguru.com/info/faqs/firfaq.htm.*
12. DSP Guru: Infinite Impulse Response FAQ *http://www.dspguru.com/info/faqs/iirfaq.htm.*

A Versatile Pipelined Hardware Implementation for Encryption and Decryption Using Advanced Encryption Standard

Nadia Nedjah[1] and Luiza de Macedo Mourelle[2]

[1] Department of Electronics Engineering and Telecommunications,
Faculty of Engineering, State University of Rio de Janeiro, Brazil
nadia@eng.uerj.br
[2] Department of Systems Engineering and Computation,
Faculty of Engineering, State University of Rio de Janeiro, Brazil
ldmm@eng.uerj.br

Abstract. The Advanced Encryption System – AES is now used in almost all network-based applications to ensure security. In this paper, we propose a very efficient pipelined hardware implementation of AES-128. The design is versatile as it allows both encryption and decryption. The core computation of AES, which is performed on data blocks of 128 bits, is iterated for several rounds, depending on the key size. The security strength of AES has been proven proportional to the number of rounds applied. we show that if the required number of rounds must increase to defeat attackers, the proposed implementation stays efficient.

1   Introduction

Cryptographic algorithms used by nowadays cryptosystems fall into two main categories: symmetric key and asymmetric-key algorithms [8]. Symmetric-key ciphers use the same key for encryption and decryption, or to be more precise, the key used for decryption is computationally easy to compute given the key used for encryption. In turn, symmetric-key ciphers, fall into two categories: block ciphers and stream ciphers. Stream ciphers encrypt the plaintext one bit at a time, in contrast to block ciphers, which operate on a block of bits of a predefined length. Most popular block ciphers are DES, IDEA [7] and AES, and most popular stream cipher is RC6 [9].

The Advanced Encryption System – AES is a block cipher, adopted as the new encryption standard in substitution to its predecessor Data Encryption Standard – DES [2]. AES main scrambling computation is performed on a fixed block size of 128 bits with a key size of 128, 192 or 256 bits. This core computation is iterated for many rounds. The number of rounds depends on the key size. Currently, it is set to 10, 12 and 14 for the cited keys sizes respectively. The resistance of AES against breaking attacks depends entirely on the number of rounds used. So far, the best known attacks are on 7 rounds for 128-bit keys, 8 rounds for

M. Daydé et al. (Eds.): VECPAR 2006, LNCS 4395, pp. 249–259, 2007.
© Springer-Verlag Berlin Heidelberg 2007

192-bit keys, and 9 rounds for 256-bit keys [5]. The small margin between these round numbers and the actual ones is very worrying for the cryptographer's community.

In this paper, we propose a novel hardware implementation of AES-128. The architecture allows one to perform the core computation of the algorithm is a pipelined manner. The throughput of the cryptographic hardware is 1Gbits per second. A unique hardware is used for encryption and decryption. The pipelined encryption and decryption allows an increase of the number of rounds without much loss of efficiency. Recall that increasing the number of rounds applied, increases the resistance of the AES algorithm.

This rest of this paper is organised in 4 subsequent sections. First, in Section 2, we give a brief description of the AES encryption and decryption algorithms as well as the modified version of these two algorithms, which are the basis of the proposed hardware architecture. Thereafter, in Section 3, we describe in a structured manner, the pipelined hardware architecture of AES-128 for encryption and decryption. Subsequently, in Section 4, we present some experimental result and compare our implementation to existing ones. Last but not least, in Section 5, we draw some conclusions and introduce some directions for future work.

2   Advanced Encryption Standard

AES is an elegant and a so-far-secure cipher. Encryption using AES proceeds as described in Algorithm 1, wherein functions *SubBytes*, *ShiftRows*, *MixColumns* and *AddroundKey* are defined as follows:

- Function *SubBytes* yields a new state simply by substituting each of the 16 bytes of *state* using a substitution box. The four most significant bits of the byte in question is used as the S-box row index while the remaining four bits are used as the S-box column index.
- Function *ShiftRows* obtains a new state by cyclically shifting the state rows. The bytes of row i are shifted i times, where $0 \leq i \leq 4$.
- Function *MixColumns* operates on the states columns. The bytes of a given column are used as coefficients of a polynomial over $GF(2^8)$. The formed polynomial is multiplied by a fixed polynomial $P(x)$ *modulo* $x^4 + 1$, wherein $P(x) = \{03\}x^3 + \{01\}x^2 + \{01\}x + \{02\}$. The details of the multiplication operation can be found in [3], [1].
- Function *AddRoundKey* computes the new state using a XOR of the columns bytes and the key schedule of the current round.

Before the cipher operation takes place, a key schedule is generated. Four subkeys are required for each round of the cipher algorithm. The subkeys for the first round are the private cipher key. For a given round, the first subkey is obtained by first rotating once the last subkey form the previous round, then substituting each of byte using the S-box used by function *subBytes*, thereafter XORing the result with a given constant and finally XORing the result with first subkey of the previous round. The subsequent subkeys of the current round are computed using a XOR of the previous key in the current round and the one inversely respective from the previous round.

Algorithm 1. AES-Cipher
input: Byte $T[4 \times nb]$, Word $K[nb \times (nr + 1)]$;
output: Byte $C[4 \times nb]$,
   Byte $state[4, nb]$;
   $state := T$;
   AddRoundKey($state$, $K[0, nb - 1]$;
   for $round := 1$ to $nr - 1$ do
      SubBytes($state$);
      ShiftRows($state$);
      MixColumns($state$);
      AddRoundKey($state$, $K[round \times nb, nb(round + 1) - 1]$);
   SubBytes($state$);
   ShiftRows($state$);
   AddRoundKey($state$, $K[nr \times nb, nb(nr + 1) - 1]$);
   $C := state$;
   return C;
end

For hardware efficiency reasons, we modified the AES cipher algorithm as in Algorithm 2. Note that Algorithm 1 and Algorithm 2 are equivalent and yield the same output.

Algorithm 2. Modified-AES-Cipher
input: Byte $C[4 \times nb]$, Word $K[nb \times (nr + 1)]$;
output: Byte $T[4 \times nb]$,
   Byte $state[4, nb]$;
   $state := C$;
   for $round := 0$ to $nr - 1$ do
      AddRoundKey($state$, $K[round \times nb, nb(round + 1) - 1]$);
      SubBytes($state$);
      ShiftRows($state$);
      if $round < nr - 1$ then MixColumns($state$);
   AddRoundKey($state$, $K[nr \times nb, nb(nr + 1) - 1]$);
   $T := state$;
   return T;
end

The decryption of a text that was ciphered using AES can be performed by Algorithm 3. Comparing Algorithm 1 and Algorithm 3, one can note that each function was replaced by its inverse. However, the application sequence of these functions is slightly different. In order to have a unique versatile hardware for encryption and decryption, this algorithm was modified as in Algorithm 4.

Algorithm 3. AES-Decipher
input: Byte $C[4 \times nb]$, Word $K[nb \times (nr + 1)]$;
output: Byte $T[4 \times nb]$,
  Byte $state[4, nb]$;
  $state := C$;
  AddRoundKey($state$, $K[round \times nb, nb(nr + 1) - 1]$);
  for $round := nr - 1$ downto 1 do
    InvShiftRows($state$); InvSubBytes($state$);
    AddRoundKey($state$, $K[nr \times nb, nb(nr + 1) - 1]$);
    InvMixColumns($state$);
  InvShiftRows($state$);
  InvSubBytes($state$);
  AddRoundKey($state$, $K[0, nb(nr + 1) - 1]$);
  $T := state$;
  return T;
end

Algorithm 3 and Algorithm 4 are equivalent as operations *InvSubBytes* and *InvShiftRows* commute. Moreover, function *InvMixColumns* is linear so we have expression *InvMixColumns*(x XOR y) is equivalent to *InvMixColumns*(x) XOR *InvMixColumns*(y). Recall that operation *AddRoundKey* is a XOR of its arguments. Using these two facts, we can swap operations *AddRoundKey* and *InvMixColumns*, provided that the columns of the decryption key schedule are modified using operation *InvMixColumns*. Note that functions *SubBytes* and *InvSusbytes* perm the same process but using distinct S-Boxes.

Algorithm 4. Modified-AES-Decipher
input: Byte $C[4 \times nb]$, Word $K[nb \times (nr + 1)]$;
output: Byte $T[4 \times nb]$,
  Byte $state[4, nb]$;
  $state := C$;
  for $round := nr - 1$ to 0 do
    AddRoundKey($state$, $K[round \times nb, nb(round + 1) - 1]$);
    InvSubBytes($state$); InvShiftRows($state$);
    if $round < nr - 1$ then InvMixColumns($state$);
  AddRoundKey($state$, $K[nr \times nb, nb(nr + 1) - 1]$);
  $T := state$;
  return T;
end

3   Pipelined Hardware Implementation of AES

The overall architecture of the AES hardware mirrors the structure of Algorithm 2 and Algorithm 4. It is a synchronous implementation of both the processes of cipher and decipher. It uses four 128-registers. Every clock transition,

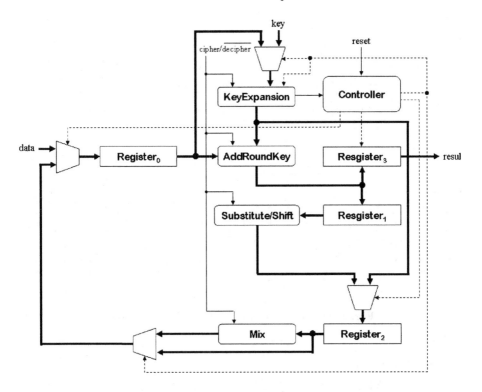

Fig. 1. Overall hardware architecture for the AES cipher/decipher

these registers are loaded, except $Register_3$, which is loaded when an input state is completely ciphered. In the encryption/decryption process, $Register_0$ is loaded with the input data or the partially encrypted/decrypted plaintext/ciphertext; $Register_1$ with the result of the *AddRoundKey* component; $Register_2$ with the state after applying functions *SubBytes* (using the appropriate S-Box) and subsequently *ShiftRows/InvShiftRows*. The block architecture of the AES cipher and decipher hardware is shown in Fig. 1.

The component that implements function *AddRoundKey* is simply a net of XOR gates that adds in $GF(2^8)$ the key schedule to the current state. The component implementing function *SubBytes* uses 16 S-boxes (8 for ciphering and 8 for deciphering) stored in a Read-Only Memory (ROM). The obtained state is row-shifted before its storage in $Register_2$. The component architecture is given in Fig. 2.

Function *MixColumns* is implemented by a massively parallel component that computes all the bytes of the new state in a single clock. It uses four components of the same architecture. This basic component produces one column os the new state. Its architecture is described in Fig. 3, wherein component *mult* yields the a special product of a given byte from the state times {01}, {02}, {03}, {09}, {0B}, {0D} or {0E} (see [3], [1] for details on the operation). The architecture of component *mult* is presented in Fig. 4. Component *xtime* computes the *xtime* operation as defined in [3] and its architecture is given in Fig. 5.

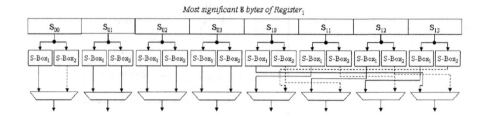

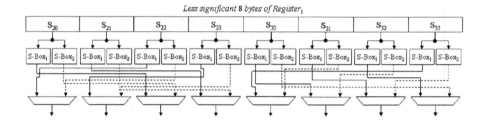

Fig. 2. The structure of *Substitute/Shif* component

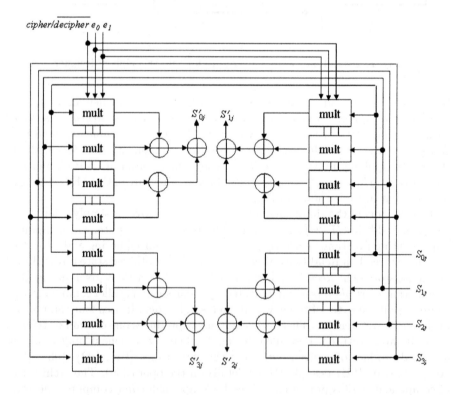

Fig. 3. Basic component in *Mix* component

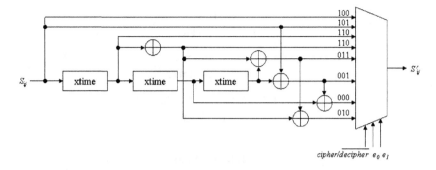

Fig. 4. Architecture of the *mult* component

For component synchronisation purposes, the architecture includes a controller. Among other actions, the controller determines when to reset the cipher hardware, accept input data, to register output results. As the excution of function *MixColumn/InvMixColumn* is conditional (see Algorithm 2), the controller decides when the result obtained by associated component can be used or must be ignored. Recall the hardware allows both encryption and decryption. When data is being deciphered, the key schedule generated by component *KeyExpansion* must be ordered differently [3]. The AES hardware of Fig. 1 takes advantage of component *MixColumn* to schedule the subkeys in the required order. The controller also controls this operation.

The controller is structured as in Fig. 6. The included combinational logic permits the conversion of the 5-bit count to a single bit that triggers state transition. The sate machine includes six states. As long as control signal *keyExpand* is set, the current state is kept unchanged in S_0. As soon as this signal is reset by the *keyExpansion* component, which means that the step of key schedule generation is complete, the machine transits to state S_1, wherein it stays for 3 clock cycles, which is the required time to complete the processing of one 128-bit state. Also, during this period of time, the data input signal is active, which

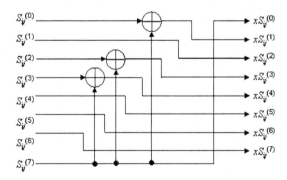

Fig. 5. Architecture of the *xtime* component

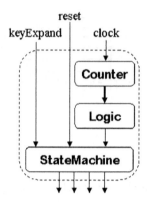

Fig. 6. Controller architecture

allows the hardware to accept the three states that will be ciphered/deciphered in pipelined manner. Synchronously with the fourth clock transition, the machine transits to state S_2 allowing to deactivate the data input signal and wait for the three accepted states are almost processed as only the last *AddRound-Key* is yet to be performed to complete the encryption/decryption process. At the 30th. clock transition, the machine state changes to S_3 to activate output result signal, which is maintained for the two subsequent clock periods. A the 33rd. clock transition, the encryption/decryption of the three accepted states is completed and therefore, the control is returned to state S_1, where in data input signal is reactivated to allow more date to be entered and processed. The state machine transition diagram is shown in Fig. 7.

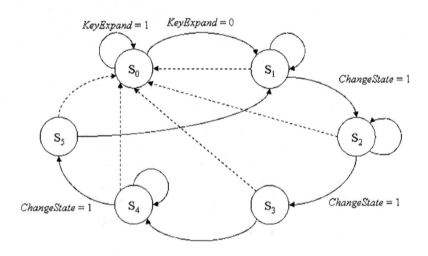

Fig. 7. State machine transition diagram

4   Experimental Results

The pipelined execution of the AES cipher using the architecture of Fig. 1 is illustrated in Fig. 8. We implemented the hardware described throughout this paper using reconfigurable hardware. The FPGA family used is VIRTEX-II. Component *KeyExpansion* introduces a delay of 78.3ns. The clock cycle is 10.44ns. Every 33 clock cycles, the hardware can yield an encrypted datastream of 3×128 bits. The throughput, say *tp* can then be calculated as in (1). The throughput is a little more than 1Gbps.

$$Tp = \frac{3 \times 128}{33 \times clockcycle} = \frac{128}{11 \times 10.44} = 1062.9 Mbs \qquad (1)$$

As far as the authors know, the versatile hardware implementation of AES algorithm that performs both encryption and decryption is novel. We compared our implementation to the ones from [6] and [10]. Note that these implementations are for the cipher algorithm only while our implementation ciphers and deciphers. One may think that the implementation proposed and those from [6] and [10] are incomparable. They are cited here for reference only. The throughput, expressed in Mbps, as well as the hardware area required, expressed in number of CLBs, are given in Table 1.

Table 1. Performance comparison

Implementation	Throughput	Area	CLB/Mbs
Our's: cipher& decipher	1063	9937	9.35
[6]: cipher only	1911	8767	4.59
[10]: cipher only	1450	542	0.37

Recall that the resistance of AES-based encryption against cryptanalysis attacks depends entirely on the number of rounds used. The pipelined implementation we propose throughout this paper can be easily adapted to a higher round number. The chart of Fig. 9 shows that this can be done without much loss in efficiency and with much gain of security strength. To be able to increase the number of round, component *KeyExpansion* needs to generate more key schedules and therefore the delay introduced by it increases with the number of rounds. The throughput, say *tp*, can be expressed in terms of the round number, say *rn*, is as in (2). The security strength, say *st* is proportional to the number of rounds applied. So, considering the security strength provided by applying 10 rounds as a reference, *st* would be defined as in (3).

$$Tp(rn) = \frac{128}{(rn+1) \times clockcycle} \qquad (2)$$

$$St(rn) = \frac{rn}{10} \qquad (3)$$

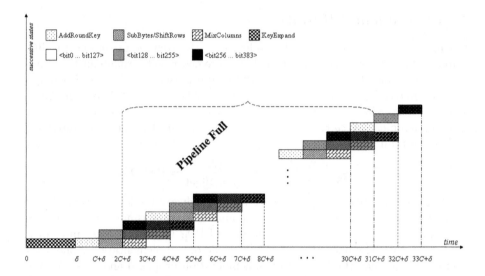

Fig. 8. Pipelined execution of the AES algorithm using the hardware of Fig. 1

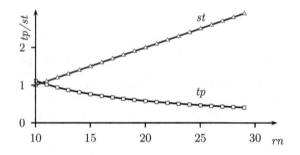

Fig. 9. The impact of increase in the round number

5   Conclusion

In this paper, we propose a novel pipelined hardware implementation of AES-128 that can be used for both encryption and decryption. Besides, we show that if the required number of rounds must increase to defeat attackers, the proposed implementation stays efficient. The hardware proposed is massively parallel and executes the four main steps of the algorithm in a pipelined manner, which allows a reasonable throughput fo a little more of 1Gbs. Compared to existing implementations of the cipher algorithm, this kind of throughput may be considered somehow low. However, considering the 2-in-1 aspect of the hardware as it allows encryption and decryption, it comes handy for devices with restricted hardware area with a not too bad throughput of 1Gbs.

In future research work, we intend to investigate further the proposed implementation, with the hope to improve the throughput without much increase in required hardware area.

References

1. J. Daemen and V. Rijmen, *The Design of Rijndael: AES – The Advanced Encryption Standard*, Springer-Verlag, 2002.
2. National Institute of Standard and Technology, *Data Encryption Standard*, Federal Information Processing Standards 46, November 1977.
3. National Institute of Standard and Technology, *Advanced Encryption Standard*, Federal Information Processing Standards 197, November 2001.
4. Nicolas Courtois, Josef Pieprzyk, *Cryptanalysis of Block Ciphers with Overdefined Systems of Equations*, Proceedings of ASIACRYPT 2002, pp 267-287, 2002.
5. N. Ferguson, J. Kelsey, S. Lucks, B. Schneier, M. Stay, D. Wagner and D. Whiting, *Improved Cryptanalysis of Rijndael*, Proceedings of FSE 2000, pp. 213-230, 2000.
6. A. Labbe, A. Perez, *AES Implementation on FPGA: Time and Flexibility Tradeoff*, in Proceedings of FPL, pp. 836-844, 2002.
7. X. Lai, J. L. Massey, *A Proposal for a New Block Encryption Standard*, EUROCRYPT'90, pp. 389–404, 1990.
8. A.J. Menezes, S.A. Vanstone and P.J. Van Oorschot, *Handbook of Applied Cryprography*, CRC Press, USA, 1997.
9. R. Rivest, M. Robshaw, R. Sidney, and Y.L. Yin. *The RC6 block cipher*, First AES Candidate Conference, 1998.
10. F. Standaert, G. Rouvroy, J. Quisquater, J. Legat, *A Methodology to Implement Block Ciphers in Reconfigurable Hardware and its Application to Fast and Compact AES RIJNDAEL*, in Proceedings of FPGA, 2003.

Combinatorial Scientific Computing: The Enabling Power of Discrete Algorithms in Computational Science

Bruce Hendrickson[1] and Alex Pothen[2]

[1] Discrete Math & Algorithms Dept., Sandia National Labs, Albuquerque, NM, USA
bah@sandia.gov
http://www.cs.sandia.gov/~bahendr
[2] Computer Science Department and Center for Computational Science,
Old Dominion University, Norfolk, VA, USA
pothen@cs.odu.edu
http://www.cs.odu.edu/~pothen

Abstract. Combinatorial algorithms have long played a crucial, albeit under-recognized role in scientific computing. This impact ranges well beyond the familiar applications of graph algorithms in sparse matrices to include mesh generation, optimization, computational biology and chemistry, data analysis and parallelization. Trends in science and in computing suggest strongly that the importance of discrete algorithms in computational science will continue to grow. This paper reviews some of these many past successes and highlights emerging areas of promise and opportunity.

1 Introduction

Combinatorial scientific computing (CSC) is a new name for research in an inter-disciplinary field that spans scientific computing and algorithmic computer science. Research in CSC comprises three key components. The first component involves identifying a problem in scientific computing and building an appropriate combinatorial model of the problem, in order to make the computation feasible or efficient. Developing the right combinatorial model is often critical to the computation of an efficient solution, and this step could be the most time-consuming of all. The second component involves the design, analysis, and implementation of algorithms to solve the combinatorial subproblem. The emphasis in this step is on practical algorithms that are efficient for large-scale problems; an algorithm with a time complexity quadratic in the input size could be too slow to be useful, if the worst-case behavior is realized. The algorithm could compute an exact, approximate, or heuristic solution to the problem, and it should run quickly within the context of the other computational steps in the scientific computation. The third component involves developing software, evaluating its performance on a collection of test problems, making it publicly available, and perhaps integrating with a larger software library. These

M. Daydé et al. (Eds.): VECPAR 2006, LNCS 4395, pp. 260–280, 2007.
© Springer-Verlag Berlin Heidelberg 2007

three components are illustrated in several examples of research activity in CSC included in this paper. Work in CSC is multi-disciplinary in its orientation, and has the twin emphases of theoretical rigor and practical impact.

While work in CSC has been ongoing for more than three decades, the myriad roles of combinatorial algorithms are scattered in standard taxonomies of scientific computing. This historical fragmentation has obscured the broad impact of combinatorial algorithms in scientific computing. Algorithmic researchers in one niche are often unaware of ongoing work in another, perhaps related niche. Yet, a developer of computational geometry algorithms for mesh generation is likely to have more in common esthetically and intellectually with a developer of sparse matrix algorithms than with a user of the meshes he or she develops. The CSC community was founded to address this fragmentation and to facilitate closer interactions among researchers in this field.

The purpose of this article is to briefly highlight the role that combinatorial algorithms have played in various fields of scientific computing, and to point to emerging opportunities for the future. The topics discussed include the role of CSC in parallel computing; differential equations, sparse linear algebra, and numerical optimization; statistical physics, computational chemistry, bioinformatics, and information sciences.

We are aware of two articles with somewhat similar goals as ours in the related fields of scientific computing and theoretical computer science. The central role of algorithms in numerical analysis (we would call it scientific computing) has been surveyed recently by Trefethen [63]. A report on challenges for theoretical computer science, as emerging from an NSF-funded workshop circa 2000, was drafted by Johnson [34].

We view this document as a work in progress, and invite suggestions and feedback from other researchers both within and outside the CSC community.

2   Parallel Computing

In recent years, parallel computing has become central to scientific and engineering simulations. The efficient parallelization of scientific computations requires the solution to a variety of combinatorial problems. Perhaps best known is the need to partition the data (and attendant work) of a problem amongst the processors of a parallel machine. For the past decade, this problem has been commonly cast in terms of graph partitioning. Vertices of the graph represent units of computation, and the edges describe data dependencies. The goal of graph partitioning is to divide the vertices into sets of approximately equal cardinality (or weight) while cutting as few edges as possible [31]. Several widely used serial and parallel graph partitioning tools have been developed for this purpose [32,36,37].

Unfortunately, the number of graph edges cut by a partitioning is only an approximation to the actual communication volume in a parallel calculation. So minimizing the number of cut edges doesn't actually minimize communication. More recently, alternative *hypergraph* partitioning models has been devised in which the number of cut *hyperedges* exactly corresponds to communication volume [8]. A hypergraph is a generalization of a graph in which a *hyper*edge can connect of two *or more* vertices. As above, vertices represent computation. But now a hyperedge joins all the vertices that consume a value with the vertex that produces it. Work is divided amongst processors by partitioning the vertices in such a way that a minimum number of hyperedges is cut. Based upon this insight, serial and parallel hypergraph partitioning tools have been developed to facilitate parallel computations [9,17].

Graph coloring is another important kernel for parallelizing some scientific operations. A *coloring* is an assignment of a color to each vertex in such a way that adjacent vertices have different colors. The goal is to label all the vertices while using only a small number of colors. Coloring is a useful tool in parallelizing applications in which an operation on a vertex has side effects on adjacent vertices. In this case, an operation cannot be simultaneously performed on neighboring vertices. The coloring identifies sets of vertices (those with the same color) that can be operated on at the same time. These operations can all be performed in parallel. Thus, a small number of colors facilitates a fast parallel calculation. As one example of this idea, Jones and Plassmann use coloring to identify elements that can be simultaneously refined in an adaptive meshing application [35].

Parallel computing also requires efficient algorithms for interprocessor communication. This challenge leads to a set of problems that can be addressed with discrete algorithms. Consider the common situation in which each processor needs to send information to a few other processors. This is a recurring kernel in many scientific problems. The network of wires that carries messages in a parallel computer is generally sparse, but regular. If the logical communication pattern in the application can be embedded well into the physical network, then communication will be efficient. A good embedding is one in which no physical link is expected to transmit a disproportionate amount of data. Once a computation is broken into P pieces, there is freedom in mapping the pieces to P physical processors. This freedom can be exploited to ensure that the communication patterns in the application map well onto the physical network. This problem can be described in terms of graph embeddings. Given two graphs G and H, an *embedding* is an assignment of vertices from G to vertices of H, with a corresponding assignment of edges of G to paths in H. In the parallel computing setting, the vertices of G are the work partitions and we want to assign them to processors (vertices of H). But we want to do this is a way such a way that communication operations (edges of G) don't overwhelm the network interconnect (edges of H). Graph embedding techniques provide a way to address this problem.

Combinatorial algorithms arise in a wide assortment of other parallel computing scenarios. Alternative load balancing models use space-filling curves or network flow algorithms. Graph matching is used to reduce data remapping costs in dynamic load balancing. A new linear time approximation algorithm for maximum weighted matching with approximation ratio 1/2 was developed by Preis [52] in the context of partitioning graphs for parallel computation. His work has spurred work on algorithms with better approximation ratios. Graph techniques are used to block or reorder operations to improve the utilization of memory hierarchies. As these and many similar examples illustrate, combinatorial scientific computing is a thriving and critical enabler for parallel computing.

3 Mesh Generation

Many methods for solving partial differential equations require the geometric space to be decomposed into simple shapes. In scientific or engineering simulations with complex geometries, this *mesh generation* problem can be extremely challenging. In many settings, more time is spent generating the mesh than in any subsequent step of simulation and analysis.

Criteria to evaluate the quality of a mesh continue to evolve. But generally speaking, a good mesh is one with *well shaped* elements, and as few of them as possible. A well shaped element is one in which angles and lengths don't vary too much from being isotropic. (When the physics being modeled is dramatically skewed, as near surfaces in fluid flow calculations, a carefully skewed element may be desirable, and meshing routines must be adjusted appropriately.)

A rich collection of geometric algorithms are employed in mesh generation. One common technique is the use of Delaunay triangulation to produce triangular meshes in two dimensions [43,58]. Delaunay triangularization is an elegant algorithm for joining a set of points with triangles in a such a way that unnecessarily badly shaped triangles are avoided. However, in the mesh generation problem the locations of mesh points in the interior of the object are generally unspecified. A wide variety of algorithms have been proposed to initialize and optimize point locations. These algorithms use geometric techniques like quadtrees (or oct-trees in three dimensions) and various point insertion algorithms. Near geometric boundaries, constrained triangulations may be required, which adds complexity to standard Delaunay algorithms.

Tetrahedral meshing in three dimensions is considerably more challenging. Unlike in two dimensions, three dimensional Delaunay meshes are not guaranteed to consist of only well shaped elements. A variety of heuristics have been proposed to avoid badly shaped *sliver* elements, and this continues to be an active area of research.

For some applications, quadrilateral (in 2D) or hexahedral (in 3D) elements are preferred to triangles and tetrahedra. Quad and hex meshing are quite challenging, and are considerably less mature than triangular or tetrahedral meshing. A quad or hex mesh has considerable topological structure. For instance, consider the path of quadrilateral elements that is constructed by entering an element on one side and departing it on the opposite side. In three dimensions, one can work with paths constructed in a similar manner, or sheets grown by expanding in two of the three pairs of opposing faces of an element. The topology of these structures greatly constrains the space of possible meshes and can be used to facilitate the mesh generation process [61].

4   Solving Sparse Linear Systems

4.1   Direct Methods

Matrix factorizations are at the core of modern numerical linear algebra. Solving systems of linear equations, least-squares data fitting, eigenvector and singular vector computations can all be described in terms of factoring a given matrix into a product of 'simpler' matrices, such as diagonal, triangular, or orthogonal matrices. When a matrix is sparse, i.e., there are many zero elements in it, the factors can be computed with fewer operations and reduced storage. As an instance, the Cholesky factorization of an $n \times n$ dense matrix requires $O(n^3)$ operations and $O(n^2)$ space, whereas if the matrix can be represented by a planar computational graph, the operations are bounded by $O(n^{3/2})$ and space by $O(n \log n)$. Appropriate graph models and sophisticated algorithms designed using these models are necessary to realize these gains.

We begin by discussing combinatorial issues associated with solving symmetric positive definite systems of equations, since this is the archetypal problem. After that we will sketch the modifications required for unsymmetric systems of equations.

The graph model for Cholesky factorization (Gaussian elimination of a symmetric positive definite matrix A) was introduced by Parter [45], and further studied by Rose [55]. The appropriate graph here is the adjacency graph of the symmetric matrix that has a vertex v_i representing the ith row and column, and an undirected edge (i, j) representing each nonzero a_{ij} and its symmetric counterpart a_{ji}. During factorization, multiples of a row of the current matrix are added to a subset of the higher-numbered rows with the goal of transforming nonzeros below the diagonal in a column into zeros. During this process, a zero element in a higher-numbered row could become nonzero, and such newly created nonzero elements are called *fill elements*. Parter described the graph transformation that models the kth step of the factorization: add edges as needed to make all higher-numbered neighbors of vertex v_k a clique, and then mark v_k and all edges incident on it as deleted. The edges added correspond to fill elements. Rose showed that the *filled graph* obtained by taking the union of all the added edges with the edges corresponding to the original matrix is a *chordal graph*. (A *chord* in a cycle is an edge that joins two non-consecutive vertices on the cycle. A chordal graph has a chord joining every cycle of length greater than or equal to four.) A *fill path* in a graph is a path joining two vertices v_i and v_j in which every interior vertex in the path is numbered lower than both end vertices. Rose, Tarjan, and Lueker [56] obtained a static characterization of fill in Cholesky factorization: A fill edge (v_i, v_j) is created during the factorization if and only if a fill path joins the vertices v_i and v_j in the adjacency graph of the original matrix A.

Sparse matrix factorizations require the computation of data structures for the factor matrices, i.e., identification of the nonzero elements and their row and column indices. Once this information is available, most of the non-numerical operations can be removed from the inner-most loop of numerical computations, so that the latter can be performed in time proportional to the number of numerical operations. Hence one of the requirements for algorithms for computing the various data structures associated with sparse factorizations is that they run as far as is possible, in time proportional to arithmetic operations or faster. This necessitates the development of graph models, efficient algorithms and high-performance implementations.

The Elimination Tree. A data structure called the *elimination tree* [41] plays center stage in determining the control flow during the factorization, and in designing efficient algorithms for computing data structures for the Cholesky factors. The elimination tree has the vertices of the adjacency graph for its nodes, and the parent of a node v is the next vertex to be eliminated from the clique created when v is eliminated. In other words, the parent of a node in the etree is the lowest-numbered node among all of its higher-numbered neighbors. The elimination tree is also the transitive reduction of the filled graph in which every edge is directed from its lower-numbered to its higher-numbered endpoint.

The elimination tree is a minimal representation of the control dependences in the Cholesky factorization in the following sense: if an edge (i, j) exists in the filled graph then j is an ancestor of i in the elimination tree; and if i and j belong to vertex-disjoint subtrees in the elimination tree then no edge joins i and j in the filled graph. In particular this implies that the computation of a column of the factor cannot be completed unless all of the columns corresponding to its children nodes has been computed. Hence the elimination tree can be used to schedule the numerical computations associated with Cholesky factorization.

The elimination tree is also useful in designing efficient algorithms for computing the nonzero structures of the rows and columns of the Cholesky factor. The nonzero row indices in the jth column of the Cholesky factor can be obtained by unioning the row indices of the jth column of the matrix A and the row indices of the columns of the Cholesky factor corresponding to the children of node j in the elimination tree, where we consider only rows from j to n in each column. Without the elimination tree, the union would have to be taken over a larger set of columns. The structure of the ith row of the factor can be obtained as a pruned subtree of the elimination tree rooted at node i.

Another important feature of modern sparse factorization algorithms is the identification of dense submatrices within the sparse factors to obtain high performance on modern multiprocessors through register and cache reuse. The dense submatrices are obtained by grouping adjacent vertices with identical sets of higher-numbered neighbors into supernodes. The occurrence of supernodes stems from the fact that eliminating a vertex creates a clique of its higher-numbered neighbors.

A survey of the several data structures employed in sparse matrix factorizations is provided in [50].

Ordering Algorithms. An important component of modern sparse matrix solvers is an algorithm that orders columns (and rows) of the initial matrix to reduce the work and storage needed for computing the factors. These orderings also influence the effectiveness of preconditioners and the convergence of iterative solvers, and can often reduce the work needed by an order of magnitude or more. Two major classes of ordering algorithms have emerged thus far: algorithms based on the divide and conquer paradigm, exemplified by nested dissection, which is a top-down algorithm that computes the ordering from n (the number of vertices) to 1. The computational graph is recursively separated into two or more connected components by removing a small set of vertices called a vertex separator at each step. The separator vertices are ordered last, and the remainder of the vertices in the graph are ordered by recursively computing separators in the subgraphs and giving them the next lower available numbers. The second class of ordering algorithms is a greedy, bottom-up algorithm that orders vertices to locally reduce fill. This class is exemplified by the minimum-degree algorithm, which chooses a vertex of minimum degree in the current graph to eliminate next.

In practice, minimum degree algorithms are implemented in a space-efficient manner such that the filled graph can be implicitly represented in the same space as the original graph, even as fill edges are created during the elimination [24]. One way to do this is to use a clique cover, i.e., a set of cliques that includes every edge in the current graph. Initially each edge in the original graph is a clique, and as a vertex is eliminated, all cliques containing that vertex are merged into a new clique. Since this union operation does not increase the size of the clique cover, we are guaranteed that the filled graph can be represented in no more space than the original graph. There is one difficulty associated with the clique cover representation though: it costs $O(n^2)$ operations to compute the degree of a vertex in the course of the ordering. Hence approximations for the degree measure which can be computed fast, in $O(n)$ time, have been developed, the most popular of which is the approximate minimum degree (AMD) algorithm [1], due to Amestoy, Davis and Duff.

Nested dissection, which was discovered by Alan George [23], has spurred much research into computing vertex separators in graphs. Important classes of graphs that occur in various applications have separators of bounded size. Planar graphs have $O(n^{1/2})$ separators [40], and this implies that systems of equations from finite element meshes of 2-dimensional problems can be solved in $O(n^{3/2})$ operations and $O(n \log n)$ space. This result explains why direct methods are often the solvers of choice for 2-dimensional problems. For 3-dimensional meshes in which each element is well shaped, the corresponding bounds are $O(n^2)$ operations and $O(n^{4/3})$ space. Spectral, geometric, and multilevel algorithms have been developed for computing separators in such graphs. Currently software for graph partitioning employs multilevel algorithms due to the good quality of the ordering and the fast computation they offer.

Unsymmetric Problems. For unsymmetric (and symmetric indefinite) matrices, algorithms for sparse Gaussian elimination cannot neatly separate combinatorial concerns from numerical concerns as in the symmetric positive definite case. These problems require pivoting based on the actual numerical values in the partially factored matrix for numerical stability. In these problems, the combinatorial task of data structure computation has to be interleaved with numerical computation on a group of columns.

Nevertheless, combinatorial algorithms for computing data structures for the factors and for determining the control flow have been designed. One of the important differences for unsymmetric problems is that instead of an elimination tree, the control flow is determined by directed acyclic graphs (DAGs) that minimally represent the directed graphs corresponding to the factors [25]. These DAGs could be used to speed up the computation of the data structures for the factors.

Another combinatorial task that arises in these problems is matchings in graphs. An unsymmetric matrix can be represented by a bipartite graph with vertices corresponding to rows and columns, and each nonzero represented by an edge joining a row vertex and a column vertex. The magnitude of a nonzero can be represented by an edge weight, and then a matching of maximum cardinality with the maximum weight can be used to permute rows and columns so as to place large elements on the diagonal [18]. This reduces the need for numerical pivoting in direct solvers, and improves the quality of incomplete factorization preconditioners.

A maximum cardinality matching in a bipartite graph can also be used to compute a canonical decomposition of bipartite graphs called the Dulmage-Mendelsohn decomposition (the ear decomposition for bipartite graphs), which corresponds to a block triangular form for reducible matrices [49]. Only the diagonal blocks in a block triangular form need to be factored, potentially leading to significant savings in work and storage for reducible matrices that arise in applications such as circuit simulations. The diagonal blocks are also called strong Hall components, since they have the property that every set of k columns has nonzeros in at least $k + 1$ rows. The strong Hall property is useful in many structure prediction algorithms for unsymmetric Gaussian elimination and orthogonal-triangular factorization [26].

4.2   Iterative Methods

For solving large systems of equations, iterative methods are often preferred to sparse direct solvers since iterative solvers require less memory and are easier to parallelize. The runtime of an iterative method depends upon the cost of each iteration and the number of iterations required to achieve convergence. Combinatorial algorithms contribute to both of these considerations.

Consider the product $c = Ab$ where A is sparse. This operation typically involves a doubly-nested loop in which the outer loop is over rows and the inner loop is over nonzero entries in a row of A. If the nonzeros in a row are not consecutive, then non-consecutive entries of c will need to be accessed. The needed elements of c will often not be in cache, and so will be comparatively expensive to access. But if nonzeros in a row happen to be adjacent, then this will not only improve access to c, but can also be exploited to reduce the number of memory indirections required to access the elements of A.

One way to improve the performance of this operation is to reorder the columns of the matrix to increase the number of consecutive nonzeros in the rows. Pınar and Heath show that this problem can be recast as a graph problem in which the objective is to order the vertices to maximize the sum of weights connecting adjacent vertices [47]. They propose a heuristic approach to this NP-Hard problem that borrows techniques from literature on the the Traveling Salesman problem. They report that this reordering improves the performance of sparse matrix-vector multiplication by more than 20%.

The second factor in the cost of an iterative solver is the speed with which the method converges. The number of iterations can be dramatically reduced by effective preconditioning. Preconditioning is a transformation of a linear system so that $Ax = b$ is replaced by $M^{-1}Ax = M^{-1}b$, where the operator $M^{-1}A$ has better numerical properties than A alone. Generally speaking, this modification should reduce the condition number or increase the degree of clustering of the eigenvalues. For a preconditioner M to be effective, it must be easy to solve systems of the form $My = z$, and the construction of M must be efficient in both time and space.

A number of preconditioning strategies have been proposed, and several classes of preconditioners have combinatorial aspects. Incomplete factorization preconditioners follow the steps of a sparse direct solver, but discard many of the fill elements. Their construction involves many of the same operations as sparse direct solvers including graph-based reordering and fill monitoring [57].

Algebraic multigrid preconditioners approximate a matrix by a sequence of smaller and smaller matrices. The construction of smaller matrices can involve graph matching [38] or independent set computations [15].

Support theory preconditioners exploit an equivalence between the numerical properties of diagonally dominant matrices and graph embedding concepts of congestion and dilation [7]. The (symmetric positive-definite) matrix A is represented as a graph, and the preconditioner is constructed via graph operations to create an approximation to A that is easy to factor [27]. Spielman and Teng have used this approach to propose preconditioners that are provably near optimal for all diagonally dominant matrices, no matter how irregularly structured or poorly conditioned [59].

5   Optimization, Derivatives, and Coloring

5.1   Overview

Many algorithms that solve nonlinear optimization problems and differential equations require the computation of derivative matrices of vector functions. When the derivative matrices are large and sparse, sparsity and matrix symmetry can be exploited to compute their nonzero entries efficiently. The problem of minimizing the number of function evaluations needed to compute a sparse derivative matrix can be formulated as a matrix partitioning problem.

Graph coloring is an abstraction for partitioning a set of objects into groups according to certain rules. Hence it is natural that the matrix partitioning problems in derivative matrix computations can be modeled as specialized graph coloring problems. Remarkably, the techniques for exploiting sparsity here are essentially the same whether derivatives are computed using the older method of finite differences or the comparatively recent method of automatic differentiation. In formulating, analyzing, and designing algorithms for these matrix partitioning problems, graph coloring has proven to be a powerful tool. Indeed, modern software for computing large, sparse Jacobians and Hessians rely on graph coloring algorithms to make the computations feasible.

5.2   A Jacobian Computation Problem

Let $F(x)$ denote a vector function of a vector variable x, and let J denote the derivative matrix of F with respect to x (the *Jacobian*). We assume that the nonzero structure of J is known or can be computed. From the approximation $\frac{1}{\epsilon}[F(x+\epsilon e_k) - F(x)] \approx J(x)e_k$, by differencing the function along the co-ordinate vector e_k, we can estimate the kth column of J through function evaluations at $F(x)$ and $F(x+\epsilon e_k)$, where ϵ is a small step size. Thus, if sparsity is not exploited, the estimation of a Jacobian matrix with n columns would require n additional function evaluations.

Now consider a subset of the columns of the Jacobian such that no two columns have a nonzero in a common row; such a subset of columns is *structurally orthogonal*. In a group of structurally orthogonal columns, the columns are pairwise orthogonal to each other independent of the numerical values of the nonzeros. Choose a column vector d with 1's in components corresponding to the indices of columns in a structurally orthogonal group of columns, and zeros in all other components. By differencing the function F along the vector d, one can simultaneously determine the nonzero elements in *all* of these columns through the function evaluations at $F(x)$ and $F(x + \epsilon d)$. Further, by partitioning the columns of the Jacobian into the fewest groups, each consisting of structurally orthogonal columns, the number of (vector) function evaluations needed to estimate the Jacobian matrix is minimized.

Curtis, Powell, and Reid [14] observed in 1974 that sparsity can be employed in this way to reduce the number of function evaluations needed to estimate the Jacobian. In 1983 Coleman and Moré [11] modeled this matrix partitioning problem as a distance-1 graph coloring problem. The model uses the *column intersection graph* of a matrix where columns correspond to vertices and two vertices are joined by an edge whenever the corresponding columns have nonzeros in a common row (i.e., the columns are structurally non-orthogonal). A distance-1 coloring of a column intersection graph, partitions the columns into groups of structurally orthogonal columns. Since the distance-1 graph coloring problem is known to be NP-hard, the work of Coleman and Moré showed that it is unlikely that there is a polynomial time algorithm for partitioning the columns of a matrix into the fewest groups of structurally orthogonal columns. Meanwhile, they developed several practically effective heuristics for the problem. More recently, Gebremedhin et al. [21] have used a different graph coloring model for the same matrix partitioning problem. This coloring formulation uses a *bipartite graph* to represent a Jacobian matrix. The vertex set V_1 in the bipartite graph corresponds to the rows of the matrix and the vertex set V_2 corresponds to the columns. An edge joins a row vertex r_k to a column vertex c_ℓ if the matrix element $j_{k\ell}$ of the Jacobian is nonzero.

Two columns in the Jacobian matrix are structurally orthogonal if and only if they are at a distance greater than two from each other in the corresponding bipartite graph. Thus, a distance-2 coloring of the set of column vertices V_2 is equivalent to a partitioning of the columns of the matrix into groups of structurally orthogonal columns. A distance-2 coloring of the vertex set V_2 is an assignment of colors to these vertices such that every pair of column vertices at a distance of exactly two edges from each other receives distinct colors. More precisely, this coloring is a *partial* distance-2 coloring of the bipartite graph since the row vertex set V_1 is left uncolored.

5.3   Variations on Matrix Computation

Depending on the type of derivative matrix being computed and the specifics of the method being applied, there exist several variant matrix partitioning problems. Specifically, the nature of a particular problem in our context depends on: whether the matrix to be computed is nonsymmetric, a *Jacobian*; or symmetric, a *Hessian*; whether the evaluation scheme employed is *direct* or *substitution*-based (a direct method requires solving a diagonal system and a substitution method relies on solving a triangular system of equations); whether a *unidirectional* (1d) partition or a *bidirectional* (2d) partition is used (a unidirectional partition involves only columns or rows whereas a bidirectional one involves both columns and rows); and whether *all* of the nonzero entries of the matrix or only a *subset* need to be determined; we refer to these as *full* and *partial* matrix computation. Each of these matrix partitioning problems can be modeled as a specialized graph coloring problem.

Hessians. In 1979 Powell and Toint [51] extended the approach of Curtis, Powell, and Reid to compute sparse Hessians. McCormick [42] introduced a distance-2 graph coloring model for the computation of Hessians in 1983. Independently, in 1984, Coleman and Moré [12] gave a more precise coloring model that exploits symmetry. Their model satisfies the two conditions: (1) every pair of adjacent vertices receives distinct colors (a distance-1 coloring), and (2) every path on four vertices uses at least three colors. This variant of coloring is called *star* coloring, since in such a coloring every subgraph induced by vertices assigned any two colors is a collection of stars.

Substitution-Based Evaluation. In a substitution-based evaluation scheme, the unknown matrix elements are determined by solving a triangular system of equations. A substitution-based evaluation is often effectively combined with the exploitation of symmetry, and hence is used in computing the Hessian. Based on the work of Powell and Toint [51], Coleman and Moré [12] found a coloring model for a restricted substitution method for evaluating a Hessian called triangular coloring. Triangular coloring exploits symmetry only to a limited extent. A more accurate model for a substitution method to compute a Hessian leads to an *acyclic coloring* problem in which the requirements are that (1) the coloring corresponds to a distance-1 coloring, and (2) vertices in every cycle of the graph are assigned at least three distinct colors. This variant of coloring is called acyclic since every subgraph induced by vertices assigned any two colors is a forest, and is due to Coleman and Cai [10]. Recently Gebremedhin et al [22] have developed the first practical heuristic algorithm for acyclic coloring and a new efficient algorithm for star coloring; they have shown that a substitution method based on acyclic coloring leads to faster Hessian computations than a direct method based on star coloring.

Bidirectional Partition. If the matrix contains a few dense columns *and* rows, it may be advantageous to consider partitioning subsets of *both* columns and rows. A partition that involves both columns and rows is called *bidirectional*. Due to symmetry, there is no advantage in considering a bidirectional partition of the Hessian, i.e., a symmetry-exploiting unidirectional partition suffices. In the context of automatic differentiation, bidirectional partitions arise when the Jacobian is computed by using the forward and reverse modes simultaneously.

Bidirectional partitioning of the Jacobian leads to specialized *bicoloring* problems in the bipartite graph, i.e., a coloring of subsets of both the row vertices and the column vertices with disjoint sets of colors. When bidirectional partitioning is used within a direct evaluation scheme for Jacobians, the coloring problem is that of *star bicoloring*; the corresponding model within a substitution-based scheme is the *acyclic bicoloring* problem. Bidirectional partitioning problems and their graph coloring formulations were studied by Hossain and Steihaug [33] and Coleman and Verma [13].

Partial Computation. The final variation within the classification scheme is whether all elements of the Jacobian and the Hessian are required, or only a subset that would be needed for preconditioning purposes. We refer to these variations as *full* and *partial* matrix computation. The latter would be useful in 'matrix-free' methods for large-scale problems, where the Jacobian is too large to be explicitly estimated, but a coarser representation of the Jacobian is used as a preconditioner. Partial matrix computation problems lead to restricted coloring problems where only a specified subset of the vertices need to be colored; however, one still needs to pay attention to the remaining vertices, since they could interfere with the estimation of the required matrix elements.

All of these variations lead to a rich collection of graph coloring problems. Table 1 shows the collection of five coloring problems that arise when we consider the computation of all nonzero entries of Jacobians and Hessians. Partial matrix computation problems lead to another set of five coloring problems, of which graph models have been formulated for direct methods by Gebremedhin et al. These authors provide a recent survey of graph coloring for computing derivatives in [21].

Table 1. Graph coloring formulations for computing *all* nonzero entries of derivative matrices. The Jacobian is represented by its bipartite graph, and the Hessian by its adjacency graph. NA stands for not applicable.

	1d partition	2d partition	
Jacobian	distance-2 coloring	star bicoloring	Direct
Hessian	star coloring	NA	Direct
Jacobian	NA	acyclic bicoloring	Substitution
Hessian	acyclic coloring	NA	Substitution

Graph and hypergraph coloring have been used in a wide collection of application areas in addition to optimization: register allocation in compilers, radio and wireless networks, scientific computing, data movement in distributed and parallel computing, facility location problems, cache-efficient algorithms, etc. Parallel computers make it feasible to solve large-scale problems in many of these application areas, especially optimization, and hence there is currently increased interest in efficient algorithms and software for coloring graphs with millions of vertices.

6   Statistical Physics

The inherent complexity of the physical world has led physicists to investigate simplified, idealized models. The hope is that these idealized models capture some of the most interesting features of reality, but their simplification allows for more detailed analysis and simulation. In many cases, these models have rich and exploitable combinatorial structure.

The best known example of this approach is the Ising model for magnetic materials. Bulk magnetism is caused by the alignment of atomic spins. The spin of each atom influences the spin of its near neighbors, leading to very complex dynamics. In real magnets, the complex, three-dimensional geometry of atomic locations and the subtlety of the interactions makes analysis quite difficult. In the 1930s, long before computational simulation was even an option, the Ising model was proposed as a simple tool for studying magnetism.

In the Ising model atoms are placed on the lattice points of a regular 1-, 2- or 3-dimensional grid. Each atom only interacts with its nearest neighbors. Various initial and boundary conditions can be applied to the problem, and many questions can be asked about its statistical dynamics or energetics. Some of these questions can be addressed via combinatorial optimization techniques involving matchings and counting of subgraphs [5].

The success of the Ising model has led to a vast array of variants and generalizations, many of which have combinatorial features of their own. It has also led to philosophically related models of very different phenomena.

One of these alternatives was proposed by Thorpe to model mechanical properties of materials [62]. Instead of trying to explicitly model the detailed bond structure of a complex composite, Thorpe's model places atoms in space and then connects them to near neighbors randomly. The number of connections is chosen to reflect the statistical properties of a specific composite material. These bonds are then treated as rigid bars, and the mechanical rigidity of the resulting structure can be analyzed. Fast algorithms for analyzing these structures have been developed which build upon concepts in graph matching and graph rigidity [3]. The application of combinatorial optimization techniques to simplified physical models continues to be a a very active area of research.

A very different class of idealized models of physical reality are provided by *cellular automata*, of which Conway's Game of Life is a prototypical example. In cellular automata, a set of entities interact via very simple rules, but in some circumstances complex collective behavior can be observed. The analysis of cellular automata is richly combinatorial [64].

7   Computational Chemistry

There is a natural correspondence between the structure of molecules consisting of atoms and bonds, and the vertices and edges of a graph. This relationship has led to a wide range of graph theoretic techniques in chemistry. In fact, the term "graph" as used here was first coined by J. J. Sylvester in his studies of molecular structure [60].

The graphs that describe molecules have special properties that sometimes allow for more efficient algorithms. Since an atom can be bonded to at most a few other atoms (typically four), the corresponding graph has a small maximum degree. Also, each atom is of a particular type (carbon, oxygen, etc.), so each vertex in the graph can be assigned a corresponding type value.

If a drug company discovers a molecule that exhibits an interesting biological effect, they will want to test similar molecules. But in a very large universe of molecules, how does one determine which ones are *similar*? A common technique is to use a set of graph properties or invariants to characterize molecules. Molecules of interest are then those that have similar graphical properties. This seemingly simplistic approach actually works quite well, and a vast collection of graphical invariants have been proposed to characterize molecules [54].

Another way to search for drug candidates is to identify a piece of a molecule, perhaps a small portion of a large protein, that displays the desired activity, and then to search for other molecules that possess the same molecular fragment. This problem can phrased in terms of *subgraph isomorphism*. Given graph G and a smaller target graph T, the subgraph isomorphism problem is a search for a subset of vertices and edges in G that comprise an exact match for T. Although subgraph isomorphism is known to be NP-complete, several aspects of this application make it solvable in practice. First, the vertex types (i.e., chemical species) constrain the search space. And the goal here is not to look in a single large graph G, but rather to scan a large library of smaller graphs, each of which corresponds to a molecule.

8 Bioinformatics

Bioinformatics has seen spectacular growth since the 1990's, and combinatorial problems abound in biological applications, so that a section of this length has to be necessarily incomplete. A few books discussing algorithms in bioinformatics include: Gusfield [28], Durbin et al. [19], Pevzner [46], and Eidhammer et al. [20]. All that we hope to do here is to highlight a few select areas of current research interest.

Algorithms on strings are used in local and multiple alignment of DNA and protein sequences. Dynamic programming is used to compute optimal pairwise alignments of sequences, but due to its quadratic time complexity, it is impractical for searching a large database of sequences against a query sequence to find the best local or global alignment. Faster heuristic algorithms such as BLAST and FASTA have been developed for this problem, and BLAST represents one of the most widely used bioinformatics tools. Optimal multiple sequence alignments are NP-hard to compute, and various approximation algorithms have been developed for this problem. Hidden Markov models are used to build probabilistic models of protein families and to answer queries about whether a specific protein belongs to the family or not [19]. Sequence data of specific proteins have been used to construct phylogenetic trees, and have provided an alternative to classifying organisms based on phenotypes. Constructing an optimal phylogenetic tree that represents given sequence data for multiple organisms is computationally intensive due to the super-exponential growth in the number of trees that must be examined as a function of the number of species.

Aligning RNA sequences is a computationally more intensive task since secondary structure needs to be taken into account to compute alignments. A recent approach to this problem involves a graph-theoretic formulation that uses weighted matchings in graphs and integer linear programs [4].

The Gene Ontology (GO) project (URL: www.geneontology.org) addresses the need to provide consistent descriptions of the proteins in multiple databases. Structured and controlled vocabularies are being developed for proteins in terms of the biological processes they are associated with, the cellular components they belong to, and their molecular functions, independent of the species. The data structure underlying GO may be viewed as partially ordered set (poset), and answering queries efficiently in GO leads to several combinatorial problems on posets. Efficient algorithms for computational problems on posets remain to be developed.

Proteomic experiments such as the yeast 2-hybrid system yield protein-protein interaction graphs at the organism-scale, and such graphs are now available for many model organisms as well as humans. Because these in-vitro experiments have high error rates, Bayesian networks have been used to integrate this data with other proteomic and genomic data to improve the reliability of the interaction graph. A functional module is a group of proteins involved in a common biological process [29]. A key computational task is to decompose a protein-protein interaction graph into functional modules, to annotate the biological process that each module is involved in, and to identify "cross-talk" between the modules, i.e., the proteins that are involved in linking different biological processes. This task can be modeled as a clustering of the graph in which the clusters can overlap; computing a clustering in these networks is challenging since these graphs are small-world networks (the average distance between any two vertices in the network is $O(\log n)$, where n is the number of vertices); hence the distance between two clusters is quite small. The degree distributions of vertices in these networks obey a power-law, and there are a few vertices of high degree that tend to confound the clustering.

Early work on clustering these networks has involved searching for cliques of small size, or local clustering approaches that grow clusters from seed vertices [2]. Spectral and multi-level clustering algorithms, similar to their analogues in graph partitioning algorithms, have been developed to compute such clusterings [53]. The emerging field of computational systems biology is rich in combinatorial problems that arise from the characterization of biological networks and knowledge discovery in such networks. Effective methods to text-mine the literature to build proteomic and genomic networks are essential as the number of publications in these emerging fields continues to grow.

9   Information Processing

Like many aspects of society, science is being transformed by the explosive growth of available information and the rapidly evolving tools for search and analysis. There are many ways to represent information, but several of them have rich combinatorial underpinnings.

Perhaps the most familiar graph in informatics is the graph in which web pages are vertices and hyperlinks become (directed) edges. The structure of this graph is a critical aspect of Google's PageRank algorithm for ranking web pages [44], and of Kleinberg's related HITS algorithm [39] which is used by other search engines. Both of these algorithms construct a matrix from the web graph and then compute rankings with eigenvectors or singular vectors of this matrix.

The ideas in PageRank were actually proposed several decades earlier in citation analysis [48]. In this field, scientific papers are vertices and their citation links are edges. A calculation analogous to PageRank is used to determine which papers (or journals) are most significant.

Many types of information can be naturally represented as graphs, including social or communication networks of all sorts. The study of these *complex networks* has recently grown to become a very active area of research amongst sociologists, statistical physicists and computer scientists. Needless to say, graph algorithms are central to this field.

Text analysis is another area of informatics in which graph models are important. Consider a set of documents and the union of all their keywords. A matrix can be used to encode the set of documents that use a particular keyword. Equivalently, this relationship can be encoded by a bipartite graph in which documents are one set of vertices and keywords are another. An edge connects a document to a keyword that the document contains. This structure can be used to analyze and organize the information contained in the corpus of documents using a variety of graph and linear algebra techniques. For instance, this structure can be used in a query processing system to identify the documents that best match a user's query [6,16,30].

10   The Future

This paper has tried to introduce some of the many ways in which abstractions and algorithmic advances in computer science have played a role in scientific computing. The critical enabling role that these algorithms play is often overlooked. One important reason for this is that the combinatorial kernel is often just one piece of a larger tool or body of work (e.g., an ordering code within a linear solver, or a Delaunay triangulation within a mesh generator). But we believe there are other, cultural factors involved as well.

Computational science is usually marketed with an emphasis on the scientific impact of the work – e.g., the insight into global warming or the design of a more efficient chemical plant. The vast collection of enabling technologies underpinning these applications often get short shrift. When these algorithms are emphasized, those that are most accessible to computational scientists are the ones that are most likely to be lauded. The training of a computational scientist often involves exposure to numerical methods or to finite elements, so these technologies are likely to be appreciated and acknowledged. But few computational scientists have taken courses in graph algorithms, and so the importance of discrete algorithms is less likely to be recognized.

As we have tried to argue in this paper, discrete algorithms have long played a crucial enabling role in science and engineering. We expect their importance to continue to grow for several reasons. Fundamentally, as the data sets we analyze and the computations we perform continue to grow in size and complexity, optimal algorithmic efficiency becomes of paramount concern. This driving force will continue to create opportunities for new research into advanced algorithms (and approximation algorithms). In addition, as the recognition of the value of CSC becomes more widespread in the scientific and engineering communities, we are already witnessing a growing receptiveness and interest in discrete algorithms. We are also seeing a growth in educational programs that expose students to a range of topics necessary to contribute to CSC. Finally, we foresee rapid growth in several areas of science that are particularly rich in combinatorial problems. Among these are biology and informatics. And the broad transition to parallel computing for scientific and engineering computations also increases the importance of combinatorial algorithms.

As with any interdisciplinary subject, the growth of CSC raises challenges on several fronts. Education is a key issue. Computational science requires training in a scientific discipline combined with training in numerical methods and software engineering. We feel strongly that it should also include exposure to basic algorithms and data structures, with a particular focus on graph algorithms. Publication venues are a challenge in CSC since its work often falls into the cracks between theory and applications. CSC sits on the periphery of several scientific communities, but is central to none of them. Visibility and recognition is particularly important for young researchers. The tenure process can be difficult for academics whose work spans traditional communities. Professional societies and funding agencies can play an important role in nurturing and supporting this field.

Many scientific breakthroughs occur at the boundaries between fields where ideas and techniques can fruitfully cross-fertilize each other. We believe that combinatorial scientific computing lies on one of these fruitful boundaries. For researchers trained in computer science algorithms, scientific applications offer a rich assortment of interesting problems with high impact. For computational scientists trained in numerical methods or in an application discipline, combinatorial techniques offer the potential for dramatic advances in simulation capability. This mutual benefit will continue to motivate and inspire important work for long into the future.

Acknowledgments

We are indebted to many of our colleagues who have contributed to our understanding of the diverse areas touched upon in this paper. We also thank Assefaw Gebremedhin and Florin Dobrian for comments on an earlier draft of this manuscript. Sandia National Laboratories is a multiprogram laboratory operated by Sandia Corporation, a Lockheed-Martin Company, for the U.S. DOE under contract number DE-AC-94AL85000.

References

1. P. R. Amestoy, T. A. Davis, and I. S. Duff. An approximate minimum degree ordering algorithm. *SIAM Journal on Matrix Analysis and Applications*, 17(4):886–905, 1996.
2. G. D. Bader and C. W. Hogue. An automated method for finding molecular complexes in large protein interaction networks. *BMC Bioinformatics*, 4(2):27 pp., 2003.
3. S. Bastea, A. Burkov, C. Moukarzel, and P. M. Duxbury. Combinatorial optimization methods in disordered systems. *Computer Phys. Comm.*, 121:199–205, 1999.
4. M. Bauer, G. W. Klau, and K. Reinert. Fast and accurate structural RNA alignment by progressive Langrangian optimization. In M. R. Berthold et al, editor, *Computational Life Sciences, Lecture Notes in Bioinformatics*, volume 3695, pages 217–228. Springer Verlag, 2005.
5. R. J. Baxter. *Exactly solved models in statistical mechanics*. Academic Press, 1982.
6. M. Berry and M. Browne. *Understanding Search Engines: Mathematical Modeling and Text Retrieval*. SIAM, Philadelphia, 1999.
7. E. G. Boman and B. Hendrickson. Support theory for preconditioning. *SIAM J. Matrix Anal. Appl.*, 25(3):694–717, 2003.

8. Ü. Çatalyürek and C. Aykanat. Hypergraph-partitioning based decomposition for parallel sparse-matrix vector multiplication. *IEEE Trans. Parallel Distrib. Syst.*, 10(7):673–693, 1999.

9. Ü. Çatalyürek and C. Aykanat. PaToH: a multilevel hypergraph partitioning tool for decomposing sparse matrices and partitioning VLSI circuits. Technical Report BU–CEIS–9902, Dept. Computer Engineering and Information Science, Bilkent Univ., Turkey, 1999.

10. T. F. Coleman and J. Cai. The cyclic coloring problem and estimation of sparse Hessian matrices. *SIAM J. Alg. Disc. Meth.*, 7(2):221–235, 1986.

11. T. F. Coleman and J. J. Moré. Estimation of sparse Jacobian matrices and graph coloring problems. *SIAM J. Numer. Anal.*, 20(1):187–209, February 1983.

12. T. F. Coleman and J. J. Moré. Estimation of sparse Hessian matrices and graph coloring problems. *Math. Program.*, 28:243–270, 1984.

13. T. F. Coleman and A. Verma. The efficient computation of sparse Jacobian matrices using automatic differentiation. *SIAM J. Sci. Comput.*, 19(4):1210–1233, 1998.

14. A. R. Curtis, M. J. D. Powell, and J. K. Reid. On the estimation of sparse Jacobian matrices. *J. Inst. Math. Appl.*, 13:117–119, 1974.

15. H. De Sterck, U. M. Yang, and J. J. Heys. Reducing complexity in parallel algebraic multigrid preconditioners. *SIAM J. Matrix Anal. Appl.*, 27:1019–1039, 2006.

16. S. Deerwester, S. Dumais, G. Furnas, T. Landauer, and R. Harshman. Indexing by latent semantic analysis. *J. Amer. Soc. Information Sci.*, 41(6):391–407, 1990.

17. K. Devine, E. Boman, R. Heaphy, R. Bisseling, and U. Catalyurek. Parallel hypergraph partitioning for scientific computing. In *Proc. IPDPS'06*. IEEE, 2006.

18. I. S. Duff and J. K. Koster. The design and use of algorithms for permuting large entries to the diagonal of sparse matrices. *SIAM Journal on Matrix Analysis and Applications*, 20(4):889–901, July 1999.

19. R. Durbin, S. Eddy, A. Krogh, and G. Mitchison. *Biological Sequence Analysis*. Cambdridge University Press, 1998.

20. I. Eidhammer, I. Jonassen, and W. R. Taylor. *Protein Bioinformatics: An algorithmic approach to sequence and structure analysis*. Wiley, 2004.

21. A. Gebremedhin, F. Manne, and A. Pothen. What color is your Jacobian? Graph coloring for computing derivatives. *SIAM Review*, 47(4):629–705, Dec. 2005.

22. A. Gebremedhin, F. Manne, A. Pothen, and A. Tarafdar. New acyclic and star coloring algorithms with application to Hessian computations. Technical report, Old Dominion University, Norfolk, VA, March 2005.

23. A. George. Nested dissection of a regular finite element mesh. *SIAM Journal on Numerical Analysis*, 10:345–363, 1973.

24. A. George and J. W. H. Liu. The evolution of the minimum-degree ordering algorithm. *SIAM Review*, 31:1–19, 1989.

25. J. R. Gilbert and J. W. H. Liu. Elimination structures for unsymmetric sparse *LU* factors. *SIAM Journal on Matrix Analysis and Applications*, 14:334–352, 1993.

26. J. R. Gilbert and E. G. Ng. Predicting structure in nonsymmetric sparse matrix factorizations. In A. George, J. R. Gilbert, and J. W. H. Liu, editors, *Graph Theory and Sparse Matrix Computation*, pages 107–139. Springer-Verlag, 1993.

27. K. Gremban. *Combinatorial Preconditioners for Sparse, Symmetric, Diagonally Dominant Linear Systems*. PhD thesis, School of Computer Science, Carnegie-Mellon University, 1996. Available as Tech. Report CMU-CS-96-123.

28. D. Gusfield. *Algorithms on Strings, Trees and Sequences*. Cambdridge University Press, 1997.

29. L. H. Hartwell, J. J. Hopfeld, and A. W. Murray. From molecular to modular cell biology. *Nature*, 402:C47–C52, 1999.
30. B. Hendrickson. Latent semantic analysis and Fiedler retrieval. Submitted for publication to Lin. Alg. Appl. Earlier version in Proc. SIAM Workshop on Text Mining'06, 2006.
31. B. Hendrickson and T. Kolda. Graph partitioning models for parallel computing. *Parallel Comput.*, 26:1519–1534, 2000.
32. B. Hendrickson and R. Leland. The Chaco user's guide: Version 2.0. Technical Report SAND94–2692, Sandia National Labs, Albuquerque, NM, June 1995.
33. S. Hossain and T. Steihaug. Computing a sparse Jacobian matrix by rows and columns. *Optimization Methods and Software*, 10:33–48, 1998.
34. D. S. Johnson. Challenges for theoretical computer science: Draft. URL: www.research.att.com/~dsj/nsflist.html, 2000.
35. M. T. Jones and P. E. Plassmann. Parallel algorithms for adaptive mesh refinement. *SIAM J. Scientific Computing*, 18:686–708, 1997.
36. G. Karypis and V. Kumar. A fast and high quality multilevel scheme for partitioning irregular graphs. Technical Report CORR 95–035, University of Minnesota, Dept. Computer Science, Minneapolis, MN, June 1995.
37. G. Karypis and V. Kumar. Parmetis: Parallel graph partitioning and sparse matrix ordering library. Technical Report 97-060, Department of Computer Science, University of Minnesota, 1997.
38. H. Kim, J. Zu, and L. Zikatanov. A multigrid method based on graph matching for convection-diffusion equations. *Numerical Lin. Alg. Appl.*, 10:181–195, 2002.
39. J. Kleinberg. Authoritative sources in a hyperlinked environment. *J. ACM*, 46(5):604–632, 1999.
40. R. J. Lipton and R. E. Tarjan. A separator theorem for planar graphs. *SIAM Journal on Applied Mathematics*, 36:177–189, 1979.
41. J. W. H. Liu. The role of elimination trees in sparse factorization. *SIAM Journal on Matrix Analysis and Applications*, 11:134–172, 1990.
42. S. T. McCormick. Optimal approximation of sparse Hessians and its equivalence to a graph coloring problem. *Math. Program.*, 26:153–171, 1983.
43. S. J. Owen. A survey of unstructured mesh generation technology. In *Proc. 7th Intl. Meshing Roundtable*, 1998.
44. L. Page, S. Brin, R. Motwani, and T. Winograd. The PageRank citation ranking: Bringing order to the web. Technical report, Stanford Digital Library Technologies Project, 1998.
45. S. V. Parter. The use of linear graphs in Gaussian elimination. *SIAM Review*, 3:119–130, 1961.
46. P. A. Pevzner. *Computational Molecular Biology: An algorithmic approach*. MIT Press, 2000.
47. A. Pınar and M. T. Heath. Improving performance of sparse matrix–vector multiplication. In *Proc. ACM and IEEE International Conference on Supercomputing (SC99)*, 1999.
48. G. Pinski and F. Narin. Citation influence for journal aggregates of scientific publications: Theory, with applications to the literature of physics. *Inf. Proc. and Management*, 12:297–312, 1957.
49. A. Pothen and C.-J. Fan. Computing the block triangular form of a sparse matrix. *ACM Transactions on Mathematical Software*, 16:303–324, Dec. 1990.
50. A. Pothen and S. Toledo. Elimination structures in scientific computing. In D. Mehta and S. Sahni, editors, *Handbook on Data Structures and Applications*, pages 59.1–59.29. CRC Press, 2004.

51. M. J. D. Powell and P. L. Toint. On the estimation of sparse Hessian matrices. *SIAM J. Numer. Anal.*, 16(6):1060–1074, 1979.
52. R. Preis. Linear-time 1/2- approximation algorithm for maximum weighted matching in general graphs. In C. Meinel and S. Tison, editors, *Symposium on Theoretical Aspects of Computer Science (STACS)*, volume 1563 of *LNCS*, pages 259–269. Springer Verlag, 1999.
53. E. Ramadan, C. Osgood, and A. Pothen. The architecture of a proteomic network in the yeast. In M. R. Berthold et al, editor, *Computational Life Sciences, Lecture Notes in Bioinformatics*, volume 3695, pages 265–276. Springer Verlag, 2005.
54. M. Randic and J. Zupan. On interpretation of well-known topological indices. *J. Chem. Inf. Comput. Sci.*, 41(3):550–560, 2001.
55. D. J. Rose. A graph-theoretic study of the numerical solution of sparse positive definite systems of linear equations. In R. C. Read, editor, *Graph Theory and Computing*, pages 183–217. Academic Press, New York, 1972.
56. D. J. Rose, R. E. Tarjan, and G. S. Lueker. Algorithmic aspects of vertex elimination on graphs. *SIAM Journal on Computing*, 5:266–283, 1976.
57. Y. Saad. *Iterative methods for sparse linear systems (2nd edition)*. SIAM, Philadelphia, 2003.
58. J. Shewchuk. Triangle: Engineering a 2D quality mesh generator and Delaunay triangulator. In M. C. Lin and D. Manocha, editors, *Applied Computational Geometry: Towards Geometric Engineering*, volume 1148 of *Lecture Notes in Computer Science*, pages 203–222. Springer-Verlag, Berlin, 1996.
59. D. Spielman and S.-H. Teng. Nearly-linear time algorithms for graph partitioning, graph sparsification, and solving linear systems. In *Proc. 36th ACM Symp. Theory of Comput.* ACM, 2004.
60. J. J. Sylvester. On an application of the new atomic theory to the graphical representation of the invariants and covariants of binary quantics: With three appendices. *Amer. J. Mathematics*, 1:64–128, 1878.
61. T. G. Tautges, T. Blacker, and S. A. Mitchell. The whisker weaving algorithm: A connectivity-based method for constructing all-hexahedral finite element meshes. *Intl. J. Numerical Methods Engng.*, 39:3327–3349, 1996.
62. M. F. Thorpe. Continuous deformations in random networks. *J. Non-Cryst. Solids*, 57:355–370, 1983.
63. L. N. Trefethen. Numerical analysis. In Timothy Gowers with June Barrow-Green, editor, *Princeton Companion to Mathematics*. Princeton University Press, 2006. To appear.
64. S. Wolfram. *A new kind of science*. Wolfram Media, 2002.

Improving the Numerical Simulation of an Airflow Problem with the BlockCGSI Algorithm

C. Balsa[1], M. Braza[2], M. Daydé[3], J. Palma[1], and D. Ruiz[3]

[1] FEUP, Porto, Portugal
{cbalsa, jpalma}@fe.up.pt
[2] IMFT–CNRS, Toulouse, France
braza@imft.fr
[3] ENSEEIHT–IRIT, Toulouse, France
{dayde, ruiz}@enseeiht.fr

Abstract. Partial spectral information associated with the smallest eigenvalues can be used to improve the solution of successive linear systems of equations, namely in the simulation of time-dependent partial differential equations, where at each time step there are several systems with the same spectral properties to be solved. We propose to perform a partial spectral decomposition with the BlockCGSI algorithm in the first time step, and exploit this information to improve the convergence of the Conjugate Gradient algorithm in the solution of the following linear systems. We describe in summary the BlockCGSI algorithm, that is a combination of the block Conjugate Gradient (blockCG) with the Inverse Subspace Iteration. Then, we validate the accelerating strategy in the simulation of the flow around an airplane wing, where the Conjugate Gradient is accelerated through the deflation of the starting residual.

1 Introduction

Partial spectral information associated with the smallest eigenvalues can be used to improve the solution of successive linear systems of equations, namely in the simulation of time-dependent partial differential equations, where at each global iteration there are several systems with the same spectral properties to be solved. We propose a two-phase acceleration technique, where in the first phase we perform a partial spectral decomposition of the system solved in the first global iteration, with the BlockCGSI algorithm. In a second phase we exploit this information to improve the convergence of the Conjugate Gradient algorithm in the solution of the following linear systems.

The two-phase acceleration strategy has been initially proposed in the experimental work in Ref. [1]. In the first phase, the BlockCGSI algorithm computes a *near*-invariant subspace associated with the smallest eigenvalues, and in the second phase this spectral information is used to deflate the eigencomponents associated with the smallest eigenvalues with an appropriate starting guess. We concluded [2] that this strategy has a good potential to reduce the computing time of a fluid flow simulation algorithm, and the success of this approach depends on the appropriate monitoring of the BlockCGSI algorithm that combines

M. Daydé et al. (Eds.): VECPAR 2006, LNCS 4395, pp. 281–291, 2007.
© Springer-Verlag Berlin Heidelberg 2007

the blockCG iterative solver with the Subspace Iteration. In [3], we analyzed the convergence of the BlockCGSI algorithm from an inner-outer iteration point of view. We establish how the eigenvalue error of the Subspace Iteration varies along the inner iteration, when the system is solved with the blockCG algorithm. In agreement with these results, we proposed an appropriate stopping criterion for each level of iteration, that enables to reduce the computational costs.

In the present work we validate the two phase accelerating strategy in the simulation of a flow around an airplane wing (see [4]). Firstly, in section 2, we describe in short, the BlockCGSI algorithm, that is a combination of the block Conjugate Gradient (blockCG) [5,1] with the Subspace Iteration [6]. We analyze the computational costs involved in the two phases, as a function of the dimension of the computed *near*-invariant subspace, and compute the *a posteriori* optimal dimension (section 3.1). Based on these results, we propose a strategy for choosing dynamically the dimension of the basis, that does not need *a priori* informations about the spectrum of the coefficient matrix. We conclude (section 3.2), showing the benefits resulting from the application of the two-phase approach to present airflow problem.

2   The BlockCGSI Algorithm

The BlockCGSI algorithm is used to compute an M-orthonormal basis W of a *near*-invariant subspace associated with the smallest eigenvalues in the preconditioned matrix $M^{-1}A$. If this basis incorporates, for instance, all the eigenvalues of $M^{-1}A$ in the range $[0, \mu]$, we can expect, when using it later as a second level of preconditioning, that the condition number of the coefficient matrix will be reduced to about $\kappa = \lambda_{\max}/\mu$, where $\lambda_{\max}$ is the largest eigenvalue in $M^{-1}A$. In Algorithm 1, $\lambda_{\max}$ and μ are considered as input parameters (a rough upper bound on $\lambda_{\max}$ is usually enough). Another input concerns the choice of the block size s that defines the dimension of the working subspace at each inverse iteration; it also gives the number of right-hand sides and solutions vectors of the multiple linear systems solved by the blockCG algorithm, and consequently the amount of memory required as working space.

As a starting point, the algorithm requires the generation of an M-orthonormal basis of size s. The closer are these vectors to the targeted *near*-invariant subspace, the faster will be the convergence of the inverse iteration. The scope of steps 1 to 4 in Algorithm 1, is to generate an initial M-orthonormal set $V^{(0)}$ of s vectors with eigencomponents corresponding to eigenvalues in the range $[\mu_f, \lambda_{\max}]$ below some predetermined value $\xi \ll 1$ (denoted as the "filtering level"). This filtering technique is based on Chebyshev polynomials (step 3) and details about it can be found in [3]. The idea behind the use of these Chebyshev filters at the starting point is to put the inverse subspace iteration in the situation of working in the orthogonal complement of a large number of eigenvectors, e.g. all those associated with the eigenvalues in the range $[\mu_f, \lambda_{\max}]$. We can also expect that the resulting filtered right-hand sides will present more favorable spectral properties that can improve the convergence behavior of the

---

ALGORITHM 1. BLOCKCGSI ALGORITHM

Inputs: $A, M = R^T R \in \mathbb{R}^{n \times n}, \mu, \lambda_{max} \in \mathbb{R}, s \in \mathbb{N}$
Output: a *near*-invariant subspace W associated with all
   eigenvalues in the range $]0, \mu]$

Begin
 Generate the initial subspace (with filtering)
 1. $Z^{(0)} =$ RANDOM(n, s)
 2. $Y^{(0)} = R^{-1} Z^{(0)} \Psi$ such that $Y^{(0)^T} M Y^{(0)} = I_{s \times s}$
 3. $Q^{(0)} =$ Chebyshev-Filter$(Y^{(0)}, \xi, [\mu_f, \lambda_{\max}], A, R)$
 4. $V^{(0)} = Q^{(0)} \Gamma$ such that $V^{(0)^T} M V^{(0)} = I_{s \times s}$
 5. $W^{(0)} =$ empty
 6. **For** $k = 1, ...,$ until convergence **Do:**

   Orthogonal iteration
   i. Solve $M^{-1} A Z^{(k)} = V^{(k-1)}$ with blockCG
   ii. $P^{(k)} = Z^{(k)} - W^{(k-1)} W^{(k-1)^T} M Z^{(k)}$
   iii. $Q^{(k)} \Gamma_k = P^{(k)}$ such that $Q^{(k)^T} M Q^{(k)} = I_{s \times s}$
   iv. $Q^{(k)} = [W^{(k-1)} \; Q^{(k)}]$

   Ritz acceleration
   v. $\beta_k = Q^{(k)^T} A Q^{(k)}$
   vi. Diagonalize $\beta_k = U_k \Delta_k U_k^T$
    where $U_k^T = U_k^{-1}$
    and $\Delta_k =$ Diag$(\delta_1, ..., \delta_{p+s})$ (Ritz Values)
   vii. $V^{(k)} = Q^{(k)} U_k$ (Ritz Vectors)

   Update the computational window
   viii. $W^{(k)} =$ converged columns of $V^{(k)}$
   ix. $V^{(k)} =$ non-converged columns of $V^{(k)}$
   x. $(n, p) =$ size$(W^{(k)})$
   xi. Incorporate new vectors in $(V^{(k)})$
 7. **EndDo**
End

---

blockCG. Obviously, there is some compromise to achieve, in the sense that very small values of μ_f and ξ will minimize the number of inverse and blockCG iterations, but will also increase the computational efforts in the Chebyshev initial filtering step.

 The essence of the inverse subspace iteration is the orthogonal iteration. It consists in multiplying a set of vectors by $A^{-1} M$ and M-ortonormalizing it in turn. In step i, the multiplication by $A^{-1} M$ is performed implicitly through the iterative solution of the system $M^{-1}AZ^{(k)} = V^{(k-1)}$ via the blockCG. In order to reduce the computational costs, this system is solved with an accuracy determined by an appropriate residual threshold ε (for details see [3]). In step ii, the approximate solution vectors $Z^{(k)}$ are then projected onto the orthogonal complement of the converged vectors $W^{(k-1)}$, in order to remove the influence of eigencomponents associated with the converged eigenvalues. The set of projected vectors $P^{(k)}$ is then M-orthonormalized (step iii), and gathered together with $W^{(k-1)}$.

To improve the rate of convergence of the subspace iteration, the orthogonal iteration is followed by the Ritz acceleration (steps v to vii), as suggested by [6]. The spectral information contained in $Q^{(k)}$ is thus redistributed in the column vectors of $V^{(k)}$, that will contain each better approximations of individual eigenvectors. Steps v, vi, and vii, yield the Ritz values, $diag(\Delta) = \delta_1, ..., \delta_{p+s}$, ranged in increasing order, and the associated Ritz vectors, $[v_1, v_2, ..., v_p, ..., v_{p+s}]$, where p is the dimension of $W^{(k-1)}$ and s is the current block size.

The end of the BlockCGSI algorithm consists in testing the convergence and updating the computational window. In step viii, all the Ritz vectors that are considered as *near*-invariant (with respect to the given accuracy) are assigned to $W^{(k)}$ (more details are given in [3]). Step xi consists in incorporating new vectors in the current set of vectors $V^{(k)}$. The operation that consists in introducing a set of ℓ new vectors, after some of the Ritz vectors have converged, is detailed in Algorithm 2. We denote this algorithmic issue in the BlockCGSI algorithm as *"sliding window"*. Its purpose is to enable the approximation of a number of eigenvectors greater than the block size s. Basically, we generate randomly a linear combination of the filtered vectors $V^{(0)}$, generated in the starting steps (1 and 2) of the BlockCGSI algorithm. Then, these vectors are projected in the M-orthogonal complement of the converged ones, in order to remove the corresponding eigencomponents. Note that we can also opt to reduce or enlarge the block size s at this stage, when setting the value of ℓ (i.e. the number of newly incorporated vectors).

ALGORITHM 2: INCORPORATE NEW VECTORS

Inputs: $\ell \in \mathbb{N}, M = R^T R \in \mathbb{R}^{n \times n}, V^{(0)} \in \mathbb{R}^{n \times s}, W^{(k)} \in \mathbb{R}^{n \times p}, V^{(k)} \in \mathbb{R}^{n \times (s-\ell)}$
Begin
    a) $Y = \text{RANDOM}(s, \ell)$
    b) $P = V^{(0)} Y$
    c) $P = Q\Gamma$ such that $Q^T M Q = I_{\ell \times \ell}$
    d) $P = Q - W^{(k)} W^{(k)^T} M Q$
    e) $V^{(k)} = [V^{(k)} P]$
End

---

3   Some Numerical Experiments in an Airflow Problem

We present some numerical results concerning the exploitation of a *near*-invariant subspace W, with dimension q, associated with the eigenvalues of

$M^{-1}A$ in the range $]0, \mu[$. The spectral information, computed with BlockCGSI algorithm, is used to improve an airflow simulation code of the flow around a wing (see details in [7,4,8]). The Navier-Stokes equations are solved by finite elements in a 2D field, through a prediction-correction algorithm and a semi-implicit time discretization scheme. To obtain all the physical structures of the flow, long periods of simulation ($T = 10$ or $T = 20$) are required. In each time step (typically 0.01 s) iteration, we need to solve a system of linear equation (Poisson type), with the same coefficient matrix and changing right-hand sides, of size $n = 27283$ and $nz = 187487$ non-zero elements.

3.1 Optimal Dimension of the Basis

After preconditioning, by means of the classical Incomplete Cholesky ($M = R^T R = IC(0)$), the spectrum is distributed from $\lambda_{\min} = 6.5e - 05$ to $\lambda_{\max} = 1.7e + 00$, which corresponds to a spectral condition number κ of order $2.6e + 04$. After the preconditioning, there are still few eigenvalues on the left of the spectrum that are responsible for the non-linear convergence of the Conjugate Gradient. To remove these problems we propose to deflate this part of the spectrum through an initial projection on the CG algorithm. The spectral projector is built with the basis W of the *near*-invariant subspace computed with the BlockCGSI algorithm. We denote the technique that combines the Conjugate Gradient with the initial deflation as the INIT-CG algorithm (see details in [3]). One open question is how many eigenvalues we must compute to improve the convergence of the INIT-CG algorithm. For instance, if we want to reduce the condition number to 100, we need to cancel the effect of the 48 smallest eigenvalues ($\mu \approx 1.7e - 02$). The desirable choice is that μ falls between two clusters.

The optimal dimension of the basis W will be the one that minimizes the total cost when solving all the systems during the simulation, with our two-phase approach. This cost is given by

$$\text{Total cost} = \mathcal{C}_{BCGSI} + \mathcal{C}_{InitCG} \times \text{NGits}, \qquad (1)$$

where NGits is the number of global iteration (or time steps), i.e. the total number of systems to be solved. The cost of pre-computing the spectral information with the BlockCGSI algorithm is given by $\mathcal{C}_{BCGSI}$ and the cost of solving one system with INIT-CG is given by $\mathcal{C}_{InitCG}$.

3.1.1 Pre-computational Cost
The cost of pre-computing the spectral information depends on the dimension q of the basis W and on working parameters in the BlockCGSI algorithm, as for instance the block size s, the filtering level ξ, and the cut-off value μ_f for the filtering. In our experiments with the current test problem, the value of μ_f was automatically set as $\mu_f = \mu$, and ξ was fixed to $1e - 10$. As we have seen in [3], the filtering level is important but does not need to be very small to reduce sub-stancialy the costs in the BlockCGSI algorithm. In an efficient implementation of the BlockCGSI algorithm, Level 3 BLAS kernels can be incorporated in order

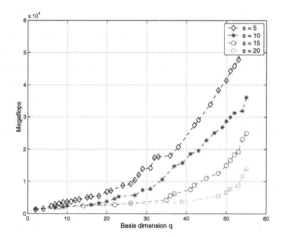

Fig. 1. Costs of pre-computing the *near*-invariant subspace ($\mathcal{C}_{BCGSI}$) for different block sizes s

to maximize the Megaflops rate, and the value of s can also be determined only on the bais of such computer aspects, keeping in mind that the *sliding window* technique adjusts the dimension of the basis W automaticaly.

Figure 1 displays the values of $\mathcal{C}_{BCGSI}$ for a block size s equal to 5, 10, 15 and 20. We can see that lower pre-computational costs are obtained with larger block sizes, specially for high dimensions of q. The principal reasons for that are the *guard vector* effect [3] and the costs of incorporating new vectors by Algorithm 2. As indicated in step b of Algorithm 2, we inject a random linear combination of the filtered starting vectors $V^{(0)}$ generated in step 4 of Algorithm 1. With this practical simplification we call the Chebyshev filtering routine only once and do not need to filter the newly incorporated vectors. The idea behind that, is that the starting vectors include already some information concerning all the eigenvalues in the range $]0, \mu[$, and to recover it we just need to redistribute this information over each of the newly incorporated vectors. In some cases, specially if we want an accurate spectral information, a breakdown can occur due to the *near*-collinearity of these new vectors relatively to the converged ones in $W^{(k)}$, which can be avoided if we force the blockCG to do a minimum number of iterations (for instance $i_{min} = 4$).

3.1.2   Solution Cost

If we analyze now the behavior of the solution costs $\mathcal{C}_{InitCG}$, as shown in figure 2. At the beginning a large decrease of the costs until q reaches approximately the dimension 20, above which the improvements are minimal ($q_{sol} \approx 20$). This occurs because until q is lower than 20 we are interpolating the extremal eigenvalues of the cluster and after that we are in the middle of the cluster. As the basis is enlarged from $q = 20$ to $q = 30$ the value of μ is shifted from $6.63e - 3$ to $1.02e - 2$, which corresponds to a small reduction of the condition number from $\kappa = 2.54e + 02$ to $\kappa = 1.68e + 02$. Additionally, the costs of the initial and restarted oblique projection in the INIT-CG algorithm also contribute to a constant level of solution costs $\mathcal{C}_{InitCG}$ when q is larger than approximately 20. This does not mean that we could not use the proposed two-phase approach with dimension larger than $q = 20$. As we will show, the large number of times that the system, with the same coefficient matrix, is solved enables the computation of larger dimension basis W. In the right of figure 2, we plot the costs of pre-computing the spectral information in terms of number of right-hand side computed through the formula

$$\text{Amor. rhs} = \left\lfloor \frac{\mathcal{C}_{BCGSI}}{\mathcal{C}_{InitCG} - \mathcal{C}_{CG}} \right\rfloor + 1,$$

where $\mathcal{C}_{CG}$ is the cost of classical CG algorithm, and $\mathcal{C}_{BCGSI}$ is the cost of pre-computing the spectral information with the BlockCGSI algorithm using a block size $s = 10$ (see figure 1). In the case of 100 systems to be solved (NGits = 100) the pre-computational costs are amortized until the basis reaches a critical value of $q = 50$. Under these conditions, if a larger number of times-steps is needed (for instance NGits = 1000) larger will be the critical value of q.

3.1.3   Minimizing the Total Cost Function

Much more important than the critical values of q is the optimal value of q that minimizes the total cost function given by (1). As we have seen, the parameter Total cost is the addition of two other cost functions that are inversely proportional, namely $\mathcal{C}_{BCGSI}$ and $\mathcal{C}_{InitCG} \times$ NGits. We have seen that the cost of pre-computing the spectral information $\mathcal{C}_{BCGSI}$ increases with q, while the solution costs $\mathcal{C}_{InitCG} \times$ NGits decrease as q increases. In figure 3, we plot all these two costs as well as the sum of two (the Total cost). The cost $\mathcal{C}_{BCGSI}$ was computed with block size $s = 10$, and the plot on the left corresponds to a simulation time of $T = 1$ where NGits $= T/\Delta t = 100$, and the plot on the right to $T = 10$ with NGits $= T/\Delta t = 1000$. The minimal value of Total cost occurs before $\mathcal{C}_{BCGSI} \approx \mathcal{C}_{InitCG} \times$ NGits because the solution cost $\mathcal{C}_{InitCG} \times$ NGits decrease very slightly when q is greater than 20 and the cost $\mathcal{C}_{BCGSI}$ grows in a larger scale. The optimal value of q (q_{opt}) is near 20 when NGits $= 100$, and near 30 when NGits $= 1000$.

The optimal value of q confirms that we must stop the BlockCGSI algorithm when the Ritz values are very close to each other. There is no benefit in approximating all the 48 eigenvalues corresponding to the targeted condition number $\kappa = 100$. The cost of solving all the systems given by $\mathcal{C}_{BCGSI}$ increases if we continue the subspace iteration when q is larger than 20. Even if the total number of systems to be solved NGits is large, as for instance 1000, there is no effective reduction of the total costs when we compute a basis of higher dimension. As shown in figure 3b, the Total cost corresponding to $q = 20$ is nearly $1.5e + 05$ Mflops and the value corresponding to $q = 30$ is nearly $1.3e + 05$ Mflops. We confirm the idea that if the spectrum is very clustered (as in the present case) the two-phase accelerating strategy is more effective if we compute only the *near*-invariant subspace associated with the extremal eigenvalues.

(a) Solution costs

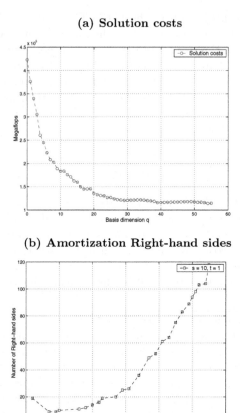

(b) Amortization Right-hand sides

Fig. 2. Solution costs and amortization right-hand sides with INIT-CG (b)

3.2   Costs-Benefits of the Two-Phase Approach

Table 1 shows the cost-benefits of accelerating strategy. We consider that the INIT-CG algorithm has converged when the backward error is below 10^{-8}. In this case the classical Conjugate Gradient (INIT-CG with $q = 0$) performs 423 Mflops. As before, we indicate the number of floating-point operations in Megaflops (Mflops) and the number of amortization right-hand sides by Amor. rhs. The pre-computational costs $\mathcal{C}_{BCGSI}$ are obtained with a block size of $s = 15$.

Firstly we can observe that when NGits $= 100$ the minimum value of Total cost, obtained with $q = 25$, is 15925 Mflops. The optimal value of q is greater than in the previous section ($q = 20$) because we run the BlockCGSI with $s = 15$ instead of $s = 10$. The same occurs when NGits $= 1000$ where the optimal value of q is equal to 40 instead of 30. This indicates that, if there were no computational restrictions, the BlockCGSI algorithm would run with a block size s as large as possible.

(a) Total costs, NGits = 100

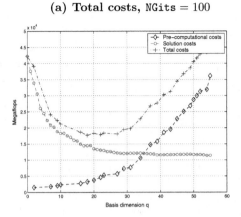

(b) Total costs, NGits = 1000

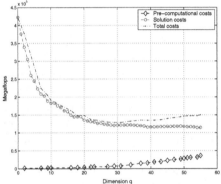

Fig. 3. Total costs of the two-phase approach

If we apply our accelerating technique to the current problem, and compute a basis W with optimal dimension ($q = 25$), 3225 Mflops are needed for the spectral pre-computation, out of which the INIT-CG achieves convergence in 127 Mflops, i.e. a reduction of 70% compared with the work needed to solve one system with the classical Conjugate Gradient (423 Mflops). Therefore, the 3225 extra Mflops are paid back after 11 consecutive global iterations of the simulation code. And, in the case of NGits = 100, the value of Total cost is reduced from 42300 to 15925 Mflops, which corresponds to a reduction of 62% of the total amount of work required to solve 100 consecutive linear systems. In the case of NGits = 1000, if a basis of size $q = 30$ is used, the value of Total cost is reduced from 423000 to 125242 Mflops, which is a reduction of order 70% over all the computational work needed to solve the 1000 systems.

Table 1. Cost-benefits of the two-phase accelerating technique

Spectral fact.		INIT-CG	Amor. rhs	Total cost	
q	C_{BCGSI}	Mflops		NGIts = 100	NGIts = 1000
0	–	423	–	42300	423000
5	2041	261	13	28141	263041
10	2377	189	11	21277	191377
15	2767	162	11	18967	164767
20	2767	145	10	17267	147767
25	3225	127	11	15925	130225
30	4242	121	15	16342	125242
35	4242	122	15	16442	126242
40	7473	116	25	19073	123473
45	11015	117	36	22715	128015
50	14972	118	50	26772	132972
55	24950	114	81	36350	138950

4   Conclusions

A two-phase approach was suggested to improve the numerical simulation of
an airflow problem. In the first phase we have computed with the BlockCGSI
algorithm a *near*-invariant subspace linked to the smallest eigenvalues. In the
second phase, the basis of this subspace is used in each run of the Conjugate
Gradient to deflate the starting residual (INIT-CG) and improve the consecutive
solutions of the linear systems with the same coefficient matrix and changing
right-hand sides.

The key question of this strategy is the dimension q of the *near*-invariant subspace to be computed. The optimal dimension depends on a compromise between
the pre-computational (first phase) costs and the solution (second phase) costs.
The cost of pre-computing the spectral information, which increases with the
dimension of the basis, depends also on the block size s used on the BlockCGSI
algorithm. The results show that larger block sizes reduce the pre-computational
costs. On the other hand, the solution costs decrease with the increasing dimension of the *near*-invariant subspace q, until a q value (q_{sol}) is reached, above
which the solution costs stagnates. The optimal dimension (q_{opt}) is thus, the one
that minimizes the sum of the two costs (pre-computing and solution) over all the
systems to be solved. The results showed that q_{opt} and q_{sol} are close to each other.

The stagnation of the solution costs occur because the convergence rate of
CG is not sufficiently improved, since the effect of all the extremal eigenvalues,
separated from the main cluster, was removed. As the remaining eigenvalues
are very close to each other, their deflation yields only a low reduction on the
condition number that governs the convergence rate of the CG.

As a consequence of the previous results, we suggest a dynamical strategy to
set up the dimension q of the *near*-invariant subspace associated with the smallest eigenvalues without *a priori* knowledge of the spectrum. At the beginning,
after setting a reduced condition number $\kappa = \lambda_{\max}/\mu$ in agreement with the

convergence rate of the CG, we request to approximate all the eigenvectors corresponding to the eigenvalues smaller that μ. At each iteration of the BlockCGSI algorithm, if the request is not satisfied, we compute the gaps between the approximated eigenvalues (Ritz values). As soon as the larger gap is below a given preset tolerance (which means that we are in the middle of a cluster), we stop the BlockCGSI algorithm and switch to the CG improved with precomputed basis (INIT-CG), to compute the remaining system solutions of the simulation problem.

References

1. Arioli, M., Ruiz, D.: Block conjugate gradient with subspace iteration for solving linear systems. In: Iterative Methods in Linear Algebra, Second IMACS Symposium on Iterative Metohds in Linear Algebra, Blagoevgrad, Bulgaria, S. Margenov and P. Vassilevski (eds.) (June, 1995) pp. 64–79
2. Balsa, C., Palma, J., Ruiz, D.: Partial spectral information from linear systems to speed-up numerical simulations in computational fluid dynamics. In Daydé, M., Dongarra, J., Hernandez, V., Palma, J., eds.: High Performance Computing for Computational Science, 6th Int. Meeting, VECPAR'04. LNCS 3402, Berlin, Springer-Verlag (2005) pp. 699–715
3. Balsa, C., Daydé, M., Palma, J., Ruiz, D.: Inexact subspace iteration to exploit partial spectral information. Technical Report TR/TLSE/05/09, Institut National Polytechnique de Toulouse, LIMA-IRIT (2005)
4. Bergmann, M.: Analyse physique de la Transition Laminaire-Turbulent 2D dans des écoulements Cisaillés a l'Aide d'un Code de Navier-Stokes en éléments Finis. Rapport de stage de DEA, Toulouse (2001)
5. O'Leary, D.P.: The block conjugate gradient algorithm and related methods. Linear Algebra and its Applications (1980) 293–322
6. Parlett, B.N.: The Symmetric Eigenvalue Problem. SIAM, Philadelphia (1998)
7. Braza, M.: Analyse Physique du Comportement Dynamique d'un Écoulement Externe, Décollé, Instationnaire en Transition Laminaire-Turbulent. Application: Cylindre Circulaire. Thse d'état, INPT, Toulouse (1986)
8. Martinat, G.: Analyse physique de la Transition Laminaire-Turbulent sous l'Effect de la Rotation par un Code en Élement Finis. Rapport de stage de DEA, Toulouse (2003)

EdgePack: A Parallel Vertex and Node Reordering Package for Optimizing Edge-Based Computations in Unstructured Grids

Marcos Martins, Renato Elias, and Alvaro Coutinho

Center for Parallel Computations and Department of Civil Engineering
Federal University of Rio de Janeiro, P. O. Box 68506,
RJ 21945-970 – Rio de Janeiro, Brazil
{marcos, renato, alvaro}@nacad.ufrj.br
http://www.nacad.ufrj.br

Abstract. A new and simple method is proposed to choose the best data configuration in terms of processing phase time according to previous probing of edge-based matrix-vector products for codes using iterative solvers in unstructured grid problems. This method is realized as a suite of routines named EdgePack, acting during both pre-solution and solution phase, based on data locality optimization techniques and variations of matrix-vector product algorithm. Results have been demonstrating the great flexibility and simplicity of this method, which is suitable for distributed memory platforms in which different data configurations can coexist.

1 Introduction

The performance optimization of codes based on iterative solvers for unstructured grids based problems has the matrix-vector product algorithm as a key issue. It is well known that data reordering techniques comprise an effective solution for good performance results during matrix-vector product computations due to data locality optimization [1]-[12]. Despite these techniques, the overall performance can be further increased by the migration from element based to edge based data structure, verified in the last decade [11], [13]-[18]. This data structure is quite suitable for reordering manipulations, strengthening even more the overall performance [19].

The combination of data reordering techniques with edge-based data structures generates lots of possibilities which can be or not the best performance solution for one or another hardware and software platform [19]. The code performance depends on many factors namely, computational architecture, compiler options, data configuration, number of degrees of freedom per node (algorithm complexity) and algorithm structure itself. According to previous results [19], there is no an ultimate data combination known prior to processing phase unless a probe is performed, since the many combinations possible may produce unexpected results.

This work presents a method to determine which data combination for a given computer platform is the best one in terms of processing time. This method is simply based on the choice of the best results after probing the several possibilities more adequate to such platform. Considering a particular parallel platform, as clusters

M. Daydé et al. (Eds.): VECPAR 2006, LNCS 4395, pp. 292–304, 2007.
© Springer-Verlag Berlin Heidelberg 2007

(heterogeneous or not), the best data configuration choice can be different from each other and thus, different data structures can be used together for the same model. Concerning distributed parallel processing, the standard approach is to partition data before performing data reordering on each processor. Data partition is performed by Metis library [20].

Based on the nodal renumbering algorithms and concepts proposed in [4], [21], [22] and edge renumbering algorithms proposed in [17], algorithms for matrix-vector product composed by 1, 3 and 4 degrees of freedom per node (hereafter, referred as *dof*), for symmetric and non-symmetric matrices, were implemented. The techniques employed try to define the most suitable data reordering according to the computational system at hand, even without any prior knowledge about the processors architecture. Among them, a sorting of edges, in increasing order by the edge first node number (namely hereafter, reduced i/a), halves the indirect addressing operations of the edge-based matrix-vector product algorithm [21].

Additionally, data locality algorithms were used in association with special edge groupings, which improve the edge-based matrix-vector product algorithm. This was implemented for tetrahedra, grouped into 3 and 6 edges, named respectively superedge3 and superedge6 for both incompressible fluid flow and geomechanics problems [17]. For the latter, it was implemented an edge-based interface element with special groupings named superedge4 and superedge9, comprising groups of 4 and 9 edges respectively [23].

Related to memory dependency, lists of edges are built where no pair of edges in the same edge list shares the same node. This arrangement is referred here as nodal disjointing and is responsible for a significant lack of performance. A Reverse Cuthill McKee (RCM) [24] algorithm in conjunction with edge and element sorting according to node numbering are further employed to diminish the negative effect provided by the nodal disjointing ordering.

The method developed is realized as a suite of routines built in Fortran90, comprising the necessary tasks for both pre-solution and solution phases of finite element codes based on iterative solution methods, for serial or parallel platform (shared, distributed or hybrid), named EdgePack. The remainder of this work is organized as follows. In the next section, we revise the edge-based data structures and the data reordering algorithms. In Section 3 we describe EdgePack in detail. In the following section, we show some numerical experiments exploring the main characteristics of EdgePack. The paper ends with a summary of our conclusions and remarks.

2   Edge-Based Structure and Data Reordering

Edge based finite element data structures have been introduced for explicit computations of compressible flow in unstructured grid finite element and finite volume computations [13], [14], [25]. It was observed in these works that residual computations with edge-based data structures were faster and required less memory than standard element-based residual evaluations. Following these ideas, Coutinho *et al* [15], Catabriga and Coutinho [16], Sydenstricker *et al* [23], Elias *et al* [18] derived edge-based finite element implementations respectively for elasto-plasticity, the

SUPG finite element formulation with shock-capturing for inviscid compressible flows, interface elements for modeling joints and faults and the SUPG/PSPG solution of incompressible flows. Differently from the previous finite volume/finite element implementations, all these works used the concept of algebraically disassembling the finite element matrices to build the edge matrices introduced by Catabriga and Coutinho [16]. This procedure makes the edge-operators construction independent of the underlying finite element formulation. To illustrate this procedure, consider three dimensional problems on unstructured meshes composed by tetrahedra. Thus, the element matrice $\mathbf{A}^e$ can be disassembled into their edge contributions as

$$\mathbf{A}^e = \sum_{s=1}^{m} \mathbf{T}_s^e \, . \tag{1}$$

where $\mathbf{T}_s^e$ is the contribution of edge s to $\mathbf{A}^e$ and m is the number of edges per element. The contributions of all elements sharing a given edge s is given by the following edge matrix,

$$\mathbf{A}_s = \sum_{s \in \mathrm{E}} \mathbf{T}_s^e \, . \tag{2}$$

where E is the set of all elements sharing a given edge s.

When working with iterative solvers, it is necessary to compute sparse matrix-vector products. These matrix-vector products are responsible for a good share of the overall computational effort. A straightforward way to implement the edge-by-edge, similar to popular element-by-element matrix-vector product, is

$$\mathbf{A}\mathbf{p} = \sum_{l=1}^{ne} \mathbf{A}^l \mathbf{p}^l \, . \tag{3}$$

where ne is the total number of local structures (edges or elements) in the mesh and $\mathbf{p}^l$ is the restriction of $\mathbf{p}$ to the edge or element degrees-of-freedom. As a prototype of such procedure, the edge-based Laplacian loop algorithm (comprising 1 dof per node) as proposed in [21], is:

Laplacian loop for a single edge sparse matrix-vector product

```
do edge = edge_begin, edge_end
    eq_1 = lm(1,edge)
    eq_2 = lm(2,edge)
    ap = a(edge) * (u(eq_2) - u(eq_1))
    p(eq_1) = p(eq_1) + ap
    p(eq_2) = p(eq_2) + ap
end do
```

where array lm stores edge equation numbers, a stores the edge coefficient and u stores the unknown values.

In order to achieve a good balance between memory accesses and floating point operations (*flops*), reordering techniques are suggested in the literature such as *superedges* [17], [26] and reduced indirect addressing (i/a) edges [21]. The *superedge* scheme reaches good balance of i/a and *flops* [15] without complex preprocessing codes [17]. In this case, computer costs are just related to the new order of the edge list, considering the edges agglomerated, in geometric sense, for example, in tetrahedral shape, swept by stride of 6 edges.

An alternative to reduce i/a is to convert an edge-based loop into a vertex-based loop [4] in which the edges are arranged in such a way that the first node always has the lower number and increases as the edge number increases with stride one. This loop reuses vertex-based data items in most or all of the accesses several times before discarding it. This approach increases flops but reduces i/a operations, whereas the edge has to be processed twice. Table 1 shows a comparison of computation parameters for the matrix-vector multiplication algorithm for reduced i/a and superedge schemes, considering 3 *dof* and symmetric operators.

Table 1. Comparison of computational parameters for matrix-vector multiplication algorithm for reduced i/a and *superedges* for 3 *dofs*

Group/Parameter	flops	i/a	Flops/(i/a)
Simple Edge	36	18	2.0
Superedge6	268	36	7.4
Superedge3	130	27	4.8
Superedge9	436	54	8.1
Superedge4	190	36	5.3
Reduced i/a	39	9	4.3

3 EdgePack

EdgePack is a suite of routines built in Fortran90 to optimize computations on unstructured grids. EdgePack is divided into two sets of routines: the first set is a preprocessing phase and aims to reorder finite element meshes composed by tetrahedra and prisms to improve the performance of iterative solvers for any number of degrees of freedom per node. The main concern is to optimize data locality and data reuse for serial or parallel and shared or distributed memory computers. The second set is composed by a series of optimized edge-based data routines for matrix-vector products and element matrix disassembling into edges.

The main tasks performed by EdgePack during the preprocessing phase comprise the edge connectivity assembly, based on building fast hash-tables, edge groupings into *superedges* [17], [26], nodal and edge reordering into reduced indirect addressing edge mode [4], [21], nodal reordering to minimize bandwidth based on Reverse Cuthill McKee (RCM) algorithm [24], nodal disjointing reordering for pipelined processing providing data with no memory dependencies, edge reordering driven by equation map and element reordering according edge connectivity.

EdgePack runs either on serial or parallel distributed memory systems. Targeting on parallel processing, the mesh partitioning is performed by Metis library [20], in weighted or non-weighted mode, and for all subdomains data reordering is accomplished locally on each processor node. This method provides the possibility of achieving the best data structure and reordering choice for each processing node, as in the case of heterogeneous clusters. EdgePack probes timing results for matrix-vector products based on equation map and element matrix topology, taking into consideration the number of degrees of freedom per node and if the matrix is symmetric or not, and chooses the data configuration from the best matrix-vector product timing result. Based on this probe, it is possible to determine which data structure will suit best on a given hardware and software configuration and automatically decide which element, edge and node data structure and order best fit on, without neither the user intervention nor concern. However, the user can set directly which data structure to use without probing. EdgePack can be used as either a stand-alone program or library.

Communication among processors is another issue treated by EdgePack. The subdomain interfaces, inherent in distributed parallel processing, can either be done by simply indexing shared nodes among subdomains – thus preserving local data order – or by ordering shared nodes sequentially for optimizing communications tasks.

For processing phase, EdgePack provides optimized edge-by-edge matrix-vector product routines for 1 up to 4 *dof* per node, for symmetric or non-symmetric operators, which account for i/a and *flops* reduction, and data reuse strategy based on data locality and agglomeration into registers. Besides data configuration paradigm, EdgePack probes the matrix-vector product routines based on typical vector lengths for chunkwise and nodal disjointed loops, alternative right hand side evaluation (RHS) [21] and loop unrolling into edges. The matrix-vector product routines are ready to run under serial and parallel (hybrid or not) mode and are set according to pre-compiler directives.

Table 2. Keywords for data and matrix-vector algorithm configuration

Keyword #	Description	Option
1	Nodal Disjointing	Yes / No
2	Chunks	Yes (List Length) / No
2.1	List Length	Starting from 64 up to 2048 and free (mandatory for reduced 1 and 2)
3	Node Order	RCM / Reduced
4	Edge Order	Reduced (0/1/2) / Simple or *Superedge*
4.1	Prescribed Edge	Yes / No (for simple and *superedge* only)
4.2	*Superedge* Omission	-Sx (for *superedge* only, x = 3, 4, 6 and 9)
5	Shared Nodes (for distributed processing)	Indexed / Sequential
6	Alternative RHS	Yes / No
7	Loop Unrolling	Yes (2/3/6) / No

The various data configurations and matrix-vector product modes available can be set by a combination of 7 keywords as presented by Table 2. The first five keywords are related to the preprocessing phase and the remaining ones set the matrix-vector product algorithm type.

4 Preliminary Results

This section presents some preliminary results from three models comprising geomechanics and incompressible fluid flow. The first two ones are models of true sedimentary basins with faults. The last model comprises the transient incompressible fluid flow around a cylinder. The first model illustrates the main data orderings available on EdgePack, for serial processing, through edge connectivity graphs, characterizing data locality for each case. In the second model we show time probing results for some data ordering available on EdgePack for serial mode, fancying the fastest ones for each case to glance the various data configuration possibilities and related results. The third model presents edge connectivity graphs, timing results for the best data configuration selection after probing, besides some validation results for parallel processing.

4.1 Sedimentary Basin – Model 1

This model represents a sedimentary basin which geometry and materials correspond to a region at Colombia, South America. A fault is present in the model, crossing it completely. The model surface is approximately 80×80 km^2 and 23 km deep, and comprising 141,766 tetrahedra, 5,133 interface elements and 28,897 nodes. Fig. 1 presents the surface of the of sedimentary basin mesh – model 1.

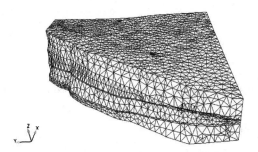

Fig. 1. Surface mesh of sedimentary basin – model 1

The model will be employed to illustrate the main data orderings available in EdgePack. Fig. 2 presents the main edge and vertex ordering generated by EdgePack for this model by a representation of its edge connectivity, where the first edge node is in black line and the second one in gray line. The edge sequence is highlighted through lines connecting the nodes. In this figure, the original vertex and edge orderings are presented in letter (a); the reduced edge order is presented in letter (b) in which the monotonic order of the first nodes can be clearly noted by the ramps. Letter (c) presents the *simpleedge* order and letter (d), the *superedge* one. It can be noted on these orderings the good data locality configuration. In letter (d), the five ramps represent the *superedges* employed as *superedge*6 (S6), *superedge*3 (S3), *superedge*9 (S9), *superedge*4 (S4) and *simpleedge* respectively.

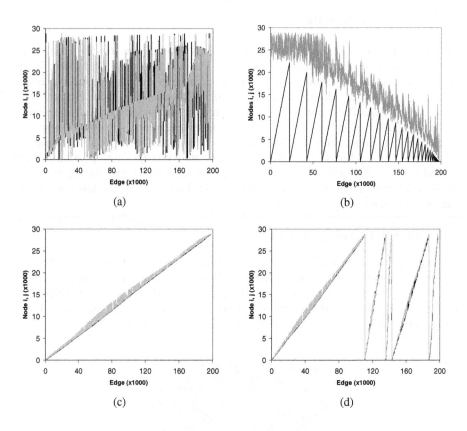

Fig. 2. Edge connectivity orders generated by EdgePack for sedimentary basin – model 1

4.2   Sedimentary Basin – Model 2

This model represents a portion of a sedimentary basin which geometry and material corresponds to a region at northeast of Brazil, South America. The model is constituted of four blocks separated by three geological faults. The model is 9.1×6 km^2 and 1.5 km deep, comprising 371,244 nodes, 2,064,940 linear tetrahedral elements and 17,317 interface elements, as shown in Fig. 3. The boundary conditions impose compression along the major dimension, normal to faults and normal displacements nullified over entire surface, besides overburden from upper layers and self-weight. Analysis comprises 3 *dofs* per node as displacements.

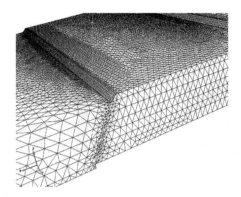

Fig. 3. Surface mesh of sedimentary basin - model 2

Fig. 4 presents the most important time probing results out of 208 jobs on Itanium 2 platform, under pipelined serial mode. Fig. 4(a) presents some time probing results where the clear benefit of simpleedge over reduced schemes for this case is represented by 45% gain in time. The labels correspondence is presented by Table 3 in which results were obtained for non-unrolled matrix-vector loops.

Table 3. Data configuration for results presented by Fig. 4(a)

Label	Nodal Order	Chunk Length (min/max)
Simpleedge	RCM	64/2048
Superedge	RCM	64/ 512
Reduced 0	Reduced	64/ 64
Reduced 1	Reduced	64/ 64
Reduced 2	Reduced	64/ 128

Fig. 4(b) presents the percentage of occurrences of nodal order configuration for all 208 jobs. The percentage is referred for each legend individually. It is clear the advantage of RCM nodal ordering over reduced one for this combination of model and platform. However, only 4% of all cases with nodes ordered by RCM attain the best results.

For the reduced edge scheme, Fig. 4(c) presents the percentage of occurrences related to each one individually. This picture shows the slight advantage of reduced 0 and 1 scheme over reduced 2, for this case. However, the distribution tends to be uniform for three modes.

Fig. 4(d) pictures time probing results for nodes ordered by RCM and edges arranged as *simpleedges* and *superedges*. In the latter, all *superedges* available were used. The supremacy of *simpleedge* over *superedge* is notorious since *superedge* only occurs in the third time scale. However, only 7% of *simpleedge* data combinations appear as best results.

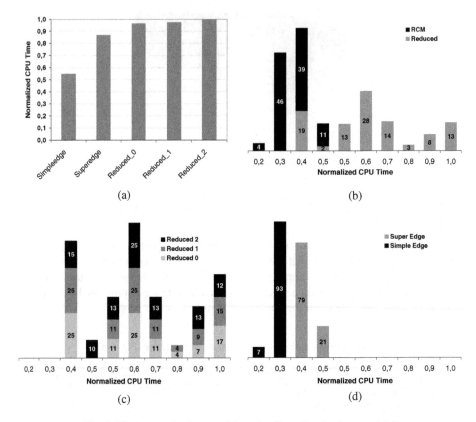

Fig. 4. Timing results from probing of sedimentary basin – model 2

4.3 Incompressible Fluid Flow

The problem of a fluid flowing around a circular cylinder is considered as an application of the EdgePack's data improvements to incompressible fluid flow codes. Time probing for this case was performed and the data configuration comprising edge ordering by *simpleedge* scheme, nodal ordering by RCM and chunk length of 4096 was chosen. For this problem an extension for transient flows of the edge-based stabilized finite element implementation described in [18], is applied to solve the three dimensional *u-p* fully coupled (4 *dofs* per node) problem arisen from the Navier-Stokes discretization. The computational domain follows the dimensions described in [27] and the mesh is formed by 446,662 linear tetrahedra elements, 1,010,367 edges, 81,991 nodes, summarizing 174,008 equations.

The results, for a Reynolds 100, are assessed and compared with those presented by [28] and [29] showing a good agreement for the time evolution of the drag and lift forces on the cylinder surface as depicted in Fig. 5 for lift coefficient. Baranyi [29] reported a Strouhal number of 0.163 and drag and lift coefficients of 1.346 and 0.228 respectively for Reynolds 100 while Williamson [28], employing an experimental correlation for the Strouhal-Reynolds pair, estimated the value of 0.1643. In this work we have found S_t = 0.16 and 1.313 and 0.225 for drag and lift coefficients respectively, which compares well with the results presented by those authors.

A typical computation of this problem considering 10,000 fixed time steps, which corresponds to 500 time units, spent 6.48 hours running in MPI mode with four processors of a SGI Altix 350 system equipped with Intel Itanium-2 1.5 GHz processors.

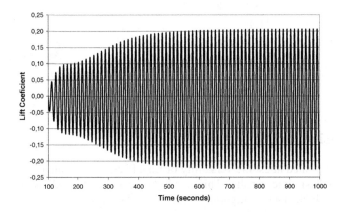

Fig. 5. Lift coefficient for flow around a cylinder problem

Fig. 6. Snapshot of cylinder surface mesh and vorticity, showing the development of the von Karman vortex streets

Fig. 7 presents nodal edge connectivity for partition 2 out of 4 partitions. For this figure, letter (a) presents the original node order for edges ordered in chunks for pipeline processing and letter (b) corresponds to the combination of *simpleedge* order in chunks for pipelined processing with nodes renumbered by RCM.

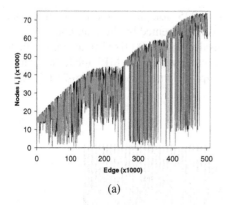
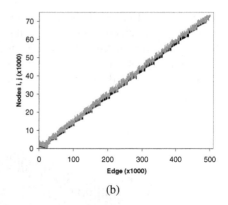

(a)   (b)

Fig. 7. Nodal edge connectivity of *simpleedge* for partition 2

5 Conclusions and Incoming Work

The results presented attest the versatility of EdgePack in choosing data configuration according to best time results. This versatility can be derived from the need of determining which data configuration produces the best time performance among hundreds of possibilities besides computational platform effects.

The model presented in Section 0 presented the main data configuration possibilities and its choice range. The mesh partitioning was also exploited for distributed parallel processing. The good effect over data locality was clearly shown by nodal edge connectivity graphs.

An example of time probing for serial run was done in 0, where performance results were undetermined a priori and probing demonstrated that a wrong choice could represent about 45% loss in time processing. The presented graphs glimpsed the various possibilities and their unexpected results strengthening the EdgePack flexibility in setting data configuration without user intervention.

Section 0 presented an example of distributed parallel processing over partitions locally ordered by EdgePack. In this example, data was prepared to run under hybrid mode, comprising data distribution, memory dependency and good data locality.

As next step in EdgePack development, the work goes towards distributed parallel processing in heterogeneous clusters and grids, where in each node, EdgePack can determine different data configurations, to set the best performance individually providing the coexistence of different data structures during same analysis and exploiting the most adequate data configuration for each processor.

References

1. Burgess, D. A. and Giles, M. B.: Renumbering unstructured grids to improve the performance of codes on hierarchical memory machines, Advances in Engineering Software 28 (1997) 189-201
2. Carey, G. F., Swift, S. and McLey, R. T.: Maximizing sparse matrix-vector product performance on RISC based MIMD computers. Journal of Parallel and Distributed Computing, v.37, p.146-158, 1996

3. Douglas, C. C, Hu, J., Kowarschik, M., Rude, U. and Weiss, C.: Cache Optimization for Structured and Unstructured Grid Multigrid. Electronic Transactions on Numerical Analysis, v.10, p.21-40, 2000

4. Gropp, W. D., Kaushik, D. K., Keyes, D. E. and Smith, B. F.: Performance Modeling and Tuning of an Unstructured Mesh CFD Application, Proceedings of SC 2000, IEEE Computer Society, 2000, Dallas, Texas, United States, Article No. 34, ISBN:0-7803-9802-5

5. Löhner, R.: Renumbering Strategies for unstructured-grid solvers operating on shared-memory, cache-based parallel machines, Computer Methods in Applied Mechanics and Engineering 163 (1998) 95-109

6. Oliker, L., Canning, A., Carter, J., Shalf, J. and Skinner, D.: Evaluation of cache-based superscalar and cacheless vector architectures for scientific computations, Proceedings of the 18th Annual International Conference on Supercomputing, Malo, France, 2004, ISBN:1-58113-839-3

7. Oliker, L., Li, X., Heber G. and Biswas, R.: Parallel Conjugate Gradient: Effects of Ordering Strategies, Programming Paradigms, and Architectural Platforms, IEEE Transactions on Parallel and Distributed Systems, 11(9):931-940, 2000

8. Oliker, L., Li, X., Heber, G. and Biswas, R.: Ordering Unstructured Meshes for Sparse Matrix Computations on Leading Parallel Systems, Lecture Notes In Computer Science, Vol. 1800, pp. 497-503, 2000

9. Oliker, L., Li, X., Husbands, P. and Biswas, R.: Effects of Ordering Strategies and Programming Paradigms on Sparse Matrix Computations, SIAM Review, Vol. 44, No. 3, pp 373-393

10. Pinar, A. and Heath, M. T.: Improving Performance of Sparse Matrix-Vector Multiplication, Conference on High Performance Networking and Computing, Proceedings of the 1999 ACM/IEEE Conference on Supercomputing (CDROM), Portland, Oregon, United States, Article No. 30, 1999, ISBN:1-58113-091-0

11. Ribeiro, F. L. B and Coutinho, A. L. G. A.: Comparison between element, edge and compressed storage schemes for iterative solutions in finite element analyses. International Journal for Numerical Methods in Engineering, Volume 63(4): 569-588, 2005

12. Vuduc, R., Demmel, J. W., Yelick, K. A., Kamil, S., Nishtala, R., and Lee, B.: Performance Optimizations and Bounds for Sparse Matrix-Vector Multiply. Conference on High Performance Networking and Computing, Proceedings of the 2002 ACM/IEEE Conference on Supercomputing, Baltimore, Maryland, Pages: 1 – 35, 2002

13. Peraire J, Peiro J, Morgan K, 1993. Multigrid solution of the 3d-compressible Euler equations on unstructured grids. Int. J. Num. Meth. Engrg.. 36(6): 1029-1044

14. Luo H, Baum JD, Löhner R, 1994. Edge-based finite element scheme for the Euler equations, AIAA Journal, 32(6):1183-1190

15. Coutinho ALGA, Martins MAD, Alves JLD, Landau L and Moraes A, 2001. Edge-based finite element techniques for non-linear solid mechanics problems. Int. J. for Num. Meth. in Engrg, 50(9):2053-2068

16. Catabriga L and Coutinho ALGA. , 2002. Implicit SUPG solution of Euler equations using edge-based data structures. Computer Methods in Applied Mechanics and Engineering, 32:3477-3490

17. Martins, M.A.D., Alves, J.L.D., Coutinho, A.L.G.A.: Parallel Edged-Based Finite Techniques for Nonlinear Solid Mechanics. Lecture Notes on Computer Science, Vol. 1981, Springer-Verlag, Berlin Heidelberg (2001), pp 506-518

18. R. N. Elias, M. A. D. Martins, A. L. G. A. Coutinho, Parallel Edge-Based Inexact Newton Solution of Steady Incompressible 3D Navier-Stokes Equations, J.C. Cunha and P.D. Medeiros (Eds.): Euro-Par 2005, LNCS 3648, pp. 1237–1245, 2005.

19. Coutinho ALGA, Martins MAD, Sydenstricker R and Elias RN. Performance comparison of data reordering algorithms for sparse matrix-vector multiplication in edge-based unstructured grid computations, Int. J. Num. Meth. Engng, accepted.
20. Karypis G. and Kumar V., Metis 4.0: Unstructured Graph Partitioning and Sparse Matrix Ordering System. Technical report, Department of Computer Science, University of Minnesota, Minneapolis, (1998). http://www.users.cs.umn.edu/~karypis/metis.
21. Löhner, R., Galle, M.: Minimization of indirect addressing for edge-based field solvers, Communications in Numerical Methods in Engineering, 18 (2002) 335-343
22. Löhner, R.: Some useful renumbering strategies for unstructured grids, International Journal for Numerical Methods in Engineering, Vol. 36, (1993) 3259-3270
23. Sydenstricker, R.M., Martins, M.A.D., Coutinho, A. L. G. A., Alves, J.L.D.: Edge-Based Interface Elements for Solution of Three-Dimensional Geomechanical Problems. Lecture Notes in Computer Science, v.2565, p.53 - 64, 2003
24. Cuthill, E. and McKee, J.: Reducing the bandwidth of sparse symmetric matrices. In Proc. ACM Nat. Conf., pp 157-172, 1969
25. Barth, T. J., "Numerical Aspects of Computing Viscous High Reynolds Number Flows on Unstructured Meshes", AIAA, 29th Aerospace Sciences Meeting, January 7-10, AIAA 91-0721, Reno, Nevada, 1991
26. Löhner, R.: Edges, Stars, Superedges and Chains; Comp. Meth. Appl. Mech. Eng. 111, 255-263 (1994)
27. Kalro V. and Tezduyar T.E., Parallel 3D Computation of Unsteady Flows around Circular Cylinders, Parallel Computing 23 (1997) 1235-1248
28. Williamson, CHK, Defining a Universal and Continuous Strouhal-Reynolds Number Relationship for the Laminar Vortex Shedding of a Circular Cylinder, Phys Fluids 31 (1988) 2742-2744
29. Baranyi, L, Computation of Unsteady Momentum and Heat Transfer from a Fixed Circular Cylinder in Laminar Flow, Journal of Computational and Applied Mechanics, vol 4, no. 1, (2003) pp. 13-25

Parallel Processing of Matrix Multiplication in a CPU and GPU Heterogeneous Environment

Satoshi Ohshima, Kenji Kise, Takahiro Katagiri, and Toshitsugu Yuba

Graduate School of Information Systems
The University of Electro-Communications
1-5-1, Chofugaoka, Chofu-shi, Tokyo, Japan
Tel.: +81-42-443-5644; Fax: +81-42-443-5644
ohshima@hpc.is.uec.ac.jp, {kis, katagiri, yuba}@is.uec.ac.jp

Abstract. GPUs for numerical computations are becoming an attractive alternative in research. In this paper, we propose a new parallel processing environment for matrix multiplications by using both CPUs and GPUs. The execution time of matrix multiplications can be decreased to 40.1% by our method, compared with using the fastest of either CPU only case or GPU only case. Our method performs well when matrix sizes are large.

1 Introduction

The performance of Graphics Processing Units (GPU) has been significantly improved in recent years. Compared with the CPU, the GPU is better suited for parallel processing and vector processing and has evolved to perform various types of computation, in addition to graphics processing, including numerical computations. General-purpose computations on GPUs (GPGPU) have been examined for various applications[1,2,3].

A high-performance computing environment is necessary for numerical computations like physics and earth environment simulations which require enormous computational power. Matrix multiplication is an important operation in numerical computation. Speeding up matrix multiplication results in a corresponding speed up increase in various numerical computations.

Basic Linear Algebra Subprograms (BLAS)[4] is frequently used as a basic numerical calculation library. Automatically Tuned Linear Algebra Software (ATLAS)[5], is a fast implementation of BLAS in CPUs. These libraries have succeeded in exploiting performance enhancing features of a CPU.

In BLAS, matrix multiplication is treated as a computation of $C = \alpha \times A \times B + \beta \times C$ where A, B, and C are matrices, and α and β are scalars. Improving performance of such computations will speedup of various numerical calculations.

We propose a heterogeneous computing environment for parallel processing using both CPUs and GPUs for numerical computations. First we divide the larger problem into two partial problems and assign one to the CPU and the other to the GPU. Ideally, this results in achieving high performance of both the

M. Daydé et al. (Eds.): VECPAR 2006, LNCS 4395, pp. 305–318, 2007.
© Springer-Verlag Berlin Heidelberg 2007

CPU and the GPU. We evaluate this method of parallel processing using the NVIDIA GeForce7800GTX and the 6600GT as our GPUs.

Section 2, discusses the background and related work. Section 3, proposes a parallel processing method using 1-CPU and 1-GPU. Implementation and analysis of our method for matrix multiplication are described in Section 4. Section 5 describes the experimental results measured on a real heterogeneous environment and section 6 discusses about future research issues.

2   Background and Related Work

Graphics processors generate large number of polygons at a very high speed. In generating polygons, vector and matrix computations are frequently used. Many computations can be executed in parallel on a GPU. GPUs have evolved rapidly with hardware suited for both vector and highly parallel computations compared with a CPU. In addition, the programmable shader, controls processor's behavior in software level, has become popular in newer GPUs. Since floating point arithmetic of a GPU is advanced these GPUs can efficiently execute various computations rather than generating polygons[2,3,6].

The GPGPU aims at resolving target calculations utilizing the computational power of a GPU. The main scope of GPGPU includes the computation of high-level shading and lighting in creating real images[7], various simulations and visualizations[8,9]. These are examples related to graphics computations, the original use of GPUs. Besides these graphics computations, utilization for numerical computations is a new application domain of the GPU[10,11,12]. Floating-point computations of GPUs have a lower precision than CPUs[13]. Therefore, further evaluation and improvement of precision are necessary because of the very high arithmetic precision required in numerical calculations.

Matrix multiplication is a popular GPGPU application. Current research includes: efficient utilization of GPU for matrix multiplication, decreasing execution time by using vector computation and programmable shader and an effective utilization of GPU inner cache[2,14,15,16,17]. However, effective performance evaluation results have not yet been obtained, because of the issues related to memory and bandwidth in inner GPU.

Task parallelization has been used to increase performance in systems having both CPUs and GPUs. For example, in a real-time movie, the CPUs calculate the position of numerous objects and the GPUs calculate the shades and high lights of these objects. However, data parallelization in both the CPUs and GPUs is rare. We propose data parallelization with the CPUs and GPUs for numerical calculation.

The research on parallel processing in heterogeneous environments includes multiple CPUs with different performance. The problems addressed are: scheduling for effective utilization of all processors, load balancing in a dynamically changing environment, and resolving differences in arithmetic precision[18]. We try to overcome such issues using a new domain as CPU and GPU complex heterogeneous system.

3   Parallel Processing in a CPU and GPU Heterogeneous Environment

3.1   Execution Time Analysis of Parallel Processing

Conventional approaches for execution time analysis for both CPU and GPU include processor speedup, increased processor utilization (various proposals and implementations have been investigated for approaching theoretical performance), and parallel processing with multiple processors.

First, we formulate the execution time for CPUs. The execution time T_{CPU_ALL} is defined in equation (1). We denote the number of operations in the target computation as R. The execution time required for solution using peak CPU performance is denoted as a function of R, or $f_{CPU}(R)$. The effective execution time increases because the CPU cannot always attain peak performance. We denote the increase of execution time relative to the ideal execution time (execution time at effective performance / execution time at peak performance) as a ($a \geq 1$). Ideally, the execution time is divided by n ($n \geq 1$), the number of CPUs used in parallel processing. The execution time is increased by the parallelization overhead when more than two CPUs are used. We neglect this time for simplification.

$$T_{CPU\_ALL} = \frac{f_{CPU}(R) \times a}{n} \tag{1}$$

Similarly, we formulate the execution time for GPUs. The execution time T_{GPU_ALL} is defined in equation (2). In this equation, the execution time required for a solution using the peak GPU performance is $f_{GPU}(R)$, the increase of execution time relative to the ideal execution time is b ($b \geq 1$), and the number of GPUs is m ($m \geq 1$).

$$T_{GPU\_ALL} = \frac{f_{GPU}(R) \times b}{m} \tag{2}$$

In previous research on numerical computations using GPGPUs, the execution time of a GPU system was compared to that of a CPU system, as shown by (1) and (2). However, GPGPU systems often have both CPUs and GPUs. Therefore, we propose a parallel processing method to obtain the overall CPUs and GPUs performance. We divide a target computation into a parts, and assign them to CPUs and GPUs to perform.

Assume that the target computation is divided into two partial computations. One partial computation with the assignment ratio r ($0 \leq r \leq 1$) of the computation is assigned to CPUs. The other, with assignment ratio $1 - r$ of the computation, is assigned to the GPUs. Then, the CPUs' execution time T_{CPU} defined in equation (1) becomes equation (3). Similarly, the GPUs' execution time T_{GPU} defined in equation (2) becomes equation (4).

$$T_{CPU} = \frac{f_{CPU}(R \times r) \times a}{n} \tag{3}$$

$$T_{GPU} = \frac{f_{GPU}(R \times (1 - r)) \times b}{m} \tag{4}$$

The execution time for a parallel system is defined as $T_{Parallel}$ in (5). Because the target computation ends when both the CPUs and the GPUs finish computations, the execution time is obtained as the maximum of either T_{CPU} or T_{GPU}. The parallelization overhead. is omitted for simplification.

$$T_{Parallel} = max(T_{CPU}, T_{GPU}) \tag{5}$$

To attain optimal performance, the execution time $T_{Parallel}$, defined by equation (5), must be minimized.

3.2   Case of One CPU and One GPU

For simplicity, we evaluate the proposed method for the case of one CPU and one GPU. If the parameters $a, n, b, and\ m$ in equations (3) and (4) are constants, then $T_{Parallel}$ is a function of only r as an input parameter. We propose a method by which to achieve high performance by properly estimating the parameter r. For one CPU and one GPU, we have $n = 1 and\ m = 1$ in equation (3) and (4). Therefore, equation (3) can be simplified to (6), and equation (4) can be simplified to (7).

$$T_{CPU} = f_{CPU}(R \times r) \times a \tag{6}$$

$$T_{GPU} = f_{GPU}(R \times (1 - r)) \times b \tag{7}$$

3.3   Parallelization of Matrix Multiplication

The interface of matrix multiplication in BLAS is denoted in (8). In this equation, $A, B, and\ C$ are matrices, and $\alpha\ and\ \beta$ are scalars. This function updates the matrix C.

$$C = \alpha \times A \times B + \beta \times C \tag{8}$$

In this equation, matrix size is assigned the three values of $M, N, and\ K$, as shown in Fig.1(a). When a certain element of matrix C is updated, only the updated element of matrix C is referenced. We divide matrices A and C into $Mc and Mg$, where Mc and Mg denote the sizes of matrices allocated to CPU and GPU respectively. Then we assign the partial matrices to the CPU and the GPU as shown in Fig.1(b). Thus, matrix multiplication can be executed in parallel without synchronization, and the assignment ratio $r = Mc/Mc + Mg$ is obtained. A matrix can be divided into any assignment ratio, and the value of r can be changed freely. So, optimal division, i.e. static load balancing, is easy to achieve.

Parallel processing on 1-CPU and 1-GPU uses two threads. One is a thread handling the CPU, this thread performs the SGEMM function (single precision floating-point GEMM function) using ATLAS. The other is a thread handling the GPU, this thread performs data transfer between the CPU and the GPU,

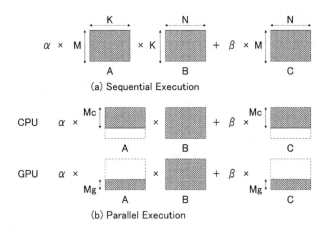

Fig. 1. Assignment of computation in Matrix Multiplication

and issues instructions to the GPU. Matrix multiplication on the GPU is implemented as follows. We use DirectX as a graphics API and HLSL as a shading language for creating programs[19]. Vector calculations are used because the GPU can handle vector data and vector operations efficiently. Although a GPU has both vertex and pixel processing units (fragment processing units), the implementation herein uses only pixel processing units.

In the next section, we measure the performance of matrix multiplication using either only one CPU or only one GPU. Based on this measurement, we can predict the performance of parallel execution using both 1-CPU and a 1-GPU heterogeneous environment.

4 Preliminary Performance Experiments

4.1 Performance of the 1-CPU System

The personal computer we used has a Pentium4 3.0GHz processor as the CPU and NVIDIA GeForce7800GTX as the GPU. Specifications of these processors are given by Table 1. First, we examine the performance of the 1-CPU system in the execution of matrix multiplication.

As described in Section 3.3, the SGEMM function of ATLAS is executed on the CPU in parallel processing in a 1-CPU and 1-GPU heterogeneous environment. We execute the SGEMM function and measure the execution time. The matrix size is 2,048, which means that $r = 1.0$ when $M = N = K = 2,048$. We examine the relationship between the value of r and the execution time by changing the vertical size Mc of matrices A and C.

The results obtained are shown in Fig.2. The horizontal axis denotes r, and the vertical axis denotes the execution time. This is a the graph of equation (6). We observe that the execution time of the SGEMM function is proportional to

Table 1. Experimental Environment of Experiments

CPU	Pentium4 3.0GHz
Memory	1.00GB
OS	Windows XP
GPU	GeForce7800GTX
Graphics Bus	PCI-Express x16
VRAM	256MB
GPU's core clock	430MHz
GPU's memory clock	1.20GHz
amount of vertex shader unit	8
amount of pixel(fragment) shader unit	24

the computation size, and the amount of computation in matrix multiplication is proportional to r. Measurements are obtained by changing the assignment ratio of matrices of the size of multiples of 64.

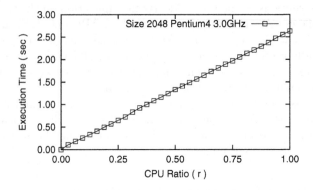

Fig. 2. CPU execution time for one Pentium4 3.0GHz

4.2   Performance of the 1-GPU System

Next, we examine the performance of the 1-GPU system. The matrix size is defined as 2,048, and the relationship between the matrix size and the execution time is examined in the same manner as the 1-CPU system. The execution time is measured from the beginning of data transfer from the CPU to the GPU to the end of data read back from the GPU to the CPU. We exclude the time required to initialize DirectX and load the HLSL program from the scope of measurement.

   The results obtained are shown in Fig.3. The horizontal axis denotes the value of $1 - r$, and the vertical axis denotes the execution time. This is a graph of equation (7). As a result, the execution time of matrix multiplication using 1-GPU is also proportional to the matrix size. This is the result of changing the assignment ratio as we did with the 1-CPU.

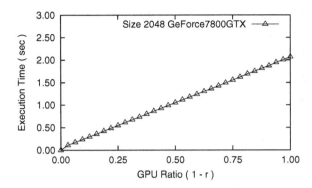

Fig. 3. GPU execution time for one NVIDIA GeForce7800GTX

4.3   Performance Prediction of the Heterogeneous Environment

We can predict the execution time on a parallel heterogeneous environment based on 1-CPU and 1-GPU execution times using the following process: we first put the 1-GPU graph (Fig.3) over the 1-CPU graph (Fig.2), while adjusting the horizontal edge. We obtain Fig.4, which depicts both T_{CPU} and T_{GPU} for the assignment ratio r. As mentioned above, the larger value of T_{CPU} and T_{GPU} is the predicted time for parallel execution of each r, because parallel execution finishes when both the CPU and the GPU complete the calculations. Figure 5 shows a graph of the prediction time obtained from Fig.4. This is a graph of equation (5). Matrix multiplication is executed at the fastest speed at the lowest point of the execution time on this graph, and its assignment ratio is optimal, that is, the value of r is minimized equation (5).

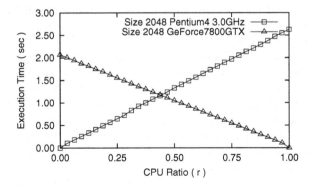

Fig. 4. CPU and GPU execution time

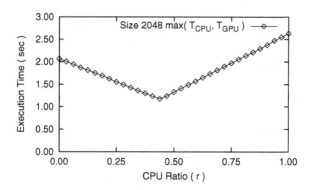

Fig. 5. Predicted execution time for parallel processing with 1-CPU and 1-GPU

The result of the above prediction is that, in this environment, the execution time is expected to be the minimum when the CPU assignment is 43.8% of the computation. The execution time is expected to be reduced 44.1% compared with the 1-CPU only case and by 59.5% compared with the 1-GPU only case.

5   Performance Evaluation of the CPU and GPU Heterogeneous Environment

5.1   Performance Evaluation of the Heterogeneous System

In this section, we measure the execution time required in parallel execution on a heterogeneous environment. We implemented a parallel program using both the thread handling CPU and GPU, as described in Section 3.3. The SGEMM function of ATLAS is executed in the CPU thread, and data transfer between the CPU and the GPU and computations using the programmable shader are executed in the GPU thread. These experiments were carried out by changing the matrix size by 64 intervals.

The results obtained are shown in Fig.6. The center of the graph is lower compared with either side. Therefore, the execution time is decreased by parallel execution. The execution time is minimum when the CPU does 40.6% of the computation. The execution time is decreased by 45.1% compared with the CPU only case, and by 60.8% compared with the GPU only case. Figure 7 shows the performance ratio when the higher performance of the 1-CPU only case and the 1-GPU only case is defined as 1.0. As a result, we got a performance improvement for the parallel case of 1.64 times.

We compare the experimental result with our prediction result. We first confirm that the assignment ratio minimizing the execution time. The execution time is predicted to be decreased the most when the CPU does 43.8% of the computation. Correspondingly, the experimental result also indicated that the execution time is minimum when the CPU does 43.8% of the computation.

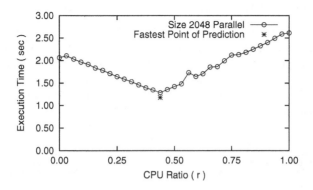

Fig. 6. Execution time measured on the CPU and GPU heterogeneous environment

In addition, these values mean that execution time is minimum when the CPU does 1, 152 of the total 2, 048 size.

Next, we confirm the minimal execution time by using the optimal assignment ratio of the computation. In the prediction, the minimal execution time is 1.23 sec. Here, the ratio of the execution time to the CPU only case is 44.1%, and the ratio of the execution time to the GPU only case is 59.6%. In the experiment, the minimal experimental execution time is 1.26 sec. At this time, the ratio of the execution time to the 1-CPU only case is 45.1%, and the ratio of the execution time to the 1-GPU only case is 60.8%. The ratio of the minimal experimental execution time to the minimal prediction time is 102.4%.

We can conclude that we obtained high performance using parallel processing method. Moreover, we can predict with high precision both the execution time of matrix multiplication on a heterogeneous environment and the assignment ratio of computation for the minimal execution time.

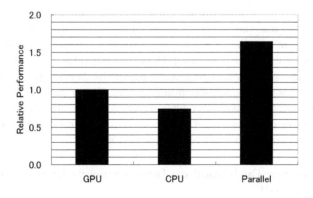

Fig. 7. Relative performance of parallel execution. (The higher performance of the CPU only case and the GPU only case is defined as 1.0.).

5.2   Performance Evaluation with Varying Matrix Sizes

We examine the performance in other matrix sizes of matrix multiplication to confirm whether our method is affected by the size of matrix in the computation. So, we evaluate performance by varying the matrix size of the matrix of computation. The results obtained are shown in Figs.8 and 9. Figure 8 shows the result of the 1-CPU only case and the 1-GPU only case with varying the matrix sizes: 512, 1024, 1536 and 2560.

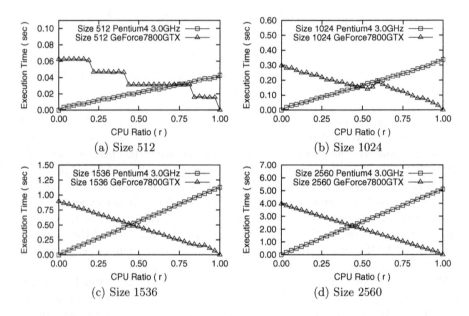

Fig. 8. 1-CPU only execution time and 1-GPU only execution time (GeForce7800GTX)

Graphs for the small computation sizes were unstable, but the tendency in the larger matrix size was the same as the figures we have already shown. Figure 9 shows the result in parallel execution. As a result, this method didn't work well when the computation size was small. However, when the computation size was large enough, the execution time was decreased by parallel processing. At this time, the assignment ratio for the minimal time in parallel execution was near the prediction point. In this heterogeneous environment, maximum speedup was obtained at the size of 2048. Further research is required to analyze the reason why the best performance was obtained in this size.

5.3   Performance Evaluation on Heterogeneous Environment with Different GPU

We tried to evaluate performance with another GPU of a different type to evaluate the effectiveness of the proposed method. We use the GeForce6600GT instead

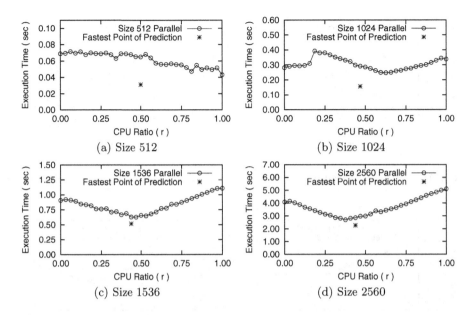

Fig. 9. Execution time measured on the implemented on parallel system with a CPU and a GPU, and fastest point of prediction (GeForce7800GTX)

of the GeForce7800GTX. The differences for each GPU are shown in Table 2. The results obtained are shown in Figs.10 and 11. Figure 10 is a graph for the 1-CPU only case and the 1-GPU only case with changing matrix sizes of 1024 and 2048. Figure 11 is a graph in parallel execution. The environment in this experiment is the same as for the GeForce7800GTX.

Table 2. Comparison of GPUs

GPU	GeForce7800GTX	GeForce6600GT
Graphics Bus	PCI-Express x16	PCI-Express x16
VRAM	256MB	128MB
GPU's core clock	430MHz	300MHz
GPU's memory clock	1.20GHz	1.00GHz
amount of vertex shader unit	8	3
amount of pixel(fragment) shader unit	24	8

Our method didn't work well when the computation size was small, but the execution time was decreased by parallel processing when the computation size was large enough. The difference between the optimal assignment ratio of computation in CPU and GPU was small. The tendencies were almost same as the case of the GeForce7800GTX. The highest rate of speedup is shown for the size

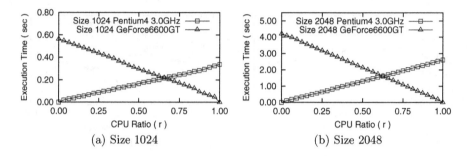

Fig. 10. CPU only execution time and GPU only execution time (GeForce6600GT)

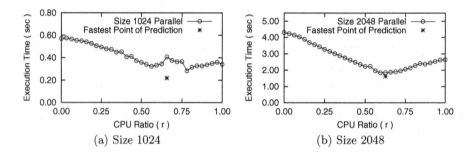

Fig. 11. Execution time measured on the implemented on parallel system with a CPU and a GPU, and fastest point of prediction (GeForce6600GT)

2048, and the ratio of the execution time for the 1-CPU only case was 70.5%. These results show that our method is useful when computation size is large. and doesn't work well when computation size is too small.

6 Conclusion and Future Work

In this paper, we proposed a method for dividing a large target computation into partial computations and executing them in parallel using 1-CPU and 1-GPU. In addition, we proposed a load balancing method for minimizing the execution time. Using the method we proved by experiment that the execution time was reduced to 44.1% for the CPU and 59.5% of that for the GPU. In addition, we demonstrated that the proposed method could be used to predict the optimal assignment ratio to the CPU and GPU according to each execution time.

Future research work is required in the following problem areas. First, we must check how useful our method is in heterogeneous environments. Therefore, it is necessary to evaluate performance in various environments. Secondly, it is necessary to evaluate the arithmetic precision of the computation using our method. In particular, we have to evaluate the differences in precision for calculated results between the CPU and GPU. Thirdly, a library of parallel programming must be developed for CPU and GPU research. We have to write CPU and GPU programs independently when we want to execute parallel programs using CPUs and GPUs. However, it is desirable that users are not concerned about whether they use CPUs or GPUs.

We are considering developing such an automatic performance tuning library. Various applications are speeded up when a library automatically assigns computations to CPUs and GPUs using our load balancing method in parallel processing. For example, if we make a library with an interface of BLAS, it can automatically assign computations to the CPU and GPU, and many applications using BLAS can be speeded up easily.

The method we proposed can be applied for more complex environments having multiple CPUs and GPUs. Utilizing such a multiple processor environment will become a new trend in GPU technology for the benefit of many CPUs and GPUs. In such environments, new approaches for realizing optimal load balancing are required to achieve the maximal speed up in the high-performance computing field.

References

1. gpgpu.org: General-Purpose computation on GPUs(GPGPU), http://gpgpu.org/
2. Thompson, C.J., Hahn, S., Oskin, M.: Using Modern Graphics Architectures for General-Purpose Computing: A Framework and Analysis. In: Proceedings of the 35th annual ACM/IEEE International Symposium on Microarchitecture. (2002) 306–317
3. Owens, J.D., Luebke, D., Govindaraju, N., Harris, M., Krüger, J., Lefohn, A.E., Purcell, T.J.: A Survey of General-Purpose Computation on Graphics Hardware. In: Eurographics 2005, State of the Art Reports. (2005) 21–51
4. Higham, N.J.: Exploiting Fast Matrix Multiplication Within the Level 3 BLAS. ACM Transactions on Mathematical Software **16** (1990) 352–368
5. Whaley, R.C., Petitet, A., Dongarra, J.J.: Automated Empirical Optimization of Software and the ATLAS Project. Parallel Computing **27** (2001) 3–35
6. John Montrym, H.M.: THE GEFORCE 6800. IEEE MICRO 2005, Vol.25, No.2 (2005)
7. Fernando, R.: GPU Gems: Programming Techniques, Tips and Tricks for Real-Time Graphics. Addison-Wesley Pub (Sd) (2004)
8. SHINOMOTO, Y., MIWA, S., SHIMADA, H., MORI, S.I., NAKASHIMA, Y., TOMITA, S.: Consideration for Speculative Rendering in PVR. In: IPSJ SIG Technical Reports. 2005-ARC-164 (2005) 145–150
9. T.Amada, M.Imura, Y.Yasumuro, Y.Manabe, K.Chihara: Partivle-Based Fluid Simulation on GPU. In: ACM Workshop on General-Purpose Computing on Graphics Processors. (2004)
10. Ádám Moravánszky: Dense Matrix Algebra on the GPU, ShaderX2 (2003)
11. Krüger, J., Westermann, R.: Linear Algebra Operators for GPU Implementation of Numerical Algorithms. In: Proceedings of ACM SIGGRAPH 2003. (2003) 908–916
12. Moreland, K., Angel, E.: The FFT on a GPU. In: Proc. SIGGRAPH / EUROGRAPHICS Workshop Graphics Hardware. (2003) 112–119
13. Hillesland, K., Lastra, A.: GPU floating-point paranoia. In: Proceedings of GP2. (2004)

14. Larsen, E., McAllister, D.: Fast matrix multiplies using graphics hardware. In: Proceedings of the 2001 ACM/IEEE conference on Supercomputing. (2001)
15. K.Fatahalian, J.Sugerman, P.Hanrahan: Understanding the Efficiency of GPU Algorithms for Matrix-Matrix Multiplication. In: Graphics Hardware 2004. (2004)
16. Jesse D. Hall, Nathan A. Carr, J.C.H.: Cache and Bandwidth Aware Matrix Multiplication on the GPU . Technical report, University of Illinois Dept. of Computer Science (2003)
17. Jiang, C., Snir, M.: Automatic Tuning Matrix Multiplication Performance on Graphics Hardware. In: Proceedings of the 14th International Conference on Parallel Architectures and Compilation Techniques (PACT'05). (2005) 185–196
18. Blackford, L.S., Hammarling, S., Cleary, A., Petitet, A., Whaley, R.C., Demmel, J., Dhillon, I., Ren, H., Stanley, K., Dongarra, J.: Practical experience in the numerical dangers of heterogeneous computing. ACM Transactions on Mathematical Software (TOMS) **23** (1997) 133–147
19. Microsoft: DirectX Developer Center, http://msdn.microsoft.com/directx/

Robust Two-Level Lower-Order Preconditioners for a Higher-Order Stokes Discretization with Highly Discontinuous Viscosities

Duilio Conceição[1,*], Paulo Goldfeld[2], and Marcus Sarkis[1]

[1] IMPA, Rio de Janeiro, Brazil
dtadeu@fluid.impa.br, msarkis@fluid.impa.br
[2] UFRJ, Instituto de Matemática, Rio de Janeiro, Brazil
goldfeld@ufrj.br

Abstract. The main goal of this paper is to present new robust and scalable preconditioned conjugate gradient algorithms for solving Stokes equations with large viscosity jumps across subregion interfaces and discretized on non-structured meshes. The proposed algorithms do not require the construction of a coarse mesh and avoid expensive communications between coarse and fine levels. The algorithms belong to the family of preconditioners based on non-overlapping decomposition of subregions known as balancing domain decomposition methods. The local problems employ two-level element-wise/subdomain-wise direct factorizations to reduce the size and the cost of the local Dirichlet and Neumann Stokes solvers. The Stokes coarse problem is based on subdomain constant pressures and on connected subdomain interface flux functions and rigid body motions. This guarantees scalability and solvability of the local Neumann problems. Estimates on the condition numbers and numerical experiments based on a parallel implementation for unstructured meshes are also discussed.

1 Introduction

The *core-flow* technique is a technology in research that can highly improve the efficiency of the production/transportation of heavy oil through a pipe. The numerical simulators available nowadays are inefficient for solving large scale problems with large jumps in viscosity such as the core-flow model. In order to develop an efficient parallel code to solve such model, we develop a preconditioner for the Stokes problem that is robust with respect to high jumps in viscosity and are suitable for unstructured meshes.

Balancing Domain Decomposition (BDD) methods are preconditioners based on non-overlapping decomposition of subregions. They have been tested successfully on several challenging large scale applications [4,7,6] and its first scalable version was developed by Mandel [6] for the Poisson equation with the introduction of a coarse problem based on the kernel of the Laplace operator. Extensions

---

* This work was supported by ANP/PRH-32. The author is candidate to the best student paper award.

M. Daydé et al. (Eds.): VECPAR 2006, LNCS 4395, pp. 319–333, 2007.
© Springer-Verlag Berlin Heidelberg 2007

of the BDD preconditioner for elliptic problems with possibly large jumps on
coefficients were treated subsequently in [2,9,10]. The extension of the BDD preconditioner for the Stokes equations had its debut only recently by Pavarino and
Widlund [7]. For the Stokes problem, the local Neumann problems are singular
and the boundary values of the local Dirichlet problems should satisfy the zero
flux condition on the boundary of the subregions. Such issues are discussed in
detail in [7] and on this paper.

The goal of this paper is to introduce several improvements of the Pavarino
and Widlund method which are essential for its efficient application. We are
particularly concerned with aspects associated to unstructured mesh parallel
implementation and the high cost of the subdomain solvers when high-order
Stokes discretizations are considered. We introduce several possible choices for
unstructured coarse spaces and discuss their advantages in terms of scalability,
implementation efforts and robustness with respect to coefficient jumps. With
regards to the high cost of the subdomain solvers, we explore how the inf-sup
condition of Stokes discretization are checked in order to perform proper element-
wise static condensation and decrease the number of interior unknowns. We
show that the computational complexity of the two discretizations, the higher-
order $(\mathbf{P_2} + \mathbf{Bubbles})/\mathrm{P}_1$ and the lower-order $\mathbf{P_2}/\mathrm{P}_0$, have comparable computational costs. The paper is organized as follows. Sections 2 and 3 present the
Stokes equations and the variational formulation, respectively, while in Section
4 we introduce the discretizations used in the numerical experiments. Section 5
is devoted to the BDD preconditioner for the Stokes equations and the coarse
spaces. In Section 6 we present some of the implementation issues, and in Section 7 we provide the numerical results. Section 8 closes the paper with the
conclusions.

2   The Stokes Model

Let $\Omega \subset \mathbb{R}^2$ be a domain with a polygonal boundary. We consider the Stokes
equations:

$$\begin{cases} -2\nabla \cdot (\nu\varepsilon(\mathbf{u})) + \nabla p = \mathbf{f} & \text{in } \Omega \\ -\nabla \cdot \mathbf{u} = g & \text{in } \Omega \\ \mathbf{u} = \mathbf{u}_d \text{ on } \partial\Omega \end{cases} \qquad (1)$$

where $\nu > 0$ is the kinematic viscosity and $\varepsilon(\mathbf{u}) = \frac{1}{2}[\nabla\mathbf{u} + \nabla\mathbf{u}^T]$ denotes the
symmetric stress tensor. In this paper, we assume only Dirichlet boundary con-
dition with the compatibility condition $\int_\Omega -g\, dx = \int_{\partial\Omega} \mathbf{u}_d \cdot \mathbf{n}\, ds$. The treatment
of natural boundary condition is similar and does not bring any extra difficulties;
see also Remark 3.

Remark 1. Since we are assuming Dirichlet boundary condition on all $\partial\Omega$, the
velocity solution is unique and the pressure is unique up to a constant. To make
the pressure unique, we impose the additional condition of zero average pressure
on Ω, i.e., $\int_\Omega p\, dx = 0$.

3   Variational Formulation

The variational formulation is introduced as follows. Let us define the space of velocities $\mathbf{X} = H_0^1(\Omega)^2$ and the space of pressures $M = L_0^2(\Omega)$, where $L_0^2(\Omega)$ stands for $L^2(\Omega)$ functions with zero average in Ω. Given $\mathbf{f} \in H^{-1}(\Omega)^2$ and $g \in L^2(\Omega)$, the variational formulation of the Stokes equations is given by:

Find $\mathbf{u} \in \mathbf{X}$ and $p \in M$ such that

$$\begin{cases} a(\mathbf{u}, \mathbf{v}) + b(\mathbf{v}, p) = F(\mathbf{v}) & \forall \mathbf{v} \in \mathbf{X}, \\ b(\mathbf{u}, q) \qquad\qquad = G(q) & \forall q \in M, \end{cases} \tag{2}$$

where $a(\mathbf{u}, \mathbf{v}) = 2\nu(\varepsilon(\mathbf{u}) : \varepsilon(\mathbf{v}))_\Omega$, $b(\mathbf{v}, p) = -(\nabla \cdot \mathbf{v}, p)_\Omega$, $F(\mathbf{v}) = (\mathbf{f}, \mathbf{v})_\Omega$, and $G(q) = (g, q)_\Omega$. The solution $(\mathbf{u}, p) \in \mathbf{X} \times M$ of (2) exists and is unique; see [3].

4   Discretization

Let $\mathcal{T}_h$ be a regular triangulation of Ω. We consider the mixed finite elements $\mathbf{P_2}/\mathrm{P_0}$ and $(\mathbf{P_2} + \mathbf{Bubbles})/\mathrm{P_1}$, where the velocity is taken continuous and the pressure discontinuous.

The $\mathbf{P_2}/\mathrm{P_0}$ mixed finite elements is described as follows: the velocity space is given by $\mathbf{X}_h = \{\mathbf{v} \in \mathbf{X}; \ \mathbf{v}|_K \in P_2(K)^2, \ \forall K \in \mathcal{T}_h\}$, while the pressure space is comprised of discontinuous piecewise constant functions $M_h = \{q \in M; \ q|_K \in P_0(K), \ \forall K \in \mathcal{T}_h\}$. To obtain more accurate results we introduce the $(\mathbf{P_2} + \mathbf{Bubbles})/\mathrm{P_1}$ mixed finite element space. This space can be considered as a stabilization of the unstable space $\mathbf{P_2}/\mathrm{P_1}$. We take the bubble function as $\hat{b}(\hat{x}, \hat{y}) = \hat{x}\hat{y}(1 - \hat{x} - \hat{y})$ defined on the element of reference $\widehat{K}$, and then for each element K in $\mathcal{T}_h$ define $b_K(x, y) = \hat{b}(F_K^{-1}(x, y))$, where F_K is the affine mapping from $\widehat{K}$ to K. The velocity space $\mathbf{X}_h$ is then given as

$$\mathbf{X}_h = \{\mathbf{v} \in \mathbf{X}; \ \mathbf{v} = \mathbf{v}_P + \mathbf{v}_B, \ \text{s.t.} \ \mathbf{v}_{P|K} \in P_2(K)^2, \ \mathbf{v}_{B|K} \in \mathbf{X}_B(K), \forall K \in \mathcal{T}_h\},$$

where for each element $K \in \mathcal{T}_h$

$$\mathbf{X}_B(K) = \left\{\mathbf{v}_B \in H_0^1(K)^2; \ \mathbf{v}_B = \begin{pmatrix} \alpha_1 b_K \\ \alpha_2 b_K \end{pmatrix} \text{ and } \alpha_1, \ \alpha_2 \in \mathbb{R}\right\}.$$

The discrete pressure space consists of discontinuous piecewise linear functions denoted by $\mathrm{P_1}$ given as $M_h = \{p \in M; \ p|_K \in P_1(K), \ \forall K \in \mathcal{T}_h\}$.

The two discretizations above satisfy a uniform *inf-sup* condition [3], i.e., there exists a constant β (independent of h) such that

$$\sup_{\substack{\mathbf{v} \in \mathbf{X}_h \\ \mathbf{v} \neq 0}} \frac{(\nabla \cdot \mathbf{v}, q)}{\|\mathbf{v}\|_{\mathbf{H^1}}} \geq \beta \|q\|_0 \quad \forall q \in M_h. \tag{3}$$

The discrete variational formulation of the Stokes problem (1) is given by:

Find $\boldsymbol{u} \in \mathbf{X}_h$ and $p \in M_h$ such that

$$\begin{cases} a(\boldsymbol{u}, \boldsymbol{v}) + b(\boldsymbol{v}, p) = F(\boldsymbol{v}) \ \ \forall \boldsymbol{v} \in \boldsymbol{X}_h, \\ b(\boldsymbol{u}, q) \qquad\quad = G(q) \ \ \forall q \in M_h. \end{cases} \tag{4}$$

The inf-sup stability of the mixed finite element spaces guarantees the existence and uniqueness of the solution of (4) (see [3]). In matricial form, the discrete linear system (4) is of the form

$$\begin{pmatrix} A & B^T \\ B & 0 \end{pmatrix} \begin{pmatrix} u \\ p \end{pmatrix} = \begin{pmatrix} f \\ g \end{pmatrix}. \tag{5}$$

5   BDD for Stokes Problem

In this section we present the matrix form of the preconditioner. Decompose the domain Ω into N non-overlapping connected subdomains Ω_i and let $\Gamma = (\cup_{i=1}^N \partial\Omega_i)\backslash\partial\Omega$, then we have $\Omega = \cup_{i=1}^N \Omega_i \cup \Gamma$. We denote the nodes inside Ω_i by Ω_i^h, the nodes on Γ by Γ_h and the nodes on $\partial\Omega_i \cap \Gamma$ by $\Gamma_h^{(i)}$.

5.1   Schur Complement System

In order to perform a static condensation of the interior variables on Ω_i we reorder and denote the variables as follows: $\boldsymbol{u}_I$ (the interior velocities), p_I (pressures with zero average in each subdomain Ω_i), $\boldsymbol{u}_\Gamma$ (interface velocities) and p_0 (constant pressure in each Ω_i and with zero average in Ω). Using this reordering, the matrix of the discrete system (5) can be written as:

$$K = \begin{pmatrix} K_{II} & K_{I\Gamma} \\ K_{\Gamma I} & K_{\Gamma\Gamma} \end{pmatrix} = \begin{pmatrix} A_{II} & B_{II}^T & A_{I\Gamma} & B_{0I}^T \\ B_{II} & 0 & B_{I\Gamma} & 0 \\ A_{I\Gamma}^T & B_{I\Gamma}^T & A_{\Gamma\Gamma} & B_0^T \\ B_{0I} & 0 & B_0 & 0 \end{pmatrix}.$$

The submatrix B_{0I} is null since by the divergence theorem, $\int_{\Omega_i} \nabla \cdot \boldsymbol{u}_I \ dx = 0$. Eliminating the interior variables $\boldsymbol{u}_I$ and p_I by static condensation we obtain the following Schur complement system:

$$S \begin{pmatrix} \boldsymbol{u}_\Gamma \\ p_0 \end{pmatrix} = \begin{pmatrix} \tilde{\boldsymbol{f}}_\Gamma \\ \tilde{g}_0 \end{pmatrix}, \tag{6}$$

where

$$S = K_{\Gamma\Gamma} - K_{\Gamma I} K_{II}^{-1} K_{I\Gamma} = \begin{pmatrix} S_\Gamma & B_0^T \\ B_0 & 0 \end{pmatrix} \text{ and } \begin{pmatrix} \tilde{\boldsymbol{f}}_\Gamma \\ \tilde{g}_0 \end{pmatrix} = \begin{pmatrix} \boldsymbol{f}_\Gamma \\ g_0 \end{pmatrix} - K_{\Gamma I} K_{II}^{-1} \begin{pmatrix} \boldsymbol{f}_I \\ g_I \end{pmatrix}.$$

Remark 2. Since A_{II} is positive definite (by Korn's inequality) and B_{II} has full row rank, the K_{II} is invertible. We note also that since B_{0I} is null, it is not possible to eliminate p_0.

Having solved the linear system (6), we can obtain the solutions u_I and p_I by solving $\begin{pmatrix} u_I \\ p_I \end{pmatrix} = \begin{pmatrix} A_{II} & B_{II}^T \\ B_{II} & 0 \end{pmatrix}^{-1} \left[\begin{pmatrix} f_I \\ g_I \end{pmatrix} - \begin{pmatrix} A_{I\Gamma} & 0 \\ B_{I\Gamma} & 0 \end{pmatrix} \begin{pmatrix} u_\Gamma \\ p_0 \end{pmatrix} \right]$, where we observe that u_I and p_I do not depend on p_0. After a reordering of the interior variables by subdomain we obtain that K_{II} is the block-diagonal matrix $K_{II} = \mathrm{diag}\{K_{II}^{(1)}, \cdots, K_{II}^{(N)}\}$. This shows that the subdomain matrices $K_{II}^{(i)}$ are decoupled and then to apply K_{II}^{-1} to a vector is equivalent to solve N decoupled saddle point problems in parallel. Notice that the multiplication by $K_{II}^{(i)-1}$ represents a discrete Stokes problem with Dirichlet velocity data on $\Gamma_h^{(i)}$. This solution exists and is unique since we consider the space of pressure and test functions q_I with zero average on Ω_i. The velocity component of $K_{II}^{(i)-1}$, denoted by $\mathcal{SH}^{(i)}$, is known as the local discrete Stokes harmonic extension operator with velocity Dirichlet boundary condition prescribed on $\Gamma_h^{(i)}$.

Our goal is to solve the linear system (6) by a preconditioned conjugate gradient method. This method does not require assembling the matrix S of the linear system, but only applying S to vectors. By definition of S, applying S to a vector w is equivalent to applying matrices $K_{\Gamma\Gamma}$, $K_{I\Gamma}$, $K_{\Gamma I}$ and K_{II}^{-1} to subvectors of w. Among those applications, the K_{II}^{-1} is the most expensive one. As we have mentioned, it can be done in parallel.

5.2 BDD Preconditioning

Let us decompose the space $\mathbf{X}_h \times M_h = \left(\oplus_{i=1}^N \mathbf{X}_{i,h} \times M_{i,h} \right) \oplus \left(\mathbf{V}_{\Gamma,h} \times M_0 \right)$ where $\mathbf{X}_{i,h} = \mathbf{X}_h \cap H_0^1(\Omega_i)^2$, $M_{i,h} = M_h \cap L_0^2(\Omega_i)$, $\mathbf{V}_{\Gamma,h} = \{v \in \mathbf{X}_h; \ v|_{\Omega_i} = \mathcal{SH}^{(i)}(v|_{\partial\Omega_i}), \ i = 1, \ldots, N\}$, and $M_0 = \{q \in M_h; \ q|_{\Omega_i} = \text{constant}, \ i = 1, \ldots, N\}$. We observe that the function $v \in \mathbf{V}_{\Gamma,h}$ is uniquely defined by its value on the interface Γ_h.

We now construct a parallel preconditioner M^{-1} for S in order to make the linear system scalable and well conditioned.

An initial attempt would be to use an additive Schwarz like preconditioner of the form

$$M^{-1} = \sum_{i=1}^N R_i^T D_i^T S^{(i)-1} D_i R_i, \tag{7}$$

where $S^{(i)}$ is the Schur complement of the local stiffness matrix $K^{(i)}$, $R_i : \Gamma_h \to \Gamma_h^{(i)}$ is the discrete restriction operator, and D_i is a diagonal matrix defining a partition of unity on Γ_h, i.e., $\sum_{i=1}^N R_i^T D_i R_i = I$ on Γ_h. The partition of unity may be defined through the *counting functions*, which can be defined for each subdomain as $\delta_i : \Gamma_h^{(i)} \to \mathbb{R}$ such that $\delta_i(x)$ equals the number of subdomains sharing the node $x \in \Gamma_h^{(i)}$. Thus, define D_i as $D_i = \mathrm{diag}\{\delta_i^{-1}\}$. When the problem has piecewise constant viscosity ν_i in each subdomain, and discontinuous across the interface Γ, then a better choice is to set

$$\delta_i = \frac{\sum_{j \in \mathcal{N}_x} \nu_j^\gamma(x)}{\nu_i^\gamma(x)}, \tag{8}$$

where $\gamma \in [1/2, \infty)$, and $\mathcal{N}_x$ is the set of indices of the subdomains that have the node x on their boundaries (see [9,10]).

Remark 3. The local problems $S^{(i)^{-1}}$ in (7) use natural boundary conditions $\nu_i \nabla \boldsymbol{u} \cdot \boldsymbol{n} - pn = r$ on $\Gamma_h^{(i)}$. In this case the pressure is uniquely determined and therefore the pressure spaces are now taken on $L^2(\Omega)$.

The preconditioner (7) is not as good as it appears to be. When the boundary of a subdomain Ω_i does not intersect the boundary of the domain $\partial\Omega$, we have a *floating subdomain* Ω_i. The problem

$$S^{(i)} \begin{pmatrix} \boldsymbol{u}_\Gamma^{(i)} \\ p_0^{(i)} \end{pmatrix} = \begin{pmatrix} \widetilde{\boldsymbol{f}}_\Gamma^{(i)} \\ \widetilde{g}_0^{(i)} \end{pmatrix} \tag{9}$$

is equivalent to solving $\begin{pmatrix} K_{II}^{(i)} & K_{I\Gamma}^{(i)} \\ K_{\Gamma I}^{(i)} & K_{\Gamma\Gamma}^{(i)} \end{pmatrix} \begin{pmatrix} \boldsymbol{u}_I^{(i)} \\ p_I^{(i)} \\ \boldsymbol{u}_\Gamma^{(i)} \\ p_0^{(i)} \end{pmatrix} = \begin{pmatrix} 0 \\ 0 \\ \widetilde{\boldsymbol{f}}_\Gamma^{(i)} \\ \widetilde{g}_0^{(i)} \end{pmatrix}$. Hence when Ω_i is a float-

ing subdomain, $S^{(i)}$ has a kernel spanned by the *rigid body motions* (RBM) and therefore the linear system (9) might not have a solution. In the two dimensional case the kernel basis three-dimensional, spanned by two translations and one rotation. To avoid the issue of existence of solution, we introduce a coarse space $\mathbf{V}_0 \subset \mathbf{V}_{\Gamma,h}$ to enforce that when solving the linear system (9) the right hand side (RHS) is on the image of $S^{(i)}$, and since $S^{(i)}$ is symmetric, this is equivalent to have the RHS in $\mathrm{Ker}^\perp(S^{(i)})$. In addition we will require that the space $\mathbf{V}_0$ must be chosen so that the pairing $(\mathbf{V}_0, M_0)$ be stable, i.e., satisfies the inf-sup condition. We discuss possible choices of coarse spaces in Subsection 5.4.

5.3   Matrix Form of Preconditioner

Let $L_0 : \mathbf{V}_0 \to \Gamma_h$ be the matrix whose columns are the basis of the space $\mathbf{V}_0$. Then define the restriction operator $R_0 = \begin{pmatrix} L_0^T & 0 \\ 0 & I \end{pmatrix}$, where I is the identity matrix whose size is the number of subdomains. To define a coarse problem Q_0, we set $S_0 = R_0 S R_0^T = \begin{pmatrix} L_0^T S_\Gamma L_0 & L_0^T B_0^T \\ B_0 L_0 & 0 \end{pmatrix}$, and $Q_0 = R_0^T S_0^{-1} R_0$. The BDD preconditioner is then given by

$$M^{-1} = Q_0 + (I - Q_0 S) \sum_{i=1}^N Q_i (I - SQ_0),$$

and the preconditioned operator by $T = M^{-1}S = P_0 + (I - P_0)\sum_{i=1}^N P_i(I - P_0)$, where $P_0 = Q_0 S$, $P_i = Q_i S$ and

$$Q_i = \begin{pmatrix} R_i^T D_i & 0 \\ 0 & 0 \end{pmatrix} \begin{pmatrix} S_\Gamma^{(i)} & B_0^{(i)^T} \\ B_0^{(i)} & 0 \end{pmatrix}^{-1} \begin{pmatrix} D_i R_i & 0 \\ 0 & 0 \end{pmatrix}.$$

The minimal size coarse space $\mathbf{V}_0$ must be related to the local RBM associated to each subdomain Ω_i. Since the local problems are scaled by D_i, we also scale the local RBM basis associated to Ω_i by D_i to define a coarse space so that the local problems (9) are compatible, i.e., for any $\boldsymbol{w} \in \mathbf{V}_{\Gamma,h}$

$$\left\langle \begin{pmatrix} D_i R_i & 0 \\ 0 & * \end{pmatrix} S(I - P_0)\boldsymbol{w}, \boldsymbol{v}_i \right\rangle_{\Gamma_i} = 0 \quad \forall \boldsymbol{v}_i \in \text{Ker}(S^{(i)}). \tag{10}$$

A desirable property of any parallel preconditioner is the scalability. To obtain that, the coarse space must also satisfy the following inf-sup condition

$$\sup_{\substack{\boldsymbol{v}_\Gamma \in \mathbf{V}_{\Gamma,h} \\ \boldsymbol{v}_\Gamma \neq 0}} \frac{(\nabla \cdot \mathcal{SH}(\boldsymbol{v}_\Gamma), q_0)^2}{a(\mathcal{SH}\boldsymbol{v}_\Gamma, \mathcal{SH}\boldsymbol{v}_\Gamma)} \geq \beta_0 \|q_0\|_{L^2}^2 \quad \forall q_0 \in M_0. \tag{11}$$

When that is the case, as in [7], we can show that the bound for the condition of the preconditioned operator in S-norm is

$$\text{cond}_{S^{1/2}}(T) \leq C(1 + \frac{1}{\beta_0})\frac{1}{\beta^2}(1 + \log(\frac{H}{h}))^2 \tag{12}$$

where β is the inf-sup constant of the original problem (3).

5.4   The Coarse Space

The coarse space $\mathbf{V}_0$ plays an important role in the BDD preconditioning. This space must guarantee solvability for the local Neumann problems and scalability for the preconditioner. The minimum coarse space $\mathbf{V}_0$ for solvability is

$\mathbf{V}_0^{(0)} = $ Rigid Body Motion of each subdomain Ω_i scaled by $\text{diag}\{D_i\}$ on $\Gamma_{h,i}$ and zero on the remaining nodes on Γ_h.

Thus, in the two dimensional case $\mathbf{V}_0^{(0)}$ has dimension $3\times$(number of subdomains). As we will see in the numerical results, the associated preconditioner T is not going to be scalable, therefore $\mathbf{V}_0^{(0)}$ must not satisfy the uniform inf-sup stability (11). This indicates that the coarse space should be enriched. Since our objective is unstructured mesh discretization, we need to design coarse space enrichments suitable for such discretizations. We enrich $\mathbf{V}_0^{(0)}$ with one coarse function per interface $\mathcal{E}_k$, i.e., connected components of an interface $\partial\Omega_i \cap \partial\Omega_j$.

Let $\mathcal{E}_k$ be an interface ordered by a sequence of vertices $(v_0, \ldots, v_{n_k})$ connected by fine edges on $T_h(\partial\Omega_i \cap \partial\Omega_j)$. We define unity normal vectors $\boldsymbol{n}_j$ (for $j = 1, \ldots, (n_k - 1)$), by using the coordinates of v_j and its two neighboring vertices v_{j-1} and v_{j+1} on $T_h(\partial\Omega_i \cap \partial\Omega_j)$. Let $\boldsymbol{\eta}_{j-1/2}$ and $l_{j-1/2}$ ($\boldsymbol{\eta}_{j+1/2}$ and $l_{j+1/2}$) be the unity normal and the length of the interval $[v_{j-1}, v_j]$ ($[v_j, v_{j+1}]$), respectively. Define

$$\boldsymbol{n}_j = (l_{j-1/2}\boldsymbol{\eta}_{j-1/2} + l_{j+1/2}\boldsymbol{\eta}_{j+1/2})/\|l_{j-1/2}\boldsymbol{\eta}_{j-1/2} + l_{j+1/2}\boldsymbol{\eta}_{j+1/2}\|_2.$$

To define the different coarse space enrichments we first define the weight functions w_k on each interface $\mathcal{E}_k$. We consider the following weight functions on $\mathcal{E}_k$ (see Fig. 1):

- for defining $\mathbf{V}_0^{(1)}$ let $w_k^{(1)} \equiv 1$
- for defining $\mathbf{V}_0^{(2)}$ let $w_k^{(2)}(v_j) = 0$ for j even and 1 for j odd
- for defining $\mathbf{V}_0^{(3)}$ let $w_k^{(3)}(v_j) = \min\{d_{(j)}^1, d_{(j)}^2\}/\text{max\_dist}$
- for defining $\mathbf{V}_0^{(4)}$ let $w_k^{(4)}(v_j) = d_{(j)}^1 d_{(j)}^2/(\text{max\_dist})^2$, where $d_{(j)}^1$ and $d_{(j)}^2$ are defined as the l_2 distances to the boundary vertices v_0 and v_{n_k}, respectively, and let $\text{max\_dist} = \max_j\{d_{(j)}^1, d_{(j)}^2\}$.

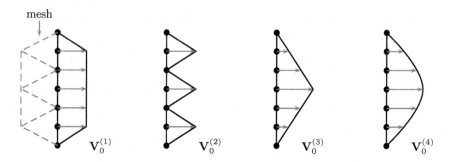

Fig. 1. Sketch of the edge enrichment functions

For each interface $\mathcal{E}_k$, we define the coarse function as

$$\boldsymbol{U}_k^{(r)}(v_j) = \begin{cases} w_k^{(r)}(v_j)\boldsymbol{n}_j & \text{for } j = 1,\ldots,(n_k - 1) \\ 0 & \text{for } j = 0, n_k \end{cases}$$

and then define the enriched coarse spaces $\mathbf{V}_0^{(r)}$, $r = 1,\ldots,4$, as the space spanned by $\mathbf{V}_0^{(0)}$ and the coarse functions $\boldsymbol{U}_k^{(r)}$. The spaces $\mathbf{V}_0^{(1)}$ and $\mathbf{V}_0^{(2)}$ are quite easy to implement, even for the tridimensional case, since their implementation depend only on the normal vector at the vertices. Since the enrichment of $\mathbf{V}_0^{(1)}$ is already a basis of the RBM for structured meshes, we do not consider $\mathbf{V}_0^{(1)}$ on the numerical tests.

6   Implementation Aspects

In this section we discuss some of the implementation details of the code. A parallel software was developed in C using the PETSc library [1] for unstructured meshes. The unstructured meshes are generated using the 2D mesh generator EMC2 from INRIA [8]. The partitioning of the mesh is by elements and it is performed using the ParMETIS library [5].

6.1   BDD Implementation

To assemble the matrix B_0 and the right hand side g_0, we define a vector $e^{(i)}$ in order to recover the constant pressure function in the subdomain Ω_i; in the case of P_0 functions, $e^{(i)}$ is the vector of ones. The matrix $B_0^{(i)}$ is computed as $B_0^{(i)^T} = B^{(i)^T} e^{(i)}$, while the vector components of the vector g_0 are computed as $g_0^{(i)} = e^{(i)^T} g^{(i)}$. Since the discrete local pressure spaces are subspaces of $L_0^2(\Omega_i)$ and the global pressure space is a subspace of $L_0^2(\Omega)$, we employ Lagrange multipliers $\lambda^{(i)}$ to enforce zero average on each $p_I^{(i)}$ in Ω_i and another Lagrange multiplier μ to enforce zero average of p_0 in Ω.

For applying the BDD preconditioner it remains to deal with another issue when solving (9): the uniqueness of the Neumann solution for the floating subdomains. The natural way of dealing with such difficulty is to search for a solution $u_I^{(i)}$ which is orthogonal to the kernel of $S^{(i)}$, i.e., orthogonal to the local RBM. This is done by introducing three Lagrange multipliers per subdomain, i.e., one for each local RBM basis function.

6.2   A Higher Order Method

Having implemented the $\mathbf{P_2}/P_0$ discretization in PETSc we reuse all the *index sets* and *local to global mappings* defined for the $\mathbf{P_2}/P_0$ to implement the $(\mathbf{P_2 + Bubbles})/P_1$. We add the bubble velocities and the linear average zero pressures on each element $K \in \mathcal{T}_h$, and then, through a static condensation at the element level, we eliminate the bubble functions and the two average zero pressures, resulting in a sort of stabilized $\mathbf{P_2}/P_0$ finite elements. After solving the linear system we can recover the P_1 discontinuous pressure solution at element level.

7   Numerical Results

A parallel software was developed in C using the PETSc library [1]. In order to study the scalability of the coarse space enrichments without the influence of the mesh partitioning, which may lead to irregular interface between subdomains, we consider in Subsections 7.1 and 7.2 a structured mesh in the domain $[0, 1] \times [0, 1]$ partitioned into $\sqrt{N} \times \sqrt{N}$ square subdomains. In Subsection 7.3 we consider an unstructured mesh example to study the parallel performance.

For the numerical experiments on Subsections 7.1 and 7.2 we impose Dirichlet boundary condition with the exact solution

$$\begin{cases} u_1(x, y) = x(1 - x) \cos(x + y) \cos(x + 3y) \\ u_2(x, y) = y(1 - y) \sin(x + y) \sin(x + y) \\ p(x, y) = xy \exp(x + 2y) \sin(x - y) \cos(y - x), \end{cases}$$

where we point out that $\nabla \cdot \boldsymbol{u}$ is non-null. Since the preconditioned operator T in (12) is symmetric positive definite with respect to S (see [7]), we use the preconditioned conjugated gradient (PCG) with the stopping criterion $\|r_k\|_2/\|r_0\|_2 \leq 10^{-6}$, where r_k is the residual at the iteration k. For solving the local problems we use the PETSc's LU with nested dissection reordering. The minimum eigenvalue is not presented in the tables since it is equal to one.

For the numerical experiments reported here we use a cluster of Linux PCs composed of 8 nodes with two Opteron processors each, where each node has 8Gbytes of shared memory among it processors. Each processor is scored at 4.8Gflops.

7.1   Constant Viscosity Tests

In this section all the numerical experiments are performed with a constant viscosity $\nu = 1$ and using the discretization $(\mathbf{P_2} + \mathbf{Bubble})/\mathbf{P_1}$. In Table 1 we fix the mesh of the subdomains to 32×32 and increase the number of subdomains. In Table 2 we fix the number of subdomains to 4×4 and refine the mesh of the subdomains. These tables show the number of PCG iterations and the maximum eigenvalue (in parenthesis) for the different coarse spaces. The minimum eigenvalue is always very close to 1.0 and is not reported. We conclude from Table 1 that the coarse spaces $\mathbf{V}_0^{(0)}$ and $\mathbf{V}_0^{(2)}$ do not satisfy the uniform inf-sup stability (11), while the coarse spaces $\mathbf{V}_0^{(3)}$ and $\mathbf{V}_0^{(4)}$ provide scalable algorithms. From Table 2, we see that the iteration counts of all the preconditioners depend very weakly on the size of the local problems. This result is expected due to (12).

Table 1. The PCG iteration counts and the largest eigenvalues of the preconditioned operator T (within parenthesis) for different coarse spaces. We fix the local mesh to 32×32.

Subdomains	$V_0^{(0)}$	$V_0^{(2)}$	$V_0^{(3)}$	$V_0^{(4)}$
3×3	19 (10.3)	19 (8.49)	17 (7.23)	16 (7.22)
4×4	23 (12.0)	22 (9.42)	20 (7.56)	20 (7.54)
5×5	27 (23.5)	25 (13.5)	20 (7.70)	20 (7.68)
6×6	28 (24.1)	24 (13.7)	20 (7.80)	20 (7.78)
7×7	30 (43.2)	26 (17.2)	20 (7.87)	20 (7.84)
8×8	35 (41.2)	27 (17.0)	21 (7.91)	20 (7.88)

Table 2. The PCG iteration counts and the largest eigenvalues of the preconditioned operator T (within parenthesis) for different coarse spaces. We fix the number of subdomains to 4×4.

Local mesh	$V_0^{(0)}$	$V_0^{(2)}$	$V_0^{(3)}$	$V_0^{(4)}$
8×8	17 (7.87)	16 (4.72)	15 (4.30)	14 (4.27)
16×16	20 (9.83)	19 (6.80)	17 (5.82)	17 (5.79)
32×32	23 (12.0)	22 (9.42)	20 (7.56)	20 (5.74)

In the sequel numerical experiments we consider only the space $\mathbf{V}_0^{(4)}$ since it shows to be the most effective coarse space tested.

On Table 3 we compare the discretization errors of the $(\mathbf{P_2} + \mathbf{Bubbles})/\mathbf{P_1}$ and the $\mathbf{P_2}/\mathbf{P_0}$ (in parenthesis). We see that the $(\mathbf{P_2} + \mathbf{Bubbles})/\mathbf{P_1}$ discretization is much more accurate than the $\mathbf{P_2}/\mathbf{P_0}$. The convergence error rates for the

$(\mathbf{P_2} + \mathbf{Bubbles})/\mathrm{P}_1$ are 10, 4, 4 for the velocity in the L^2, H^1, div norms, and 4 for the pressure in the L^2 norm, respectively. For the $\mathbf{P_2}/\mathrm{P}_0$ discretization the rates are 4, 2, 2 for the velocity in the L^2, H^1, div norms, and 2 for the pressure in the L^2 norm, respectively.

On Table 4 we compare the discretizations $(\mathbf{P_2} + \mathbf{Bubbles})/\mathrm{P}_1$ and $\mathbf{P_2}/\mathrm{P}_0$ with respect to iteration counts, conditioning, execution and assembling times (given in seconds). The table shows that the overall CPU time for the discretization $(\mathbf{P_2} + \mathbf{Bubble})/\mathrm{P}_1$ is not much larger than the one for $\mathbf{P_2}/\mathrm{P}_0$. Also we can see that the number of PCG iterations and the condition number are approximately the same for both discretizations. The high CPU time in the case of the local mesh 64×64 will be discussed in Subsection 7.3.

Table 3. The discretization errors of velocity for $(\mathbf{P_2} + \mathbf{Bubbles})/\mathrm{P}_1$ and $\mathbf{P_2}/\mathrm{P}_0$ (within parenthesis). The number of subdomains is fixed to 4×4.

Local mesh	$\|u - u_h\|_0$	$|u - u_h|_1$	$|u - u_h|_{div}$	$\|p - p_h\|_0$
4×4	3.64e-5 (5.73e-4)	3.88e-3 (3.63e-2)	2.82e-3 (3.31e-2)	1.39e-2 (7.42e-2)
8×8	3.71e-6 (1.47e-4)	9.13e-4 (1.84e-2)	6.93e-3 (1.69e-2)	3.81e-3 (3.72e-2)
16×16	4.13e-7 (3.73e-5)	2.18e-4 (9.26e-3)	1.71e-4 (8.52e-3)	9.78e-4 (1.86e-2)
32×32	4.97e-8 (9.40e-6)	5.39e-5 (4.64e-3)	4.27e-5 (4.27e-3)	2.46e-4 (9.31e-3)
64×64	6.60e-9 (2.36e-6)	1.34e-5 (2.33e-3)	1.07e-5 (2.14e-3)	4.65e-5 (4.65e-3)

Table 4. PCG iteration counts (Its.), largest eigenvalue of the preconditioned operator T ($\lambda_{\max}$), CPU time for assembling the matrix and CPU times for all the running (T_2) for the discretizations $(\mathbf{P_2} + \mathbf{Bubbles})/\mathrm{P}_1$ and $\mathbf{P_2}/\mathrm{P}_0$ (within parenthesis). The number of subdomains is fixed to 4×4.

Local mesh	Its.	$\lambda_{\max}$	$T_1(s)$	$T_2(s)$
4×4	11 (13)	2.98 (3.42)	0.08 (0.06)	2.35 (2.30)
8×8	14 (14)	4.27 (4.57)	0.10 (0.07)	3.12 (2.90)
16×16	17 (16)	5.79 (5.96)	0.16 (0.10)	8.65 (8.53)
32×32	20 (18)	7.53 (7.61)	0.58 (0.34)	108.6 (107.1)
64×64	22 (21)	9.52 (9.51)	1.80 (0.93)	5687.1 (5682.6)

7.2 Discontinuous Viscosities

In this section we assume that the viscosity is constant in each subdomain, however with a jump across the subdomains. We study the case where the viscosity is given by two constant values ν_1 and ν_2, in such a way that it has a checker board pattern.

We consider the discretization $(\mathbf{P_2} + \mathbf{Bubbles})/\mathrm{P}_1$ and fix $\nu_1 = 1$. On Table 5 we provide the number of iterations and the maximum eigenvalue (in parenthesis), for different values of the exponent γ; see (8). The best result is obtained when $\gamma = 1$, although for $\gamma > 1$ the condition numbers present similar behavior. In addition, as predicted in [9,10], we confirm the strong deterioration of the performance of the algorithms when γ is less than $1/2$ and ν_2 is large.

Table 5. PCG iteration counts and larget eigenvalue within parenthesis, for different viscosities ν_2 and exponent γ (see 8). The number of subdomains is fixed to 4×4.

γ	local mesh	$\nu_2 = 10$	$\nu_2 = 100$	$\nu_2 = 1000$
	8×8	19 (11.2)	25 (44.5)	26 (172)
$\gamma = 0.25$	16×16	23 (16.0)	31 (65.3)	35 (254)
	32×32	25 (22.0)	35 (90.5)	43 (352)
	8×8	15 (5.72)	17 (7.71)	17 (8.70)
$\gamma = 0.5$	16×16	18 (7.93)	19 (10.7)	19 (12.1)
	32×32	20 (10.6)	22 (14.3)	22 (16.1)
	8×8	13 (4.42)	11 (4.09)	11 (4.04)
$\gamma = 1$	16×16	14 (5.72)	13 (5.13)	12 (5.04)
	32×32	16 (7.08)	15 (6.17)	13 (6.03)
	8×8	13 (5.05)	11 (4.15)	11 (4.05)
$\gamma = 2$	16×16	15 (6.57)	13 (5.21)	12 (5.05)
	32×32	17 (8.17)	15 (6.26)	13 (6.04)

7.3   Parallel Performance

In order to analyze the parallel performance of the code we consider the discretization $(\mathbf{P_2} + \mathbf{Bubble})/\mathrm{P}_1$ and the coarse space enrichment $\mathbf{V}_0^{(4)}$ in the preconditioner. We also consider the domain Ω as in Figure (2) with an unstructured mesh. We impose the following Dirichlet boundary conditions

$$u(x, y) = \begin{cases} y(1-y); & \text{for } x = 0 \text{ (inflow)} \\ y(1-y); & \text{for } x = 6 \text{ (outflow)} \\ 0; & \text{otherwise (no-slip condition)} \end{cases}$$

On Table 6 we run problems with a mesh of 23008 elements (116283 dofs). In order to study the scalability we solve a problem in one processor only with LU using nested dissection reordering. The speedup in N processors (S_N) is calculated as the ratio of total execution time in 1 processor (T_1) and in N processors (T_N) as $S_N = T_1/T_N$. The efficiency in N processors is computed as the ratio of the speedup in N processors and the number of processors, i.e., S_N/N. The CPU times show that the proposed preconditioner is more effective when the size of the local problems is small. This is due to the high cost of the local LU factorizations of the Dirichlet and Neumann matrices. These LU factorization times leads to the high CPU time on Table 4 in the case of a local mesh 64×64. The CPU time in assembling and in LU factorization of the coarse matrix is very small. The decreasing of T_C as the number os subdomains increases is due to the assembling time of the coarse problem. This assembling is performed by inner product of vectors defined on the edges and hence the time is proportional to the size of the edges. The speedup factor grows super linearly, when we increase the number of processors, due to the smaller size of the local factorizations. The efficiency of the method grows due to the same reason; we point out that in the last case of 32 subdomains there is an overload of the processors. We also mention that a postprocessing of the mesh partition can improve a little the iteration counts by smoothing the interface between the subdomains.

On Table 7 we fix the local mesh to 3222 elements. We point out that to setup the preconditioner for more than one subdomain it is required two LU factorizations, while in one subdomain we need just one. We remark that the band of the matrix in the one subdomain case is smaller than in the cases with more subdomains, due to the shape of the domain. Thus, the execution time for one subdomain is more than twice faster than the 4 subdomains case, however, from the case of 4 subdomains to 16 subdomains the increase in the execution

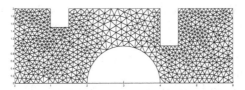

Fig. 2. Domain for parallel performance test and sketch of an unstructured mesh

Table 6. This table shows the iteration counts (It.), total execution time (T_{tot}), the speedup factors, the efficiency, and the CPU times to solve iteratively the linear system (T_S), to compute the LU factorizations of the local problems (T_F) and to compute the coarse matrix (this includes the LU factorization of the coarse matrix and is denoted T_C). The cases of 32 subdomains is performed by overloading some processors. The global mesh is fixed to 23008 elements.

Subs.	Its.	T_{tot} (s)	Speedup	Efficiency	T_S(s)	T_F(s)	T_C(s)
1 (LU)	–	4.91e+4	–	–	–	–	–
2	10	1.67e+4	2.94	1.47	1.06e+2	1.65e+4	1.15e+1
4	13	2.11e+3	23.3	5.82	3.85e+1	2.06e+3	5.83e+0
8	17	3.21e+2	153	19.1	2.17e+1	2.95e+2	3.49e+0
12	22	1.17e+2	420	35.0	2.56e+1	8.65e+1	2.52e+0
16	28	6.48e+1	758	47.4	2.09e+1	4.01e+1	1.84e+0
32	31	3.47e+1	1420	44.4	1.55e+1	1.13e+1	7.57e-1

Table 7. This table shows the iteration counts (It.), total execution time (T_{tot}) and the CPU times to solve iteratively the linear system (T_S), to compute the LU factorizations of the local problems (T_F) and to compute the coarse matrix (this includes the LU factorization of the coarse matrix and is denoted T_C). The local mesh is fixed to 3222 elements.

Subs.	Its.	T_{tot}(s)	T_S(s)	T_F(s)	T_C(s)
1 (LU)	–	1.41e+2	–	–	–
4	11	4.06e+2	1.32e+1	3.88e+2	2.80e+0
16	25	4.71e+2	5.43e+1	4.06e+2	5.38e+0

time is almost all due to the iterative solver, that takes 15 more iterations than in the 4 subdomains case. Hence, by comparing the 4 and 16 subdomain cases, the scalability is obtained. We expect that the iteration counts will stabilize for large number of subdomains due to the theory and Table 1.

8 Conclusions

We propose four coarse spaces suitable for BDD preconditioning on unstructured meshes. It is verified that the coarse spaces $\mathbf{V}_0^{(0)}$ and $\mathbf{V}_0^{(2)}$ are not stable, while the coarse spaces $\mathbf{V}_0^{(3)}$ and $\mathbf{V}_0^{(4)}$ are stable and scalable. We show that the discretization $(\mathbf{P_2} + \mathbf{Bubble})/\mathrm{P}_1$ is much more accurate than the $\mathbf{P_2}/\mathrm{P}_0$, with no significant extra computational cost. We have numerically confirmed that the choice $\gamma \geq 1$ in the definition of the diagonal scaling (8) is a robust choice for highly discontinuous viscosities.

We developed a code based on PETSc library for 2D unstructured meshes, extensible to 3D meshes, with very good efficiency and speedup factors. In addition, as indicated by the numerical results, we can increase the performance of the local LU factorizations with the use of better reorderings.

Acknowledgements. D. Conceição is gratefull to the PETSc team for valuable suggestions, and to IMPA's cluster support staff, S. Pilotto and D. Albuquerque.

References

1. Balay, S., Buschelman, K., Gropp, W.D., Kaushik, D., Knepley, M.G., McInnes, L.C., Smith, B.F., Zhang, H.: PETSc Web page : http://www.mcs.anl.gov/petsc (2001).
2. Dryja, M., Widlund, O.: Schwarz Methods of Neumann-Neumann type for three-dimensional elliptic finite element problems. Comm. Pure Appl. Math. **48** (1995), no.2, 121–155.
3. Girault, V., Raviart, P.-A.: Finite element methods for Navier-Stokes equations: Theory and algorithms. Springer Series in Computational Mathematics, 5. Springer, Berlin, (1986).
4. Goldfeld, P., Pavarino, L.F., Widlund, O.: Balancing Neumann-Neumann preconditioners for mixed approximations of heterogeneous problems in linear elasticity. Numer. Math. **95** (2003), no. 2, 283–324.
5. Karypis, G., Schloegel, K., Kumar, V.: ParMETIS – Parallel Graph Partitioning and Sparse Matrix Ordering Library. Version 3.1. Web page: http://glaros.dtc.umn.edu/gkhome/metis/parmetis/overview
6. Mandel, J,.: Balancing domain decomposition, Comm. Appl. Numer. Methods **9** (1993) 233–241.
7. Pavarino, L.F., Widlund, O.: Balancing Neumann-Neumann methods for incompressible Stokes equations. Comm. Pure Appl. Math. **55** (2002), no.3, 302–335.
8. Saltel, E., Hecht, F.: EMC2 Wysiwyg 2D finite elements mesh generator. INRIA. EMC2 web page: http://www-rocq1.inria.fr/gamma/cdrom/www/emc2/eng.htm

9. Sarkis, M.: Two-level Schwarz methods for nonconforming finite elements and discontinuous coefficients, Proceedings of the Sixth Copper Mountain Conference on Multigrid Methods, N.D. Melson, T.A. Manteuffel and S.F. McCormick, eds., Vol. 2, no. 3224, 543–566, NASA, Hampton VA, 1993.
10. Sarkis, M.: Nonstandard coarse spaces and Schwarz methods for elliptic problems with discontinuous coefficients using non-conforming element, Num. Math., **77** (1997) 383–406.

The Impact of Parallel Programming Models on the Performance of Iterative Linear Solvers for Finite Element Applications

Kengo Nakajima

Department of Earth and Planetary Science, The University of Tokyo
7-3-1 Hongo, Bunkyo-ku, Tokyo 112-0002, Japan
nakajima@eps.s.u-tokyo.ac.jp
http://www-solid.eps.s.u-tokyo.ac.jp/~nakajima

Abstract. Parallel iterative linear solvers for unstructured grids in FEM applications, originally developed for the Earth Simulator (ES), are ported to various types of parallel computer. The performance of flat MPI and hybrid parallel programming models is compared for the ES, Hitachi SR8000, IBM SP-3 and IBM p5-model 595 supercomputers. The effect of coloring and of different storage methods for coefficient matrices are evaluated in various types of application. Performance for more than 10^4 processors is estimated using measured data for up to 10^3 processors.

1 Introduction

1.1 Parallel Programming Models on SMP Cluster Architectures

Recently, symmetric multiprocessor (SMP) cluster architectures have become very popular in teraflop-scale parallel computers, such as the DOE-ASC (Advanced Simulation & Computing, formerly *ASCI*) [1] machines and the Earth Simulator (ES) [2].

In order to achieve minimal parallelization overhead, a multi-level *hybrid* programming model is often employed for these architectures (Fig. 1). In this method, coarse-grained parallelism is achieved through domain decomposition by message passing among SMP nodes, and fine-grained parallelism is obtained via loop-level parallelism inside each SMP node using compiler-based thread parallelization such as OpenMP.

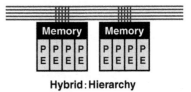

Hybrid: Hierarchy

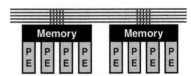

Flat MPI: Each PE -> Independent

Fig. 1. Parallel programming models for SMP cluster architectures

Another often-used programming model is the single-level *flat MPI* model (Fig. 1), in which separate single-threaded MPI processes are executed on each processing element (PE). The efficiency of each model depends on hardware performance (CPU speed, communication bandwidth, memory bandwidth, and the balance between these), application features, and problem size [3].

M. Daydé et al. (Eds.): VECPAR 2006, LNCS 4395, pp. 334–348, 2007.
© Springer-Verlag Berlin Heidelberg 2007

1.2 Previous Work

In previous work [4, 5], the author developed parallel iterative linear solvers for un-structured grids in finite element applications using GeoFEM [6] on the ES, using both the flat MPI and hybrid parallel programming models. Multicolor and reverse Cuthill-McKee (RCM) ordering methods [7, 8] provide excellent parallel and vector performance on the ES for iterative solvers with ILU/IC-type preconditioning. The performances of the flat MPI and hybrid parallel programming models are similar in most cases. The hybrid model outperforms the flat MPI model when the number of SMP nodes is large and the problem size is not too large. This is probably due to the effect of communication latency in MPI processes [9]. The effect of the number of colors processed has also been investigated.

1.3 Present Work

In this paper, parallel iterative solvers for unstructured grids, developed in [4, 5], are implemented on three more supercomputers: the Hitachi SR8000/MPP (University of Tokyo) [10], the IBM SP-3 (NERSC/LBNL) [11], and the IBM p5-model 595 (Kyushu University) [12]. The effect on performance of the number of colors and the storage method for coefficient matrices is evaluated through benchmarks based on real applications. Single processing element (PE), single SMP node, and multiple nodes are used for both the flat MPI and hybrid parallel programming models. Performance for more than 10^4 processors is estimated by extrapolating from measured data for up to 1,000 PEs.

Recently, several reports have been published relating to the performance of applications with unstructured grids on SMP cluster architectures (e.g. used in the finite element method (FEM)) [13, 14]. However, these are focused mainly on the flat MPI programming model.

2 Overview of Hardware and Software Environments

2.1 Hardware

Table 1 summarizes the architectures of the four supercomputers studied in this paper.

Table 1. Summary of architectures of Earth Simulator, Hitachi SR8000, IBM SP-3, and IBM p5-595 platforms

	Earth Simulator[2]	Hitachi SR8000 [10]	IBM SP-3 [11]	IBM p5-595 [12]
PE#/node	8	8	16	16
Clock rate (MHz)	500	450	375	1,900
Peak performance/PE (GFLOPS)	8.00	1.80	1.50	7.60
Memory/node (GB)	16	16	16~64	64~128
Memory BW (GB/sec)	32	4	1	6.4
Network BW (GB/sec/node)	12.3	1.60	1.00	4.00
MPI Latency (μsec)	5.6-7.7 [15]	6-20 [16]	16.3 [14]	3.9 [12]

The Earth Simulator [2] is a parallel vector system based on the NEC SX-6, with 640 SMP nodes, 5,120 vector processors, and 10 TB memory. The total peak performance is 40 TFLOPS. Each node is connected through a single-stage crossbar network. The Hitachi SR8000/MPP (Hitachi SR8000) at the University of Tokyo, based on the Hitachi SR8000 model G1 [10], has a very similar architecture to that of the ES. The entire system has 128 SMP nodes, 1,024 Power3-based processors and 2 TB memory. The total peak performance is 1.84 TFLOPS. Each PE is a scalar processor, but displays excellent performance with codes for vector processors through its pseudo-vector capability [10]. Each SMP node is connected through a three-dimensional crossbar network. The IBM SP-3 at NERSC/LBNL (Seaborg) [11] is a Power3-based superscalar system, with 380 SMP nodes, 6,080 processors, and 7.3 TB memory. Its total peak performance is 9.12 TFLOPS. Each PE has a 64 KB Level 1 data cache and an 8 MB Level 2 cache. Multi-node configurations are networked via the Colony switch. In this study, only 8 of 16 PEs on each SMP node are used for comparison with the ES and Hitachi SR8000. The IBM p5-model 595 (IBM p5-595) at Kyushu University [12] is a Power5-based superscalar system, with 26 SMP nodes, 416 processors, and 2.0 TB memory. The total peak performance is 3.16 TFLOPS. In this study, only one SMP node is used, and 8 of 16 PEs on an SMP node are used. Each PE has an 18 MB Level 3 cache.

2.2 Software

This paper contains evaluations of parallel iterative solvers with preconditioning for various types of applications on unstructured grids, originally developed for the ES. The following three types of preconditioning methods are considered [5, 6]:

I.   Localized block ILU(0) method for 3D solid mechanics
II.   Selective blocking method for 3D solid mechanics with contact conditions [5, 17]
III. Parallel multigrid method for 3D Poisson equations derived from incompressible Navier-Stokes solvers with adaptive grids [5]

The GeoFEM local data structure is applied. A proper definition of the distributed data structure is an important factor determining the efficiency of parallel computations with unstructured grids. The local data structures in GeoFEM are node-based with overlapping elements, and are appropriate for the preconditioned iterative solvers used in GeoFEM [6]. In FEM-type applications, most communication between processors occurs via information exchange at domain boundaries (Fig. 2). The ratio of communication to computation is usually small [4, 5].

In order to achieve efficient parallel/vector computation for applications with unstructured grids, there must be: (1) local operations and no global dependency, (2) continuous memory access, and (3) sufficiently long innermost loops for vector computation [4, 5, 8]. For unstructured grids, in which data and memory access patterns are very irregular, reordering methods are very effective in achieving highly parallel and vector performance, especially for factorization operations in ILU/IC preconditioning. The most popular reordering methods are RCM and multicoloring (MC) [7]. RCM is a typical level set reordering method with much less fill-in than for Gaussian elimination. MC is based on the concept that no two adjacent nodes have the same

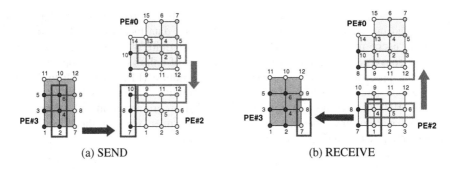

Fig. 2. Communication between processors in parallel FEM [4,5]

color. In both methods, elements having the same color are independent. Therefore, parallel operation is possible for elements with the same color.

In the hybrid parallel programming model, the following three levels of parallelism are considered: (1) MPI for inter-SMP node communication, (2) OpenMP for intra-node parallelization, and (3) compiler directives for vectorization of each PE.

Coefficient matrices are stored in descending-order jagged diagonal manner (DJDS) (Fig. 3(a)) in the original code [4, 5]. This method provides long innermost loops, and is suitable for vector processors. In this study, descending-order compressed row storage (DCRS) (Fig. 3(b)) is also tested for the Hitachi SR8000, IBM

```
do j= 1, NJmax
  do i= 1, Imax(j)
    k=(j-1)*N+i; kk=IA(k)
    Y(i)= Y(i)+A(k)*X(kk)
    ...
  enddo
enddo
```

```
do i= 1, N
  do k= IND(i-1)+1, IND(i)
    kk=IA(k)
    Y(i)= Y(i)+A(k)*X(kk)
    ...
  enddo
enddo
```

(a) DJDS   (b) DCRS

Fig. 3. Storage scheme and loop organization for matrix operations

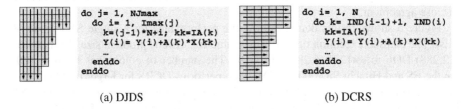

Fig. 4. Forward/backward substitution procedure using OpenMP and vectorization directives during ILU(0)/IC(0) preconditioning [4, 5]

SP-3, and IBM p5-595 machines. DCRS provides rather shorter loops than DJDS, but the reduced innermost loops of DCRS achieve good data locality, which is advantageous for cache utilization [18]. DJDS and DCRS require the same number of iterations for convergence, as long as the same number of colors has been applied. Fig. 4 shows the procedure for forward/backward substitution (FBS) using OpenMP and vectorization directives during ILU(0)/IC(0) preconditioning using DJDS/MC ordering (i.e. DJDS with multicoloring). In the flat MPI programming model, **PEsmpTOT** is set to 1 without any OpenMP options for the compiler, while in the hybrid model **PEsmpTOT** is set to the number of OpenMP threads.

3 Single PE/SMP Node Performance

Table 2 shows the single PE performance of a conjugate gradient solver with incomplete Cholesky preconditioning (ICCG) using DJDS with multicolor ordering, for a simple FEM application. The application concerned 3D linear elastic solid mechanics for a simple cubic geometry [4, 5], with homogeneous isotropic material properties and boundary conditions. The measured performance was computed from the results of 8 PEs for each system. The estimated performance was computed based on peak performance and measured memory bandwidth from STREAM benchmarks [19] using the estimation method described in [20]. The effect of the scalar processors' caches is not considered here. The problem size was 6,291,456 (3×128^3) DOF (degrees of freedom), and the number of colors was set to 100 for sufficiently long innermost loops. For each supercomputer, the measured and estimated performances are in close agreement.

Figure 5 shows results demonstrating the performance for a single SMP node (8 PEs). The elapsed execution time was measured for various problem sizes from 3×16^3 (12,288) DOF to 3×128^3 (6,291,456) DOF. The number of colors was fixed at 100. On the ES and Hitachi SR8000, the DJDS outperforms DCRS for larger problems due to the longer innermost loops. On the ES, the DJDS performance improved from 3.81 GFLOPS to 22.7 GFLOPS with problem size (i.e. from 6.0% to 35.5% of the peak performance). The pseudo-vector capability of the Hitachi compiler showed good performance. For the IBM SP-3 and IBM p5-595, the difference between DJDS and DCRS is not significant, and performance is better for a small problem size due to the effect of the cache. The DCRS method performs better than DJDS for a small problem size because DCRS utilizes cache memory more effectively. On the ES, the flat MPI and hybrid models are closely matched, though the flat MPI type has slightly better performance for the DJDS method. On the Hitachi SR8000, the hybrid model performs much better. The IBM SP-3 and IBM p5-595 machines display similar performance if the problem size is large, but the flat MPI model is much better for small problems, especially for the IBM SP-3. The cache on each processor is utilized more efficiently in the flat MPI parallel programming model. Reasons for the difference between the flat MPI and hybrid models on the Hitachi SR8000 with DJDS are not clear. According to the investigation in [21], the pseudo-vector capability does not seem to work efficiently in the flat MPI model.

Table 2. Single CPU performance for finite element type applications for each architecture

	Earth Simulator	Hitachi SR8000	IBM SP-3	IBM p5-595
Peak performance/PE (GFLOPS)	8.00	1.80	1.50	7.60
Measured Memory BW (GB/sec/PE) [19]	26.6	2.85	.623	3.65
Estimated performance (GFLOPS (% of peak))	2.31-3.24 (28.8-40.5)	.291-.347 (16.1-19.3)	.072-.076 (4.80-5.05)	.419-.444 (5.52-5.84)
Measured performance	2.93 (36.6)	.335 (18.6)	.122 (8.11)	.461 (6.07)

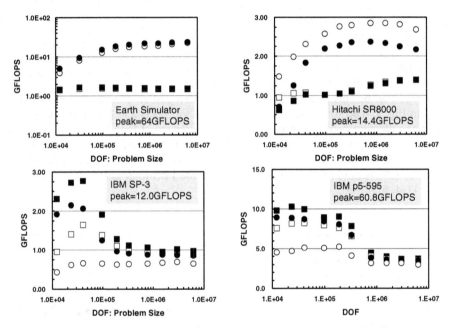

Fig. 5. Effect of coefficient matrix storage method and flat MPI/hybrid model type for a 3D linear elastic problem with simple cubic geometry for various problem sizes on a single SMP node (8 PEs) (100 colors)

● Flat-MPI DJDS
○ Hybrid DJDS
■ Flat-MPI DCRS
□ Hybrid DCRS

4 Effect of the Number of Colors

The convergence of iterative solvers using a multicolor reordering method can be improved by increasing the number of colors, because of fewer incompatible graphs [8]. However, this reduces the number of elements of each color, which means shorter innermost loops for vectorization [4, 5, 8]. In this section, this effect is investigated for both the flat MPI and hybrid programming models using a single SMP node (8 PEs) for each supercomputer for various types of applications and geometries.

4.1 Elastic Solid Mechanics

The first example is the simple 3D linear elastic problem for a cube in [4, 5] with 3×10^6 DOF (3×100^3). Figure 6 shows the effect of the number of colors on convergence of ICCG solvers using DJDS and DCRS with multicolor ordering. The number of iterations for convergence decreases as the number of colors increases in both the flat MPI and hybrid models. The hybrid programming model requires slightly fewer numbers of iterations for convergence. Figure 7 shows the effect of the number of colors on performance of the ES for DJDS. For both the flat MPI and hybrid models, the GFLOPS value decreases as the number of colors increases. Therefore, the elapsed time for computation is longer for 1,000 colors, even though the number of iterations decreases, as

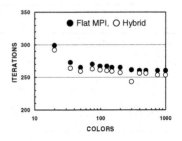

Fig. 6. Effect of number of colors in multicolor reordering: number of iterations for convergence in 3D linear elastic problem for cube (3×10^6 DOF) (3×100^3) using 1 SMP node

shown in Fig. 6. This phenomenon is much more significant in the hybrid model, as seen in Fig. 7. The size of the vector register in the ES is 256 [2]. In this case, with 10^6 finite element nodes on 8 PEs, the average innermost loop length is 256 for the case with 488 colors. However, Fig. 7 shows that the performance of the hybrid model worsens when the number of colors is more than about 100. This is mainly due to the synchronization overhead of OpenMP in the FBS loop in ILU/IC factorization (Fig. 4) [4, 5].

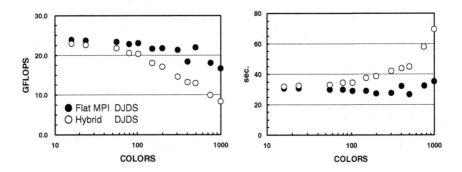

Fig. 7. Effect of number of colors in DJDS ordering on ES with 1 SMP node (8 PEs) for 3D linear elastic problem for cube (problem size=3×10^6 DOF) (3×100^3), peak=64 GFLOPS

On the Hitachi SR8000, the performance of the hybrid programming model decreases slightly when many colors are used, as shown in Fig. 8. However this is not as significant as on the ES. In the flat MPI model with DJDS, the performance improves as the number of colors increases. For the IBM SP-3 and IBM p5-595 machines, the effect of the number of colors on performance is not clear (Fig. 8). The DCRS performance is slightly better than for DJDS, and the DJDS method seems more sensitive to the number of colors than DCRS. The performance of the flat MPI model with DJDS improves as the number of colors increases. The performance of the hybrid model with DJDS also improves as the number of colors increases from 10 to 100, but then the performance deteriorates from 100 to 1,000 colors due to the OpenMP overhead. On the IBM p5-595, this drop is not so significant.

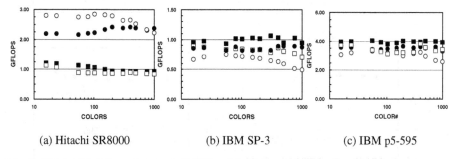

(a) Hitachi SR8000   (b) IBM SP-3   (c) IBM p5-595

Fig. 8. Effect of number of colors in DJDS and DCRS ordering on Hitachi SR8000, IBM-SP3 and IBM p5-595 with 1 SMP node (8 PEs) for 3D linear elastic problem for cube (problem size=3×10^6 DOF) (3×100^3)

● Flat-MPI DJDS
○ Hybrid   DJDS
■ Flat-MPI DCRS
□ Hybrid   DCRS

4.2 Selective Blocking Preconditioning for Contact Problems

Selective blocking is a special preconditioning method for contact problems with penalty constraints developed by the author. The target application is a simulation of processes of stress accumulation at plate boundaries around the Japanese islands (Fig.9) [4, 5, 6, 17]. In the selective blocking method, finite element nodes in the same contact group coupled through penalty constraints are placed into a

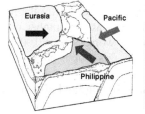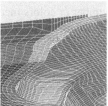

Fig. 9. Description of the Southwest Japan model with crust (dark gray) and subduction plate (light gray) [4, 5, 6, 17]

large block (the *selective block* or *super node*). For symmetric positive definite matrices, incomplete block Cholesky factorization without inter-block fill-in using selective blocking (SB-BIC(0)) shows excellent performance and robustness for a wide range of penalty parameter values [4, 5, 17]. Figure 10 shows the results for the South West Japan model with 784,000 tril-inear hexahedral elements, 823,813 nodes, and 2,471,439 DOF on a single SMP node (8 PEs). Only DJDS ordering was evaluated. With this geometry, the relationship between the number of colors and the performance has more marked characteristics. On the IBM SP-3 and IBM p5-595, the effect of the number of colors on performance is small. However it improves slightly as the

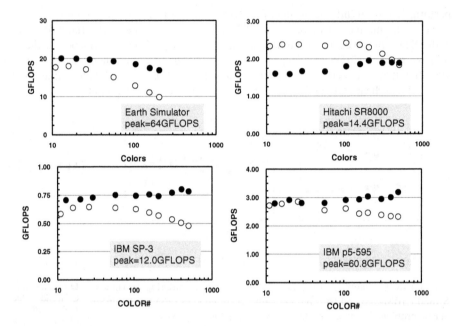

Fig. 10. Effect of number of colors for DJDS ordering on a single SMP node ● Flat MPI DJDS
(8 PEs) for 3D contact problem in Fig. 9   ○ Hybrid   DJDS

number of colors increases, especially for the flat MPI model. The performance of the
hybrid model also improves as the number of colors increases from 10 to 100, but
thereafter the performance worsens from 100 to 500 colors due to the OpenMP overhead. On the IBM p5-595, this drop is not so significant.

4.3 Multigrid Preconditioning for Poisson Equations

The next example is a multigrid preconditioned conjugate gradient iterative method
(MGCG) for Poisson equations, as described in [5]. The target application is 3D incompressible thermal convection in the region between dual spherical surfaces. This
type of geometry often appears in the simulations of earth sciences for both fluid earth
(atmosphere and ocean) and solid earth (mantle and outer core). Semi-unstructured
prismatic grids generated from triangles on spherical surfaces are used. Grids start as
icosahedrons and are then globally refined recursively as shown in Fig. 11 [5]. The
grid hierarchy resulting from recursive refinement can be used to generate coarse
grids. As stated in the author's previous work [5], the drop in performance for the
many colors case in the hybrid parallel programming model on the ES was very significant, because of shorter loop length and greater overhead. In this study, the same
problem is applied to different hardware. Figure 12 shows the performance of MGCG
cycles on the Poisson equations with 6,144,000 DOF on a single SMP node (8 PEs).

Here, only DJDS ordering was evaluated. For calculations with many colors, fewer iterations are required for convergence, however the performance is worse due to the smaller loop length and greater overhead. Performance of the ES is greatly affected by loop length. Moreover, the hybrid parallel programming model is much more sensitive to the number of colors and innermost vector length than the flat MPI model. Results for the Hitachi SR8000, IBM SP-3 and IBM p5-595 show similar features. On the IBM p5-595, the deterioration in performance for the hybrid model with many colors is not so great.

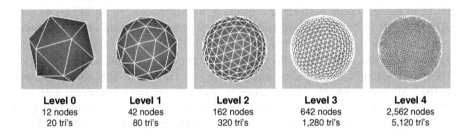

Level 0	Level 1	Level 2	Level 3	Level 4
12 nodes	42 nodes	162 nodes	642 nodes	2,562 nodes
20 tri's	80 tri's	320 tri's	1,280 tri's	5,120 tri's

Fig. 11. Surface triangle grids generated from an icosahedron

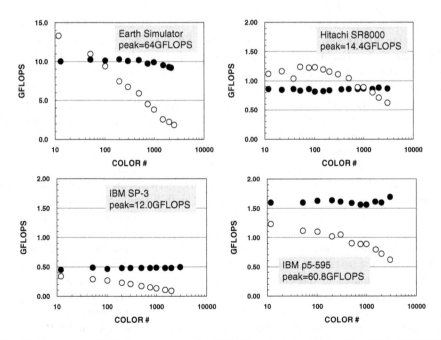

Fig. 12. Performance of Poisson solvers with MGCG on a single SMP node   ● Flat MPI DJDS
(8 PEs) with 6,144,000 prisms   ○ Hybrid DJDS

5 Multiple Nodes

Finally, a large-scale 3D simple elastic application with simple cubic geometry [4, 5] was solved, using more than 100 SMP nodes on the ES, Hitachi SR8000 and IBM SP-3 machines. The performance of both the flat MPI and hybrid models was evaluated. The problem size for one SMP node was fixed, and the number of nodes was varied between 1 and 176 (1,408 PEs) for the ES, and between 1 and 128 (1,024 PEs) for the Hitachi SR8000 and IBM SP-3. On the ES, the largest problem size was 2.21×10^9 DOF, for which the performance was about 3.80 TFLOPS, corresponding to 33.7 % of the total peak performance of the 176 SMP nodes (10.24 TFLOPS) with DJDS (Fig. 13) [4, 5]. The hybrid and flat MPI programming models display similar performance, but the hybrid outperforms the flat MPI when a large number of SMP nodes are involved, especially if the problem size per node is small, as shown in Fig. 13. Figures 14 and 15 show results obtained using the Hitachi SR8000 with DJDS and the IBM SP-3 with DCRS. On the Hitachi SR8000, the largest problem size was 8.05×10^8 DOF, for which the performance was about 335 GFLOPS, corresponding to 18.2 % of the total peak performance of the 128 SMP nodes. On the IBM SP-3, the largest problem size was 3.84×10^8 DOF, yielding a performance of around 110 GFLOPS, which corresponds to 7.16 % of the total peak performance of the 128 SMP nodes. In both cases, the deterioration in performance of the flat MPI model as seen on the ES was not observed.

For these applications, the sustained GFLOPS rate for a single SMP node of the ES is 20 to 30 times as large as that of the IBM SP-3, as shown in Table 2 and Fig. 5. The network bandwidth is also 10 times faster. However, the rate of MPI latency is very similar. According to [21], if there are 32^3 FEM nodes on a PE (=98,304 unknowns/PE), the computation time for one matrix-vector multiplication procedure (*mat-vec*) for 3D solid mechanics is about 6 msec on the ES if the performance is 2.80 GFLOPS/PE (35% of peak). The MPI latency of the ES is 6-8 μsec, as shown in Table 1, therefore the effect of MPI latency could be very significant in cases with more than 1,000 PEs on the ES.

Figure 16 shows the communication overhead measured for the ES and IBM SP-3 with 3×10^6 DOF/SMP node. The difference between the elapsed computation time per iteration for each case and the result with a single SMP node (8 PEs) is considered to be the communication overhead per iteration. Generally, the communication overhead is smaller for the hybrid programming model. On the ES, the communication overhead of the flat MPI model increases constantly, while for the other cases the overhead saturates for many PEs. The relative communication overhead compared to the elapsed computation time with a single SMP node (8 PEs) has been estimated for cases with more than 1,000 PEs according to experimental data. The resulting estimated regression curves are displayed in Fig. 17. The measured overhead ratios with 1,024 PEs are 27.7 % (ES/flat MPI), 10.8 % (ES/hybrid), 12.7 % (IBM SP-3/flat MPI), and 10.1 % (IBM SP-3/hybrid), respectively. The estimated ratios with 10^4 PEs are 110-280 % (ES/flat MPI), 16 % (ES/hybrid), 18-31 % (IBM SP-3/flat MPI), and 16-21 % (IBM SP-3/hybrid),

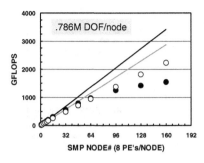
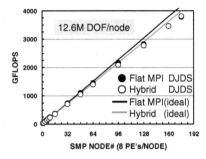

Fig. 13. Parallel performance on the ES for the 3D linear elastic problem using between 1 and 176 SMP nodes (1,408 PEs) with DJDS/MC ordering

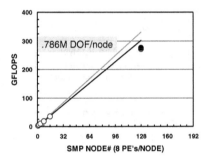
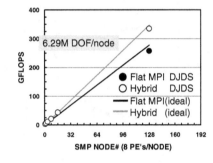

Fig. 14. Parallel performance on the Hitachi SR8000 for the 3D linear elastic problem using between 1 and 128 SMP nodes (1,024 PEs) with DJDS/MC ordering

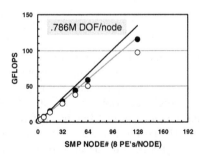
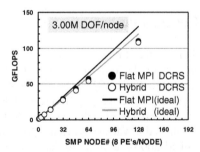

Fig. 15. Parallel performance on the IBM SP-3 for the 3D linear elastic problem using between 1 and 128 SMP nodes (1,024 PEs) with DCRS/MC ordering

respectively. Tests with multiple nodes on the IBM p5-595 have not been conducted, but the results would be expected to be similar to those of the IBM SP-3, going by the performance parameters in Table 1 and the single PE/node performance data in Table 2 and Fig. 5.

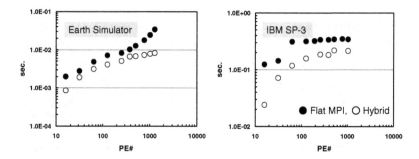

Fig. 16. Measured communication overhead per iteration of the ES and IBM SP-3 with 3×10^6 DOF/SMP node for the 3D linear elastic problem

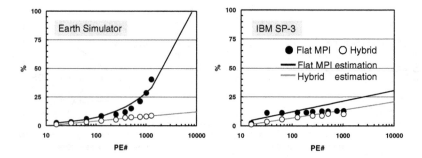

Fig. 17. Relative communication overhead of ES and IBM SP-3 with 3×10^6 DOF/SMP node for 3D linear elastic problem. Estimate based on extrapolation of measured results. Ratio based on elapsed computation time with an SMP node (8 PEs).

6 Conclusions

Parallel iterative linear solvers for unstructured grids in FEM applications, originally developed for the ES, were ported to three other SMP cluster supercomputers: the Hitachi SR8000, the IBM SP-3 and the IBM p5-595. The performance of the flat MPI and hybrid parallel programming models was compared using more than 100 SMP nodes. The effects of coloring and of storage method for coefficient matrices were also evaluated in various types of applications. The performance characteristics of the Hitachi SR8000 are very similar to those of the ES, mainly because of its pseudo-vector capability. Performance degradation for larger numbers of colors is not so significant as for the ES. The IBM SP-3 exhibits better performance for small problems. The combination of the DCRS and the flat MPI model gives the best performance, because this utilizes cache memory most efficiently. In the DJDS with flat MPI combination, increasing the number of colors gives improved performance due to data locality. The performance characteristics of the IBM p5-595 are similar to those of the IBM SP-3, but the performance of the hybrid parallel programming model with OpenMP is much improved in the IBM p5-595. The flat MPI and hybrid parallel programming models show similar performance in most cases for each supercomputer. On the ES, the hybrid outperforms the flat MPI when the number of SMP nodes is large and the problem size is small. This phenomenon was not observed on the other computers. This is because of the relatively high MPI latency of the ES. Generally, communication overhead with many PE's is higher in the flat MPI model than in the hybrid programming model. The performance of parallel FEM on massively parallel computers strongly depends on the balance between single PE performance, communication latency, and communication bandwidth.

Acknowledgements

This work is supported by the 21st Century Earth Science COE Program at the University of Tokyo, and CREST/Japan Science and Technology Agency. The author would like to thank the Earth Simulator Center, the Information Technology Center at the University of Tokyo, the National Energy Research Scientific Computing Center at Lawrence Berkeley National Laboratory, and the Computing and Communication Center at Kyushu University, for use of their computer resources.

References

1. ASCI: http://www.llnl.gov/asci/
2. Earth Simulator Center: http://www.es.jamstec.go.jp/
3. Rabenseifner, R. (2002), Communication Bandwidth of Parallel Programming Models on Hybrid Architectures, Lecture Notes in Computer Science 2327, 437-448.
4. Nakajima, K. (2003), Parallel Iterative Solvers of GeoFEM with Selective Blocking Preconditioning for Nonlinear Contact Problems on the Earth Simulator, ACM/IEEE Proceedings of SC2003.
5. Nakajima, K. (2004), Preconditioned Iterative Linear Solvers for Unstructured Grids on the Earth Simulator, IEEE Proceedings of HPC Asia 2004, 150-169.
6. GeoFEM: http://geofem.tokyo.rist.or.jp/
7. Saad, Y. (2003), Iterative Methods for Sparse Linear Systems (2nd Edition), SIAM.
8. Doi, S. and Washio, T. (1999), Using Multicolor Ordering with Many Colors to Strike a Better Balance between Parallelism and Convergence, Proceedings of RIKEN Symposium on Linear Algebra and its Applications, 19-26.
9. Kerbyson, D.J., Hoisie, A. and Wasserman, H. (2002), A Comparison Between the Earth Simulator and AlphaServer Systems using Predictive Application Performance Models, LA-UR-02-5222, Los Alamos National Laboratory.
10. Information Technology Center, The University of Tokyo: http://www.cc.u-tokyo.ac.jp/
11. National Energy Research Scientific Computing Center, Lawrence Berkeley National Laboratory: http://www.nersc.gov/
12. Computing and Communication Center, Kyushu University: http://www.cc.kyushu-u.ac.jp/
13. Adams, M.F., Bayraktar, H.H., Keaveny, T.M. and Papadopoulos, P. (2003), Applications of Algebraic Multigrid to Large-Scale Finite Element Analysis of Whole Bone Micro-Mechanics on the IBM SP, ACM/IEEE Proceedings of SC2003.
14. Oliker, L., Canning, A., Carter, J., Shalf, J., and Ethier, S. (2004), Scientific Computations on Modern Parallel Vector Systems, ACM/IEEE Proceedings of SC2004.

15. Uehara, H., Tamura, M., Itakura, K. and Yokokawa, M. (2003), MPI Performance Evaluation on the Earth Simulator (in Japanese), IPSJ Transactions on High-Performance Computing System, 44 SIG 1 (HPS 6), 24-34.
16. HLRS (High Performance Computing Center Stuttgart): http://www.hlrs.de/
17. Nakajima, K. and Okuda, H. (2004), Parallel Iterative Solvers with Selective Blocking Preconditioning for Simulations of Fault Zone Contact, Journal of Numerical Algebra with Applications, 11, 831-852.
18. Hatazaki, T. (2004), Lessons from porting vector computer applications onto Non-Uniform Memory Access scalar machines, IEEE Proceedings of HPC Asia 2004, 236-243.
19. STREAM (Sustainable Memory Bandwidth in High Performance Computers): http://www.cs.virginia.edu/stream/
20. Nakajima, K. (2005), Three-Level Hybrid vs. Flat MPI on the Earth Simulator: Parallel Iterative Solvers for Finite-Element Method, Applied Numerical Mathematics, 54, 237-255.
21. Nakajima, K. (2005), Parallel programming models for finite-element method using preconditioned iterative solvers with multicolor ordering on various types of SMP cluster supercomputers, IEEE Proceedings of HPC Asia 2005, 83-90.

Efficient Parallel Algorithm for Constructing a Unit Triangular Matrix with Prescribed Singular Values

Georgina Flores-Becerra[1,2], Victor M. Garcia[1], and Antonio M. Vidal[1]

[1] Departamento de Sistemas Informáticos y Computación
Universidad Politécnica de Valencia
Camino de Vera s/n, 46022 Valencia, España
{gflores, vmgarcia, avidal}@dsic.upv.es
[2] Departamento de Sistemas y Computación. Instituto Tecnológico de Puebla
Av. Tecnológico 420, Col. Maravillas, C.P. 72220, Puebla, México

Abstract. The problem tackled in this paper is the parallel construction of a unit triangular matrix with prescribed singular values, when these fulfill Weyl's conditions [9]; this is a particular case of the Inverse Singular Value Problem. A sequential algorithm for this problem was proposed in [10] by Kosowsky and Smoktunowicz. In this paper parallel versions of this algorithm will be described, both for shared memory and distributed memory architectures. The proposed parallel implementation is better suited for the shared memory paradigm; this is confirmed by the numerical experiments; the shared memory version, reaches an efficiency over 90%, and reduces substantially the execution times compared with the sequential algorithm.

1 Introduction

Inverse problems can be found in many branches of Science and Engineering, such as simulation of mechanical systems, geophysics, tomography, and many others [3,6,11,12]. A particular instance of this family of problems is the Inverse Singular Value Problem (ISVP), which can be defined as:

Given a set of n positive real numbers $S^* = \{s_1^*, s_2^*, ..., s_n^*\}$, where $s_1^* \geq s_2^* \geq ... \geq s_n^*$, find a matrix $A \in \Re^{n \times n}$, with a certain structure, whose singular values are S^*.

There exist several algorithms to solve this problem, such as MI, MIII, EP and FB [5], which are iterative Newton-like algorithms, with high computational cost ($O(n^4)$ for MI, MIII and EP; and $O(n^6)$ for FB). If the desired matrix must have a certain structure, the computational costs can be drastically reduced. As an example, the ISVP problem for Toeplitz matrices can be solved with Newton algorithms with cost $O(n^3)$.

The problem of the construction of a unit lower triangular matrix $A \in R^{n \times n}$, such that the singular values of A are $s_1^* \geq s_2^* \geq ... \geq s_n^*$, was proposed by Kosowski and Smoktunowicz in [10]. It can be seen as a special case of the

M. Daydé et al. (Eds.): VECPAR 2006, LNCS 4395, pp. 349–362, 2007.
© Springer-Verlag Berlin Heidelberg 2007

ISVP which could be named Inverse Unit Triangular Singular Value Problem (IUTSVP). The existence of solution was given by Horn [9], who proved that such matrix A exists if and only if the following conditions are fulfilled (these are called Weyl conditions):

$$s_1^* s_2^* s_3^* ... s_i^* \geq 1, \quad (i = 2 : n) \quad and \quad s_1^* s_2^* s_3^* ... s_n^* = 1.$$

Kosowski and Smoktunowicz proposed an $O(n^2)$ algorithm (based on Horn's proof) to solve the IUTSVP. It is a direct algorithm (that is, it solves the problem in a finite number of steps), in contrast with the iterative methods needed to solve the general ISVP.

This paper is focused on the design of a parallel version of the algorithm proposed by Kosowski and Smoktunowicz, and its implementation for shared memory and distributed memory computers; of course, the primary goal is the reduction of the time needed to solve this problem. Both implementations are compared from different points of view.

This paper is organized as follows: The theoretical background, along with the sequential algorithm are shown in the Section 2. In Section 3 the distributed memory parallel algorithm is introduced and discussed, and the shared memory algorithm is discussed in Section 4. In these three sections numerical results are given. Finally, in Section 5 the results obtained are compared and analyzed, offering the conclusions of the study.

2   Method Based in Weyl's Conditions(WE Method)

The method proposed by Kosowski and Smoktunowicz to solve IUTSVP is based on the construction of a sequence of unit lower triangular matrices $A^{(i)}$ $(i = 1 : n)$ equivalent to the diagonal matrix $diag(s_1^*, s_2^*, ..., s_n^*)^1$. To build the matrices of this sequence the following lemma is applied:

Lemma 1. *Two real numbers $s_i^*, s_j^* > 0$ such that $s_i^* \geq 1 \geq s_j^*$ or $s_j^* \geq 1 \geq s_i^*$, are the singular values of the matrix*

$$\begin{bmatrix} 1 & 0 \\ \sqrt{(s_i^{*2} - 1)(1 - s_j^{*2})} & s_i^* s_j^* \end{bmatrix}.$$

This lemma leads to take submatrices 2×2 of $A^{(i)}$ $(i = 1 : n)$ in the form $diag(d_i, d_j)$ such that d_i, d_j fulfill

$$d_i \geq 1 \geq d_j \quad or \quad d_j \geq 1 \geq d_i. \tag{1}$$

To ensure that (1) is fulfilled, Kosowski et.al. apply the next lemma:

---

[1] Two matrices M and N are unitarily equivalent if exist unitary matrices U, V such that $M = UNV^t$; under these conditions M and N shall have the same singular values.

Lemma 2. *If the real numbers $s_1^* \geq s_2^* \geq ... \geq s_n^* > 0$ satisfy Weyl's conditions, then there exists a permutation $\{d_1, d_2, ..., d_n\}$ of $\{s_1^*, s_2^*, ..., s_n^*\}$ such that*

$$d_1 d_2 ... d_{i-1} \geq 1 \geq d_i \quad or \quad d_i \geq 1 \geq d_1 d_2 ... d_{i-1} \quad (i = 2 : n). \qquad (2)$$

Given the matrix $A^{(1)} = diag(d_1, d_2, ..., d_n)$, and if d_1 and d_2 satisfy (1), then the following matrix exists:

$$L^{(2)} = \begin{bmatrix} 1 & 0 \\ \sqrt{(d_1^2 - 1)(1 - d_2^2)} & d_1 d_2 \end{bmatrix}$$

with singular values d_1, d_2. Then, we can build $A^{(2)} = diag(L^{(2)}, D_{n-2 \times n-2}^{(1)})$, where $D_{n-2 \times n-2}^{(1)} = diag(d_3, d_4, ..., d_n)$. $A^{(2)}$ is equivalent to $A^{(1)}$ because there exist 2×2 unitary matrices $U^{(2)}$, $V^{(2)}$ such that $L^{(2)} = U^{(2)} diag(d_1, d_2) V^{(2)T}$.

Once $A^{(2)}$ has been built, starts an iterative process to build $A^{(3)}, A^{(4)}, ... A^{(n)}$. For example, the construction of $A^{(3)}$ is based on the singular value decomposition (SVD) of the 2×2 matrix $L^{(3)}$:

$$L^{(3)} = \begin{bmatrix} 1 & 0 \\ \sqrt{(d_1^2 d_2^2 - 1)(1 - d_3^2)} & d_1 d_2 d_3 \end{bmatrix} = U^{(3)} \ diag(d_1 d_2, \ d_3) \ V^{(3)T} \qquad (3)$$

and the $A^{(3)}$ can be written as $A^{(3)} = Q^{(3)} A^{(2)} Z^{(3)T}$, where:

$$Q^{(3)} = diag(I_{1 \times 1}, U^{(3)}, I_{n-3 \times n-3}), \quad Z^{(3)T} = diag(I_{1 \times 1}, V^{(3)T}, I_{n-3 \times n-3}), \qquad (4)$$

$$A^{(2)} = \begin{bmatrix} 1 & 0 & 0 & 0 & ... & 0 \\ \sqrt{(d_1^2 - 1)(1 - d_2^2)} & d_1 d_2 & 0 & 0 & ... & 0 \\ 0 & 0 & d_3 & 0 & ... & 0 \\ \hline 0 & 0 & 0 & d_4 & ... & 0 \\ \vdots & \vdots & \vdots & \vdots & \ddots & \vdots \\ 0 & 0 & 0 & 0 & ... & d_n \end{bmatrix} = \begin{bmatrix} B_{1 \times 1}^{(2)} & & \\ \hline C_{2 \times 1}^{(2)} & diag(d_1 d_2, d_3) & \\ \hline & & D_{n-3 \times n-3}^{(2)} \end{bmatrix}$$

$$(5)$$

(See eq. (8) for the definition of B) and, performing the matrix multiplications, $A^{(3)}$ can be written as:

$$A^{(3)} = \begin{bmatrix} B_{1 \times 1}^{(2)} & & \\ \hline U^{(3)} C_{2 \times 1}^{(2)} & L^{(3)} & \\ \hline & & D_{n-3 \times n-3}^{(2)} \end{bmatrix}. \qquad (6)$$

The same procedure is followed to compute $A^{(4)}, A^{(5)}, ..., A^{(n)}$. The final result will be the unit lower triangular matrix $A^{(n)}$, whose singular values are S^*. Therefore, if the numbers p_i, z_i are defined as follows:

$$p_i = d_1 d_2 ... d_i, \ (i = 1 : n); \quad and \quad z_i = \sqrt{(p_{i-1}^2 - 1)(1 - d_i^2)}, \ (i = 2 : n); \quad (7)$$

$A^{(n)}$ has the form:

$$A^{(n)} = \left[\begin{array}{c|c} B^{(n-1)}_{n-2\times n-2} & \\ \hline U^{(n)}_{2\times 2}C^{(n-1)}_{2\times n-2} & L^{(n)}_{2\times 2} \end{array} \right],$$

where $B^{(n-1)}_{n-2\times n-2}$, $U^{(n)}_{2\times 2}C^{(n-1)}_{2\times n-2}$ and $L^{(n)}_{2\times 2}$ are:

$$B^{(n-1)}_{n-2\times n-2} = \left[\begin{array}{cccc|cccc|c|c} 1 & & & & 0 & & & & \cdots & 0 \\ u^{(3)}_{11}z_2 & & & & 1 & & & & \cdots & 0 \\ u^{(4)}_{11}u^{(3)}_{21}z_2 & & & & u^{(4)}_{11}z_3 & & & & \cdots & 0 \\ \vdots & & & & \vdots & & & & \ddots & \vdots \\ u^{(n-1)}_{11}u^{(n-2)}_{21}...u^{(3)}_{21}z_2 & & & & u^{(n-1)}_{11}u^{(n-2)}_{21}...u^{(4)}_{21}z_3 & & & & \cdots & 1 \end{array} \right] \quad (8)$$

$$U^{(n)}_{2\times 2}C^{(n-1)}_{2\times n-2} = \left[\begin{array}{c|c|c|c} u^{(n)}_{11}u^{(n-1)}_{21}...u^{(3)}_{21}z_2 & u^{(n)}_{11}u^{(n-1)}_{21}...u^{(4)}_{21}z_3 & \cdots & u^{(n)}_{11}z_{n-1} \\ u^{(n)}_{21}u^{(n-1)}_{21}...u^{(3)}_{21}z_2 & u^{(n)}_{21}u^{(n-1)}_{21}...u^{(4)}_{21}z_3 & \cdots & u^{(n)}_{21}z_{n-1} \end{array} \right] \quad (9)$$

$$L^{(n)}_{2\times 2} = \left[\begin{array}{cc} 1 & 0 \\ z_n & p_n \end{array} \right] = \left[\begin{array}{cc} 1 & 0 \\ z_n & 1 \end{array} \right]. \quad (10)$$

From these equations (8), (9) and (10), it becomes clear that $A^{(n)}$ can be built with the entries of the $U^{(i)}$ ($i = 3 : n$) matrices and the p_i ($i = 1 : n$), z_i ($i = 2 : n$) and d_i ($i = 1 : n$) values. $U^{(i)}$ is the matrix of the left singular vectors of

$$L^{(i)} = \left[\begin{array}{cc} 1 & 0 \\ z_i & p_i \end{array} \right]. \quad (11)$$

The algorithm to compute $A^{(n)}$ (called WE), must start by computing d_i ($i = 1 : n$), since p_i ($i = 1 : n$) and z_i ($i = 2 : n$) depend on d_i; recall that these d_is are a permutation of S^* that can be built using the Lemma 2; the algorithm that performs this permutation is taken from [10].

The WE algorithm can be written as follows:

```
Algorithm Sequential_WE
    1: build d (as mentioned above, taken from [10])
    2: compute p_i (i = 1:n) and z_i (i = 2:n), in accordance with (7)
    3: build L^(i) (i = 3:n), in accordance with (11)
    4: compute SVD(L^(i)) to obtain U^(i) (i = 3:n), using LAPACK_dgesvd
    5: build A^(n) as shown in (8-10), using BLAS_[dgemm/dscal]
```

It was proved in [10] that the time complexity of the WE Algorithm is

$$T(n) = \left\{ n^2 + \frac{328n}{3} \right\} t_f,$$

where t_f is the execution time for a single floating point operation.

Table 1 shows the results of some numerical experiments with the WE algorithm, where S denotes the singular values of the computed lower triangular matrix. In all the cases the results are quite good.

Table 1. Experimental Results (accuracy) with WE algorithm

n	4	5	8	30	50	100	150	300	500
$\|S^* - S\|_2$	4e-16	4e-16	5e-16	1e-15	1e-15	2e-14	4e-14	8e-15	1e-14
$\frac{\|S^*-S\|_2}{\|S^*\|_2}$	1e-16	3e-17	1e-16	1e-16	1e-16	3e-16	2e-16	3e-16	5e-16

3 Parallel Algorithm for Distributed Memory Model

The tools to implement the distributed memory version were standard linear algebra subroutines and libraries, such as LAPACK [1], BLAS [8] and the communications library BLACS [4] over MPI [7].

Recall that $U^{(i)}$ $(i = 3 : n)$ depends on the SVD of $L^{(i)}$, and $L^{(i)}$ (see (11)) depends on p_i and z_i (defined in (7)). Each element of p can be computed independently of the others; as z_i depends on p_{i-1} and d_i, z_i can also be computed in parallel with z_j $(j = 2 : n, j \neq i)$. Then, the SVD of the matrices $L^{(i)}$ can also be computed in parallel and the $U^{(i)}$ matrices can be computed at the same time.

On the other hand, in the former section it was proved that $A \equiv A^{(n)}$ can be computed without explicitly computing $A^{(1)}, A^{(2)}, ..., A^{(n-1)}$. In order to avoid unnecessary floating point operations, we can order the products of each row of A as in the following scheme (suppose the 5-th row of a 6×6 matrix A):

with $U^{(6)}$: $\Rightarrow A_{5,1:6} = \boxed{\;\;\;\;\;\; u_{11}^{(6)} \;\; 1 \;\;}$

with $U^{(5)}$: $A_{5,4}u_{21}^{(5)} \Rightarrow A_{5,1:6} = \boxed{\;\;\;\; u_{11}^{(6)} \;\; u_{21}^{(5)} \;\; u_{11}^{(6)} \;\; 1 \;\;}$

with $U^{(4)}$: $A_{5,3}u_{21}^{(4)} \Rightarrow A_{5,1:6} = \boxed{\;\; u_{11}^{(6)} \;\; u_{21}^{(5)} \;\; u_{21}^{(4)} \;\; u_{11}^{(6)} \;\; u_{21}^{(5)} \;\; u_{11}^{(6)} \;\; 1 \;\;}$

with $U^{(3)}$: $A_{5,2}u_{21}^{(3)} \Rightarrow A_{5,1:6} = \boxed{ u_{11}^{(6)} u_{21}^{(5)} u_{21}^{(4)} u_{21}^{(3)} \;\; u_{11}^{(6)} u_{21}^{(5)} u_{21}^{(4)} \;\; u_{11}^{(6)} u_{21}^{(5)} \;\; u_{11}^{(6)} \;\; 1 }$

The general expressions of this procedure are:

$$A_{i,i-2} = u_{21}^{(i-2)}; \quad A_{i,j} = A_{i,j+1}u_{21}^{(j)}; \quad (i = n; \; j = i - 3, i - 2, ..., 1); \qquad (12)$$

$$A_{i,i-1} = u_{11}^{(i-1)}; \quad A_{i,j} = A_{i,j+1}u_{21}^{(j)}; \quad (i = 2 : n - 1; \; j = i - 2, i - 3, ..., 1). \; (13)$$

To finish the construction of A, the columns of the lower triangular of A, (except the diagonal), are multiplied with the values of z; this operation can be expresed by:

$$A_{ij} = z_j A_{ij}; \quad (i = 2 : n; \; j = 1 : i - 1). \qquad (14)$$

Then, the rows of A can be obtained simultaneously if the values of U and z are available.

Therefore, there are three sections of the WE algorithm amenable for parallelization: the computing of the U matrices, the z components and the A rows.

In this work, the parallelization consists in the distribution of the $n-1$ components of z $(z_2,...,z_n)$, $(n-2)$ U^i matrices $(U^{(3)},...,U^{(n)})$ and $n-1$ rows of A $(A_{2,1:1}, A_{3,1:2}, A_{4,1:3},...,A_{n,1:n-1})$ among P processors.

To control the distribution of the work among the processors two indexes have been used, called *low* and *up*, which give the limits of the subinterval of components of U and z which each processor must compute.

The distribution of the work needed to obtain A is controlled through the data structures *Rows* and *CountRows*; *Rows* controls which rows of A belongs to each processor, and *CountRows* gives the number of rows in each processor. The distribution of pairs of rows is made trying to equilibrate the computational work. For example, if $n = 20$ and $P = 7$, the following pairs of rows can be formed:

pairs	(2,20)	(3,19)	(4,18)	(5,17)	(6,16)	(7,15)	(8,14)	(9,13)	(10,12)	(11,-)
flops	36	38	38	38	38	38	38	38	38	19

The pairs are formed picking rows from both extremes, so that the total number of flops is approximately the same for every processor. In the example, each processor owns a pair, and the rest of the pairs are distributed among the processors:

Proc	0	1	2	3	4	5	6
Rows	2, 20, 9	3, 19, 13	4, 18, 10	5, 17, 12	6, 16, 11	7, 15	8, 14
CountRows	3	3	3	3	3	2	2

To implement this idea in a distributed memory computer, all the arrays *Rows*, *CountRows*, U and z must be available in all the processors; therefore, the algorithm must contain at least two communication stages. The following diagram outlines how this is scheduled for the case $n = 19$, $P = 4$:

	$Proc_0$	$Proc_1$	$Proc_2$	$Proc_3$
Build	$d_{1:n}$	$d_{1:n}$	$d_{1:n}$	$d_{1:n}$
Compute	$z_{3:7}$	$z_{8:11}$	$z_{12:15}$	$z_{16:19,2}$
Compute	$U^{(3)}...U^{(7)}$	$U^{(8)}...U^{(11)}$	$U^{(12)}...U^{(15)}$	$U^{(16)}...U^{(19)}$
	$\Leftarrow$ - $\Rightarrow$			
	All–to–All Broadcast of U and z			
	$\Leftarrow$ - $\Rightarrow$			
Compute	$A_{2,1}, A_{19,1:18}$ $A_{16,1:15}, A_{15,1:14}$ $A_{11,1:10}$	$A_{3,1:2}, A_{18,1:17}$ $A_{7,1:6}, A_{14,1:13}$ $A_{10,1:9}$	$A_{4,1:3}, A_{17,1:16}$ $A_{8,1:7}, A_{13,1:12}$	$A_{5,1:4}, A_{16,1:15}$ $A_{9,1:8}, A_{12,1:11}$
	$\Leftarrow$ -			
	All–to–One communication to build A in $Proc_0$			
	$\Leftarrow$ -			

The algorithm for the distributed computation of z, U and A is described below.

```
Algorithm Parallel_WE
   In Parallel For Proc = 0, 1, ..., P - 1
      1: build d
      2: compute pᵢ, zᵢ (i = low : up), in accordance with (7)
      3: build L⁽ⁱ⁾ (i = low : up), in accordance with (11)
      4: compute SVD(L⁽ⁱ⁾) to obtain U⁽ⁱ⁾ (i = low : up),
                                    using LAPACK_dgesvd
      5: All-to-all broadcast of z and U, using BLACS_dgeb[r/s]2d
      6. compute Aᵢⱼ (i = Rows₁, Rows_countRows; j = 1 : i - 2),
                                    in accordance with (12-14)
      7: All-to-One Reduction of A to Proc = 0,
                                    using MPI_[Pack/Send/Recv]
   EndParallelFor
```

3.1   Theoretical and Experimental Costs

The code described above has been tested in a cluster of 2GHz biprocessor Intel Xeons (Kefren[2]) composed of 20 nodes, each one with 1 Gbyte of RAM, disposed in a 4×5 mesh with 2D torus topology and interconnected through a SCI network.

The theoretical analysis of the Parallel WE algorithm shows that its speedup is severely affected by the construction of A in the processor $Proc_0$, since the volume of data to be transferred is $O\left(\frac{n^2}{\sqrt{P}}\right)$ (assuming that the diameter of the network is the square root of P), as can be seen in the theoretical execution time:

$$T_{WE}(n, P) = \left\{\frac{n^2}{P} + \frac{322n}{3P} + 2n - \frac{656}{3P}\right\} t_f + 5\sqrt{P}t_m + \frac{n^2 + 4n}{\sqrt{P}}t_v,$$

where t_m is the network latency and t_v is the inverse of the bandwidth; therefore, the WE speedup does not reach the theoretical optimum, according with the following expression:

$$\lim_{n \to \infty} S_{WE}(n, P) = \frac{P}{1 + \sqrt{P}\frac{t_v}{t_f}}.$$

Some experiments performed in the Kefren cluster confirm this behaviour. The execution times are recorded in the Table 2, where it is quite clear that the execution times do not decrease when the number of processors increases.

The theoretical cost of the same algorithm without the final construction of A in the processor $Proc_0$ is:

$$T_{WE}(n, P) = \left\{\frac{n^2}{P} + \frac{322n}{3P} + 2n - \frac{656}{3P}\right\} t_f + 4\sqrt{P}t_m + \frac{6n}{\sqrt{P}}t_v,$$

[2] http://www.grycap.upv.es/usuario/kefren.htm

this shows that the degree of parallelism reached during the computation of U, z and A is theoretically good, since the speedup reaches the optimum asymptotically:

$$\lim_{n \to \infty} S_{WE}(n, P) = P.$$

The experiments with the algorithm without building A can be seen in the Table 3. In these experiments there are execution times reductions when we use more than one processor, except in $n = 500$, that is a case efficiently solved by the sequential algorithm. The efficiency of these experiments are in Figure 1. The efficiency curve at $n = \{2000, 2500, 3000\}$ represents a typical case where the use of the processor cache influences the execution times; however, this phenomenon tends to disappear when the problem size increases.

The reconstruction of a matrix with prescribed singular values is usually a part of a larger problem to be solved in parallel. So, the gathering of the matrix in a single processor or the redistribution of the matrix among processors may be necessary or not, depending on the larger problem. Therefore, the analysis

Table 2. WE Execution Times in Kefren, building A in $Proc_0$

P	Seconds							
1	0.41	5.08	12.05	22.28	35.76	66.08	191	270
2	2.61	10.27	21.92	39.31	73.79	125	294	456
4	2.59	10.29	22.02	39.66	72.60	113	258	400
6	2.66	10.60	23.26	42.15	72.71	111	244	388
8	2.81	11.63	26.67	45.55	75.08	114	240	328
10	3.00	11.82	26.90	47.61	77.52	115	237	379
12	3.22	11.90	26.74	47.28	77.27	115	232	375
14	3.51	12.13	27.69	48.67	79.58	118	230	370
16	3.69	12.60	28.51	50.44	81.27	120	230	368
n	500	1000	1500	2000	2500	3000	4000	5000

Table 3. WE Execution Times in Kefren, without building A in $Proc_0$

P	Seconds							
1	0.41	5.08	12.05	22.28	35.76	66.08	191	270
2	2.20	4.68	7.85	12.51	21.30	42.13	104	136
4	1.19	2.56	4.39	7.10	12.23	23.12	53	69
6	0.69	1.89	3.27	5.29	8.96	16.08	35	47
8	0.55	1.55	2.76	4.37	7.18	12.90	27	35
10	0.48	1.39	2.43	3.87	6.21	10.52	21	28
12	0.49	1.26	2.20	3.34	5.43	8.94	18	24
14	0.45	1.18	2.00	3.08	4.92	8.02	16	20
16	0.44	1.10	1.89	2.89	4.49	7.21	14	18
n	500	1000	1500	2000	2500	3000	4000	5000

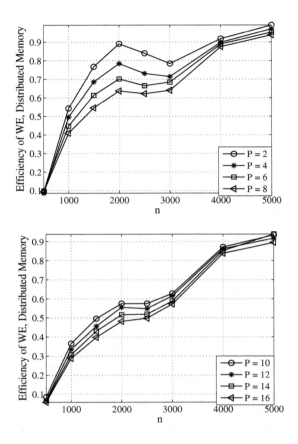

Fig. 1. Experimental WE Efficiency in Kefren, without building A in $Proc_0$

above shows that, leaving aside the final communications needed to recover the matrix in a single processor, the resolution of the problem has been reasonably parallelized. The remaining pitfall will be adressed in the next section.

4 Parallel Algorithm for Shared Memory Model

The analysis of the costs of the distributed memory code shows that the communications needed to collect the results and put it in a single processor damage seriously the performance of the code. In a Shared Memory Machine, this last step would be not needed, so that it is a natural way to improve the overall speed of the code. This implementation has been carried out using OpenMP [2] compiler directives, such as *omp parallel do* (to parallelize a loop), *omp parallel* (to parallelize a section of code), *omp barrier* (to synchronize execution threads) and *omp do* (to define shared work in a cycle).

Using H process threads and assuming that each one is executed in one processor $(P = H)$, each thread shall compute a subset of components of z, U and A; it was seen in the last section that there are no data dependency problems.

The data distribution of z, U and A might be performed by the programmer as in the distributed memory code (through indexes low, up, $Rows$ and $CountRows$). However, in this case it is more efficient to let the openMP compiler do the job; the directives $omp\ parallel\ do$ and $omp\ do$ split the work among processors automatically. The following diagram shows schematically how would proceed the computation:

Thread	Th_0	Th_1	Th_2	Th_3
Build	$d_{1:n}$	$- - - - -$	$- - - - -$	$- - - - -$
Compute	$z_{3:7}$	$z_{8:11}$	$z_{12:15}$	$z_{16:19,2}$
Compute	$U^{(3)}...U^{(7)}$	$U^{(8)}...U^{(11)}$	$U^{(12)}...U^{(15)}$	$U^{(16)}...U^{(19)}$
	$\Leftarrow - \Rightarrow$			
	Synchronisation Barrier			
	$\Leftarrow - \Rightarrow$			
Compute	$A_{2,1}, A_{3,1:2}, A_{4,1:3}$ $A_{5,1:4}, A_{18,1:17}$	$A_{6,1:5}, A_{7,1:8}$ $A_{8,1:7}, A_{9,1:8}$	$A_{10,1:9}, A_{11,1:10}$ $A_{12,1:11}, A_{13,1:12}$	$A_{14,1:13}, A_{15,1:14}$ $A_{16,1:15}, A_{17,1:16}, A_{19,1:18}$

Comparing this diagram with the Distributed Memory diagram, it is clear that the communication stages disappear, so that the efficiency is expected to increase. This process is written with detail in the ParallelSh_WE algorithm, where we have used $omp\ parallel$ in order to create a team of threads $(Th_1, ..., Th_{H-1})$, to run in parallel a code segment, and in ParallelSh_A and ParallelSh_zU algorithms, where the directive $omp\ do$ was used to divide the iterations (of the "for" loop) among the the threads created with $omp\ parallel$.

```
Algorithm ParallelSh_WE
  1: build d /* executed by the main thread */
  2: !$omp parallel private(Th) /* slave threads created by
       3: call ParallelSh_zU                the main thread */
       4: !$omp barrier
       5: call ParallelSh_A
  6: !$omp end parallel /* slave threads finished by
                                        the main thread */
```

```
Algorithm ParallelSh_A
  1: !$omp do /* The iterations are divided among the threads */
     2: For i = 2, 3, ..., n
        3: compute A_{i,1:i-1}, in accordance with (12-14)
     4: EndFor
  5: !$omp enddo
```

```
Algorithm ParallelSh_zU
  1: !$omp do /* The iterations are divided among the threads */
     2: For i = 3, 4, ..., n
        3: compute z_{i-1}, p_i, in accordance with (7)
        4: build L^{(i)}, in accordance with (11)
        5: compute SVD(L^{(i)}) to obtain U^{(i)}, using LAPACK_dgesvd
     6: EndFor
  7: !$omp enddo
  8: If Th = Th_{H-1} compute z_1, in accordance with (7)
```

4.1 Experimental Tests

The shared memory code has been tested in a multiprocessor (Aldebaran[3]) SGI Altix 3700 with 48 processors Intel 1.5 GHz Itanium 2, each one with 16 Gbytes of RAM; these are connected with a SGI NumaLink network, with hypercube topology. Although from the programmer's point of view it is a shared memory multiprocessor, actually it is a distributed memory cluster as well, though with a very fast interconnection network. The execution times in this machine are summarized in the Table 4.

The efficiency corresponding to these experiments (Figure 2) is good even with relatively small problem sizes. With $n = 1000$ and two threads the efficiency is 81%; from $n = 2000$ the efficiency is very good with up to 4 threads and acceptable for 6. For the largest case tested in this work ($n = 5000$) the efficiency with 6 threads is also good.

The Scaled Speedup is computed increasing in the same proportion the size of the problem and the number of threads; as WE is $O(n^2)$ we have taken its Speedup with $n = \{1000, 1400, 2000, 2800, 3400, 4000\}$ respectively with $H = \{1, 2, 4, 8, 12, 16\}$ in Figure 3. This figure shows an acceptable scalability.

Table 4. WE execution times in Aldebaran (Shared memory)

H	Seconds							
1	1.3	4.9	10.1	22	26	37	68	95
2	1.0	3.1	6.1	12	14	20	35	48
3	0.64	2.1	4.3	9.0	10	14	23	31
4	0.55	1.6	3.4	7.4	7.1	10	18	23
6	0.45	1.1	2.7	5.0	6.2	9	16	18
8	0.44	1.1	2.6	4.3	5.5	8	13	17
10	0.33	1.2	2.3	3.7	4.8	7	10.3	16
12	0.35	1.0	2.6	3.7	4.6	6.1	10.2	13
14	0.32	0.9	2.2	3.3	4.2	6.0	9.8	11
16	0.25	0.8	2.1	3.2	4.1	5.4	9.6	10
n	500	1000	1500	2000	2500	3000	4000	5000

---

[3] http://www.asic.upv.es

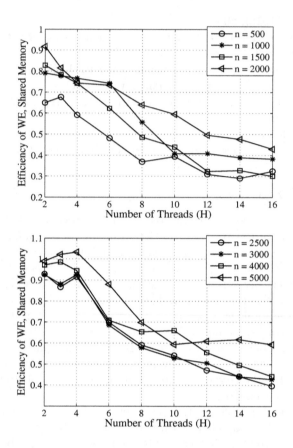

Fig. 2. Experimental WE Efficiency in Aldebaran (Shared memory)

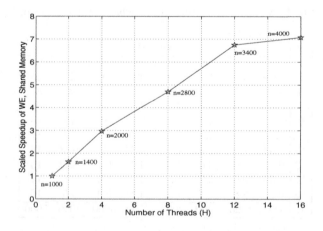

Fig. 3. WE Scaled Speedup in Aldebaran. Shared memory

5   Conclusions

The parallel code written for distributed memory reaches a good level of parallelism, as far as the computation phase is concerned. However, once the matrix is computed, it could be necessary to bring it back to a single processor or redistribute it among processors, by depending of the design of the parallel algorithm that uses this matrix; these communications spoil all the gains obtained with the parallel code. This trouble can be overcome if the same algorithm is adapted to a shared memory machine, where these final communications would not be needed. Furthermore, some communications that would happen in the distributed memory code (replicating U, z) would not be necessary either in the shared memory code.

We can establish the following comparisons between both approaches:

Shared Memory	Distributed Memory
* Easy Implementation	* Complex Implementation
* No extra data structures	* Require extra data structures
* Load Distribution through compiler directives	* Load Distribution by the programmer
* Efficiency larger than 90% with up to 4 threads	* Do not reach acceptable performance due to the last communications stage

Therefore, the shared memory implementation decreases the execution times of the sequential WE code, and gives good performances with up to 4 threads. It reaches an efficiency $> 90\%$ (Figure 2), obtaining as well an acceptable scalability (Figure 3). The performance is damaged when more than 4 threads are used; this is due to the fact that the machine Aldebaran is a multiprocessor with logically shared but physically distributed memory, so that at the end there is a large (transparent to the programmer) traffic of messages.

It seems clear that the nature of this problem makes it more addequate to be processed in a shared memory environment, rather than in a distributed memory cluster.

Acknowledgement

This work has been supported by Spanish MEC and FEDER under Grant TIC2003-08238-C02-02 and SEIT-DGEST-SUPERA-ANUIES (México).

References

1. Anderson E., Bai Z., Bishof C., Demmel J., Dongarra J.: LAPACK User Guide; Second edition. SIAM (1995)
2. Chandra, R., Dagum L., Kohr D., Maydan D., McDonald J., Menon R.: Parallel Programming in OpenMP. Morgan Kaufmann Publishers (2001)
3. Chu, M.T.: Inverse Eigenvalue Problems. SIAM, Review, Vol. 40 (1998)

4. Dongarra J., Van de Geijn R.: Two dimensional basic linear algebra comunications subprograms. Tecnical report st−cs−91−138, Department of Computer Science, University of Tennessee (1991)
5. Flores-Becerra G., García V. M., Vidal A. M.: Numerical Experiments on the Solution of the Inverse Additive Singular Value Problem. Lecture Notes in Computer Science, Vol. 3514, (2005) 17-24
6. Groetsch, C.W.: Inverse Problems. Activities for Undergraduates. The mathematical association of America (1999)
7. Groupp W., Lusk E., Skjellum A.: Using MPI: Portable Parallel Programming with Message Passing Interface. MIT Press (1994)
8. Hammarling S., Dongarra J., Du Croz J., Hanson R.J.: An extended set of fortran basic linear algebra subroutines. ACM Trans. Mathemathical Software (1988)
9. Horn A.: On the eigenvalues of a matrix with prescribed singular values. Proc. Amer. Math. Soc., Vol. 5, (1954) 4-7
10. Kosowski P., Smoktunowicz A.: On Constructing Unit Triangular Matrices with Prescribed Singular Values. Computing, Vol. 64, No. 3 (2000) 279-285
11. Neittaanmki, P., Rudnicki, M., Savini, A.: Inverse Problems and Optimal Design in Electricity and Magnetism. Oxford: Clarendon Press (1996)
12. Sun, N.: Inverse Problems in Groundwater Modeling. Kluwer Academic (1994)

A Rewriting System for the Vectorization of Signal Transforms

Franz Franchetti, Yevgen Voronenko, and Markus Püschel[*]

Electrical and Computer Engineering,
Carnegie Mellon University,
5000 Forbes Avenue, Pittsburgh, PA 15213
{franzf, yvoronen, pueschel}@ece.cmu.edu
http://www.spiral.net

Abstract. We present a rewriting system that automatically vector-izes signal transform algorithms at a high level of abstraction. The input to the system is a transform algorithm given as a formula in the well-known Kronecker product formalism. The output is a "vectorized" formula, which means it consists exclusively of constructs that can be directly mapped into short vector code. This approach obviates compiler vectorization, which is known to be limited in this domain. We included the formula vectorization into the Spiral program generator for signal transforms, which enables us to generate vectorized code and further optimize for the memory hierarchy through search over alternative al-gorithms. Benchmarks for the discrete Fourier transform (DFT) show that our generated floating-point code is competitive with and that our fixed-point code clearly outperforms the best available libraries.

1 Introduction

Most recent architectures feature short vector SIMD instructions that provide data types and instructions for the parallel execution of scalar operations in short vectors of length ν (called ν-way). For example, Intel's SSE family pro-vides $\nu = 2$ for double precision and $\nu = 4$ for single precision floating point arithmetic as well as $\nu = 8$ for 16-bit and $\nu = 16$ for 8-bit integer arithmetic. The potential speed-up offered by these instructions makes them attractive in domains were high performance is crucial, but they come at a price: Compilers often cannot make optimal use of vector instructions, since the necessary pro-gram transformations are not well understood. This moves the burden to the programmer, who is required to leave the standard C programming model, for example by using so-called intrinsics interfaces to the instruction set.

In [1] we have argued that for the specific domain of signal transforms such as the discrete Fourier transform (DFT) there is an attractive solution to this

---

[*] This work was supported by NSF through awards 0234293, 0325687, and by DARPA through the Department of Interior grant NBCH1050009. Franz Franchetti was sup-ported by the Austrian Science Fund FWF's Erwin Schroedinger Fellowship J2322.

M. Daydé et al. (Eds.): VECPAR 2006, LNCS 4395, pp. 363–377, 2007.
© Springer-Verlag Berlin Heidelberg 2007

problem, namely to perform the vectorization at a higher level of abstraction by manipulating Kronecker product expressions through mathematical identities. The Kronecker product formalism has been known to be useful for the representation and derivation for DFT algorithms [2] but also in the derivation of parallel algorithms [3].

In this paper we describe an implementation of this formal vectorization in the form of a rewriting system [4], the common tool used in symbolic computation. We then include the rewriting system into the Spiral program generator [5], which uses the Kronecker product formalism as internal algorithm representation. This enables additional optimization through Spiral search mechanism. Namely, for the desired transform, Spiral will generate alternative algorithms, each of which is vectorized using the rewriting system, and return the fastest for the given platform. In effect this heuristic procedure optimizes for the platform's memory hierarchy.

We show that our approach works for the DFT (speed-up in parenthesis) on a Pentium 4 for 2-way (1.5 times) and 4-way floating point (3 times), and 8-way (5 times) and 16-way (6 times) integer code. Benchmarks of our generated code against the Intel MKL and IPP libraries and FFTW [6] show that our generated floating-point code is competitive and our generated fixed-point code is at least a factor of 2 faster than the vendor library.

We have used the rewriting approach to the related but different problems of parallelization for shared and distributed memory computers [7,8].

Related work. The Intel C++ compiler includes a vectorizer based on loop vectorization and translation of complex operations into two-way vector code. FFTW 3.0.1 combines loop vectorization and an approach based on the linearity of the DFT to obtain 4-way single-precision SSE code [6]. This combines a hardcoded vectorization approach with FFTW's capability to automatically tune for the memory hierarchy. An approach to designing embedded processors with vector SIMD instructions and for designing software for these processors is presented in [9].

Organization of the paper. In Section 2 we provide background on SIMD vector instructions, signal transforms and their fast algorithms, and the Spiral program generator. The rewriting system is explained in Section 3 including examples of vectorization rules and vectorized formulas. Section 4 shows a number of runtime benchmarks for the DFT. We offer conclusions in Section 5.

2   Background

SIMD vector instructions. Recently, major vendors of general purpose microprocessors have included short vector SIMD (single instruction, multiple data) extensions into their instruction set architecture. Examples of SIMD extensions include Intel's MMX and SSE family, AMD's 3DNow! family, and Motorola's AltiVec extension. SIMD extensions have the potential to speed up implementations in areas where the relevant algorithms exhibit fine grain parallelism but are a major challenge to software developers.

In this paper we denote the vector length with ν. For example, SSE2 provides 2-way ($\nu = 2$) double and 4-way ($\nu = 4$) single precision floating point as well as 8-way ($\nu = 8$) 16-bit and 16-way ($\nu = 16$) 8-bit integer vector instructions.

Signal transforms. A (linear) signal transform is a matrix-vector multiplication $x \mapsto y = Mx$, where x is a real or complex input vector, M the transform matrix, and y the result. Examples of transforms include the discrete Fourier transform (DFT), multi-dimensional DFTs (MDDFT), the Walsh-Hadamard transform (WHT), and discrete Wavelet transforms (DWT) like the Haar wavelet. For example, for an input vector $x \in \mathbb{C}^n$, the DFT is defined by the matrix

$$\mathrm{DFT}_n = [\omega_n^{k\ell}]_{0 \le k, \ell < n}, \quad \omega_n = \exp(-2\pi i / n).$$

Algorithms for transforms can be written using the Kronecker product formalism [2,3,5] in the form of structured sparse matrix factorizations. In the following, we use I_n to denote an $n \times n$ identity matrix.

$$A \otimes B = [a_{k\ell} B] \quad \text{for } A = [a_{k\ell}]$$

is the tensor product of matrices. Further we introduce the stride permutation matrix defined, for $m | n$, by

$$\mathrm{L}_m^n : jk + i \mapsto im + j, \quad 0 \le i < k,\ 0 \le j < m.$$

Equations (1)–(6) show examples of recursive, divide-and-conquer, transform algorithms, written in the form of rules:

$$\mathrm{DFT}_{mn} \to \left(\mathrm{DFT}_m \otimes \mathrm{I}_n \right) D_{m,n} \left(\mathrm{I}_m \otimes \mathrm{DFT}_n \right) \mathrm{L}_m^{nm} \tag{1}$$

$$\mathrm{DFT}_n \to X_n \,\mathrm{RDFT}_n \tag{2}$$

$$\mathrm{WHT}_{mn} \to \mathrm{WHT}_m \otimes \mathrm{WHT}_n \tag{3}$$

$$\mathrm{MDDFT}_{n_1 \times \cdots \times n_k} \to \mathrm{MDDFT}_{n_1 \times \cdots \times n_r} \otimes \mathrm{MDDFT}_{n_{r+1} \times \cdots \times n_k} \tag{4}$$

$$\mathrm{MDDFT}_n \to \mathrm{DFT}_n \tag{5}$$

$$\mathrm{Haar}_n \to \mathrm{L}_2^n \left(\mathrm{I}_{n/2} \otimes \mathrm{DFT}_2 \right) \tag{6}$$

In (1), $D_{m,n}$ is a complex diagonal matrix [2]. In (2), RDFT is the real version (i.e., for real input) of the DFT and X_n is an X-shaped matrix containing only the entries $0, \pm 1, \pm i$ [2]. The WHT is a real matrix and exists only for two-power size. $\mathrm{WHT}_2 = \mathrm{DFT}_2$ together with (3) defines the transform. In (4), the transform takes as input a $n_1 \times \cdots \times n_k$ array, stored linearized in a vector.

Spiral. Recursive application of rules like (1)–(6) yields many different algorithms for a given transform. Spiral [5] uses this fact to search for the fastest one on a given platform. A user-specified transform (like DFT_{256}) is expanded by Spiral using rules into a formula, which is then translated into a C program by a special formula compiler. The runtime of the program is measured and fed into a search module, which triggers, in a feedback loop, the generation of a modified formula based on a search strategy. Upon termination, Spiral outputs the fastest program found.

In this paper, we explain a crucial module in Spiral: A rewriting system that manipulates formulas to enable their direct compilation into SIMD vector code, which obviates the need for compiler vectorization.

Complex arithmetic. To describe complex transforms in terms of real arithmetic, we represent complex data vectors as real vectors using the interleaved complex format (alternating real and imaginary parts of the complex entries). Since the complex multiplication $(u + iv)(y + iz)$ is equivalent to the real multiplication $\begin{bmatrix} u & -v \\ v & u \end{bmatrix}\begin{bmatrix} y \\ z \end{bmatrix}$, we can write the complex matrix-vector multiplication $Mx \in \mathbb{C}^n$ as $\overline{M}x' \in \mathbb{R}^{2n}$, where we define $\overline{M}$ by replacing every entry $u + iv$ as $\begin{bmatrix} u & -v \\ v & u \end{bmatrix}$, and x' is x in the interleaved complex format.

3   Vectorization Through Rewriting

Our goal is to take formulas obtained by the recursive application of rules like (1)–(6) and automatically manipulate them into a form that enables a direct mapping into SIMD vector code. Further, we also want to explore different vectorizations for the same formula. The solution is a suitably designed rewriting system that implements our previous ideas for formula-based vectorization in [1,10].

We have used a very similar approach, but with different rewriting rules, for the related but different problems of parallelization for shared and distributed memory computers [7,8].

Formula vectorization: The basic idea. The central formula construct that can be implemented on all ν-way short vector extensions is

$$A \otimes I_\nu, \tag{7}$$

where A is an arbitrary real matrix. Vector code is obtained by generating scalar code for A (i.e., for $x \mapsto Ax$) and replacing all scalar operations by their respective ν-way vector operations. For example, c=a+b is replaced by c=vadd(a,b).

Of course, most formulas do not match (7). In these cases we manipulate the formula using rewriting rules to consist of components that either match (7) or are among a small set of base cases. It turns out that for a large class of formulas the only base cases needed are

$$\mathrm{L}_2^{2\nu}, \ \mathrm{L}_\nu^{2\nu}, \ \mathrm{L}_\nu^{\nu^2}, \ (\mathrm{I}_{n/\nu} \otimes \mathrm{L}_2^{2\nu})\overline{D_n}(\mathrm{I}_{n/\nu} \otimes \mathrm{L}_\nu^{2\nu}), \tag{8}$$

where D_n is any complex diagonal matrix. For $\nu = 2$ we also need the additional base case

$$\overline{[1, i]} = \begin{bmatrix} 1 & 0 & 0 & -1 \\ 0 & 1 & 1 & 0 \end{bmatrix}. \tag{9}$$

We assume that vectorized implementations of (8) are available. Note that $I_m \otimes L_2^{2\nu}$ converts a real vector $x' \in \mathbb{R}^{2m\nu}$ (which originates from a complex vector $x \in \mathbb{C}^{m\nu}$) from interleaved complex format (alternating real and imaginary part) into a block-interleaved complex format with block size ν (alternating ν real and ν imaginary parts). Analogously, $I_m \otimes L_\nu^{2\nu}$ converts back from block-interleaved into the interleaved complex format.

Definition 1. *We call a formula ν-way vectorized if it is either of the form (7) or one of the forms in (8) and (9), or of the form*

$$I_m \otimes A \text{ or } AB, \tag{10}$$

where A and B are ν-way vectorized.

Formula manipulation. We vectorize formulas through formula manipulation using well-known mathematical identities such as the following; we assume that A is $n \times n$ and B is $m \times m$:

$$I_{mn} = I_m \otimes I_n, \tag{11}$$
$$I_m \otimes A = L_m^{mn}(A \otimes I_m) L_n^{mn}, \tag{12}$$
$$L_n^{kmn} = (L_n^{kn} \otimes I_m)(I_k \otimes L_n^{mn}), \tag{13}$$
$$L_{km}^{kmn} = (I_k \otimes L_m^{mn})(L_k^{kn} \otimes I_m), \tag{14}$$
$$A \otimes (BC) = (A \otimes B)(A \otimes C), \tag{15}$$
$$A \otimes B = (A \otimes I_m)(I_n \otimes B) = (I_n \otimes B)(A \otimes I_m). \tag{16}$$

As a small example, we assume A is a real $n \times n$ matrix, and vectorize

$$I_m \otimes A \tag{17}$$

for a ν-way vector instruction set.

We first apply (11) to obtain

$$I_{m/\nu} \otimes I_\nu \otimes A$$

and then apply (12) to $I_\nu \otimes A$ to get

$$I_{m/\nu} \otimes \left(L_\nu^{n\nu}(A \otimes I_\nu) L_n^{n\nu} \right). \tag{18}$$

Note that in this formula $A \otimes I_\nu$ is already vectorized, but the stride permutations are not. To vectorize the stride permutations, we apply (13) and (14) to get

$$I_{m/\nu} \otimes \left((L_\nu^n \otimes I_\nu)(I_{n/\nu} \otimes L_\nu^{\nu^2})(A \otimes I_\nu)(I_{n/\nu} \otimes L_\nu^{\nu^2})(L_{n/\nu}^n \otimes I_\nu) \right). \tag{19}$$

Inspection shows that this formula is vectorized in the sense of Definition 1.

3.1   Rewriting System

Our goal is to *automatically* apply formula identities like (12)–(16) to transform given formulas into vectorized formulas. Note that the order and actual parameters chosen for each of the applied identities is a nontrivial choice. Only the correct choice will lead to vectorized formulas. Thus, automatic formula manipulation requires an appropriately designed rewriting system [4]. Specifically, it is a difficult problem to identify the right objects and rules in the system to guarantee that it is confluent and converges to fully vectorized formulas, when possible.

Vector tags. We introduce a set of tags to propagate vectorization information through the formulas and to perform algebraic simplification of permutations. Note that all objects remain matrices.

We tag a formula construct A to be translated into vector code for vector length ν by

$$\underbrace{A}_{\text{vec}(\nu)} = A.$$

Further, we write

$$A \,\bar{\otimes}\, \mathrm{I}_\nu = A \otimes \mathrm{I}_\nu$$

to stipulate that the tensor product is to be mapped into vector code as explained in Section 3. We use a tag "base" to mark the base cases defined in (8):

$$\underbrace{\mathrm{L}_2^{2\nu}}_{\text{base}}, \ \underbrace{\mathrm{L}_\nu^{2\nu}}_{\text{base}}, \ \underbrace{\mathrm{L}_\nu^{\nu^2}}_{\text{base}}, \ \underbrace{\overline{D_n}^{\nu}}_{\text{base}}, \ \underbrace{\overline{[1, i]}}_{\text{base}}.$$

Constructs marked with $\bar{\otimes}$ and "base" are final, i.e., they will not be changed by rewriting rules.

In addition we need variants of the operator $\overline{(.)}$ to handle the vectorization of complex formulas (A is assumed to be $n \times n$):

$$\overleftrightarrow{A}^{\nu} = \overline{A}, \tag{20}$$

$$\overrightarrow{A}^{\nu} = (\mathrm{I}_{n/\nu} \otimes \mathrm{L}_2^{2\nu})\overline{A}, \tag{21}$$

$$\overleftarrow{A}^{\nu} = \overline{A}(\mathrm{I}_{n/\nu} \otimes \mathrm{L}_\nu^{2\nu}), \tag{22}$$

$$\overline{A}^{\nu} = (\mathrm{I}_{n/\nu} \otimes \mathrm{L}_2^{2\nu})\overline{A}(\mathrm{I}_{n/\nu} \otimes \mathrm{L}_\nu^{2\nu}). \tag{23}$$

(20)–(23) are the four variants of $\overline{A}$ that have either interleaved or block-interleaved input and output format. The format conversions introduce the building blocks $\mathrm{L}_2^{2\nu}$ and $\mathrm{L}_\nu^{2\nu}$ defined in (8). A key idea in our rewriting system is to minimize these format conversions by applying the identity

$$\mathrm{L}_2^{2\nu} \mathrm{L}_\nu^{2\nu} = \mathrm{I}_{2\nu} .$$

To facilitate this simplification we introduce (20)–(23) as objects into our rewriting system. The rules (31)–(45) introduced below then operate on these objects and encode the knowledge where to introduce $L_2^{2\nu}$ and $L_\nu^{2\nu}$ to minimize format conversion overhead.

Rules. The goal is to vectorize a given formula. In our rewriting system, this is done by tagging the formula with $\text{vec}(\nu)$ and applying rules that vectorize the formula in the sense of Definition 1. Rules are applied by matching the left side of a rule against a given expression, extracting the parameters defined in the left side, and replacing it with one of the choices in the right side parameterized by the extracted parameters. Most of our rewriting rules are shown in Tables 1–3.

Table 1. Stride permutation rules

$$\underbrace{L_n^{n\nu}}_{\text{vec}(\nu)} \rightarrow \left(I_{n/\nu} \otimes \underbrace{L_\nu^{\nu^2}}_{\text{base}} \right) \left(L_{n/\nu}^n \bar{\otimes} I_\nu \right) \tag{24}$$

$$\underbrace{L_\nu^{n\nu}}_{\text{vec}(\nu)} \rightarrow \left(L_\nu^n \bar{\otimes} I_\nu \right) \left(I_{n/\nu} \otimes \underbrace{L_\nu^{\nu^2}}_{\text{base}} \right) \tag{25}$$

$$\underbrace{L_m^{mn}}_{\text{vec}(\nu)} \rightarrow \left(L_m^{mn/\nu} \bar{\otimes} I_\nu \right) \left(I_{mn/\nu^2} \otimes \underbrace{L_\nu^{\nu^2}}_{\text{base}} \right) \left((I_{n/\nu} \otimes L_{m/\nu}^m) \bar{\otimes} I_\nu \right) \tag{26}$$

Table 2. Tensor product rules. A is an $n \times n$ matrix.

$$\left(\underbrace{A \otimes I_m}_{\text{vec}(\nu)} \right) \rightarrow \left(A \otimes I_{m/\nu} \right) \bar{\otimes} I_\nu \tag{27}$$

$$\left(\underbrace{I_m \otimes A}_{\text{vec}(\nu)} \right) \rightarrow \begin{cases} I_{m/\nu} \otimes \underbrace{\left(I_\nu \otimes A \right)}_{\text{vec}(\nu)} \\ \underbrace{L_m^{mn}}_{\text{vec}(\nu)} \underbrace{\left(A \otimes I_m \right)}_{\text{vec}(\nu)} \underbrace{L_n^{mn}}_{\text{vec}(\nu)} \end{cases} \tag{28}$$

$$\underbrace{\left(I_m \otimes A \right) L_m^{mn}}_{\text{vec}(\nu)} \rightarrow \begin{cases} \underbrace{L_m^{mn}}_{\text{vec}(\nu)} \underbrace{\left(A \otimes I_m \right)}_{\text{vec}(\nu)} \\ \left(I_{m/\nu} \otimes \underbrace{L_\nu^{n\nu}}_{\text{vec}(\nu)} \left(A \bar{\otimes} I_\nu \right) \right) \left(L_{m/\nu}^{mn/\nu} \bar{\otimes} I_\nu \right) \end{cases} \tag{29}$$

$$\underbrace{\left(I_k \otimes \left(I_m \otimes A^{n \times n} \right) L_m^{mn} \right) L_k^{kmn}}_{\text{vec}(\nu)} \rightarrow \left(\underbrace{L_k^{km} \otimes I_n}_{\text{vec}(\nu)} \right) \left(I_m \otimes \underbrace{\left(I_k \otimes A^{n \times n} \right) L_k^{kn}}_{\text{vec}(\nu)} \right) \left(\underbrace{L_m^{mn} \otimes I_k}_{\text{vec}(\nu)} \right) \tag{30}$$

Table 3. Bar operator rules (left: recursive; right: base cases). A is an $n \times n$ matrix.

$$\underset{\text{vec}(\nu)}{\underbrace{(\overline{A})}} \rightarrow \overleftrightarrow{(\underset{\text{vec}(\nu)}{\underbrace{A}})}^{\nu} \quad (31)$$

$$\overleftrightarrow{\overline{AB}}^{\nu} \rightarrow \overleftarrow{A}^{\nu}\overrightarrow{B}^{\nu} \quad (32)$$

$$\overleftarrow{\overline{AB}}^{\nu} \rightarrow \overleftarrow{A}^{\nu}\overleftarrow{B}^{\nu} \quad (33)$$

$$\overrightarrow{\overline{AB}}^{\nu} \rightarrow \overrightarrow{A}^{\nu}\overrightarrow{B}^{\nu} \quad (34)$$

$$\overline{\overline{AB}}^{\nu} \rightarrow \overline{A}^{\nu}\overline{B}^{\nu} \quad (35)$$

$$\overleftrightarrow{\overline{I_m \otimes A}}^{\nu} \rightarrow I_m \otimes \overleftrightarrow{A}^{\nu} \quad (36)$$

$$\overleftarrow{\overline{I_m \otimes A}}^{\nu} \rightarrow I_m \otimes \overleftarrow{A}^{\nu} \quad (37)$$

$$\overrightarrow{\overline{I_m \otimes A}}^{\nu} \rightarrow I_m \otimes \overrightarrow{A}^{\nu} \quad (38)$$

$$\overline{\overline{I_m \otimes A}}^{\nu} \rightarrow I_m \otimes \overrightarrow{A}^{\nu} \quad (39)$$

$$\overleftarrow{\overline{A \otimes I_\nu}}^{\nu} \rightarrow (I_{n/\nu} \otimes \underset{\text{base}}{\underbrace{L_\nu^{2\nu}}})(\overline{A} \bar{\otimes} I_\nu) \quad (40)$$

$$\overleftarrow{\overline{A \otimes I_\nu}}^{\nu} \rightarrow \overline{A} \bar{\otimes} I_\nu \quad (41)$$

$$\overrightarrow{\overline{A \bar{\otimes} I_\nu}}^{\nu} \rightarrow (\overline{A} \bar{\otimes} I_\nu)(I_{n/\nu} \otimes \underset{\text{base}}{\underbrace{L_2^{2\nu}}}) \quad (42)$$

$$\overline{\overline{L_m^{mn}}} \rightarrow L_m^{mn} \otimes I_2 \quad (43)$$

$$\overleftarrow{(\underset{\text{vec}(\nu)}{\underbrace{L_\nu^{\nu^2}}})}^{\nu} \rightarrow (L_\nu^{2\nu} \bar{\otimes} I_\nu)(I_2 \otimes \underset{\text{base}}{\underbrace{L_\nu^{\nu^2}}})(L_2^{2\nu} \bar{\otimes} I_\nu) \quad (44)$$

$$\overleftarrow{(\underset{\text{vec}(\nu)}{\underbrace{D_n}})}^{\nu} \rightarrow \underset{\text{base}}{\underbrace{\overline{D_n}}}^{\nu} \quad (45)$$

The important difference between identities like (11)–(16) and rules like (24)–(45) is that the rules encode the decisions *how* to apply the identities, i.e., fix the choice of parameters. For instance, both identities (13) and (14) can be applied to $L_n^{n\nu}$ for composite n and a two-power ν; however, only (14) with the specific choice $k = m/\nu$, $m = \nu$, and $n = \nu$ leads to a vectorized result:

$$L_n^{n\nu} = (I_{n/\nu} \otimes L_\nu^{\nu^2})(L_{n/\nu}^n \otimes I_\nu).$$

This knowledge is encoded in rule (24) which chooses the right parameters.

Similarly, rule (27) and the first alternative of rule (28) are based on identity (11) but also encode the knowledge how to apply (11) depending on the vector length ν.

Simple example. We return to our previous example (17) and explain how it is handled by our rewriting system. We start with the tagged formula

$$\underset{\text{vec}(\nu)}{\underbrace{I_m \otimes A}},$$

which means "$I_m \otimes A$ is to be vectorized." The system can only apply one of the alternatives of rule (28). Suppose it chooses the first alternative, which yields

$$I_{m/\nu} \bar{\otimes} \underset{\text{vec}(\nu)}{\underbrace{(I_\nu \otimes A)}}$$

and then applies the second alternative of (28) to $(I_\nu \otimes A)$, which leads to

$$I_{m/\nu} \bar{\otimes} \left(\underset{\text{vec}(\nu)}{\underbrace{L_\nu^{n\nu}}} (A \bar{\otimes} I_\nu) \underset{\text{vec}(\nu)}{\underbrace{L_n^{n\nu}}} \right).$$

Next, only rules (24) and (25) match, which yields

$$I_{m/\nu} \otimes \Big((L_\nu^n \bar{\otimes} I_\nu)(I_{n/\nu} \otimes \underbrace{L_\nu^{\nu^2}}_{\text{base}})(A \bar{\otimes} I_\nu)(I_{n/\nu} \otimes \underbrace{L_\nu^{\nu^2}}_{\text{base}})(L_{n/\nu}^n \bar{\otimes} I_\nu) \Big).$$

This is the properly tagged version of the vectorized formula (19).

Example: DFT. We now show how our rewriting system vectorizes DFT_{mn} with $\nu^2 \mid mn$. The vectorization process has to overcome three crucial problems for an arbitrary two-power ν: 1) handle the interleaved complex format, 2) vectorize the stride permutation, and 3) vectorize the complex diagonal matrix. Our example shows how these problems are solved and how to obtain the short-vector FFT algorithm [10] as a result.

The DFT is a complex transform, but vector instructions operate on real vectors. Thus, we have to start with $\overline{DFT_{mn}}$, tagged for vectorization. First the system commutes the vector tag using (31) and the $\overleftrightarrow{(.)}^\nu$ operator and breaks the product using (32)–(39):

$$\underbrace{(\overline{DFT_{mn}})}_{\text{vec}(\nu)} \to \underbrace{\Big(\overline{(DFT_m \otimes I_n)D_{m,n}(I_m \otimes DFT_n)L_m^{mn}} \Big)}_{\text{vec}(\nu)}$$

$$\to \Big(\underbrace{\overleftarrow{(DFT_m \otimes I_n)D_{m,n}(I_m \otimes DFT_n)L_m^{mn}}^{\nu}}_{\text{vec}(\nu)} \Big)$$

$$\to \Big(\underbrace{\overleftarrow{(DFT_m \otimes I_n)}}_{\text{vec}(\nu)} \underbrace{\overleftarrow{D_{m,n}}}_{\text{vec}(\nu)} \underbrace{\overleftarrow{(I_m \otimes DFT_n)L_m^{mn}}^{\nu}}_{\text{vec}(\nu)} \Big)$$

$$\to \underbrace{\overleftarrow{(DFT_m \otimes I_n)}^{\nu}}_{\text{vec}(\nu)} \underbrace{\overleftarrow{(D_{m,n})}^{\nu}}_{\text{vec}(\nu)} \underbrace{\overleftarrow{(I_m \otimes DFT_n)L_m^{mn}}^{\nu}}_{\text{vec}(\nu)}$$

We now continue with the three factors separately. The system applies (27) and (40) to the first factor

$$\underbrace{\overleftarrow{(DFT_m \otimes I_n)}^{\nu}}_{\text{vec}(\nu)} \to \overleftarrow{(DFT_m \otimes I_{n/\nu}) \bar{\otimes} I_\nu}^{\nu}$$

$$\to (I_{mn/\nu} \otimes \underbrace{L_\nu^{2\nu}}_{\text{base}})(\overline{DFT_m \otimes I_{n/\nu}} \bar{\otimes} I_\nu)$$

which is vectorized. The second factor is already vectorized. For the third factor, suppose the system chooses the second alternative of (29) and then breaks and propagates $\overrightarrow{(.)}^\nu$ using (32)–(39):

$$\underbrace{\overrightarrow{(I_m \otimes DFT_n)\, L_m^{mn}}^{\nu}}_{\text{vec}(\nu)} \to \overrightarrow{\left(I_{m/\nu} \otimes \underbrace{L_\nu^{n\nu}}_{\text{vec}(\nu)} (DFT_n \bar{\otimes} I_\nu)\right)(L_m^{mn} \bar{\otimes} I_\nu)}^{\nu}$$

$$\to \left(I_{m/\nu} \otimes \underbrace{\overline{L_\nu^{n\nu}\,(DFT_n \bar{\otimes} I_\nu)}^{\nu}}_{\text{vec}(\nu)}\right)\overrightarrow{(L_m^{mn} \bar{\otimes} I_\nu)}^{\nu}$$

$$\to \left(I_{m/\nu} \otimes \underbrace{\overline{L_\nu^{n\nu}}^{\nu} \overline{(DFT_n \bar{\otimes} I_\nu)}^{\nu}}_{\text{vec}(\nu)}\right)\overrightarrow{(L_m^{mn} \bar{\otimes} I_\nu)}^{\nu}$$

We now continue with the factors of the tensor product. One difficult part of the vectorization is the interplay of $\overline{(.)}$ and the stride permutation $L_\nu^{n\nu}$. First, rule (25) factors the stride permutation, then rules (32)–(39) handle $\overline{(.)}^{\nu}$. Finally, rules (40)–(44) encode the rather involved manipulation required to completely vectorize:

$$\underbrace{\overline{L_\nu^{n\nu}}^{\nu}}_{\text{vec}(\nu)} \to \overline{\underbrace{(L_\nu^{n} \otimes I_\nu)}_{\text{vec}(\nu)}\underbrace{(I_{n/\nu} \otimes L_\nu^{\nu^2})}_{\text{vec}(\nu)}}$$

$$\to \overline{(L_\nu^{n} \bar{\otimes} I_\nu)}^{\nu}\left(I_{n/\nu} \otimes \underbrace{\overline{(L_\nu^{\nu^2})}^{\nu}}_{\text{vec}(\nu)}\right)$$

$$\to (\overline{L_\nu^{n}} \bar{\otimes} I_\nu)\left(I_{n/\nu} \otimes (L_\nu^{2\nu} \bar{\otimes} I_\nu)(I_2 \otimes \underbrace{L_\nu^{\nu^2}}_{\text{base}})(L_2^{2\nu} \bar{\otimes} I_\nu)\right)$$

The vectorization of the remaining constructs is straightforward using rules (40)–(44):

$$\overline{(DFT_n \bar{\otimes} I_\nu)}^{\nu} \to (\overline{DFT_n} \bar{\otimes} I_\nu)$$

and

$$\overrightarrow{(L_m^{mn} \bar{\otimes} I_\nu)}^{\nu} \to (\overline{L_m^{mn}} \bar{\otimes} I_\nu)\left(I_{mn/\nu} \otimes \underbrace{L_2^{2\nu}}_{\text{base}}\right)$$

$$\to \left((L_m^{mn} \otimes I_2)\bar{\otimes} I_\nu\right)\left(I_{mn/\nu} \otimes \underbrace{L_2^{2\nu}}_{\text{base}}\right).$$

Collecting the vectorized formulas and applying (45) yields a completely vectorized FFT:

$$(I_{mn/\nu} \otimes \underbrace{L_\nu^{2\nu}}_{\text{base}})(\overline{DFT_m \otimes I_{n/\nu}} \bar{\otimes} I_\nu)\overline{D_{m,n}}^{\nu}$$

$$\left(I_{m/\nu} \otimes (\overline{L_\nu^{n}} \bar{\otimes} I_\nu)(I_{n/\nu} \otimes (L_\nu^{2\nu} \bar{\otimes} I_\nu)(I_2 \otimes \underbrace{L_\nu^{\nu^2}}_{\text{base}})(L_2^{2\nu} \bar{\otimes} I_\nu))(\overline{DFT_n} \bar{\otimes} I_\nu)\right)$$

$$((L_m^{mn} \otimes I_2)\bar{\otimes} I_\nu)(I_{mn/\nu} \otimes \underbrace{L_2^{2\nu}}_{\text{base}}).$$

Inspection shows that this formula is indeed vectorized in the sense of Definition 1.

Note that there are degrees of freedom in applying our rule set, which thus yields different vectorizations. The search in the Spiral system will select the best for the given platform.

4   Experimental Results

We incorporated our rewriting system into the Spiral code generator to automatically generate vector code and search over alternative algorithms or formulas. We show runtime benchmarks on a 3 GHz Intel Pentium 4 running Windows XP and a 3.6 GHz Intel Pentium 4 running Linux kernel 2.6. We used the Intel C++ compiler 9.0 with options "/QxKW /O3 /G7 /Qc99 /Qrestrict" for vector code and "/O3 /G7 /Qc99 /Qrestrict" for scalar x86 and x87 code. These options turned out to produce the fastest code. DFT_n performance is measured in pseudo Mflop/s for floating-point code and in pseudo Mfpop/s for fixed-point code, both computed as $5n \log_2 n/(\text{runtime [ms]})$. The Haar_n wavelet performance is measured in Mfpop/s $= 2N/(\text{runtime [ms]})$. For all performance results higher is better. We compare our generated code with the Intel MKL 8.0 (DFTI functions) and IPP 5.0 library and with FFTW 3.1 [6] for both two-powers and multiples of ν.

2-way double-precision. Figure 1 evaluates our approach for two-way vectorization. We compare two-power DFTs of sizes $2^7 \le n \le 2^{14}$: 1) Spiral generated scalar x87 code; 2) Spiral generated SSE2 code; 3) FFTW 3.1 with enabled SSE2 support; and 4) Intel IPP 5.0 using SSE2. Spiral and FFTW achieve similar performance with FFTW being slightly faster, and IPP is between 5% and 15% faster than both. For Spiral generated code, SSE2 vectorization provides around 50% speed-up over scalar code.

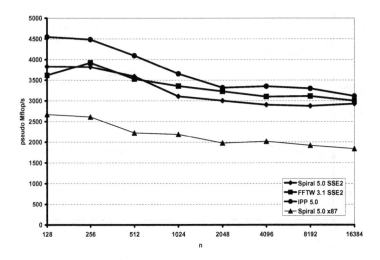

Fig. 1. Performance of DFT_n with $n = 2^k$, implemented in double-precision on a 3.6 GHz Pentium 4

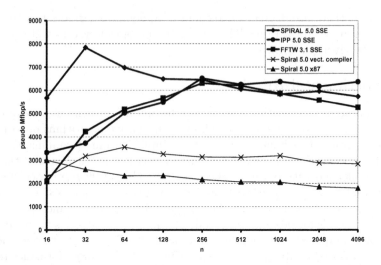

Fig. 2. Performance of DFT$_n$ with $n = 2^k$, implemented in single-precision on a 3 GHz Pentium 4

4-way single-precision. Figure 2 compares DFT code for two-power sizes $2^4 \leq n \leq 2^{12}$: 1) Spiral generated scalar x87 code; 2) Spiral generated scalar x87 code vectorized by Intel's compiler (option "/QxKW"); 3) Spiral generated 4-way SSE code; 4) FFTW 3.1 with enabled SSE support; and 5) Intel IPP 5.0 using SSE. Spiral generated SSE code is up to 3 times faster as Spiral generated scalar x87 code. Using the Intel C++ compiler to vectorize code leads only to

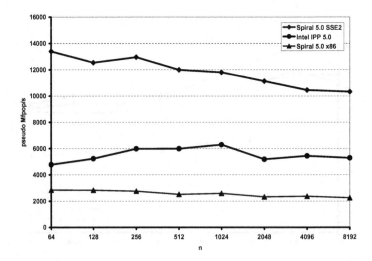

Fig. 3. Performance of DFT$_n$ with $n = 2^k$, implemented in 16-bit fixed-point on a 3.6 GHz Pentium 4

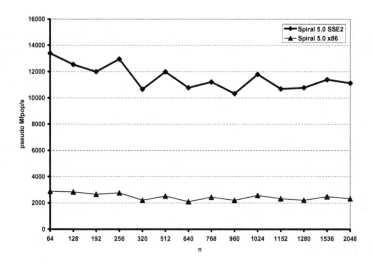

Fig. 4. Performance of DFT_n with $n = 64 \times 2^k 3^\ell 5^m$, implemented in 16-bit fixed-point on a 3.6 GHz Pentium 4

around 50% speed-up. For $16 \leq n \leq 128$, Spiral generated SSE code is clearly the fastest. For $n=256$ both FFTW and IPP are slightly faster as Spiral generated SSE code. For $512 \leq n \leq 2048$, Spiral generated SSE code is within 10% of IPP.

8-way 16-bit fixed-point. Figure 3 compares two-power DFT fixed-point code. It shows 1) Spiral generated SSE2 code (8-way, 16-bit), 2) scalar 16-bit

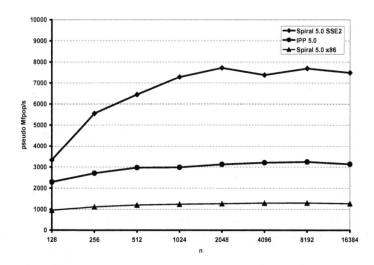

Fig. 5. Performance of the wavelet $Haar_n$ with $n = 2^k$, implemented in 8-bit fixed-point on a 3.6 GHz Pentium 4

x86 code, and 3) Intel IPP 5.0 (16-bit). Spiral's SSE2 vectorization consistently provides speed-up of 5 times over scalar x86 code. Spiral generated SSE2 code is 2 to 2.5 times faster as the IPP and 5 to 6 times faster than Spiral generated scalar code. Figure 4 shows that for DFTs of size $n = 64 \times 2^k 3^\ell 5^m$ SSE2 code generated by Spiral maintains the speed-up of 5 times over scalar x86 code. IPP does only provide two-power FFTs for 16-bit fixed-point. FFTW does not provide fixed-point code.

16-way 8-bit fixed-point. Figure 5 compares implementations of the Haar wavelet: 1) Spiral generated SSE2 code, 2) Spiral generated scalar 8-bit x86 code (16-way, 8-bit), and 3) the Intel IPP 5.0 (8-bit). For Spiral, SSE2 vectorization provides a speed-up of up to 6 times over scalar x86 code. Spiral generated SSE2 code is 2 to 2.5 times faster than the IPP. FFTW does not implement Haar wavelets.

5   Conclusion

SIMD vector instructions have a huge potential to speed up performance critical computational kernels with fine-grain parallelism. However, compiler support is limited and typically programmers have to resort to low-level C extensions or to assembly language programming to realize the potential of SIMD extensions. To overcome these problems for the domain of signal transforms, we presented a domain-specific vectorization framework for signal transform algorithms and in particular FFTs. The basic idea is to vectorize at a high mathematical level of abstraction, where more structural information is available as in the corresponding C code. The suitable tool for implementing this technique is a rule based rewriting system, which we included in Spiral to enable search in tandem with vectorization. The viability of the approach in this domain is demonstrated by both the high performance of the generated code and by the applicability of the approach to the related and equally difficult problem of parallelization.

References

1. Franchetti, F., Püschel, M.: A SIMD vectorizing compiler for digital signal processing algorithms. In: Proc. IEEE Int'l Parallel and Distributed Processing Symposium (IPDPS). (2002) 20–26
2. Van Loan, C.: Computational Framework of the Fast Fourier Transform. SIAM (1992)
3. Johnson, J.R., Johnson, R.W., Rodriguez, D., Tolimieri, R.: A methodology for designing, modifying, and implementing Fourier transform algorithms on various architectures. Circuits, Systems, and Signal Processing **9**(4) (1990) 449–500
4. Dershowitz, N., Plaisted, D.A.: Rewriting. In Robinson, A., Voronkov, A., eds.: Handbook of Automated Reasoning. Volume 1. Elsevier (2001) 535–610
5. Püschel, M., Moura, J.M.F., Johnson, J., Padua, D., Veloso, M., Singer, B.W., Xiong, J., Franchetti, F., Gačić, A., Voronenko, Y., Chen, K., Johnson, R.W., Rizzolo, N.: SPIRAL: Code generation for DSP transforms. Proc. of the IEEE **93**(2) (2005) 232–275 Special issue on *Program Generation, Optimization, and Adaptation.*

6. Frigo, M., Johnson, S.G.: The design and implementation of FFTW3. Proceedings of the IEEE **93**(2) (2005) 216–231 Special issue on "Program Generation, Optimization, and Adaptation".
7. Franchetti, F., Voronenko, Y., Püschel, M.: FFT program generation for shared memory: SMP and multicore. In: Proc. Supercomputing. (2006)
8. Bonelli, A., Franchetti, F., Lorenz, J., Püschel, M., Ueberhuber, C.W.: Automatic performance optimization of the discrete Fourier transform on distributed memory computers. In: Proc. International Symposium on Parallel and Distributed Processing and Applications (ISPA). (2006)
9. Robelly, J., Cichon, G., Seidel, H., Fettweis, G.: A HW/SW design methodology for embedded SIMD vector signal processors. International Journal of Embedded Systems (2005)
10. Franchetti, F., Püschel, M.: Short vector code generation for the discrete Fourier transform. In: Proc. IEEE Int'l Parallel and Distributed Processing Symposium (IPDPS). (2003) 58–67

High Order Fourier-Spectral Solutions to Self Adjoint Elliptic Equations

Moshe Israeli[1] and Alexander Sherman[2]

[1] Department of Computer Science, Technion, Technion City, Haifa 32000, Israel
`israeli@cs.technion.ac.il`
[2] Department of Applied Mathematics, Technion, Technion City, Haifa 32000, Israel
`asherman@tx.technion.ac.il`

Abstract. We develop a High Order Fourier solver for nonseparable, selfadjoint elliptic equations with variable (diffusion) coefficients. The solution of an auxiliary constant coefficient equation, serves in a transformation of the dependent variable. There results a "modified Helmholtz" elliptic equation with almost constant coefficients. The small deviations from constancy are treated as correction terms. We developed a highly accurate, fast, Fourier-spectral algorithm to solve such constant coefficient equations. A small number of correction steps is required in order to achieve very high accuracy. This is achieved by optimization of the coefficients in the auxiliary equation. For given coefficients the approximation error becomes smaller as the domain decreases. A highly parallelizable hierarchical procedure allows a decomposition into smaller sub-domains where the solution is efficiently computed. This step is followed by hierarchical matching to reconstruct the global solution. Numerical experiments illustrate the high accuracy of the approach even at coarse resolutions.

1 Introduction

Variable coefficient elliptic equations are ubiquitous in many scientific and engineering applications the most important case being that of the self-adjoint operator appearing for example in diffusion processes in non uniform media. Many repeated solutions of such problems are required when solving variable coefficient or non linear time dependent problems by implicit marching methods.

Application of high-order (pseudo) spectral methods, which are based on global expansions into orthogonal polynomials (Chebyshev or Legendre polynomials), to the solution of elliptic equations, results in full (dense) matrix problems. The spectral element method allows for some sparsity. On the other hand the Fourier spectral method for the solution of the Poisson equation gives rise to diagonal matrices and has an exponential rate of convergence but looses accuracy for non-periodic boundary conditions due to the Gibbs phenomenon.

Our method to resolve the Gibbs phenomenon represents the RHS as a sum of a smooth periodic function and another function which can be integrated analytically. This approach is sometimes called "subtraction".

M. Daydé et al. (Eds.): VECPAR 2006, LNCS 4395, pp. 378–390, 2007.
© Springer-Verlag Berlin Heidelberg 2007

The subtraction technique for the reduction of the Gibbs phenomenon in the Fourier series solution of the Poisson equation goes back to Sköllermo [2] who considered

$$\Delta u = f$$

in the rectangle $[0, 1] \times [0, 1]$ with non periodic boundary conditions. We note that the subtraction algorithm in [2] was of limited applicability. We develop in section 4 a high order generalization for the case of the modified Helmholtz equation. The Poisson equation case is just a particular case.

The subtraction technique (in the physical space) has the following advantages:

a) After subtraction, the Fast Fourier Transform can be applied to the remaining part of RHS with a high convergence rate.
b) The algorithm preserves the diagonal representation of the Laplace operator.
c) The computation of the subtraction functions inexpensive.

In the framework of the present paper we solve the elliptic equation:

$$\nabla \cdot (a(x, y) \, \nabla u(x, y)) - c(x, y)u(x, y) = f(x, y), \quad (x, y) \in \mathbf{D}, \tag{1}$$

where $\mathbf{D}$ is a rectangular domain, with Dirichlet boundary conditions

$$u(x, y) = g(x, y), \quad (x, y) \in \partial \mathbf{D}. \tag{2}$$

We assume $a(x, y) > 0$ for any $(x, y) \in \mathbf{D}$.

1. We develop first a fast direct algorithm for the solution of Eq. (1) for any function $a(x, y)$, such that $a(x, y)^{1/2}$ is equal to the solution $w(x, y)$ of a certain, appropriately chosen, constant coefficient equation (see below). The algorithm is based on our improvement of the fast direct solver of [1] and a transformation described in [3].
2. If $a(x, y)^{1/2}$ is not equal to $w(x, y)$, we substitute $w(x, y)^2$ for $a(x, y)$ and transfer the difference to the right hand side. The solution is found in a short sequence of correction steps.
3. An adaptive hierarchical domain decomposition approach allows improved approximation for any function $a(x, y)$.

2   Outline of the Algorithm

Following [3] we make the following change of variable in Eq. (1):

$$w(x, y) = a(x, y)^{1/2}u(x, y), \tag{3}$$

then Eq. (1) takes the form

$$\Delta w(x, y) - p(x, y)w(x, y) = q(x, y), \tag{4}$$

where

$$p(x, y) = \Delta(a(x, y)^{1/2}) \cdot a(x, y)^{-1/2} + c(x, y) \cdot a(x, y)^{-1},$$
$$q(x, y) = f(x, y) \cdot a(x, y)^{-1/2}. \tag{5}$$

If $p(x, y)$ happens to be a constant we have achieved a reduction to a constant coefficient case. As $a(x, y)$ and $c(x, y)$ are prescribed in the formulation of the problem we have no control over $p(x, y)$, nevertheless we will show that a constant approximation to $p(x, y)$ is achievable. We note in particular that if $c(x, y)$ vanishes and $a(x, y)^{1/2}$ is a harmonic function, Eq. (14) becomes a Poisson equation for w:

$$\Delta w(x, y) = q(x, y) \tag{6}$$

Otherwise let

$$P = \Delta(\tilde{a}(x, y))^{1/2} \cdot \tilde{a}(x, y)^{-1/2} + c(x, y) \cdot \tilde{a}(x, y)^{-1/2} \cdot a(x, y)^{-1/2} \tag{7}$$

Which can be rearranged as,

$$\Delta(\tilde{a}(x, y))^{1/2} - P \cdot \tilde{a}(x, y)^{1/2} = -c(x, y) \cdot a(x, y)^{1/2} \tag{8}$$

For any constant P, the above PDE becomes a Constant Coefficients Modified Helmholtz Equation (CCMHE) for the unknown function $\tilde{a}(x, y))^{1/2}$ with Dirichlet boundary conditions $\tilde{a}(x, y)^{1/2} = a(x, y)^{1/2}$ on $\partial \mathbf{D}$. P serves as an optimization parameter in order to get the best fit of $\tilde{a}(x, y)^{1/2}$ to $a(x, y)^{1/2}$. We introduce

$$\varepsilon(x, y) = a(x, y) - \tilde{a}(x, y)$$
$$\tilde{u}(x, y) = u(x, y) - u_0(x, y)$$

where $u_0(x, y)$ is the solution of Eq. (1), where $a(x, y)$ is replaced by $\tilde{a}(x, y)$. Then Eq. (1) could be rewritten in the following form:

$$\nabla \cdot ((\tilde{a}(x, y) + \varepsilon(x, y)) \, \nabla(u_0(x, y) + \tilde{u}(x, y)) -$$
$$c(x, y)(u_0(x, y) + \tilde{u}(x, y)) = f(x, y) \tag{9}$$

One can verify that

$$\nabla \cdot (\tilde{a}(x, y) \, \nabla u_0(x, y)) +$$
$$c(x, y) \, (\tilde{a}^{1/2}(x, y) \, a^{-1/2}(x, y)) \, u_0(x, y) = f(x, y) \tag{10}$$

can be reduced to a MH equation for $u_0(x, y)$. Thus we are left with

$$\nabla \cdot (\tilde{a} \, \nabla \tilde{u}) + c(\tilde{a}^{1/2} a^{-1/2}) \tilde{u} =$$
$$- \left[\nabla \cdot \varepsilon \nabla + c\varepsilon(a + (a\tilde{a})^{1/2})^{-1} \right] (u_0 + \tilde{u}) \tag{11}$$

Therefore, the following iterative procedure for $\tilde{u}(x, y)$ can be set up:

$$\left[\nabla \cdot \tilde{a} \, \nabla + c(\tilde{a}^{1/2} a^{-1/2})\right] u_{n+1} =$$

$$- \left[\nabla \cdot \varepsilon \nabla + c\varepsilon(a + (a\tilde{a})^{1/2})^{-1}\right] u_n, \quad n \geqslant 0 \qquad (12)$$

By defining

$$u^n \triangleq u_0 + u_n, \quad n \geqslant 1$$

we can show that $\|\varepsilon\| \leqslant s\|a\|$ in a certain Sobolev semi-norm, where s is a small constant and it follows that the error decreases according to:

$$\|u^{n+1} - u\| \leqslant s\|u^n - u\| \qquad (13)$$

3   The Auxiliary Equation

If p in Eq. (5) is not zero but a constant (larger then the first eigenvalue of the Laplacian) we have an elliptic constant coefficient partial differential equation of Helmholtz or modified Helmholtz type. Such equations can be easily solved by the subtraction technique as illustrated in section 4. By assumption, $a(x, y)^{1/2}$ is positive and does not vanish. Consider for example a region R, the values of $a(x, y)^{1/2}$ on its boundary are positive which is tantamount to positive Dirichlet boundary conditions for our approximation which should satisfy also the equation:

$$\Delta w(x, y) - Pw(x, y) = c(x, y)a(x, y)^{-1/2}, \qquad (14)$$

where P is a constant to be chosen so that $w(x, y)$ gives the best approximation to $a(x, y)^{1/2}$. If $a(x, y)^{1/2}$ is constant on the boundaries and dome shaped, and c(x,y) vanishes, the harmonic approximation will be a horizontal plane. On the other hand a negative P will give rise to a dome shaped approximation, and P can be chosen so that the function $w(x, y)$ will match the height of the dome. As we take more negative P (but larger then the lowest eigenvalue of the Laplacian) we get higher and higher domes. Conversely, if $a(x, y)^{1/2}$ is bowl shaped, a positive P will give rise to deeper and deeper bowls. For large P we will get values close to zero in most of the interior of R.

4   Solution of Modified Helmholtz Equation in a Box

In this section, we will describe a method for the solution of Modified Helmholtz equation with arbitrary order accuracy. We will start with an algorithm of $O(N^{-4})$ order of accuracy, then we construct the algorithm for $O(N^{-6})$ and generalize it to the arbitrary order of accuracy.

4.1   Problem Formulation

We are interested in the solution of the two-dimensional Modified Helmholtz (MH) equation in the rectangular region $\Omega = [0\ 1] \times [0\ 1]$ with Dirichlet boundary conditions.

$$\begin{cases} \triangle u(x,y) - k^2 u(x,y) = f(x,y) & \text{in } \Omega \\ u(x,y) = \Phi(x,y) & \text{on } \partial\Omega \end{cases} \tag{15}$$

The boundary functions

$$\phi_1(x) \triangleq \Phi(x,0), \quad \phi_3(x) \triangleq \Phi(x,1)$$
$$\phi_2(y) \triangleq \Phi(0,y), \quad \phi_4(y) \triangleq \Phi(1,y)$$

are assumed to be smooth and continuous at the corners . In addition, $f(x,y)$ is supposed to be known on $\partial\Omega$. We introduce the following notations:

$$f^{(p)}(x) \triangleq \frac{\partial^p f(x)}{\partial x^p}, \quad f^{(p,q)}(x,y) \triangleq \frac{\partial^{p+q} f(x,y)}{\partial x^p \partial y^q}$$

$$\text{Vandermonde}(\lambda_1, \lambda_2, \ldots, \lambda_n) \triangleq \begin{pmatrix} 1 & 1 & \cdots & 1 \\ \lambda_1 & \lambda_2 & \cdots & \lambda_n \\ \lambda_1^2 & \lambda_2^2 & \cdots & \lambda_n^2 \\ \vdots & \vdots & \ddots & \vdots \\ \lambda_1^{n-1} & \lambda_2^{n-1} & \cdots & \lambda_n^{n-1} \end{pmatrix}$$

$$\mathcal{H}_M \triangleq \triangle - k^2$$

Let $I = \{1,2,3,4\}$ be an index set of corner points or edges. Denote by $p_j, j \in I$ the four corner points of $\partial\Omega$:

$$p_1 = (0,0), \ p_2 = (0,1), \ p_3 = (1,1), \ p_4 = (1,0)$$

and by $E_j, j \in I$ the four edges of $\partial\Omega$:

$$E_1 = \{(x,y)|y=0\}, \quad E_2 = \{(x,y)|x=0\},$$
$$E_3 = \{(x,y)|y=1\}, \quad E_4 = \{(x,y)|x=1\},$$

and define

$$\partial\Omega_C \triangleq \{p_j \,|\, j \in I\}, \quad \partial\Omega_E \triangleq \partial\Omega \setminus \partial\Omega_C$$

4.2   Constructions of Auxiliary Function

In order to apply the subtraction technique, we construct a family of functions $q_{2r}(x), r \geqslant 0$ with the following property:

$$q_{2r}(1) = 1, \quad q_{2r}(0) = 0 \qquad\qquad \text{if } r = 0$$
$$\begin{cases} q_{2r}^{(2s)}(0) = 0, \quad q_{2r}^{(2s)}(1) = 0, \quad 0 \leqslant s \leqslant r-1 \\ q_{2r}^{(2r)}(0) = 1, \quad q_{2r}^{(2r)}(1) = 0 \end{cases} \quad \text{if } r \geqslant 1 \tag{16}$$

We look for a function $q_{2r}(x)$ as the linear combination

$$q_{2r}(x) = \sum_{i=1}^{r+1} \alpha_{2r,i} \frac{\sinh(\lambda_{2r,i}(1-x))}{\sinh(\lambda_{2r,i})}, \qquad \text{where} \quad \forall i: \lambda_{2r,i} > 0 \qquad (17)$$

Lemma 1. *For any* $r \geqslant 0$ *we can find constants* $\alpha_i \in \mathbb{R}$ *and* $0 < \lambda_i \in \mathbb{R}$ *such that the function* $q_{2r}(x)$ *takes form of (17)*

Lemma 2. *Let* $\lambda, \mu > 0$ *and define* $f(x, y)$ *as follows*

$$f(x,y) \triangleq \frac{\sinh(\lambda(1-x))}{\sinh(\lambda)} \frac{\sinh(\mu(1-y))}{\sinh(\mu)} \qquad (18)$$

If in addition $\lambda^2 + \mu^2 = k^2$, *where* k *is defined in Eq. (15), then* $f(x,y) \in Ker(\mathcal{H}_M)$.

Definition 1. *We say that boundary function* $\Phi(x, y)$ *is compatible with RHS* $f(x, y)$ *of Eq. (15) with respect to operator* $\mathcal{H}_M$ *if*

$$\forall p \in \partial\Omega_C, \quad \mathcal{H}_M(\Phi(p)) = f(p) \qquad (19)$$

4.3   Solution of the Modified Helmholtz Equation with Homogeneous RHS

As an intermediate stage in the solution of Eq. (15), we solve the Modified Helmholtz equation with zero RHS. We are interested in the solution of

$$\begin{cases} \triangle u_0(x,y) - k^2 u_0(x,y) = 0 & \text{in } \Omega \\ u_0(x,y) = \Phi_0(x,y) & \text{on } \partial\Omega \end{cases} \qquad (20)$$

The boundary function $\Phi(x, y)$ is assumed to be smooth and compatible with respect to $\mathcal{H}_M$. In order to utilize a rapidly convergent series expansions (see [1]), the boundary functions ϕ_j, $j \in I$ should vanish at the p_j along with a number of even derivatives. For each function $q_{2r}(x)$ in the form (17), define four functions $Q_{2r,j}(x, y)$, $j \in I$ as follows:

$$Q_{2r,1}(x,y) = \sum_{i=1}^{r+1} \alpha_{2r,i} \frac{\sinh(\lambda_{2r,i}(1-x))}{\sinh(\lambda_{2r,i})} \frac{\sinh(\mu_i(1-y))}{\sinh(\mu_{2r,i})},$$

$$Q_{2r,2}(x,y) = Q_{2r,1}(x, 1-y), \qquad (21)$$
$$Q_{2r,3}(x,y) = Q_{2r,1}(1-x, 1-y),$$
$$Q_{2r,4}(x,y) = Q_{2r,1}(1-x, y)$$

where $\forall i: \lambda_{2r,i}, \mu_{2r,i} > 0$ and $\lambda_{2r,i}^2 + \mu_{2r,i}^2 = k^2$.
By virtue of Lemma (2), $\forall j \in I: Q_{2r,j}(x,y) \in Ker(\mathcal{H}_M)$.

We define $w_0(x,y)$ and $\Phi_2(x,y)$ as follows

$$w_0(x,y) = \phi_1(0)\,Q_{0,1}(x,y) + \phi_3(0)\,Q_{0,2}(x,y)$$
$$+ \phi_3(1)\,Q_{0,3}(x,y) + \phi_1(1)\,Q_{0,4}(x,y), \tag{22}$$
$$\Phi_2(x,y) = \Phi_0(x,y) - w_0(x,y)|_{\partial\Omega}$$

$\Phi_2(x,y)$ has the following property: $\forall p \in \partial\Omega_C,\ \Phi_2(p) = 0$.
By solving a new equation

$$\begin{cases} \triangle u_2(x,y) - k^2 u_2(x,y) = 0 & \text{in } \Omega \\ \qquad\qquad u_2(x,y) = \Phi_2(x,y) & \text{on } \partial\Omega \end{cases} \tag{23}$$

we obtain that $u_0(x,y) = u_2(x,y) + w_0(x,y)$.

The subtraction procedure can be continued. In general, for $r \geqslant 1$, we define

$$w_{2(r-1)}(x,y) = \sum_{j\in I} \Phi_{2(r-1)}^{(2r,0)}(x,y)\big|_{P_j}\, Q_{2(r-1),j}(x,y)$$
$$\Phi_{2r}(x,y) = \Phi_{2(r-1)}(x,y) - w_{2(r-1)}(x,y)|_{\partial\Omega} \tag{24}$$

Lemma 3. *For any $r \geqslant 1$ and any s, $0 \leqslant s \leqslant r-1$ the function $\Phi_{2r}(x,y)$ defined in (24) has the following property:*

$$\Phi_{2r}^{(2s,0)}(p) = \Phi_{2r}^{(0,2s)}(p) = 0,\ \forall p \in \partial\Omega_C \tag{25}$$

Thus, by solving

$$\begin{cases} \triangle u_{2r}(x,y) - k^2 u_{2r}(x,y) = 0 & \text{in } \Omega \\ \qquad\qquad u_{2r}(x,y) = \Phi_{2r}(x,y) & \text{on } \partial\Omega \end{cases} \tag{26}$$

using rapidly convergent series (as suggested in [1]) we can achieve any prescribed (depending on $r \in \mathbb{N}$) order of accuracy. For $r \geqslant 1$, the general formula for the sought solution of Eq. (20) is

$$u_0(x,y) = u_{2r}(x,y) + \sum_{s=0}^{r-1} w_{2s}(x,y) \tag{27}$$

It is worthwhile to mention that all the functions $w_{2s}(x,y)$, $0 \leqslant s \leqslant r-1$ are explicitly known.

4.4   Solution of the Modified Helmholtz Equation with Nonhomogeneous RHS

We are interested in the solution of

$$\begin{cases} \triangle u_0(x,y) - k^2 u_0(x,y) = f_0(x,y) & \text{in } \Omega \\ \qquad\qquad u_0(x,y) = \Psi(x,y) & \text{on } \partial\Omega \end{cases} \tag{28}$$

In addition to the assumptions made in (20), we assume that $f_0(x, y)$ is smooth and $\Phi(x, y)$ is compatible with $f(x, y)$ with respect to $\mathcal{H}_M$. We extend further the technique developed in [2]. In order to solve Eq. (28) with high accuracy, $f_0(x, y)$ should satisfy the conditions stated in the next theorem which where obtained in ([2]).

Theorem 1. *Assume $f_0(x, y)$ is smooth and $p \geqslant 2$. If $\forall s, \ 0 \leqslant s \leqslant p - 2$*

$$f_0^{(2s, 2s)}(x, y) = 0, \quad \forall p \in \partial \Omega \tag{29}$$

then the direct Fourier method applied to (28) with $\Psi(x, y) = 0$ is of order of accuracy $O(N^{-2p})$.

We look for a function $f(x, y)$ that is an eigenfunction of the operator $\mathcal{H}_M$.

Lemma 4. *Let $\lambda, \mu > 0$ and $f(x, y)$ defined as in (18). If in addition $\lambda^2 + \mu^2 = 1 + k^2$, where k is defined in Eq. (15), then $\mathcal{H}_M(f(x, y)) = f(x, y)$.*

Define four functions $\widetilde{Q}_{2r, j}(x, y)$, $j \in I$ as follows:

$$
\begin{aligned}
\widetilde{Q}_{2r,1}(x, y) &= \sum_{i=1}^{r+1} \alpha_{2r,i} \frac{\sinh(\lambda_{2r,i}(1 - x))}{\sinh(\lambda_{2r,i})} \frac{\sinh(\mu_i(1 - y))}{\sinh(\mu_{2r,i})}, \\
\widetilde{Q}_{2r,2}(x, y) &= \widetilde{Q}_{2r,1}(x, 1 - y), \\
\widetilde{Q}_{2r,3}(x, y) &= \widetilde{Q}_{2r,1}(1 - x, 1 - y), \\
\widetilde{Q}_{2r,4}(x, y) &= \widetilde{Q}_{2r,1}(1 - x, y)
\end{aligned}
\tag{30}
$$

where $\forall i : \lambda_{2r,i}, \mu_{2r,i} > 0$ and $\lambda_{2r,i}^2 + \mu_{2r,i}^2 = 1 + k^2$.

By virtue of Lemma (4), $\forall j \in I : \mathcal{H}_M(\widetilde{Q}_{2r,j}(x, y)) = \widetilde{Q}_{2r,j}(x, y)$.

We split Eq. (28) to two equations one with homogeneous and one with non-homogeneous R.H.S. The main idea is to solve Eq. (28) with carefully constructed boundary conditions such that we can achieve any prescribed order of accuracy.

Define $h_0(x, y)$ and $f_1(x, y)$ as follows

$$
\begin{aligned}
h_0(x, y) &= \sum_{j \in I} f_0(p_j)\, \widetilde{Q}_{0,j}(x, y) \\
f_1(x, y) &= f_0(x, y) - h_0(x, y)
\end{aligned}
\tag{31}
$$

Obviously, $\mathcal{H}_M(h_0(x, y)) = h_0(x, y)$ and $f_1(x, y)$ has the following property: $f_1(p) = 0, \forall p \in \partial \Omega_C$. In order to apply Theorem 1, we need that $f_1(x, y)$ will vanish on the boundaries, that is: $f_1(p) = 0, \ \forall p \in \partial \Omega_E$.

For $q_0(x)$ as defined in (17), define

$$
\begin{aligned}
\tilde{q}_{0,1}(y) &= q_0(y), & \tilde{q}_{2r,2}(x) &= q_{2r}(x), \\
\tilde{q}_{0,3}(y) &= q_0(1 - y), & \tilde{q}_{2r,4}(x) &= q_{2r}(1 - x)
\end{aligned}
$$

and also (where $\zeta \equiv y$ for $j = 1, 3$ and $\zeta \equiv x$ for $j = 2, 4$)

$$h_{1,j}(x, y) = f_1(x, y)|_{E_j} \, \tilde{q}_{0,j}(\zeta), \quad h_1(x, y) = \sum_{j \in I} h_{1,j}(x, y),$$

$$f_2(x, y) \triangleq f_1(x, y) - h_1(x, y)$$

We introduce the following problems:

$$\forall j \in I : \begin{cases} \triangle w_{0,j}(x, y) - k^2 w_{0,j}(x, y) = h_1(x, y) & \text{in } \Omega \\ w_{0,j}(x, y) = 0 & \text{on } \partial\Omega \end{cases}$$

$$\begin{cases} \triangle u_2(x, y) - k^2 u_2(x, y) = f_2(x, y) & \text{in } \Omega \\ u_2(x, y) = 0 & \text{on } \partial\Omega \end{cases} \tag{32}$$

Using the technique of [2] for the error estimates, it can be shown that each equation in (32) can be solved with $O(N^{-4})$ order of accuracy and therefore, Eq. (28) with $\tilde{\Psi}(x, y) = h_0(x, y)|_{\partial\Omega}$ can be also solved with $O(N^{-4})$ accuracy. In addition, we need to solve Eq. (20) with $\Phi(x, y) = \Psi(x, y) - \tilde{\Psi}(x, y)$. We can proceed further and by constructing $h_2(x, y)$ and $h_3(x, y)$ obtain $O(N^{-6})$ accuracy etc.

4.5   Solution of the Modified Helmholtz Equation in the Non-compatible Case

In the formulation of the original problem, the boundary function $\Phi(x, y)$ is not necessary compatible with the R.H.S. with respect to $\mathcal{H}_M$. We utilize the idea that by changing the boundary function $\Phi(x, y)$ along with $f(x, y)$ in (15) we can achieve compatibility of the boundary function and the R.H.S. For this purpose we can use functions of the form

$$\tau_{2k}(x, y) = \mathcal{R}e\{c_{2k} z^{2k} log(z)\} \tag{33}$$

where $c_{2k} = a_{2k} + ib_{2k}$ and where $a_{2k} = 0$ while $b_{2k} = (-1)^k \dfrac{2}{\pi(2k)!}$.

As an example, assume that compatibility doesn't hold at p_1, that is:

$$\phi_1''(0) + \phi_2''(0) - k^2\phi_1(0) = f(p_1) + A$$

Let $v(x, y) = u(x, y) - A\tau_2(x, y)$. For $v(x, y)$ compatibility holds at p_1. Also, assume that compatibility already holds for the other corners. Thus, if we define

$$\tilde{f}(x, y) \triangleq f(x, y) + Ak^2\tau_2(x, y)$$

$$\tilde{\Phi}(x, y) \triangleq \Phi(x, y) - A\tau_2(x, y)|_{\partial\Omega}$$

then for

$$\begin{cases} \triangle v(x, y) - k^2 v(x, y) = \tilde{f}(x, y) & \text{in } \Omega \\ v(x, y) = \tilde{\Phi}(x, y) & \text{on } \partial\Omega \end{cases} \tag{34}$$

compatibility of $\tilde{\Phi}(x, y)$ with $\tilde{f}(x, y)$ holds. After solution of Eq. (34) we return back to $u(x, y)$.

5   Domain Decomposition

The present algorithm incorporates the following novel elements:

1. It extends our previous fast Poisson solvers [1] as it provides an essentially direct solution for equations (1) where $a(x,y)^{1/2}$ is an arbitrary harmonic function.
2. In the case where $a(x,y)^{1/2}$ is not harmonic, we approximate it by $\widetilde{a}(x,y)^{1/2}$ (which is a superposition of harmonic functions) and apply several correction steps to improve the accuracy.
3. In the case where $a(x,y)^{1/2}$ is dome shaped or bowl shaped, we approximate it by $\widetilde{a}(x,y)^{1/2}$ which is now a solution of Eq. (??) and apply several correction steps to improve the accuracy. The value of P is determined to match the average Gaussian curvature of $a(x,y)^{1/2}$.

However high accuracy for the solution of (1) requires an accurate approximation of $a(x,y)^{1/2}$ by the functions discussed above. Such an approximation is not always easy to derive in the global domain, however it can be achieved readily in smaller subdomains. In this case we suggest the following Domain Decomposition algorithm.

1. The domain is decomposed into smaller rectangular subdomains. Where the boundary of the subdomains coincides with full domain boundary we take on the original boundary conditions. For other interfaces we introduce some initial boundary conditions which do not contradict the equation at the corners, where the left hand side of (1) can be computed. The function $a(x,y)$ is approximated by $\tilde{a}(x,y)^{1/2}$ in each subdomain such that $\tilde{a}(x,y)^{1/2}$ is harmonic(or subharmonic or superharmonic). An auxiliary equation (12) is solved in each subdomain.
2. The collection of solutions obtained at Step 1 is continuous but doesn't have continuous derivatives at domain interfaces. To further match subdomains, a hierarchical procedure can be applied similar to the one described in [4]. For example, if we have four subdomains 1,2,3 and 4, then 1 can be matched with 2, 3 with 4, while at the final step the merged domain 1,2 is matched with 3,4.

We illustrate the effectiveness of the domain decomposition approach by solving a one dimensional variable coefficient equation where the coefficient function is not harmonic namely $a(x) = (2x + 3 + \sin(\pi x))^2$ with exact solution $u(x) = \sin(\pi x)$. We change the number of domains from 1 to 8. The correction procedure works much better when the subdomains become smaller. With 4 domains and with only 2 correction steps we reach an error of order 10^{-6}, with 8 domains we get 10^{-8}. Thus the present approach behaves essentially as a direct fast method. The Domain Decomposition of course has the further advantage of easy parallelization on massively parallel computers.

6   Numerical Results

First let us demonstrate the rate of convergence of the improved subtraction algorithm. Assume that u is the exact solution of Eq. (1) and u' is the computed solution. We will use the following measure to estimate the errors:

$$\varepsilon_{MAX} = \max |u'_i - u_i| \qquad (35)$$

Example 1. Consider the Modified Helmholtz equation with $f(x, y) = -k^2(x^2-y^2)$, where k is defined in (15); the right hand side and the boundary conditions correspond to the exact solution $u(x, y) = x^2 - y^2$ in the domain $[0, 1] \times [0, 1]$. The results are presented in Table 1.

Table 1. MAX error for the fourth order subtraction methods with $k = 1$

$$f(x, y) = -k^2(x^2 - y^2)$$
$$u(x, y) = x^2 - y^2$$

$N_x \times N_y$	$\varepsilon_{MAX}(4)$	ratio
8×8	1.29e-6	-
16×16	1.23e-7	10.5
32×32	9.81e-9	12.5
64×64	7.04e-10	13.93
128×64	4.81e-11	14.66

Example 2. Consider the same equation as in the previous example but with $k = 10$. The results are presented in Table 2.

Table 2. MAX error for the fourth order subtraction methods with $k = 10$

$$f(x, y) = -k^2(x^2 - y^2)$$
$$u(x, y) = x^2 - y^2$$

$N_x \times N_y$	$\varepsilon_{MAX}(4)$	ratio
8×8	3.38e-3	-
16×16	2.89e-4	11.7
32×32	2.73e-5	10.58
64×64	2.16e-6	12.64
128×64	1.54e-7	14.02

Example 3. Here we consider the elliptic equation of Eq. (1), there are four tables where we change a parameter, b, from 1.5 to $3, 6$ and 12. As b increases both the solution and the coefficient $a(x, y)$ become more and more oscillatory. We denote the number of the iteration by j and the resolution by N.

Table 3. $b = 1.5$ and $m = 4$

$$u(x,y) = \sin(b\,x) + \sin(b\,y)$$
$$a(x,y) = \left(1.5 + \sin(0.5\,b\,x) \cdot \sin(0.5\,b\,y)\right)^2$$
$$c(x,y) = ma(x,y)$$

$j\backslash N$	7	14	28	56	112	224
0	1.12 e-3	1.13 e-1	1.13 e-3	1.13 e-3	1.13 e-3	1.13 e-3
1	3.03 e-5	2.80 e-5	2.44 e-6	2.42 e-6	2.42 e-6	2.42 e-6
2	3.00 e-5	2.60 e-6	1.95 e-7	1.35 e-8	7.61 e-9	7.51 e-9
3	3.00 e-5	2.60 e-6	1.95 e-7	1.35 e-8	9.03 e-10	6.45 e-11

Table 4. $b = 3$ and $m = 4$

$$u(x,y) = \sin(b\,x) + \sin(b\,y)$$
$$a(x,y) = \left(1.5 + \sin(0.5\,b\,x) \cdot \sin(0.5\,b\,y)\right)^2$$
$$c(x,y) = ma(x,y)$$

$j\backslash N$	7	14	28	56	112	224
0	7.87 e-3	8.42 e-3	8.41 e-3	8.42 e-3	8.43 e-3	8.43 e-3
1	1.42 e-4	1.51 e-4	1.50 e-4	1.50 e-4	1.51 e-4	1.51 e-4
2	7.79 e-5	8.44 e-6	2.58 e-6	2.76 e-6	2.77 e-6	2.77 e-6
3	7.81 e-5	8.41 e-6	7.10 e-7	6.85 e-8	6.12 e-8	6.12 e-8
4	—	—	—	—	9.32 e-9	1.34 e-9
5	—	—	—	—	—	1.32 e-9

Table 5. $b = 6$ and $m = 6$

$$u(x,y) = \sin(b\,x) + \sin(b\,y)$$
$$a(x,y) = \left(1.5 + \sin(0.5\,b\,x) \cdot \sin(0.5\,b\,y)\right)^2$$
$$c(x,y) = ma(x,y)$$

$j\backslash N$	7	14	28	56	112	224
0	3.13 e-1	3.14 e-2	3.21 e-2	3.22 e-2	3.22 e-2	3.22 e-2
1	9.10 e-4	1.48 e-3	1.49 e-3	1.49 e-3	1.15 e-3	1.15 e-3
2	1.58 e-4	1.63 e-4	5.79 e-5	6.47 e-5	6.51 e-5	6.51 e-5
3	1.57 e-4	1.63 e-4	1.37 e-5	3.10 e-6	3.43 e-6	3.46 e-6
4	—	—	—	1.00 e-6	1.57 e-7	1.79 e-7
5	—	—	—	9.98 e-7	1.04 e-7	1.29 e-8

We see that it takes 1 to 5 iterations to converge to the accuracy which is determined by the resolution, we see that the accuracy behavior is like a fourth order method as expected. For the most oscillatory case, $b = 12$, there is no

Table 6. $b = 12$ and $m = 4$

$$u(x,y) = \sin(bx) + \sin(by)$$
$$a(x,y) = \left(1.5 + \sin(0.5\,bx) \cdot \sin(0.5\,by)\right)^2$$
$$c(x,y) = ma(x,y)$$

$j\backslash N$	7	14	28	56	112	224
0	8.37 e-1	9.54 e-1	1.00	1.00	1.00	1.00
1	1.10	6.90 e-1	6.90 e-1	7.16 e-1	7.20 e-1	7.20 e-1
2	4.44 e-1	5.47 e-1	5.45 e-1	5.62 e-1	5.63 e-1	5.62 e-1
3	5.75 e-1	4.65 e-1	4.87 e-1	4.82 e-1	4.91 e-1	4.90 e-1
4	3.39 e-1	3.98 e-1	4.31 e-1	4.29 e-1	4.30 e-1	4.31 e-1
5	3.92 e-1	3.52 e-1	4.01 e-1	4.03 e-1	4.03 e-1	4.03 e-1

convergence. Here we must resort to domain decomposition, by dividing the domain to four sub domains we would essentially reduce this case to that of b=6 from the point of view of convergence. The results are presented in Tables 3-6.

References

1. Averbuch, A., Vozovoi, L., Israeli, M.: On a Fast Direct Elliptic Solver by a Modified Fourier Method, Numerical Algorithms, Vol. **15** (1997) 287–313
2. Skölermo, G.: A Fourier method for numerical solution of Poisson's equation, Mathematics of Computation, Vol. **29**, No. 131 (Jul., 1975) 697–711
3. Concus P., Golub G.H.: Use of fast direct methods for the efficient numerical solution of nonseparable elliptic equations, SIAM J. Numer. Anal. **10** (1973), No. 6, 1103–1120.
4. Israeli M., Braverman E., Averbuch A.: A hierarchical domain decomposition method with low communication overhead, Domain decomposition methods in science and engineering, (Lyon, 2000), 395-403, Theory Eng. Appl. Comput. Methods, Internat. Center Numer. Methods Eng. (CIMNE), Barcelona, 2002.

Multiresolution Simulations Using Particles

Michael Bergdorf and Petros Koumoutsakos

Computational Science & Engineering Laboratory
ETH Zürich
CH-8092, Switzerland
{petros,bergdorf}@inf.ethz.ch

Abstract. We present novel multiresolution particle methods with extended dynamic adaptivity in areas where increased resolution is required. In the framework of smooth particle methods we present two adaptive approaches: one based on globally adaptive mappings and one employing a wavelet-based multiresolution analysis to guide the allocation of computational elements. Preliminary results are presented from the application of these methods to problems involving the development of sharp vorticity gradients. The present particle methods are employed in large scale parallel computer architectures demonstrating a high degree of parallelization and enabling state of the art large scale simulations of continuum systems using particles.

1 Approximations Using particles

The development of particle methods is based on the integral representation of functions and differential operators. The integrals are discretized using particles as quadrature points.

1.1 Function Approximation

The approximation of continuous functions by particle methods starts with the equality

$$q(\boldsymbol{x}) \equiv \int q(\boldsymbol{x} - \boldsymbol{y})\, \delta(\boldsymbol{y})\, d\boldsymbol{y}\,. \tag{1}$$

Using N particles we discretize above equality by numerical quadrature and get the "point-particle" approximation of q:

$$q^h(\boldsymbol{x}) = \sum_p Q_p\, \delta(\boldsymbol{x} - \boldsymbol{x}_p)\,. \tag{2}$$

Point particle methods based on the approximation (2) yield exact weak solutions of conservation laws. A drawback of point particle approximations is that the function q^h can only be reconstructed on particle locations $\boldsymbol{x}_p$.

This shortcoming is addressed by mollifying the Dirac delta function in (1) resulting on a mollified approximation:

$$q^\varepsilon(\boldsymbol{x}) = \int q(\boldsymbol{x} - \boldsymbol{y})\, \zeta^\varepsilon(\boldsymbol{y})\, d\boldsymbol{y}\,, \tag{3}$$

M. Daydé et al. (Eds.): VECPAR 2006, LNCS 4395, pp. 391–402, 2007.
© Springer-Verlag Berlin Heidelberg 2007

where $\zeta^\varepsilon = \varepsilon^{-d}\zeta(x/\varepsilon)$, $x \in \mathbb{R}^d$, and ε being a characteristic length scale of the kernel. For consistency of the approximation the kernel ζ has to fulfill the following moment conditons:

$$\int \zeta \, x^\alpha \, dx = 0^\alpha \text{ for } 0 \le |\alpha| < r \,. \tag{4}$$

The kernel ζ is of order r and the following error bound holds:

$$\|q - q^\varepsilon\| \le C\varepsilon^r \|q\|_\infty \,. \tag{5}$$

Now again, we get a discrete but smooth function approximation by approximating the integral in (3) by a midpoint quadrature rule yielding

$$q^{\varepsilon,h}(x) = \sum_p Q_p \, \zeta^\varepsilon(x - x_p) \,, \tag{6}$$

The error of (6) can be assessed by splitting $\|q - q^{\varepsilon,h}\|$ into

$$
\begin{aligned}
\|q - q^{\varepsilon,h}\| &\le \quad \|q - q^\varepsilon\| + \qquad \|q^\varepsilon - q^{\varepsilon,h}\| \\
&\le \quad C_1 \varepsilon^r \|q\|_\infty + \quad C_2 \left(\tfrac{h}{\varepsilon}\right)^m \|q\|_\infty \,.
\end{aligned} \tag{7}
$$

We conclude from this, that (h/ε) must be smaller than 1, *i.e.* smooth particles *must* overlap[1].

1.2   Differential Operator Approximation

In smooth particle methods differential operators can be approximated by discrete integral operators. Degond & MasGallic developed an integral representation of the diffusion operator - isotropic and anisotropic - which was later extended to differential operators of arbitrary degree in [6]. The integral operator for the 1D Laplacian for instance takes the form

$$\Delta^\varepsilon q = \frac{1}{\varepsilon^2} \int [q(y) - q(x)] \, \eta^\varepsilon(x - y) \, dy \,, \tag{8}$$

where the kernel $\eta(x)$ has to fulfill $\int x^2 \, \eta(x) \, dx = 2$. This integral is discretized by particles using their locations as quadrature points:

$$\left(\Delta^{\varepsilon,h} q\right)(x_{p'}) = \varepsilon^{-2} \sum_p [q_p - q_{p'}] \, \eta^\varepsilon(x_{p'} - x_p) \, v_p \,. \tag{9}$$

2   Solving Transport Problems with Particle Methods

Particle methods discretize the Lagrangian form of the governing equation,

$$\frac{\partial q}{\partial t} + \nabla \cdot (\mathbf{u} \, q) = \mathcal{L}(q, x, t) \,, \tag{10}$$

---

[1] For certain kernels, an r-th order approximation can be achieved even with $\varepsilon = h$ [17].

resulting in the following set of ODEs:

$$\frac{d\boldsymbol{x}_p}{dt} = \mathbf{u}(\boldsymbol{x}_p, t), \qquad \text{positions}$$

$$\frac{dv_p}{dt} = v_p \left(\nabla \cdot \mathbf{u} \right)(\boldsymbol{x}_p, t), \quad \text{volumes} \qquad (11)$$

$$\frac{dQ_p}{dt} = v_p \, \mathcal{L}^{\varepsilon,h}(q, \boldsymbol{x}_p, t). \quad \text{weights}$$

Particle positions are usually initialized as a regular lattice with spacing h, volumes are thus initially set to $v_p = h^d$ and $Q_p = q_o(\boldsymbol{x}_p) \, h^d$. The ODES (11), are now advanced using a standard explicit time stepper and the transported quantity q can be reconstructed as

$$q(\boldsymbol{x}, t) = \sum_p Q_p(t) \, \zeta^\varepsilon \left(\boldsymbol{x} - \boldsymbol{x}_p(t) \right). \qquad (12)$$

However, as the particles follow the flow map $\mathbf{u}(\boldsymbol{x}, t)$ their positions eventually become irregular and distorted, and the function approximation (12) ceases to be well-sampled. To ascertain convergence, it is therefore necessary to periodically regularize the particle locations; this process is called "remeshing".

2.1   Remeshing

Remeshing involves interpolation of particle weights from irregular particle locations onto a regular lattice. New particles are then created on the lattice, replacing the old particles. This interpolation process takes the form

$$Q_{p'}^{\text{new}} = \sum_p W(\boldsymbol{x}_{p'} - \boldsymbol{x}_p) \, Q_p^{\text{old}}, \qquad (13)$$

where $Q_{p'}^{\text{new}}$ are the new particle weights, and $\boldsymbol{x}_{p'}$ are located on a regular lattice. The interpolation functions $W(\boldsymbol{x})$ is commonly chosen to be a tensor product of one-dimensional interpolation function which for accuracy have to be sufficiently smooth and moment-conserving. The M_4' function [14] is commonly used in the context of particle methods; it is in $C^1(\mathbb{R})$ and of third order.

The introduction of a grid clearly detracts from the meshless character of particle methods. The use of a grid in the context of particle methods does not restrict the adaptive character of the method and provides the basis for a new class of "hybrid" particle methods with several computational and methodological benefits

2.2   Hybrid Particle Methods

The introduction of a grid enables fast evaluation of differential operators using compact PSE kernels, enables the use of fast grid-based Poisson solvers [8], facilitates parallelization and is a key component in adaptive particle methods,

which we will present in section 3. Hybrid particle methods make heavy use of
these computational advantages [19,3,11].

Recently, we have developed a generic hybrid particle method framework [16],
enabling efficient, parallel simulations of large-scale transport problems as diverse as the DNS of turbulent flows and diffusion processes in complex biological
organelles. Figure 1 shows visualizations of the Crow instability and the elliptic instability of two counter-rotating vortex tubes employing a maximum of 33
million particles. The simulations were performed on a 16 cpu Opteron cluster.
One time step for 1 million particles took less than 30 seconds. Current implementations using the fast multipole method which retain the meshless character
of the particle method require approximately 2400 seconds per time step [20].
This clearly demonstrates the advantages of hybrid methods.

Fig. 1. Crow (left) and short-wave or elliptic instability (right)

3   Adaptive Particle Methods

The accuracy of smooth particle methods is determined by the core size ε of
the kernel $\zeta^\varepsilon(\boldsymbol{x})$. For computational efficiency this core size needs to be spatially
variable to resolve small scales in different parts of the flow, such as the boundary
layer and the wake of bluff body flows. As particles need to overlap, varying core
sizes imply spatially varying particle spacings. This can be achieved in two ways:

- remeshing particles on a regular grid corresponding to variable size particles
  in a mapped using a global (adaptive) mapping
- remeshing particles by combining several simple local mappings in a domain
  decomposition frame.

In the context of vortex methods, Hou [10] first introduced spatially varying
particle sizes and proved the convergence of the method in the case of the 2D
Euler equations. This proof was extended in [15] to the viscous case and the
method was used for the simulation of wakes with stetched particle resolution.
In [2] Cottet, Koumoutsakos, and Ould-Salihi formulated a convergent variable
core size vortex method for the Navier-Stokes equations by using mappings from
a reference space $\hat{\Omega} \subseteq \mathbb{R}^d$ with uniform core size $\hat{\varepsilon}$ to the "physical" space
$\Omega \subseteq \mathbb{R}^d$ with cores of varying size $\varepsilon(\boldsymbol{x})$ in conjunction with an anisotropic
diffusion operator, *i.e.*

$$\boldsymbol{x} = \boldsymbol{f}(\hat{\boldsymbol{x}}), \qquad \hat{\boldsymbol{x}} = \mathbf{g}(\boldsymbol{x}), \qquad \{\boldsymbol{\Phi}\}_{ij} = \frac{\partial \hat{x}_i}{\partial x_j} \qquad \text{and} \qquad |\boldsymbol{\Phi}| = \det \boldsymbol{\Phi} \qquad (14)$$

Like in the uniform core size method (11), we convect the particles in physical space, but diffusion is performed in reference space, so that with N particles, located in $\{\boldsymbol{x}_j(t)\}_{j=1}^{N} = \{\boldsymbol{f}(\hat{\boldsymbol{x}}_j)\}_{j=1}^{N}$ we find an approximate solution to (10) by integrating the following set of ODEs:

$$\frac{d\boldsymbol{x}_j}{dt} = \mathbf{u}(\boldsymbol{x}_j, t),$$

$$\frac{dQ_j}{dt} = \frac{\nu}{\hat{\epsilon}^2} \sum_k \psi_{pq}^{\hat{\epsilon}}(\hat{\boldsymbol{x}}_j - \hat{\boldsymbol{x}}_k) \left(\frac{m_{pq}(\hat{\boldsymbol{x}}_j) + m_{pq}(\hat{\boldsymbol{x}}_k)}{2} \right) [\hat{v}_j \hat{Q}_k - \hat{v}_k \hat{Q}_j], \qquad (15)$$

$$\frac{d\hat{v}_j}{dt} = \hat{\nabla} \cdot (\underline{\boldsymbol{\Phi}}\mathbf{u})(\boldsymbol{x}_j, t) \, \hat{v}_j.$$

In the above equation Q_j and $\hat{Q}_j$ denote the particle strength in physical and reference space, respectively, related by

$$\hat{Q}_j = Q_j |\boldsymbol{\Phi}|(\boldsymbol{x}_j),$$

and $m_{pq} = b_{pq} - \frac{1}{d+2}\delta_{pq}\delta_{p'q'}b_{p'q'}$, with

$$b_{p'q'} = \frac{1}{\boldsymbol{\Phi}} \frac{\partial(\hat{\boldsymbol{x}})_{p'}}{\partial(\boldsymbol{x})_r} \frac{\partial(\hat{\boldsymbol{x}})_{q'}}{\partial(\boldsymbol{x})_r}$$

and $\psi_{pq}(\boldsymbol{x}) = (\boldsymbol{x})_p (\boldsymbol{x})_q \rho(\boldsymbol{x})$, $\rho(\boldsymbol{x})$ being a radially symmetric kernel with suitable moment properties [4].

In [2] analytic, invertible mappings have been employed. Albeit being a simple and robust way to efficiently resolve the range of length scales in the flow, this method requires prior knowledge about the flow physics. In [1] we extended this method by introducing two different approaches to dynamical adaptivity in particle methods; One approach makes use of a global adaptive mapping (AGM, see section 3.1), and one employing dynamically placed patches of smaller sized particles, reminiscent of adaptive mesh refinement in finite volume methods (AMR).

3.1 Particle Method with Adaptive Global Mappings

We introduce a transient smooth map $\mathbf{f} : \hat{\Omega} \times [0, T] \to \Omega$ represented by particles:

$$\boldsymbol{x}(\hat{\boldsymbol{x}}, t) = \mathbf{f}(\hat{\boldsymbol{x}}, t) = \sum_j \boldsymbol{\chi}_j(t) \zeta^{\hat{\epsilon}}(\hat{\boldsymbol{x}} - \boldsymbol{\xi}_j), \qquad (16)$$

where $\boldsymbol{\xi}_p$ are fixed at grid point locations. The parameters in the map that are changed in the process of adaptation are the node values $\boldsymbol{\chi}_j$. As the map (16) is not easily invertible, we require it to be smooth in both space and time. Given this property, the governing equation (10) can be entirely cast into reference space, again yielding a transport equation:

$$\frac{\partial \hat{q}'}{\partial t} + \hat{\nabla} \cdot (\hat{q}' \, \tilde{\boldsymbol{u}}) = \hat{\mathcal{L}}(\hat{q}', \hat{\boldsymbol{x}}, t), \qquad (17)$$

where $\hat{q}' = (|\Phi|)^{-1}\hat{q}$ and

$$\tilde{u} = \Phi(\hat{u} - \mathcal{U}), \qquad \text{and} \qquad \mathcal{U} = \frac{\partial \mathbf{f}}{\partial t} = \sum_j \frac{\partial \chi_j}{\partial t} \zeta^{\hat{e}}(\hat{x} - \xi_j). \qquad (18)$$

What remains is to chose a $\mathcal{U}$, such that particle core sizes in physical space are small where small scale features are present in the flow. In [1] this was accomplished by setting $\mathcal{U}$ to be the solution of a moving mesh partial differential equation (MMPDE),

$$\mathcal{U} = \hat{\nabla} \cdot \left(M(\hat{x}, t)\hat{\nabla}\mathbf{f}(\hat{x}, t)^{\mathrm{T}} \right), \qquad (19)$$

where $M(\hat{x}, t)$ is a so-called monitor function: a positive measure which takes great values where numerical resolution should be increased, e.g.

$$M(\hat{x}, t) = \sqrt{1 + \alpha|\mathcal{B}\hat{q}|^2}, \qquad (20)$$

$\mathcal{B}$ being a high-pass filter. We applied this method in [1] to the evolution of an elliptical vortex governed by the 2D Euler equations. Figure 2 depicts the adaptation of the underlying grid, and thus the particle core sizes $\varepsilon(x)$.

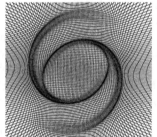

Fig. 2. Simulation of the evolution of an inviscid elliptical vortex using the AGM particle method: vorticity (left), particle sizes (middle, dark areas represent coarse particle sizes) and grid (right)

3.2   Wavelet-Based Multiresolution Particle Method

We employ a wavelet-based multiresolution analysis (MRA) using $L + 1$ levels of refinement to guide the creation of particles on the grid. The function $q(x, t)$ can be represented as

$$q^L = \sum_k q_k^0 \zeta_k^0 + \sum_{0 \leq l < L} \sum_k d_k^l \psi_k^l, \qquad (21)$$

where q_k^0 are the weights, ζ_k^0 are the scaling functions (or kernels) on the coarsest level, d_k^l are the so-called detail coefficients and ψ_k^l are the wavelets. The MRA here is based on an iterative interpolation scheme as introduced by Deslauriers and Dubuc [5], thus we do not have explicit scaling functions ζ and wavelets ψ. In this scheme the scaling coefficients of two subsequent levels are related through

$$q_{2k}^{l+1} = q_k^l$$
$$q_{2k+1}^{l+1} = d_k^l + \sum_j w_{j-k}^l \, q_{j-k}^l \,, \tag{22}$$

where w_j^l are coefficients related to the polynomial interpolation of the scheme [5].

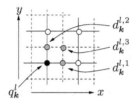

Fig. 3. Each detail coefficient $d_k^{l,m}$, with $m = 1, \ldots, 2^d - 1$ corresponds to a specific grid point on the next higher level

As illustrated in Figure 3 each detail coefficient is associated with a grid point on the next finer grid. Let $child(\boldsymbol{k}, m)$ be the grid point associated with $d_k^{l,m}$ and let $ancs(\boldsymbol{k})$ denote the set of grid points $\boldsymbol{k}'$ needed to interpolate the value q_k^l from values $q_{\boldsymbol{k}'}^{l-1}$ and detail coefficients $\boldsymbol{d}_{\boldsymbol{k}'}^{l-1}$ of the next coarser level. Then an adapted grid is constructed by discarding all those grid points whose $|d_k^{l,m}|$ are smaller as a prescribed threshold, i.e.

$$\mathcal{K}_> = \mathcal{K}^0 \cup \left\{ \boldsymbol{k}' = child(\boldsymbol{k}, m) \cup ancs(\boldsymbol{k}') \,\Big|\, |d_k^{l,m}| > \varepsilon \,, \, l \in [0, L-1] \right\}. \tag{23}$$

Note that $\mathcal{K}_>^0 \equiv \mathcal{K}^0$ and that $ancs(\boldsymbol{k}')$ are added to maintain proper nestedness of the grids (see for instance [18] for details). In order to be able to capture small scales that may emerge between two subsequent MRAs we follow the conservative approach of Liandrat and Tchamitchian [13] and additionally activate all children of the active grid points, *i.e.*

$$\mathcal{K}_>^l \leftarrow \mathcal{K}_>^l \cup \{\, child(\boldsymbol{k}, m) \mid \boldsymbol{k} \in \mathcal{K}_>^{l-1} \,,\, m = 1, 2, 3 \,\} \text{ for } l = L, \ldots, 1. \tag{24}$$

Multilevel remeshing interpolates particles created on a set of grid points $\mathcal{K}_>^l$ onto a set of grid points $\mathcal{K}_\times^l$. This is accomplished in the following way: let M denote the kernel used for remeshing the particles, then (i) horizontally extend the set of source grid points $\mathcal{K}_>^l$ by $\boldsymbol{\mathcal{B}}^l$, where

$$\boldsymbol{\mathcal{B}}^l = \left\{ \boldsymbol{k}' \,\Big|\, \min_{\boldsymbol{k} \in \mathcal{K}_>^l} |\boldsymbol{k}' - \boldsymbol{k}| \le \lceil \tfrac{1}{2} supp(M) + \text{LCFL} \rceil \right\}, \tag{25}$$

where the "Lagrangian CFL" LCFL $\equiv \delta t \, \|\nabla \otimes \boldsymbol{u}\|_\infty$, (ii) create particles on $\mathcal{K}_>^l \cup \mathcal{B}_1^l$, *i.e.*

$$Q_p^l = c_{\boldsymbol{k}}^l \, (h^l)^d \,, \qquad v_p^l = (h^l)^d \,, \qquad \boldsymbol{x}_p^l = \boldsymbol{x}_{\boldsymbol{k}}^l \,,$$

(iii) after convection, interpolate these particles onto a new set of grid points $\mathcal{K}_\times^l$. Clearly, for consistency $\mathcal{K}_\times^l$ cannot be chosen arbitrarily. We propose the following method: Introduce and indicator function χ^l defined as

$$\chi_{\boldsymbol{k}}^l = \begin{cases} 1 \,, & \boldsymbol{k} \in \mathcal{K}_>^l \\ 0 \,, & \boldsymbol{k} \in \mathcal{B}^l \,, \end{cases} \qquad (26)$$

and convect the particles, *i.e.* solve the following set of equations

$$\frac{dQ_p^l}{dt} = \mathcal{L}(q, \boldsymbol{x}, t) \,, \qquad \frac{d\chi_p^l}{dt} = 0 \,, \qquad \frac{d\boldsymbol{x}_p^l}{dt} = \boldsymbol{u}(\boldsymbol{x}_p^l, t) \,, \qquad \frac{dv_p^l}{dt} = v_p^l \, (\nabla \cdot \boldsymbol{u}) \, (\boldsymbol{x}_p^l, t) \,. \tag{27}$$

The particle weights and the indicator are then interpolated onto the grid and grid points with $\tilde{\chi}_{\boldsymbol{k}}^l > 0$ are selected to consitute $\mathcal{K}_\times^l$, where $\tilde{\chi}^l$ denotes the remeshed indicator function. Using this technique, the scale distribution $\{\mathcal{K}_>^l\}_{l=0}^L$ is naturally convected with the flow and we obtain and adaptation mechanism which is independent of the CFL number.

To demonstrate the Lagrangian character of the adaptation we considered the convection of a passive scalar in 2D, subject to a vortical velocity field [12]. The problem involves strong deformation of a initial circular "blob" which at the end of the simulation returns to the initial condition. The remeshing function and particle kernel were both chosen as

$$W(\boldsymbol{x}) = \zeta(\boldsymbol{x}) = \prod_{l=1}^d M_6''' ((\boldsymbol{x})_l) \,,$$

where the fourth-order accurate interpolating function M_6''' is of higher order than the M_4' function at the expense of a larger support. The wavelets employed were also fourth-order accurate. Figure 4 illustrates the adaptation of the grid/particles at two different times. We measure the L^2 and L^∞ error of the final solution for different choices of ϵ and observe second order convergence, corresponding to fourth order convergence in h, as depicted in Figure 6. The maximum CFL measured during the course of the simulation was 40.7.

We also applied the presented method to the simulation of a propagating interface using a level set formulation. A "narrow band" formulation is easily accomplished with the present method by truncating the detail coefficients that are far from the interface. We consider the well-established 2D deformation test case which amounts to the propagation of a circle subject to the same velocity field as above. Figure 5 depicts the grid adaptation and comparing to Figure 4, one can clearly see the restriction of the refinement to a small neighborhood around the interface. We measure the error of the area encompassed by the

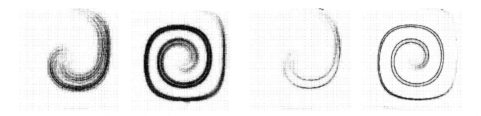

Fig. 4. Active grid points/particles at two different times of the simulation of a passive scalar subject to a single vortex velocity field

Fig. 5. Active grid points/particles at two different times of the simulation of a propagating interface subject to a single vortex velocity field

interface at the final time and compare it against a non-adaptive particle level set method [9] and against the "hybrid particle level set method" of Enright *et al.* [7]. Figure 7 displays this comparison and we find that our adaptive approach performs favorably, which may be attributed in part to the adaptive character and in part to the high order of the method.

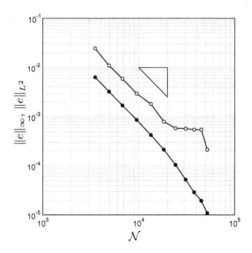

Fig. 6. ε-refinement study; the data points correspond to $\varepsilon = 2^{-p} \times 10^{-3}$ for $p = 0, \ldots, 10$. The triangle represents 2nd-order convergence. $\mathcal{N}$ is the number of active grid points/particles.

3.3   Parallelization

Recently we have developed a Parallel Particle-Mesh (PPM) software library [16] that facilitates large-scale calculations of transport and related problems using particles. The library provides the mechanisms necessary to achieve good parallel efficiency and load balancing in these situations where both meshes and particles operate as computational elements. Figure 8 and Figure 9 depict

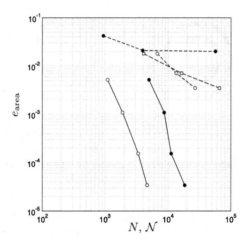

Fig. 7. Plot of relative error of the area enclosed by the interface against degrees of freedom: Hieber & Koumoutsakos [9] (•--•, particles at time t=0), Enright *et al.*[7] (o--o, auxiliary particles at time t=0 and □--□, grid points) and present method (o—o, active grid points at time t=0, •—•, active grid points at the final time)

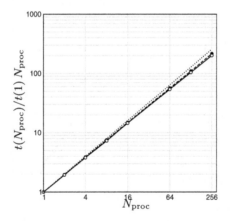

Fig. 8. Parallel scaling of a particle-based Navier-Stokes solver built using the PPM library. The calculations were performed on the Cray XT3, with 524,288 particles per CPU. Curves denote double precision (o—o), and single-precision (•--•) results, respectively.

the parallel performance of the library for a Navier-Stokes solver based on the vortex method. The calculations were run on the Cray XT3 at the Swiss National Supercomputing Centre (CSCS). Our current work aims at implementing the adaptive techniques described herein into the parallel framework of the PPM library.

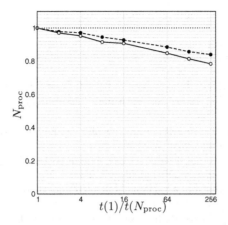

Fig. 9. Parallel Efficiency of a PPM-based Navier-Stokes solver. Curves denote double precision (○—○), and single-precision (●--●) results, respectively.

4   Conclusions

We present multiresolution particle-mesh methods for simulating transport equations. We outline two methods introducing enhanced dynamic adaptivity and multiresolution capabilities for particle methods. The first method is based on an adaptive global mapping from a reference space to physical space for the particle locations; it has been successfully applied to the evolution of an elliptical vortex in an inviscid incompressible fluid. The second method is based on a wavelet multiresolution decomposition of the particle function representation. It is equipped with a Lagrangian adaptation mechanism that enables the simulation of transport problems and interface capturing problems independent of the CFL number. We have presented results of an interface tracking problem where the method has shown to have superior volume conservation properties. We are currently working on the application of this method to the Navier-Stokes equations.

References

1. M. Bergdorf, G.-H. Cottet, and P. Koumoutsakos. Multilevel adaptive particle methods for convection-diffusion equations. *Multiscale Modeling and Simulation*, 4(1):328–357, 2005.
2. G.-H. Cottet, P. Koumoutsakos, and M. L. Ould Salihi. Vortex methods with spatially varying cores. *Journal of Computational Physics*, 162:164–185, 2000.
3. G.-H. Cottet and P. Poncet. Advances in direct numerical simulations of 3d wall-bounded flows by vortex-in-cell methods. *Journal of Computational Physics*, 193:136–158, 2003.
4. P. Degond and S. Mas-Gallic. The weighted particle method for convection-diffusion equations. part 2: The anisotropic case. *Mathematics of Computation*, 53(188):509–525, 1989.

5. G. Deslauriers and S. Dubuc. Symmetric iterative interpolation processes. *Constructive Approximation*, 5:49–68, 1989.
6. J. D. Eldredge, A. Leonard, and T. Colonius. A general deterministic treatment of derivatives in particle methods. *Journal of Computational Physics*, 180:686–709, 2002.
7. D. Enright, R. Fedkiw, J. Ferziger, and I. Mitchell. A hybrid particle level set method for improved interface capturing. *Journal of Computational Physics*, 183(1):83–116, 2002.
8. F. H. Harlow. Particle-in-cell computing method for fluid dynamics. *Methods in Computational Physics*, 3:319–343, 1964.
9. S. E. Hieber and P. Koumoutsakos. A Lagrangian particle level set method. *Journal of Computational Physics*, 210(1):342–367, 2005.
10. T. Y. Hou. Convergence of a variable blob vortex method for the euler and navier-stokes equations. *SIAM Journal on Numerical Analysis*, 27(6):1387–1404, 1990.
11. P. Koumoutsakos. Multiscale flow simulations using particles. *Annual Review of Fluid Mechanics*, 37(1):457–487, 2005.
12. R. J. Leveque. High-resolution conservative algorithms for advection in incompressible flow. *SIAM Journal on Numerical Analysis*, 33(2):627–665, 1996.
13. J. Liandrat and P. Tchamitchian. Resolution of the 1D regularized burgers equation using a spatial wavelet approximation. ICASE Report 90-83, NASA Langley Research Center, 1990.
14. J. J. Monaghan. Extrapolating b-splines for interpolation. *Journal of Computational Physics*, 60:253–262, 1985.
15. P. Ploumhans and G. S. Winckelmans. Vortex methods for high-resolution simulations of viscous flow past bluff bodies of general geometry. *Journal of Computational Physics*, 165:354–406, 2000.
16. I. F. Sbalzarini, J. H. Walther, M. Bergdorf, S. E. Hieber, E. M. Kotsalis, and P. Koumoutsakos. PPM – a highly efficient parallel particle-mesh library. *Journal of Computational Physics*, 215(2):566–588, 2006.
17. A.-K. Tornberg and B. Engquist. Numerical approximations of singular source terms in differential equations. *Journal of Computational Physics*, 200:462–488, 2004.
18. O. V. Vasilyev. Solving multi-dimensional evolution problems with localized structures using second-generation wavelets. *International Journal of Computational Fluid Dynamics*, 17(2):151–168, 2003.
19. J. H. Walther and P. Koumoutsakos. Three-dimensional particle methods for particle laden flows with two-way coupling. *Journal of Computational Physics*, 167:39–71, 2001.
20. Q. X. Wang. Variable order revised binary treecode. *Journal of Computational Physics*, 200(1):192 –210, 2004.

Evaluation of Several Variants of Explicitly Restarted Lanczos Eigensolvers and Their Parallel Implementations*

V. Hernandez, J.E. Roman, and A. Tomas

D. Sistemas Informáticos y Computación, Universidad Politécnica de Valencia,
Camino de Vera, s/n, E-46022 Valencia, Spain
Tel.: +34-963877356; Fax: +34-963877359
{vhernand,jroman,atomas}@itaca.upv.es

Abstract. It is well known that the Lanczos process suffers from loss of orthogonality in the case of finite-precision arithmetic. Several approaches have been proposed in order to address this issue, thus enabling the successful computation of approximate eigensolutions. However, these techniques have been studied mainly in the context of long Lanczos runs, but not for restarted Lanczos eigensolvers. Several variants of the explicitly restarted Lanczos algorithm employing different reorthogonalization strategies have been implemented in SLEPc, the Scalable Library for Eigenvalue Computations. The aim of this work is to assess the numerical robustness of the proposed implementations as well as to study the impact of reorthogonalization in parallel efficiency.

Topics: Numerical methods, parallel and distributed computing.

1 Introduction

The Lanczos method [1] is one of the most successful methods for approximating a few eigenvalues of a large real symmetric (or complex Hermitian) matrix, A. It computes a sequence of Lanczos vectors, v_j, and scalars α_j, β_j as follows

Choose a unit-norm vector v_1 and set $\beta_1 = 0$
For $j = 1, 2, \ldots$
$\quad u_{j+1} = Av_j - \beta_j v_{j-1}$
$\quad \alpha_j = v_j^* u_{j+1}$
$\quad u_{j+1} = u_{j+1} - \alpha_j v_j$
$\quad \beta_{j+1} = \|u_{j+1}\|_2 \quad$ (if $\beta_{j+1} = 0$, stop)
$\quad v_{j+1} = u_{j+1}/\beta_{j+1}$
end

Every iteration of the loop computes the following three-term recurrence

$$\beta_{j+1} v_{j+1} = Av_j - \alpha_j v_j - \beta_j v_{j-1} \,. \tag{1}$$

---

* This work was supported in part by the Valencian Regional Administration, Directorate of Research and Technology Transfer, under grant number GV06/091.

M. Daydé et al. (Eds.): VECPAR 2006, LNCS 4395, pp. 403–416, 2007.
© Springer-Verlag Berlin Heidelberg 2007

The first m iterations can be summarized in matrix notation as follows

$$AV_m - V_m T_m = \beta_{m+1} v_{m+1} e_m^*, \tag{2}$$

where $V_m = [v_1, v_2, \ldots, v_m]$, $e_m^* = [0, 0, \ldots, 1]$, and

$$T_m = \begin{bmatrix} \alpha_1 & \beta_2 & & & \\ \beta_2 & \alpha_2 & \beta_3 & & \\ & \ddots & \ddots & \ddots & \\ & & \beta_{m-1} & \alpha_{m-1} & \beta_m \\ & & & \beta_m & \alpha_m \end{bmatrix} \tag{3}$$

It can be shown that the Lanczos vectors are mutually orthonormal, i.e. $V_m^* V_m = I_m$, where I_m is the $m \times m$ identity matrix. As described in section 2, the above procedure can be used as a basis for implementing a solver for symmetric eigenvalue problems, because eigenvalues of T_m approximate eigenvalues of A. However, practical implementations have to deal with issues such as:

1. The loss of orthogonality of Lanczos vectors in finite-precision arithmetic.
2. The convenience of eventually restarting the recurrence.

If loss of orthogonality is not treated appropriately, then duplicate eigenvalues appear in the spectrum of T_j as the iteration progresses. This effect is well understood since the work by Paige [2], who showed that orthogonality is lost as soon as an eigenvalue has converged. This helped researchers devise effective reorthogonalization strategies for preserving (semi-) orthogonality, as described in section 2. These techniques make use of previously computed Lanczos vectors, thus increasing the storage needs and computational cost with respect to the original algorithm, growing as the iteration proceeds. For this reason, practical implementations of Lanczos must generally be restarted, especially in the case of very large-scale sparse problems. In a restarted version, the number of Lanczos steps is limited to a maximum allowed value, after which a new recurrence begins. The simplest form of restart, usually called explicit restart, consists in computing a new starting vector, v_1, from the spectral information available before the restart. Although this strategy is typically less effective than other techniques such as implicit restart [3], it can still be competitive in some cases.

The motivation of this work is to provide a robust and efficient parallel implementation of an explicitly restarted symmetric Lanczos eigensolver in SLEPc, the Scalable Library for Eigenvalue Problem Computations [4]. The main goal is to analyze how different reorthogonalization techniques behave in this context, both from the stability and efficiency viewpoints. We focus on single-vector Lanczos variants, in contrast to block variants such as that proposed in [5].

The text is organized as follows. In section 2, the Lanczos method is described in more detail, including the different strategies for coping with loss of orthogonality. Implementation details such as how to efficiently parallelize the orthogonalization operation are discussed as well. In section 3, the particular implementations available in SLEPc are described. Sections 4 and 5 show the analysis results with respect to numerical stability and parallel performance, respectively. Finally, in section 6 some conclusions are given.

2 Description of the Method

This section provides an overview of the Lanczos method and some of its variations, including techniques for avoiding loss of orthogonality. For more detailed background material the reader is referred to [6,7].

2.1 Basic Lanczos Algorithm

Apart from viewing the Lanczos process from the perspective of the three-term recurrence described in the previous section, it can also be seen as the computation of the orthogonal projection of matrix A onto the Krylov subspace $\mathcal{K}_m(A, v_1) \equiv \text{span}\{v_1, Av_1, \ldots, A^{m-1}v_1\}$. From this perspective, the Lanczos method is equivalent to the Arnoldi method, and can be described as follows.

Algorithm 1. Basic Lanczos
Input: Matrix A, number of steps m, and initial vector v_1 of norm 1
Output: $(V_m, T_m, v_{m+1}, \beta_{m+1})$ so that $AV_m - V_mT_m = \beta_{m+1}v_{m+1}e_m^*$
 For $j = 1, 2, \ldots, m$
  $u_{j+1} = Av_j$
  Orthogonalize u_{j+1} with respect to V_j (obtaining α_j)
  $\beta_{j+1} = \|u_{j+1}\|_2$
  $v_{j+1} = u_{j+1}/\beta_{j+1}$
 end

In the above algorithm, the second line in the loop performs a Gram-Schmidt process in order to orthogonalize vector u_{j+1} with respect to the columns of V_j, that is, the vectors $v_1, v_2, \ldots, v_j$ (see subsection 2.4 for details about Gram-Schmidt). In this operation, j Fourier coefficients are computed. In exact arithmetic, the first $j - 2$ coefficients are zero, and therefore the corresponding operations need not be carried out (orthogonality with respect to the first $j - 2$ vectors is automatic). The other two coefficients are β_j and α_j. According to Paige [8], the β_j computed in this operation should be discarded and, instead, use the value $\|u_j\|_2$ computed in the previous iteration. As we will see in subsection 2.2, orthogonalization will be a key aspect of robust Lanczos variants that cope with loss of orthogonality.

Since $V_m^* v_{m+1} = 0$ by construction, then by premultiplying Eq. 2 by V_m^*

$$V_m^* AV_m = T_m, \tag{4}$$

that is, matrix T_m represents the orthogonal projection of A onto the Krylov subspace spanned by the columns of V_m, and this fact allows us to compute Rayleigh-Ritz approximations of the eigenpairs of A. Let (λ_i, y_i) be an eigenpair of matrix T_m, then the Ritz value, λ_i, and the Ritz vector, $x_i = V_m y_i$, can be taken as approximations of an eigenpair of A. Typically, only a small percentage of the m approximations are good. This can be assessed by means of the residual norm for the Ritz pair, which turns out to be very easy to compute:

$$\|Ax_i - \lambda_i x_i\|_2 = \|AV_m y_i - \lambda_i V_m y_i\|_2 = \|(AV_m - V_m T_m)y_i\|_2 = \beta_{m+1}|e_m^* y_i|. \tag{5}$$

2.2   Lanczos in Finite Precision Arithmetic

When implemented in finite precision arithmetic, the Lanczos algorithm does not behave as expected. The eigenvalues of the tridiagonal matrix T_j (the Ritz values) converge very rapidly to well-separated eigenvalues of matrix A, typically those in the extreme of the spectrum. However, if enough iterations of the algorithm are carried out, then multiple copies of these Ritz values appear, beyond the multiplicity of the corresponding eigenvalue in A. In addition, the process gives wrong Ritz values as converged, which are usually called spurious eigenvalues. It can be observed that this unwanted behavior appears at the same time that the Lanczos vectors start to lose mutual orthogonality. Lanczos himself was already aware of this problem and suggested to explicitly orthogonalize the new Lanczos vector with respect to all the previous ones at each step. Although effective, this costly operation seems to invalidate all the appealing properties of the algorithm. Other alternatives, discussed below, have been proposed in order to be able to deal with loss of orthogonality at less cost.

Full Orthogonalization. The simplest cure to loss of orthogonality is to orthogonalize vector u_{j+1} explicitly with respect to all the previously computed Lanczos vectors. That is, performing the computation for all vectors, including the first $j - 2$ ones for which the Fourier coefficient is zero in exact arithmetic.

The main advantage of full orthogonalization is its robustness, since orthogonality is maintained to full machine precision. (Note that for this to be true it may be necessary to resort to double orthogonalization, see subsection 2.4.) The main drawback of this technique is that the cost of orthogonalization is high and grows as more Lanczos steps are carried out. This recommends a restarted version, in which the number of Lanczos vectors is bounded, see section 2.3.

Local Orthogonalization. The quest for more efficient solutions to the problem of loss of orthogonality started with a better theoretical understanding of the Lanczos process in finite precision arithmetic, unveiled by Paige's work [8,2,9]. One key aspect of Paige's analysis is that Lanczos vectors start to lose orthogonality as soon as an eigenvalue of T_j stabilizes or, in other words, when a Ritz value is close to convergence, causing the subsequent Lanczos vectors to contain a non-negligible component in the direction of the corresponding Ritz vector. Until this situation occurs, the Lanczos algorithm with local orthogonalization (that is, if vector u_{j+1} is orthogonalized only with respect to v_j and v_{j-1}) computes the same quantities as the variant with full orthogonalization. This fact suggests that an algorithm could proceed with local orthogonalization until an eigenvalue of T_j has stabilized, then either start a new Lanczos process with a different initial vector, or continue the Lanczos process with the introduction of some kind of reorthogonalization. The latter approach gave way to the development of semiorthogonal Lanczos methods, discussed below.

A completely different approach is to simply ignore loss of orthogonality and perform only local orthogonalization at every Lanczos step. This technique is obviously the cheapest one, but has several important drawbacks. For one thing, convergence of new Ritz values is much slower since multiple copies of already converged ones keep on appearing again and again. This makes it necessary to carry out many Lanczos steps to obtain the desired eigenvalues. On the other hand, there is the problem of determining the correct multiplicity of the computed eigenvalues as well as discarding those which are spurious. A clever technique for doing this was proposed in [10]. An eigenvalue of T_j is identified as being spurious if it is also an eigenvalue of the matrix T_j', which is constructed by deleting the first row and column of T_j. Furthermore, good eigenvalues are accepted only after they have been replicated at least once.

Semiorthogonal Techniques. As mentioned above, the idea of these techniques is to perform explicit orthogonalization only when loss of orthogonality is detected. Two aspects are basic in this context:

1. How to carry out the orthogonalization so that the overall cost is small.
2. How to determine when an eigenvalue has stabilized or, in other words, how to monitor loss of orthogonality, without incurring a high cost.

With respect to the first aspect, several different approaches have been proposed: selective [11], periodic [12], and partial [13] reorthogonalization. In brief, they consist, respectively, in: orthogonalizing every Lanczos vectors with respect to all nearly converged Ritz vectors; orthogonalizing u_{j+1} and u_{j+2} with respect to all the Lanczos vectors; and orthogonalizing u_{j+1} and u_{j+2} with respect to a subset of the Lanczos vectors. The second aspect can be addressed in two ways, basically. One is to compute the error bounds associated to the Ritz pairs (Eq. 5) at each iteration, and the other is to use a recurrence for estimating a bound of the level of orthogonality, such as the one proposed by Simon in [13]. If we define the level of orthogonality at the j-th Lanczos step as

$$\omega_j \equiv \max_{1 \leq k < j} |\omega_{j,k}|, \quad \text{with} \quad \omega_{j,k} \equiv v_j^* v_k, \tag{6}$$

then the full orthogonalization technique keeps it at roundoff level in each step, $\omega_j \approx \epsilon_M$. However, all that effort is not necessary since, as shown in [13,14], maintaining semiorthogonality, i.e. $\omega_j \approx \sqrt{\epsilon_M}$, is sufficient so that properties of the Rayleigh-Ritz projection are still valid.

2.3 Explicit Restart

As mentioned above, restarting is intended for reducing the storage requirements and, more importantly, reducing the computational cost of orthogonalization, which grows as more Lanczos vectors become available. Restart can be accomplished in several ways. The idea of explicit restart is to iteratively compute different m-step Lanczos factorizations (Eq. 2) with successively "better" initial vectors. The initial vector for the next Lanczos run is computed from the information available in the most recent factorization. The simplest way to select the new initial vector is to take the Ritz vector associated to the first wanted, non-converged Ritz value.

In order for a restarted method to be effective, it is necessary to keep track of already converged eigenpairs and perform a deflation, by *locking* converged Ritz vectors. Suppose that after a Lanczos run, the first k eigenpairs have already converged to the desired accuracy, and write V_m as

$$V_m = \left[V_{1:k}^{(l)} \middle| V_{k+1:m}^{(a)} \right],\tag{7}$$

where the (l) and (a) superscripts indicate locked and active vectors, respectively. In the next Lanczos run, only $m - k$ Lanczos vectors must be computed, the active ones, and in doing this the first k vectors have to be deflated. This can be done simply by orthogonalizing every new Lanczos vector also with respect to the locked ones. Therefore, deflation can be incorporated to Algorithm 1 simply by explicitly including locked vectors in the orthogonalization operation. With this change, a restarted Lanczos method can be described as in Algorithm 2.

Algorithm 2. Explicitly Restarted Lanczos
Input: Matrix A, initial vector v_1 of norm 1, and dimension of the subspace m
Output: A partial eigendecomposition $AV_k = V_k\Theta_k$, with $\Theta_k = \mathrm{diag}(\theta_1,\ldots,\theta_k)$
    Initialize $k = 0$
    Restart loop
        Perform $m - k$ steps of Lanczos (Algorithm 1) with initial vector v_{k+1}
        Compute eigenpairs of T_m, $T_m y_i = y_i \theta_i$
        Compute residual norm estimates, $\tau_i = \beta_{m+1}|e_m^* y_i|$
        Lock converged eigenpairs
        $V_m = V_m Y$
    end

In a restarted Lanczos method, it is also necessary to deal with loss of orthogonality. In the case of the simple explicit restart scheme, it is safe to use any of the techniques described in the previous subsection, since full orthogonality of the Lanczos vectors is not required for the restart to work correctly. Only in the case of local orthogonalization, the following considerations should be made:

- The restart vector has to be orthogonalized with respect to locked vectors.
- Since the value of m (the largest allowable subspace dimension) is usually very small compared to n (the matrix dimension), then the heuristics suggested in [10] cannot be applied. Therefore, another technique should be used in order to discard spurious eigenvalues as well as redundant duplicates.

2.4   Gram-Schmidt Orthogonalization

Gram-Schmidt procedures are used for orthogonalizing a vector u_{j+1} with respect to a set of vectors V_j. In finite precision arithmetic, simple versions such as Classical Gram-Schmidt (CGS) or Modified Gram-Schmidt (MGS) will not be reliable enough in some cases, and may produce numerical instability. This problem can be solved by introducing *refinement*, that is, to take the resulting vector and perform a second orthogonalization. In exact arithmetic, the Fourier coefficients of the second orthogonalization ($c_{1:j,j}$) are zero and therefore it has no effect. However, this is not the case in finite precision arithmetic, where those coefficients can be thought of as a correction to coefficients of the first orthogonalization ($h_{1:j,j}$), which is not necessarily small.

In cases where large rounding errors have not occurred in first place, refinement is superfluous and could be avoided. In order to determine whether the computed vector is good enough or requires a refinement, it is possible to use a simple criterion such as

$$\beta_{j+1} < \eta\,\rho \tag{8}$$

for some constant parameter $\eta < 1$ (a safe value is $\eta = 1/\sqrt{2}$). This criterion compares the norm of u_{j+1} before (ρ) and after (β_{j+1}) orthogonalization. An orthogonalization procedure based on this scheme is illustrated in Algorithm 3. For further details about iterative Gram-Schmidt procedures, see [15,16].

Algorithm 3. CGS with selective refinement and estimated norm

$$h_{1:j,j} = V_j^* u_{j+1}$$
$$\rho = \|u_{j+1}\|_2$$
$$u_{j+1} = u_{j+1} - V_j h_{1:j,j}$$
$$\beta_{j+1} = \sqrt{\rho^2 - \sum_{i=1}^{j} h_{i,j}^2}$$
if $\beta_{j+1} < \eta\,\rho$
$$\qquad c_{1:j,j} = V_j^* u_{j+1}$$
$$\qquad \sigma = \|u_{j+1}\|_2$$
$$\qquad u_{j+1} = u_{j+1} - V_j c_{1:j,j}$$
$$\qquad h_{1:j,j} = h_{1:j,j} + c_{1:j,j}$$
$$\qquad \beta_{j+1} = \sqrt{\sigma^2 - \sum_{i=1}^{j} c_{i,j}^2}$$
end

A similar orthogonalization scheme might be considered for the MGS variant. However, the resulting numerical quality is about the same, as pointed out in [15]. In this work, we do not consider MGS variants since they lead to poor parallel performance.

In the context of parallel implementations, orthogonalization is usually the operation that introduces more performance penalty. In order to reduce this effect, the number of synchronizations should be reduced whenever possible. One possibility for this is to defer the normalization of the vector to the next Lanczos step, as proposed in [17]. Algorithm 3 tries to optimize parallel performance by means of estimation of the norm. The main objective of this technique is to avoid the explicit computation of the Euclidean norm of u_{j+1} and, instead, use an estimation based on the original norm (prior to the orthogonalization), by simply applying the Pythagorean theorem. The original norm is already available, since it is required for the selective refinement criterion, and can be computed more efficiently in parallel since its associated reduction is susceptible of being integrated in a previous reduction (that is, with one synchronization less). More details about these techniques can be found in [18].

3   Lanczos Methods in SLEPc

SLEPc, the Scalable Library for Eigenvalue Problem Computations [4], is a software library for the solution of large, sparse eigenvalue problems on parallel computers. It can be used for the solution of problems formulated in either standard or generalized form, both Hermitian and non-Hermitian, with either real or complex arithmetic. SLEPc provides a collection of eigensolvers such as Arnoldi, Lanczos, Subspace Iteration and Power/RQI. It also provides built-in support for different types of problems and spectral transformations such as the shift-and-invert technique.

SLEPc is built on top of PETSc (Portable, Extensible Toolkit for Scientific Computation, [19]), a parallel framework for the numerical solution of partial differential equations, whose approach is to encapsulate mathematical algorithms using object-oriented programming techniques in order to be able to manage the complexity of efficient numerical message-passing codes. In PETSc, the application programmer works directly with objects such as vectors and matrices, rather than concentrating on the underlying data structures. Built on top of this foundation are various classes of solver objects, including linear, nonlinear and time-stepping solvers. SLEPc extends PETSc with all the functionality necessary for the solution of eigenvalue problems, and thus inherits all the good properties of PETSc, including portability, scalability, efficiency and flexibility. SLEPc also leverages well-established eigensolver packages such as ARPACK, integrating them seamlessly.

As of version 2.3.0, SLEPc provides a symmetric Lanczos eigensolver, which is based on explicit restart and allows the user to select among several types of reorthogonalization strategies. In SLEPc, these strategies are referred to as `local`, `full`, `selective`, `periodic`, and `partial`, and are related to the techniques described in section 2. However, the implementations slightly differ from the originally proposed techniques due to the implications of having a restarted algorithm.

In the case of local orthogonalization, the post-processing technique proposed in [10] cannot be used in the context of a restarted method, because of the relatively small size of matrix T_m. The approach taken in SLEPc is to explicitly compute the residual norm for every converged eigenpair, then from the correct values accept only the first replica in each restart. Although this may seem a very costly strategy, results show that the incurred overhead is small (see section 5).

In the case of selective orthogonalization, there are two possible approaches in the context of a restarted method. The first one, used in [20], is to exit the Lanczos loop as soon as an eigenvalue has converged. This provokes the computation of the corresponding Ritz vector which will be used for deflation in subsequent restarts. The other approach is to run the Lanczos process completely up to the maximum subspace dimension, managing the loss of orthogonality with the selective orthogonalization technique. The latter is the approach implemented in SLEPc since, to our experience, it is faster in terms of overall convergence.

With respect to periodic and partial reorthogonalization, in both cases we use Simon's recurrence for monitoring loss of orthogonality [13]. The difference with a non-restarted method is that in our implementation in all Lanczos steps the current vector is orthogonalized explicitly with respect to locked vectors, that is, the vectors used for deflation are not considered in the recurrence for monitoring loss of orthogonality. A better approach would be to do this explicit deflation only when necessary, as in [21]. This issue is proposed as future work.

All these orthogonalization strategies employ CGS with selective refinement and estimated norm, as described in section 2. Additionally, we allow a second refinement that may slightly improve numerical robustness in some cases and makes it possible to detect linear dependence of the Lanczos vectors.

4 Numerical Results

In this section, we consider an empirical test with a battery of real-problem matrices using the implementation referred to in section 3 with standard double precision arithmetic. The analysis consists in measuring the level of orthogonality and the residual norm when computing the 10 largest eigenvalues of every symmetric matrix from the Harwell-Boeing collection [22]. These 67 matrices come from a variety of real problems and some of them are particularly challenging for eigenvalue computation. For this test, the solvers are configured with tolerance equal to 10^{-7} and a maximum of 50 basis vectors.

The level of orthogonality is defined as the maximum value of $\|I - V_m^* V_m\|_F$ at each restart, and the residual norm is computed as the maximum of $\|Ax - \lambda x\|_2 / \|\lambda x\|_2$ for every converged eigenvalue.

The results of these tests are shown in Figures 1 and 2, where each dot corresponds to one matrix from the collection. As expected, the algorithm with full orthogonalization maintains the orthogonality level close to full machine precision and the algorithm with local orthogonalization does not guarantee the orthogonality among Lanczos vectors. The semiorthogonal methods (selective, periodic and partial) have a good level of orthogonality, in all cases between full and half machine precision. Another remarkable conclusion that can be drawn from these results is that the Gram-Schmidt procedure with iterative refinement and estimated norm described in subsection 2.4 is a well-suited orthogonalization scheme for these algorithms.

5 Performance Analysis

In order to compare the parallel efficiency of the proposed Lanczos variants, several test cases were analyzed in a cluster platform. This machine consists of 55 biprocessor nodes with Pentium Xeon processors at 2.8 GHz interconnected with an SCI network in a 2-D torus configuration. Only one processor per node was used in the tests reported in this section. The solver was requested to compute 10 eigenvalues with tolerance equal to 10^{-7} using a maximum of 50 basis vectors.

Two types of tests were considered. On one hand, matrices arising from real applications were used for measuring the parallel speed-up. These matrices are listed in Table 1 and are taken from the University of Florida Sparse Matrix

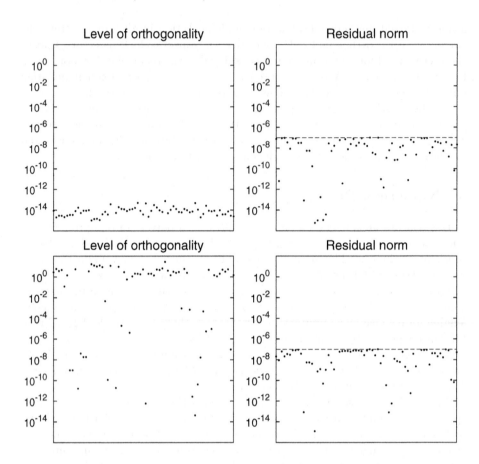

Fig. 1. Level of orthogonality and residual norm for Lanczos with full (top) and local (bottom) orthogonalization. In the horizontal axis, each point represents a test case.

Collection [23]. This speed-up is calculated as the ratio of elapsed time with p processors to the elapsed time with one processor corresponding to the fastest algorithm. This latter time always corresponds to the local orthogonalization variant (including the post-process) as Table 1 shows. On the other hand, a synthetic test case was used for analyzing the scalability of the algorithms, measuring the scaled speed-up (with variable problem size) and Mflop/s per processor. For this analysis, a tridiagonal matrix was used, with a dimension of $10,000 \times p$, where p is the number of processors.

As Figure 3 illustrates, all algorithms show overall good parallel performance. However, the selective reorthogonalization algorithm has poorer performance with a large number of processor than the rest. This is due to the extra synchronizations needed to perform the deflation against converged Ritz vectors in each iteration, which cannot be combined with the other orthogonalizations. Also, because of the deflation against locked vectors needed by the explicit restart

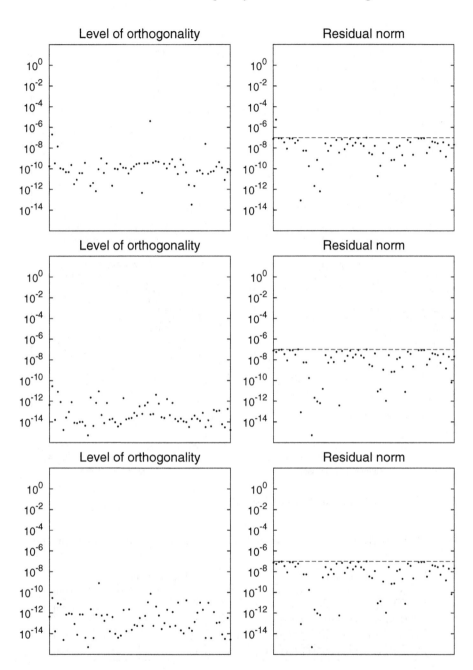

Fig. 2. Level of orthogonality and residual norm for the Lanczos algorithm with selective (top), periodic (center) and partial (bottom) reorthogonalization. In the horizontal axis, each point represents a test case.

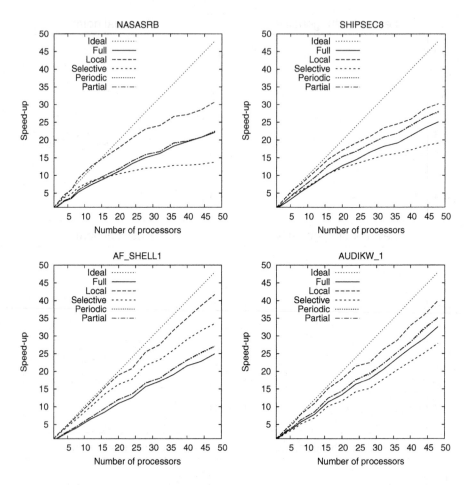

Fig. 3. Measured speed-up for test matrices

Table 1. Properties of test matrices and elapsed time in seconds with one processor

Test matrix			Elapsed time				
Name	Order	Non-zeros	Full	Local	Selective	Periodic	Partial
NASASRB	54,870	2,677,324	29.19	17.49	18.64	27.52	26.85
SHIPSEC8	114,919	3,303,553	6.79	4.28	5.49	5.63	5.58
AF_SHELL1	504,855	17,562,051	126.89	76.03	82.08	117.18	116.97
AUDIKW_1	943,695	77,651,847	55.87	43.90	61.29	51.21	51.03

scheme, the partial and periodic algorithms have no practical advantage over the full orthogonalization variant. The local orthogonalization scheme has the best performance in these tests in spite of having a post-processing phase and an additional reorthogonalization for the restart vector.

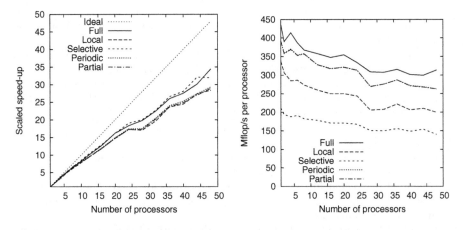

Fig. 4. Measured scaled speed-up and Mflop/s for synthetic matrix

The results in Figure 4 show overall good speed-up in all alternatives. The selective reorthogonalization algorithm has the lowest Mflops/s rate due to the extra synchronizations needed. In the rest of algorithms, the Mflops/s rate improves as the average number of vectors involved in the orthogonalization grows.

6 Conclusions

In this work, an explicit restarting scheme has been applied to different Lanczos variants. The orthogonalization of vectors in these algorithms is done with an optimized version of CGS with selective refinement. All the implemented algorithms are numerically robust for the considered test cases.

The performance results presented in section 5 show that the algorithms achieve good parallel efficiency in all the test cases analyzed, and scale well when increasing the number of processors. The best algorithm will depend on the application, so testing different alternatives is often useful. Thanks to the object-oriented structure of SLEPc, the user can try different Lanczos variants without even having to recompile the application.

References

1. Lanczos, C.: An iteration method for the solution of the eigenvalue problem of linear differential and integral operators. J. Res. Nat. Bur. Standards **45** (1950) 255–282
2. Paige, C.C.: Error analysis of the Lanczos algorithm for tridiagonalizing a symmetric matrix. J. Inst. Math. Appl. **18**(3) (1976) 341–349
3. Sorensen, D.C.: Implicit application of polynomial filters in a k-step Arnoldi method. SIAM J. Matrix Anal. Appl. **13** (1992) 357–385
4. Hernandez, V., Roman, J.E., Vidal, V.: SLEPc: A scalable and flexible toolkit for the solution of eigenvalue problems. ACM Trans. Math. Software **31**(3) (2005) 351–362

5. Grimes, R.G., Lewis, J.G., Simon, H.D.: A shifted block Lanczos algorithm for solving sparse symmetric generalized eigenproblems. SIAM J. Matrix Anal. Appl. **15**(1) (1994) 228–272
6. Parlett, B.N.: The Symmetric Eigenvalue Problem. Prentice-Hall, Englewood Cliffs, NJ (1980) Reissued with revisions by SIAM, Philadelphia, 1998.
7. Nour-Omid, B.: The Lanczos algorithm for solution of large generalized eigenproblem. In Hughes, T.J.R., ed.: The Finite Element Method. Prentice-Hall, Englewood Cliffs, USA (1987) 582–630
8. Paige, C.C.: Computational variants of the Lanczos method for the eigenproblem. J. Inst. Math. Appl. **10** (1972) 373–381
9. Paige, C.C.: Accuracy and effectiveness of the Lanczos algorithm for the symmetric eigenproblem. Linear Algebra Appl. **34** (1980) 235–258
10. Cullum, J.K., Willoughby, R.A.: Lanczos Algorithms for Large Symmetric Eigenvalue Computations. Vol. 1: Theory. Birkhaüser, Boston, MA (1985) Reissued by SIAM, Philadelphia, 2002.
11. Parlett, B.N., Scott, D.S.: The Lanczos algorithm with selective orthogonalization. Math. Comp. **33** (1979) 217–238
12. Grcar, J.F.: Analyses of the Lanczos algorithm and of the approximation problem in Richardson's method. Technical Report 1074, Department of Computer Science, University of Illinois at Urbana-Champaign, Urbana, Illinois (1981)
13. Simon, H.D.: The Lanczos algorithm with partial reorthogonalization. Math. Comp. **42**(165) (1984) 115–142
14. Simon, H.D.: Analysis of the symmetric Lanczos algorithm with reorthogonalization methods. Linear Algebra Appl. **61** (1984) 101–132
15. Hoffmann, W.: Iterative algorithms for Gram-Schmidt orthogonalization. Computing **41**(4) (1989) 335–348
16. Björck, Å.: Numerical Methods for Least Squares Problems. Society for Industrial and Applied Mathematics, Philadelphia (1996)
17. Kim, S.K., Chronopoulos, A.T.: A class of Lanczos-like algorithms implemented on parallel computers. Parallel Comput. **17**(6–7) (1991) 763–778
18. Hernandez, V., Roman, J.E., Tomas, A.: Parallel Arnoldi eigensolvers with enhanced scalability via global communications rearrangement. submitted (2006)
19. Balay, S., Buschelman, K., Gropp, W., Kaushik, D., Knepley, M., McInnes, L.C., Smith, B., Zhang, H.: PETSc users manual. Technical Report ANL-95/11 - Revision 2.3.1, Argonne National Laboratory (2006)
20. Szularz, M., Weston, J., Clint, M.: Explicitly restarted Lanczos algorithms in an MPP environment. Parallel Comput. **25**(5) (1999) 613–631
21. Cooper, A., Szularz, M., Weston, J.: External selective orthogonalization for the Lanczos algorithm in distributed memory environments. Parallel Comput. **27**(7) (2001) 913–923
22. Duff, I.S., Grimes, R.G., Lewis, J.G.: Sparse matrix test problems. ACM Trans. Math. Software **15**(1) (1989) 1–14
23. Davis, T.: University of Florida Sparse Matrix Collection. NA Digest (1992) Available at http://www.cise.ufl.edu/research/sparse/matrices.

PyACTS: A High-Level Framework for Fast Development of High Performance Applications

L.A. Drummond[1], V. Galiano[2], O. Marques[1], V. Migallón[3], and J. Penadés[3]

[1] Lawrence Berkeley National Laboratory
One Cyclotron Road, MS 50F-1650
Berkeley, California 94720, USA
LADrummond@lbl.gov, OAMarques@lbl.gov
[2] Departamento de Física y Arquitectura de Computadores
Universidad Miguel Hernández
03202 Elche, Alicante, Spain
vgaliano@umh.es
[3] Departamento de Ciencia de la Computación e Inteligencia Artificial
Universidad de Alicante, 03071 Alicante, Spain
violeta@dccia.ua.es, jpenades@dccia.ua.es

Abstract. Software reusability has proven to be an effective practice to speed-up the development of complex high-performance scientific and engineering applications. We promote the reuse of high quality software and general purpose libraries through the Advance CompuTational Software (ACTS) Collection. ACTS tools have continued to provide solutions to many of today's computational problems. In addition, ACTS tools have been successfully ported to a variety of computer platforms; therefore tremendously facilitating the porting of applications that rely on ACTS functionalities. In this contribution we discuss a high-level user interface that provides a faster code prototype and user familiarization with ACTS tools. The high-level user interfaces have been built using Python. Here we focus on Python based interfaces to ScaLAPACK, the PyScaLAPACK component of PyACTS. We briefly introduce their use, functionalities, and benefits. We illustrate a few simple example of their use, as well as exemplar utilization inside large scientific applications. We also comment on existing Python interfaces to other ACTS tools. We present some comparative performance results of PyACTS based versus direct LAPACK and ScaLAPACK code implementations.

1 Introduction

The development of high performance engineering and scientific applications is an expensive process that often requires specialized support and adequate information about the available computational resources and software development tools. The development effort is increased by the complexity of the phenomena that can be addressed by numerical simulation, along with the increase and evolution of computing resources. We promote high-quality and general purpose software tools that provide a plethora of computational services to the growing computational sciences community.

M. Daydé et al. (Eds.): VECPAR 2006, LNCS 4395, pp. 417–425, 2007.
© Springer-Verlag Berlin Heidelberg 2007

The Advanced CompuTational Software (ACTS) Collection [1, 2, 3] is a set of computational tools developed primarily at DOE laboratories and is aimed at simplifying the solution of common and important computational problems. The use of the tools reduces the development time for new codes and the tools provide functionality that might not otherwise be available. All this potential cannot be achieved, however, if the tools are not used effectively or not used at all. For this reason, we look at creating didactic frameworks to help scientists and engineers deploy the ACTS functionality. Thus, our intent with PyACTS is not to substitute tool interfaces but rather provide a self-learning mechanisms for tool users to familiarize themselves with ACTS tools, their interfaces and functionality.

In this article, we will focus on PyScaLAPACK by introducing its use in simple ScaLAPACK calls, and also in large scientific applications. These examples are followed by some performance results. We later reference other Python implementations that interface tools in the ACTS Collection.

2 Why Python?

Python [4] is an interpreted, interactive, object-oriented programming language. Python combines remarkable power with very clear syntax. It has modules, classes, exceptions, very high level dynamic data types, and dynamic typing. There are interfaces to many system calls and libraries, as well as to various windowing managing systems. New built-in modules are easily written in C or C++. Early performance numbers on some components of PyACTS [5] have demonstrated a low overhead induced by the use of the Python-based high-level interface. However, there is a substantial gain in the simplification of tool interfaces because call to the PyACTS interfaces contain only high-level objects that are familiar to the user, for instance a matrix A associated with a linear system $Ax = b$, rather than the matrix A and all the computational parameters associated with the performance of the algorithm. In fact, the PyACTS interfaces generate all the other extra information necessary to actually call the ACTS tool and exploit the tools high-performance capabilities. This extra information includes data pertaining to the parallel environment, tool optimization, specific data distributions and storage techniques, etc.

Notice that the use of Python allows not only for a user friendlier environment but also to easily implement interoperable interfaces between these tools, and easily maintain different versions of the tools as the tools continue to evolve independently. In summary, PyACTS aims at easing the learning curve, hide performance and tuning parameters from beginner users while supporting interoperable interfaces as individual tools continue to evolve.

3 Software Tools

ACTS tools tackle a number of common computational issues found in many applications, mainly implementation of numerical algorithms, and support for code development, execution and optimization. The ever increasing number of users of Python has motivated tool developers to include Python interfaces. In

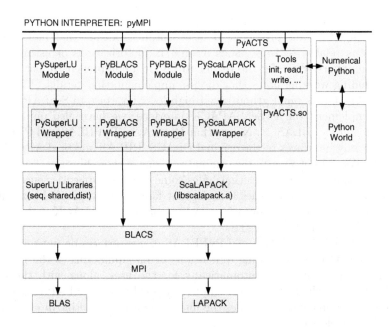

Fig. 1. Main components of PyACTS. The flexible infrastructure allows for easily addition of new modules or versions of the different ACTS Tools.

this article we focus on the PyACTS interface to ScaLAPACK [6]. Figure 1 illustrates the overall structure of the PyACTS framework which includes interfaces to other tools in the Collection (the reader is referred to *http://acts.nersc.gov* for a full list of tools in the ACTS Collection).

In its current implementation, PyACTS uses **Numeric** and **RandomArray** from Numpy [7] to implement and handle array objects. In addition, we use pyMPI [8] to implement and handle the parallelism.

Another relevant aspect of PyACTS is that it facilitates high-level interoperable interfaces between the different tools in the ACTS Collection since objects (e.g., a given matrix in a particular storage format) from one library can be internally converted to objects that are used by another library. Some developers of ACTS tools have also implemented their own Python interfaces, and in the future PyACTS will interface with them. Instances of such implementations include the Python interface to PETSc [9], PyTrilinos [10], a Python based interface to selected packages in the Trilinos framework, and a Python interface to ODE solvers in SUNDIALS [11].

3.1   Introduction to PyScaLAPACK

ScaLAPACK is a library of high-performance linear algebra routines for distributed memory message-passing computers. It complements the LAPACK library [12], which provides analogous software for workstations, vector supercomputers, and shared-memory parallel computers. ScaLAPACK contains routines for solving systems of linear equations, least squares, eigenvalue problems and singular value problems. It also contains routines that handle many computations related to those, such as matrix factorizations or estimation of condition numbers. We refer the reader to [6] for a comprehensive list of references, including working notes that discuss implementation details.

PyScaLAPACK is our Python based high-level interface to ScaLAPACK. In order to implement the PyScaLAPACK interface, we have also implemented PyBLACS and PyPBLAS [5]. Notice that PyBLACS, PyPBLAS and PyScaLA-PACK are only interfaces to the original BLACS, PBLAS and ScaLAPACK, respectively. We did not rewrite the original versions of these libraries, but instead aggregated high-level interfaces that hides some of the complexities encountered by users of the original libraries that are not familiar with parallel computing, linear algebra or matrix computations. Additionally, ScaLAPACK users can call other ACTS Tools using PyACTS.

PyScaLAPACK, PyBLACS and PyPBLAS user interfaces do not directly include arguments like the leading dimensions or manipulations to the processor grids. These are generated automatically for the user, along with the corresponding block-cyclic distributions and then passed to the ScaLAPACK library. Therefore, it significantly simplifies the interface for the scientific or engineering application developer.

4   Examples of PyScaLAPACK Utilization

In this section we will look at how to use the PyScaLAPACK interface through a set of simple calls to ScaLAPACK (we assume the reader is familiar with the ScaLAPACK library or refer to [6] for more information).

We begin with our simple example in ScaLAPACK. To show the performance, we present in Figure 2 the results for the routines PSGESV and PDGESVD. We have used the routine PSGESV to compute the solution of a simple precision system of linear equations $Ax = b$, where $A \in \mathbb{R}^{n \times n}$ and $x, b \in \mathbb{R}^n$. The routine PDGESVD has been used to compute the singular value decomposition (SVD) of a square double precision matrix. ScaLAPACK users need to define the different variables and descriptors that are associated with the parallel data layout and environment used by ScaLAPACK. Then, there is a sequence of calls to BLACS and ScaLAPACK to initialize the environment. In PyScaLAPACK this is simplified by the use of

```
PyACTS.gridinit()            # Initializes the process grid.
ACTS_LIB = 1                 # 1 Identifies ScaLAPACK in PyACTS,
                             # 2 is SuperLU, and so on ...
A = num2PyACTS(A, ACTS_lib)  # Converts a Numeric Array into
                             # PyScaLAPACK array; A was previously
                             # defined as 2D NumArray.
```

The call to PyACTS.gridinit resolves automatically to the corresponding ScaLA-PACK and BLACS routines. Parameters are taken from the input data (e.g., number of processor, and command line argument) entered by the user. The invocation to **num2PyACTS** resolves in the creation of the descriptors associated with A, and they handled internally by PyACTS, and this includes all the data distribution. Thus the actual call to PDGESVD using ScaLAPACK and PyScaLAPACK are as follow:

```
CALL PDGESV(N,NRHS,A,IA,JA,DESCA,IPIV,B,IB,JB,DESCB,INFO)
```

and

```
x,info = PyScaLAPACK.pvgesv(A,B)
```

Figures 2(a) and 2(b) present an example of performance results obtained in a Linux Cluster consisting of Pentium IV processors and connected through a 1 Gigabit network switch. In both graphs, we compared the straight Fortran 77 version of ScaLAPACK vs PyScaLAPACK. As it can be seen in the graph, the overhead induced by the Python infrastructure is rather nominal. Thus, PyScaLAPACK does not hinder the performance deliverance of ScaLAPACK.

We introduce a few examples of real scientific application codes that can be easily prototyped or extend its functionality with the use of PyACTS. For each application we present a summary of the highlights of the PyACTS implementation and performance results.

4.1 PyClimate: A Set Climate Analysis Tools

PyClimate [13] is a Python based package that provides support to common tasks during the analysis of climate variability data. It provides functions that range from simple IO operations and operations with COARDS-compliant netCDF files to Empirical Orthogonal Function (EOF) analysis, Canonical Correlation

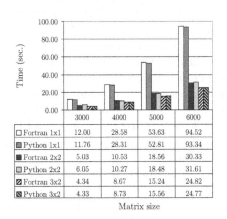

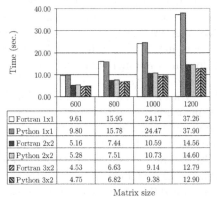

(a) psgesv   (b) pdgesvd

Fig. 2. Performance of PyScaLAPACK vs ScaLAPACK for the ScaLAPACK routines PSGESV and PDGESVD. (a) uses REAL arithmetic and (b) uses DOUBLE PRECISION.

Analysis (CCA) and Singular Value Decomposition (SVD) analysis of coupled data sets, some linear digital filters, kernel based probability-density function estimation and access to DCDFLIB.C library from Python. PyClimate uses functionality available in LAPACK.

There has been a growing need for PyClimate to scale-up its functionality to support parallel and scalable algorithms. Rather than implementing these new functions from scratch, we collaborate with the PyClimate team to provide PyScaLAPACK. In this case, PyScaLAPACK integrates well with all the PyClimate development and application environment. Here we present an example concerning a meteorological study by means of the EOF and SVD analysis.

The EOF analysis is widely used to decompose a long-term time series of spatially observed data set into orthogonal spatial and temporal modes. EOF can be calculated in a single step using singular value decomposition (SVD) without constructing either version of the covariance matrix as shown by Kelly in [14]. Concretely, EOFs can be computed, via SVD, after removing the spatial (i.e., column) mean from the data matrix at each time step. In this case, the EOFs decompose the variability of the spatial property gradients rather than variability of the property itself. These spatial variance EOFs are useful when the purpose is to investigate the variance associated with features that do not vary strongly over time. In practical terms, EOFs are a means of reducing the size of a data set while retaining a large fraction of the variability present in the original data.

PyClimate implements these features in the routine "svdeofs". An example of use of this routine can be found in *www.pyclimate.org* (script: example 1). This script performs the SVD decomposition after removing the column mean from the data matrix at each time step. Some other computations are accomplished after the SVD decomposition. We have parallelized the aforementioned script by using both PyScaLAPACK (for the SVD) and PyPBLAS (for the extra computation). We would like to emphasize that the parallel PyScaLAPACK script resembles the coding structure of the serial PyClimate one. Moreover, the PyScaLAPACK version is semantically the parallel implementation of the serial version, just like the relationships between ScaLAPACK and LAPACK. All this is done in a manner that is almost-transparent to the user.

The data sets used in our experiments correspond to air temperature in a 2.5° latitude × 2.5° longitude global grid with 144 × 73 points. The first data set measures the air temperature over 365 days (referred as air.day), and the second one contains measures over the mean of 694 months (referred as air.mon), obtained both from the Climate Diagnostics Center.

In Figure 3 we show the computational times obtained for the sequential version (using PyClimate) and for the parallel version (using our proposed interfaces) for different number of processors. Figure 3(a) corresponds to the air.day data set, and Figure 3(b) corresponds to the air.month. These results have been obtained on a cluster of 28 nodes with two Intel Xeon processors (2.4 GHz, 1 GB DDR RAM, 512 KB L2 cache) per node connected via a Myrinet network (2.0 Gigabit/s). As it can be appreciated, we obtain a substantial reduction of time when our proposed parallel interfaces are used.

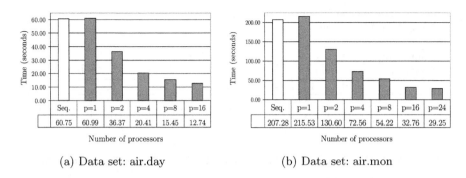

(a) Data set: air.day   (b) Data set: air.mon

Fig. 3. EOF analysis using PyClimate and its parallel version

4.2 Large Inverse Problems in Geo-Physics

In this application we are interested in using singular values and singular vectors in the solution of large inverse problems that arise in the study of physical models for the internal structure of the Earth [15, 16]. The Earth is discretized into layers and the layers into cells, and travel times of sound waves generated by earthquakes are used to construct the corresponding physical models. Basically, we deal with an idealized linear equation relating arrival time deviations to perturbations in Earth's structure. The underlying discretization lead to very large sparse matrices whose singular values and singular vectors are then computed and used in the solution of the associated inverse problems. They are also used to estimate uncertainties. In one phase of these calculations we need to solve a (block) symmetric tridiagonal eigenvalue problem that arises in the context of a (block) Lanczos-based algorithm. This is done in a post processing phase using ScaLAPACK, which requires the block-cyclic distribution of the tridiagonal matrix and the corresponding eigenvectors.

First, Table 1 presents some results of interactive runs comparing the PyScaLAPACK version of the code against the original Fortran implementation that uses the original ScaLAPACK. The runs were performed in an IBM SP Power 3, 350 MHz per processors, and each node has 16 processors. In the tests shown in Table 1, we noticed the slight influence of the Python interpreter in the timings. In this example we have called the ScaLAPACK subroutine PSSYEV, which computes the eigenvalues and corresponding eigenvectors of a symmetric matrix A. In Table 1, we show the results for three different sizes of A, 1000, 5000 and 7500. The overhead introduced by Python is currently under study and we will try to use a different MPI-Python implementation on the IBM system. Nevertheless, if we take a look at the original calls to ScaLAPACK from the Fortran code versus the PyScaLAPACK, we observe a significant simplification of the user interface. As in the previous examples, there is already a simplification at the level of declarations of variables, data distribution and initialization of the parallel environment. The Fortran call to PSSYEV looks like this:

Table 1. Performance Results Earth Science applications using ScaLAPACK (left) and PyScaLAPACK (right)

Number of	Matrix Size					
Processors	1000		5000		7500	
4	9.00	10.22	671.79	816.67	-	-
8	6.72	9.05	339.39	428.72	-	-
16	6.41	8.19	188.37	195.10	713.05	850.12

```
CALL PSSYEV('V','L',N,A,1,1,DESCA,S,X,1,1,DESCY,WORK,LWORK,INFO)
```

and the PyScaLAPACK version:

```
s,x,info= PyScaLAPACK.pvsyev(a_ACTS,jobz='V')
```

Further, there are two calls to PSSYEV in the original Fortran code. The first one precomputes the size of the work array. In the PyScaLAPACK all these details are hidden from the user and performed internally by PyScaLAPACK.

Comparing the PyClimate and the Earth Sciences application we notice that in the case of PyClimate, not only we obtain a simplified and friendlier interface but also a parallel version of the code. In the Earth Sciences case, the Fortran code already calls ScaLAPACK and as shown in Table 1 the code shows some speed ups even for the small problem sizes. The benefit in the latter application is seen at the level of the interface.

5   Conclusions

PyACTS aims at easing the learning curve, hide performance and tuning parameters from beginner users while supporting interoperable interfaces as individual tools continue to evolve. In addition, PyACTS reduces the time users spend prototyping and deploying high-end software tools like the ones in the ACTS Collection. The results shown in the previous graphs and examples show not only the many advantages of using the simplified interfaces but also that there are not major performance degradations by using the PyScaLAPACK interface.

One item of our future work will consist in replacing the MPI interface with a more scalable version of MPI for Python. Furthermore, PyACTS will guide its users via a scriber that produces Fortran or C language code from the PyACTS high-level commands. Then, the user will can use the PyACTS generated Fortran or C language code pieces for generating production codes in either language.

The Python interfaces PyACTS, as well as some examples, a user guide reference and the PyACTS prerequisites are available at *www.pyacts.org*.

Acknowledgements

This research was partially supported by the Spanish Ministry of Science and Education under grant number TIN2005-093070-C02-02, and by Universidad de Alicante under grant number VIGROB-020.

References

[1] Drummond, L.A., Marques, O.A.: An overview of the Advanced CompuTational Software (ACTS) Collection. ACM Transactions on Mathematical Software **31** (2005) 282–301

[2] Boisvert, R.F., Drummond, L.A., Marques, O.A.: Introduction to the special issue on the Advanced CompuTational Software (ACTS) Collection. ACM Transactions on Mathematical Software **31** (2005) 281

[3] Drummond, L.A., Marques, O.: The Advanced Computational Testing and Simulation Toolkit (ACTS): *What can ACTS do for you?* Technical Report LBNL-50414, Lawrence Berkeley National Laboratory (2002)

[4] G. van Rossum, F.D.J.: An Introduction to Python. Network Theory Ltd (2003)

[5] Drummond, L.A., Galiano, V., Migallón, V., Penadés, J.: Improving ease of use in BLACS and PBLAS with Python. In: Joubert G.R., Nagel W.E., Peters F.J., Plata O., Tirado P., Zapata E., editors. Parallel Computing: Current & Future Issues of High-End Computing (Proceedings of the International Conference ParCo 2005), NIC Series Volume 33 ISBN 3-00-017352-8.

[6] Blackford, L.S., Choi, J., Cleary, A., D'Azevedo, E., Demmel, J.W., Dhillon, I., Dongarra, J.J., Hammarling, S., Henry, G., Petitet, A., Stanley, K., Walker, D., Whaley, R.C.: ScaLAPACK User's Guide. SIAM, Philadelphia, Pennsylvania (1997)

[7] Ascher, D., Dubois, P.F., Hinsen, K., Hugunin, J., Oliphant, T.: An Open Source Project: Numerical Python. Technical Report http://numeric.scipy.org/numpydoc/numpy.html, Lawrence Livermore National Laboratory (2001)

[8] Miller, P.: An Open Source Project: Numerical Python. Technical Report UCRL-WEB-150152, Lawrence Livermore National Laboratory (2002)

[9] Balay, S., Buschelman, K., Gropp, W.D., Kaushik, D., Knepley, M., McInnes, L.C., Smith, B.F., Zhang, H.: PETSc users manual. Technical Report ANL-95/11 - Revision 2.1.5, Argonne National Laboratory (2002)

[10] Sala, M.: Distributed Sparse Linear Algebra with PyTrilinos. Technical Report SAND2005-3835, Sandia National Laboratories (2005)

[11] Gates, M., Lee, S., Miller, P.: User-friendly Python Interface to ODE Solvers. Technical Report www.ews.uiuc.edu/~mrgates2/python-ode-small.pdf, University of Illinois, Urbana-Champaign (2005)

[12] Anderson, E., Bai, Z., Bischof, C., Blackford, S., Demmel, J.W., Dongarra, J.J., Croz, J.D., Greenbaum, A., Hammarling, S., McKenney, A., Sorensen, D.C.: LAPACK User's Guide. third edn. SIAM, Philadelphia, Pennsylvania (1999)

[13] Saenz, J., Zubillaga, J., Fernandez, J.: Geophysical data analysis using Python. Computers and Geosciences **28/4** (2002) 457–465

[14] Kelly, K.: Comment on "Empirical orthogonal function analysis of advanced very high resolution radiometer surface temperature patterns in Santa Barbara Channel" by G.S.E. Lagerloef and R.L. Bernstein. Journal of Geohysical Research **93** (1988) 15753–15754

[15] Vasco, D.W., Johnson, L.R., Marques, O.: Global Earth Structure: Inference and Assessment. Geophysical Journal International **137** (1999) 381–407

[16] Marques, O., Drummond, L.A., Vasco, D.W.: A Computational Strategy for the Solution of Large Linear Inverse Problems in Geophysics. In: International Parallel and Distributed Processing Symposium (IPDPS), Nice, France (2003)

Sequential and Parallel Resolution of the Two-Group Transient Neutron Diffusion Equation Using Second-Degree Iterative Methods

Omar Flores-Sánchez[1,2], Vicente E. Vidal[1], Victor M. García[1], and Pedro Flores-Sánchez[3]

[1] Departamento de Sistemas Informáticos y Computación
Universidad Politécnica de Valencia
Camino de Vera s/n, 46022 Valencia, España
{oflores, vvidal, vmgarcia}@dsic.upv.es
[2] Departamento de Sistemas y Computación
Instituto Tecnológico de Tuxtepec
Av. Dr. Victor Bravo Ahuja, Col. 5 de Mayo, C.P. 68300, Tuxtepec, Oaxaca, México
oflores70@hotmail.com
[3] Telebachillerato "El Recreo"
Tierra Blanca, Veracruz, México
pedrofs080374@hotmail.com

Abstract. We present an experimental study of two versions of a second-degree iterative method applied to the resolution of the sparse linear systems related to the 3D multi-group time-dependent Neutron Diffusion Equation (TNDE), which is important for studies of stability and security of nuclear reactors. In addition, the second-degree iterative methods have been combined with an adaptable technique, in order to improve their convergence and accuracy. The authors consider that second-degree iterative methods can be applied and extended to the study of transient analysis with more than two energy groups and they might represent a saving in spatial cost for nuclear core simulations. These methods have been coded in PETSc [1][2][3].

1 Introduction

For design and safety reasons, nuclear power plants need fast and accurate plant simulators. The centre point of concern in the simulation of a nuclear power plant is the reactor core. Since it is the source of the energy that is produced in the reactor, a very accurate model of the constituent processes is needed. The neutron population into the reactor core is modeled using the Boltzmann transport equation. This three-dimensional problem is modeled as a system of coupled partial differential equations, the multigroup neutron diffusion equations[4][5], that have been discretised using a nodal collocation method in space and one-step Backward-Difference Method in time. The solution of these equations can involve very intensive computing. Therefore, it is necessary to find effective

M. Daydé et al. (Eds.): VECPAR 2006, LNCS 4395, pp. 426–438, 2007.
© Springer-Verlag Berlin Heidelberg 2007

algorithms for the solution of the three-dimensional model. The progress in the area of multiprocessor technology suggests the application of High-Performance Computing to enable engineers to perform faster and more accurate safety analysis of Nuclear Reactors [6].

Bru et al in [7] apply two Second-Degree methods [8] to solve the linear system of equations related to a 2D Neutron-Diffusion equation case. Thus, the main goal of this paper is the application of those methods and some modifications that we have proposed to decrease the computational work, but applied to a 3D real test case.

The outline of the paper is as follows. The mathematical model of the Time-dependent Neutron Diffusion Equation and its discretisation are described in Section 2. The second-degree iterative methods are introduced at Section 3. Section 4 describes hardware and software platform used. The test case is presented in Section 5. Section 6 presents a sequential study of the second-degree iterative methods and the modifications proposed. In Section 7 numerical parallel results are presented. Finally, we will draw some conclusions in Section 8.

2   Problem Description

Plant simulators mainly consist of two different modules which account for the basic physical phenomena taking place in the plant: a neutronic module which simulates the neutron balance in the reactor core, and the evaporation and condensation processes. In this paper, we will focus on the neutronic module. The balance of neutrons in the reactor core can be approximately modeled by the time-dependent two energy group neutron diffusion equation, which is written using standard matrix notation as follows[9]:

$$[v^{-1}]\dot{\phi} + \mathcal{L}\phi = (1 - \beta)\mathcal{M}\phi + \chi \sum_{k=1}^{K} \lambda_k \mathcal{C}_k \tag{1}$$

$$\dot{\mathcal{C}}_k = \beta_k [\nu \Sigma_{f_1} \nu \Sigma_{f_2}]\phi - \lambda_k \mathcal{C}_k, \qquad k = 1, \dots, K \tag{2}$$

where

$$\mathcal{L} = \begin{bmatrix} -\nabla \cdot (D_1 \nabla) + \Sigma_{a1} + \Sigma_{12} & 0 \\ -\Sigma_{12} & -\nabla \cdot (D_2 \nabla) + \Sigma_{a2} \end{bmatrix}, [v^{-1}] = \begin{bmatrix} \frac{1}{v_1} & 0 \\ 0 & \frac{1}{v_2} \end{bmatrix},$$

and

$$\mathcal{M} = \begin{bmatrix} \nu \Sigma_{f1} & \nu \Sigma_{f2} \\ 0 & 0 \end{bmatrix}, \phi = \begin{bmatrix} \phi_f \\ \phi_t \end{bmatrix}, \chi = \begin{bmatrix} 1 \\ 0 \end{bmatrix},$$

where

- ϕ is the neutron flux on each point of the reactor; so, it is a function of time and position.
- $\mathcal{C}_k$ is the concentration of the k-th neutron precursor on each point of the reactor (it is as well a function of time and position). $\lambda_k \mathcal{C}_k$ is the decay rate of the k-th neutron precursor.

- K is the number of neutron precursors. β_k is the proportion of fission neutrons given by transformation of the k-th neutron precursor; $\beta = \sum_{k=1}^{K} \beta_k$.
- $\mathcal{L}$ models the diffusion $(-\nabla \cdot (D_1 \nabla))$, absorption $(\sum_{a1}, \sum_{a2})$ and transfer from fast group to thermal group $(\sum_{12})$.
- $\mathcal{M}$ models the generation of neutrons by fission.
- $\nu \sum_{fg}$ gives the amount of neutrons obtained by fission in group g.
- v^{-1} gives the time constants of each group.

To study rapid transients of neutronic power and other space and time phenomena related to neutron flux variations, fast codes for solving these equations are needed. The first step to obtain a numerical solution of these equations consists of choosing a spatial discretization for equation (1). For this , the reactor is divided in cells or nodes and a nodal collocation method is applied[10][11]. In this collocation method, neutron flux is expressed as a series of Legendre Polynomials.

After a relatively standard process (setting boundary conditions, making use of the orthonormality conditions, using continuity conditions between cells) we obtain the following systems of ordinary differential equations:

$$[v^{-1}]\dot{\psi} + L\psi = (1 - \beta)M\psi + X \sum_{k=1}^{K} \lambda_k C_k, \tag{3}$$

$$\dot{C}_k = \beta_k [M_{11} M_{12}]\psi - \lambda_k C_k, \qquad k = 1, \ldots, K, \tag{4}$$

where unknowns ψ and C_k are vectors whose components are the Legendre coefficients of ϕ and $\mathcal{C}_k$ in each cell, and L, M, $[v^{-1}]$ are matrices with the following block structure:

$$L = \begin{bmatrix} L_{11} & 0 \\ -L_{21} & L_{22} \end{bmatrix}, M = \begin{bmatrix} M_{11} & M_{12} \\ 0 & 0 \end{bmatrix}, v^{-1} = \begin{bmatrix} v^{-1} & 0 \\ 0 & v^{-1} \end{bmatrix}, X = \begin{bmatrix} I \\ 0 \end{bmatrix}.$$

Depending on flux continuity conditions imposed among the discretisation cells of the nuclear reactor, the blocks L_{11} and L_{22} can be symmetric or not. For our test case, these blocks are symmetric positive definite matrices[12], while blocks L_{21}, M_{11} and M_{12} are diagonal.

The next step consists of integrating the above ordinary differential equations over a series of time interval, $[t_n, t_{n+1}]$. Equation (4) is integrated under the assumption that the term $[M_{11}M_{12}]\psi$ varies linearly from t_n to t_{n+1}, obtaining the solution C_k at t_{n+1} expressed as

$$C_k^{n+1} = C_k^n e^{\lambda_k h} + \beta_k (a_k [M_{11}M_{12}]^n \psi^n + b_k [M_{11}M_{12}]^{n+1} \psi^{n+1} \tag{5}$$

where $h = t_{n+1} - t_n$ is a fixed time step size, and the coefficients a_k and b_k are given by

$$a_k = \frac{(1 + \lambda_k h)(1 - e^{\lambda_k h})}{\lambda_k^2 h} - \frac{1}{\lambda_k}, b_k = \frac{\lambda_k h - 1 + e^{\lambda_k h}}{\lambda_k^2 h}.$$

To integrate (3), we must take into account that it constitutes a system of stiff differential equations, mainly due to the elements of the diagonal matrix $[v^{-1}]$. Hence, for its integration, it is convenient to use an implicit backward difference formula (BDF). A stable one-step BDF to integrate (3) is given by

$$\frac{[v^{-1}]}{h}(\psi^{n+1} - \psi^n) + L^{n+1}\psi^{n+1} = (1 - \beta)M^{n+1}\psi^{n+1} + X\sum_{k=1}^{K}\lambda_k C_k^{n+1} \quad (6)$$

Taking into account equation (5) and the structure of matrices L and M, we rewrite (6) as the system of linear equations

$$\begin{bmatrix} T_{11} & T_{12} \\ T_{21} & T_{22} \end{bmatrix} \begin{bmatrix} \psi_1^{n+1} \\ \psi_2^{n+1} \end{bmatrix} = \begin{bmatrix} R_{11} & R_{12} \\ 0 & R_{22} \end{bmatrix} \begin{bmatrix} \psi_1^n \\ \psi_2^n \end{bmatrix} + \sum_{k=1}^{K}\lambda_k e^{-\lambda_k h} \begin{bmatrix} C_k^n \\ 0 \end{bmatrix}, \quad (7)$$

where

$$T_{11} = \frac{1}{h}v_1^{-1} + L_{11}^{n+1} - (1 - \beta)M_{11}^{n+1} - \sum_{k=1}^{K}\lambda_k\beta_k b_k M_{11}^{n+1},$$

$$T_{21} = -L_{21}^{n+1},$$

$$T_{12} = -(1 - \beta)M_{12}^{n+1} - \sum_{k=1}^{K}\lambda_k\beta_k b_k M_{12}^{n+1},$$

$$T_{22} = \frac{1}{h}v_2^{-1} + L_{22}^{n+1},$$

$$R_{11} = \frac{1}{h}v_1^{-1} + \sum_{k=1}^{K}\lambda_k\beta_k a_k M_{11}^n,$$

$$R_{12} = \sum_{k=1}^{K}\lambda_k\beta_k a_k M_{12}^n, \qquad R_{22} = \frac{1}{h}v_2^{-1}.$$

Thus, for each time step it is necessary to solve a large and sparse system of linear equations, with the following block structure:

$$\begin{bmatrix} T_{11} & T_{12} \\ T_{21} & T_{22} \end{bmatrix} \begin{bmatrix} \psi_1 \\ \psi_2 \end{bmatrix} = \begin{bmatrix} e_1 \\ e_2 \end{bmatrix} \quad (8)$$

where the right-hand side depends on both the solution in previous time steps and the backward difference method used. Usually, the coefficients matrix of system (8) has similar properties as the matrices L and M in equation (3), namely blocks T_{11}, T_{22} are symmetric positive definite matrices, and blocks T_{12}, T_{21} are singular diagonal matrices. System (8) will be also denoted as

$$T\psi = e. \quad (9)$$

3   Second-Degree Iterative Methods

We begin this section with a brief introduction to the second-degree methods presented and applied to a 2D neutron diffusion equation case in [7].

3.1   Second Degree Method A

Consider the coefficient matrix T of the linear system (9) and the Jacobi splitting, $T = M - N$, with matrices M and N given by

$$M = \begin{bmatrix} T_{11} & 0 \\ 0 & T_{22} \end{bmatrix}, N = \begin{bmatrix} 0 & -T_{12} \\ -T_{21} & 0 \end{bmatrix}$$

where iteration matrix B_J is represented by

$$B_J = M^{-1}N = \begin{bmatrix} 0 & -T_{11}^{-1}T_{12} \\ -T_{22}^{-1}T_{21} & 0 \end{bmatrix}$$

Now, considering the matrices $G_1 = \omega B_J$, $G_0 = (1 - \omega)B_J$, where ω is an extrapolation factor, and the vector $k = M^{-1}e$, we can write the following second degree method based on the Jacobi Over-relaxation (JOR) splitting

$$\psi^{(n+1)} = G_1\psi^{(n)} + G_0\psi^{(n-1)} + k = B_J(\omega\psi^{(n)} + (1 - \omega)\psi^{(n-1)}) + k,$$

which corresponds to the following operations

$$\begin{aligned} T_{11}\psi_1^{l+1} &= e_1 - T_{12}(\omega\psi_2^l + (1 - \omega)\psi_2^{l-1}), \\ T_{22}\psi_2^{l+1} &= e_2 - T_{21}(\omega\psi_1^l + (1 - \omega)\psi_1^{l-1}). \end{aligned} \tag{10}$$

Let us identify these operations as Method A.

3.2   Second Degree Method B

In the same manner, we can construct another method based on the accelerated Gauss-Seidel splitting, whose iteration matrix B_{GS} is given by

$$B_{GS} = M^{-1}N = \begin{bmatrix} T_{11} & 0 \\ T_{21} & T_{22} \end{bmatrix}^{-1} \begin{bmatrix} 0 & -T_{12} \\ 0 & 0 \end{bmatrix},$$

The operations that correspond to this method are represented by

$$\begin{aligned} T_{11}\psi_1^{l+1} &= e_1 - T_{12}(\omega\psi_2^l + (1 - \omega)\psi_2^{l-1}), \\ T_{22}\psi_2^{l+1} &= e_2 - T_{21}(\omega\psi_1^{l+1} + (1 - \omega)\psi_1^l). \end{aligned} \tag{11}$$

Methods A and B, can be described by the following algorithmic scheme

Second-Degree Iterative Method Algorithm

(1) Set ψ_2^0; $\{\psi_2^0 := \psi_2^*\}$
(2) Solve $T_{11}\psi_1^1 = e_1 - T_{12}\psi_2^0$
(3) Solve $T_{22}\psi_2^1 = e_2 - T_{21}\psi_1^1$
(4) Do $l = 1, 2, \ldots$

(4a) Solve ψ_1^{l+1} in accordance with A or B method.
(4b) Solve ψ_2^{l+1} in accordance with A or B method.
until $\|\psi_1^{l+1} - \psi_1^l\| < tol$ and $\|\psi_2^{l+1} - \psi_2^l\| < tol$

where, ψ_2^* represents the solution of a previous time step.

As we can see, the main difference between methods A and B, is that in method B, the new solution for ψ_1 is used as soon as it is available to compute ψ_2. Therefore, a faster convergence rate may be expected. In both methods, we distinguish between outer and inner iterations. The outer iterations are identified by the Step (4), and inner iterations are represented by Step (4a) and (4b), which correspond to iterations needed for solving the linear systems with matrices T_{11} and T_{22} respectively. Since these blocks are symmetric positive-definite matrices the *Conjugate-Gradient method*[16] was applied. General theorems about the convergence of second-degree methods can be found in [8].

The next section presents the hardware and software tools that we have used to carry out the numerical experiments.

4   Hardware and Software Platform

Sequential and parallel experiments have been performed on a 12-node biprocessor cluster with Red Hat 8.0 operating system, using only one CPU per node at the Polytechnic University of Valencia. Each CPU is a 2 GHz Intel Xeon processor and has 1 GB of RAM memory. All nodes are connected by a SCI network with a Torus 2D topology in a 4x5 mesh.

The Portable, Extensible Toolkit for Scientific Computation (PETSc)[1][2][3], is a suite of data structures and routines that provide the building blocks for the implementation of large-scale application codes on parallel (and serial) computers. PETSc uses the MPI Standard for all message-passing communication. Some of the PETSc modules deal with vectors, matrices (generally sparse), distributed arrays, Krylov subspace methods, preconditioners including multigrid and sparse direct solvers, etc.

Figure 1 illustrates the PETSc library hierarchical organization, which enables users to employ the level of abstraction that is most appropriate for a particular problem.

PETSc uses the message-passing model for parallel programming and employs MPI for all interprocessor communication. In PETSc the user is free to employ MPI routines as needed throught an application code. However, by default the user is shielded from many of the details of message passing within PETSc, since these are hidden within parallel objects, such as vectors, matrices, and solvers. In addition, PETSc provides tools such as generalized vector scatter/gathers and distributed arrays to assist in the management of parallel data.

PETSc provides a variety of matrix implementations because no single matrix format is appropriate for all problems. Currently PETSc supports dense storage and compressed sparse row storage, as well as several specialized formats. There are sequential and parallel AIJ sparse matrix format in PETSc. In the sequential AIJ sparse matrix, the nonzero elements are stored by rows, along with an array

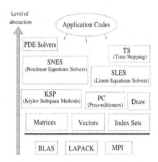

Fig. 1. Organization of PETSc library

of corresponding column numbers and an array of pointers to the beginning of each row. Parallel sparse matrices with the AIJ format can be created with the command

```
MatCreateMPIAIJ(MPI_Comm comm,int m, int n, int M,int N,
                int d_nz,int *d_nnz,int o_nz,int *o_nnz, Mat *A);
```

A is the newly created matrix, while the arguments m, M and N, indicate the number of local rows and the number of global rows and columns, respectively. In the PETSc partitioning scheme, all the matrix columns are local and n is the number of columns corresponding to local part of a parallel vector. Either the local o global parameters can be replaced with PETSC_DECIDE, so that PETSc will determine them. The matrix is stored with a number of rows on each process, given by m, or determined by PETSc if m is PETSC_DECIDE. If PETSC_DECIDE is not used for the arguments m and n, then the user must ensure that they are chosen to be compatible with the vectors. To do this, one first considers the matrix-vector product $y = Ax$. The m that is used in the matrix creation routine MatCreateMPIAIJ() must match the local size used in the vector routine VecCreateMPI() for y. Likewise, the n used must match that used as the local size in VecCreateMPI() for x. For example, the PETSc partitioning scheme using the parallel sparse matrix AIJ format for operation Ax, must be as follows

$$
\begin{array}{c}
p_0 \\ \\ \\ p_1 \\ \\ \\ p_2 \\
\end{array}
\quad Ax =
\left(
\begin{array}{ccc|cc|cc|c}
1 & 2 & 0 & 0 & 3 & 0 & 0 & 4 \\
0 & 5 & 6 & 7 & 0 & 0 & 8 & 0 \\
9 & 0 & 10 & 11 & 0 & 0 & 12 & 0 \\
\hline
13 & 0 & 14 & 15 & 16 & 17 & 0 & 0 \\
0 & 18 & 0 & 19 & 20 & 21 & 0 & 0 \\
0 & 0 & 0 & 22 & 23 & 0 & 24 & 0 \\
\hline
25 & 26 & 27 & 0 & 0 & 28 & 29 & 0 \\
30 & 0 & 0 & 31 & 32 & 33 & 0 & 34 \\
\end{array}
\right)
\left(
\begin{array}{c}
1 \\ 0 \\ 5 \\ \hline 7 \\ 9 \\ 0 \\ \hline 10 \\ 11 \\
\end{array}
\right)
\quad
\begin{array}{c}
p_0 \\ \\ \\ p_1 \\ \\ \\ p_2 \\
\end{array}
$$

where the local parts of matrix A and vector x stored in processor p_0 are

$$A_{p_0} = \begin{pmatrix} 1 & 2 & 0 & | & 0 & 3 & 0 & | & 0 & 4 \\ 0 & 5 & 6 & | & 7 & 0 & 0 & | & 8 & 0 \\ 9 & 0 & 10 & | & 11 & 0 & 0 & | & 12 & 0 \end{pmatrix} \text{ and } \begin{pmatrix} 1 \\ 0 \\ 5 \end{pmatrix} = x_{p_0},$$

respectively.

In PETSc, user must specify a communicator upon creation of any PETSc object (such as a vector, matrix or solver) to indicate the processors over which the object is to be distributed.

Among the most popular Krylov subspace iterative methods contained in PETSc are *Conjugate Gradient, Bi-Conjugate Gradient, Stabilized BCG, Transpose Free Quasi-Minimal Residual, Generalized-Minimal Residual* and so on[16]. PETSc offers preconditioners as *Additive Schwarz, Block Jacobi, Jacobi, ILU, ICC*, etc. We do not apply preconditioning for our test case due to the good spectral properties of T_{11} and T_{22} blocks, as we can see in the convergence curves represented in the Figure 2.

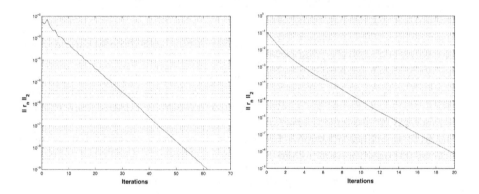

Fig. 2. CG convergence curves for T_{11} and T_{22} blocks

Timing is obtained through the use of real-time (wall-clock time) clock function available in PETSc library. For all methods, we have verified their accuracy and precision with regard to the global system $T\psi = e$ using the Matlab software.

5   Test Case

The test case chosen is the comercial reactor of Leibstadt[13], which has been discretised in a 3D form. The spatial discretisation has 32*32*27 cells, so that the total number of equations and cells is quite large: 157248 equations and 796080 nonzero elements in the Jacobian matrix.

We have applied all methods presented here to the set of matrices belongs to time step $t = 0$, which corresponds to a stability test carried out in 1990 where the reactor oscillates out of phase, to test their robustness and efficiency.

In the next section, the sequential performance is analysed, and some variations to A and B methods are introduced.

6   Sequential Study of Methods A and B and Some Variations

In order to identify the optimum ω for A and B methods in our test case, we have carry out a heuristic study. Some results of this study are described in Table 1.

Table 1. Sequential execution times (secs) for method A and B with different ω values. The symbol † indicates that convergence was not attained.

ω	0.1	0.5	0.8	0.9	1.0	1.3	1.5	1.9
A	252.46	223.97	196.75	187.39	**138.08**	†	†	†
B	211.08	175.44	143.61	130.40	116.28	**63.98**	290.85	†

In accordance with CPU execution times of this table, the optimum ω for method A is 1.0 and for method B is 1.3.

The errors attained with methods A and B are represented in Table 2, where r_{l+1} and $\|\cdot\|$ represent the residual $e - T\psi^{l+1}$ and the Euclidean norm, respectively.

From Tables 1 and 2, we can observe that method B is twice more efficient than method A as we had expected.

In order to reduce even more the computational work of method B, we have modified the operations as follows

$$T_{11}\psi_1^{l+1} = e_1 - T_{12}(\omega_1\psi_2^l + (1 - \omega_1)\psi_2^{l-1}),$$
$$T_{22}\psi_2^{l+1} = e_2 - T_{21}(\omega_2\psi_1^{l+1} + (1 - \omega_2)\psi_1^l). \tag{12}$$

Under this scheme, we have added two different parameters ω_1 and ω_2. Let us identify (12) as method C.

From application of method C to the test case, the optimum value of ω_1 is 1.0 and for ω_2 is 1.9. Table 3 shows a comparison of the number of iterations and execution times registered by method B and C, and we can see that the goal of decrease the computational work has been reached without lost of accuracy. From Table 3 we can observe a time reduction of 43% with regard to method B.

Since the precision of method C is good as A and B methods, we have experimented with an 'adaptable' precision technique, achieving some improvements in the efficiency of the process. This technique solves T_{11} and T_{22} blocks with a cheap precision (ϵ_{ρ_i}) at initial stages of the method. Then, this precision is 'adapted' or 'improved' towards a more demanding one ($\epsilon_{\rho_{i+1}}$) in successive iterations. Application of this technique to method C give rise to the following algorithm (method D in this work).

Table 2. Precision of methods A and B with optimum ω value

	$\|r_{l+1}\|_2$	$\|r_{l+1}\|_2/\|e\|_2$	Its.
A	8.83e-6	7.58e-5	394
B	4.91e-6	4.21e-5	183

Table 3. Performance comparison between method B and C

	Its.	CPU Time (secs.)	$\|r_{l+1}\|_2/\|e\|_2$
B	183	63.98	4.21e-5
C	**89**	**36.29**	5.42e-5

Second-Degree Iterative Algorithm (Adaptable version)

(1) Set ψ_2^0; $\{\psi_2^0 := \psi_2^*\}$
(2) Set $\epsilon_\rho = \{\epsilon_{\rho_1}, \epsilon_{\rho_2}, \ldots, \epsilon_{\rho_n}\}$ where $\epsilon_{\rho_i} > \epsilon_{\rho_{i+1}}$;
(2) Solve $T_{11}\psi_1^1 = e_1 - T_{12}\psi_2^0$
(3) Solve $T_{22}\psi_2^1 = e_2 - T_{21}\psi_1^1$
(4) Do $l = 1, 2, \ldots$
           (4a) Solve for ψ_1^{l+1} with tolerance ϵ_{ρ_i}
           (4b) Solve for ψ_2^{l+1} with tolerance ϵ_{ρ_i}
           (4c) if *precision of* $r_{l+1} \leq r_l$
                  $i := i + 1$
            end if
    until $\|\psi_1^{l+1} - \psi_1^l\| < tol$ and $\|\psi_2^{l+1} - \psi_2^l\| < tol$

Numerical experiments have shown that Method D is 25% more efficient than method C. Also, it is as exact as the rest of methods for the test case (See Table 4).

Table 4. Performance comparison between method C and D

	Its.	CPU Time (secs.)	$\|r_{l+1}\|_2/\|e\|_2$
C	89	36.29	5.42e-5
D	90	**27.31**	5.42e-5

Next section presents the parallel numerical results for all methods.

7   Parallel Numerical Results

All methods have been coded using the following PETSc operations facilites:

- VecNorm Computes the vector norm.
- VecPointwiseMult Computes the componentwise multiplication w = x*y.
- VecAYPX Computes y = x + alpha y.

- VecCopy Copies a vector.
- KSPSolve Solves a linear system. Steps (4a) and (4b) are carry out through the use of this operation.

Method A presents a good parallelism degree because the different linear systems of equations in (10) can be simultaneously solved by different groups of processors, and then interchange their solutions. For that reason, we have implemented two different parallel versions of this method, based on two different MPI communication routines: *gather/scatter* and *send/recv*. Also, we use the MPI facility to manage groups of processes through the use of communicators. For example, for the case of use $p = 2$ processors in method A, processor p_0 is dedicated to solve system T_{11} and processor p_1 is dedicated to solve system T_{22}. For the case of use $p = 4$ processors, $\frac{p}{2}$ processors are dedicated to solve system T_{11} and the rest dedicated to solve T_{22}, and so on.

The timing results for different number of processors (p) are registered in Table 5. As we can see, version based on *send/recv* is slightly more efficient than version based on *gather/scatter* primitives.

Table 5. Parallel execution times (secs) with method A

p	1	2	4	6	8	12
gather/scatter	138.08	89.33	54.90	40.51	33.86	25.47
send/recv	138.08	**88.27**	**53.07**	**38.05**	**31.30**	**22.60**

We have attempted to implement an asynchronous version of method A, but this was not possible, due to the strong dependency between T_{11} and T_{22} blocks.

We present a summary of parallel execution times with all methods in Table 6, where we can observe that the use of High-Performance Computing has decreased the sequential execution times for all methods. For example, for A and B methods the sequential execution times have been reduced until 16% of the original time value when we use $p = 12$ processors. For C and D methods, the execution time was reduce to 14% for the same number of processors.

Table 6. Parallel execution times (secs) for all methods

Method	$p = 1$	$p = 2$	$p = 4$	$p = 6$	$p = 8$	$p = 12$
A	138.08	88.27	53.07	38.05	31.30	22.60
B	63.98	36.83	28.78	20.26	11.94	10.41
C	36.29	21.11	12.01	8.68	6.91	5.23
D	**27.31**	**15.83**	**8.90**	**6.51**	**5.17**	**3.90**

From Table 6 we observe that method D offers the best execution time. The *speedup* and *efficiency*[15] of method D are represented in Figure 3, where for all values of p, parallel efficiency remains above of 50%.

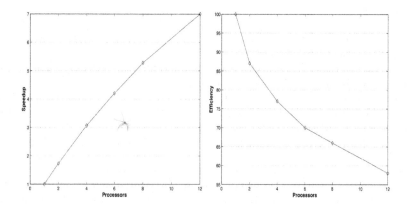

Fig. 3. Speedup and efficiency of method D

8   Conclusions

We have presented the application and parallelisation of two second-degree methods (A and B methods) to solve the sparse linear system related to a 3D Neutron-Diffusion equation of a real nuclear reactor using the numerical parallel library of PETSc.

In addition, we have modified A and B methods in order to reduce the computational work. For this, we have implemented two versions: the first one, based on two different relaxation parameters for each energy group obtaining a great performance which we call method C; and a second one, named method D, which is based on an *adaptable* technique that improves even more the performance with regard to the others methods.

We have carry out a heuristic study of the optimum relaxation parameter for each one of the methods presented and for our particular test case. These parameters have helped to accelerate the methods, specially C and D methods.

The main advantage of the second-degree methods presented in this work, is that matrix T do not need be formed explicitly; thus, simulation with more than 2 energy groups can be feasible.

It is important to emphasize that the application of High Performance Computing has reduced the sequential time of the different methods presented in this work.

Future works will contain the integration of these methods to DDASPK and FCVODE routines and the simulation of a full transient.

Acknowledgement

This work has been supported by Spanish MEC and FEDER under Grant ENE2005-09219-C02-02 and SEIT-SUPERA-ANUIES (México).

438   O. Flores-Sánchez et al.

References

1. Balay S., Gropp W.D., McInnes L.C., Smith B.F.: PETSc home page http://www.mcs.anl.gov/petsc (2002)
2. Balay S., Gropp W.D., McInnes L.C., Smith B.F.: PETSc Users Manual. ANL-95/11 - Revision 2.1.5, Argonne National Laboratory (1997)
3. Balay S., Gropp W.D., McInnes L.C., Smith B.F.: Efficient Management of Parallelism in Object Oriented Numerical Software Libraries. Modern Software Tools in Scientific Computing (1997) 163-202
4. Weston J.R., Stacey M.: SpaceTime Nuclear Reactor Kinetics. Academic Press (1970)
5. Henry A.F.: Nuclear Reactor Analysis. The M.I.T Press (1975)
6. García V.M., Vidal V., Verdú G., Miró R.: Sequential and Parallel Resolution of the 3D Transient Neutron Diffusion Equation. Mathematics and Computation, Supercomputing, Reactor Physics and Nuclear and Biological Applications, on CD-ROM, American Nuclear Society (2005)
7. Bru R., Ginestar D., Marín J., Verdú G., Mas J., Manteuffel T.: Iterative Schemes for the Neutron Diffusion Equation. Computers and Mathematics with Applications, Vol.44, (2002) 1307-1323
8. D.M. Young.: Iterative Solution of Large Linear Systems. Academic Press Inc.,New York, N.Y. (1971)
9. Stacey W.M.: Space-Time Nuclear Reactor Kinetics. Academic Press, New York (1969)
10. Verdú G., Ginestar D., Vidal. V., Muñoz-Cobo J.L.: A Consistent Multidimensional Nodal Method for Transient Calculation. Ann. Nucl. Energy, 22(6), (1995) 395-410
11. Ginestar D., Verdú G., Vidal V., Bru R., Marín J., Muñoz J.L.: High order backward discretization of the neutron diffusion equation. Ann. Nucl. Energy, 25(1-3), (1998) 47-64
12. Hébert A.: Development of the Nodal Collocation Method for Solving the Neutron Diffusion Equation. Ann. Nucl. Energy, 14(10), (1987) 527-541
13. Blomstrand J.: The KKL Core Stability Test, conducted in September 1990. ABB Report, BR91-245 , (1992)
14. D. Ginestar and J. Marín and G. Verdú.: Multilevel methods to solve the neutron diffusion equation. Applied Mathematical Modelling, 25, (2001) 463-477
15. V. Kumar and A. Grama and A. Gupta and G. Karypis.: Introduction to parallel computing:design and analysis of parallel algorithms. The Benjamin/Cummings Publishing Company, Inc.,Redwood City, CA (1994)
16. Y. Saad. Iterative Methods for Sparse Linear Systems PWS Publishing Company, Boston, MA (1996)

Enhancing the Performance of Multigrid Smoothers in Simultaneous Multithreading Architectures*

Carlos García, Manuel Prieto, Javier Setoain, and Francisco Tirado

Dto. Arquitectura de Computadores y Automática
Universidad Complutense de Madrid
Avd. Complutense s/n, 28040 Madrid, Spain
{garsanca,mpmatias,jsetoain,ptirado}@dacya.ucm.es

Abstract. We have addressed in this paper the implementation of red-black multigrid smoothers on high-end microprocessors. Most of the previous work about this topic has been focused on cache memory issues due to its tremendous impact on performance. In this paper, we have extended these studies taking *Simultaneous Multithreading (SMT)* into account. With the introduction of *SMT*, new possibilities arise, which makes a revision of the different alternatives highly advisable. A new strategy is proposed that focuses on inter-thread sharing to tolerate the increasing penalties caused by memory accesses. Performance results on an *IBM's Power5* based system reveal that our alternative scheme can compete with and even improve sophisticated schemes based on tailored loop fusion and tiling transformations aimed at improving temporal locality.

1 Introduction

Multigrid methods are regarded as being the *fastest* iterative methods for the solution of the linear systems associated with elliptic partial differential equations, and as amongst the *fastest* methods for other types of integral and partial differential equations [16]. *Fastest* refers to the ability of Multigrid methods to attain the solution in a computational work which is a small multiple of the operation counts associated with discretizing the system. Such efficiency is known as *textbook multigrid efficiency* (TME) [15] and has made multigrid one of the most popular solvers on the niche of large-scale problems, where performance is critical.

Nowadays, however, the number of executed operations is only one of the factors that influences the actual performance of a given method. With the advent of parallel computers and superscalar microprocessors, other factors such as *inherent parallelism* or *data locality* (i.e. the memory access behavior of the algorithm) have also become relevant. In fact, recent evolution of hardware has exacerbated this trend since:

---

* This work has been supported by the Spanish research grants TIC 2002-750 and TIN 2005-5619.

M. Daydé et al. (Eds.): VECPAR 2006, LNCS 4395, pp. 439–451, 2007.
© Springer-Verlag Berlin Heidelberg 2007

- The disparity between processor and memory speeds continues to grow despite the integration of large caches.
- Parallelism is becoming the key of performance even on high-end microprocessors, where multiple cores and multiple threads per core are becoming mainstream due to clock frequency and power limitations.

In the multigrid context, these trends have prompted the the development of specialized multigrid-like methods [1,2,10,5], and the adoption of new schemes that try to bridge the processor/memory gap by improving locality [14,18,7,4,8]. Our focus in this paper is the extension of this cache-aware schemes to *Simultaneous Multithreading (SMT)* processors.

As its name suggests, *SMT* architectures allows several independent threads to issue instructions simultaneously in a single cycle [17]. Its main goal is to yield better use of the processor's resources, hiding the inefficiencies caused by long operational latencies such as memory accesses. At first glance, these processors can be seen as a set of logical processors that share some resources. With HT, the Intel Pentium 4 behaves as two logical processors sharing some resources (Functional Units, Memory Hierarchy, etc). The exploitation of this additional level of parallelism has been performed in this work by means of OpenMP directives, which are directly supported by the Intel ICC compiler consequently, one may think that optimizations targeted for *Symmetric Multiprocessors (SMP)* systems are also good candidates for *SMT*. However, unlike *SMP* systems, *SMT* provides and benefits from fine-grained sharing of processor and memory resources. On the other hand, unlike conventional superscalar architectures, *SMT* exposes and benefits from thread level parallelism hiding latencies. Therefore, optimizations that are appropriate for these conventional machines may be inappropriate or less effective for *SMT* [9].

Unfortunately, *SMT* potentials are not yet fully exploited in most applications due to the relative underdevelopment of compilers, which despite many improvements still lag far behind. Due to this gap between compiler and processor technology, applications cannot benefit from *SMT* hardware unless they are explicitly aware of thread interactions. In this paper, we have revisited the implementation of multigrid smoothers in this light. The popularity of multigrid makes this study of great practical interest. In addition, it also provides certain insights about the potential benefits of this relatively new capability and how to take advantage of it, which could ideally helps to develop more efficient compiler schemes.

The organization of this paper is as follows. We begin in Sections 2 and Section 3 by briefly introducing multigrid methods and describing the main characteristics of our target computing platform respectively. In Section 4 we describe the baseline codes used in our study for validation and assessment. They are based on the *DIME* project (*DIME* stands for *Data Local Iterative Methods For The Efficient Solution of Partial Differential Equations*) [3,14,18,7,4,8], which is one of the most outstanding and systematic studies about the optimization of multigrid smoothers. Afterwards, in Section 5, we discuss our *SMT*-aware implementation. Performance results are discussed in Section 6. Finally, the paper ends with some conclusions and hints for future research.

2   Multigrid Introduction

This section provides a brief introduction to multigrid, defining basic terms and describing the most relevant aspects of these methods so that we have a basis on which to discuss some of the performance issues.

The fundamental idea behind Multigrid methods [16] is to capture errors by utilizing multiple length scales (multiple grids). They consist of the following complementary components:

- *Relaxation.* The relaxation procedure, also called smoother in multigrid lingo, is basically a simple (and inexpensive) iterative method like *Gauß-Seidel*, damped *Jacobi* or block *Jacobi*. Its election depends on the target problem, but if well chosen, it is able to reduce the high-frequency or oscillatory components of the error in relatively few steps.
- *Coarse-Grid Correction.* Smoothers are ineffectual in attenuating low-frequency content of the error, but since the error after relaxation should eliminate the oscillatory components, it can be well-approximated using a coarser grid. On that grid, errors appear more oscillatory and thus the smoother can be applied effectively. New values are transferred afterwards to the target grid to update the solution.

The *Coarse-Grid Correction* can be applied recursively in different ways, constructing different cycling strategies. Algorithm 1. shows the pseudo-code of one of the most popular choices, known as V-cycle due to its pattern. This algorithm telescopes down to a given coarsest grid, and then works its way back to the target finest grid. The transfer operators I_h^{2h} and I_{2h}^h connect the grids levels: I_h^{2h} is known as the restriction operator and transfers values from a finer to a coarser level, whereas I_{2h}^h is known as the prolongation operator and maps from a coarser to a finer level.

Algorithm 1. V-cycle(ν_1,ν_2,v_h,b_h) multigrid V-cycle applied to the system $A_h u_h = b_h$ defined on a grid Ω_h.

---

  if h == Coarsest **then**
    Return $u_H \leftarrow$ Solve(A_H,v_H,b_H)
  else
    $v_h \leftarrow$ Smooth(ν_1,v_h,b_h)
    $b_{2h} \leftarrow I_h^{2h}(b_h - A_h v_h)$
    $v_{2h} \leftarrow$ V-cycle($\nu_1,\nu_2,0_{2h},b_{2h}$)
    $v_h \leftarrow v_h + I_{2h}^h(v_{2h}$
    Return $u_n \leftarrow$ Smooth(ν_2,v_h,b_h)
  end if

The most time-consuming part of a multigrid method is the smoother and hence is the primary parameter in optimizing the performance. In this initial study we have focused on point-wise smoothers. Block smoothers [12,11] are more efficient in certain problems but are beyond the scope of this paper and will not be addressed at this time. The discussion about the implementation of point-wise smoothers is taken up in Section 4.

3   Experimental Platform

Our experimental platform consists in an *IBM's Power5* processor running under Linux, the main features of which are summarized in Table 1.

Table 1. Main features of the target computing platform

		IBM 2-way 1.5GHz Power5 (2 way core SMP)
Processor	L1 DataCache	32 KB 4-way associative, LRU
	L2 Unified Cache	1.9MB 10-way associative, LRU
	L3 Unified Cache (*off-chip*)	36MB shared per processor pair 10-way associative, LRU
Memory		2048 MBytes (4x512) DIMMS 266 MHz DDR SDRAM
Operating System		GNU Debian Linux kernel 2.6.14-SMP for 64 bits
IBM XL Fortran Switches (Advance Ed. v9.1)		-O5 -qarch=pwr5 -qtune=pwr5 -q64 -qhot -qcache=auto Parallelization with OpenMP: -qsmp=omp

This processor has introduced *SMT* to the *IBM's Power* family [6]. With this design, each core of this dual-core processor appears to software as two logical CPUs, usually denoted as threads, that share some resources such as functional units or the memory hierarchy.

Apart from *SMT*, we should also highlight the impressive memory subsystem of the *Power5*. The memory controller is moved on chip and the main memory is directly connected to the processor via three buses: the address/command bus, the unidirectional write data bus, and the unidirectional read data bus. The 36-MB off-chip L3 has been removed from the path between the processor to the memory controller and operates as a victim cache for the L2. This means that data is transferred to the L3 only when it is replaced from the L2.

Finally, it is worth mentioning that the exploitation of *SMT* has been performed in this work by means of *OpenMP* directives, which are directly supported by the IBM's FORTRAN compiler. Single thread performance has been measured using a sequential code and enabling the *Power5 single-threaded* mode of the *Power5*, which gives all the resources of the chip to one of the logical CPUs.

4   Cache-Aware Red-Black Smoothers

Gauß-Seidel has long been the smoother of choice within multigrid on both structured and unstructured grids [1]. Although, it is inherently sequential in its natural form (the lexicographic ordering), it is possible to expose parallelism by applying multi-coloring, i.e. splitting grid nodes into disjoint sets, with each set having a different color, and updating simultaneously all nodes of the same color.

The best known example of this approach is the red-black Gauß-Seidel for the 5-point Laplace stencil, which is schematically illustrated in Figure 1. For the 9-point Laplacian, a red-black ordering may lead to a race condition (depending on the implementation), and at least a four color ordering of the grid space is needed to decouple the grid nodes completely.

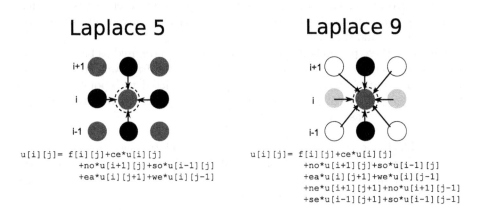

Fig. 1. 2D red-black Gauß-Seidel for the 5-point (on left-hand side) and 9-point (on right-hand side) Laplace stencils

Apart from exposing parallelism, multi-coloring also impacts on the convergence rate, but unlike other techniques such as block Gauß-Seidel (i.e. applying Gauß-Seidel locally on every processor), the overall multigrid convergence rates remain satisfactory.

Unfortunately, multi-coloring deteriorate the memory access and may lead to poor performance. Algorithm 4 shows the pseudo-code of a red-black smoother, denoted as *rb1* by the *DIME* project. This naïve implementation performs a complete sweep through the grid for updating all the red nodes, and then another complete sweep for updating all the black nodes. Therefore, *rb1* exhibits lower spatial locality than a lexicographic ordering. Furthermore, if the target grid is large enough, temporal locality is not fully exploited.

Algorithm 2. Red-Black Gauß-Seidel naïve implementation

```
for it=1,nIter do
  // red nodes:
  for i = 1; n-1 do
    for j = 1+(i+1)%2; n-1; j=j+2 do
      Relax_point( i,j )
    end for
  end for
  // black nodes:
  for i = 1, n-1 do
    for j = 1+i%2; n-1; j=j+2 do
      Relax_point( i,j )
    end for
  end for
end for
```

Alternatively, some authors have successfully improved cache reuse (locality) using loop reordering and data layout transformations that were able to improve both temporal and spatial data locality [13,18].

Following these previous studies, in this paper we have used as baseline codes the different red-black smoothers developed within the framework of the *DIME* project. To simplify matters, these codes are restricted to 5-point as well as 9-point discretization of the Laplacian operator. Figures 2-4 illustrate some of them, which are based on the following observations:

– The black nodes of a given row $i - 1$ can be updated once the red nodes of the i row has been updated. This is the idea behind the *DIME's rb2* (see figure 2) and *rb3* schemes, which improve both temporal and spatial locality fusing the *red* and *black* sweeps.

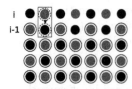

Fig. 2. *DIME's* rb2. The update of red and black nodes is fused to improve temporal locality.

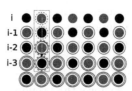

Fig. 3. *DIME's* rb5. Data within a tile is reused as much as possible before moving to the next tile.

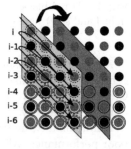

Fig. 4. *DIME's* rb9. Data within a tile is reused as much as possible before moving to the next tile.

– If several successive relaxation have to be performed, additional improvements can be achieved transforming the iteration transversal so that the operations are performed on small 1D or 2D tiles of the whole array. Data within a tile is used as many times as possible before moving to the next tile.

Table 2. MFlops achieved by the different *DIME's* variants of the red-black Gauß-Seidel for a 5-point Laplace stencil. MELT denotes the number of successive iterations of this smoother.

	16		32		64	
	no PAD.	PADDING	no PAD.	PADDING	no PAD.	PADDING
rb1 (MELT=1)	1220.41	1328.02	1554.55	1886.03	1846.53	2151.69
rb2 (MELT=1)	1654.44	1799.24	1590.53	2353.55	2396.09	2464.36
rb3 (MELT=1)	1655.28	1800.82	1588.90	2364.93	2387.07	2474.08
rb4 (MELT=2)	1317.74	1353.41	1418.75	1435.21	1468.09	1471.15
rb4 (MELT=3)	1375.39	1410.50	1447.29	1489.77	1368.22	1384.25
rb5 (MELT=2)	1315.34	1355.12	1418.45	1433.10	1462.85	1472.53
rb5 (MELT=3)	1371.65	1411.75	1446.33	1490.16	1364.34	1386.79
rb6 (MELT=2)	1834.07	2055.36	2139.22	2743.28	2961.18	3101.63
rb6 (MELT=3)	1851.40	2121.46	2137.32	2908.11	3138.46	3235.82
rb7 (MELT=2)	2079.85	2091.77	2406.82	2826.85	3052.28	3128.04
rb7 (MELT=3)	1728.44	1768.35	1916.17	2066.12	2030.17	2124.82
rb8 (MELT=2)	2061.58	2087.47	2511.27	2833.20	3052.79	3128.73
rb8 (MELT=3)	1726.98	1766.26	1871.81	2038.05	1998.76	2069.93
rb9 (MELT=4)	1603.72	1672.13	2075.09	2349.55	2453.04	2643.53

	128		256		512	
	no PAD.	PADDING	no PAD.	PADDING	no PAD.	PADDING
rb1 (MELT=1)	2637.98	2673.35	2801.20	2823.32	2139.93	2412.95
rb2 (MELT=1)	2642.36	2667.40	2794.10	2794.10	2178.04	2495.04
rb3 (MELT=1)	2661.24	2664.53	2840.70	2840.70	2150.20	2500.19
rb4 (MELT=2)	1368.15	1380.44	1370.40	1375.65	1248.74	1292.01
rb4 (MELT=3)	1386.74	1393.14	1359.37	1368.66	1377.90	1427.34
rb5 (MELT=2)	1367.39	1379.36	1371.12	1373.43	1248.87	1292.34
rb5 (MELT=3)	1388.05	1398.70	1359.08	1369.72	1386.58	1425.09
rb6 (MELT=2)	3362.78	3445.60	3453.84	3513.04	2009.28	2561.72
rb6 (MELT=3)	3677.55	3710.23	3376.04	3487.90	1372.93	2390.08
rb7 (MELT=2)	3461.82	3553.01	3597.04	3625.99	1976.21	2619.95
rb7 (MELT=3)	2190.06	2207.63	2038.09	2109.38	1332.34	1833.58
rb8 (MELT=2)	3488.76	3549.89	3553.04	3618.60	1995.86	2620.77
rb8 (MELT=3)	2124.59	2152.84	2033.73	2054.91	1355.92	1788.43
rb9 (MELT=4)	2785.56	2965.07	2262.49	2981.07	864.31	2079.22

	1024		2048		4096	
	no PAD.	PADDING	no PAD.	PADDING	no PAD.	PADDING
rb1 (MELT=1)	1530.51	1904.80	1012.56	1130.23	827.73	908.38
rb2 (MELT=1)	1135.85	2012.24	1108.30	1322.43	909.35	1060.96
rb3 (MELT=1)	1138.72	1995.31	1108.83	1311.42	909.32	1060.65
rb4 (MELT=2)	691.73	1366.76	645.31	1198.68	607.49	1048.54
rb4 (MELT=3)	699.54	1405.66	664.97	1283.75	625.90	1096.99
rb5 (MELT=2)	691.23	1364.57	644.51	1191.35	607.49	1047.26
rb5 (MELT=3)	700.39	1406.21	664.47	1283.65	625.21	1093.09
rb6 (MELT=2)	797.33	2405.67	424.95	1228.52	425.91	1202.53
rb6 (MELT=3)	740.90	2385.51	377.16	1046.89	359.09	985.14
rb7 (MELT=2)	769.53	2319.02	490.97	1336.46	465.97	1248.94
rb7 (MELT=3)	706.90	1658.36	598.65	1132.00	555.64	988.34
rb8 (MELT=2)	888.24	2507.34	460.97	1087.26	474.43	1193.49
rb8 (MELT=3)	713.10	1670.59	551.20	1201.43	464.93	954.55
rb9 (MELT=4)	444.54	1928.21	353.77	1248.75	348.30	1058.10

DIME's rb4-rb9 schemes perform different 1D and 2D tiling transformations. Figures 3 and 4 illustrate the *rb5* and *rb9* schemes respectively.

Tables 2 and 3 show the MFlops achieved by *DIME's rb1-9* codes on our target platform. The speedup of the best transformation ranges from 1.2 to 1.8

Table 3. MFlops achieved by the different *DIME's* variants of the red-black Gauß-Seidel for a 9-point Laplace stencil. MELT denotes the number of successive iterations of this smoother.

	16		32		64	
	no PAD.	PADDING	no PAD.	PADDING	no PAD.	PADDING
rb1 (MELT=1)	846.24	845.91	1134.27	1172.41	1321.73	1348.14
rb2 (MELT=1)	950.62	1034.2	1042.20	1414.47	1126.24	1656.04
rb3 (MELT=1)	962.94	964.20	1248.63	1278.89	1371.60	1464.29
rb4 (MELT=2)	746.65	747.42	760.19	764.49	740.99	748.75
rb4 (MELT=3)	780.77	783.89	782.92	793.98	752.32	770.89
rb5 (MELT=2)	623.14	627.04	602.06	607.46	561.58	564.05
rb5 (MELT=3)	662.97	670.53	624.34	636.21	586.11	598.19
rb6 (MELT=2)	796.40	802.01	837.04	848.68	835.16	848.87
rb6 (MELT=3)	777.82	977.49	1094.88	1159.87	1125.33	1173.56
rb7 (MELT=2)	791.24	819.63	809.41	852.67	813.76	821.87
rb7 (MELT=3)	916.94	932.90	1007.62	1038.20	1029.50	1066.33
rb8 (MELT=2)	818.48	846.73	857.17	883.96	858.65	868.43
rb8 (MELT=3)	919.03	936.28	1007.72	1038.38	1047.92	1084.02
rb9 (MELT=4)	817.50	837.66	937.08	983.44	1039.88	1101.31

	128		256		512	
	no PAD.	PADDING	no PAD.	PADDING	no PAD.	PADDING
rb1 (MELT=1)	1667.04	1670.97	1729.36	1759.88	1562.05	1554.85
rb2 (MELT=1)	1831.58	1925.38	2018.26	2067.22	1951.36	1941.23
rb3 (MELT=1)	1727.06	1747.52	1862.76	1857.99	1742.83	1715.75
rb4 (MELT=2)	698.13	699.38	696.54	696.83	719.67	731.25
rb4 (MELT=3)	765.17	766.85	755.06	753.95	727.05	737.16
rb5 (MELT=2)	564.48	564.48	558.92	560.02	545.81	548.25
rb5 (MELT=3)	585.55	585.71	575.29	576.74	560.88	564.15
rb6 (MELT=2)	864.13	863.54	868.21	868.07	675.26	836.98
rb6 (MELT=3)	1284.45	1286.39	1280.24	1293.26	765.69	1135.30
rb7 (MELT=2)	831.41	831.08	821.43	830.45	725.22	794.60
rb7 (MELT=3)	1083.47	1083.95	1076.12	1080.96	903.37	913.06
rb8 (MELT=2)	885.39	878.57	885.83	892.54	776.17	859.16
rb8 (MELT=3)	1104.18	1105.80	1099.24	1100.55	961.44	976.59
rb9 (MELT=4)	1130.08	1146.71	871.69	1147.63	532.22	868.32

	1024		2048		4096	
	no PAD.	PADDING	no PAD.	PADDING	no PAD.	PADDING
rb1 (MELT=1)	1209.56	1249.72	497.73	554.38	504.74	571.47
rb2 (MELT=1)	932.55	1656.93	374.89	521.46	420.05	574.23
rb3 (MELT=1)	1295.14	1468.16	583.47	671.95	662.89	632.52
rb4 (MELT=2)	679.67	717.89	593.96	659.83	593.07	663.24
rb4 (MELT=3)	691.21	726.90	611.44	681.95	609.99	684.88
rb5 (MELT=2)	386.53	554.37	366.12	535.87	367.82	533.03
rb5 (MELT=3)	388.97	561.67	373.21	542.73	374.81	546.64
rb6 (MELT=2)	361.31	594.64	274.79	566.75	273.87	567.97
rb6 (MELT=3)	317.80	552.96	285.76	449.79	274.89	445.53
rb7 (MELT=2)	436.52	653.60	340.72	685.85	359.81	570.45
rb7 (MELT=3)	618.65	1053.75	762.01	754.74	471.25	701.31
rb8 (MELT=2)	438.12	697.92	333.05	617.26	384.65	613.42
rb8 (MELT=3)	641.91	1073.55	760.33	795.94	497.73	663.63
rb9 (MELT=4)	339.64	896.80	290.39	617.66	278.66	573.70

for the 5-point stencil, and from 1.15 to 1.25 for the 9-point version. Our first insight is that these gains are lower than on other architectures. For instance, the improvements on a DEC PWS 500au reported on *DIME's* website reach a factor of 4 [3]. Furthermore, the sophisticated two-dimensional blocking transformation *DIME's rb9* does not provide additional improvements, being *DIME's rb7* and sometimes *DIME's rb2* the most effective transformations.

The main reason behind this difference in behavior is the relatively large amount of on-chip and off-chip caches included in the *IBM's Power5*, as well as their higher degree of associativity.

5   SMT-Aware Red-Black Smoothers

The availability of *SMT* introduces a new scenario in which thread-level parallelism can also be applied to hide memory accesses. As mentioned above, *SMT* processors can be seen as a set of logical processors that share execution units, systems buses and the memory hierarchy. This logical view suggests the application of the general principles of data partitioning to get the multithreaded versions of the different *DIME* variants of the red-black Gauß-Seidel smoother. This strategy, which can be easily expressed with *OpenMP* directives, is suitable for shared memory multiprocessor. However, in a *SMT* microprocessor, the similarities amongst the different threads (they execute the same code with just a different input dataset) may cause contention since they have to compete for the same resources.

Alternatively, we have employed a dynamic partitioning where computations are broken down into different tasks with are assigned to the pool of available threads. Intuitively, the smoothing of the different colors is interleaved by assigning the relaxation of each color to a different thread. This interleaving is controlled by a scheduler, which avoids race conditions and guarantees a deterministic ordering.

Algorithm 3. shows a pseudo-code of this approach for red-black smoothing. Our actual implementation is based on the *OpenMP's parallel* and *critical* directives. The critical sections introduce some overhead but are necessary to avoid race-conditions. However, the interleaving prompted by the scheduling allows the *black thread* to take advantage of some sort of data prefetching since it processes grid nodes that have just been processed by the *red thread*, i.e. the *red thread* acts as a helper thread that performs data prefetching for the *black* one.

This interleaved approach can also be combined with *DIME's rb4-8* variants. If two successive iterations have to be performed, the intuitive idea is that one thread performs the first relaxation step whereas the other performs the second one. The scheduler guarantees again a deterministic ordering.

In the next section we compare the performance of this novel approach over traditional block-outer, cyclic-outer and cyclic-inner distributions of the relaxation nested loop. For the naïve implementation, all of them are straightforward. However, for *DIME's rb2-8* variants, the cyclic-outer version is

---

Algorithm 3. Interleaved implementation of a red-black Gauß-Seidel

---

```
#pragma omp parallel private(task,more_tasks) shared(control_variables)
more_tasks = true
while more_tasks do

    #pragma omp critical
    Scheduler.next_task(&task);

    if (task.type == RED) then
        Relax_RED_line(task);
    end if

    if (task.type == BLACK) then
        Relax_BLACK_line(task);
    end if

    #pragma omp critical
    more_task=Scheduler.commit(task);

end while
```

---

non-deterministic, whereas the block-outer requires the processing of block boundaries in advance.

We have omitted a parallel version of *DIME's rb9* since even in the sequential setting, that version does not provide superior performance over *DIME's rb5-8*. We have also omitted block-outer and cyclic-outer distributions of the red-black smoother for the 9-point stencil, since they are also non-deterministic. Note, however, that both the cyclic-inner and our interleaved approach avoid race-conditions.

6   Performance Results

Figure 5 shows the speedup achieved by the different parallel strategies over the baseline code (with the best DIME's transformation) for the the 5-point stencil.

As can be noticed, the election of the most suitable strategy depends on the grid size:

– For small and medium grid sizes block and cyclic distributions outperform our approach, although for the smallest sizes none of them is able to improve performance. This is the expected behavior given that for small and medium working sets, memory bandwidth and data cache exploitation are not a key issue and traditional strategies beat our approach on performance due to the overheads introduced by the dynamic task scheduling.

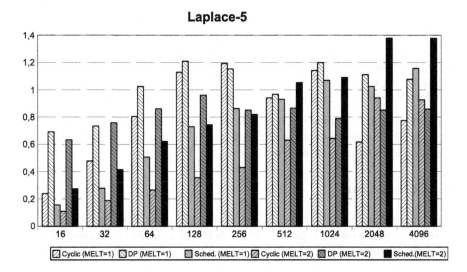

Fig. 5. Speedup achieved by different parallel implementations of a red-black Gauß-Seidel smoother for a 5-point Laplace stencil. Sched denotes our strategy, whereas DP and Cyclic denote the best block and a cyclic distribution of the smoother's outer loop respectively. MELT is the number of successive relaxations that have been applied.

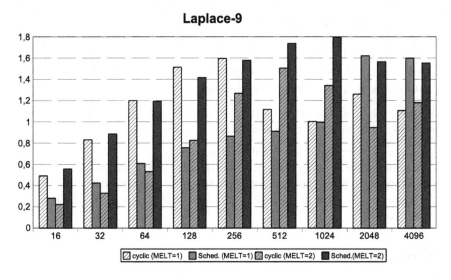

Fig. 6. Speedup achieved by different parallel implementations of a red-black Gauß-Seidel smoother for a 9-point Laplace stencil. Sched denotes our strategy, whereas Cyclic denotes the best cyclic distribution of the smoother's inner loop. MELT is the number of successive relaxations that have been applied.

– For large sizes we observe the opposite behavior given that the overheads involved in task scheduling become negligible, whereas the competition for memory resources becomes a bottleneck in the other versions. In fact, we should highlight that the block and cyclic distributions become clearly inefficient for large grids.

- The break-even point between the static distributions and our interleaved approach is a relative large grid due to the impressive L3 cache (36 MB) of the *Power5*.

Figure 6 confirms some of these observations for the the 9-point stencil. Furthermore, the improvements over *DIME's* variants are higher in this case, since this is a more demanding problem.

7   Conclusions

In this paper, we have introduced a new implementation of red-black Gauß-Seidel Smoothers, which on *SMT* processors fits better than other traditional strategies. From the results presented above, we can draw the following conclusions:

- Our alternative strategy, which implicitly introduces some sort of tiling amongst threads, provides noticeable speed-ups that match or even outperform the results obtained with the different *DIME's rb2-9* variants for large grid sizes. Notice that instead of improving *intra-thread locality*, our strategy improves locality taking advantage of fine-grain thread sharing.
- For large grid sizes, competition amongst threads for memory bandwidth and data cache works against traditional block distributions. Our interleaved approach performs better in this case, but suffers important penalties for small grids, since its scheduling overheads does not compensate its better exploitation of the temporal locality. Given that multigrid solvers process multiple scales, we advocate hybrid approaches.

We are encouraged by these results, and based on what we have learned in this initial study we are proceeding with:

- Analyzing more elaborated multigrid solvers.
- Combining interleaving with grid partitioning distributions to scale beyond two threads. The idea is to use grid partitioning to distribute data amongst a large scale system, and interleaving to exploit thread level parallelism inside their cores.

References

1. M. F. Adams, M. Brezina, J. J. Hu, and R. S. Tuminaro. Parallel multigrid smoothing: polynomial versus Gauss-Seidel. *J. Comp. Phys.*, 188(2):593–610, 2003.
2. Edmond Chow, Robert D. Falgout, Jonathan J. Hu, Raymond S. Tuminaro, and Ulrike Meier Yang. A survey of parallelization techniques for multigrid solvers,. Technical report, 2004.
3. Friedrich-Alexander University Erlangen-Nuremberg. Department of Computer Science 10. DIME project. Available at http://www10.informatik.uni-erlangen.de/Research/Projects/DiME-new.

4. C.C. Douglas, J. Hu, M. Kowarschik, U. Rüde, and C. Weiß. Cache Optimization for Structured and Unstructured Grid Multigrid. *Electronic Transactions on Numerical Analysis (ETNA)*, 10:21–40, 2000.
5. F. Hülsemann, M. Kowarschik, M. Mohr, and U. Rde. Parallel geometric multigrid. *Lecture Notes in Computer Science and Engineering*, 51:165–208, 2005.
6. Ronald N. Kalla, Balaram Sinharoy, and Joel M. Tendler. IBM Power5 chip: A dual-core multithreaded processor. *IEEE Micro*, 24(2):40–47, 2004.
7. M. Kowarschik, U. Rüde, C. Weiß, and W. Karl. Cache-Aware Multigrid Methods for Solving Poisson's Equation in Two Dimensions. *Computing*, 64:381–399, 2000.
8. M. Kowarschik, C. Weiß, and U. Rüde. Data Layout Optimizations for Variable Coefficient Multigrid. In P. Sloot, C. Tan, J. Dongarra, and A. Hoekstra, editors, *Proc. of the 2002 Int. Conf. on Computational Science (ICCS 2002), Part III*, volume 2331 of *Lecture Notes in Computer Science (LNCS)*, pages 642–651, Amsterdam, The Netherlands, 2002. Springer.
9. Jack L. Lo, Susan J. Eggers, Henry M. Levy, Sujay S. Parekh, and Dean M. Tullsen. Tuning compiler optimizations for simultaneous multithreading. In *International Symposium on Microarchitecture*, pages 114–124, 1997.
10. W. Mitchell. Parallel adaptive multilevel methods with full domain partitions. *App. Num. Anal. and Comp. Math*, 1:36–48, 2004.
11. Manuel Prieto, Rubén S. Montero, Ignacio Martín Llorente, and Francisco Tirado. A parallel multigrid solver for viscous flows on anisotropic structured grids. *Parallel Computing*, 29(7):907–923, 2003.
12. Manuel Prieto, R. Santiago, David Espadas, Ignacio Martín Llorente, and Francisco Tirado. Parallel multigrid for anisotropic elliptic equations. *J. Parallel Distrib. Comput.*, 61(1):96–114, 2001.
13. D. Quinlan, F. Bassetti, and D. Keyes. Temporal locality optimizations for stencil operations within parallel object-oriented scientific frameworks on cache-based architectures. In *Proceedings of the PDCS'98 Conference*, July 1998.
14. U. Rüde. Iterative Algorithms on High Performance Architectures. In *Proc. of the EuroPar-97 Conf.*, Lecture Notes in Computer Science (LNCS), pages 26–29. Springer, 1997.
15. James L. Thomas, Boris Diskin, and Achi Brandt. Textbook multigrid efficiency for fluid simulations. *Annual Review of Fluid Mechanics*, 35:317–340, 2003.
16. U. Trottenberg, C. Oosterlee, and A. Schller. *Multigrid*. Academic Press, 2000.
17. Dean M. Tullsen, Susan J. Eggers, and Henry M. Levy. Simultaneous multithreading: Maximizing on-chip parallelism. In *25 Years ISCA: Retrospectives and Reprints*, pages 533–544, 1998.
18. C. Weiß, W. Karl, M. Kowarschik, and U. Rüde. Memory Characteristics of Iterative Methods. In *Proc. of the ACM/IEEE Supercomputing Conf. (SC99)*, Portland, Oregon, USA, 1999.

Block Iterative Algorithms for the Solution of Parabolic Optimal Control Problems

Christian E. Schaerer[1], Tarek Mathew[1], and Marcus Sarkis[1,2]

[1] Instituto de Matemática Pura e Aplicada-IMPA,
Estrada Dona Castorina 110, Rio de Janeiro, RJ 22460-320, Brazil
[2] Department of Mathematical Sciences-WPI,
100 Institute Road, Worcester, MA 01609, USA
cschaer@fluid.impa.br, tmathew@fluid.impa.br, msarkis@impa.br

Abstract. In this paper, we describe block matrix algorithms for the iterative solution of *large scale* linear-quadratic optimal control problems arising from the control of parabolic partial differential equations over a finite control horizon. After spatial discretization, by finite element or finite difference methods, the original problem reduces to an optimal control problem for n coupled ordinary differential equations, where n can be quite *large*. As a result, its solution by conventional control algorithms can be prohibitively expensive in terms of computational cost and memory requirements.

We describe two iterative algorithms. The first algorithm employs a CG method to solve a symmetric positive definite reduced linear system for the unknown control variable. A preconditioner is described, which we prove has a rate of convergence independent of the space and time discretization parameters, however, double iteration is required. The second algorithm is designed to avoid double iteration by introducing an auxiliary variable. It yields a symmetric indefinite system, and for this system a positive definite block preconditioner is described. We prove that the resulting rate of convergence is independent of the space and time discretization parameters, when MINRES acceleration is used. Numerical results are presented for test problems.

1 Introduction

Systems governed by parabolic partial differential equations arise in models of various processes in the oil industry. An instance is the model of a production strategy whose main objective is the displacement of a resident fluid (oil) by the injection of another fluid (gas) [15]. The associated partial differential equation for the pressure is parabolic. In this context, recent work has demonstrated that control strategies based on Optimal Control Theory (OCT) can potentially increase the production in oil and gas fields [15]. In addition, the efficiency of the OCT model makes it suitable for application to real reservoirs simulated using large scale models, in contrast to many existing techniques [12].

M. Daydé et al. (Eds.): VECPAR 2006, LNCS 4395, pp. 452–465, 2007.
© Springer-Verlag Berlin Heidelberg 2007

The main bottleneck in this approach, however, is the need for a fast simulator to test all the necessary scenarios to decide upon an adequate strategy for each reservoir.

Our purpose in this paper, is to study iterative algorithms for the solution of finite time *linear-quadratic optimal control* problems governed by a parabolic partial differential equation. Such problems are computationally intensive and require the minimization of some quadratic objective functional $J(\cdot)$ (representing some cost to be minimized over time), subject to linear constraints given by a *stiff* system of n ordinary differential equations, where n is typically quite large. An application of the Pontryagin maximum principle to determine the optimal solution, see [9], results in a Hamiltonian system of ordinary differential equations, with initial and final conditions. This system is traditionally solved by reduction to a matrix Riccati equation for an unknown matrix function $P(t)$ of size n, on an interval $[0, T]$, see [9,7,11]. Solving the Riccati equation, and storing matrix $P(t)$ of size n for each $t \in [0, T]$ can become prohibitively expensive for large n. Instead, motivated by the *parareal* algorithm (of Lions, Maday and Turinici [6]) and iterative shooting methods in the control context [4,13], we propose iterative algorithms for such control problems; see also [14].

The iterative algorithms we formulate for parabolic optimal control problems are based on a *saddle point* formulation [11]. We consider a finite difference (or finite element) discretization of the parabolic equation in space, and a θ-scheme discretization in time. The cost functional is discretized in time using a trapezoidal or midpoint rule, and the control variable is approximated by piecewise constant finite element functions in time and space. Lagrange multipliers (adjoint) variables are introduced to enforce the constraints, and a saddle point linear system is formulated for the optimal solution. Inspired by the reduction approach employed in [11] for elliptic control problems, we develop two algorithms whose rate of convergence does not deteriorate as the mesh parameters become small. The first algorithm uses a CG method to solve a symmetric positive definite reduced linear system for determining the unknown *control variable*. We show under specific assumptions that the resulting system has a condition number independent of the mesh parameters. For the second algorithm, we expand the reduced system consistently by introducing an auxiliary variable. We describe a block preconditioned algorithm using a MINRES method on the auxiliary and control variables. We analyze the convergence rates of these two proposed iterative algorithms.

Our discussion is organized as follows. In Section 2, we introduce the optimal control problem for the parabolic problem. In Section 3, we describe the finite dimensional linear-quadratic optimal control problem. We then describe the saddle point system obtained by a stable discretization of the parabolic control problem. In Section 4, we describe the preconditioners and theoretical results that justify the efficiency of the proposed methods. Finally, in Section 5, numerical results are presented which show that the rate of convergence of both proposed algorithms are independent of the space and time step parameters.

2   The Optimal Control Problem

Let (t_0, t_f) denote a time interval and let $\mathcal{A}$ denote an operator from a Hilbert space $L^2(t_0, t_f; Y)$ to $L^2(t_0, t_f; Y')$, where $Y = H_0^1(\Omega)$ in our applications, and $L^2(t_0, t_f; Y)$ is endowed with its standard norm [7]. Given $z_0 \in Y$, we consider the following state equation on (t_0, t_f) with $z(t) \in Y$:

$$\begin{cases} z_t + \mathcal{A}z = \mathcal{B}v, & \text{for } t_0 < t < t_f \\ \quad z(0) = z_o, \end{cases} \tag{1}$$

where $z(\cdot) \in Y$ is referred to as a state variable and operator $\mathcal{A}$ is coercive. The distributed control $v(\cdot)$ belongs to an admissible space $\mathcal{U} = L^2(t_0, t_f; \Omega)$ and $\mathcal{B}$ is an operator in $\mathcal{L}(\mathcal{U}, L^2(t_0, t_f; Y'))$. We assume that for each $v(\cdot)$, this problem is well posed, and indicate the dependence of z on $v \in \mathcal{U}$ using the notation $z(v)$. Given parameters $q \geq 0$, $r \geq 0$, $s \geq 0$, we shall employ the following cost function, which we associate with the state equation (1):

$$J(z(v), v) := \tfrac{q}{2} \int_{t_0}^{t_f} \|z(v)(t, .) - z_*(t, .)\|_{L^2(\Omega)}^2 \, dt + \tfrac{r}{2} \int_{t_0}^{t_f} \|v(t, .)\|_{L^2(\Omega)}^2 \, dt$$
$$+ \tfrac{s}{2} \|z(v)(t_f, .) - z_*(t_f, .)\|_{L^2(\Omega)}^2,$$

where $z_*(., .)$ is a given target. The optimal control problem for equation (1) consists of finding a controller $u \in \mathcal{U}$ which *minimizes* the cost function (2):

$$J(z(u), u) = \min_{v \in \mathcal{U}} J(z(v), v). \tag{2}$$

Since $\tfrac{r}{2} \int_{t_0}^{t_f} \|v(t, .)\|_{L^2(\Omega)}^2 \, dt > 0$ and $\tfrac{q}{2} \int_{t_0}^{t_f} \|z(v)(t, .) - z_*(t, .)\|_{L^2(\Omega)}^2 \geq 0$ for $r > 0$, $q > 0$ and $v \neq 0$ in (2), the optimal control problem (2) will be well posed, see [7]. Let $(., .)$ denote the $L^2(\Omega)$ inner product, then the weak formulation of (1) will seek $z \in L^2(t_0, t_f; Y)$ with $\dot{z} \in L^2(t_0, t_f; Y')$ satisfying:

$$(\dot{z}(t), \eta) + (\mathcal{A}z(t), \eta) = (\mathcal{B}u(t), \eta), \quad \forall \eta \in Y, \text{ for } t_0 < t < t_f. \tag{3}$$

The bilinear form $(\mathcal{A}z, \eta)$ will be assumed to be continuous on $Y \times Y$ and Y-elliptic, and bilinear form $(\mathcal{B}u, \eta)$ will be assumed to be continuous on $U \times Y$. To discretize the state equation (1), we apply the finite element method to its weak formulation for each fixed $t \in (t_0, t_f)$. Let $Y_h \subset Y = H_0^1(\Omega)$ denote a finite element space for approximating $z(t, .)$ and let $U_h(\Omega) \subset U$ denote a finite element space for approximating u. If $z_{ho} \in Y_h$ is a good approximation of $z(t_o)$ (for instance, a $L^2(\Omega)$-projection), then a semi-discretization of (1) will be:

$$(\dot{z}_h(t), \eta_h) + (\mathcal{A}z_h(t), \eta_h) = (\mathcal{B}u_h(t), \eta_h), \quad \forall \eta_h \in Y_h, \text{ for } t_o < t < t_f, \tag{4}$$
$$z_h(t_o) = z_{ho}. \tag{5}$$

Let $\{\phi_1(x), ..., \phi_n(x)\}$ denote a basis for Y_h and $\{\varphi_1(x), ..., \varphi_m(x)\}$ a basis for U_h, so that we can represent $z_h(t) = \sum_{j=1}^n \phi_j(x)\xi_j(t)$ and $u_h(t) = \sum_{j=1}^m \varphi_j(x)\mu_j(t)$. Then, for any $t \in (t_o, t_f)$, the discrete variational equality (4) is equivalent to:

$$\sum_{j=1}^n (\phi_j, \phi_i) \dot{\xi}_j(t) + \sum_{j=1}^n (\mathcal{A}\phi_j, \phi_i) \xi_j(t) = \sum_{j=1}^m (\mathcal{B}\varphi_j, \phi_i) \mu_j(t) \quad \text{for all } i \in \{1, .., n\}.$$

Define the matrices $(\hat{A}_h)_{ij} := (\mathcal{A}\phi_j, \phi_i)$, $(\hat{M}_h)_{ij} := (\phi_j, \phi_i)$ and $(\hat{B}_h)_{ij} := (\mathcal{B}\varphi_j, \phi_i)$, and the vectors $\xi := (\xi_j(t))_j$, $\mu := (\mu_j(t))_j$ and $\xi_o := \xi(t_0)$. Then, the preceding equations correspond to the following system of ordinary differential equations:

$$\hat{M}_h\dot{\xi} + \hat{A}_h\xi = \hat{B}_h\mu, \ \ t \in (t_o, t_f) \ \text{ and } \xi(t_o) = \xi_o. \tag{6}$$

We discretize the functional (2) as follows:

$$\begin{aligned} J_h(\xi, u) = \ & \tfrac{q}{2} \int_{t_o}^{t_f} (\xi - \xi_*)^T(t)\hat{M}_h(\xi - \xi_*)(t) + \tfrac{r}{2} \int_{t_o}^{t_f} u^T(t)R_h u(t) \\ & + \tfrac{s}{2}(\xi - \xi_*)^T(t_f)\hat{M}_h(\xi - \xi_*)(t_f), \end{aligned} \tag{7}$$

where both R_h and $\hat{M}_h$ are mass matrices. In our applications, Y_h will be piecewise linear finite elements and $\mathcal{U}_h$ will be piecewise constant finite elements. Since matrix $\hat{M}_h$ will be symmetric positive definite, we factorize $\hat{M}_h = U_h^T U_h$ and introduce new variables $y = U_h\xi$ and $u = \mu$. Then, functional (7) will be:

$$\begin{aligned} J_h(y, u) = \ & \tfrac{q}{2} \int_{t_o}^{t_f} (y - y_*)^T(y - y_*) + \tfrac{r}{2} \int_{t_o}^{t_f} u^T R_h u \\ & + \tfrac{s}{2}(y - y_*)^T(y - y_*)(t_f), \end{aligned} \tag{8}$$

and the state equation (6) can be reduced to:

$$\begin{cases} \dot{y} = Ay + Bu, \ \ t \in (0, t_f) \\ y(t_o) = y_0, \end{cases} \tag{9}$$

where $A := U_h^{-T}\hat{A}_h U_h^{-1}$ and $B := U_h^{-T}\hat{B}_h$.

In summary, spatial discretization of (1) transforms the constraints into a system of n linear ordinary differential equations (9), where $y(\cdot) \in \mathbb{R}^n$ denotes the discrete state space variables having initial value y_0, while $u(\cdot) \in \mathbb{R}^m$ denotes the discrete control variables. Although, A, B can be $n \times n$ and $n \times m$ matrix functions, respectively, we shall only consider the time-invariant case with A being a symmetric and *negative* definite matrix of size n, where n is large. When $\mathcal{A} = -\Delta$ and the finite element space Y_h is defined on a triangulation with mesh size h, matrix A will correspond to a discrete Laplacian, and its eigenvalues will lie in an interval $[-c, -d]$ where $c = O(h^{-2})$ and $d = O(1)$.

A general *linear-quadratic* optimal control problem seeks $y(\cdot) \in \mathbb{R}^n$ and $u(\cdot) \in \mathbb{R}^m$ satisfying (9) and *minimizing* a non-negative quadratic cost functional $J(.,.)$, more general than (8), given by:

$$\begin{cases} J(y, u) \equiv \int_{t_o}^{t_f} l(y, u)\, dt + \psi(y(t_f)), \ \ \text{where} \\ l(y, u) \equiv \tfrac{1}{2}\left(e(t)^T Q(t)e(t) + u(t)^T R(t)u(t)\right), \\ \psi(y(t_f)) \equiv \tfrac{1}{2}\left(y(t_f) - y_*(t_f)\right)^T C \left(y(t_f) - y_*(t_f)\right), \end{cases} \tag{10}$$

where $e(t) := y(t) - y_*(t)$ and $Q(.)$ is an $n \times n$ symmetric positive semi-definite matrix function, $y_*(\cdot) \in \mathbb{R}^n$ is a given *tracking* function, C is an $n \times n$ symmetric positive semidefinite matrix, and $R(.)$ is an $m \times m$ symmetric positive definite matrix function. The linear-quadratic optimal control problem seeks the minimum of $J(\cdot)$ in (10) subject to the constraints (9). Given the tracking function $y_*(\cdot)$, the optimal control $u(\cdot)$ must ideally yield $y(\cdot)$ "close" to $y_*(\cdot)$.

3   The Basic Saddle Point System

We now consider a stable discretization of the optimal control problem:

$$J(\hat{y}, \hat{u}) = \min_{(y,u) \in \mathcal{K}} J(y, u), \tag{11}$$

where the constraint set $\mathcal{K}$ consists of (y, u) satisfying:

$$\begin{cases} \dot{y} = A\,y + B\,u, & \text{for } t_o < t < t_f \\ y(t_o) = y_0, \end{cases} \tag{12}$$

where $J(y, u)$ is defined in (10) and matrices Q, R and S are time-invariant. We discretize the time domain $t \in [t_o, t_f]$ using $(l-1)$ interior grid points, so that the time step is $\tau = (t_f - t_o)/l$ with $t_i = i\,\tau$. The state variable y at the time t_i is denoted by $y_i := y(t_i)$. We assume that the discrete controller u is constant on each interval $(t_i, t_{i+1}]$ with the value $u_{i+1/2} = u(t_{i+1/2})$. A stable discretization of equation (12) using the θ-scheme can be written as:

$$F_1\,y_{i+1} = F_0\,y_i + \tau\,B\,u_{i+1/2}, \quad \text{with} \quad y_0 = y(t_o), \quad i = 0, 1, ..., l-1, \tag{13}$$

where matrices F_1, $F_0 \in \Re^{n \times n}$ are given by $F_0 := I + \tau(1-\theta)A$ and $F_1 := I - \tau\theta A$. The preceding discretization of equation (12) takes the matrix form:

$$E\,\mathbf{y} + N\,\mathbf{u} = \mathbf{f}, \tag{14}$$

where the discrete state vector $\mathbf{y} \in \Re^{nl}$ and control vector $\mathbf{u} \in \Re^{ml}$ are:

$$\mathbf{y} := [y_1, \dots, y_l]^T \quad \text{and} \quad \mathbf{u} := [u_{1/2}, \dots, u_{l-1/2}]^T, \tag{15}$$

respectively, the input vector $\mathbf{f} \in \Re^{nl}$ is given by $\mathbf{f} := [-F_0 y_0, 0, ..., 0]^T$, and the matrices $E \in \Re^{(nl) \times (nl)}$ and $N \in \Re^{(ml) \times (ml)}$ have the following block structure:

$$E := \begin{bmatrix} -F_1 & & & \\ F_0 & -F_1 & & \\ & \ddots & \ddots & \\ & & F_0 & -F_1 \end{bmatrix} \quad \text{and} \quad N := \tau \begin{bmatrix} B & & \\ & \ddots & \\ & & B \end{bmatrix}. \tag{16}$$

The discretization of the performance functional $J(y, u)$ takes the form:

$$J_h(y, u) \equiv \frac{1}{2}(\mathbf{u}^T G \mathbf{u}^T + \mathbf{e}^T Z \mathbf{e} + \mathbf{e}^T C e(t_f)), \tag{17}$$

where the discrete error vector $\mathbf{e} := [e_1^T, \dots, e_l^T]^T \in \Re^{nl}$ is defined in terms of the discrete errors at time t_i as follows $e_i := y(i\tau) - y_*(i\tau)$ for $i = 1, ..., l$. In the numerical experiments, we consider matrix G to be diagonal since we approximate the controller using piecewise constant functions in time and also in space. The finite element element functions representing the tracking error are piecewise linear in both time and space, and hence matrix Z is block tri-diagonal where each non-zero block is matrix $\hat{M}_h$. The discrete Lagrangian $\mathcal{L}_h(\mathbf{y}, \mathbf{u}, \mathbf{p})$ associated with the constrained minimization problem has the matrix form:

$$\mathcal{L}_h(\mathbf{y}, \mathbf{u}, \mathbf{p}) = \frac{1}{2}(\mathbf{u}^T G \mathbf{u}^T + \mathbf{e}^T K \mathbf{e}) + \mathbf{p}^T (E\mathbf{y} + N\mathbf{u} - \mathbf{f}), \qquad (18)$$

where K is defined as $K := Z + \Gamma$ and $\Gamma = \text{blockdiag}(0, 0, ..., 0, C)$. A fully discrete version of the optimal control problem (11) will seek the minimum of (17) subject to the constraints (14). To obtain a saddle point formulation, we require the first derivatives of $\mathcal{L}_h(\mathbf{y}, \mathbf{u}, \mathbf{p})$ with respect to $\mathbf{y}$, $\mathbf{u}$ and $\mathbf{p}$ to be zero at the saddle point. This yields the linear system:

$$\begin{bmatrix} K & 0 & E^T \\ 0 & G & N^T \\ E & N & 0 \end{bmatrix} \begin{bmatrix} \mathbf{y} \\ \mathbf{u} \\ \mathbf{p} \end{bmatrix} = \begin{bmatrix} K\mathbf{g} \\ 0 \\ \mathbf{f} \end{bmatrix} \qquad (19)$$

where $\mathbf{g} := [g_i]$ for $g_i = y_*(i\tau)$. In the following, we shall estimate the condition number of matrix EE^T, where E is the evolution matrix.

Theorem 1. *Let A be a $n \times n$ symmetric negative definite matrix. Denote its eigenvalues as $\lambda_i(A)$ for $1 \leq i \leq n$. Let the evolution matrix E be as defined in (16) with matrices F_0 and F_1 defined for $0 \leq \theta \leq 1$ as follows:*

$$F_0 := I + \tau(1-\theta)A \quad and \quad F_1 := I - \tau\theta A \qquad (20)$$

Then, for $\theta \geq \frac{1}{2}$, scheme (13) will be stable for all $\tau > 0$, while for $\theta < \frac{1}{2}$, scheme (13) will be stable only if $\tau \leq 2/((1-2\theta)\rho_{max})$. The following bound:

$$\text{cond}(EE^T) \leq \frac{4(1 + \tau\theta\rho_{\max})^2}{(\tau\rho_{\min})^2}, \qquad (21)$$

will hold, where $\rho_{\max} := \max |\lambda_i|$ and $\rho_{\min} := \min |\lambda_i|$.

Proof. Part 1. Consider the marching scheme for equation (1) given by:

$$y_{k+1} = \Phi y_k + F_1^{-1} \tau B u \qquad (22)$$

where Φ is the marching matrix given by

$$\Phi := (I - \tau \theta A)^{-1}(I + \tau(1-\theta)A). \qquad (23)$$

The stability condition for (22) is given by

$$|(1 - \tau\theta\lambda_i)^{-1}(1 + \tau(1-\theta)\lambda_i)| \leq 1 \qquad (24)$$

or equivalently,

$$\begin{cases} 1 + \tau(1-\theta)\lambda_i \leq 1 - \tau\theta\lambda_i \\ -1 - \tau(1-\theta)\lambda_i \leq 1 - \tau\theta\lambda_i. \end{cases} \qquad (25)$$

From (25), we obtain $\tau\lambda_i \leq 0$ and $\tau \mid \lambda_i \mid (1 - 2\theta) \leq 2$ since $\lambda_i < 0$. In the case $\theta \geq 1/2$, there is no restriction on τ, consequently the marching scheme is unconditionally stable. On other hand, if $\theta < 1/2$ then $0 < (1 - 2\theta)$ and in order for the scheme to be stable it is necessary that $\tau \leq 2/\left((1 - 2\theta)\rho_{\max}\right)$. In this case, the marching scheme is conditionally stable.

Part 2. To estimate cond(EE^T), we shall diagonalize the blocks of EE^T:

$$EE^T = \begin{bmatrix} F_1F_1^T & -F_1F_0^T & & & \\ -F_0F_1^T & F_0F_0^T + F_1F_1^T & -F_1F_0^T & & \\ & -F_0F_1^T & F_0F_0^T + F_1F_1^T & -F_1F_0^T & \\ & & \ddots & \ddots & \ddots \\ & & & -F_0F_1^T & F_0F_0^T + F_1F_1^T \end{bmatrix}.$$

$$(26)$$

Let $Q^TAQ = \Lambda = \operatorname{diag}(\lambda_i)$ denote the diagonalization of A where $Q = [q_1, \ldots, q_n]$ is orthogonal. Then, F_0 and F_1 will also be diagonalized by Q, yielding that $\Lambda_0 = Q^TF_0Q = Q^T(I - \tau\theta A)Q$ and $\Lambda_1 = Q^TF_1Q = Q^T(I + \tau(1 - \theta)A)Q$. If $\mathcal{Q} := \operatorname{blockdiag}(Q, \ldots, Q)$, then $\mathcal{Q}EE^T\mathcal{Q}^T$ will have diagonal blocks:

$$\mathcal{Q}EE^T\mathcal{Q}^T = \begin{bmatrix} \Lambda_1^2 & -\Lambda_0\Lambda_1 & & & \\ -\Lambda_0\Lambda_1 & \Lambda_0^2 + \Lambda_1^2 & -\Lambda_1\Lambda_0 & & \\ & -\Lambda_0\Lambda_1 & \Lambda_0^2 + \Lambda_1^2 & -\Lambda_1\Lambda_0 & \\ & & \ddots & \ddots & \ddots \\ & & & -\Lambda_0\Lambda_1 & \Lambda_0^2 + \Lambda_1^2 \end{bmatrix}.$$

$$(27)$$

Next, we permute the rows and columns of the block tridiagonal matrix (27) using a permutation matrix P, so that $P(\mathcal{Q}EE^T\mathcal{Q}^T)P^T = \operatorname{blockdiag}(\Theta_1, \ldots, \Theta_l)$ where each block submatrix Θ_i is a tridiagonal matrix with entries:

$$\Theta_i := (P\mathcal{Q}EE^T\mathcal{Q}^TP^T)_i = \begin{bmatrix} a_i^2 & -a_ib_i & & & \\ -a_ib_i & a_i^2 + b_i^2 & -a_ib_i & & \\ & -a_ib_i & a_i^2 + b_i^2 & -a_ib_i & \\ & & \ddots & \ddots & \ddots \\ & & & -a_ib_i & a_i^2 + b_i^2 \end{bmatrix}, \quad (28)$$

where $b_i := (1 + \tau(1 - \theta)\lambda_i)$ and $a_i := (1 - \tau\theta\lambda_i)$. Let $\mu(\Theta_i)$ denote an eigenvalue of submatrix Θ_i (and hence also of Θ). Then, Gershgorin's Theorem [2] yields:

$$\mid \mu(\Theta_i) - a_i^2 \mid \leq \mid a_ib_i \mid \quad \text{or} \quad \mid \mu(\Theta_i) - a_i^2 - b_i^2 \mid \leq 2 \mid a_ib_i \mid \qquad (29)$$

Using condition (24), we guarantee stability when $\mid b_i \mid \leq \mid a_i \mid$ obtaining

$$\mu(\Theta_i) \leq \max\left(\mid a_i \mid (\mid a_i \mid + \mid b_i \mid), (\mid a_i \mid + \mid b_i \mid)^2\right) \leq \max 4 \mid a_i \mid^2 \qquad (30)$$

and

$$\mu(\Theta_i) \geq \min\left((\mid a_i \mid^2 - \mid a_i \mid\mid b_i \mid), (\mid a_i \mid - \mid b_i \mid)^2\right) \geq \min(\mid a_i \mid - \mid b_i \mid)^2. \qquad (31)$$

To obtain an upper bound for $\mu(\Theta_i)$ from (30), we define $\rho_{max} := \max |\lambda_i|$, therefore we have $\mu(\Theta_i) \leq 4(1 + \tau\theta\rho_{max})^2$. To obtain a lower bound for $\mu(\Theta_i)$, from (31) we define $\rho_{min} := \min |\lambda_i|$ obtaining $\mu(\Theta_i) \geq (\tau\rho_{min})^2$. Therefore, the condition number of the matrix EE^T will satisfy the bound:

$$\text{cond}(EE^T) \leq 4\left(\frac{1 + \tau\theta\rho_{max}}{\tau\rho_{min}}\right)^2. \tag{32}$$

This completes the proof.

Remark. For finite difference and finite element discretizations on a domain of size $O(1)$, the eigenvalues $\lambda_i(A)$ of the scaled matrix A will satisfy the bounds $\alpha_1 \leq |\lambda_i(A)| \leq \alpha_2 h^{-2}$ Then, using (32) we obtain:

$$\text{cond}(EE^T) \approx \left(\frac{1 + \tau\theta\alpha_2 h^{-2}}{\tau\alpha_1}\right)^2. \tag{33}$$

Thus, matrix EE^T will be ill-conditioned with a condition number that can grow as $O(h^{-4})$ depending on τ and h. If system (19) is solved using Uzawa's method, it will be necessary to solve $-(EK^{-1}E^T + NG^{-1}N^T)\mathbf{p} = \mathbf{f} - E\mathbf{g}$, where matrix $S := (EK^{-1}E^T + NG^{-1}N^T)$ is the Schur complement of system (19) with respect to the Lagrange multiplier $\mathbf{p}$.

Next, we analyze the condition number of S. Notice that due to the positive semi-definiteness of matrix C in (17), we obtain in the sense of quadratic forms that $K^{-1} = (Z + \Gamma)^{-1} \leq Z^{-1}$ and apply it in the following estimate for the condition number of the Schur complement S. Henceforth, we normalize $q = 1$.

Lemma 1. *Let the upper and lower bound for the singular values of EE^T be given by $4(1 + \tau\theta\rho_{max})^2$ and $(\tau\rho_{min})^2$, respectively. Let us assume, using (8) and (9), that the mass matrices Z, G, N, and Γ satisfy*

$$c_1\tau\mathbf{y}^T\mathbf{y} \leq \mathbf{y}^T Z\mathbf{y} \leq c_2\tau\mathbf{y}^T\mathbf{y} \tag{34}$$

$$c_3 r\tau h^d\mathbf{u}^T\mathbf{u} \leq \mathbf{u}^T G\mathbf{u} \leq c_4 r\tau h^d\mathbf{u}^T\mathbf{u}, \tag{35}$$

$$c_5\tau^2 h^d\mathbf{p}^T\mathbf{p} \leq \mathbf{p}^T NN^T\mathbf{p} \leq c_6\tau^2 h^d\mathbf{p}^T\mathbf{p} \quad and \tag{36}$$

$$0 \leq \mathbf{y}^T\Gamma\mathbf{y} \leq c_7 s\mathbf{y}^T\mathbf{y}. \tag{37}$$

Then, the condition number of matrix S will satisfy the bound:

$$\text{cond}(S) \leq \left(\frac{c_4 r(c_5\tau + c_7 s)}{c_1\tau c_3 r}\right)\left(\frac{4 c_3 r(1 + \rho_{max}\tau\theta)^2 + c_6\tau^2 c_1}{c_4 r(\tau\rho_{min})^2 + c_5\tau(c_2\tau + c_7 s)}\right) \tag{38}$$

where $S := EK^{-1}E^T + NG^{-1}N^T$ denotes the Schur complement.

Proof. Using the upper and lower bounds for K, EE^T, NN^T and G we obtain:

Upper bound:

$$\mathbf{p}^T S \mathbf{p} = \mathbf{p}^T E K^{-1} E^T \mathbf{p} + \mathbf{p}^T N G^{-1} N^T \mathbf{p} \tag{39}$$

$$\leq \mathbf{p}^T E Z^{-1} E^T \mathbf{p} + \mathbf{p}^T N G^{-1} N^T \mathbf{p} \tag{40}$$

$$\leq \frac{1}{c_1 \tau} \mathbf{p}^T E E^T \mathbf{p} + \frac{1}{c_3 r \tau h^d} \mathbf{p}^T N N^T \mathbf{p} \tag{41}$$

$$\leq \left(\frac{4}{c_1 \tau} (1 + \tau \theta \rho_{\max})^2 + \frac{c_6 \tau^2 h^d}{c_3 r \tau h^d} \right) \mathbf{p}^T \mathbf{p} \tag{42}$$

$$= \left(\frac{4}{c_1 \tau} (1 + \tau \theta \rho_{\max})^2 + \frac{c_6 \tau}{c_3 r} \right) \mathbf{p}^T \mathbf{p}. \tag{43}$$

Lower bound:

$$\mathbf{p}^T S \mathbf{p} \geq \frac{1}{(c_2 \tau + c_7)} \mathbf{p}^T E E^T \mathbf{p} + \frac{1}{c_4 r \tau h^d} \mathbf{p}^T N N^T \mathbf{p} \tag{44}$$

$$\geq \left(\frac{(\tau \rho_{\min})^2}{(c_2 \tau + c_7 s)} + \frac{c_5 \tau^2 h^d}{c_4 r \tau h^d} \right) \mathbf{p}^T \mathbf{p} \tag{45}$$

$$= \left(\frac{(\tau \rho_{\min})^2}{(c_2 \tau + c_7 s)} + \frac{c_5 \tau}{c_4 r} \right) \mathbf{p}^T \mathbf{p}. \tag{46}$$

Therefore, the condition number of matrix S can be estimated by:

$$\text{cond}(S) = \frac{c_4 r (c_5 \tau + c_7 s)}{c_1 \tau c_3 r} \frac{4 c_3 r (1 + \rho_{\max} \tau \theta)^2 + c_6 \tau^2 c_1}{c_4 r (\tau \rho_{\min})^2 + c_5 \tau (c_2 \tau + c_7 s)}. \tag{47}$$

Remark. The estimate given in (47) shows that matrix S is ill-conditioned. Indeed, let all the constants $c_i = 1$. Then the expression (47) reduces to:

$$\text{cond}(S) \approx \left(\frac{\tau + s}{\tau} \right) \left(\frac{r (1 + h^{-2} \tau \theta)^2 + \tau^2}{r \tau^2 + \tau^2 + \tau s} \right). \tag{48}$$

Choosing $\theta = 1$ and $h \approx \tau$, and using the reasonable assumption:

$$0 < O(h^4) \leq r \leq O(s/\tau),$$

yields that $\text{cond}(S) \approx O(r h^{-4})$.

4   The Reduced System for u

We shall now describe an algorithm to solve the saddle point system (19) based on the solution of a reduced Schur complement system for the control variable $\mathbf{u}$. Assuming that $G \neq 0$ and solving the first and third block rows in (19) yields $\mathbf{p} = -E^{-T} K \mathbf{y} + E^{-T} K \mathbf{g}$ and $\mathbf{y} = -E^{-1} N \mathbf{u} + E^{-1} \mathbf{f}$, respectively. System (19) can then be reduced to the following Schur complement system for $\mathbf{u}$:

$$(G + N^T E^{-T} K E^{-1} N) \mathbf{u} = N^T E^{-T} K E^{-1} \mathbf{f} - N^T E^{-T} K \mathbf{g}, \tag{49}$$

where matrix $(G + N^T E^{-T} K E^{-1} N)$ is symmetric and positive definite. In the next Lemma, we show that $(G + N^T E^{-T} K E^{-1} N)$ is spectrally equivalent to G.

Lemma 2. *Let the bounds for G, E, K, N, Γ presented in Lemma 1 hold. Then, there exists $\mu_{mim} > 0$ and $\mu_{max} > 0$, independent of h and u, such that:*

$$\mu_{min}\left(\mathbf{u}^T G \mathbf{u}\right) \leq \mathbf{u}^T (N^T E^{-T} K E^{-1} N)\mathbf{u} \leq \mu_{max}\left(\mathbf{u}^T G \mathbf{u}\right) \tag{50}$$

Proof. Using the upper and lower bounds for K, EE^T, NN^T and G we obtain:

Upper bound:

$$\mathbf{u}^T N^T E^{-T} K E^{-1} N \mathbf{u} \leq (c_2\,\tau + c_7\,s)\,\mathbf{u}^T N^T E^{-T} E^{-1} N \mathbf{u} \tag{51}$$

$$\leq \frac{(c_2\,\tau + c_7\,s)}{(\tau\,\rho_{\min})^2}\,\mathbf{u}^T N^T N \mathbf{u} \tag{52}$$

$$\leq \frac{(c_2\,\tau + c_7\,s)\,c_6\,\tau^2\,h^d}{(\tau\,\rho_{\min})^2}\,\mathbf{u}^T \mathbf{u} \tag{53}$$

$$= \frac{(c_2\,\tau + c_7\,s)c_6\,h^d}{(\rho_{\min})^2}\,\mathbf{u}^T \mathbf{u} \tag{54}$$

$$\leq \frac{(c_2\,\tau + c_7\,s)\,c_6}{(\rho_{\min})^2\,c_3\,r\,\tau}\,\mathbf{u}^T G \mathbf{u} \tag{55}$$

$$= \mu_{\max}\,\mathbf{u}^T G \mathbf{u}. \tag{56}$$

Lower bound:

$$\mathbf{u}^T N^T E^{-T} K E^{-1} N \mathbf{u} \geq (c_1\,\tau)\,\mathbf{u}^T N^T E^{-T} E^{-1} N \mathbf{u} \tag{57}$$

$$\geq \frac{c_1\,\tau}{4\,(1 + \tau\,\rho_{\max}\,\theta)^2}\,\mathbf{u}^T N^T N \mathbf{u} \tag{58}$$

$$\geq \frac{c_1\,c_5\,\tau^3\,h^d}{4\,(1 + \tau\,\rho_{\max}\,\theta)^2}\,\mathbf{u}^T \mathbf{u} \tag{59}$$

$$\geq \frac{c_1\,c_5\,\tau^2\,h^d}{4\,(1 + \tau\,\rho_{\max}\,\theta)^2\,c_4\,r}\,\mathbf{u}^T G \mathbf{u} \tag{60}$$

$$= \mu_{\min}\,\mathbf{u}^T G \mathbf{u}. \tag{61}$$

This completes the proof.

First Algorithm. The Schur complement system (49) can be solved using a CG algorithm (conjugate gradient) using G as a preconditioner. Note that:

$$\mathbf{u}^T G \mathbf{u} \leq \mathbf{u}^T (G + N^T E^{-T} K E^{-1} N)\mathbf{u} \leq (1 + \mu_{max})\,\mathbf{u}^T G \mathbf{u}. \tag{62}$$

Since $\rho_{\min}$ is $O(1)$ and $\rho_{\max}$ is $O(h^{-4})$, it is easy to see that $\mu_{\min} = O(\frac{h^4}{r})$ and $\mu_{\max} = O(\frac{1+s/\tau}{r})$. Hence, the rate of convergence of this algorithm will be independent of h, with a condition number estimate bounded by $O(1 + \frac{1+\frac{s}{\tau}}{r})$. This algorithm is simple to implement however has two drawbacks. It has *inner* and *outer* iterations, and requires applications of E^{-1} (and E^{-T}) which are not directly parallelizable.

Second Algorithm. Our second algorithm avoids double iteration. Define:

$$\hat{\mathbf{b}} := -N^T E^{-T} K E^{-1}\mathbf{f} + N^T E^{-T} K \mathbf{g} \quad \text{and} \quad \mathbf{w} := -E^{-T} K E^{-1} N \mathbf{u}.$$

Then, the solution to system (49) can be obtained by solving the system:

$$\begin{bmatrix} EK^{-1}E^T & N \\ N^T & -G \end{bmatrix} \begin{bmatrix} \mathbf{w} \\ \mathbf{u} \end{bmatrix} = \begin{bmatrix} \mathbf{0} \\ \hat{\mathbf{b}} \end{bmatrix}, \tag{63}$$

which is symmetric and *indefinite*. The action of E^{-1} is required only in a pre-computed step to assemble the right hand side input vector $\hat{\mathbf{b}}$. Since system (63) is symmetric indefinite, it can be solved iteratively using the MINRES algorithm with a positive definite block diagonal *preconditioner* $\mathrm{diag}(E_o K_o^{-1} E_o^T, G_o)$, where K_o is any matrix spectrally equivalent to the mass matrix K, matrix E_o is any matrix spectrally equivalent to (or a preconditioner for) the evolution matrix E, see [6,13,3], and matrix G_o is a preconditioner for matrix G. The following Theorem estimates the number when $E_0 = E$, $K_0 = K$ and $G_0 = G$.

Theorem 2. *Let the bounds for matrices G, E, K, N and Γ hold as presented in lemma 1. Let $\mathcal{P} := blockdiag(EK^{-1}E^T, G)$ denote a block diagonal precondi-tioner for the coefficient matrix $\mathcal{H}$ of system (63). Then, the condition number of the preconditioned system will satisfy the bound:*

$$\kappa(\mathcal{P}^{-1}\mathcal{H}) \leq O\left(\left(1 + \frac{1 + s/\tau}{r} \right)^{1/2} \right). \tag{64}$$

Proof. Since the preconditioner $\mathcal{P}$ is positive definite, we consider the generalized eigenvalue problem given by:

$$\begin{bmatrix} EK^{-1}E^T & N \\ N^T & -G \end{bmatrix} \begin{bmatrix} \mathbf{w} \\ \mathbf{u} \end{bmatrix} = \lambda \begin{bmatrix} EK^{-1}E^T & \\ & G \end{bmatrix} \begin{bmatrix} \mathbf{w} \\ \mathbf{u} \end{bmatrix}, \tag{65}$$

We obtain the equations:

$$(\lambda - 1)EK^{-1}E^T\mathbf{w} = N\mathbf{u} \quad \text{and} \quad (\lambda + 1)G\mathbf{u} = N^T\mathbf{w}. \tag{66}$$

These equations yield $N^T E^{-T} K E^{-1} N u = (\lambda^2 - 1)Gu$ where $(\lambda^2 - 1)$ is the generalized eigenvalue of $N^T E^{-T} K E^{-1} N$ with respect to G. Using Lemma 2, we obtain bounds for λ as follows:

$$max|\lambda| \leq (1 + \mu_{max})^{1/2} = O\left(\left(1 + \frac{1 + s/\tau}{r} \right)^{1/2} \right) \tag{67}$$

$$min|\lambda| \geq (1 + \mu_{min})^{1/2} = O(1). \tag{68}$$

The desired result now follows, since:

$$\kappa(\mathcal{P}^{-1}\mathcal{H}) \leq \frac{max|\lambda|}{min|\lambda|}. \tag{69}$$

This completes the proof.

Remark. Generalization of this theorem for matrices G_o and $E_o K_o^{-1} E_o^T$ (spec-trally equivalent to G and $EK^{-1}E^T$ respectively) follows directly from [5].

Table 1. Number of CG iterations for Algorithm 1. The parameters $s = 0$ ($s = 1$).

Nx \ Nt	32	64	128	256	512
32	36 (41)	38 (44)	40 (46)	40 (47)	40 (48)
64	36 (41)	38 (44)	40 (46)	40 (47)	40 (48)
128	36 (41)	38 (44)	40 (46)	40 (47)	40 (48)
256	36 (41)	38 (44)	40 (46)	40 (47)	40 (48)
512	36 (41)	38 (44)	40 (46)	40 (47)	40 (48)

Remark. Applying matrix E is highly unstable, but applying E^{-1} is stable. The algorithms presented here do not require application of E or E^T since:

$$\mathcal{P}\mathcal{H} = \begin{bmatrix} I & E^{-T}KE^{-1}N \\ G^{-1}N^T & -I \end{bmatrix}. \tag{70}$$

5 Numerical Experiments

In this section, we consider the numerical solution of an optimal control problem involving the 1D-heat equation. In this case, the constraints are given by:

$$z_t - z_{xx} = v, \ \ 0 < x < 1, \ \ t > 0$$

with boundary conditions $z(t,0) = 0$ and $z(t,1) = 0$ for $t \geq 0$, with initial data $z(0,x) = 0$ for $x \in [0,1]$, and with the performance function $z_* = x(1-x)e^{-x}$ for all $t \in [0,1]$. Following [8], we take $q = 1$ and $r = 0.0001$. The backward Euler discretization ($\theta = 1$) is considered in the numerical experiments. As a stopping criteria for the iterative solvers, we take $\|\mathbf{r_k}\|/\|\mathbf{r_0}\| \leq 10^{-9}$ where $\mathbf{r_k}$ is the residual at each iteration k.

Table 2. Condition number of the preconditioned matrix of Algorithm 1. The parameters are $q = 1$, $r = 0.0001$, $t_f = 1$, $h = 1/32$ and $\tau = 1/64$.

r \ s	10^4	10^2	1	10^{-2}	0
10^{-2}	4.9 10^4	5.0 10^2	6.2	1.9	1.9
10^{-4}	4.7 10^6	4.8 10^4	5.2 10^2	93	93
10^{-6}	4.2 10^8	4.8 10^6	5.1 10^4	9.2 10^3	9.1 10^3

Algorithm 1: Reduction to u. We consider matrix G as a preconditioner for system (49) and use the CG method to solve the resulting preconditioned system. For the case where $s = 0$ and in parenthesis $s = 1$, Table 1 presents the number of iterations for different time and space meshes. As predicted by the theory in Section 4, the number of iterations remains constant as h is refined. Table 1 also shows that the number of iterations deteriorates very weakly when the time discretization τ gets finer. As expected from the analysis, this

Table 3. Number of CG iterations for Algorithm 1 for different values of τ and s. The space discretization is $h = 1/32$.

Nt \ s	0	1	10	100
32	36	41	50	68
64	38	44	52	73
128	40	46	57	83
256	40	47	62	89
512	40	48	64	96

deterioration is more noticeable for larger s, (see Tables 1 and 3). Table 2 shows that the condition number estimates in Section 4 are sharp for different values of parameters r and s.

Algorithm 2. Table 4 presents the number of iterations required to solve system (63) using MINRES acceleration when both time and space grid sizes are refined. As predicted from the analysis, as the space grid is refined, the number of iterations remains bounded. As before, for larger s, a slight deterioration in the number of iterations is observed.

Table 4. Number of MINRES iterations for algorithm 2. Parameter $s = 0$ ($s = 1$).

Nx \ Nt	32	64	128	256	512
32	56 (60)	58 (62)	60 (64)	60 (66)	60 (68)
64	56 (60)	58 (62)	60 (64)	60 (66)	60 (68)
128	56 (60)	58 (62)	60 (64)	60 (66)	60 (68)
256	56 (60)	58 (62)	60 (64)	60 (66)	60 (68)
512	56 (60)	58 (62)	60 (64)	60 (66)	60 (68)

6   Concluding Remarks

In this paper we have described two approaches for iteratively solving the linear quadratic parabolic optimal control problem. The first method is based on the CG solution of a Schur complement. This is obtained by reducing the saddle point system to the system associated with the control variable. This method is simple to implement but requires double iteration. The second method avoids double iteration by introducing an auxiliary variable. The resulting system is symmetric and indefinite, so that MINRES can be used. The structure of this method also allows parallel block preconditioners. The preconditioners described yield a rate of convergence independent of the time and space parameters.

Acknowledgments. C.E.S. acknowledges Etereldes Gonçalves Junior for a fruitful discussion on topics of this work.

References

1. G. BIROS AND O. GATTAS, *Parallel Lagrange-Newton-Krylov-Schur Methods for PDE-Constrained Optimization. Part I: The Krylov-Schur Solver*, SIAM Journal on Scientific Computing, 27(2), pp. 687–713, 2005.
2. J. W. DEMMEL, *Applied Numerical Linear ALgebra*, SIAM, Philadelphia, 1997.
3. M. J. GANDER AND S. VANDEWALLE, *On the super linear and linear convergence of the parareal algorithm*, Proceedings of the 16th International Conference on Domain Decomposition Methods, 2005.
4. M. HEINKENSCHLOSS, *A time-domain decomposition iterative method for the solution of distributed linear quadratic optimal control problems*, Journal of Computational and Applied Mathematics, Volume 173, Issue 1, No 1, Pages 169-198, 2005.
5. A. KLAWON, *Preconditioners for Indefinite Problems*, Reports CS-NYU, TR1996-716, March, 1996.
6. J. L. LIONS AND Y. MADAY AND G. TURINICI, *Résolution d'EDP par un schéma en Temps Pararéel*, C.R. Acad. Sci. Paris, t. 332, Série I, pp. 661–668, 2001.
7. J.L. LIONS *Optimal Control of Systems Governed by Partial Differential Equations problems*, Springer, Berlin, 1971.
8. A. LOCATELLI, *Otimal Control: An Introduction*, Birkhäuser, Berlin, 2001.
9. D. LUENBERGER, *Introduction to Dynamic Systems: Theory, Models and Applications*, Wiley, New York, 1979.
10. Y. MADAY AND G. TURINICI, *A parareal in time procedure for the control of partial differential equations*, C.R.Acad. Sci. Paris, t. 335, Ser. I, pp. 387–392, 2002.
11. T. MATHEW AND M. SARKIS AND C.E. SCHAERER , *Block matrix preconditioners for elliptic optimal control problems*, accepted in Numerical Linear Algebra with Applications, 2006.
12. P. SARMA, K. AZIZ AND L.J. DURLOFSKY, *Implementation of adjoint solution for optimal control of smart well*, SPE reservoir Simulation Symposium, Texas-USA, January 31- February 2, 2005.
13. C.E. SCHAERER AND E. KASZKUREWICZ, *The shooting method for the numerical solution of ordinary differential equations: a control theoretical perspective*, International Journal of Systems Science, Vol. 32, No. 8, pp. 1047-1053, 2001.
14. C.E. SCHAERER AND T. MATHEW AND M. SARKIS, *Parareal-block matrix preconditioners for the control of PDEs of parabolic type*, Proceedings of the 17th International Conference on Domain Decomposition Methods. Lecture Notes in Computational Science and Engineering, Springer Verlag, 2007. Submitted.
15. B. SUDARYANTO AND Y. C. YORTSOS, *Optimization of fluid fronts dynamics in porous media using rate control. I. Equal mobility fluids*, Physics of Fluids, Vol. 12, No 7, pp. 1656–1670, 2000.

Evaluation of Linear Solvers for Astrophysics Transfer Problems[*]

Osni Marques[1] and Paulo B. Vasconcelos[2]

[1] Lawrence Berkeley National Laboratory,
1 Cyclotron Road, MS 50F-1650, Berkeley, CA 94720-8139, USA
oamarques@lbl.gov
[2] Faculdade de Economia da Universidade do Porto,
Rua Dr. Roberto Frias s/n, 4200-464 Porto, Portugal
pjv@fep.up.pt

Abstract. In this work we consider the numerical solution of a radiative transfer equation for modeling the emission of photons in stellar atmospheres. Mathematically, the problem is formulated in terms of a weakly singular Fredholm integral equation defined on a Banach space. Computational approaches to solve the problem are discussed, using direct and iterative strategies that are implemented in open source packages.

Keywords: High performance computing, Fredholm integral equation, weakly singular kernel, projection approximation, numerical methods.

AMS subject classification: 32A55, 45B05, 65D20, 65R20, 68W10.

1 Introduction and Problem Overview

The emission of photons in stellar atmospheres can be modeled by a strongly coupled system of nonlinear equations. In this work we consider a restriction of the system by taking into account the temperature and pressure. We refer the reader to [2] and [13] for details on the model. The resulting integral equation, a radiative transfer problem, is expressed by

$$T\varphi - z\varphi = f, \quad \varphi \in L^1(I), \quad I = [0, \tau^*], \tag{1}$$

defined on a Banach space $L^1(I)$, where the integral operator T is defined as

$$(T\varphi)(\tau) = \frac{\varpi}{2} \int_0^{\tau^*} E_1(|\tau - \tau'|) \varphi(\tau') d\tau'. \tag{2}$$

The variable τ represents the optical depth, τ^* is the optical thickness of a stellar atmosphere, z is in the resolvent set of T and $\varpi \in]0, 1[$ is the albedo (assumed

[*] This work was partly supported by the Centro de Matemática da Universidade do Porto, through the programs POCTI and POSI, and partly by the Director, Office of Science, Division of Mathematical, Information, and Computational Sciences of the U.S. Department of Energy under contract No. DE-AC02-05CH11231.

M. Daydé et al. (Eds.): VECPAR 2006, LNCS 4395, pp. 466–475, 2007.
© Springer-Verlag Berlin Heidelberg 2007

to be constant in the present work). The free term f is taken to be $f(\tau) = -1$ if $0 \le \tau \le \tau^*/2$, and $f(\tau) = 0$ if $\tau^*/2 < \tau \le \tau^*$. The first exponential-integral function E_1, defined by

$$E_1(\tau) = \int_1^\infty \frac{\exp(-\tau\mu)}{\mu} d\mu, \quad \tau > 0, \tag{3}$$

has a logarithmic behavior in the neighborhood of 0.

The numerical approach used to solve this problem is based on the projection of the integral operator into a finite dimensional subspace. By evaluating the projected problem on a specific basis function we obtain a linear system of equations whose coefficient matrix is banded, sparse and nonsymmetric. In order to obtain a good accuracy for the solution it is necessary to use a large dimension for the space where the problem is projected into. One possible approach is to compute an approximate initial solution in a subspace of moderate (small) size and then iteratively refine it by a Newton-type method. This approach was adopted with success in [17]. Alternatively, one can discretize the problem on a finer grid and then solve a large banded sparse algebraic linear system. In this case, depending on the dimension of the problem, we can employ either direct or iterative methods.

This work aims to explore the second approach mentioned above. Due to the large dimensional cases of interest to the astrophysicists and due to the memory limitation of computers, one needs to have access to scalable parallel versions of the code. Scalability is crucial either for the generation phase of the matrix coefficients as well as for the solution phase. In [17], scalability of the solution phase was not achieved because the (moderate size) systems were not solved in a distributed way. The parallelization of the solution phase was limited to the iterative refinement process. MPI was used for this purpose.

A large number of computational models and simulations that are analyzed and solved on nowadays high end computers benefit from the use of advanced and promptly available software tools and libraries to achieve performance, scalability and portability. In these lines, we are interested in investigating the trade offs and capabilities implemented in several packages, in particular the ones that are available in the DOE Advanced CompuTational Software (ACTS) Collection [9]. In the following sections we outline the projection and matrix formulation that we use to tackle the integral operator. Next, we give a brief description of the ACTS Collection, and the tools that are pertinent to our application. Finally, we present some numerical results and drawn up some conclusions.

2 Projection Phase and Matrix Formulation

Integral equations as the one described in the previous section are usually solved by discretization mechanisms, for instance by projection into a finite dimensional subspace. The operator T is thus approximated by T_n, with its projection into the finite dimensional subspace given by $X_n = \text{span}\{e_{n,j}, j = 1, \ldots, n\}$ (spanned by n linearly independent functions in X). In this case, we will take for X_n the

basis $e_n = [e_{n,1} \ldots e_{n,n}]$ of piecewise constant functions on each subinterval of $[0, \tau^*]$ determined by a grid of $n+1$ points $0 = \tau_{n,0} < \tau_{n,1} < \ldots < \tau_{n,n} = \tau^*$.
For $x \in X$ let

$$\langle x, e_{n,j}^* \rangle = e_{n,j}^*(x) = \frac{1}{\tau_{n,j} - \tau_{n,j-1}} \int_{\tau_{n,j-1}}^{\tau_{n,j}} x(\tau) d\tau,$$

and define

$$T_n x = \pi_n T x = \sum_{j=1}^{n} \langle x, T^* e_{n,j}^* \rangle e_{n,j}, \tag{4}$$

where $\pi_n x = \sum_{j=1}^{n} \langle x, e_{n,j}^* \rangle e_{n,j}$ and $T^* e_n^* \in X^*$(the adjoint space of X). The approximate problem

$$(T_n - zI)\varphi_n = f \tag{5}$$

is then solved by means of an algebraic linear system of equations

$$(A - zI)x = b, \tag{6}$$

where A is a non singular matrix of order n, and $A(i,j) = \langle e_{n,j}, T^* e_{n,i}^* \rangle$, $b(i) = \langle f, T^* e_{n,i}^* \rangle$, $x(j) = \langle \varphi_n, T^* e_{n,j}^* \rangle$ (see [2]). The relation between x and φ_n is given by

$$\varphi_n = \frac{1}{z} \left(\sum_{j=1}^{n} x(j) e_{n,j} - f \right).$$

In order to achieve an approximate solution φ_n with good accuracy by this method it may be necessary to use a very large dimensional linear system.
To obtain the elements of A we need to compute

$$A(i,j) = \frac{\varpi}{2(\tau_{n,i} - \tau_{n,i-1})} \int_{\tau_{n,i-1}}^{\tau_{n,i}} \int_{0}^{\tau^*} E_1(|\tau - \tau'|) e_{n,j}(\tau') d\tau' d\tau$$

for $i, j = 1, \ldots, n$. Using the fact that $E_3(0) = 1/2$, we obtain

$$A(i,j) = \begin{cases} \frac{\varpi}{2(\tau_{n,i} - \tau_{n,i-1})} [-E_3(\tau_{n,i} - \tau_{n,j}) + E_3(\tau_{n,i-1} - \tau_{n,j}) + \\ \qquad + E_3(\tau_{n,i} - \tau_{n,j-1}) + E_3(\tau_{n,i-1} - \tau_{n,j-1})] & \text{if } i \neq j \\ \\ \varpi \left[1 + \frac{1}{\tau_{n,i} - \tau_{n,i-1}} (-E_3(\tau_{n,i} - \tau_{n,i-1}) - 1) \right] & \text{if } i = j \end{cases}, \tag{7}$$

where

$$E_3(\tau) = \int_{1}^{\infty} \frac{\exp(-\tau\mu)}{\mu^3} d\mu. \tag{8}$$

For computational purposes, this function is evaluated according to [1].

3 ACTS: Tools of the Trade

The ACTS Collection consists of a set of computational tools for the solution of common and important computational problems. The tools were developed in various laboratories and universities and have allowed a wide spectrum of important computational problems to be solved to content [10]. We refer the reader to [9] for an overview of the project and available numerical tools, and also to the ACTS Information Center [12] for details about all tools available in the Collection.

In this paper we are interested in solving equation (1) on a fine mesh. ACTS incorporates the packages ScaLAPACK [6], SuperLU [7], PETSc [5] and Trilinos [11]. ScaLAPACK provides routines for distributed-memory message-passing MIMD architectures, in particular routines for solving systems of linear equations, least squares, eigenvalue problems and singular value problems. SuperLU is a library for the direct solution of large, sparse, nonsymmetric systems of linear equations, but that can also be applied efficiently to many symmetric systems. Working precision iterative refinement subroutines are provided for improved backward stability. PETSc provides a number of functionalities for the numerical solution of PDEs that require solving large-scale, sparse linear and nonlinear systems of equations. It includes nonlinear and linear equation solvers that employ a variety of Newton techniques and Krylov subspace methods. Trilinos is one the the last additions to ACTS. It targets the development of parallel solver algorithms and libraries within an object-oriented software framework. It provides self-contained packages, each one with its own set of requirements. One of this packages is AztecOO, which superseded the widely used package Aztec.

In order to solve the problem for larger values of $\tau^\star$ we need to use high performance computers as well a scalable software. Taking into account the characteristics of the coefficient matrix here we will focus on SuperLU and PETSc, for the direct and iterative solution, respectively, of a large, sparse, nonsymmetric system of linear equations.

4 Numerical Results

The numerical results showed in this section were obtained on an SGI Altix 350, an AMD Opteron cluster, and an IBM SP. The Altix is configured with 32 64-bit 1.4 GHz Intel Itanium-2 processors, with 192 GBytes of shared memory. The cluster is configured with 356 dual-processor nodes, each processor running at a clock speed of 2.2 GHz, with 6 GB of memory per node, interconnected with a high-speed InfiniBand network. The IBM SP is configured with 380 compute nodes with 16 Power 3+ processors per node. Most nodes have 16 GB of memory. These three systems are located at the National Energy Research Scientific Computing Center (NERSC), Lawrence Berkeley National Laboratory, of the US Department of Energy. To validade our implementation, our experiments used only a small fraction of the computer power provided by those systems. For the physical problem we considered $\varpi = 0.75$ and $\varpi = 0.90$, and explored the band and sparse characteristics of the coefficient matrix obtained for this particular kernel as mentioned earlier.

Table 1. Normalized times for the generation of the matrix and solution of the system of equations with SuperLU, for various matrix sizes (m) and $\varpi = 0.75$, on the SGI Altix

	generation	solution	
m		factor	solve
1000	3.26E+03	6.95E+01	1.00E+00
2000	2.12E+04	1.65E+02	3.00E+00
4000	9.71E+04	3.59E+01	6.00E+00
8000	4.26E+05	7.51E+02	1.80E+01
16000	1.80E+06	1.54E+03	3.00E+01
32000	7.36E+06	3.12E+03	5.35E+01

Table 2. Normalized times for the generation of matrices of various sizes (m), with the corresponding number of nonzeros in the matrix (nnz) for $\varpi = 0.75$, and solution with two distinct solvers, on one processor of the IBM SP

		generation	SuperLU	GMRES
m	nnz		(factor+solve)	(22 iterations)
1000	104918	5.79E+01	1.14E+00	1.00E+00
2000	215718	2.27E+02	2.75E+00	1.90E+00
4000	445264	8.83E+02	6.11E+00	3.36E+00
8000	880518	3.46E+03	1.26E+01	6.96E+00

In Table 1 we show normalized times required for the generation of the matrix (and right-hand side) and for the solution of problem with SuperLU on one processor of the SGI Altix, for $\varpi = 0.75$. As can be seen in the table, the most time consuming part of the simulation is the generation of the matrix, due to the large number of exponential evaluations. This phase is orders of magnitude more expensive that the other calculations and grows exponentially. The factor phase of the solution is then the second most time consuming part. One of the main advantages of the LU factorization is the potential gain that we can achieve if there is a need to solve several linear systems with the same coefficient matrix. However, this is not the case here. In addition, for higher dimensional problems direct methods usually becomes less competitive.

Table 2 shows normalized times required for the generation of the matrix (and right-hand side), and for the solution of problem with SuperLU and GMRES with Jacobi preconditioner on one processor of the IBM SP. The tolerance for the iterative method was set to 10^{-10}. We notice that for the parameters we have used in defining the problem an iterative method is very adequate in the sequential case.

The numbers in Tables 1 and 2 stress the need for parallelization in order to solve the problem for higher dimensions. It turns out that the terms in (7) can be analytically developed such that their generation becomes embarrassing parallel, that is, without any communication needed among the processors [17]. As a result, the data distribution can also be done accordingly to the solver used.

Table 3. Normalized times and number of iterations for various matrix sizes (m) and for $\varpi = 0.75$ on up to 64 processors (p) of the IBM SP

m	p	generation	GMRES (22 iterations)	BiCGstab (14 iterations)
10000	2	7.70E+03	8.77E+00	9.16E+00
	4	3.87E+03	6.32E+00	6.11E+00
	8	1.93E+03	3.06E+00	3.24E+00
	16	9.72E+02	1.90E+00	2.43E+00
	32	4.87E+02	1.33E+00	1.00E+00
25000	16	6.13E+03	4.95E+00	3.62E+00
	32	3.00E+03	2.47E+00	2.04E+00
	64	1.49E+03	1.76E+00	1.36E+00

Table 4. Normalized times and number of iterations for various matrix sizes (m) and for $\varpi = 0.75$ on up to 32 processors (p) on the Opteron cluster

m	p	generation	GMRES (22 iterations)	BiCGstab (14 iterations)
10000	1	5.40E+03	7.54E+00	7.95E+00
	2	2.67E+03	4.02E+00	4.58E+00
	4	1.39E+03	2.32E+00	2.56E+00
	8	6.90E+02	1.80E+00	1.97E+00
	16	3.51E+02	1.15E+00	1.25E+00
	32	1.79E+02	1.15E+00	1.36E+00
25000	4	8.41E+03	5.42E+00	5.61E+00
	8	4.28E+03	3.02E+00	3.15E+00
	16	2.16E+03	2.05E+00	1.83E+00
	32	1.07E+03	1.00E+00	1.15E+00
50000	16	8.57E+03	3.14E+00	3.20E+00
	32	4.24E+03	1.53E+00	1.86E+00

In Table 3 we list (normalized) times for the generation phase and for two preconditioned iterative methods implemented in PETSc on the IBM SP, for up to 64 processors, for $\varpi = 0.75$. We observe that there are significant gains by using the parallel version of the code. The generation phase is still the most time consuming part of the algorithm. The two sparse iterative solvers show similar times, although for the parameters chosen BICGstab requires fewer iterations.

In Table 4 we list (normalized) times for the generation phase and for two preconditioned iterative methods implemented in PETSc on the Opteron cluster, for $\varpi = 0.75$. Once more, we observe that there are significant gains by using the parallel version of the code. The generation phase dominates the computational costs. The two sparse iterative solvers show similar times, although, as before, for the parameters chosen BICGstab requires fewer iterations. The achieved speedup is not ideal but together with the generation phase the performance of the code is almost linear, see Figure 1. The stagnation of the speedup curve for the linear

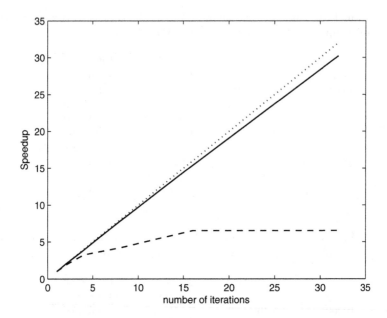

Fig. 1. Speedup for $m = 10^4$, solid line: ideal, dashed line: preconditioned GMRES

Table 5. Normalized times and number of iterations for various matrix sizes (m) and for $\varpi = 0.90$ on up to 32 processors (p) on the Opteron cluster

m	p	generation	GMRES (24 iterations)	BiCGstab (37 iterations)
10000	1	2.83E+03	6.14E+00	6.99E+00
	2	1.36E+03	3.62E+00	4.12E+00
	4	7.20E+03	2.20E+00	2.31E+00
	8	3.59E+02	1.67E+00	1.76E+00
	16	1.80E+02	1.11E+00	1.30E+00
	32	9.19E+02	1.02E+00	1.00E+00
25000	4	2.83E+03	4.96E+00	5.37E+00
	8	2.22E+03	2.96E+00	3.39E+00
	16	1.11E+03	1.78E+00	2.11E+00
	32	5.55E+03	1.33E+00	1.46E+00
50000	16	4.36E+03	2.82E+00	3.14E+00
	32	2.21E+03	2.15E+00	2.05E+00

solver in Figure 1 only indicates that for $m = 10000$ it is not worthy to use more than 16 processors. In fact, the speedup gets better for larger values of m. The time required to generate the matrix for $m = 2.5 \times 10^4$ on 8 processors was similar to the time to generate the matrix for $m = 5 \times 10^4$ on 32 processors. For the solver, the time to solve the linear system for $m = 2.5 \times 10^4$ on 8 processors

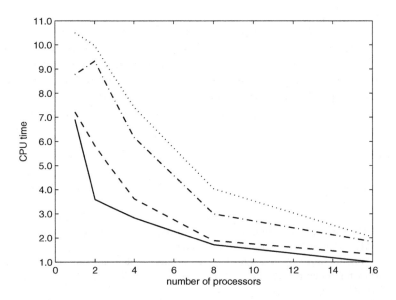

Fig. 2. CPU time in seconds for $m = 5 \times 10^4$, solid line: GMRES/Jacobi, dashed line: BiCGstab/Jacobi, dashdotted line: GMRES/blockJacobi, dotted line: BiCGstab/blockJacobi

was similar to the time required for the solution of a linear system for $m = 5 \times 10^4$ on 16 processors, showing a good relative speedup.

In Table 5 we list, as in Table 4, the normalized times on the Operon cluster but now for $\varpi = 0.90$. The time for the generation phase was the same but, as expected, for higher values of the albedo the linear system becomes more difficult to solve. In fact, for $\varpi = 0.90$ GMRES performed better than BiCGstab, requiring only a few more iterations than for $\varpi = 0.75$. That was not the case of BiCGstab, which required almost three times as many iterations and therefore a degradation in performance for this system.

The entries of the coefficient matrices show a high decay in magnitude from the diagonal. This property allowed us to successfully employ the highly parallel Jacobi preconditioner for the iterative methods. As we can see in Fig. 2, both iterative solvers performed better with the simple Jacobi preconditioner than with the block-Jacobi.

5 Conclusions and Future Work

In this contribution we discussed the numerical solution of a radiative transfer equation for modeling the emission of photons in stellar atmospheres, in particular mechanisms that we have implemented to enable the solution of large systems. This is necessary because the generation of the matrix associated to the model requires a significant amount of time. The parallelization of generation phase, as dicussed in the previous section, dramatically reduces the time to solution. At the same time, it is also important to select an appropriate solver for the resulting system of linear equations. Here we focused on tools available in the DOE ACTS Collection, and in particular (the sequential version of) SuperLU and iterative methods implemented in PETSc. These tools have delivered capability and portability and thus have been very useful in the development of a number of applications, including the one discussed here.

References

1. M. Abramowitz and I.A. Stegun. *Handbook of Mathematical Functions.* Dover, New York, 1960.
2. M. Ahues, F.D. d'Almeida, A. Largillier, O. Titaud and P. Vasconcelos. An L^1 Refined Projection Approximate Solution of the Radiation Transfer Equation in Stellar Atmospheres. *J. Comput. Appl. Math.*, 140:13–26, 2002.
3. F.D. d'Almeida and P.B. Vasconcelos, A Parallel Implementation of the Atkinson Algorithm for Solving a Fredholm Equation. *Lecture Notes in Computer Science*, 2565:368–376, 2003.
4. E. Anderson and Z. Bai and C. Bischof and S. Blackford and J. W. Demmel and J. J. Dongarra and J. Du Croz and A. Greenbaum and S. Hammarling and A. McKenney and D. C. Sorensen, *LAPACK User's Guide*, SIAM, Philadelphia, USA, 1999.
5. S. Balay and K. Buschelman and V. Eijkhout and W. D. Gropp and D. Kaushik and M. G. Knepley and L. Curfman McInnes and B. F. Smith and H. Zhang, PETSc Users Manual, Technical Report ANL-95/11, Revision 2.1.5, Argonne National Laboratory, 2004.
6. L.S. Blackford, J. Choi, A. Cleary, E. D'Azevedo, J. Demmel, I. Dhillon, J. Dongarra, S. Hammarling, G. Henry, A. Petitet, K. Stanley, D. Walker, and R.C. Whaley. *ScaLAPACK Users' Guide.* SIAM, Philadelphia, PA, 1997.
7. J. Demmel and J. Gilbert and X. Li. SuperLU Users' Guide, LBNL-44289, 1999.
8. J.J. Dongarra, I.S. Duff, D.C. Sorensen and H.A. van der Vorst. *Numerical Linear Algebra for High-Performance Computers.* Society for Industrial and Applied Mathematics, Philadelphia, 1998.
9. L.A. Drummond and O. Marques, An Overview of the Advanced CompuTational Software (ACTS) Collection. *ACM TOMS*, 31:282–301, 2005.
10. L.A. Drummond, V. Hernandez, O. Marques, J.E. Roman, and V. Vidal, A Survey of High-Quality Computational Libraries and their Impact in Science and Engineering Applications Collection. *Lecture Notes in Computer Science*, Springer Verlag, 3402:37–50, 2005.
11. M. Heroux and J. Willenbring, Trilinos Users Guide, Technical Report SAND2003-2952, Sandia national Laboratories, 2003.
12. O.A. Marques and L.A. Drummond, The DOE ACTS Information Center. http://acts.nersc.gov.
13. B. Rutily. Multiple Scattering Theoretical and Integral Equations. *Integral Methods in Science and Engineering: Analytic and Numerical Techniques*, Birkhauser, 211–231, 2004.
14. Y. Saad, SPARSKIT: A basic tool kit for sparse matrix computations, Technical Report RIACS-90-20, NASA Ames Research Center, Moffett Field, CA, 1990.

15. Y. Saad. *Iterative Methods for Sparse Linear Systems.* PWS Publishing Company, 1996.
16. M. Snir, S. Otto, S. Huss-Lederman, D. Walker, J.J. Dongarra. *MPI: The Complete Reference.* The MIT Press, 1996.
17. P.B. Vasconcelos and F.D. d'Almeida, Performance evaluation of a parallel algorithm for a radiative transfer problem. *Lecture Notes in Computer Science,* Springer Verlag, 3732: 864–871, 2006.

Scalable Cosmological Simulations
on Parallel Machines

Filippo Gioachin[1], Amit Sharma[1], Sayantan Chakravorty[1], Celso L. Mendes[1],
Laxmikant V. Kalé[1], and Thomas Quinn[2]

[1] Dept. of Computer Science, University of Illinois at Urbana-Champaign
Urbana, IL 61801, USA
{gioachin, asharma6, schkrvrt, cmendes, kale}@uiuc.edu
[2] Dept. of Astronomy, University of Washington
Seattle, WA 98105, USA
trq@astro.washington.edu

Abstract. Cosmological simulators are currently an important component in the study of the formation of galaxies and planetary systems.
However, existing simulators do not scale effectively on more recent machines containing thousands of processors. In this paper, we introduce
a new parallel simulator called ChaNGa (Charm N-body Gravity). This
simulator is based on the CHARM++ infrastructure, which provides a
powerful runtime system that automatically maps computation to physical processors. Using CHARM++ features, in particular its measurement-
based load balancers, we were able to scale the gravitational force calculation of ChaNGa on up to one thousand processors, with astronomical
datasets containing millions of particles. As we pursue the completion
of a production version of the code, our current experimental results
show that ChaNGa may become a powerful resource for the astronomy
community.

1 Introduction

Cosmological simulators are currently an important component in the study of
the formation of galaxies and planetary systems. Galaxies are the most distinctive objects in the universe, containing almost all the luminous material. They
are remarkable dynamical systems, formed by non-linear collapse and a drawn-
out series of mergers and encounters. Galaxy formation is indeed a challenging
computational problem, requiring high resolutions and dynamic timescales. For
example, to form a stable Milky Way-like galaxy, tens of millions of resolution
elements must be simulated to the current epoch. Locally adaptive timesteps
may reduce the CPU work by orders of magnitude, but not evenly throughout
the computational volume, thus posing a considerable challenge for parallel load
balancing. No existing N-body/Hydro solver can handle this regime efficiently.

The scientific payback from such studies can be enormous. There are a number
of outstanding fundamental questions about the origins of planetary systems
which these simulations would be able to answer.

M. Daydé et al. (Eds.): VECPAR 2006, LNCS 4395, pp. 476–489, 2007.
© Springer-Verlag Berlin Heidelberg 2007

To address these issues, various cosmological simulators have been created recently. PKDGRAV [1], developed at the University of Washington, can be considered among the state-of-the-art in that area. However, PKDGRAV does not scale efficiently on newer machines with thousands of processors. In this work, we present a new N-body cosmological simulator that utilizes the Barnes-Hut tree topology to compute gravitational forces. Our new simulator, named ChaNGa, is based on the CHARM++ runtime system [2]. We leverage the object based virtualization [3] inherent in the CHARM++ runtime system to obtain automatic overlapping of communication and computation time, as well as to perform automatic runtime measurement-based load balancing. ChaNGa advances the state-of-the-art in N-Body simulations by allowing the programmer to achieve higher levels of resource utilization with moderate programming effort. In addition, as confirmed by our experimental results, the use of CHARM++ has enabled ChaNGa to efficiently scale on large machine configurations.

The remainder of this paper is organized as follows. Section 2 presents an overview of previous work in the development of parallel simulators for cosmology. Section 3 describes the major components of ChaNGa. Section 4 presents the various optimizations that we have applied to ChaNGa, with the resulting improvement in performance measured for each optimization. Finally, Section 5 contains our conclusions and the future directions of our work.

2   Related Work

There have been numerous studies on the N-Body problem, which involves the evolution of interacting particles that are under the effects of Newtonian gravitational forces. One of the most widely used methods was proposed by Barnes and Hut [4]. In their scheme, the particles are associated to a hierarchical structure comprising a tree. This tree is traversed and the forces between particles are computed exactly or by approximations, depending on the distance between the given particles. With N particles, this approach achieves reduction in the complexity of the problem from the original $O(N^2)$ to $O(N \log N)$.

Given the power of hierarchical methods for N-Body simulations, such methods have been adopted for quite some time by the astronomy community [5]. One of the most popular simulators currently is PKDGRAV [1], a parallel hierarchical tree-structured code used to conduct cosmological simulations on shared-memory and distributed-memory systems. It is portable across different communication substrates (e.g. MPI, PVM, etc.), and supports adaptive decomposition of work among the processors. PKDGRAV has been used in production simulations of systems with millions of particles, and has been shown to scale well on up to hundreds of processors. One restriction in PKDGRAV's current version, however, arises from its limited load-balancing capability. This effectively prevents scaling the code efficiently on newer machines with thousands of processors.

Other cosmological simulators have been in use as well. Among these, two of the major codes are GADGET [6], developed in Germany, and falcON [7], developed at the University of Maryland. However, despite claiming a good scalability with the number of particles, falcON is a serial simulator. Meanwhile, GADGET originally had some of the same limitations of PKDGRAV when scaling to a large number of processors. This has been addressed in a more recent version of their code (GADGET-2), but there are not yet results reported with more than around one hundred processors [8].

3   New ChaNGa Code

In order to leverage the features that the CHARM++ runtime system offers, we decided to develop a new cosmological simulator called ChaNGa (formerly ParallelGravity). Our goal in developing this new application is to create a full production cosmological simulator that scales to thousands of processors.

This new simulator is capable of computing gravitational forces generated by the interaction of a very large number of particles, integrating those forces over time to calculate the movement of each particle. Since most of the running time of the application is devoted to force computation, our focus has been in optimizing this aspect of the code. The integration over time is typically easier to parallelize, and is not the focus of our analysis in this paper.

Since the gravitation field is a long range force, the total force applied to a given particle has contributions from all the other particles in the entire space. The algorithm we applied is based on a Barnes-Hut tree topology [4], which enables achieving an algorithmic performance of $O(N \log N)$. The tree generated by this algorithm is constructed globally over all the particles, and distributed across elements that are named *TreePieces*. This distribution is done according to the particles contained in each internal tree node. Figure 1 shows an example of such distribution. In this scheme, some internal nodes are replicated in more than one element. The particles are at the leaves of the tree, and are grouped by spatial proximity into *buckets* of a user-defined size. While walking the tree to compute forces, a single walk is performed for all the particles contained in a given bucket. Mass moments needed for the gravity calculation are evaluated as the tree is built. For upper parts of the tree, requests are made to neighboring *TreePieces* for their contributions to the moments. The code allows the user to choose between different available tree distributions. Currently, two types of distributions are implemented: *SFC*, where a Morton-ordered Space Filling Curve is used to impose a total ordering on the particles, with a contiguous portion of the curve being assigned to each TreePiece; and *Oct*, where particles are divided based on the nodes of an Octree covering the entire space, with each TreePiece assigned a complete subtree rooted at an internal node.

3.1   Charm++ Infrastructure

Our new ChaNGa code is based on the CHARM++ [2] infrastructure. CHARM++ is a parallel C++ library that implements the concept of *processor virtualization*: an application programmer decomposes her problem into a large number of components, or objects, and the interactions among those objects. The objects, called *chares* in CHARM++ nomenclature, are automatically mapped to

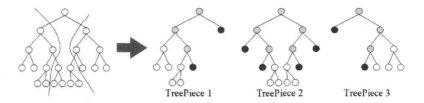

Fig. 1. Distribution of a tree across TreePieces (top levels). White nodes are owned by one TreePiece, black nodes are placeholders for remote nodes, gray nodes are shared among multiple TreePieces.

physical processors by the CHARM++ runtime system. Typically, the number of chares is much higher than the number of processors. By making the number of chares independent of the number of existing processors, CHARM++ enables execution of the same code on different machine configurations. This separation between logical and physical abstractions provides higher programmer productivity, and has allowed the creation of parallel applications that scale efficiently to thousands of processors, such as the molecular dynamics NAMD code [9].

The CHARM++ runtime system has the ability to migrate chares across processors during execution of an application. This migration capability is used by the powerful measurement-based load-balancing mechanism of CHARM++ [10]. The runtime system can measure various parameters in the chares, such as computational load or communication frequency and volume. CHARM++ provides a family of load balancers, targeting optimization of a variety of metrics. The user simply needs to select her desired balancers at application launch. During execution, the selected balancers will collect the measured chare values for the appropriate metrics, and dynamically remap chares across the available processors in a way that execution performance is optimized. This dynamic optimization capability is critical for applications such as particle system simulators, where particles can move in space and cause overloading on a given processor as the simulation progresses, while other processors become underutilized.

3.2   Major ChaNGa Features

An early decision in the design of ChaNGa was to select where to compute the forces applied to a particle. Historically, two methods have been used: (a) distributing the computation of the forces on that particle across all processors, with each processor computing the portion of the forces given by its subtrees, or (b) gathering at the processor owning that particle all the data needed to compute the forces on it. We opted for the second scheme, since the CHARM++ capabilities could be better exploited, as explained later in this section.

In our implementation of ChaNGa, each TreePiece in Figure 1 is a CHARM++ chare. Thus, TreePieces are dynamically mapped to physical processors by the CHARM++ runtime system. The overall structure of how the code works is shown in Figure 2, and described in the next paragraphs.

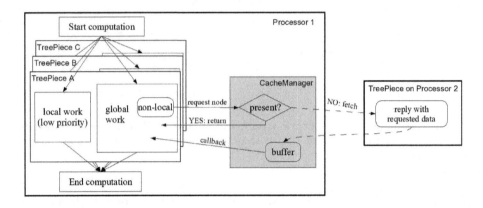

Fig. 2. Control flow of the execution of an iteration of force calculation

To perform the computation of the forces on its particles, a TreePiece processes its buckets independently. For each bucket, the TreePiece must walk the overall tree and compute the forces due to all other particles. During the walk, visited nodes may be local (i.e. owned by this TreePiece) or non-local. For local nodes, the force computation can proceed immediately. For non-local nodes, a retrieval must be carried out, to bring the corresponding data into the TreePiece. A non-local node may reside either at another TreePiece of the same processor, or at a remote processor. In the first case, we use a direct data transfer between chares. In the second case, data must be requested to the remote processor. While waiting for remote data to arrive, the TreePiece can process other buckets.

Instead of repeating fetches of the same remote data for different bucket walks, we can use the property that buckets close in space will require similar remote portions of data. Thus, we can buffer the imported data and have it used by all buckets in the TreePiece before discarding it. Because in CHARM++ we may have multiple chares in a single processor, we implemented this optimization at the processor level using a CHARM++ *group*, which we call *CacheManager*.

The purpose of the CacheManager is to serve all requests made by the TreePieces, and provide a caching mechanism to hide the latency of interprocessor data fetching. The CacheManager implements a random access to the cached data through the use of a hash table. To reduce the overhead of table lookup, the imported data is reconstructed into a local tree. Thus, once entering a subtree, TreePieces can iterate over direct pointers, until another cache miss occurs. Upon detecting a miss, the CacheManager will fetch the remote data and use callbacks to notify the requesting TreePiece when the data arrives. More advanced features provided by the CacheManager are presented in the next section, together with the observed experimental results.

Because CHARM++ executes chare methods in a non-preemptive fashion, a long sequence of consecutive tree walks might potentially prevent a processor from serving incoming data requests from other processors. In order to provide good responsiveness to incoming requests, we partitioned the processing of tree

Table 1. Characteristics of the parallel systems used in the experiments

System Name	Location	Number of Processors	Processors per Node	CPU Type	CPU Clock	Memory per Node	Type of Network
Tungsten	NCSA	2,560	2	Xeon	3.2 GHz	3 GB	Myrinet
BlueGene/L	EPCC	2,048	2	Power440	700 MHz	512 MB	Torus
HPCx	HPC-UK	1,536	16	Power5	1.5 GHz	32 GB	Federation

walks with a fine granularity. The grainsize is a runtime option, and corresponds to the number of buckets that will walk the tree without interruption. After that number of walks is performed, the TreePiece will yield the processor, enabling the handling of existing incoming data requests.

While dividing the computation into fine grains, we also distinguish between *local* and *global* computation. Local computation is defined as the interaction with the particles present in the same TreePiece. In contrast, global computation is defined as the interaction with the rest of the tree, i.e. the computation that involves non-local nodes. In particular, because this global computation is performed on the imported sections of the tree, it is on the more critical path. To express this different criticality, we utilized the prioritization mechanism embedded into CHARM++. This mechanism allows establishing a total order of priority for the different operations performed by a TreePiece: the highest priority is assigned to accepting requests arriving from other processors, followed by sending replies to such requests, and finally the two types of computation (local and global), with the local one having the lowest priority. The CHARM++ runtime system will schedule these operations according to such priorities.

4 Optimizations and Experimental Evaluation

After having a basic version of ChaNGa in place, we studied its performance and added a number of optimizations to the code. Some of these optimizations were designed to exploit CHARM++ aspects that enable high performance, whereas others were aimed at specific characteristics of particle codes. In this section, we describe the various optimizations that we have added, and present, in each case, the performance improvement that we obtained by applying such techniques to real cosmological datasets. Although the following subsections describe the effect of each optimization technique separately, our integrated version of ChaNGa contains all the optimizations. It is this integrated, fully optimized version that we use in the last subsection, to show how the current code scales with increasing system size. In our experiments, we used the parallel systems described in Table 1, and the following particle datasets:

lambs: Final state of a simulation of a $71Mpc^3$ volume of the Universe with 30% dark matter and 70% dark energy. Nearly three million particles are used (3M). This dataset is highly clustered on scales less than 5 Mpc, but becomes uniform on scales approaching the total volume. Three subsets of this dataset are obtained by taking random subsamples of size thirty thousand (30K), three hundred thousand (300K), and one million (1M) particles, respectively.

dwarf: Snapshot at $z = .3$ of a multi-resolution simulation of a dwarf galaxy forming in a $28.5Mpc^3$ volume of the Universe with 30% dark matter and 70% dark energy. The *mass* distribution in this dataset is uniform, but the *particle* distribution is very centrally concentrated and therefore highly clustered. The central regions have equivalent resolution of 2048^3 particles in the entire volume. The total dataset size is nearly five million particles.

4.1   Uniprocessor Performance

While developing a parallel application like ChaNGa, we are concerned not only with scalability, but also with performance (i.e. execution time). Hence, it is important to evaluate the single processor performance as well. To do this, we compared the serial performances of ChaNGa and PKDGRAV, on different subsets of the *lambs* dataset.

Table 2 shows the execution times for the gravitational force calculation phase of the two simulators, running on one Xeon processor of Tungsten. As the table shows, ChaNGa's serial performance is comparable to that of PKDGRAV, even for the larger datasets. The slightly greater times for ChaNGa (increase of less than 6%) are caused by optimizations aimed at improving parallel performance. As the next subsections will demonstrate, this is a very small price to pay in view of the large gains achievable with those optimizations in the parallel case.

4.2   Software Cache Mechanism

As mentioned in Section 3.2, the CacheManager not only reduces the number of messages exchanged to fetch remote data, but also hides the latency of fetching data from other processors. We evaluated the effectiveness of the CacheManager

Table 2. Time, in seconds, for one step of force calculation in serial execution

	Number of Particles			
Simulator	30,000	300,000	1 million	3 million
PKDGRAV	0.83	12.0	48.5	170.0
ChaNGa	0.83	13.2	53.6	180.6

Table 3. CacheManager effects in terms of number of messages and iteration time

		Number of Processors				
		4	8	16	32	64
Number of Messages	No Cache	48,723	59,115	59,116	68,937	78,086
($\times 10^3$)	With Cache	72	115	169	265	397
Time	No Cache	730.7	453.9	289.1	67.4	42.1
(seconds)	With Cache	39.0	20.4	11.3	6.0	3.3

on the 1 million lambs subset running on varying numbers of HPCx processors. Table 3 shows that the CacheManager dramatically reduces the number of messages exchanged. The performance improvement due to sending a much lower number of messages, combined with the latency-hiding effects of the CacheManager, produces a sharp reduction in the execution time, as seen in Table 3. Thus, the software cache mechanism is absolutely necessary to obtain good parallel performance.

4.3   Data Prefetching

As in PKDGRAV, we can take the principle of the software cache one step further by fetching not only the node requested by a TreePiece, but proactively also part of the subtree rooted at that node. The user can specify the *cache depth* (analogous to the concept of cache line in hardware) as the number of levels in the tree to recursively prefetch. The rationale for this is that if a node is visited, most probably its children will be visited as well. This mechanism of prefetching more data than initially requested helps to reduce the total number of messages exchanged during the computation. Since every message has both a fixed and a variable cost, prefetching reduces the total fixed cost of communication. On the other hand, a cache depth of more than zero might cause some data to be transferred but never used, thus increasing the variable part of the cost.

If a TreePiece requested data to the CacheManager only when required by the tree-walk computation, the CacheManager might not have it. This would trigger a fetch of the data from the remote node, but at the same time it would suspend the computation for the requesting bucket until the moment of data arrival. Both the interruption of the tree walk and the notification from the CacheManager incur an overhead. To limit this effect, we developed a *prefetching phase* which precedes the real tree-walk computation. During this phase, we traverse the tree and prefetch all the data that will be later used during the computation in the regular tree walk. This prefetching phase can work with different cache depths.

We used the lambs dataset on 64 processors of Tungsten to evaluate the impact of cache depth and the prefetching phase. Figure 3(a) shows the execution

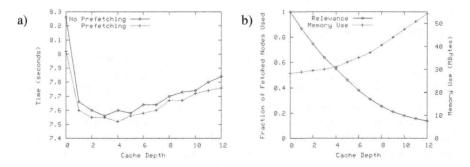

Fig. 3. Impact of cache depth and prefetching on (a) iteration time, and (b) relevance and memory use

time for different cache depths with and without the prefetching phase. In both cases there is an optimal value of cache depth, at which the execution time is minimal. The optimal point is achieved when the fixed cost associated with every message and the variable cost of transferring data over the network are in balance. According to our results in Figure 3(a), the optimal cache depth seems to vary between 3 and 5.

We can also see that the prefetching phase improves performance for all considered values of cache depth. This is due to the increased hit rate of the cache. While executions without the prefetching phase generate a cache hit rate of about 90%, with the prefetching active the hit rate rises to 95-97% for SFC tree decomposition, and 100% for Oct decomposition. The greater accuracy in prefetching for Oct decomposition is due to the better prefetching algorithm we developed, given the constraint that prefetching must be lightweight. Although Oct decomposition provides a clear benefit in terms of cache hit rate over SFC, the full effects on the entire execution time are more complex and will require more detailed studies to be fully characterized.

We define the *relevance* as the ratio between the number of nodes fetched and used, and the total number of nodes fetched. Ratios closer to 1.0 represent a better relevance. In Figure 3(b), we plot the relevance on the left vertical axis. The observed relevance decreases with increasing cache depth, leading to unnecessarily higher memory consumption, as plotted on the right vertical axis of the same graph. Nevertheless, this higher memory consumption due to caching is limited to a fraction of the total memory footprint for moderate values of cache depth. At a very low value of relevance, the cost of fetching a large amount of extra data is not offset by the benefit of having the data already present in the software cache when it is requested. This is why the execution time rises for large values of cache depth in Figure 3(a). The prefetching phase does not affect the relevance, since it does not change which data items are transferred. Prefetching simply causes those data transfers to occur earlier.

Thus, we see that using the prefetching phase along with a small but non-zero value of cache depth improves performance. In the following subsections, we will assume that the prefetching phase is active, and a reasonable value of cache depth is used.

4.4   Tree-in-Cache Effects

In Section 3 we introduced the concept of local and global computation. We pointed that the global work is on the critical path, and that the local work can be used to hide the latency of data transfers. From this, it is clear that we should have as much local work as possible. One point to notice is that in the CHARM++ environment we fragment the particle dataset in more TreePieces than the number of physical processors available. This over-decomposition reduces the amount of local work per TreePiece. In some of our experiments, when increasing the number of processors beyond one hundred, the local work became insufficient to maintain the processor busy during the entire computation.

Table 4. Distribution of work between local and global computation

	Local	Global
Original Code	16%	84%
Code with Tree-in-cache	42%	58%

By noticing that during the force computation there is no migration of TreePieces, we can consider collectively all the TreePieces residing on a given processor. We can attribute to local computation not only the work related to nodes/particles present in the same TreePiece, but also the work related to particles and nodes present in other TreePieces in the same processor. This is implemented by having each TreePiece registering to the CacheManager at the beginning of the computation step. The CacheManager will then create a superset tree including all the trees belonging to the registered TreePieces. Each TreePiece will now consider as local work this entire tree. During this operation, only the nodes closest to the root of the tree will be duplicated. According to our tests with datasets of a few million particles, less than one hundred nodes were duplicated.

Table 4 summarizes the percentage of local and global work for a simulation on 64 Tungsten processors with the lambs-300K subset. The percentages changed considerably before and after this optimization. In our tests, this new scheme enabled scaling the computation up to hundreds of processors. However, when reaching the limit of one thousand processors, even the extra work from co-resident TreePieces becomes insufficient. A solution that we are investigating is to split the global walk into multiple sub-walks.

4.5 Interaction Lists

After having preceded the computation with a prefetching phase, and verifying that it is accurate, we explored a faster algorithm for gravitational force computation similar to the Fast Multipole Method [11]. This algorithm is centered on the concept of *interaction lists*, which we describe in this subsection. The new algorithm is based on the same principle of the CacheManager: two buckets close in space will tend to interact similarly with a given remote node.

In the regular ChaNGa algorithm, whenever a bucket walk visits a tree node, a fundamental test is carried out. In this test, we check the spatial position of the bucket in respect to the particles in that node. If the bucket is sufficiently *far* from the node, the forces on the bucket due to the entire subtree rooted at that node are immediately computed, using the subtree's center of mass. Otherwise, ChaNGa *opens* the node, i.e. it recursively traverses the subtree rooted at that node. Thus, the threshold used to decide if a node is close enough to the bucket represents the *opening criteria* for deciding whether the visited node must be opened or not.

Instead of checking the opening criteria at a given node for each bucket independently, we can modify the algorithm and do that check for various local

Table 5. Number of checks for opening criteria, in millions

	lambs 1M	dwarf 5M
Original algorithm	120	1,108
Modified algorithm	66	440

buckets at once. We can do this collective check using the buckets' ancestors in the local tree. These ancestors will be local nodes containing particles which are close in space. If an ancestor needs to open a visited node, that node will be opened for every bucket that is a descendent of such ancestor. On the other hand, if a node is far enough for that ancestor, this node will be far enough for all the ancestor's buckets too. In this second case, we can directly compute the interaction between the node and all these local buckets.

By grouping the checking for various local buckets, we can reduce the total number of checks for opening nodes. As an example, Table 5 shows the number of checks that are observed with the two algorithms, executing on the HPCx system with our two datasets. A potential problem in this modified algorithm is that it may cause less effective usage of the hardware cache: because the computation of interactions proceeds for various local buckets, one bucket's data may flush another bucket's data from the hardware cache. We can reduce the number of hardware-cache misses by storing all the nodes that interact with a given bucket in a bucket's *interaction list*, and perform the entire computation of forces on that bucket at the end of the tree walk. Performance is improved even further with interaction lists because compilers may keep a particle's data inside CPU registers while computing interactions with the nodes in the list.

Figure 4(a) plots the execution time for both the regular algorithm and the new algorithm employing interaction lists, showing also cases where load balancing was employed (load balancing is the subject of our analysis in the next subsection). The new algorithm shows a performance improvement over the entire range considered. This improvement varies between 7% and 10%. We used ChaNGa with interaction lists for the uniprocessor tests of Section 4.1.

4.6 Load Balancer Importance

After describing all the optimizations applied to the basic ChaNGa code, we assess the importance of the CHARM++ automatic load balancing framework in improving the performance of our simulations. In particular, we emphasize the fact that the code instrumentation and the migration of chares in the system are totally automated, and do not require any programmer intervention.

Figure 4(a) shows the effect of load balancing on both versions of ChaNGa, one with the regular algorithm and the other with the interaction-list implementation. The improvement from load balancing is similar in both algorithms. We see that, before load balancing, the behavior of the algorithms is somewhat random and determined only by the particle decomposition. This happens because different particles in space require different amounts of computation. TreePieces

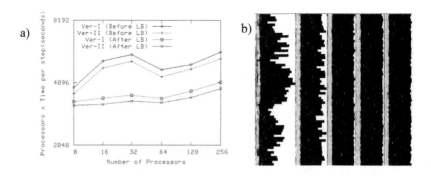

Fig. 4. (a) Comparison between regular ChaNGa (Ver-I) and the one with interaction lists (Ver-II) before (1st) and after load balancing (5th iter.) on BlueGene/L for dwarf dataset. (b) Effect of Load Balancer for dwarf dataset on 64 BlueGene/L processors.

owning heavy particles will be overloaded, hence cause bad performance. After load balancing, performance improves between 15% and 35%.

To further analyze the improvements from the load balancer, Figure 4(b) displays a view from our PROJECTIONS performance analysis tool, a component of CHARM++. This view corresponds to five timesteps of a simulation on 64 Blue-Gene/L processors. The horizontal axis represents time, while each horizontal bar represents a processor. Darker colors represent higher utilization, with black as full utilization and white as idleness. One can see that even starting from a very unbalanced situation on the first timestep, after two timesteps the load balancer improves performance quite significantly, approaching almost perfect balance. The gray region at the beginning of each timestep, where utilization is lower, corresponds to the communication overhead due to prefetching. The time spent by the application in load balancing and in domain decomposition is hardly visible in the figure. It corresponds to the period between the end of the longest black bar in one timestep and the beginning of the gray region of the next timestep. That time is negligible.

It is relevant to notice that the input dataset (dwarf) is highly clustered at the center of the simulation space, and its spatial distribution of particles is very uneven. This non-uniform particle distribution is reflected by the varying processor utilization in the first timestep of the simulation. Situations like this present the biggest challenge to obtain load balance across processors. Nevertheless, the CHARM++ load balancers achieved very good balance.

4.7 Scalability with Number of Processors

By applying all the optimizations described in the previous subsections, and making use of various CHARM++ features, we obtain our best performing version of ChaNGa. We used this version to conduct scaling tests on large machine configurations, and to make scaling comparisons with PKDGRAV.

Figure 5 shows the scaling of ChaNGa on BlueGene/L, HPCx and Tungsten for five iterations. The vertical axis is the product of the time per iteration and

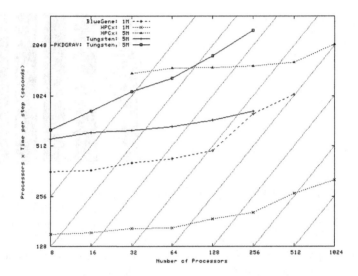

Fig. 5. ChaNGa scaling on various systems and comparison with PKDGRAV

the number of processors in the simulation. Horizontal lines represent perfect scalability, while the diagonal lines represent no gain in scaling.

For the lambs1M dataset, the algorithm scales well up to 128 processors on BlueGene/L and 256 processors on HPCx. Beyond these points, there is not adequate work available for each processor, and the gain is reduced. BlueGene/L has the most problems, and there is almost no advantage from the increased number of processors. The dwarf dataset, being larger with 5 million particles, allows good scaling up to 1024 processors of HPCx.

Figure 5 also presents the scaling comparison between ChaNGa and PKD-GRAV on Tungsten. We can see that ChaNGa scales much better than PKD-GRAV, maintaining a good performance over the entire range considered. Due to machine unavailability, we ran tests only up to 256 Tungsten processors.

5   Conclusions and Future Work

In this paper, we have presented a new cosmological simulator named ChaNGa. Our design was guided by the goal of achieving good scalability on modern parallel machines, with thousands of processors. Our experimental results show that ChaNGa's serial performance is comparable to that of one of the top-level simulators existing today. Meanwhile, by employing various optimizations enabled by the CHARM++ runtime system, the gravity calculation phase in ChaNGa was shown to scale very well up to one thousand processors with real astronomical datasets. This level of scalability places ChaNGa as a potentially powerful resource for the astronomy community.

Despite ChaNGa's good observed scalability, we intend to study other load balancing schemes and parallelization techniques that may provide even further benefits. Moreover, to become a production-level simulator, ChaNGa still needs a few more features. We are adding support for more physics, such as fluid-dynamics and periodic boundaries, as well as providing multiple timestepping. In addition, as we start our tests on thousands of processors, we are also analyzing the performance of other phases of the simulation, such as the construction of the particle tree. Given the support for various types of trees already present in the code, we will conduct a detailed study of their effects on the simulation.

Acknowledgments. This work was supported in part by the National Science Foundation, under grant number NSF ITR 0205611. We are thankful for the access to parallel systems at Edinburgh's HPCx Consortium and EPCC Center, and Illinois' NCSA.

References

1. M. D. Dikaiakos and J. Stadel, "A performance study of cosmological simulations on message-passing and shared-memory multiprocessors," in *Proceedings of the International Conference on Supercomputing - ICS'96*, (Philadelphia, PA), pp. 94–101, December 1996.
2. L. V. Kale and S. Krishnan, "Charm++: Parallel Programming with Message-Driven Objects," in *Parallel Programming using C++* (G. V. Wilson and P. Lu, eds.), pp. 175–213, MIT Press, 1996.
3. L. V. Kalé, "Performance and productivity in parallel programming via processor virtualization," in *Proc. of the First Intl. Workshop on Productivity and Performance in High-End Computing (at HPCA 10)*, (Madrid, Spain), February 2004.
4. J. Barnes and P. Hut, "A hierarchical $O(N \log N)$ force-calculation algorithm," *Nature*, vol. 324, pp. 446–449, December 1986.
5. G. Lake, N. Katz, and T. Quinn, "Cosmological N-body simulation," in *Proceedings of the Seventh SIAM Conference on Parallel Processing for Scientific Computing*, (Philadelphia, PA), pp. 307–312, February 1995.
6. V. Springel, N. Yoshida, and S. White, "GADGET: A code for collisionless and gasdynamical simulations," *New Astronomy*, vol. 6, pp. 79–117, 2001.
7. W. Dehnen, "A hierarchical $O(N)$ force calculation algorithm," *Journal of Computational Physics*, vol. 179, pp. 27–42, 2002.
8. V. Springel, "The cosmological simulation code GADGET-2," *MNRAS*, vol. 364, pp. 1105–1134, 2005.
9. J. C. Phillips, G. Zheng, S. Kumar, and L. V. Kalé, "NAMD: Biomolecular simulation on thousands of processors," in *Proceedings of SC 2002*, (Baltimore, MD), September 2002.
10. G. Zheng, *Achieving High Performance on Extremely Large Parallel Machines: Performance Prediction and Load Balancing*. PhD thesis, Department of Computer Science, University of Illinois at Urbana-Champaign, 2005.
11. L. Greengard and V. Rokhlin, "A fast algorithm for particle simulations," *Journal of Computational Physics*, vol. 73, pp. 325–348, 1987.

Performance Evaluation of Scientific Applications on Modern Parallel Vector Systems

Jonathan Carter, Leonid Oliker, and John Shalf

NERSC/CRD, Lawrence Berkeley National Laboratory, Berkeley, CA 94720
{jtcarter,loliker,jshalf}@lbl.gov

Abstract. Despite their dominance of high-end computing (HEC) through the 1980's, vector systems have been gradually replaced by microprocessor-based systems. However, while peak performance of microprocessors has grown exponentially, the gradual slide in sustained performance delivered to scientific applications has become a growing concern among HEC users. Recently, the Earth Simulator and Cray X1/X1E parallel vector processor systems have spawned renewed interest in vector technology for scientific applications. In this work, we compare the performance of two Lattice-Boltzmann applications and the Cactus astrophysics package on vector based systems including the Cray X1/X1E, Earth Simulator, and NEC SX-8, with commodity-based processor clusters including the IBM SP Power3, IBM Power5, Intel Itanium2, and AMD Opteron processors. We examine these important scientific applications to quantify the effective performance and investigate if efficiency benefits are applicable to a broader array of numerical methods.

1 Introduction

Despite their dominance of high-end computing (HEC) through the 1980's, vector systems have been progressively replaced by microprocessor based systems due to the lower costs afforded by mass-market commercialization and the relentless pace of clock frequency improvements for microprocessor cores. However, while peak performance of superscalar systems has grown exponentially, the gradual slide in sustained performance delivered to scientific applications has become a growing concern among HEC users. This trend has been widely attributed to the use of superscalar-based commodity components whose architectural designs offer a balance between memory performance, network capability, and execution rate, that is poorly matched to the requirements of large-scale numerical computations. Furthermore, now that power dissipation is limiting the growth rate in clock frequency, the low sustained performance of superscalar systems has risen to the forefront of concerns. The latest generation of custom-built parallel vector systems have the potential to address these performance challenges for numerical algorithms amenable to vectorization.

M. Daydé et al. (Eds.): VECPAR 2006, LNCS 4395, pp. 490–503, 2007.
© Springer-Verlag Berlin Heidelberg 2007

The architectural complexity of superscalar cores has grown dramatically in the past decade in order to support out-of-order execution of instructions that feed an increasing number of concurrent functional units. However, there is growing evidence that despite the enormously complex control structures, typical superscalar implementations are only able to exploit a modest amount of instruction-level parallelism. Vector technology, by contrast, is well suited to problems with plenty of inherent data parallelism. For such problems, the vector approach reduces control complexity because each operation defined in the instruction stream implicitly controls numerous functional units operating in tandem, allowing memory latencies to be masked by overlapping pipelined vector operations with memory fetches.

However, when such operational parallelism cannot be found, the efficiency of the vector architecture can suffer from the properties of Amdahl's Law, where the time taken by the portions of the code that are non-vectorizable can easily dominate the execution time. In this regard, modern vector machines are quite unlike the Cray 1 [1] in that the scalar performance is well below average compared to commodity systems targeted at business applications. It is difficult for vector vendors to compete on scalar processor performance, as the enormous technology investment necessary to keep pace with the microprocessors is too great to sustain without a large market share. Thus today's vector systems have been unable to produce competitive scalar processor implementations, resulting in more significant performance penalties for non-vectorizable code portions when compared to classic vector system implementations.

In the context of these evolving architectural changes, it is important to continue the assessment of vector platforms in the face of increasing algorithm complexity. For this reason, our study focuses on full applications to get more realistic assessments of state-of-the-art supercomputing platforms. This work compares performance between the vector-based Cray X1/X1E, Earth Simulator (ES) and NEC SX-8, with commodity-based superscalar platforms: Intel Itanium2, AMD Opteron, and the IBM Power3 and Power5 systems. We study the behavior of three scientific codes with the potential to run at ultra-scale: Lattice-Boltzmann (LB) simulations of magnetohydrodynamics and fluid dynamics (LBMHD3D and ELBM3D), and astrophysics (CACTUS). Our work builds on our previous efforts [2,3] and makes the contribution of adding recently acquired performance data for the SX-8, and the latest generation of superscalar processors. Additionally, we explore improved vectorization techniques Cactus boundary conditions, and the effects of cache-bypass pragmas for the LB applications. Overall results show that the SX-8 attains unprecedented aggregate performance across our evaluated applications, continuing the trend set by the ES in our previous performance investigations. Our study also shows that the slide in sustained performance of microprocessor cores is not irreversible if microprocessor architectures are willing to invest the effort to make architectural decisions that eliminate bottlenecks for scientific applications.

2   HEC Platforms and Evaluated Applications

In this section we briefly outline the computing platforms and scientific applications examined in our study. Tables 1 and 2 present an overview of the salient features for the eight parallel HEC architectures. Observe that the vector machines have higher peak performance. Additionally, the X1, ES, SX-8, and to a much lessor extent the X1E, have high memory bandwidth (as measured by HPCC EP Stream [4]) relative to peak CPU speed (bytes/flop), allowing them to more effectively feed the arithmetic units. Note also that the NEC vector platforms utilize very high bandwidth interconnects with full crossbar topologies that minimize network congestion. However, the lower latency networks of the Thunder and Bassi systems will offer significant advantages for small point-to-point messaging.

Four superscalar commodity-based platforms are examined in our study. The IBM Power3 experiments reported were conducted on Seaborg, the 380-node pSeries system, running AIX 5.2 (xlf compiler 8.1.1) and located at Lawrence Berkeley National Laboratory (LBNL). Each SMP node consists of sixteen 375 MHz processors (1.5 Gflop/s peak) connected to main memory via the Colony switch using an omega-type topology. The Power5 Bassi system, also located at LBNL's NERSC facility, consists of 111 eight-way Power5 nodes operating at 1.9 GHz (7.6 Gflop/s peak) and interconnected by a dual-plane Federation interconnect using a fat-tree/CLOS topology. Like the Power3, this system also runs AIX 5.2, but uses the newer xlf 9.1 compiler. The AMD Opteron system, called Jacquard, is also located at LBNL and contains 320 dual nodes, running Linux 2.6.5 (PathScale 2.0 compiler). Each node contains two 2.2 GHz Opteron processors (4.4 Gflop/s peak), interconnected via Infiniband fabric in a fat-tree configuration. Finally, the Intel Itanium experiments were performed on Thunder, the 1024 node system located at Lawrence Livermore National Laboratory. Each node contains four 1.4 GHz Itanium2 processors (5.6 Gflop/s peak) and runs Linux Chaos 2.0 (Fortran version ifort 8.1). The system is interconnected using Quadrics Elan4 in a fat-tree configuration,

Table 1. CPU overview of the Power3, Power5, Itanium2, Opteron, X1/X1E, ES, and SX-8 platforms

Name/ Center	Platform	CPU/ Node	Clock (MHz)	Peak (GF/s)	Stream BW (GB/s)	Peak/Stream BW (Byte/Flop)
Seaborg	Power3	16	375	1.5	0.4	0.3
Bassi	Power5	8	1900	7.6	6.8	0.9
Thunder	Itanium2	4	1400	5.6	1.1	0.2
Jacquard	Opteron	2	2200	4.4	2.3	0.5
Phoenix	X1	4	800	12.8	14.9	1.2
Phoenix	X1E	4	1130	18.1	9.7	0.5
ESC	ES	8	1000	8.0	26.3	3.3
HLRS	SX-8	8	2000	16.0	41.0	2.6

We also examine four state-of-the-art parallel vector systems. The fundamental compute processor of the Cray X1/X1E is the multi-streaming processor (MSP). Each MSP contains a 2-way set associative 2 MB data Ecache, a unique feature for vector architectures that allows extremely high bandwidth (25–51 GB/s) for computations with temporal data locality. The MSP is comprised of four single-streaming processors (SSP), each containing two 32-stage vector pipes running at (800 MHz) 1130 MHz on (X1) X1E. Each X1E SSP operates at 4.5 Gflop/s peak for 64-bit data. The SSP also contains a two-way out-of-order superscalar processor running at 400 MHz. The X1E node consists of eight MSPs sharing a flat memory, and large system configuration are networked through a modified 2D torus interconnect. X1E nodes are partitioned into two logical 4-way SMP nodes from the application developers viewpoint. All reported X1E experiments were performed on Phoenix, the 1024-MSP system (several reserved for system services) running UNICOS/mp 3.1 (5.5 programming environment) and operated by Oak Ridge National Laboratory. The X1 experiments were performed on the 512-MSP system at ORNL prior to the upgrade to X1E.

The 1000 MHz Earth Simulator processor was the precursor to the NEC SX6, containing an 4-way replicated vector pipe with a peak performance of 8.0 Gflop/s per CPU. The system contains 640 ES nodes, 5120-processor, connected through a custom single-stage IN crossbar. The ES runs Super-UX, a 64-bit Unix operating system based on System V-R3 with BSD4.2 communication features. As remote ES access is not available, the reported experiments were performed during the authors' visit to the Earth Simulator Center located in Kanazawa-ku, Yokohama, Japan in 2004 and 2005.

Finally, we examine the NEC SX-8. The SX-8 architecture operates at 2 GHz, and contains four replicated vector pipes for a peak performance of 16 Gflop/s per processor. The SX-8 architecture has several enhancements compared with the ES/SX6 predecessor, including improved divide performance, hardware square root functionality, and in-memory caching for reducing bank conflict overheads. However, the SX-8 used in our study uses commodity DDR-SDRAM; thus, we expect higher memory overhead for irregular accesses when compared with the

Table 2. Interconnect performance of the Power3, Power5, Itanium2, Opteron, X1, ES, and SX-8 platforms

Platform	Network	MPI Lat (μsec)	MPI BW (GB/s/CPU)	Network Topology
Power3	Colony	16.3	0.13	Fat-tree
Power5	Federation	4.7	0.69	Fat-tree
Itanium2	Quadrics	3.0	0.25	Fat-tree
Opteron	InfiniBand	6.0	0.59	Fat-tree
X1	Custom	8.0	0.44	4D-Hypercube
X1E	Custom	6.4	0.15	4D-Hypercube
ES	Custom (IN)	5.6	1.5	Crossbar
SX-8	IXS	5.0	2.0	Crossbar

specialized high-speed FPLRAM (Full Pipelined RAM) of the ES. Both the ES and SX-8 processors contain 72 vector registers each holding 256 doubles, and utilize scalar units operating at the half the peak of their vector counterparts. All reported SX-8 results were run on the 72 node system located at High Performance Computer Center (HLRS) in Stuttgart, Germany. This HLRS SX-8 is interconnected with the NEC Custom IXS network and runs Super-UX (Fortran Version 2.0 Rev.313).

2.1 Scientific Applications

Three applications from two scientific computing domains were chosen to compare the performance of the vector-based and superscalar-based systems.

We examine LBMHD3D, a three-dimensional plasma physics application that uses the Lattice-Boltzmann method to study magneto-hydrodynamics [5]; ELBM3D, a a three-dimensional fluid dynamic application that uses the Lattice-Boltzmann method to study turbulent fluid flow [6]; and CACTUS, a modular framework supporting a wide variety of multi-physics applications [7], using the Arnowitt-Deser-Misner (ADM) formulation for the evolution of the Einstein equations from first principles that are augmented by the Baumgarte-Shapiro-Shibata-Nakamura (BSSN) [8] method to improve numerical stability for simulation of black holes.

These codes represent grand-challenge scientific problems that require access to ultra-scale systems and provide code bases mature enough that they have the potential to fully utilize the largest-scale computational resources available. Performance results, presented in Gflop/s per processor and percentage of peak, are used to compare the relative time to solution of our evaluated computing systems. When different algorithmic approaches are used for the vector and scalar implementations, this value is computed by dividing a baseline flop-count obtained from the ES system hardware counters by the measured wall-clock time of each platform.

3 Lattice-Boltzmann Turbulence Simulations

Lattice-Boltzmann methods (LBM) are an alternative to conventional numerical approaches for simulating fluid flows and modeling physics in fluids [9]. The basic idea is to develop a simplified kinetic model that incorporates the essential physics, and reproduces correct macroscopic averaged properties. These algorithms have been used extensively over the past ten years for simulating Navier-Stokes flows, and more recently, several groups have applied the LBM to the problem of magneto-hydrodynamics (MHD) [10,11,12] with promising results [5]. As can be expected from explicit algorithms, LBM are prone to numerical nonlinear instabilities as one pushes to higher Reynolds numbers. These numerical instabilities arise because there are no constraints imposed to enforce the distribution functions to remain non-negative. Such entropic LBM algorithms, which do preserve the non-negativity of the distribution functions—even in the limit of arbitrary small transport coefficients—have recently been developed for

Navier-Stokes turbulence [13,14]. Our LBM applications are representative of these two active research areas: the LBMHD3D code simulates the behavior of a conducting fluid evolving from simple initial conditions through the onset of turbulence; and the ELBM3D code uses the entropic LB algorithm to simulate the behavior of Navier-Stokes turbulence [15].

While LBM methods lend themselves to easy implementation of difficult boundary geometries, e.g., by the use of bounce-back to simulate no slip wall conditions, here we report on 3D simulations under periodic boundary conditions, with the spatial grid and phase space velocity lattice overlaying each other. Each lattice point is associated with a set of mesoscopic variables, whose values are stored in vectors proportional to the number of streaming directions. The lattice is partitioned onto a 3-dimensional Cartesian processor grid, and MPI is used for communication. As in most simulations of this nature, ghost cells are used to hold copies of the planes of data from neighboring processors.

In logical terms an LB simulation proceeds by a sequence of collision and stream steps. A collision step involves data local only to that spatial point, allowing concurrent, dependence-free point updates; the mesoscopic variables at each point are updated through a complex algebraic expression originally derived from appropriate conservation laws. A stream step evolves the mesoscopic variables along the streaming lattice to adjacent lattice sites. However, in an actual implementation, a key optimization described by Wellein and co-workers [16] is often carried out. The two phases of the simulation can be combined, so that either the newly calculated particle distribution function could be scattered to the correct neighbor as soon as it was calculated, or equivalently, data could be gathered from adjacent cells to calculate the updated value for the current cell. Our implementation uses the latter method.

For ELBM3D, a non-linear equation must be solved for each grid-point and at each time-step so that the collision process satisfies certain constraints. The equation is solved via Newton-Raphson iteration (5 iterations are usually enough to converge to within 10^{-8}), and as this equation involves taking the logarithm of each component of the distribution function at each iteration, the whole algorithm become heavily constrained by the performance of the `log` function.

Figure 1 shows a slice through the xy-plane in the (left) LBMHD3D and (right) ELBM3D simulation, where the vorticity profile has distorted considerably after several hundred time steps as turbulence sets in.

3.1 Vectorization Details

The basic structure of both applications consists of three nested loops over spatial grid points (typically 100s iterations per loop) with inner loops over velocity streaming vectors and, in the case of LBMHD3D, magnetic field streaming vectors (typically 10s iterations). Within these innermost loops the various macroscopic quantities and their updated values are calculated via various algebraic expressions.

For the LBMHD3D case, on both the ES and SX-8, the innermost loops were unrolled via compiler directives and the (now) innermost grid point loop was vectorized. This proved a very effective strategy, and was also followed on the X1E. In the case of the X1E, however, the compiler needed more coercing via directives to multi-stream the outer grid point loop and vectorize the inner grid point loop once the streaming loops had been unrolled. We then inserted the compiler directive NO_CACHE_ALLOC in the attempt to optimize cache use on the X1E [17]. This directive works by indicating to the compiler that certain arrays that have low reuse are not be to be cached. In this way, space is preserved for data that can be more beneficially cached, producing a speedup of 10%. For the superscalar architectures, we utilized a data layout that has been previously shown to be optimal on cache-based machines [16], but did not explicitly tune further for any architecture.

For ELBM3D, in the case the vector architectures, the compilers were able to vectorize all loops containing `log` functions. The routine containing the non-linear equation solver was rewritten to operate on an array of grid points, rather than a single point, allowing vectorization of this recursive operation. After this optimization, high performance was achieved on all the vector systems. For the X1E, two other factors are important to note. In a similar way to LBMHD3D, compiler directives to enable efficient cache use led to a modest 5% speedup. Less data is moved per gridpoint in ELBM3D as compared with LBMHD3D, so cache tuning could reasonably be expected to produce less of a speedup. Additionally, the call to the non-linear equation solving routine prevented multistreaming of the outer grid point loop on the X1E. Because of this, the innermost grid point loop is now both multistreamed and vectorized. For the tests run here, the vector length does not drop below 64, but it does lead to shorter vector lengths compared to the LBMHD3D code.

For the superscalar systems, using the rewritten non-linear equation solving routine proved to be much faster than the original approach. Presumably this is due to a reduction of routine-call overhead and better use of the functional

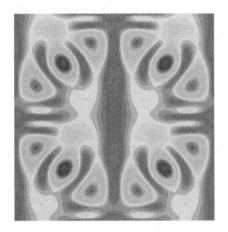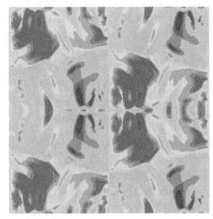

Fig. 1. Contour plot of xy-plane showing the evolution of vorticity into turbulent structures using (left) LBMHD3D and (right) ELBM3D

units. Depending on the architecture, a speedup of 20-30% is achieved on switching to the new routine. Another important optimization was to use optimized library routines to compute a vector of logarithm values per invocation. Each architecture offers an optimized math function library: MASS for IBM Power5 and Power3, MKL for Intel Itanium2; and ACML for AMD Opteron. A 15-30% speedup over the the "non-vector" log function is achieved, with the Itanium2 showing the largest speedup. In addition, the innermost grid point loop was blocked to try and improve cache reuse. A downside to this optimization is that it reduces the length of the array being passed to the log function. This produced very minor speedups for Power3 and Power5, a slowdown for the Itanium2, but a moderate improvement (roughly 15%) for the Opteron system.

3.2   Experimental Results

Tables 3 and 4 and present the performance of both LB applications across the seven architectures evaluated in our study. Cases where the memory or number of processors required exceeded that available are indicated with a dash.

For LBMHD3D, the vector architectures outperform the scalar systems by a large margin. This is largely unsurprising since our efforts at optimization had produced highly vectorized applications. Performance monitoring tools showed that the application exhibits an average vector length (AVL) very close to the maximum and a vector operation ratio (VOR) of more than 99%. In accordance with peak performance, the SX-8 is the leader by a wide margin, achieving the highest per processor performance to date for LBMHD3D; this is followed by the X1E, and then the ES. Although the SX-8 achieves the highest absolute performance, the percentage of peak is somewhat lower than that of ES. Based on previous work [18], we believe that this is related to the memory subsystem and use of DDR-SDRAM.

Turning to the superscalar architectures, the Opteron cluster outperforms the Itanium2 system by almost a factor of 2x. One source of this disparity is that the Opteron system achieves stream memory bandwidth (see Table 1) of more than twice that of the Itanium2 system. Another possible source of this degradation are the relatively high cost of inner-loop register spills on the Itanium2, since the floating point values cannot be stored in the first level of cache. Given the

Table 3. LBMHD3D performance in GFlop/s (per processor) across the studied architectures for a range of concurrencies and grid sizes. Percentage of peak is shown in parenthesis.

P	Size	Power3	Power5	Itanium2	Opteron	X1E	ES	SX-8
16	256^3	0.14 (9)	0.81 (11)	0.26 (5)	0.70 (16)	6.19 (34)	5.50 (69)	7.89 (49)
64	256^3	0.15 (10)	0.82 (11)	0.35 (6)	0.68 (15)	5.73 (32)	5.25 (66)	8.10 (51)
256	512^3	0.14 (9)	0.79 (10)	0.32 (6)	0.60 (14)	5.65 (31)	5.45 (68)	9.66 (60)
512	512^3	0.14 (9)	0.79 (10)	0.35 (6)	0.59 (13)	5.47 (30)	5.21 (65)	—

Table 4. ELBM3D performance in GFlop/s (per processor) across the studied architectures for a range of concurrencies and grid sizes. Percentage of peak is shown in parenthesis.

P	Size	Power3	Power5	Itanium2	Opteron	X1E	ES	SX-8
64	512^3	0.49 (32)	2.31 (30)	1.86 (33)	1.15 (26)	4.49 (25)	3.36 (42)	5.87 (37)
256	512^3	0.45 (30)	2.02 (27)	1.51 (27)	1.08 (25)	4.43 (25)	3.35 (42)	5.86 (37)
512	1024^3	—	2.04 (27)	1.79 (27)	1.04 (24)	4.62 (26)	3.16 (39)	—
1024	1024^3	—	—	1.54 (26)	—	—	3.12 (39)	—

age and specifications, the Power3 does quite reasonably, obtaining a higher percent of peak that the Itanium2, but falling behind the Opteron. The Power5 achieves a slightly better percentage of peak than the Power3, but somewhat disappointingly trails the Opteron.

For ELBM3D (Table 4), all superscalar architectures achieve a high percentage of peak performance. The main reason is the much higher computational intensity and less complex data access patterns of the application relative to LBMHD3D. For the vector architectures, the converse is true—all achieve a lower percentage of peak, as compared to LBMHD3D, with the ES decreasing the most. The problem is not due to a significant increase of non-vectorizable code portions, as the ELBM3D application has an AVL and VOR very close to that of LBMHD3D. Lack of arithmetic operations and data movement in the application has lessened the advantage of the fast ES memory, and the `log` function is probably a bottleneck in computation. However, although the advantage of vector over superscalar has diminished, the SX-8 still achieves the highest overall performance, followed by the X1E and ES.

4   CACTUS

Einsteins equations from theory of general relativity are among most complex in physics: Dozens of coupled nonlinear hyperbolic and elliptic equations, each with thousands of terms. The Cactus Computational ToolKit [19,8] evolves these equations to simulate gravitational waves, such as from two black holes colliding or neutron star mergers. Gravitational waves are ripples in spacetime curvature, causing distances to change. Their existence was postulated nearly 90 years ago by Albert Einstein, and constitutes the only major component of his General Theory of Relativity (GR) that has yet to be tested. If gravitational waves do exist, then an exciting new field of scientific research is about to be born that will provide fundamentally new information about the universe. The Cactus calculations aid in the experimental programs that are set to detect these phenomena using extremely sensitive laser interferometers. While Cactus is a modular framework supporting a wide variety of multi-physics applications [7], this study focuses exclusively on the GR solver, which implements the ADM-BSSN [8] method for stable evolutions of black holes. Figure 2 presents a visualization of one of the first simulations of the grazing collision of two black holes computed

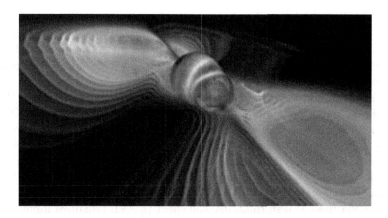

Fig. 2. Visualization of grazing collision of two black holes as computed by Cactus[1]

by the Cactus code. The merging black holes are enveloped by their "apparent horizon", which is colorized by its Gaussian curvature.

The Cactus General Relativity components solve Einstein's equations as an initial value problem that evolves partial differential equations on a regular grid using the method of finite differences. For the purpose of solving Einstein's GR equations, the ADM solver decomposes the solution into 3D spatial hypersurfaces that represent different slices of space along the time dimension. In this formalism, the equations are written as four constraint equations and 12 evolution equations. Additional stability is provided by the BSSN modifications to the standard ADM method [8]. The BSSN implementation uses the Method of Lines (MoL) to reformulate a partial differential equation (PDE) solution so that it can be solved as a coupled set of ordinary differential equations (ODEs). MoL greatly improves the numerical efficiency of the PDE solver. A "lapse" function describes the time slicing between hypersurfaces for each step in the evolution, while a "shift metric" is used to move the coordinate system at each step to avoid being drawn into a singularity. The four constraint equations are used to select different lapse functions and the related shift vectors. For parallel computation, the grid is block domain decomposed so that each processor has a section of the global grid. The standard MPI driver for Cactus solves the PDE on a local grid section and then updates the values at the ghost zones by exchanging data on the faces of its topological neighbors in the domain decomposition.

4.1   Vectorization Details

For the superscalar systems, the computations on the 3D grid are blocked in order to improve cache locality. Blocking is accomplished through the use of temporary "slice buffers", which improve cache reuse while modestly increasing the computational overhead. On vector architectures these blocking optimizations

---

[1] Visualization by Werner Benger (AEI/ZIB) using Amira [20].

were disabled, since they reduced the vector length and inhibited performance. The ES compiler misidentified some of the temporary variables in the most compute-intensive loop of the ADM-BSSN algorithm as having inter-loop dependencies. When attempts to force the loop to vectorize failed, a temporary array was created to break the phantom dependency.

Another performance bottleneck that arose on the vector systems was the cost of calculating radiation boundary conditions. The cost of boundary condition enforcement is inconsequential on the microprocessor based systems, however they unexpectedly accounted for up to 20% of the ES runtime and over 30% of the X1 overhead. The boundary conditions were vectorized using very lightweight modifications such as inline expansion of subroutine calls and replication of loops to hoist conditional statements outside of the loop. Although the boundaries were vectorized via these transformations, the effective AVL remained infinitesimally small. Obtaining longer vector lengths would have required more drastic modifications that were deemed impractical due the amount of the Cactus code that would be affected by the changes. The boundary condition modification was very effective on the X1 because the loops could be successfully multistreamed by the compiler. Multistreaming enabled an easy 3x performance improvement in the boundary calculations that reduced their runtime contribution from the most expensive part of the calculation to just under 9% of the overall wallclock time. These same modifications produced no net benefit for the ES or SX-8, however, because of the extremely short vector lengths.

4.2   Experimental Results

The full-fledged production version of the Cactus ADM-BSSN application was run on each of the architectures with results for two grid sizes shown in Table 5. The problem size was scaled with the number of processors to keep the computational load the same (weak scaling). Cactus problems are typically scaled in this manner because their science requires the highest-possible resolutions.

For the vector systems, Cactus achieves almost perfect VOR (over 99%) while the AVL is dependent on the x-dimension size of the local computational domain. Consequently, the larger problem size (250x64x64) executed with far higher efficiency on both vector machines than the smaller test case (AVL = 248 vs. 92),

Table 5. Cactus performance in GFlop/s (per processor) shown for a range of concurrencies. Percentage of peak is shown in parenthesis.

P	Size/CPU	Power3	Power5	Itanium2	Opteron	X1	ES	SX-8
16	80^3	0.31 (21)	1.12 (15)	0.60 (11)	0.98 (22)	0.54 (4)	1.47 (18)	1.86 (12)
64	80^3	0.22 (14)	1.04 (14)	0.58 (10)	0.81 (18)	0.43 (3)	1.36 (17)	1.81 (11)
256	80^3	0.22 (14)	1.12 (15)	0.58 (10)	0.76 (17)	0.41 (3)	1.35 (17)	1.75 (11)
16	$250x64^2$	0.10 (6)	1.07 (14)	0.58 (10)	0.82 (19)	0.81 (6)	2.83 (35)	4.27 (27)
64	$250x64^2$	0.08 (6)	0.95 (13)	0.57 (10)	0.92 (21)	0.72 (6)	2.70 (34)	4.04 (25)
256	$250x64^2$	0.07 (5)	0.95 (13)	0.55 (10)	0.68 (16)	0.68 (5)	2.70 (34)	3.87 (24)

achieving 34% of peak on the ES. The oddly shaped domains for the larger test case were required because the ES does not have enough memory per node to support a 250^3 domain. This rectangular grid configuration had no adverse effect on scaling efficiency despite the worse surface-to-volume ratio. Additional performance gains could be realized if the compiler was able to fuse the X and Y loop nests to form larger effective vector lengths. Also, note that for the Cactus simulations, bank conflict overheads are negligible for the chosen (non power of two) grid sizes.

Recall that the boundary condition enforcement was not vectorized on the ES and accounts for up to 20% of the execution time, compared with less than 5% on the superscalar systems. This demonstrates a different dimension of architectural balance that is specific to vector architectures: seemingly minor code portions that fail to vectorize can quickly dominate the overall execution time. The architectural imbalance between vector and scalar performance was particularly acute of the X1, which suffered a much greater impact from unvectorized code than the ES. (Cactus results are not available on the X1E due to code crashing; Cray engineers have been notified of the problem.) On the SX-8, the boundary conditions occupy approximately the same percentage of the execution time as it did on the ES, which is consistent with the fact that the performance improvements in the SX8 scalar execution unit have scaled proportionally with the vector performance improvements. The decreased execution efficiency is primarily reflected in lower efficiency in the vector execution.

The microprocessor based systems offered lower peak performance and generally lower efficiency than the NEC vector systems. The Opteron, however, offered impressive efficiency as well as peak performance in comparison to the Power3 and the Itanium2. Unlike the Power3, the Opteron maintains its performance even for the larger problem size. The relatively low scalar performance on the microprocessor-based systems is partially due to register spilling, which is caused by the large number of variables in the main loop of the BSSN calculation. However, the much lower memory latency of the Opteron and higher effective memory bandwidth relative to its peak performance allow it to maintain higher efficiency than most of the other processors. The Power5 shows much higher performance than the Power3 for the larger problem size thanks to much improved memory bandwidth and more advanced prefetch features. For the large case, it approaches the efficiency of the Opteron and achieves the highest raw performance amongst the superscalar system.

5 Conclusions

This study examined three scientific codes on the parallel vector architectures of the X1/X1E, ES and SX-8, and four superscalar platforms, Power3, Power5, Itanium2, and Opteron. Results show that the SX-8 achieves the highest performance of any architecture tested to date for our applications. However, the SX-8 could not match the computational efficiency of the ES, due in part, to a relatively higher memory latency and higher overhead for irregular data accesses. Both the SX-8 and ES also consistently achieved a significantly higher fraction of peak than the X1/X1E, due to superior scalar processor performance, memory bandwidth, and network bandwidth relative to the peak vector flop rate. Taken together, these results indicate that for applications that have a high degree of exploitable data parallelism, vector architectures have a tremendous performance capabilities.

A comparison of the superscalar platforms shows the Power5 having the best absolute performance overall, sometimes overtaking the X1. However, it is often less efficient than the Opteron processor, which in turn, consistently outperforms the Itanium2 and Power3 in terms of both raw speed and efficiency. We note that although the Itanium2 exceeds the performance of the older Power3 processor, the percentage of peak achieved often falls below that of Power3. Our study also shows that the slide in the sustained performance of microprocessor cores is not irreversible if microprocessor architects are willing to invest the effort to make architectural decisions that eliminate bottlenecks in scientific applications. For instance, the Power5 shows some improvement over its predecessors (the Power3 and Power4) in the execution efficiency for the all the applications, thanks to dramatically improved memory bandwidth and increased attention to latency hiding through advanced prefetch features. Future work will expand our study to include additional areas of computational sciences, with a focus on irregular and unstructured algorithms, while examining the latest generation of supercomputing platforms, including high-scalability experiments on BG/L and the XT3.

Acknowledgments

The authors would like to thank the staff of the Earth Simulator Center, especially Dr. T. Sato, S. Kitawaki and Y. Tsuda, for their assistance during our visit. We are also grateful for the early SX-8 system access provided by HLRS, Stuttgart, Germany. This research used the resources of several computer centers supported by the Office of Science of the U.S. Department of Energy: the National Energy Research Scientific Computing Center under Contract No. DE-AC02-05CH11231; Lawrence Livermore National Laboratory under contract No. W-7405-Eng-48; the National Center for Computational Sciences at Oak Ridge National Laboratory under Contract No. DE-AC05-00OR22725. The authors were supported by the Office of Advanced Scientific Computing Research in the Department of Energy Office of Science under contract number DE-AC02-05CH11231.

References

1. Russell, R.: The CRAY-1 Computer System. Comm. ACM **V 21, N 1** (1978)
2. Oliker, L., Canning, A., Carter, J., Shalf, J., Ethier, S.: Scientific computations on modern parallel vector systems. In: Proc. SC2004: High Performance Computing, Networking, and Storage Conference. (2004)
3. Oliker, L., et al.: Evaluation of cache-based superscalar and cacheless vector architectures for scientific computations. In: Proc. SC2003: High Performance Computing, Networking, and Storage Conference. (2003)

4. Dongarra, J., Luszczek, P.: HPC Challenge Benchmarks - Stream EP. http://icl.cs.utk.edu/hpcc/index.html (2006)
5. Carter, J., Soe, M., Oliker, L., Tsuda, Y., Vahala, G., Vahala, L., Macnab, A.: Magnetohydrodynamic turbulence simulations on the Earth Simulator using the lattice Boltzmann method. In: Proc. SC2005: High performance computing, networking, and storage conference. (2005)
6. Vahala, G., Yepez, J., Vahala, L., Soe, M., Carter, J.: 3D entropic lattice Boltzmann simulations of 3D Navier-Stokes turbulence. In: Proc. of 47th Annual Meeting of the APS Division of Plasma Physics. (2005)
7. Font, J.A., Miller, M., Suen, W.M., Tobias, M.: Three dimensional numerical general relativistic hydrodynamics: Formulations, methods, and code tests. Phys. Rev. D **61** (2000)
8. Alcubierre, M., Allen, G., Brgmann, B., Seidel, E., Suen, W.M.: Towards an understanding of the stability properties of the 3+1 evolution equations in general relativity. Phys. Rev. D **(gr-qc/9908079)** (2000)
9. Succi, S.: The lattice Boltzmann equation for fluids and beyond. Oxford Science Publ. (2001)
10. Dellar, P.: Lattice kinetic schemes for magnetohydrodynamics. J. Comput. Phys. **79** (2002)
11. Macnab, A., Vahala, G., Pavlo, P., , Vahala, L., Soe, M.: Lattice Boltzmann model for dissipative incompressible MHD. In: Proc. 28th EPS Conference on Controlled Fusion and Plasma Physics. Volume 25A. (2001)
12. Macnab, A., Vahala, G., Vahala, L., Pavlo, P.: Lattice Boltzmann model for dissipative MHD. In: Proc. 29th EPS Conference on Controlled Fusion and Plasma Physics. Volume 26B., Montreux, Switzerland (June 17-21, 2002)
13. Ansumali, S., Karlin, I.V.: Stabilization of the lattice Boltzmann method by the H theorem: A numerical test. Phys. Rev. **E62** (2000) 7999–8003
14. Ansumali, S., Karlin, I.V., Öttinger, H.C.: Minimal entropic kinetic models for hydrodynamics. Europhys. Lett. **63 (6)** (2003)
15. Keating, B., Vahala, G., Vahala, L., Soe, M., Yepez, J.: Entropic lattice boltzmann simulations of turbulence. In: Proceeding of 48th Annual Meeting of the Division of Plasma Physics. (2006)
16. Wellein, G., Zeiser, T., Donath, S., Hager, G.: On the single processor performance of simple lattice bolzmann kernels. Computers and Fluids **35** (2006) 910
17. Worley, P.: Private communication (2005)
18. Carter, J., Oliker, L.: Performance evaluation of lattice-Boltzmann magnetohydrodrodynamics simulations on modern parallel vector systems. In: High Performance Computing on Vector Systems. (2006) 41–50
19. Schnetter, E., et al.: Cactus Code Server. http://www.cactuscode.org (2006)
20. TGS Inc.: Amira - Advanced 3D Visualization and Volume Modeling. http://www.amiravis.com (2006)

Numerical Simulation of Three-Phase Flow in Heterogeneous Porous Media

Eduardo Abreu[1,*], Frederico Furtado[2], and Felipe Pereira[1]

[1] Universidade do Estado do Rio de Janeiro,
Nova Friburgo, RJ 25630-050, Brazil
eabreu@iprj.uerj.br, pereira@iprj.uerj.br
http://www.labtran.iprj.uerj.br
[2] University of Wyoming, Laramie 82071-3036, USA
furtado@uwyo.edu
http://www.uwyo.edu/furtado/

Abstract. We describe an efficient numerical simulator, based on an operator splitting technique, for three-phase flow in heterogeneous porous media that takes into account capillary forces, general relations for the relative permeability functions and variable porosity and permeability fields. Our numerical procedure combines a non-oscillatory, second order, conservative central difference scheme for the system of hyperbolic conservation laws modeling the convective transport of the fluid phases with locally conservative mixed finite elements for the approximation of the parabolic and elliptic problems associated with the diffusive transport of fluid phases and the pressure-velocity calculation. This numerical procedure has been used to investigate the existence and stability of non-classical waves (also called transitional or undercompressive waves) in heterogeneous two-dimensional flows, thereby extending previous results for one-dimensional problems.

Keywords: Three-phase flow, porous media, central difference scheme, mixed finite elements, non-classical waves, operator splitting.

1 Introduction

Three-phase flow in porous media is important in a number of scientific and technological contexts. Examples include gas injection and thermal flooding in oil reservoirs, flow of non-aqueous phase liquids in the vadose zone, and radio-nuclide migration from repositories of nuclear waste. In this paper we are concerned with the accurate numerical simulation of three-phase flow in heterogeneous porous media.

Three-phase flow in a porous medium can be modeled, using Darcy's law, in terms of the relative permeability functions of the three fluid phases (say, oil, gas, and water). Distinct empirical models have been proposed for the relative

---

* Corresponding Author.

M. Daydé et al. (Eds.): VECPAR 2006, LNCS 4395, pp. 504–517, 2007.
© Springer-Verlag Berlin Heidelberg 2007

permeability functions [10,13,26,17]. It is well known that for some of these models [10,26], which have been used extensively in petroleum engineering, the 2×2 system of conservation laws (the saturation equations) that arises when capillarity (diffusive) effects are neglected fails to be strictly hyperbolic somewhere in the interior of the saturation triangle (the phase space). This loss of strict hyperbolicity leads to the frequent occurrence of non-classical waves (also called transitional or undercompressive shock waves) in the solutions of the three-phase flow model. Crucial to calculating transitional shock waves is the correct modeling of capillarity effects [15].

We describe a numerical procedure, based on a two-level operator splitting technique, for three-phase flow that takes into account capillary pressure differences. This procedure combines a non-oscillatory, second order, conservative central difference scheme, introduced by Nessyahu-Tadmor (NT) [24], for the numerical approximation of the system of conservation laws describing the convective transport of the fluid phases with locally conservative mixed finite elements for the approximation of the parabolic and elliptic problems associated with the diffusive transport of fluid phases and the pressure-velocity calculation [23] (see also [2,3,4]). This numerical procedure has been used to indicate the existence of non-classical transitional waves in multidimensional heterogeneous flows (see [3,4] for preliminary computational results), thereby extending previous results for one-dimensional problems [22,2]. The authors are currently investigating, with the numerical procedure developed, the existence and stability (with respect to viscous fingering) of transitional waves in heterogeneous formations as a first step in the analysis of the scale-up problem for three-phase flow.

We list four distinctive aspects of our numerical scheme:

- Dimensional splitting is unnecessary. Recently, a "corrected" time-splitting method for one-dimensional nonlinear convection-diffusion problems was introduced in [18,19] to better account for the delicate balance between the focusing effects of nonlinear convection, which lead to the formation of shocks, and the smoothing effects of diffusion. As a consequence, this new method reduces considerably the error associated with viscous splitting, allowing accurate large time-steps to be taken in the computation. However, the extension of this method to multidimensional problems requires the use of dimensional splitting. It is known that in the presence of strong multidimensional effects the errors of dimensional splitting might be large (see [9]).
- Riemann solvers or approximate Riemann solvers are unnecessary.
- A CFL time-step restriction applies only to the hyperbolic part of the calculation. The parabolic part of the calculation is performed implicitly, and does not restrict the size of the time-steps for stability.
- We compute accurate velocity fields in the presence of highly variable permeability fields by discretizing the elliptic equation with mixed finite elements.

Different approaches for solving numerically the three-phase flow equations are discussed in [5,7,21].

The rest of this paper is organized as follows. In Section 2 we introduce the model for three-phase flow in heterogeneous porous media that we consider. In

Section 3 we discuss strategies for solving the hyperbolic and diffusive problems taking into account variable porosity fields. In Section 4 we present computational solutions for the model problem considered here. Conclusions appear in section 5.

2   Governing Equations for Three-Phase Flows

We consider two-dimensional, horizontal flow of three immiscible fluid phases in a porous medium. The phases will be refereed to as water, gas, and oil and indicated by the subscripts w, g, and o, respectively. We assume that there are no internal sources or sinks. Compressibility, mass transfer between phases, and thermal effects are neglected.

We assume that the three fluid phases saturate the pores; thus, with S_i denoting the saturation (local volume fraction) of phase i, $\sum_i S_i = 1$, $i = g, o, w$. Consequently, any pair of saturations inside the triangle of saturations $\triangle :=\{(S_i, S_j) : S_i, S_j \geq 0, S_i + S_j \leq 1, i \neq j\}$ can be chosen to describe the state of the fluid.

We refer the reader to [25,3] for a detailed description of the derivation of the phase formulation of the governing equations of three-phase flow. In our model we shall work with the saturations S_w and S_g of water and gas, respectively. Then, the equations governing the three-phase flow are as follows:

Saturation equations:

$$\frac{\partial}{\partial t}(\phi(\mathbf{x})S_w) + \nabla \cdot (\mathbf{v}f_w(S_w, S_g)) = \nabla \cdot \mathbf{w}_w \tag{1}$$

$$\frac{\partial}{\partial t}(\phi(\mathbf{x})S_g) + \nabla \cdot (\mathbf{v}f_g(S_w, S_g)) = \nabla \cdot \mathbf{w}_g. \tag{2}$$

The diffusion terms $\mathbf{w}_w$ and $\mathbf{w}_g$ that arise because of capillary pressure differences are given by

$$[\mathbf{w}_w, \mathbf{w}_g]^T = K(\mathbf{x})\, B(S_w, S_g)\, [\nabla S_w, \nabla S_g]^T. \tag{3}$$

Here, $[\mathbf{a}, \mathbf{b}]$ denotes the 2-by-2 matrix with column vectors $\mathbf{a}$ and $\mathbf{b}$, and $B(S_w, S_g) = QP'$, where

$$Q(S_w, S_g) = \begin{bmatrix} \lambda_w(1 - f_w) & -\lambda_w f_g \\ -\lambda_g f_w & \lambda_g(1 - f_g) \end{bmatrix}, \quad P'(S_w, S_g) = \begin{bmatrix} \dfrac{\partial p_{wo}}{\partial S_w} & \dfrac{\partial p_{wo}}{\partial S_g} \\ \dfrac{\partial p_{go}}{\partial S_w} & \dfrac{\partial p_{go}}{\partial S_g} \end{bmatrix}. \tag{4}$$

In the above, $K(\mathbf{x})$ and $\phi(\mathbf{x})$ are the absolute permeability and the rock porosity of the porous medium, respectively. $\lambda_i(S_w, S_g) = k_i/\mu_i$, $i = w, g$, denote the phase mobilities, given in terms of the phase relative permeabilities k_i and phase viscosities μ_i. The fractional flow function of phase i is given by $f_i(S_w, S_g) = \lambda_i/\lambda$.

The capillary pressures $p_{ij} = p_i - p_j$, $i \neq j$, where p_i is the pressure in phase i, are assumed to depend solely on the saturations.

Pressure-Velocity equations:

$$\nabla \cdot \mathbf{v} = 0, \tag{5}$$
$$\mathbf{v} = -K(\mathbf{x})\lambda(S_w, S_g)\nabla p_o + \mathbf{v}_{wo} + \mathbf{v}_{go}, \tag{6}$$

where $\mathbf{v}_{wo}$ and $\mathbf{v}_{go}$ are "correction velocities" defined by

$$\mathbf{v}_{ij} = -K(\mathbf{x})\lambda_i(S_w, S_g)\nabla p_{ij}. \tag{7}$$

Boundary and initial conditions for the system of equations (1)-(7) must be imposed to complete the definition of the mathematical model. In particular, S_w and S_g must be specified at the initial time $t = 0$.

3   The Numerical Simulator

We employ a two-level operator-splitting procedure for the numerical solution of the three-phase flow system (1)-(7). Operator splitting techniques constitute one of the several bridges between numerical and functional analysis. In numerical analysis, they represent algorithms intended to approximate evolution equations accurately in a computationally efficient fashion. In functional analysis, they are used to prove estimates, existence and representation theorems. The survey article [8] discusses both uses and point to a large bibliography.

The splitting technique discussed here allows for time steps for the pressure-velocity calculation that are longer than those for the diffusive calculation, which, in turn, can be longer than those for advection. Thus, we introduce three time steps: Δt_c for the solution of the hyperbolic problem for the advection, Δt_d for the solution of the parabolic problem for the diffusive calculation and Δt_p for the elliptic problem for the pressure-velocity calculation, so that $\Delta t_p \geq \Delta t_d \geq \Delta t_c$. We remark that in practice variable time steps are always useful, especially for the advection micro-steps subject dynamically to a CFL condition.

The oil pressure and the Darcy velocity, Eqs. (5)-(7), are approximated at times $t^m = m\Delta t_p$, $m = 0, 1, 2, \ldots$ using locally conservative mixed finite elements (see [3]). The linear system of algebraic equations that arises from the discretization can be solved by a preconditioned conjugate gradient procedure (PCG) or by a domain decomposition procedure [11,4,3].

The saturations S_w and S_g are approximated at times $t_n = n\Delta t_d$, $n = 1, 2, \ldots$ in the diffusive calculation; recall that they are specified at $t = 0$. For $t > 0$ these values are obtained from last solution of the hyperbolic subsystem of conservation laws modeling the convective transport of the fluid phases. In this stage the parabolic subsystem associated to the system (1)-(4) is solved. Locally conservative mixed finite elements are used to discretize the spatial operators in the diffusion system. The time discretization of the latter is performed by means of the implicit backward Euler method (see [3]).

In addition, there are values for the saturations computed at intermediate times $t_{n,\kappa} = t_n + k\Delta t_c$ for $t_n < t_{n,\kappa} \leq t_{n+1}$ that take into account the convective transport of water and gas but ignore the diffusive effects. In these intermediate times the subsystem of nonlinear conservation laws is approximated by a non-oscillatory, second order, conservative central difference scheme (see [24,3]).

We refer to [3,2] for a detailed description of the fractional-step procedure.

3.1   The NT Central Scheme for Variable Porosity Fields

In this section we discuss a possible implementation of the NT central differencing scheme for variable porosity fields (see [24] for the original scheme) and its application to the solution of the hyperbolic subsystem associated with system (1)-(4). For brevity, we only discuss the ideas for a scalar conservation law and in one space dimension. The simplicity of the extension of the ideas to systems of equations, by a component-wise application of the scalar scheme, and to multi-dimensions is one of the hallmarks of the NT scheme.

The key features of the NT scheme are: a non-oscillatory, piecewise linear (bilinear in two-space dimension) reconstruction of the solution point-values from their given cell averages and central differencing based on the *staggered* evolution of the reconstructed averages.

Consider the following scalar conservation law,

$$\frac{\partial}{\partial t}(\phi\, s) + \frac{\partial}{\partial x} f(s) = 0, \tag{8}$$

where $\phi = \phi(x)$ is the porosity and $s = s(x,t)$ is the saturation (the volume fraction of one of the fluid phases). At each time level, a piecewise constant approximate solution over cells of width $\Delta x = x_{j+\frac{1}{2}} - x_{j-\frac{1}{2}}$ (see Figure 1),

$$\bar{s}(x,t) = s_j(t), \qquad x_{j-\frac{1}{2}} \leq x \leq x_{j+\frac{1}{2}}, \tag{9}$$

is first reconstructed by a piecewise linear approximation of the form

$$L_j(x,t) = s_j(t) + (x - x_j)\frac{1}{\Delta x} s'_j(t), \qquad x_{j-\frac{1}{2}} \leq x \leq x_{j+\frac{1}{2}}, \tag{10}$$

using nonlinear MUSCL-type slope limiters (see [24] and references therein) to prevent oscillations. This reconstruction compensates the excessive numerical diffusion of central differencing. We observe that (9) and (10) can be interpreted as grid projections of solutions of successive noninteracting Riemann problems which are integrated over a staggered grid ($x_j \leq x \leq x_{j+1}$; see Figure 1). The form (10) retains conservation, i.e., (here the over-bar denotes the $[x_{j-\frac{1}{2}}, x_{j+\frac{1}{2}}]$-cell average),

$$\bar{L}_j(x,t) = \bar{s}(x,t) = s_j(t). \tag{11}$$

Second-order accuracy is guaranteed if the numerical derivatives, defined as $\frac{1}{\Delta x} s'_j$, satisfy (see [24]):

$$\frac{1}{\Delta x} s'_j(t) = \frac{\partial}{\partial x} s(x = x_j, t) + O(\Delta x). \tag{12}$$

In the second stage, the piecewise linear interpolant (10) is evolved in time through the solution of successive noninteracting Generalized Riemann (GR) problems (see Figure 1),

$$s(x, t + \Delta t_c) = GR(x, t + \Delta t_c; L_j(x, t), L_{j+1}(x, t)), \quad x_j < x < x_{j+1}. \quad (13)$$

The resulting solution (13) is then projected back into the space of staggered piecewise constant grid-functions to yield

$$s_{j+\frac{1}{2}}(t + \Delta t_c)\phi_{j+\frac{1}{2}} \equiv \frac{1}{\Delta x} \int_{x_j}^{x_{j+1}} \phi(x)s(x, t + \Delta t_c) \, dx, \quad (14)$$

where $\phi_{j+\frac{1}{2}}$ is the average value of $\phi(x)$ on the cell $[x_j, x_{j+1}]$. In view of the conservation law (8),

$$
\begin{aligned}
s_{j+\frac{1}{2}}(t + \Delta t_c)\phi_{j+\frac{1}{2}} = {} & \frac{1}{\Delta x}\left[\int_{x_j}^{x_{j+\frac{1}{2}}} \phi(x)L_j(x, t)\, dx + \int_{x_{j+\frac{1}{2}}}^{x_{j+1}} \phi(x)L_{j+1}(x, t)\, dx\right] \\
& - \frac{1}{\Delta x}\left[\int_t^{t+\Delta t_c} f(s(x_{j+1}, \tau))\, d\tau - \int_t^{t+\Delta t_c} f(s(x_j, \tau))\, d\tau\right].
\end{aligned}
$$

$$(15)$$

The first two integrands on the right of (15), $L_j(x, t)$ and $L_{j+1}(x, t)$, can be integrated exactly. We remark that the porosity is assumed to be constant on cells, $\phi(x) = \phi_j$ for $x_{j-1/2} < x < x_{j+1/2}$. Moreover, if the CFL condition

$$\frac{\Delta t_c}{\Delta x} \max_{x_j \leq x \leq x_{j+1}} \left\{ \frac{f'(s(x, t))}{\phi(x)} \right\} < \frac{1}{2}, \quad (16)$$

holds, then the last two integrands on the right of (15) are smooth functions of τ. Hence, they can be integrated approximately by the midpoint rule, at the expense of an $O(\Delta t^3)$ local truncation error, to yield the following corrector step,

$$
\begin{aligned}
s_{j+\frac{1}{2}}(t + \Delta t_c)\phi_{j+\frac{1}{2}} = {} & \frac{1}{2}[\phi_j s_j(t) + \phi_{j+1}s_{j+1}(t)] + \frac{1}{8}[\phi_j s'_j(t) - \phi_{j+1}s'_{j+1}(t)] \\
& - \alpha_x \left[f\left(s\left(x_{j+1}, t + \tfrac{\Delta t_c}{2}\right)\right) - f\left(s\left(x_j, t + \tfrac{\Delta t_c}{2}\right)\right)\right],
\end{aligned}
$$

$$(17)$$

where $\alpha_x = \Delta t_c/\Delta x$.

We observe that the spatial integration in (15) is performed over the entire Riemann fan, which consists of both left- and right-going waves. This is the distinctive feature of the NT scheme. On the one hand, this integration eliminates the need of any detailed knowledge about the exact (or approximate) generalized Riemann solver $GR(\cdot; \cdot, \cdot)$; on the other hand, it facilitates accurate computation of the numerical flux, $\int_t^{t+\Delta t_c} f(s(x_j, \tau))d\tau$, whose values are extracted from the smooth interface of two noninteracting generalized Riemann problems (see Figure 1).

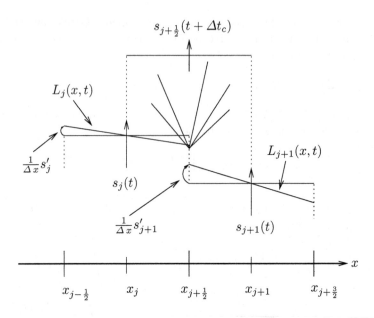

Fig. 1. Evolution from the time level t to the time level $t + \Delta t_c$. The porosity is assumed to be piecewise constant, with constant values on the cells of the original grid: $\phi(x) = \phi_j$, $x_{i-1/2} < x < x_{i+1/2}$.

By Taylor expansion and the conservation law (8),

$$s(x_j, t + \Delta t_c/2) = s_j(t) - \frac{1}{2\phi_j}\alpha_x f'_j(t), \tag{18}$$

may serve (as a predictor step) for the approximation of the saturation mid-values of the numerical fluxes that appears in (17) within the permissible second-order accuracy requirement. Here, $\frac{1}{\Delta x}f'_j$ stands for an approximate numerical derivative of the numerical flux $f(s(x = x_j, t))$,

$$\frac{1}{\Delta x}f'_j(t) = \frac{\partial}{\partial x}(f(s(x = x_j, t))) + O(\Delta x). \tag{19}$$

Next, a piecewise linear interpolant is reconstructed

$$L_{j+\frac{1}{2}}(x, t+\Delta t_c) = s_{j+\frac{1}{2}}(t+\Delta t_c) + (x - x_{j+\frac{1}{2}})\frac{1}{\Delta x}s'_{j+\frac{1}{2}}(t+\Delta t_c), \quad x_j \le x \le x_{j+1}, \tag{20}$$

again using nonlinear slope limiters, and then averaged over the original grid to yield the non-staggered cell average

$$s_j^{t+\Delta t_c} = \frac{1}{2}\left(s_{j+\frac{1}{2}}^{t+\Delta t_c} + s_{j-\frac{1}{2}}^{t+\Delta t_c}\right) + \frac{1}{8}\left(s'^{t+\Delta t_c}_{j-\frac{1}{2}} - s'^{t+\Delta t_c}_{j+\frac{1}{2}}\right). \tag{21}$$

(Here $s_j^{t+\Delta t_c} \equiv s_j(t + \Delta t_c)$.)

Remarks:

1) The NT central differencing scheme for the approximation of the hyperbolic conservation law (8) can be written in the form of three separate steps: a prediction step (18), a correction step (17), and a projection step (21).
2) The numerical derivatives that appear in equations (17), (18), and (21) should obey the accuracy constraints (12) and (19). The second-order accurate correction step (17) augments the first-order accurate prediction step (18), and results in a high-resolution second-order central difference approximation of (8).
3) To guarantee the desired non-oscillatory property of these approximations, the numerical derivatives $\frac{1}{\Delta x}s'_j$ and $\frac{1}{\Delta x}f'_j$ must be carefully chosen [24] (see [3] for our choice).
4) To solve the hyperbolic subsystem associated to the system (1)-(4) we use a component-wise extension [24] of the NT scheme for scalar equations discussed above.
5) The CFL condition for the subsystem of hyperbolic conservation laws assumes the form

$$\frac{\Delta t_c}{\Delta x} \max_{x_j \le x \le x_{j+1}} \rho\left(\frac{1}{\phi(x)}J(f_w, f_g)\right) < \frac{1}{2}, \tag{22}$$

where $\rho(A)$ denotes the spectral radius of matrix A and $J(f_w, f_g)$ is the Jacobian matrix of the fractional flow functions associated to the system (1)-(4).

3.2 Numerical Approximation of the Diffusive System with Variable Porosity Field

We discuss a numerical procedure in two space dimensions that we employ for the solution of the parabolic subsystem associated to the system (1)-(4). This procedure combines a domain decomposition technique with an implicit time backward Euler method (see [3]) in the construction of an efficient iterative method which allows for variable porosity.

We consider an element-by-element domain decomposition and require that the pairs $(S_{w_j}, \mathbf{w}_{w_j})$ and $(S_{g_j}, \mathbf{w}_{g_j})$ (where $S_{i_j} = S_i|_{\Omega_j}$, $i = w, g$.) be a solution of the subsystem associated with (1)-(4) for $\mathbf{x} \in \Omega_j$, $j = 1, \ldots, M$. It is also necessary to impose the consistency conditions,

$$\begin{aligned} S_{w_j} &= S_{w_k}, & S_{g_j} &= S_{g_k} & \mathbf{x} \in \Gamma_{jk}, \\ \mathbf{w}_{w_{jk}} \cdot \nu_j + \mathbf{w}_{w_{kj}} \cdot \nu_k &= 0, & \mathbf{w}_{g_{jk}} \cdot \nu_j + \mathbf{w}_{g_{kj}} \cdot \nu_k &= 0, & \mathbf{x} \in \Gamma_{jk}, \end{aligned} \tag{23}$$

where ν_j is a outward normal unit vector of the element Ω_j.

In order to define an iterative method to solve the above problem, it will be convenient to replace the consistency conditions in Eq. (23) by the equivalent Robin transmission boundary conditions [12]. These consistency conditions are given by

$$-\chi_{w_{jk}}\mathbf{w}_{w_j} \cdot \nu_{j_j} + S_{w_j} = -\chi_{w_{jk}}\mathbf{w}_{w_k} \cdot \nu_{j_k} + S_{w_k}, \quad \mathbf{x} \in \Gamma_{jk} \subset \partial\Omega_j, \tag{24}$$

$$-\chi_{w_{kj}}\mathbf{w}_{w_k} \cdot \nu_{j_k} + S_{w_k} = \chi_{w_{kj}}\mathbf{w}_{w_j} \cdot \nu_{j_j} + S_{w_j}, \quad \mathbf{x} \in \Gamma_{kj} \subset \partial\Omega_k, \tag{25}$$

$$- \chi_{g_{jk}} \mathbf{w}_{g_j} \cdot \nu_{j_j} + S_{g_j} = \chi_{g_{jk}} \mathbf{w}_{g_k} \cdot \nu_{j_k} + S_{g_k}, \qquad \mathbf{x} \in \Gamma_{jk} \subset \partial \Omega_j, \qquad (26)$$

$$- \chi_{g_{kj}} \mathbf{w}_{g_k} \cdot \nu_{j_k} + S_{g_k} = \chi_{g_{kj}} \mathbf{w}_{g_j} \cdot \nu_{j_j} + S_{g_j}, \qquad \mathbf{x} \in \Gamma_{kj} \subset \partial \Omega_k, \qquad (27)$$

where $\chi_{w_{jk}}$ and $\chi_{g_{jk}}$ are positive functions on Γ_{jk} (see [12]).

We consider the lowest index Raviart-Thomas space [23] over Ω_j to approximate the pairs $(S_w, \mathbf{w}_w)$ and $(S_g, \mathbf{w}_g)$. The degrees of freedom on an element Ω_j are the values S_{w_j} and S_{g_j} and the two values $w_{w_{j_\beta}}$ and $w_{g_{j_\beta}}$, $\beta = L, R, B, T$, of the diffusive fluxes across the edge of the elements. We shall also introduce the Lagrange multipliers ℓ_{w_β} and ℓ_{g_β}, $\beta = L, R, B, T$, for the water and gas saturations, respectively, on Γ_{jk}; these multipliers are constant on each edge.

So, after some standard calculations the discrete form of the parabolic subsystem can be written as (see [1,11]):

$$\phi_j \left(\frac{S_{w_j} - \bar{S}_{w_j}}{\Delta t_d} \right) - \frac{1}{h_x} (w_{w_{jR}} + w_{w_{jL}}) + \frac{1}{h_y} (w_{w_{jU}} + w_{w_{jD}}) = 0, \quad (28)$$

$$w_{w_{j_\beta}} B_{11_\beta}^{-1} + w_{g_{j_\beta}} B_{12_\beta}^{-1} = \frac{2}{h_x} (S_{w_j} - \ell_{w_{j_\beta}}), \quad \beta = L, R, \qquad (29)$$

$$w_{w_{j_\beta}} B_{11_\beta}^{-1} + w_{g_{j_\beta}} B_{12_\beta}^{-1} = \frac{2}{h_y} (S_{w_j} - \ell_{w_{j_\beta}}), \quad \beta = B, T, \qquad (30)$$

$$\phi_j \left(\frac{S_{g_j} - \bar{S}_{g_j}}{\Delta t_d} \right) - \frac{1}{h_x} (w_{g_{jR}} + w_{g_{jL}}) + \frac{1}{h_y} (w_{g_{jU}} + w_{g_{jD}}) = 0, \quad (31)$$

$$w_{w_{j_\beta}} B_{21_\beta}^{-1} + w_{g_{j_\beta}} B_{22_\beta}^{-1} = \frac{2}{h_x} (S_{g_j} - \ell_{g_{j_\beta}}), \quad \beta = L, R, \qquad (32)$$

$$w_{w_{j_\beta}} B_{21_\beta}^{-1} + w_{g_{j_\beta}} B_{22_\beta}^{-1} = \frac{2}{h_y} (S_{g_j} - \ell_{g_{j_\beta}}), \quad \beta = B, T, \qquad (33)$$

where $B_{ij_\beta}^{-1}$ are the entries of the inverse matrix $B^{-1}(\ell_{w_\beta}, \ell_{g_\beta}) = (QP')^{-1}(\ell_{w_\beta}, \ell_{g_\beta})$. Here a trapezoidal rule is used for the evaluation of the pertinent integrals in the derivation of Eqs. (28)-(30) and Eqs. (31)-(33). To simplicity of notation, in this section ϕ_j means the value of the porosity in the element Ω_j.

Define an iterative scheme for the solution of the parabolic subsystem by applying Eqs. (24)-(25) to Eqs. (29)-(30) and Eqs. (26)-(27) to Eqs. (32)-(33) to express all Lagrange multipliers in terms of Lagrange multipliers and fluxes associated with adjacent elements. This scheme, developed in [1,3] for constant porosity (see also [11]) is a natural extension for parabolic systems of the procedure introduced in [12] for scalar elliptic and parabolic problems.

The time discretization for the equations (28)-(33) is performed by means of the implicit backward Euler method (see [3]). Note that $\bar{S}_w$ and $\bar{S}_g$ are the initial conditions for the diffusive (discrete form) system (28)-(33).

4   Numerical Experiments

We consider the following Riemann problem for the numerical experimentsreported in this work:

$$S_w^L = 0.721 \quad S_g^L = 0.279 \quad \text{and} \quad S_w^R = 0.05 \quad S_g^R = 0.15. \tag{34}$$

We take the Leverett model [20] for capillary pressure which is given by

$$p_{wo} = 5\epsilon(2 - S_w)(1 - S_w) \quad \text{and} \quad p_{go} = \epsilon(2 - S_g)(1 - S_g), \tag{35}$$

where the coefficient ϵ controls the relative importance of convective and diffusive forces. We take $\epsilon = 0.001$ and fluid viscosities $\mu_o = 1.0$, $\mu_w = 0.5$, and $\mu_g = 0.3$.

We adopt two distinct sets of relative permeability functions in our numerical experiments. These sets are particular choices of the following expressions

$$k_w = S_w^2, \quad k_o = S_o^2, \quad \text{and} \quad k_g = (1 - \alpha_g)S_g^2 + \alpha_g S_g, \quad 0 \le \alpha_g \le 1. \tag{36}$$

By setting the parameter $\alpha_g = 0$ we obtain the classical immiscible Corey-type model for phase relative permeabilities. For this model, the subsystem of conservation laws modeling phase convection loses strict hyperbolicity at a particular point in the interior of the saturation triangle, whose location is determined by the fluid viscosities. It is well known that non-classical transitional shock waves typically arise in solutions of this model, and that their correct computation requires the precise modeling of capillarity effects. See [6] for some experimental evidence of the occurrence of transitional shock waves.

Following [16], any choice

$$\alpha_g > \frac{\mu_g}{\sqrt{\mu_w \, \mu_o}} \tag{37}$$

leads to a strictly hyperbolic (in the interior of the saturation triangle) subsystem of conservation laws for the convective transport of fluid phases.

The boundary conditions and injection and production specifications for three-phase flow equations (1)-(7) are as follows. For the horizontal slab geometry (Figure 2), injection is performed uniformly along the left edge ($x = 0$ m) of the reservoir (see top picture in Figure 2) and the (total) production rate is taken to be uniform along the right edge ($x = 512$ m); no flow is allowed along the edges appearing at the top and bottom of the reservoir. In the case of a five-spot geometry (Figure 3), injection takes place at one corner and production at the diametrically opposite corner; no flow is allowed across the entirety of the boundary. In the simulations reported in Figures 2, 3, and 4 (right column) the Corey-type model was used ($\alpha_g = 0$).

Note in Figure 3 that a low porosity region drives a fast finger towards this region (see top left picture in Figure 3). This finger is better resolved under refinement (see bottom left and right pictures in Figure 3).

For the study reported in Figure 2 we consider a scalar absolute permeability field $K(\mathbf{x})$ taken to be log-normal (a fractal field, see [14] and references therein for more details) with moderately large heterogeneity strength. The spatially variable permeability field is defined on a 512×128 grid with the coefficient of variation ((standard deviation)/mean) $C_v = 0.5$. The porosity field is piecewise constant with two distinct alternating values, 0.1 and 0.3 (see top picture in Figure 2).

Next we turn to a 1D comparison between the two models for phase relatives permeabilities. Figure 4 shows that the numerical solution of (1)-(7) with Riemann problem data (34) for the Corey-type model ($\alpha_g = 0$) has a transitional

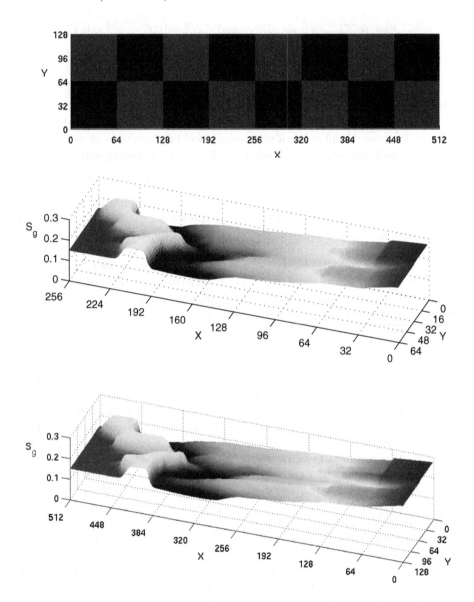

Fig. 2. Mesh refinement study for the gas saturation surface for three-phase flow after 920 days of simulation. The heterogeneous reservoir extends over a 512 m × 128 m
rectangle and has a random permeability field with coefficient of variation $C_v = 0.5$. The
porosity field is piecewise constant with two distinct alternating values 0.1 (blue) and 0.3
(red) (top picture). Computational grids: 256 × 64 (middle) and 512 × 128 (bottom).

shock wave which is not present in the solution of the model with $\alpha_g = 0.43$
(left), which is a strictly hyperbolic subsystem of conservation laws modeling
the convective transport of fluid phases in the three-phase flow region.

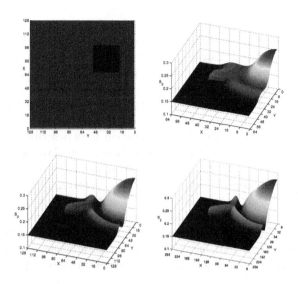

Fig. 3. Mesh refinement study in a 5-spot pattern reservoir after 250 days of simulation. The porosity field (top left) is piecewise constant with only two distinct values: 0.002 in a small rectangular region which drives the development of a finger and 0.2 elsewhere. The 128 m × 128 m reservoir is discretized with computational grid having 64 × 64 (top right), 128 × 128 (bottom left), and 256 × 256 (right left) elements.

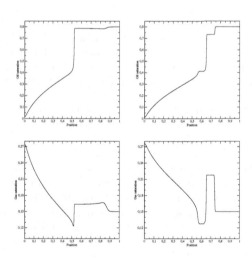

Fig. 4. Oil and gas saturation profiles are shown as functions of dimensionless distance from top to bottom at time 750 days for two models of phase relative permeabilities. We remark that for the choice $\alpha_g = 0$ a transitional (intermediate) shock wave is simulated (right) which is not present in the solution model of phase relative permeabilities with $\alpha_g = 0.43$ (left) that leads to a strictly hyperbolic subsystem of conservation laws.

5   Conclusions

We described the development of a numerical simulation tool for three-phase immiscible incompressible flow in porous media. The porous medium may be heterogeneous with variable porosity and permeability fields. General relations for the relative permeability functions may be used.

Commonly used relative permeability functions lead to the loss of strict hyperbolicity and, thus, to the existence of elliptic regions or umbilic points for the system of nonlinear hyperbolic conservation laws describing the convective transport of the fluid phases. In such situations, non-classical waves, such as transitional or undercompressive shocks, are frequently observed in solutions. We remark that such non-classical waves display a strong dependence upon the physical diffusion being modeled (see [15] and references therein). Thus their accurate computation constitutes a bona fide test for numerical simulators.

The numerical procedure described here has been used to investigate the existence and stability of non-classical waves in heterogeneous two-dimensional flows, thereby extending previous results for one-dimensional problems.

Acknowledgments. E.A. thanks CAPES/Brazil (IPRJ/UERJ) for a Ph.D. Fellowship. F.F. was supported by NSF grant INT-0104529. F.P. was supported by CNPq grants 472199/01-3, CTPetro/CNPq, 470216/2003-4, 504733/2004-4, and CNPq/NFS grant 490696/2004-0.

References

1. Abreu, E.: Numerical simulation of three-phase water-oil-gas flows in petroleum reservoirs. Universidade do Estado do Rio de Janeiro, M.Sc Thesis (2003) (in Portuguese - Available at http://www.labtran.iprj.uerj.br/Orientacoes.html)
2. Abreu, E., Furtado, F., Pereira, F.: On the Numerical Simulation of Three-Phase Reservoir Transport Problems. Transport Theory and Statistical Physics, **33** (5-7) (2004) 503–526
3. Abreu, E., Douglas, J., Furtado, F., Marchesin, D., Pereira, F.: Three-Phase Immiscible Displacement in Heterogeneous Petroleum Reservoirs. *Mathematics and Computers in Simulation*, **73**, (1-4) (2006) 2-20
4. Abreu, E., Furtado, F., Marchesin, D., Pereira, F.: Transitional Waves in Three-Phase Flows in Heterogeneous Formations. *Computational Methods for Water Resources*, Edited by C. T. Miller, M. W. Farthing, W. G. Gray and G. F. Pinder, Series: Developments in Water Science, **I**, (2004) 609–620
5. Berre, I., Dahle, H. K., Karlson, K. H., Nordhaug, H. F.: A streamline front tracking method for two- and three-phase flow including capillary forces. *Contemporary Mathematics: Fluid flow and transport in porous media: mathematical and numerical treatment*, **295** (2002) 49–61
6. Bruining, J., Duijn, C.J. van.: Uniqueness Conditions in a Hyperbolic Model for Oil Recovery by Steamdrive. Computational Geosciences (February 2000), **4**, 65–98
7. Chen, Z., Ewing, R. E.: Fully-discrete finite element analysis of multiphase flow in ground-water hydrology. *SIAM J. on Numerical Analysis*. **34**, (1997) 2228–2253
8. Chorin, A. J., Hughes, T. J. R., McCraken, M. F., and Marsden, J. E.: Product Formulas and Numerical Algorithms. *Comm. Pure Appl. Math.*. **31**, (1978) 205–256

9. Colella, P., Concus, P., and Sethian, J.: Some numerical methods for discontinuous flows in porous media. The Mathematics of Reservoir Simulation. *SIAM Frontiers in Applied Mathematics 1.* Edited by Richard E. Ewing, (1984) 161–186

10. Corey, A., Rathjens, C., Henderson, J., Wyllie, M.: Three-phase relative permeability. *Trans. AIME*, **207**, (1956) 349–351

11. Douglas, Jr. J., Furtado, F., Pereira, F.: On the numerical simulation of waterflooding of heterogeneous petroleum reservoirs. *Comput. Geosci.*, **1**, (1997) 155–190

12. Douglas, Jr. J., Paes Leme, P. J., Roberts, J. E. and Wang, J.: A parallel iterative procedure applicable to the approximate solution of second order partial differential equations by mixed finite element methods. *Numer. Math.*, **65**, (1993) 95–108

13. Dria, D. E., Pope, G.A, Sepehrnoori, K.: Three-phase gas/oil/brine relative permeabilities measured under CO_2 flooding conditions. *SPE 20184*, (1993) 143–150

14. J. Glimm, B. Lindquist, F. Pereira, and R. Peierls.: The fractal hypothesis and anomalous diffusion. *Computational and Applied Mathematics*, **11** (1992) 189–207

15. Isaacson, E., Marchesin, D., and Plohr, B.: Transitional waves for conservation laws. *SIAM J. Math. Anal.*, **21**, (1990), 837–866

16. Juanes, R., Patzek, T. W.: Relative permeabilities for strictly hyperbolic models of three-phase flow in porous media. *Transp. Porous Media*, **57** No. 2 (2004) 125-152

17. Juanes, R., Patzek, T. W.: Three-Phase Displacement Theory: An Improved Description of Relative Permeabilities. *SPE Journal.* **9**, No. 3, (2004) 302–313

18. Karlsen, K. H, Risebro, N. H.: Corrected operator splitting for nonlinear parabolic equations. SIAM Journal on Numerical Analysis, **37** No. 3, (2000) 980-1003

19. Karlsen, K. H, Lie, K.-A., Natvig, J. R., Nordhaug, H. F., and Dahle, H. K.: Operator splitting methods for systems of convection-diffusion equations: nonlinear error mechanisms and correction strategies. *J. of Computational Physics*, **173**, Issue 2, (2001) 636-663

20. Leverett, M. C., Lewis, W. B.: Steady flow of gas-oil-water mixtures through unconsolidated sands. *Trans. SPE of AIME*, **142**, (1941) 107–16

21. Li, B., Chen, Z., Huan, G.: The sequential method for the black-oil reservoir simulation on unstructured grids. *J. of Computational Physics*, **192**, (2003) 36–72

22. Marchesin, D., Plohr, B. J.: Wave structure in WAG recovery. SPE 71314, *Society of Petroleum Engineering Journal*, **6**, no. 2, (2001) 209–219

23. Raviart P-A., Thomas, J. M.: A mixed finite element method for second order elliptic problems. *Mathematical Aspects of the Finite Element Method*, Lecture Notes in Mathematics, Springer-Verlag, Berlin, New York, I. Galligani, and E. Magenes, eds. **606** (1977) 292–315

24. Nessyahu, N., Tadmor, E.: Non-oscillatory central differencing for hyperbolic conservation laws. *J. of Computational Physics*, (1990) 408–463

25. Peaceman, D. W.: Fundamentals of Numerical Reservoir Simulation. Elsevier, Amsterdam (1977)

26. Stone, H. L.: Probability model for estimating three-phase relative permeability. *Petrol. Trans. AIME, 249. JPT*, **23**(2), (1970) 214–218

Simulation of Laser Propagation in a Plasma with a Frequency Wave Equation

R. Sentis[1], S. Desroziers[1], and F. Nataf[2]

[1] CEA/Bruyeres, Service SEL, 91680 Bruyeres, France
[2] Labo J-L-Lions, Université Paris VI, 75013 Paris, France

Abstract. The aim of this work is to perform numerical simulations of the propagation of a laser beam in a plasma. At each time step, one has to solve a Helmholtz equation with variable coefficients in a domain which may contain more than hundred millions of cells.

One uses an iterative method of Krylov type to deal with this system. At each inner iteration, the preconditioning amounts essentially to solve a linear system which corresponds to the same five-diagonal symmetric non-hermitian matrix. If n_x and n_y denote the number of discretization points in each spatial direction, this matrix is block tri-diagonal and the diagonal blocks are equal to a square matrix A of dimension n_x which corresponds to the discretization form of a one-dimension wave operator. The corresponding linear system is solved by a block cyclic reduction method.

The crucial point is the product of a full square matrix Q of dimension n_x by a set of n_y vectors where Q corresponds to the basis of the n_x eigenvectors of the tri-diagonal symmetric matrix A. We show some results which are obtained on a parallel architecture. Simulations with 200 millions of cells have run on 200 processors and the results are presented.

Keywords: Cyclic reduction method, Domain Decomposition Method, Separable matrix, Non-hermitian linear solver, Helmholtz equation.

1 Introduction

The numerical simulation of propagation of high power intensity lasers in a plasma is of importance for the "NIF project" in USA and "LMJ Facility project" in France. It is a very challenging area for scientific computing indeed the wave length $2\pi/k_0$ is equal to a fraction of one micron and the simulation domain has to be much larger than 500 microns. One knows that in a plasma the index of refraction is equal to $\sqrt{1 - N_e/N_c}$, where N_e is the electron plasma density and the critical density N_c is a constant depending only on the wave length. In macroscopic simulations (where the simulation lengths are in the order of some millimeters), geometrical optics models are used and numerical solutions are based on ray tracing methods. To take into account more specific phenomena such as diffraction, autofocusing and filamentation, one generally uses models based on a paraxial approximation of the full Maxwell equations

M. Daydé et al. (Eds.): VECPAR 2006, LNCS 4395, pp. 518–529, 2007.
© Springer-Verlag Berlin Heidelberg 2007

(see for example [5], [1] or, for a new approach in a tilted frame, [6]). But this approximation is valid only if the macroscopic index of refraction is quite constant, in such a way that the wave vector is quite constant in the simulation domain.

There are situations where the macroscopic variations of the plasma density N_e are not small. Particularly if one considers a laser beam propagating in a region near the critical density, it undergoes a total change of direction near a surface called *caustic surface* and the wave vector is strongly varying near this surface. So, the paraxial approximation is no more valid and one has to deal with a model based on a frequency wave equation (obtained by time envelope of the solution of the full Maxwell equations).

Whatever propagation model is used, it is necessary to perform a coupling with the fluid dynamics system for modelling the plasma behavior. For a derivation of the models and a physical exposition of the phenomena under interest, see e.g. [13] or [7]. This paper is aiming at describing the numerical methods for solving the frequency wave equation. Notice that our simulation have been performed to take into account diffraction, refraction and auto-focusing phenomena but the Brillouin parametric instabilities which create laser backscattering are not taken into account up to now.

In the section 2, we describe the model based on the frequency wave equation. In this paper, only 2D problems are considered but the method may be extended to 3D computations. Denote by $\mathbf{x}$ the space variable and set $\mathbf{x} = (x, y)$ the two spatial coordinates. After time discretization, to find the laser field ψ at each time step, one has to solve a Helmholtz equation of the following form

$$\Delta\psi + \left(k_0^2(1 - N) + ik_0\mu\right)\psi = f \tag{1}$$

where f is a given complex function and μ a real coefficient. We assume that the gradient of the macrocsopic non-dimension density $N(\mathbf{x}) = N_e/N_c$ is parallel to the x-axis, then we set

$$N(x, y) = N_0(x) + \delta N(x, y) \tag{2}$$

where N_0 depends on the x variable only and δN is small compared to 1. To solve accurately equation (1), one considers a spatial discretization of finite difference type with a spatial step equal to a fraction of the wave length. If n_x and n_y denote the number of discretization points in each direction, it leads to solve a the linear system with $n_x n_y$ degrees of freedom (which may be equal to 10^8 for a typical 2D spatial domain). One chooses an iterative method of Krylov type with a preconditioning which amounts to solve a linear system corresponding to a five-diagonal symmetric non-hermitian matrix

$$\begin{pmatrix} \alpha + A & -T & & & \\ -T & A & -T & & \\ & -T & A & -T & \cdots \\ & & -T & A & \cdots \end{pmatrix}$$

where T is equal to a constant times the identity matrix of dimension n_x, and A corresponds to the discretization of one-dimension Helmholtz operator (α is a constant). Since this matrix is separable, the corresponding linear system may be solved by the block cyclic reduction method. This method which is derived from the classical cyclic reduction method, has been used for instance in [12] for the numerical solution of Helmholtz problems, but the problem here is a more complicated, indeed one has to deal with Perfectly Matched Layers on two sides of the simulation domain.

The crucial point is the product of a full square matrix Q of dimension n_x by the set of n_y vectors which are of length n_x, where Q corresponds to the basis of the n_x eigenvectors of the tri-diagonal matrix A.

In section 3, we describe the key points of the numerical method for solving (1). In section 4 we give some details on the parallel implementation ; for that purpose the processors are shared out according to horizontal slabs. In the last section we present numerical results in a small simulation domain with only 3 millions of cells and another case of 200 millions of cells which has run on 200 processors.

2   The Model and the Boundary Conditions

The laser beam is characterized by an electromagnetic wave with a fixed pulsation ck_0 where c is the light speed and the wave length in vacuum is equal to $2\pi/k_0$. For modelling the laser, one considers the time envelope $\psi = \psi(t, \mathbf{x})$ of the transverse electric field. It is a slowly time varying complex function. On the other hand, for modelling the plasma behavior one introduces the non-dimension electron density $N = N(t, \mathbf{x})$ and the plasma velocity $\mathbf{U} = \mathbf{U}(t, \mathbf{x})$.

Modelling of the plasma. For the plasma, the simplest model is the following one. Let $P = P(N, \mathbf{x})$ a smooth function of the density N which may depend also of the position $\mathbf{x}$, according to the variation of the electron temperature. Then one has to solve the following barotropic Euler system :

$$\frac{\partial}{\partial t} N + \nabla(N\mathbf{U}) = 0, \tag{3}$$

$$\frac{\partial}{\partial t}(N\mathbf{U}) + \nabla(N\mathbf{U}\mathbf{U}) + \nabla(P(N)) = -N\gamma_p \nabla|\psi|^2. \tag{4}$$

The term $\gamma_p \nabla|\psi|^2$ corresponds to a ponderomotive force due to a laser pressure (the coefficient γ_p is a constant depending only on the ion species).

Modelling of the laser beam. The laser field $\psi = \psi(t, \mathbf{x})$ is a solution to the following frequency wave equation (which is of Schrödinger type)

$$2i\frac{1}{c}\frac{\partial}{\partial t}\psi + \frac{1}{k_0}\Delta\psi + k_0(1 - N)\psi + i\nu\psi = 0, \tag{5}$$

where the absorption coefficient ν depends on space and the density $N = N(t, \mathbf{x})$ is solution to the fluid system stated above. Of course, the problem is interesting only in the region where $N(t, \mathbf{x}) \leq 1$.

General framework. For the numerical solution of the fluid system, we use the method described in [8] or [1] which has been implemented in a parallel platform called HERA. For solving (5), the spatial mesh has to be very fine, at least 10 cells per wave length in each direction. Generally the modulus $|\psi|$ of the electric field is slowly varying with respect to the spatial variable, one can use a crude mesh for the simulation of the Euler system (the mesh size has to be of order of the $2\pi/k_0$). If the modulus $|\psi|$ was not slowly varying in a region, one would have to solve in this region the Euler system with a fine mesh also.

So we handle a two-level mesh of finite difference type : in a 2D simulation, each cell of the fluid system is divided into $p_0 \times p_0$ cells for the Helmholtz level, with $p_0 = 10$ or 5. We assume in the whole paper that the hypothesis (2) holds, so it allows to perform a preconditioning of the global linear system by another system which is simpler since it does not take into account the perturbation $\delta N(x, y)$; this last system corresponds to a separable matrix and therefore a block cyclic reduction method may be used for its numerical solution.

Boundary conditions. The laser beam is assumed to enter in $x = 0$. Since the density N depends mainly on the $x-$variable, we may denote by N^{in} the mean value of the incoming density on the boundary and by N^{out} the mean value of the density on the outgoing boundary . Let $\mathbf{e}_b$ be the unit vector related to the direction of the incoming laser beam and set $\mathbf{K}^{in} = \mathbf{e}_b\sqrt{1 - N^{in}}$. The boundary condition on the part of the boundary ($x = 0$) reads (with $\mathbf{n} = (-1, 0)$ the outwards normal to the boundary)

$$(k_0^{-1}\mathbf{n}.\nabla + i\mathbf{K}^{in}.\mathbf{n})(\psi - \alpha^{in}e^{ik_0\mathbf{K}^{in}\mathbf{x}}) = 0. \qquad (6)$$

where $\alpha^{in} = \alpha^{in}(y)$ is a smooth function which is, roughly speaking, independent of the time. On the part of the boundary $x = x_{\max}$, there are two cases according to the value N^{out} :

i) If $N^{out} > 1$ the wave does not propagate up to the boundary and the boundary condition may read as $\partial\psi/\partial x = 0$.

ii) If $N^{out} \leq 1$ it is necessary to consider a transparent boundary condition. Here we take the simplest one, that is to say

$$(k_0^{-1}\mathbf{n}.\nabla + i\sqrt{1 - N^{out}})(\psi) = 0.$$

On the other hand, on the part of the boundary corresponding to $y = 0$ and $y = y_{\max}$, it is crucial to have a good transparent boundary condition, so we introduce perfectly matched layers (the P.M.L. of [2]). For the simple equation $-\Delta\psi - \omega^2\psi = f$, this technique amounts to replace in the neighborhood of the boundary, the operator $\frac{\partial}{\partial y}$ by $\left(1 + \frac{\sigma}{i\omega}\right)^{-1}\frac{\partial}{\partial y}$, where σ is a damping function which is not zero only on two or three wave lengths and which increases very fast up to the boundary. Notice that the feature of this method is that it is necessary to modify the discretization of the Laplace operator on a small zone near the boundary.

Time discretization. At each time step δt determined by the CFL criterion for the Euler system, one solves first the Euler system with the ponderomotive force and afterwards the frequency wave equation (5). For the time discretization of this equation, an implicit scheme is used. The length $c\delta t$ is very large compared to the spatial step therefore the time derivative term may be considered as a perturbation and one has to solve the following equation of the Helmholtz type

$$\Delta\psi + \left(k_0^2(1 - N) + ik_0(\mu_0 + \nu)\right)\psi = i\mu_0\psi^{ini} \tag{7}$$

where $\mu_0 = 2k_0/(c\delta t)$. The boundary conditions are the same as above.

3   Principle of the Numerical Methods for the Helmholtz Equation

The spatial discretization (7) is the classical one of finite difference type. Denote by n_x and n_y the number of discretization points in each direction. Beside the interior domain, there are two zones corresponding to the two PMLs near the boundary $y = 0$ and $y = y_{max}$, the width of these layers corresponds to $2p_0$ points. Then the linear system to be solved has the following form

$$\begin{pmatrix} P_1 & C_1 & 0 \\ E_1 & A_I + D & E_2 \\ 0 & C_2 & P_2 \end{pmatrix} \Psi = F, \tag{8}$$

where P_1 and P_2 are square matrices whose dimension is $2p_0n_x$, it corresponds to the discretization of the equation in the P.M.L. On the other hand C_i, E_i are coupling matrices (whose dimensions are n_xn_y, times $2p_0n_x$). The square matrix A_I whose dimension is n_xn_y, corresponds to the discretization of

$$(\Delta + k_0^2(1 - N_0) + ik_0\mu_0)\bullet \tag{9}$$

in the interior domain. Moreover, D is a diagonal matrix corresponding to the terms $\delta N(x, y) + ik_0\nu(x, y)$. Notice that the domain decomposition method is used with Robin interface conditions (see [10], [3]) which corresponds to a discretization of the condition on the interfaces between subdomains

$$\frac{\partial}{\partial n}\psi + \alpha\psi = \frac{\partial}{\partial n}\psi^{neib} + \alpha\psi^{neib}$$

(α is a complex parameter and ψ^{neib} is the value in the other subdomain).

3.1   Solution of the Linear System

To solve (8), the principle is to performe an iterative Krilov method with a preconditioning which correponds to the discretization of the Helmholtz operator (9). To save CPU time, the GMRES method has been found to be the best Krylov method ; since the number of iterations is not very high, it is not necessary to use a restart procedure. The preconditioning is performed by solving the linear system based on A_I in the interior domain and on P_1 and P_2 in the P.M.L.

domains, that is to say the main point is to solve as fast as possible a system of the following form

$$\mathcal{P}U = f, \qquad \text{where} \quad \mathcal{P} = \begin{pmatrix} P_1 & 0 & 0 \\ 0 & A_I & 0 \\ 0 & 0 & P_2 \end{pmatrix} \qquad (10)$$

where P_1 and P_2 are small matrices which may be easily factorized in the standard LU product. The symmetric non-hermitian matrix A_I has the following form

$$A_I = \begin{pmatrix} B & -T & & & \\ -T & A & -T & & \\ & \ddots & \ddots & \ddots & \\ & & -T & A & -T \\ & & & -T & B \end{pmatrix},$$

where T is equal to a constant times the identity matrix, $B = A + \alpha$ and A is a tri-diagonal matrix of dimension n_x related to the discretization of the operator $(\partial_{xx}^2 + k_0^2(1 - N_0) + ik_0\mu_0)\cdot$; that amounts to the solution of the following system

$$A_I \cdot \begin{pmatrix} u_1 \\ u_2 \\ \vdots \\ u_{n_y-1} \\ u_{n_y} \end{pmatrix} = \begin{pmatrix} f_1 \\ f_2 \\ \vdots \\ f_{n_y-1} \\ f_{n_y} \end{pmatrix} \qquad (11)$$

where the elements u_m and f_m are n_x−vectors,

3.2   The Cyclic Reduction Method

Since A_I is separable, to solve the system (11) in the central domain, we use the block cyclic reduction method. Let us recall the principle of this method. For the sake of simplicity, assume $n_y = 2^k - 1$. We know that A and T are commutative. Consider 3 successive lines of (11) for $i = 2, 4, ..., n_y - 1$:

$$\begin{cases} -Tu_{i-2} + Au_{i-1} - Tu_i & = f_{i-1} \\ \quad - Tu_{i-1} + Au_i - Tu_{i+1} & = f_i \\ \quad\quad - Tu_i + Au_{i+1} - Tu_{i+2} = f_{i+1}. \end{cases} \qquad (12)$$

After a linear combination of these lines, we get :

$$-T^2 A^{-1} u_{i-2} + \left(A - 2T^2 A^{-1}\right) u_i - T^2 A^{-1} u_{i+2} = f_i + T A^{-1} \left(f_{i-1} + f_{i+1}\right) \quad (13)$$

After this first step, the elimination procedure may be performed again by induction. That is to say, denote $A^{(0)} = A$, $B^{(0)} = B$, $T^{(0)} = T$ and $f^{(0)} = f$; after r elimination steps, the reduced system for $0 \le r \le k - 1$ owns $2^{k-r} - 1$ blocs and reads as:

$$
\begin{pmatrix}
B^{(r)} & -T^{(r)} & & & \\
-T^{(r)} & A^{(r)} & -T^{(r)} & & \\
& \ddots & \ddots & \ddots & \\
& & -T^{(r)} & A^{(r)} & -T^{(r)} \\
& & & -T^{(r)} & B^{(r)}
\end{pmatrix}
\begin{pmatrix}
u_{2^r} \\
u_{2.2^r} \\
\vdots \\
u_{(n_y-1)-2^r+1} \\
u_{n_y-2^r+1}
\end{pmatrix}
=
\begin{pmatrix}
f_{2^r}^{(r)} \\
f_{2.2^r}^{(r)} \\
\vdots \\
f_{(n_y-1)-2^r+1}^{(r)} \\
f_{n_y-2^r+1}^{(r)}
\end{pmatrix}
$$

where for $r = 1, ..., k - 2$:

$$
A^{(r)} = A^{(r-1)} - 2\left(T^{(r-1)}\right)^2 \left(A^{(r-1)}\right)^{-1}
$$

$$
B^{(r)} = A^{(r-1)} - \left(T^{(r-1)}\right)^2 \left(\left(A^{(r-1)}\right)^{-1} + \left(B^{(r-1)}\right)^{-1}\right) \tag{14}
$$

$$
T^{(r)} = \left(T^{(r-1)}\right)^2 \left(A^{(r-1)}\right)^{-1}
$$

For the right hand side, we get the induction formula :

$$
f_{i.2^r}^{(r)} = f_{i.2^r}^{(r-1)} + T^{(r-1)} \left(A^{(r-1)}\right)^{-1} \left(f_{i.2^r-2^{r-1}}^{(r-1)} + f_{i.2^r+2^{r-1}}^{(r-1)}\right) \tag{15}
$$

After all the elimination steps, it remains only one equation for finding $u_{2^{k-1}}$. Once this value is obtained, one deduces all the other values step by step by induction.

4   Parallel Implementation

Notice first that $A = ik_0\mu_0 + A^0$ where A^0 is a symmetric tri-diagonal matrix whose coefficients are real except the one in the first line and the first column (due to the boundary condition (6)). We have checked that it is possible to find a basis of eigenvectors of A^0 which are orthogonal for the pseudo scalar product $< u, v >= u^T.v$. They are computed by using the LR algorithm of Parlett (cf. [11]) although it was designed for Hermitian matrices. So denote Q the matrix whose columns are the eigenvectors of A^0, the matrix Q is orthonormal for the pseudo scalar product, that is to say

$$
QQ^T = Q^TQ = I
$$

Since T is the identity matrix up to a multiplicative constant, one can introduce the diagonal matrices $\Lambda^{(0)}$ and $\Gamma^{(0)}$

$$
A = Q\Lambda^{(0)}Q^T, \quad T = Q\Gamma^{(0)}Q^T. \tag{16}
$$

So we get

$$
A^{(r)} = Q\Lambda^{(r)}Q^T, \quad T^{(r)} = Q\Gamma^{(r)}Q^T \tag{17}
$$

with the following induction formulas

$$
\Lambda^{(r)} = \Lambda^{(r-1)} - 2\left(\Gamma^{(r-1)}\right)^2 \left(\Lambda^{(r-1)}\right)^{-1}, \quad \Gamma^{(r)} = \left(\Gamma^{(r-1)}\right)^2 \left(\Lambda^{(r-1)}\right)^{-1} \tag{18}
$$

Let us summarize the algorithm

- Introduce the vectors $\tilde{f}_i$ transformed of f_i in the eigenvector basis

$$\tilde{f}_i = Q^T f_i \text{ for } i = 1, \ldots, n_y.$$

- At each step r, the vector $\tilde{f^r}_i$ transformed of $f^r{}_i$ of the right hand side, reads

$$\tilde{f}_{i.2^r}^{(r)} = \tilde{f}_{i.2^r}^{(r-1)} + \Gamma^{(r-1)} \left(\Lambda^{(r-1)} \right)^{-1} \left(\tilde{f}_{i.2^r-2^{r-1}}^{(r-1)} + \tilde{f}_{i.2^r+2^{r-1}}^{(r-1)} \right)$$

- One computes the vectors $\tilde{u}_{2^k-1}$ by solving

$$\Lambda^{(k-1)} \tilde{u}_{2^k-1} = \tilde{f}_{2^k-1}^{(k-1)}$$

- One recursively distributes the solutions by solving sub-systems of the following type

$$\Lambda^{(r)} \tilde{u}_{j.2^{r+1}-2^r} = \tilde{g}_{j.2^{r+1}-2^r}^{(r)}$$

$$\text{where} \quad \tilde{g}_{j.2^{r+1}-2^r}^{(r)} = \tilde{f}_{j.2^{r+1}-2^r}^{(r)} + \Gamma^{(r)} \left(\tilde{u}_{(j-1).2^{r+1}} + \tilde{u}_{(j).2^{r+1}} \right)$$

- Lastly, the solution u is given by

$$u_i = Q\tilde{u}_i \text{ pour } i = 1, \ldots, n_y.$$

For the parallel implementation, the crucial point is the product of a full matrix Q of dimension $n_x \times n_x$ by a set of n_y vectors (each vector corresponds to a horizontal line of the mesh). So the processors are shared out according to horizontal slabs of the mesh. On our architecture, each node ownes four processors, but often the memory devoted to the node is not large enough to store four times the matrix Q, so multi-thread techniques are used to overcome this difficulty. The matrix Q is stored once on the local memory of the node and four threads are carried out simultaneously on the four processors on the node ; then the products of the matrix Q by the vectors are performed simultaneously for the four horizontal slabs.

Scalability. The code has run on a massively parallel architecture with HP-Compaq processors of the EV67 type. For a typical problem, with $40 \; 10^6$ complex unknowns, when the CPU time for one Krilov iteration is equal to 1 with 16 processors, it is equal to 0.98 with 32 processors and 0.96 with 64 processors, so the efficiency of the parallelism is very good.

On the other hand, consider now problems whose size is multiplied by 2 in each direction. When the number of degrees of freedom is $n_x n_y = 1.6 \; 10^6$, the CPU time with 4 processors is equal to 1 for one Krilov iteration, it is equal to 2.1 with 16 processors for $n_x n_y = 6.4 \; 10^6$, and it is equal to 4.2 with 32 processors for $n_x n_y = 25.6 \; 10^6$. That is to say the CPU is about two time larger when the number of processors and the number of degrees of freedom are 4 times

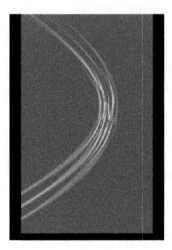

Fig. 1. Laser intensity at time 3 ps. Incoming boundary condition with 3 speckles.

larger; this is coherent with the fact that the number of operations for the cyclic reduction method grows like $n_x^2 n_y$.

5   Numerical Results

The incoming boundary condition $\alpha^{in}(y)$ is roughly speaking equal to a sum of narrow Gaussian functions depending of the y variable ; the half height width of each Gaussian function is equal to 8 wave lengths and is assumed to describe a speckle (a speckle is a light spot of high intensity). One considers first a simulation domain of 100×300 wave lengths ; the initial profile of density is a linear function increasing from 0.1 at $x = 0$ to 1.1 at $x = x_{max}$. The incoming boundary condition consists in three speckles with the same incidence angle. At the Helmholtz level, one handles only 3 millions of cells. With 32 PEs, the CPU time is about 20 seconds per time step for approximately 10 Krylov iterations at each time step. Without the coupling with the plasma, it is well known that the solution is very close to the one given by the geometrical optics ; the speckles propagate in a parallel way, undergo macroscopic refraction when the electron density increases and are tangent to a caustic line (here it is the line corresponding to $x = x_\star$ such that $N_0(x_\star) = \cos^2(\theta)$, where θ is the incidence angle of the speckles). With our model, if the laser intensity is low (which corresponds to a weak coupling with the plasma), one notices that a small digging of the plasma density occurs. This digging is more significant when the laser intensity is larger, then an autofocusing phenomenon takes place.

On figure 1, one sees the map of the laser intensity that is to say the quantity $|\psi|^2$, which corresponds to this situation after some picoseconds, knowing that the time step in about 0.05 picosecond. We notice here that the speckles undergo autofocusing phenomena and some filamentation may be observed.

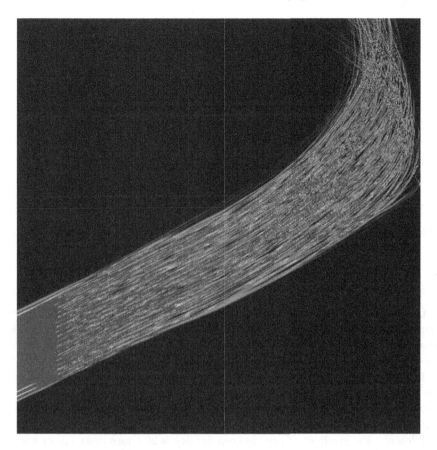

Fig. 2. Laser intensity at time 22 ps. Incoming condition with 20 speckles.

Another case is considered, corresponding to simulation domain of 2000×2000 wave lengths. In the left half of the domain, the electron density is constant and equal to 0.15 and one uses the paraxial model ; in the right half of the domain the density increases from 0.15 to 0.95 and the Helmholtz model is used. The coupling of the two method has to be performed accurately ; it is described in [4].

The incoming boundary condition consists in 20 speckles with various intensity. In this domain, one handles 200 millions of cells (whose size is 1/10 of the wave length) and the simulation have run on 200 PEs. The map of the laser intensity is shown on figure 2 after 22 ps (the time step is roughly equal to 0.02 ps). We have chosen a small absorption coefficient $\nu = 2.10^{-5}N^2$ so the problem is quite sharp. Here the digging of the plasma is locally very important since the variation of density δN reaches 0.05 in a region where $N(x) = 0.8$, see a map of the non-dimension density on the figure 3.

About 13 iterations of the Krilov method are enough to converge. The CPU time is equal to 240 s for solving the full linear system and 270 s for each time step (including the solution of the LR algoritm and the hydrodynamics of the plasma).

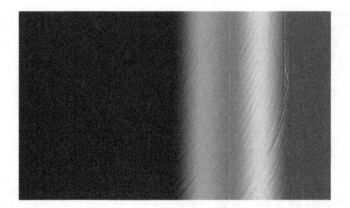

Fig. 3. Plasma density at time 22 ps (zoom). Incoming condition with 20 speckles.

6   Conclusion

In the framework of the hydrodynamics parallel platform HERA, we have developed a solver for the laser propagation based on the frequency wave equation. The assumption that the density N depends mainly on the x−variable only allows to perform a preconditioning by a domain decomposition method (two PMLs and a large Helmholtz zone) where the linear system corresponding to the Helmholtz zone is solved by the block cyclic reduction method. This kind of simulation is new. Up to our knowledge, the solution of this kind of model in a wide two-dimension domain has been published only in [9], but the framework is different : the gradient of the electron density is 20 times more larger and the simulation domain is 100 times smaller than in our problem.

Most of the computer time is spent by the product of the full matrix Q by a set a n_y vectors. In the future some CPU time may be saved if inside the inner iteration loop of the Krylov method, we do not consider the whole spatial domain that is to say all the n_y vectors but only the vectors which does not belong to some subinterval $[n_y^1, n_y^2]$ for instance the ones where the solution varies very few from an iteration to the other.

References

1. Ph. Ballereau, M.Casanova, F.Duboc, D. Dureau, H.Jourdren, P.Loiseau, J.Metral, O.Morice, R. Sentis. *Coupling Hydrodynamics with a Paraxial solver,* to appear.
2. J.-P. Berenger. A Perfectly Matched Layer for the Absorption of Electromagnetic Waves. *J. Comp. Physics* 114, p185-200 (1994).
3. B. Després, Domain decomposition method and the Helmholtz problem. II, in: *Second International Conference on Mathematical and Numerical Aspects of Wave Propagation (Newark, DE, 1993),* SIAM, (1993).
4. S. Desroziers *Modelisation de la propagation laser par résolution de l'équation d'Helmholtz,* Ph. D. dissertation, University Paris VI, (2006).

5. M.R. Dorr , F.X. Garaizar, J.A. Hittinger, *Simuation of laser-Plasma filamentation.* J. Comp. Phys. **17**, p233-263 (2002).
6. M. Doumic, F. Golse, R. Sentis. *Propagation laser paraxiale en coordonnées obliques,* Note C. R. Ac. Sciences, Paris, série I, t.**336**, p.23-28 (2003).
7. S. Hüller, Ph. Mounaix, V.T. Tikhonchuk, D. Pesme. *Interaction of two neighboring laser beams,* Phys. Plasmas, **4**, p.2670-2680, (1997).
8. H. Jourdren. HERA hydrodynamics AMR Plateform for multiphysics simulation, *Proceedings of Chicago workshop on AMR methods,* Plewa T. et al., eds., Springer Verlag, Berlin (2005).
9. Maximov A.V. and al Modeling of stimulated Brillouin .. Phys Plasmaa **11**, p.2994-3000, (2004).
10. P.-L. Lions, On the Schwarz alternating method, III. in: *Third International Symposium on Domain Decomposition Methods for PDE,* Chan and al., ed., SIAM (1990).
11. B. N.Parlett : *Acta Numerica,* p 459-491, (1995).
12. T. Rossi, J. Toivanen. *A parallel fast direct solver for block tridiagonal systems with separable matrices,* SIAM J. Sci. Comput. **20** (1999), pp. 1778-1796.
13. R. Sentis, *Mathematical Models for Laser-Plasma Interaction,* ESAIM: Math. Modelling and Num. Analysis, **39**, p275-318 (2005).

A Particle Gradient Evolutionary Algorithm Based on Statistical Mechanics and Convergence Analysis

Kangshun Li[1,2,3], Wei Li[1,3], Zhangxin Chen[4], and Feng Wang[5]

[1] School of Information Engineering, Jiangxi University of Science and Technology,
Ganzhou 341000, China
{Kangshun Li, lks}@public1.gz.jx.cn
[2] Key Laboratory of High-Performance Computing Technology of Jiangxi Province,
Jiangxi Normal University, Nanchang 330022, China
[3] Key Laboratory of Intelligent Computation and Network Measurement-Control
Technology of Jiangxi Province, Jiangxi University of Science and Technology,
Ganzhou 341000, China
[4] Center for Scientific Computation and Department of Mathematics,
Southern Methodist University, Dallas, TX 75275-0156, USA
[5] Computer School of Wuhan University, Wuhan 430072, China

Abstract. In this paper a particle gradient evolutionary algorithm is presented for solving complex single-objective optimization problems based on statistical mechanics theory, the principle of gradient descending, and the law of evolving chance ascending of particles. Numerical experiments show that we can easily solve complex single-objective optimization problems that are difficult to solve by using traditional evolutionary algorithms and avoid the premature phenomenon of these problems. In addition, a convergence analysis of the algorithm indicates that it can quickly converge to optimal solutions of the optimization problems. Hence this algorithm is more reliable and stable than traditional evolutionary algorithms.

1 Introduction

Evolutionary algorithms (EAs) are searching methods that take their inspiration from natural selection and survival of the fittest in the biological world [1,2]. EAs differ from traditional optimization techniques in that they involve a search from a "population" of solutions, not from a single point. Each iteration of an EA involves a competitive selection that weeds out poor solutions. The solutions with high "fitness" are "recombined" with other solutions by crossing parts of a solution with another. Solutions are also "mutated" by making a small change to a single element of the solutions. Recombination and mutation are used to generate new solutions that are biased toward regions of the space for which good solutions have already been seen. However, there are two main problems puzzling researches in the literature of EC (evolutionary computation) research. The first is the premature, which is one of the basic problems in EC research, and the second is the lack of a proper stopping criterion in problem solution. Previous evolutionary algorithms (we call them traditional evolutionary algorithms) are difficult to avoid the premature phenomenon, and fall into local optimal solutions; the reason is that the traditional evolutionary algorithms cannot take all the individuals of population to participate in crossing and mutating all the time.

M. Daydé et al. (Eds.): VECPAR 2006, LNCS 4395, pp. 530–543, 2007.
© Springer-Verlag Berlin Heidelberg 2007

In this paper a particle gradient evolutionary algorithm for solving complex single-objective optimization problems (SPGEA) is presented to overcome the shortcomings of the traditional evolutionary algorithms mentioned above. SPGEA adopts the method of solving the gradient of an optimization problem to construct the fitness function of the problem, which simulates the principle of energy minimizing of particles in statistical mechanics, and designs an evolving chance function of individuals as the amount of individual crossing, which simulates the law of entropy increasing of particles in statistical mechanics. Based on this construction method, the algorithm guarantees that all the particles have a chance to cross and evolve all the time and produces the global optimization solution of a problem.

This paper is organized as follows: In Section 2, theoretical foundations of statistical mechanics are discussed. The principle of gradient descending and the law of evolving chance ascending in a particle system are then analyzed theoretically in Section 3. A detailed description of a SPGEA flow is designed in Section 4. In Section 5, we perform experiments to test SPGEA by solving three complex optimization problems. The convergence of SPGEA is studied in Section 6. Finally, we draw some conclusions in Section 7.

2   Relevant Theories of Statistical Mechanics

Statistical mechanics [3,4] is to apply a statistical analysis method of applied mathematics to study the average behavior and statistical rules of a number of particles. It is an important branch of theoretical physics. The non-equilibrium statistical mechanics is to study more complex problems. Not until in the mid-20th century has the study of statistical mechanics achieved a rapid development. For a macro physical system being composed of a number of particles, the probability of the system that keeps a more disordered state exceeds the probability of the system that keeps a more ordered state. A closed physical system always trends to the disordered state from the ordered state. In thermodynamics, this is the corresponding law of entropy ascending. Therefore, the free energy theory and entropy theory of statistical mechanics are very important in the course of discussing the equilibrium and non-equilibrium particle system below.

2.1   Law of Entropy Ascending

Assume that a closed system is composed of two open subsystems that may exchange energy and particles so that the entropy of the system increases, i.e., $S = S_1 + S_2$, where S_1 and S_2 denote the entropies of the first and second systems, respectively. Furthermore, assume that the relationship between the micro-state number of the micro-canonical ensemble and the entropy function is $S = f(\Delta\Omega)$, and the two subsystems are independent of each other. As a result, the micro-state number of an isolated system is $\Delta\Omega = \Delta\Omega_1 \Delta\Omega_2$. Thus $S_1 + S_2 = f(\Delta\Omega_1 \Delta\Omega_2)$ and $S = k_B \ln \Delta\Omega$, where k_B is called the Boltzmann constant. According to the entropy equilibrium equation and Boltzman H-theorem, we see that the entropy function is a monotonically increasing function of time in a closed system; i.e., $\dfrac{dS(t)}{dt} \geq 0$. There-fore, the entropy is irreversible in the thermo-insulated system, which is the law of entropy increasing.

2.2 Principle of Energy Descending

The concept of "free energy" is a key concept to characterize physically relevant states in statistical mechanics. Given an equilibrium system of statistical mechanics with energy levels E_i of the microstates i, the Helmholtz free energy is defined as

$$F(\beta) = -\frac{1}{\beta} \log Z(\beta),$$

where

$$Z(\beta) = \sum_i e^{-\beta E_i}$$

is the partition function and β is the inverse temperature. Apparently, the Helmholtz free energy is different from the internal energy U given by

$$U = -\frac{\partial}{\partial \beta} \log Z(\beta) = \langle E_i \rangle.$$

The difference is given by the entropy times the temperature:

$$F = U - TS.$$

This equation can also be regarded as descending a Legendre transformation from U to F. Equilibrium states minimize the free energy; in this sense F is more relevant than U. The minimum of F can be achieved in two competing ways: Either by making the internal energy U small or by making the entropy S large. The basic principle underlying statistical mechanics, the maximum entropy principle, can also be formulated as a "principle of minimum free energy ".

Through the above analysis of a particle system, we know that the equilibrium state of the particle system depends on the result of the competition between free energy descending of this particle system and entropy ascending.

3 Principle of Gradient Descending and Law of Evolving Chance Ascending

We apply the principle of free energy descending and the law of entropy ascending in statistical mechanics to the SPGEA design. In the design of SPGEA, we consider individuals of a population as particles in the particle phase space, and the population of each generation as a system of particles. Our purpose is to simulate the particle system discipline in the physics system to cross and mutate individuals of the population, which tries to change its state from non-equilibrium to equilibrium, and as a result, solves for all the optimal solutions, and avoid problems' premature.

Because the establishment of a fitness function and an iterative stopping criterion of SPGEA is based on the principle of gradient descending and the law of evolving chance ascending in a physical system, which simulates the law of entropy ascending and the principle of energy descending, SPGEA is guaranteed to drive all the particles in the phase space to participate in crossing and mutating, and to speed up its convergence; in the meantime it improves its computing performance so that the probability of the phase space equals and the equilibrium state in the phase space is achieved.

4   Algorithm Flow of SPGEA

4.1 Description of Optimization Problem

We consider the optimization problem :

$$\min_{X \in D} f(X), \quad D = \{X \in S; g_k(X) \le 0, k = 1,2,\cdots,q,$$

where $S \subset R^n$ is the searching space, usually a hypercube of N dimensions, namely, $l_i \le x_i \le u_i, i = 1,2,\cdots,n$, $f : S \to R$ the objective function, n the dimension of the decision space, and D the set of feasible points.

4.2 Variation of the Objective Function

We assume that the population size is N, and the individuals $\mathbf{x}_1, \mathbf{x}_2, \cdots, \mathbf{x}_N$ as N particles in a physical system. Then we add the number t of a continuous evolving iteration into the objective function of the optimization problem, and get the new dynamical single-objective function of the optimization problem $\min_{X \in D, t > 0} f(t, X)$ related to the iteration time. We say that $\min_{X \in D, t > 0} f(t, X)$ is a SPGEA objective function.

4.3 Algorithm Process of SPGEA

According to the principle of free energy descending and the law of entropy ascending of the physical system, we give the definitions of a gradient descending equation and a evolving chance ascending equation of SPGEA as follows:

Definition 1 (SPGEA gradient descending equation): We call the difference equation $\nabla p(t, \mathbf{x}_i) = \nabla f(t, \mathbf{x}_i) - \nabla f(t-1, \mathbf{x}_i)$ as a SPGEA gradient descending equation (SPGEA free energy) of i th   particle $\mathbf{x}_i$ at time t, where $f(\mathbf{x})$ is a function on D, $\mathbf{x}_i \in D$, $i = 1,2,3\cdots\cdots$.

Definition 2 (SPGEA evolving chance ascending): We call the evolving chance counting function $\alpha(t, \mathbf{x}_i)$ of i th   particle $\mathbf{x}_i$ at time t the SPGEA evolving chance ascending (SPGEA entropy), whose value is determined as follows: When particles $\mathbf{x}_i$ participate in the evolving operation in time t,

$$\alpha(t, \mathbf{x}_i) = a(t-1, \mathbf{x}_i) + 1,$$

Otherwise,

$$\alpha(t, \mathbf{x_i}) = a(t-1, \mathbf{x_i}), \quad \mathbf{x_i} \in D, \quad i = 1, 2, 3 \cdots\cdots N \qquad \square$$

Definition 3 (SPGEA fitness function): We define the weighted function $select(t, \mathbf{x_i}) = \lambda_1 \sum_{k=0}^{t} \|\nabla p(k, \mathbf{x_i})\|_p + \lambda_2 \ln(\alpha(t, \mathbf{x_i}) + 1)$ as the SPGEA fitness function, where $\lambda_1, \lambda_2 \in [0,1]$, $\lambda_1 + \lambda_2 = 1$, and λ_1, λ_2 are called SPGEA Boltzmann constants, whose values depend on the significance of $\sum_{k=0}^{t} \|\nabla p(k, \mathbf{x_i})\|_p$ and $\ln(\alpha(t, \mathbf{x_i}) + 1)$ on the right-hand side of the fitness function equation, respectively. That is, the more significant it is, the larger the corresponding SPGEA Boltzmann constant is. This ensures the whole physical system to reach the equilibrium state from the non-equilibrium state, and hence to achieve the equal probability in the phase space; in the meantime, all the individuals in the population have a chance to take part in crossing and mutating at all the iteration times so that global optimal solutions can be achieved. In the SPGEA fitness function we can also see that the reason why SPGEA can avoid the premature phenomenon is that the SPGEA fitness function contains the SPGEA gradient descending term (SPGEA free energy) and the SPGEA evolving chance ascending term (SPGEA entropy).

Definition 4 (SPGEA stopping criterion): We define a SPGEA stopping criterion by

$$(\sum_{i=1}^{N} \|\nabla p(t, \mathbf{x_i})\|_p)/t < \varepsilon \text{ or } \sum_{i=1}^{N} \ln(\alpha(t, \mathbf{x_i}) + 1) > T,$$

where ε is a given small positive constant.

The first SPGEA stopping criterion is constructed by SPGEA free energy, and the second SPGEA stopping criterion is built by SPGEA entropy. We can easily see that the purpose of SPGEA is to minimize SPGEA free energy and maximize SPGEA entropy. These two terms are like the Helmholtz free energy and the entropy of particles in the physical system, and always compete with each other in the course of changing from non-equilibrium to equilibrium spontaneously under the same temperature.

According to the above four basic definitions of SPGEA we design the detailed algorithm of SPGEA as follows:

Step 1: Initialize particles in the physical system to generate an initial population with N individuals $\Gamma_N = \{\mathbf{x_1}, \mathbf{x_2}, \cdots, \mathbf{x_N}\}$ randomly, and set $t := 0$.

Step 2: Calculate all the function values of the particles in Γ_N and set $\nabla p(t, \mathbf{x_i}) = 0$, $\alpha(t, \mathbf{x_i}) = 0$, $\mathbf{x_i} \in \Gamma_N$; then calculate the fitness values of fitness functions $select(t, \mathbf{x_i})$, which are in the order from small to large.

Step 3: Save all the particles and their function values in the system Γ_N.

Step 4: Begin to iterate: $t := t + 1$.

Step 5: Select n particles $\mathbf{x_i'}$, $i = 1, 2, \cdots, n$ on the forefront of $select(t-1, \mathbf{x_i})$; if all the values of $select(t-1, \mathbf{x_i})$ are the same, select n particles randomly.

Step 6: Implement evolving operations on the n particles of the physical system, and generate n random numbers $\alpha_i \in [-1,1], i = 1,2,\cdots,n$ that satisfy $-0.5 \leq \sum_{i=1}^{n} \alpha_i \leq 1.5$ and $\hat{\mathbf{x}} = \sum_{i=1}^{n} \alpha_i \mathbf{x}_i' \in \mathbf{X}$; if the function value at the point $\hat{\mathbf{x}}$ is better than the worst function value at the point $\tilde{\mathbf{x}}_i$, then we replace the individual $\tilde{\mathbf{x}}_i$ by $\hat{\mathbf{x}}$; otherwise repeat this evolving operation.

Step 7: Save the best particles, and their function values and fitness values in the system Γ_N.

Step 8: Renew all the values of $select(t, \mathbf{x}_i)$ and re-sort in an ascending order.

Step 9: Calculate the stopping criterion; if $(\sum_{i=1}^{N} \|\nabla p(t,\mathbf{x}_i)\|_p)/t < \varepsilon$ or $\sum_{i=1}^{N} \ln(\alpha(t,\mathbf{x}_i)+1) > T$, stop iteration; otherwise, go to step 4.

In the above two SPGEA stopping criteria we know that the individuals which are not selected in the previous generation have more chance to be selected to take part in the evolving operation in the next generation because the fitness values added up by non-selected individuals in the previous generation are less than other individuals' fitness values calculated by the selected individuals in the previous generation. In this way, it is guaranteed that all the individuals in the population have a chance to take part in crossing and mutating all the time; this is one of the main features of SPGEA.

5 Data Experiments

In this section, three typical optimization problems that are difficult to solve using the traditional EA [5-8] will be experimented to test the performance of SPGEA. In the first experiment, we use SPGEA to solve the minimization problem of the function

$$\min_{x \in S} f(x_1, x_2) = x_1^2 + x_2^2 - 0.3\cos(3\pi x_1) - 0.4\cos(4\pi x_2) + 0.7,$$

where $-50 \leq x_1 \leq 50$ and $-50 \leq x_2 \leq 50$. From Fig.1 we can see that this optimization problem has almost an infinite number of local optimal points in the searching space, but there is only one global minimum point at $x_1 = 0$ and $x_2 = 0$ that reaches the minimum value $f^* = 0$ of the function. Only the local optimal points can be solved by using the traditional evolutionary algorithm in general.

In this experiment, we set the population size $N = 80$. The weighted coefficient $\lambda_1 = 0.8, \lambda_2 = 0.2, \varepsilon = 10^{-11}$, and the maximal value of the evolving chance function $T = 10^8$, and then select four particles (individuals) that are located in front of the fitness values of the function $select(t, \mathbf{x}_i)$ in the order from small to large to cross and mutate. According to the above configured parameters we run the SPGEA program 10 times continuously, in every iteration time we can get the optimal point that is given

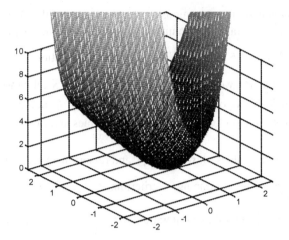

Fig. 1. The landscape of experiment 1

Table 1. The results of running SPGEA program 10 times in experiment 1

Min value f	x_1	x_2	step
0	1.53257e-010	-1.92252e-009	796
0	2.20815e-010	9.12350e-012	780
0	-1.13563e-009	-4.89725e-010	696
0	-1.92390e-009	1.30710e-010	718
0	3.66944e-010	6.67918e-010	708
0	-1.63819e-010	2.56498e-010	670
0	-2.59710e-010	1.85425e-011	831
0	1.02611e-009	-3.04113e-010	747
0	-5.06763e-011	5.91274e-010	745
0	4.64298e-010	2.06257e-010	812

in Table 1. The convergent speed by using SPGEA is faster and the results are more accurate than the traditional evolutionary algorithm in Ref. [8].

In the second experiment, we use SPGEA to test a non-convex function as follows:

$$\min_{x \in S} f(x_1, x_2) = 100(x_1^2 - x_2)^2 + (1 - x_1)^2$$

where $-2.048 \leq x_1 \leq 2.048, -2.048 \leq x_2 \leq 2.048$; this function is non-convex (see Fig.2). In running SPGEA, we set the population size $N = 60$, $\lambda_1 = 0.8, \lambda_2 = 0.2, \varepsilon = 10^{-20}$, and the maximal value of the evolving chance function $T = 10^6$; the crossing and mutating method is the same as in the first experiment. Running the SPGEA program 10 times continuously, we can get the optimal point every time which is given in Table 2.

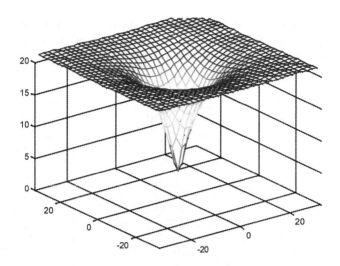

Fig. 2. The landscape of experiment 2

Table 2. The results of running SPGEA program 10 times in experiment 2

Min value f	x_1	x_2	step
3.11543E-244	1.00000E+00	1.00000E+00	9976
1.06561E-241	1.00000E+00	1.00000E+00	9999
9.01807E-241	1.00000E+00	1.00000E+00	10000
7.33725E-243	1.00000E+00	1.00000E+00	9991
8.85133E-243	1.00000E+00	1.00000E+00	9984
1.75654E-243	1.00000E+00	1.00000E+00	9989
4.45935E-242	1.00000E+00	1.00000E+00	9999
5.99649E-242	1.00000E+00	1.00000E+00	9998

In the third experiment, a complex single-objective minimization problem (Ackley function) is tested by using SPGEA, and the optimization problem is as follows:

$$\min_{x \in S} f(x_1, x_2) = -20 \exp\left(-0.2\sqrt{\frac{1}{n}\sum_{i=1}^{n} x_i^2}\right) - \exp\left(\frac{1}{n}\sum_{i=1}^{n}\cos(2\pi x_i)\right) + 20 + e$$

where $-32.768 \le x_i \le 32.768, i = 1,2, \cdots n$, $n = 2$. It is obvious that the optimal solution $f^* = 0$ is reached at the point $x_1 = 0$ and $x_2 = 0$. From Fig.3 we can also see that this function is non-convex and includes multi-local-optimal-points. So it is difficult to solve by using the traditional evolutionary algorithms. In fact, it is very easy to fall into the local solutions, i.e., the premature phenomenon of the algorithms.

We set the parameters of SPGEA: population size $N = 80$. $\lambda_1 = 0.8, \lambda_2 = 0.2, \varepsilon = 10^{-20}$, and the maximal value of the evolving chance function

$T = 10^6$; the crossing and mutating method is same as in the first experiment. Running the SPGEA program 10 times continuously, in each iteration time we also can easily get the optimal point that is given in Table 3.

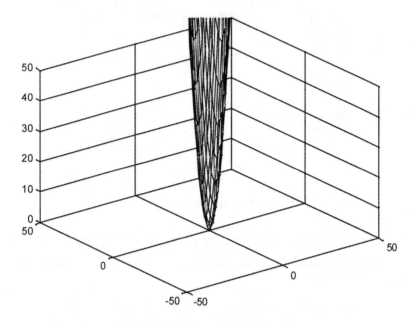

Fig. 3. The landscape of experiment 3

Table 3. The results of running SPGEA program 10 times in experiment 3

Min value f	x_1	x_2	step
1.54096e-009	1.23727e-016	2.74191e-016	913
1.54096e-009	-1.03297e-016	-7.22339e-017	975
1.54096e-009	-3.43706e-017	-8.11170e-017	855
1.54096e-009	2.35071e-016	1.00479e-016	777
1.54096e-009	2.75004e-017	2.20330e-017	871
1.54096e-009	-2.01528e-016	1.24738e-016	780
1.54096e-009	1.66060e-017	1.26505e-017	893
1.54096e-009	-7.42641e-019	7.21961e-019	822
1.54096e-009	-4.29195e-016	-1.45410e-016	938
1.54096e-009	-4.87201e-017	-1.01773e-017	902

Furthermore, we have also done many experiments to solve some well-konwn complex single-objective optimization problems like the Six Hump Camel Back Function, Axis-Parallel Hyperellipsoid Function, and Griewangk's Function by using SPGEA, and we have obtained very accurate optimal solutions, which are difficult to solve by using the traditional evolutionary algorithms.

6 Convergent Analysis of SPGEA

The convergence, time efficiency, and precision of optimal solutions are very important factors when optimization problems are solved by using evolutionary algorithms [9-12], which provide a reliable indication to an efficient algorithm. In this section the convergence of SPGEA is studied to illustrate the advantages of SPGEA according to the theory of the Markov chain and other convergent theories related to EC [13-15].

Assume the optimization problem:

$$\min\{f(x) \mid x \in S\},\tag{1}$$

where f is a function in the decision space S and $\forall \mathbf{x} \in S, f(\mathbf{x}) \geq 0$. S can be either a finite set (e.g., composition optimization problems), or a set in the real space R^n (e.g., continuous optimization problems). Then we get four definitions as follows:

Definition 5: For the SPGEA optimization problem (1), suppose that random variable $T = \min\{t \in \overline{Z}^- : F_t = f^* : \overline{Z}^- = \{0,1,\cdots,n,\cdots\}\}$ represents the time the global optimal point is found at the first time, if $P\{T < \infty\} = 1$ and independent of the initial population; then we say that the SPGEA algorithm can find the global optimal solutions of optimization problems in probability 1 in the finite time [16].

Definition 6: For a non-negative random variable sequence $\{X_n\}$, $t = 0,1,\cdots,\cdots$, which is defined in the probability space (Ω, A, P).

(1) If $\forall \varepsilon > 0$ such that $\sum_{n=0}^{\infty} P(X_n > \varepsilon)$ is convergent, then $\{X_n\}$ is called completely convergent to 0.

(2) If $\forall \varepsilon > 0$ such that $\sum_{n=0}^{\infty} P\{\omega : \lim_{n \to \infty} X_n(\omega) = 0\} = 1$, then $\{X_n\}$ is called convergent to 0 in the probability 1.

(3) If $\forall \varepsilon > 0$ such that $\lim_{n \to \infty} P\{X_n(\omega) > \varepsilon\} = 0$, then $\{X_n\}$ is called convergent to 0 in probability.

In the above three convergent forms, the completely convergent is the strongest, which implies both the convergent in the probability 1 and the convergent in probability, and the convergent in probability is the weakest [16].

Define $D_t = d(X_t) = F_t - f^*$, where f^* is the optimal solution of the optimization problem, and $F_t = f(\mathbf{x}_t)$ is the best solution of the optimization problem in the t th generation, and then set the convergence definition of SPGEA as follows:

Definition 7: We call solving optimization problem (1) the completely convergent (the convergent in the probability 1 or the convergent in probability) to the global optimal points of the problem, if the non-negative random sequence $(D_t : t \geq 0)$ produced by this optimization problem is completely convergent (convergent in the probability 1 or convergent in probability) to 0 [16].

Definition 8: We call solving optimization problem (1) which adopts the elite reservation strategy by using SPGEA the completely convergent (the convergent in the probability 1 or the convergent in probability) to the global optimal point f^*, if the non-negative random sequence $(D_t : t \geq 0)$ produced by this optimization problem is completely convergent (convergent in the probability 1 or convergent in probability) to 0 [16].

According to the above definitions, we get the next convergence theories of SPGEA.

Theorem 1: If the optimization problem (1) solved by using SPGEA satisfies the following conditions:

(1) In every evolving iteration t, if $\mathbf{x} \neq \tilde{\mathbf{x}}$ for all individuals $\mathbf{x}$ ($\mathbf{x} \in P(t)$) in the population $P(t)$ and $\forall \tilde{\mathbf{x}} \in S$, then through crossover and mutation operation once, the probability mutating $\mathbf{x}$ to $\tilde{\mathbf{x}}$ is more than or equal to $p(t)$, where $p(t)$ is a constant more than 0, and the probability is related to generation t.

(2) $\prod_{t=1}^{\infty}(1 - p(t)) = 0$.

Then SPGEA can certainly find the global optimal solution of the optimization problem in probability in finite generation times, that is,
$P\{T < \infty\} = 1$, and it has nothing to do with the distribution of the iterating initial population.

Proof: In the evolving process of tth generation, it needs to mutate to N individuals of the population by using a mutation operator independently from condition (1); we know that in tth generation of any evolving operation, through the mutation of the mutating operator, any individual in population $P(t)$ can mutate to any other individual in the search space S on the lower boundary probability $p(t)$. Therefore, in the mutation process the probability which mutates any individual $\mathbf{x} \notin \arg f^*$ to one of the global optimal solutions is no less than $p(t)$, i.e. the probability which is the first found global optimal point is at least $p(t)(> 0)$ in the evolving process of t th generation. So, after the t generations, the probability $\overline{p}(t)$ that no global optimal point found satisfies

$$\overline{p}(t) \leq \prod_{i=1}^{t}(1 - p(t)),$$

namely, $\lim_{t \to \infty} \overline{p}(t) = 0$, and then we get

$P\{T < \infty\} \geq P\{ \text{find a global optimal point in } t \text{ generations} \} = 1 - \overline{p}(t)$.

Setting $t \to \infty$ on both sides of the above equation, it reduce to $P\{T < \infty\} = 1$; that is, SPGEA can find global optimal solutions of the optimization problem in the probability 1 in the finite evolving times, and obviously, from the proof process we can see that it has nothing to do with the selecting method of an initial population in this theorem's proof.

Theorem 2: If the optimization problem (1) solved by using SPGEA satisfies the following conditions:

(1) In every evolving iteration t, if $\mathbf{x} \neq \mathbf{y}$ for all individuals $\mathbf{x}$ ($\mathbf{x} \in P(t)$) in the population $P(t)$ and $\forall \mathbf{y} \in S$, then by crossover and mutation operation once, the probability mutating $\mathbf{x}$ to $\mathbf{y}$ is more than or equal to $p(t)$, where $p(t)$ is a constant more than 0 and the probability is related to generation t.

(2) $\prod_{t=1}^{\infty}(1 - p(t)) = 0$.

(3) Adopt the strategy of the elite reservation to evolve.

Then, SPGEA certainly converges to the optimal solution of the optimization problem in probability, and it has nothing to do with the selecting method of an initial population.

If SPGEA satisfies the following additional condition:

(4) There exists a constant $p > 0$ such that $p(t) \geq p$ for all generation t, then SPGEA is completely convergent.

Proof: Assume that the global optimal point of the optimization problem is first found in the tth generation, because SPGEA is evolved according to the strategy of elite reservation. This guarantees the first found optimal solution individual to be maintained ever to the last generation in the evolving. Hence we get

$$P\{F_t > f^*\} = P\{F_t - f^* > 0\} = P\{D_t > 0\} = \prod_{i=1}^{t}(1 - p(i)) .$$

Setting $t \to \infty$, it is

$\lim_{t \to \infty} P\{D_t > 0\} = 0$; namely, SPGEA is convergent in probability.

If SPGEA satisfies condition (4), too, e.g., if we can find a constant $p > 0$ such that $p(t) \geq p$ for all t, then

$$P\{F_t > f^*\} = P\{F_t - f^* > 0\} = P\{D_t > 0\} = \prod_{i=1}^{t}(1 - p(i))$$
$$\leq \prod_{i=1}^{t}(1 - p) = (1 - p)^t$$

Because the Taylor series $\sum_{n=0}^{\infty}(1 - p)^t$ is convergent, according to definitions 6 (1) and 7, we conclude that $(D_t : t \geq 0)$ is completely convergent to 0; accordingly, SPGEA is completely convergent to 0 as well. From all the proof process we can see that the convergence of SPGEA has nothing to do with the selection method of an initial population.

7 Conclusions

Through the above theoretical and experimental analysis of SPGEA, we conclude that SPGEA has obviously more advantages than traditional EAs. Because SPGEA is based on statistical mechanics theory according to the principle of gradient descending and the law of evolving chance ascending of particles, which simulate the principle of energy minimizing and law of entropy increasing in the phase space of particles in statistical mechanics, it makes all the particles to have a chance to evolve, and drives all the particles to cross and mutate to reproduce new individuals of the next generation from the beginning to the end. Because of these reasons SPGEA can easily and quickly search for the global optimal solutions and avoid premature phenomenon of the algorithm. Meanwhile, convergent analysis of SPGEA has proved that it is reliable, stable, and secure by using SPGEA to solve complex single-objective optimization problems.

Acknowledgements

This work is supported by the National Natural Science Key Foundation of China with the Grant No.60133010, the Research Project of Science and Technology of Education Department of Jiangxi Province with the Grant No.Gan-Jiao-Ji-Zi[2005]150, and the Key Laboratory of High-Performance Computing Technology of Jiangxi Province with the Grand No.JXHC-2005-003.

References

1. Pan Zhengjun, Kang Lishan, Chen Yuping. Evolutionary Computation[M], Tsinghua University Press, 1998.
2. Lack D L. Daewin's Finches[M]. Cambridge University Press, Cambridge, England, 1947.
3. Reichl L E. A Modern Course in Statistical Mechanics[M]. University of Texas Press, Austin, Texas, 1980.
4. Radu Balescu. Statistical Dynamics[M]. Imperial College Press, London, 2000.
5. Li Yuanxiang, Zou Xiufen, Kang Lishan and Zbigniew Michalewicz, A New Dynamical Evolution Algorithm Based on Statistical Mechanics [J], Computer Science & Technology 2003, Vol. 18, No.3.
6. Michaelwicz Z.. Genetic Algorithms + Data Structures = Evolution Programs[M]. Springer-Verlag, Berlin, Her-delberg, New York., 1996.
7. Mitchell M, Forrest S, Holland J H. The royal road for genetic algorithms: Fitness landscapes and GA perform-ance[A]. In Proc. The first European Conference on Artificial Life, Varela F J, Bourgine P (eds.), MIT Press, Cambridge, Massachusetts, 1992, pp.245-254.
8. Zhijian Wu, Lishan Kang, Xiufen Zou. An Elite Subspace Evolutionary Algorithm for Solving Function Optimization Problems [J], Computer Applications, Vol.23, No.2 13-15
9. Bäck T, Fogel D B, Michalewicz Z. Handbook of Evolutionary Computation. Oxford: Oxford University Press, 1997.
10. Bäck T, Kok J N, DeGraaf J M, et al. Theory of Genetic Algorithms[J]. Bulletin of the EATCS, 1997, 63:161-192.

11. Rudolph, G. Finite Markov Chain Results in Evolutionary Computation: A Tour d'Horizon. Fundamenta in Informaticae[J], 1998, 35(1-4):67-89.
12. Eiben, A E, Rudolph G. Theory of Evolutionary Algorithms: A Bird Eye View[J]. Theoretical Computer Science, 1999,229(1/2):3-9.
13. Jun Qin, Lishan Kang. A Convergence Analysis Framework for Multi-objective Optimization Evolutionary Algorithm[J]. Computer Application and Research, 2005, No.2, pp: 68-70.
14. Zhou Yu-Ren, Yue Xishun, Zhou Ji-Xiang. The Convergence Rate and Eff iciency of Evolutionary Algorithms[J]. Chinese Journal of Computer, Vol.27 No.11, 2004, pp: 1485-1491.
15. Günter Rudolph. Local Convergence Rates of Simple Evolutionary Algorithms with Cauchy Mutations[J]. IEEE Transaction on Evolutionary Computation, Vol.1, No.4, 1997, pp: 249-258.
16. Minqiang Li, Jisong Kou, Dan Lin, Shuquan Li. Basic Theory and Applications on Genetic Algorithm[M]. Chinese Science Press, 2002.

A Computational Framework for Cardiac Modeling Based on Distributed Computing and Web Applications

D.M.S. Martins[1,2], F.O. Campos[1], L.N. Ciuffo[1], R.S. Oliveira[1], R.M. Amorim[1], V.F. Vieira[1], N.F.F. Ebecken[2], C.B. Barbosa[1], and R. Weber dos Santos[1]

[1] Department of Computer Science , Universidade Federal de Juiz de Fora,
Juiz de Fora, Minas Gerais, Brazil
rodrigo.weber@ufjf.edu.br
[2] Department of Computer Science , Universidade Federal do Rio de Janeiro,
Rio de Janeiro, Rio de Janeiro, Brazil
nelson@ntt.ufrj.br

Abstract. Cardiac modeling is here to stay. Computer models are being used in a variety of ways and support the tests of drugs, the development of new medical devices and non-invasive diagnostic techniques. Computer models have become valuable tools for the study and comprehension of the complex phenomena of cardiac electrophysiology. However, the complexity and the multidisciplinary nature of cardiac models still restrict its use to a few specialized research centers in the world. We propose a computational framework that provides support for cardiac electrophysiology modeling. This framework integrates different computer tools and allows one to bypass many complex steps during the development and use of cardiac models. The implementation of cardiac cell models is automatically provided by a tool that translates models described in CellML language to executable code that allows one to manipulate and solve the models numerically. The automatically generated cell models are integrated in an efficient 2-dimensional parallel cardiac simulator. The set up and use of the simulator is supported by a user-friendly graphical interface that offers the tasks of simulation configuration, parallel execution in a pool of connected computer clusters, storage of results and basic visualization. All these tools are being integrated in a Web portal that is connected to a pool of clusters. The Web portal allows one to develop and simulate cardiac models efficiently via this user-friendly integrated environment. As a result, the complex techniques and the know-how behind cardiac modeling are all taken care of by the web distributed applications.

1 Introduction

The phenomenon of electric propagation in the heart comprises a set of complex non-linear biophysical processes. Its multi-scale nature spans from nanometre processes such as ionic movements and protein dynamic conformation, to centimetre phenomena such as whole heart structure and contraction. Computer models [1,2] have become valuable tools for the study and comprehension of such complex phenomena, as they

M. Daydé et al. (Eds.): VECPAR 2006, LNCS 4395, pp. 544–555, 2007.
© Springer-Verlag Berlin Heidelberg 2007

allow different information acquired from different physical scales and experiments to be combined in order to generate a better picture of the whole system functionality.

Not surprisingly, the high complexity of the biophysical processes translates into complex mathematical models. The modern cardiac electrophysiology models are described by non-linear systems of partial differential equations with millions of variables and hundreds of parameters. Whereas the setup process of the simulations is time consuming and error prone, the numerical resolution demands high performance computing environments. In spite of the difficulties, the benefits and applications of these complex models justify their use. Computer models have been used during the tests of drugs [3], development of new medical devices [4], and of new techniques of non-invasive diagnosis [5] for several heart diseases.

We propose a computational framework that provides support for cardiac electrophysiology modelling. A web portal architecture which combines server and applications is presented in Figure 1.

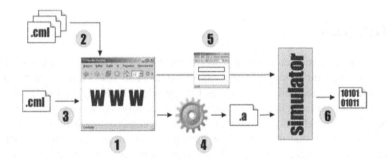

Fig. 1. The high level architecture

Through a public website (1), a user can select a biological model previously stored in the system (2), or submit its own model described in CellML meta-language (3), which has recently emerged as an international standard for the description of cell models [6]. Once established the biological model to be used, a compiler for CellML [6] will generate parallel C code based on the Message Passing Interface library (MPI [7]) (4). The result of the compilation is used as input by the simulator software which we have previously developed [8]. At the same time, the user can type the parameters in an electronic form to configure the initial states and conditions for the simulation (5). These parameters are used as simulator's data input. The simulations run on a pool of clusters and generate binary files in the end of the process (6) which can be downloaded and visualized (Figure 2).

The tools described above (XML based code generator, parallel cardiac simulators, graphical user interface environments and the web portal) provide a user-friendly environment for cardiac simulation. The complex techniques and know-how that are behind cardiac simulations, such as parallel computing, advanced numerical methods, visualization techniques and even computer code programming, are all hidden behind the integrated and easy-to-use web based framework.

The next sections describe the details of each of these components.

Fig. 2. Visualization of simulated results showing an electrical wave that propagates through the ventricles

2 AGOS Tool

There are two basic components in mathematical models of cardiac electric propagation: the cell model and the tissue model. The first component models the flow of ions across the cell membrane as first proposed by Hodgkin and Huxley [2] in their work on nerve cells. This component typically comprises of a system of Ordinary Differential Equations (ODEs). The Second component is an electrical model for the tissue that describes how currents from one region of a cell membrane interact with the neighborhood. This component is represented by a Partial Differential Equation (PDE). In this section, we present an on-line tool, the AGOS tool, aimed to help researchers in the development and solution of cell models or any other scientific model based on systems of ODEs. Special computational tools for handling the second component (the PDEs) are covered in the next sections.

AGOS stands for API (Application Program Interface) Generator for ODE Solution. Through its use one can submit a meta-model file to automatically generate a C++ API for solving first-order initial-value ODE systems.

The input data is a CellML [6] or a Content MathML [9] file, i.e., XML-based languages. CellML is an open-source mark-up language used for defining mathematical and electrophysiological models of cellular function. MathML is a W3C standard for describing mathematical notation. A CellML file includes Content MathML to provide both a human- and computer-readable representation of mathematical relationships of biological components [10]. Therefore, the AGOS tool allows the submission of a complete CellML file or just its MathML subset.

Once submitted, the XML file is translated to an API. The generated API is an object oriented C++ code. Functions are created for system initialization (initialization of parameters like the number of iterations, discretization interval and initial values of variables), numerical solution (via Explicit Euler scheme) and results storage. In addition, the API offers public reflexive functions used, for example, to restore the number of variables and their names. These reflexive functions allow the automatic creation of model-specific interfaces. This automatic generated interface enables one to set any model initial condition or parameter, displaying their actual names, as documented in the CellML or MathML input file. The AGOS tool is available at (www.fisiocomp.ufjf.br), from where it is possible to download the API source-code. AGOS can also be used online via a web application (see section 4), which uses the generated API to solve ODE systems and visualize their results.

In the next section, we present how the XML code is translated to C++ code.

2.1 The Translator

The AGOS application was implemented in C++ and makes use of basic computer structure and algorithms in order to capture the variables, parameters and equations, i.e. the ODE conceptual elements, that are embedded in a MathML file and translate these to executable C++ code, i.e. the AGOS API. The translator tool comprises of three basic components: a Preprocessor for XML format, an Extractor of ODE conceptual elements, and a Code Generator. The components are organized as a pipeline. The Preprocessor reads an XML-based file (MathML or CellML) and extracts the content into an array of tree data structures. Every tree of this array is processed by the ODE extractor that identifies the ODE elements and stores them in appropriate data formats. At the end of the pipeline, the Code Generator combines the extracted information to a code template and generates the AGOS API. The adopted strategy for code generation is largely based on code templates. The syntactical structure of code templates is described using formal grammar notation. Details related to the AGOS API and to the translator are documented in the AGOS manuals that can be found at [11].

The MathML description language uses a prefix format, i.e., the operators precedes the operands. The translator goal is achieved via the creation of a structure that supports easy identification of the operands and operators. AGOS converts the XML embedded equations in a tree-like structure. We briefly illustrate the translator tasks via a simple example. Consider the following equation:

$$t = 6.0 + \sqrt[n]{a} \tag{1}$$

The corresponding Content MathML code and the generated tree are presented in Figure 3.

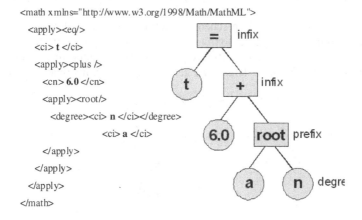

Fig. 3. Content MathML code and the extracted tree structure

The tree nodes contain information about each operand and operator, besides the equation type (if it is a differential equation or an algebraic one). The translator uses this information to include the mathematical code in the right place in the API. Using a search in depth, the following code is generated: "t = (6.0 + pow(a, 1.0/n));".

3 The Parallel Cardiac Simulator

The set of Bidomain equations [12] is currently one of the most complete mathematical models to simulate the electrical activity in cardiac tissue:

$$\nabla \cdot (\sigma_i \nabla \phi_i) = \chi \left(C_m \frac{\partial \phi}{\partial t} + f(\phi, \vec{n}) \right) \tag{2}$$

$$\nabla \cdot (\sigma_e \nabla \phi_e) = -\chi \left(C_m \frac{\partial \phi}{\partial t} + f(\phi, \vec{n}) \right) \tag{3}$$

$$\frac{\partial \vec{n}}{\partial t} = g(\phi, \vec{n}) \tag{4}$$

where ϕ_e is the extracellular potential, ϕ_i the intracellular potential and ϕ is the transmembrane potential. Eq. 4 is a system of non-linear equations that accounts for the dynamics of several ionic species and channels (proteins that cross cell membrane) and their relation to the transmembrane potential. The system of Eq. 4 typically accounts for over 20 variables, such as ionic concentrations, protein channel resistivities and other cellular features. σ_i and σ_e are the intracellular and extracellular conductivity tensors, i.e. 3x3 symmetric matrices that vary in space and describe the anisotropy of the cardiac tissue. Cm and χ are the cell membrane capacitance and the surface-to-volume ratio, respectively.

Unfortunately, a solution of this large nonlinear system of partial differential equations (PDEs) is computationally expensive. One way to solve (2)–(4) at every time step is via the operator splitting technique [13]-[15]. The numerical solution reduces to a modular three step scheme which involves the solutions of a parabolic PDE, an elliptic PDE and a nonlinear system of ordinary differential equations (ODEs) at each time step. Rewriting equations (2)–(4) using the operator splitting technique (see [16] for more details) we get the following numerical scheme:

$$\varphi^{k+1/2} = (1 + \Delta t A_i)\varphi^k + \Delta t A_i (\varphi_e)^k; \tag{5}$$

$$\varphi^{k+1} = \varphi^{k+1/2} - \Delta t f\left(\varphi^{k+1/2}, \vec{\xi}^k\right) / Cm$$
$$\vec{\xi}^{k+1} = \vec{\xi}^k + \Delta t g\left(\varphi^{k+1/2}, \vec{\xi}^k\right); \tag{6}$$

$$(A_i + A_e)(\varphi_e)^{k+1} = -A_i \varphi^{k+1}; \tag{7}$$

where φ^k, φ_e^k and ξ^k discretizes ϕ, ϕ_e and η at time k Δt; A_i and A_e are the discretizations for $\nabla.\sigma_i\nabla$ and $\nabla.\sigma_e\nabla$, respectively. Spatial discretization was done via the Finite Element Method using a uniform mesh of squares and bilinear polynomials as previously described in [16].

Steps (5), (6) and (7) are solved as independent systems. Nevertheless, (5), (6) and (7) are still computationally expensive. One way of reducing the time spent on solving these equations is via parallel computing.

3.1 The 2-Dimensional Parallel Cardiac Simulator

A solution for the Bidomain model was implemented in parallel using the MPI and PETSc [17] libraries. PETSc provides a suite of data structures and routines for the scalable solution of applications modeled by PDEs. The nonlinear ODE system was solved via the explicit forward-Euler scheme (see section 3.2 for more details). The PDEs are the most computationally expensive portion of the model and thus need a more robust algorithm. The Conjugate Gradient (CG) method combined with an appropriate preconditioner has become a standard choice for an iterative solver and was applied for the solution of the elliptic and parabolic PDEs.

The CG was parallelized via linear domain decomposition. The spatial rectangular domain was decomposed into *nproc* nonoverlapping domains of equal size, where *nproc* was the number of processors involved in the simulation. For the 2-D problem, the slice was made in the y direction.

3.2 Integration with AGOS

As described in Section 2, one can select or submit a biological model through a public web site. This model is compiled into a scalable (parallel) shared library, which can be used by the simulator.

The shared library is dynamic, thus the Simulator does not need to be compiled for every new cell model. Therefore, when it is necessary to solve the nonlinear system of ODEs, i.e. during the step described by equation (6), the Simulator makes a call to the functions of the automatic generated API library on run time, the functions are loaded to memory and executed.

The parallelization is easily obtained: each processor is responsible for a rectangular domain that has nx . ny /nproc ODE systems associated with. These are independent systems. The solution of these systems does not involve communication.

3.3 Simulator Graphic User Interface (GUI)

In order to provide an easy way for setting up simulations, a Java GUI was developed and it is here briefly presented.

The GUI allows the creation, set up, execution and visualization of simulations. First, one must create a project and set several parameters for simulating the spread of cardiac electric activity. Then, the parallel simulation can be started and its results will be stored in a folder that has the same name of the project. Visualization of the results can also be initiated via the GUI. Currently the visualization is done via an interface to FLOUNDER, a free software developed at the Calgary University [18].

Once the project is created, it is possible to select which variables should be saved for later analyzes (intra-, transmembrane, extracellular potentials, etc.). Parameters as time and space discretization, size of the bidimensional portion of tissue and number of iterations can be set. Via the selection of input files one defines the geometry of the model, cardiac fiber and sheet orientations. Stimuli type, location and intensity can also be configured. Figure 4 shows an example of a simulation set up using the simulator GUI.

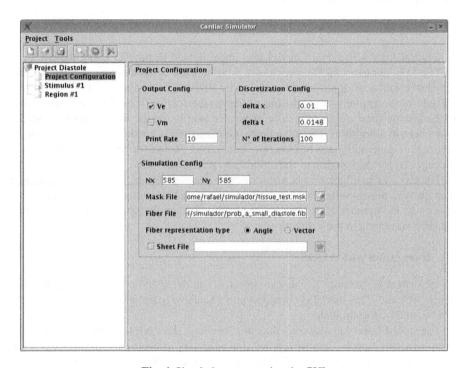

Fig. 4. Simulation set up using the GUI

3.4   The Cluster

The set up of a parallel environment based on commodity network and desktop computers can be a complex task. In order to set up a Linux cluster for scientific computing, several software packages are necessary for tasks such as installation, configuration and management. Fortunately, there are some popular kits that support these tasks.

The cluster used in this work [11], is based on NAPCI Rocks architecture and it is made up of eight nodes, Athlon 64 3000+ with 2 GB of RAM, connected by a fast Gigabit Ethernet switching device. Rocks [19] is a collection of open source software integrated to the Red Hat Linux which aims at building high performance clusters.

For monitoring the cluster status, Rocks provides a tool named Ganglia, which is a scalable distributed monitoring system for high-performance computing systems. Ganglia allows the cluster administrator to visualize historical monitoring information for cluster, host, and metric trends over different time granularities ranging from minutes to years. It generates graphics that present the historical trends of metrics versus time. Typical and useful graphics include information on network bandwidth utilization and CPU load for the whole cluster as well as for each individual node. All the monitoring in Ganglia can be done via a web browser.

In addition, Rocks is integrated to Sun Grid Engine (SGE). SGE schedules the jobs submitted to the cluster to the most appropriate nodes, based on management policies previously defined by the cluster administrator. SGE is integrated with Message Passing Interface and Parallel Virtual Machine and allows users to run parallel jobs based on these libraries. Any number of different parallel environment interfaces can be configured concurrently. Sun Grid Engine also provides dynamic scheduling and job migration via checkpoints, i.e., the procedure of storing the state of an active process. A graphical interface called QMON provides for easy control and configuration of all SGE capabilities.

3.5 Simulation Example

In this section we present an example of the use of the environment discussed in the preceding sections. We simulate the cardiac electric propagation on a 2-dimensional cut of the left ventricle obtained during the cardiac systole phase by the resonance magnetic technique of a healthful person. After segmenting the resonance image, a two-dimensional mesh of 769 X 769 points is generated, that models the cardiac tissue, blood and torso. All bidomain parameters were taken from [20]. The cardiac tissue conductivity values have been set to: $\sigma_{il} = 3$ mS/cm, $\sigma_{it} = 0.31$ mS/cm, $\sigma_{el} = 2$ mS/cm and $\sigma_{et} = 1.35$ mS/cm, where i(e) denotes intracellular (extracellular), l(t) and stands for longitudinal (transversal) to the fiber orientation. The capacitance per unit area and the surface area-to-volume ratio are set to 1 mF/cm^2 and 2000 cm^{-1}, respectively. The interface between cardiac tissue and bath is modeled as described in [21]. All the other boundaries are assumed to be electrically isolated. The spatial and temporal discretization steps of the numerical model are set to 150 μm and 10 μs, respectively. The simulation was carried out for 20 ms after a single current stimulus was introduced at a selected endocardial site.

For simulating the action potential of cardiac cells we used the human ventricular model of ten Tusscher [22]. The explicit Euler implementation described by equation (6) was generated automatically by AGOS, based on a CellMl model description downloaded from the CellML repository [6]. The linear system associated to the parabolic part of the bidomain formulation, see equation (5), is solved with the Conugate Gradient (CG) method and ILU (Incomplete LU factorization with zero fill-in) preconditioner. The linear system associated to the elliptic part, equation (7), dominates computation and is solved with CG and a parallel Algebraic Multigrid preconditioner. In this work we adopted the parallel AMG code BoomerAMG [23] with its Falgout-coarsening strategy.

Figure 5 shows the simulation result overlapped to the resonance image. The color-coded image represents the transmembrane potential distribution for a certain time instant.

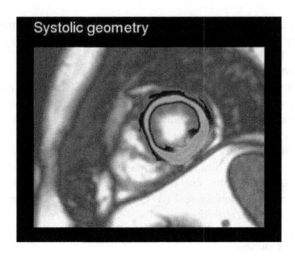

Fig. 5. Simulated electrical wave propagation overlapped to the Resonance Image

The simulation was run using one, two, four and eight processors. As presented in Figure 6, when running the simulation on eight processors the relative speedup (execution time using 1 processor / execution time with n processors) is near 5. The execution time drops from near 5 hours when running with 1 processor to less than 1 hour running with 8 processors. Linear speedups were not achieved. This is mainly due to communication overhead and to the Multigrid Preconditioner adopted. The direct method used in the coarsest grid of the preconditioner is not parallelized and thus limits scalability. Nevertheless, the results indicate the importance and benefits of cluster computing for cardiac electrophysiology modeling.

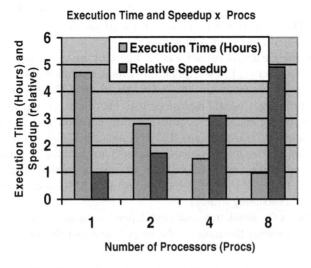

Fig. 6. Parallel speedup and execution time in hours

4   The Web Portal: Goals, Current Status and Future Work

The Web portal is under development and has three main goals:

1- To popularize the technology of cardiac simulation. The tools described above (XML based code generator, parallel cardiac simulators and user interface environments) are being combined in a single Web application which has access to a pool of clusters. The web portal supports the development and simulation of cardiac models efficiently via this user-friendly integrated environment. The complex techniques and know-how that are behind cardiac simulations, such as parallel computing, advanced numerical methods, visualization techniques and even computer code programming, are all taken care by the integrated and easy-to-use web application. Online tutorials [11] instruct the users on how to make efficient use of the integrated environment.

2- To promote the share of the computational resources among different research centers. The web application under development will allow users to execute their simulations on a pool of clusters made of clusters residing on different research centers. This will bring a parallel environment to those that do not have access to it as well as increase the computational power of the participating centers. Currently, there are three small clusters being integrated to the Web Portal (4-node cluster from Lanec-UFSJ (Neuroscience Laboratory), 5-node cluster from Labma-UFRJ (Applied Mathematics Laboratory), and the 8-node cluster from FISIOCOMP-UFJF (Computational Physiology Laboratory). This integration will be done using the Sun Grid Engine.

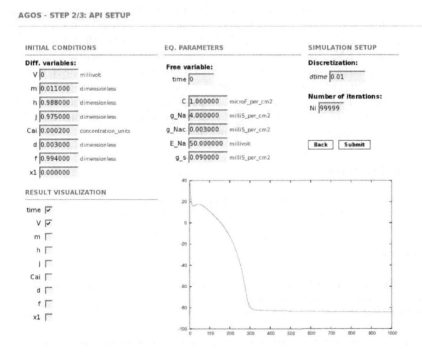

Fig. 7. Web portal usage example

3- To promote the development of cardiac modeling. The integration of the above mentioned tools in a single web portal will speed up the development of new and more realistic electro-physiological models of the heart and further integrate different research centers, promoting international collaborations.

Currently the AGOS tool is fully operational and integrated to the Web Portal. After registration, the user is granted a new account and is able to create, manage, execute and store the results of simulation projects. Associated to each registered researcher there is a folder in the server's hard disk. This folder contains one subfolder for each created project, with all the input and output files generated by the AGOS application. Those files can be downloaded and visualized at anytime. In addition, the researcher may modify the API parameters to generate new PDF and PS graphics. One of the portal screenshots is shown in the Figure 7.

We are using JSP [24] and Struts framework [25] and some web pages also use PHP [26] language. All data is being stored in a MySQL [27] database and the website runs in an Apache Tomcat server [28]. The simulator GUI described in section 3.3 is currently being integrated to the web portal. The easy-to-use computational framework composed of AGOS, the parallel cardiac simulator and the SGE grid computing tool, will efficiently support cardiac modeling in distributed environments.

5 Conclusion

In this work we presented a computational framework that supports cardiac electrophysiology modeling. The framework is made of different components and technologies and aims on simplifying the development and use of cardiac models. The combination of an XML based automatic code generator, parallel cardiac simulators, graphical user interface environments and a web portal provides an user-friendly environment for cardiac simulation. The complex techniques and know-how that are behind cardiac modeling, such as parallel computing, advanced numerical methods, visualization techniques and even computer code programming are all hidden behind the integrated and easy-to-use web based framework.

Acknowledgements

This work was supported by the Brazilian Ministry of Science and Technology, CNPq (process 506795/2004-7).

References

[1] HENRIQUEZ C. S. (1993): 'Simulating the electrical behavior of cardiac tissue using the bidomain model', Crit Rev. Biomed. Eng, 21, 1-77
[2] HODGKIN A. L., and HUXLEY A. F. (1952): 'A quantitative description of membrane current and its application to conduction and excitation in nerve', J. Physiol., 117, 500-544
[3] GIMA K., and RUDY Y. (2002): 'Ionic current basis of electrocardiographic waveforms: a model study', Circ. Res., 90, 889-896

[4] SANTOS R. W. D., STEINHOFF U., HOFER E., SANCHEZ-QUINTANA D., and KOCH H. (2003): 'Modelling the electrical propagation in cardiac tissue using detailed histological data', Biomedizinische Technik. Biomedical Engineering, 48, 476-478

[5] SANTOS R. W. D., KOSCH O., STEINHOFF U., BAUER S., TRAHMS L., and KOCH H. (2004): 'MCG to ECG source differences: measurements and a 2D computer model study', Journal Of Electrocardiology, 37 Suppl

[6] CellML biology, math, data, knowledge., Internet site address: http://www.cellml.org/

[7] MPI (Message Passing Interface), Internet site address: http://www.mpi-forum.org/

[8] SANTOS R. W. D., PLANK G., BAUER S., and VIGMOND E. J. (2004): 'Parallel Multigrid Preconditioner for the Cardiac Bidomain Model', IEEE Trans. Biomed. Eng., 51(11), 1960-1968

[9] Mathematical Markup Language (MathML) Version 2.0 (Second Edition), Internet site address: http://www.w3.org/TR/MathML2/

[10] LLOYD C. M., HALSTEAD M. D. B., and NIELSEN P. F. (2004): 'CellML: its future, present and past', in Biophysics & Molecular Biology, 85, 433-450

[11] FISIOCOMP: Laboratory of Computational Physiology, internet site address: http://www.fisiocomp.ufjf.br/

[12] SEPULVEDA N. G., ROTH B. J., and WIKSWO Jr. J. P. (1989): 'Current injection into a two-dimensional anistropic bidomain', Biophysical J., 55, 987-999

[13] VIGMOND E., AGUEL F., and TRAYANOVA N. (2002): 'Computational techniques for solving the bidomain equations in three dimensions', IEEE Trans. Biomed. Eng., 49, 1260-9

[14] SUNDNES J., LINES G., and TVEITO A. (2001): 'Efficient solution of ordinary differential equations modeling electrical activity in cardiac cells', Math. Biosci., 172, no. 2, 55-72

[15] KEENER J., and BOGAR K. (1998): 'A numerical method for the solution of the bidomain equations in cardiac tissue', Chaos, 8, no. 1, 234-241

[16] SANTOS R. W. D., PLANK G., BAUER S., and VIGMOND E. J. (2004): 'Preconditioning techniques for the bidomain equations'. Lecture Notes In Computational Science And Engineering, 40, 571-580

[17] PETSc: Portable, Extensible Toolkit for Scientific Computation, Internet site address: http://www-unix.mcs.anl.gov/petsc/petsc-as/

[18] Calgary University, Internet site address: http://www.ucalgary.ca/

[19] NAPCI Rocks, Internet site address: http://www.rocksclusters.org/Rocks/

[20] MUZIKANT, A. L. and HENRIQUEZ, C. S. (1998): 'Validation of three-dimensional conduction models using experimental mapping: are we getting closer?' Prog. Biophys. Mol. Biol. 69:205-223

[21] KRASSOWSKA, W. and NEU, J. C. (1994): 'Effective boundary conditions for syncytial tissues.' IEEE Trans. Biomed. Eng 41:143-150

[22] ten TUSSCHER, K. H. W. J., NOBLE D., NOBLE P. J., and PANFILOV A. V. (2004): 'A model for human ventricular tissue', J. Physiol., 286, 1573-1589

[23] HENSON V. E., and YANG U. M. (2000): 'BoomerAMG: a Parallel Algebraic Multigrid Solver and Preconditioner. ' Technical Report UCRL-JC-139098, Lawrence Livermore National Laboratory

[24] Sun Microsystems, Internet site address: http://java.sun.com/products/jsp/

[25] Apache Software Foundation, internet site address: http://struts.apache.org/

[26] The PHP Group, Internet site address: http://www.php.net/

[27] MySQL AB., Internet site address:http://www.mysql.com/

[28] The Apache Jakarta Project, Internet site address: http://jakarta.apache.org/tomcat/

Triangular Clique Based Multilevel Approaches to Identify Protein Functional Modules*

S. Oliveira and S.C. Seok

Department of Computer Science, 14 MLH, University of Iowa,
Iowa City IA 52242, USA
{oliveira, sseok}@cs.uiowa.edu
Phone: (319)-335-0731, (319)-353-4851
Fax: (319)-335-3624

Abstract. Identifying functional modules is believed to reveal most cellular processes. There have been many computational approaches to investigate the underlying biological structures [1,4,9,13]. A spectral clustering method plays a critical role identifying functional modules in a yeast protein-protein network in [9]. One of major obstacles clustering algorithms face and deal with is the limited information on how close two proteins with or without interactions are. We present an unweighted-graph version of a multilevel spectral algorithm which identifies more protein complexes with less computational time [8]. Existing multilevel approaches are hampered with no preliminary knowledge how many levels should be used to expect the best or near best results. While existing matching based multilevel algorithms try to merge pairs of nodes, we here present a new multilevel algorithms which merges groups of three nodes in triangular cliques. These new algorithms produce as good clustering results as previously best known matching based coarsening algorithms. Moreover, our algorithms use only one or two levels of coarsening, so we can avoid a major weakness of matching based algorithms.

Topic: Computing in Biosciences, Data Processing, Numerical Methods.

1 Introduction

Most cellular processes are carried out by groups of proteins. Identifying functional modules in protein-protein networks is considered as one of the most important and challenging research topics in computational systems biology. There has been many recent computational approaches to disclose the underlying biological structures.[1,4,9,13]

Successful approaches to clustering functional modules include partition-based algorithms [4,9]. Pothen et al. proposed a two-level architecture for a yeast proteomic network [9] and Ding et al. introduced a partitioning algorithm on a bipartite model. Pothen et al. construct a smaller network from a protein-protein interaction network by removing proteins which interact with too many or too

---

* This work was supported in part by NSF ITR grant DMS-0213305.

M. Daydé et al. (Eds.): VECPAR 2006, LNCS 4395, pp. 556–565, 2007.
© Springer-Verlag Berlin Heidelberg 2007

few proteins. And then a spectral clustering method was applied to identify functional modules in the protein-protein network in their research.

The biggest obstacle to identifying functional modules is that protein-protein interaction networks are unweighted or uniformly weighted. Unweighted graphs are considered to be harder to partition than weighted graphs because unweighted graphs provide only limited information on the strength of connection between two vertices. Multilevel (ML) algorithms have been introduced to identify more functional modules based on matching algorithms in PPI networks [8]. But these algorithms are hampered by two weaknesses. First, it is hard to find the optimal number of levels. Second, most matching based algorithms use random algorithms, so the clustering results vary from an experiment to another.

Our Triangular Clique (TC) based multilevel algorithm was inspired by Spirin et al's approach to investigate the large-scale structure of PPI networks [12]. They used the Clique idea to identify highly connected clusters of proteins in protein-protein interaction networks. They not only enumerated all cliques of size 3 and larger (complete subgraphs) but also partially complete subgraphs with high quality.

Our algorithm use only triangular cliques (cliques of size 3). These are different from matching based ML algorithms which pick pairs of nodes based on edge or node related information to merge, TC based algorithms try to merge highly connected triples of nodes. We present four different kinds of TC based algorithms according to the decision on how to deal with two TCs which share one or two nodes. These algorithms are compared and analyzed with the computational results.

We show some TC based algorithms identify as good as or better functional modules with one or two levels of coarsening than the best matching based ML algorithm we found in [8].

2 Features of Interaction Networks and Two-Level Approach

Graph theory is commonly used as a method for analyzing protein-protein interaction (PPI) networks in Computational Biology. Each vertex represents a protein, and edges correspond to experimentally identified PPIs. Proteomic networks have two important features [2]. One is that the degree distribution function $P(k)$ follows a power law (and so is considered a scale-free network). This means that, most vertices have low degrees, called low-shell proteins, and a few are highly connected, called hub proteins. The other feature is the *small world* property which is also known as *six degrees of separation*. This means the diameter of the graph is small compared with the number of nodes.

A two level approach was proposed by Pothen et al. [9] to identify functional modules in a proteomic network in yeast. The main idea is derived from the k-cores concept which was originally suggested as a tool to simplify complex graph structures by Seidman in 1983 [10]. If we repeatedly remove vertices of degree less than k from a graph until there are no longer any such vertices, the result is the k-core of the original graph. The vertices removed are called the low-shell vertices. The two-level approach pays attention to three facts in protein-protein interaction networks:

- The hub proteins have interactions with many other proteins, so it is hard to limit them to only one cluster and the computational complexity increases when they are included.
- There are many low-shell proteins, which increases the size of network. These nodes are easy to cluster when the nodes they are connected to are clustered first.
- Proteomic networks are mostly comprised of one big component and several small components.

So, disregarding hub proteins and low-shell proteins, and confining attention to the biggest component of proteomic networks leaves us to focus on the nodes which are most meaningful to cluster. We keep track of the path connecting the low-shell proteins to the others. After the clusters are created we can then add the low-shell back to the appropriate cluster.

3   Background on Multilevel Approaches and Clustering Algorithms

Let $G = (V, E)$ be a graph with vertex set V and set of undirected edges E. One of the most commonly used data structures for graphs are matrices. Matrix representations are very useful to store weights for edges and vertices. We can also use a lot of well-known computational techniques from Linear Algebra. In our matrix representations $S = (s_{ij})$, diagonal entries s_{ii} store the weights of vertices and off-diagonal entries s_{ij} represent edge weights. Our ML algorithms use this matrix representation.

3.1   Multilevel Spectral Clustering

The basic concept of "Multilevel clustering" algorithms is that when we have a large graph $G = (V, E)$ to partition, we construct a smaller graph whose vertices are groups of vertices from G. We can apply a clustering method to this smaller graph, and transfer the partition to the original graph. This idea is very useful because smaller matrices or graphs require much less time to cluster. The process of constructing the smaller matrix is called coarsening, and the process of transferring the partition is called decoarsening.

   Coarsening and decoarsening steps are implemented by multiplying a graph matrix S by a special coarsening matrix C. Each entry of C is either 0 or 1. We set $c_{ij} = 1$ if node i of the fine graph belongs to node j of the coarsened graph. A series of matrices $S_0, S_1, \cdots, S_l$ are recursively constructed using $C_1, \cdots, C_l$ in the form of $S_i = C_i' * S_{i-1} * C_i$ with $i = 1, \cdots, l$. Note that C' is the transpose of C (i.e. $c_{ij} = c_{ji}$) . A partitioning algorithm is applied to matrix S_l and we will have an initial partition Cut in the coarsest level.

Partitioning is done by using a recursive spectral bipartitioning (divisive partitioning). Recursive bipartitioning algorithms repeatedly performs two main steps. One is selecting a cluster to split, and the other is applying a two-way clustering algorithm.

The best known spectral clustering algorithms is the MinMaxCut algorithm [5]. Two-way MinMaxCut clustering algorithm aims to minimize

$$J_{MMC}(A, B) = \frac{s(A, B)}{s(A, A)} + \frac{s(A, B)}{s(B, B)} = \frac{s(A, \bar{A})}{s(A, A)} + \frac{s(B, \bar{B})}{s(B, B)}, \quad (1)$$

where $s(A, B) = \sum_{i \in A, j \in B} s_{ij}$.

In [5], a continuous approximation to this problem has the solution which is the eigenvector q_2 associated with the second smallest eigenvalue of the system $(D - S)q = \lambda Dq$, where $D = diag(d_1, d_2, \cdots, d_n)$ and $d_i = \sum_j s_{ij}$. The partition (A, B) is calculated by finding index opt such that the corresponding objective function gets optimum value with the partition, $A = \{q_2(i) \mid q_2(i) < q_2(opt)\}$ and $B = \{q_2(i) \mid q_2(i) \geq q_2(opt)\}$.

The optimum value of two-way MinMaxCut is called the cohesion of the cluster and can be an indicator to show how closely vertices of the current cluster are related [5]. This value can be used for the cluster selection algorithm. Divisive algorithms recursively choose a cluster which has the least cohesion for partitioning until we have the predefined number of clusters or until all current clusters satisfy a certain condition. On level i we have a partition (A_j) of the vertices of G_i. To represent the partition, we use a vector Cut_i on level i where $(Cut_i)_k = j$ if $k \in A_j$.

Decoarsening is how we get back to the original graph. The partition from the coarsest level is mapped into finer levels by using a proper coarsening matrix $Cut_i = C_i' \cdot Cut_{i-1}$ where i is the level number of the coarser level. Then a Kernighan-Lin (KL) type refinement algorithm is applied to improve the quality at each level [6]. KL starts with an initial partition; it iteratively searches for nodes from each cluster of the graph if moving a node to one of the other clusters leads to a better partition. For each node, there may be more than one cluster to give smaller objective function values than the current cut. So the node moves to the cluster that gives the biggest improvement. The iteration terminates when it does not find any node to improve the partition.

3.2   Matching Based Coarsening Algorithms

A matching in a graph is a set of edges in which no two of them are incident on the same node. We introduced a heuristic matching algorithm which works very well on weighted graphs in [7], called Sorted Matching (SM). SM was used earlier by us to improve clustering results for groups of documents, which compose weighted graphs. In SM, nodes are merged in order of decreasing edge weight.

The simplest matching for unweighted graphs is random matching. One node is randomly visited and one of unmatched node is randomly chosen to be merged with the node (RVRM). A drawback is that the nodes with low degrees have higher chance to be left unmerged than high degree nodes. In order to avoid this problem we can pick the lowest degree node among unmerged nodes and choose one of the unmerged nodes randomly to merge (LVRM). Thus this algorithm tends to merge more nodes than RVRM.

Our matching algorithm for unweighted graphs introduced in [8] goes as follows: we define the weights of edges as follows. The edge weights are all 1's to start with, but become the sum of the number of edges combined after a matching step. A node weight is defined as the total number of nodes merged into it.

In the PPI network, at first we have equal edge weights. We perform the first level of coarsening by combining nodes with each other, as long as they are not matched. The results are similar for any order we pick up for this step. After this matching we will have groups of edges which share the same weight (the maximum resulting edge weight will be 4 for a clique with 4 nodes/vertices). We then give the higher priority to the edge with lower combined node weights, i.e. we take the edge with maximum $1/w(n_i) + 1/w(n_j)$ as a tie-break rule, where $w(n_i)$ and $w(n_j)$ are the node weights, that is, the number of nodes, of supernodes n_i and n_j. We call this matching scheme Heavy-Edge-Small-Node (HESN). HESN was introduced and shown to outperform the other matching based algorithms in [8].

4   Coarsening with Triangular Cliques

Matching based coarsening algorithms merge groups of at most two nodes which have an edge between them. These algorithm have worked well especially on weighted graphs because all edges have different weights. These weights play a key role for picking pairs of nodes to merge. Meanwhile unweighted graphs do not provide this information. The ML algorithm with HESN works well on the unweighted graphs even though HESN is a matching based coarsening. However, in general, matching based ML algorithms have two main weaknesses. First, it is hard to find the optimal number of levels which generates the best clustering. Second, most matching based algorithms has a random component, so the clustering results vary from an experiment to another.

A clique in an undirected graph $G = (V, E)$ is a subset $V' \subseteq V$, where each pair of vertices is connected by an edge. The problem of finding the maximal size of a clique for a given graph is an NP-complete problem [11]. However, finding all cliques comprised of three vertices takes $O(|E|^2/|V|)$ time. We pay attention to that, first, if both proteins p_1 and p_2 interact with the protein p_3, p_1 and p_2 should interact and second, when three proteins p_1, p_2, p_3 forms a triangular clique (TC), the chance all three proteins are clustered in the same functional module is high. We show the quality of triangular cliques in section 5. These three nodes p_1, p_2, p_3 in a graph compose a triangular clique.

When we use TCs to form sets of vertices to be merged, we have to make two decisions. The first is because many TCs shares one or two nodes with others. Overlapping pairs of TCs fall into different cases: sharing one or two common vertices. How do we merge TCs for each case? We consider two aspects when dealing with two TCs sharing nodes. One is a criterion to decide to merge two TCs. The criterion we use in this paper is the density of the subgraph after merging two TCs. When two TCs are decided to be merged, no other consideration is necessary. However, when two TCs sharing one or two nodes are not merged, a decision should be made regarding unmerged nodes. That is, we have to decide that if we leave them for possible merging with other TCs or group them in a separate supernodes. The graph at the upper-left of Figure 1 has four TCs, from TC1 through TC4. TC1 and TC2 share two common nodes and TC3 and TC4 share one common node. We present four schemes according to the above two aspects. One way is to merge all nodes in TCs which share one or two nodes into one supernode, let us call this method TC_ALL. In this case, any group of TCs which share one or more vertices is merged into a supernode. Another simple way is to merge one of two TCs for both cases and leave other vertices unmerged, let us call this TC_ONLY. When we assume that TC1 and TC3 are chosen over TC2 and TC4 and merged into two separate supernodes by TC_ONLY, two nodes of TC4 and one node of TC2 are left unmerged. We consider two variants of TC_ONLY according to how to deal with these three unmerged nodes. The one unmerged node of TC2 forms an edge with the supernode after coarsening and looks reasonable to be grouped in the same cluster with the supernode made of TC1. So we devise a new algorithm, TC_ONE, which is basically the same as TC_ONLY but merging the left out node of TC2 and TC1 into a supernode. Similarly, the two unmerged nodes of TC4 form a TC with the supernode created. The two unmerged nodes have high chance to be merged with newly created supernode into a supernode if we have additional levels of coarsening. If we do not want to merge them with the supernode, we can have all three nodes of TC3 and two nodes of TC4 form two separate supernodes. Let us call this algorithm TC_TWO. In this case, the one unmerged node of TC2 by TC_ONLY is still left unmerged as shown in Figure 1.

Second problem we face is whether or not we keep creating more levels. If we use more than one level then the question is how many levels to create. In this paper we focus on using the same algorithm to create more levels. The performance of these four algorithms is presented and compared with a matching based coarsening algorithm, HESN, in the following section.

5 Model Networks and Computational Experiments

The budding yeast, *Saccharomyces cerevisiae*, is considered the simplest and so is the most investigated organism. Pothen's two-level architecture is applied to the CORE dataset of DIP (Database of Interacting Proteins, dip.doe-mbi.ucla.edu), where the pairs of interacting proteins identified were validated according to the criteria described in Deane et al. [3]. The network has 2610 proteins and 6236 interactions. Their idea is that removing high degree proteins (hub proteins) and low degree proteins (low-shell proteins) from the network before clustering leads to a better partitioning and then the removed nodes can be added to the partitioning. The residual network after removing hub proteins and low shell proteins has 499 proteins and 1229 interactions.

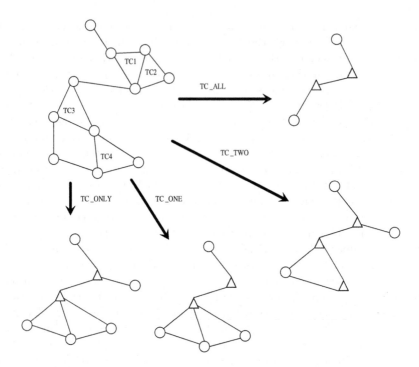

Fig. 1. Four different TC based coarsening algorithms. Each circle stands for a node and a triangle represents a supernode after coarsening.

Instead of using the small network (CORE dataset), we use the DIP network which has 4931 nodes and 17471 edges to validate our ML algorithms. Constructing a residual network starts with removing nodes that have degree 20 or more from the original network. Then low-shell proteins whose degree is 3 or less are pruned from the biggest component. The residual network has 1078 nodes and 2778 edges.

After our ML spectral algorithm is applied to this residual network, the clustering results are compared with the MIPS (mips.gsf.de) dataset as we did in [8]. Note that the residual network and MIPS dataset share 800 proteins. So the maximum number of correctly clustered nodes is 800 for any experiment.

Now we present various computational results to investigate the properties of TC based ML algorithms. Table 1 shows the number of nodes, the number of TCs, and the number of correctly grouped nodes as the number of levels increases. The sum of the maximum number of proteins which belong to the same functional module in each supernode is what we use as the number of correctly grouped proteins. Notice that, as the number of levels increases, the number of correctly grouped proteins should not increase. The original number of proteins in the residual network is 1078. As expected, TC_ALL collapses the largest number of nodes at each level of coarsening (see the first entries in row TC_ALL). But the quality of grouping worsens pretty fast (see the third entries in row TC_ALL). So with

TC_ALL as the number of level increases the size of network shrinks very fast and the quality of coarsening is becoming very bad. Meanwhile, TC_TWO merges the least after the first level, 819 proteins. TC_ONLY merges more nodes than TC_TWO at the beginning but the least after all. Even though TC_ONLY merges the least nodes after all, the quality of grouping remains good. TC_ONE merges the most nodes except when compared to TC_ALL and the quality of grouping also decreases faster than TC_ONLY and TC_TWO.

There are 1195 TCs found in the original graph. The number of TCs decreases at the first level to as few as 145 for TC_ONE or as many as 580 for TC_TWO. And then the number of TCs does not change much for the first a few levels except TC_ALL for which the number of TCs significantly continues to decrease.

Table 1. The comparison of four different TC based coarsening algorithms. Note that $xx/yy/zz$ means xx nodes, yy TCs, and zz correctly grouped nodes.

level	1	2	3	4	5
HESN	601	333	182	102	56
TC_ALL	513/272/586	303/198/265	89/39/91	35/4/47	
TC_ONE	662/145/733	550/145/647	501/342/544	468/333/460	438/316/405
TC_TWO	819/580/759	686/261/719	561/147/679	533/133/650	517/140/610
TC_ONLY	770/273/754	630/168/693	578/214/624	554/308/567	540/334/523

Finally, we see the clustering results when the ML spectral clustering algorithm is actually applied with the coarsening algorithms. Tables 2 shows the number of correctly clustered proteins with four different TC based coarsening algorithms when 1 up to 5 levels are used to form 40, 60, and 80 clusters. These results are compared with the HESN matching [8] based ML algorithm. Some algorithms, like HESN and TC_ALL, do not have big enough networks to form particular number of clusters after particular levels of coarsening. For example, there are only 56 nodes after three levels of coaresening with HESN, so we do not try to construct 60 and 80 clusters. First row has the results with HESN which is considered to work best among the matching based algorithms in [8]. Without any ML algorithm 201, 234, and 286 proteins are reported to be correctly clustered to form 40, 60, and 80 clusters respectively.

The most remarkable point from the table is the clustering results of TC_ONE and TC_ONLY with one level of coarsening are the best or almost best compared with more than one level of coarsening. As for TC_TWO, the results are improving up to some point. The quality of grouping by TC_TWO is shown well in Table 1. However, the clustering results of TC_TWO are not good compared to TC_ONLY and TC_ONE when these algorithms are actually applied to ML spectral algorithm. While TC_ONLY provides the best results with one level of coarsening, TC_ONE generates the best results with one or two levels. And the quality of clustering of TC_ONLY drops significantly with two levels. We guess that more than one level of coarsening causes overmerging, that is, the quality of grouping with two or more levels is not good enough to improve clustering.

Table 2. The numbers of correctly clustered proteins with four different TC based coarsening algorithms and one matching based algorithm to form 40, 60, and 80 clusters, when 1 up to 5 levels are used. First row has the clustering results with a matching based ML algorithm, HESN. Best results are in bold font.

	40	60	80
HESN	232/248/258/253/**267**	300/312/**329**/308	354/364/342/**367**
TC_ALL	**245**/234/235	**299**/286	**346**/321
TC_ONE	257/**261**/253/246/238	**323**/319/319/307/312	368/**370**/358/359/361
TC_TWO	203/230/249/**253**/248	283/296/300/**314**/314	336/**363**/357/355/360
TC_ONLY	**271**/248/248/243/238	**329**/306/316/298/301	**369**/361/367/354/349

6   Conclusion

In this paper we presented Triangular Clique (TC) based multi level algorithms not only to avoid problems caused by matching based algorithms but also to improve the quality of clustering. Triangular Clique based coarsening algorithms works easily by finding TCs in a given graph and then merging in the nodes to form a supernode in the next level as described in section 4. Among our four TC based algorithms TC_ONLY with one level usually gives the best results. TC_ONE also shows almost the same result as TC_ONLY with one or two levels. Our TC based ML algorithms do not rely as much on random algorithms. We believe our TC based algorithms outperforms matching based ML algorithm because TC based algorithms take advantage of the fundamental structure of unweighted graphs.

References

1. G. D. Bader and C. W. Hogue. An automated method for finding molecular complexes in large protein interaction networks. *BMC Bioinformatics*, 4(1), 2003.
2. S. Bornholdt and H. G. Schuster, editors. *Handbook of Graphs and Networks*. Wiley VCH, 2003.
3. C. M. Deane, L. Salwinski, I. Xenarios, and D. Eisenberg. Protein interactions: two methods for assessment of the reliability of high throughput observations. *Mol Cell Proteomics.*, 1(5):349–56, May 2002.
4. C. Ding, X. He, R. F. Meraz, and S. R. Holbrook. A unified representation of multiprotein complex data for modeling interaction networks. *Proteins: Structure, Function, and Bioinformatics*, 57(1):99–108, 2004.
5. C. Ding, X. He, H. Zha, M. Gu, and H. Simon. A minmaxcut spectral method for data clustering and graph partitioning. Technical Report 54111, LBNL, December 2003.
6. B. W. Kernighan and S. Lin. An efficient heuristic procedure for partitioning graphs. *The Bell System Technical Journal*, 1970.
7. S. Oliveira and S. C. Seok. A multi-level approach for document clustering. *Lecture Notes in Computer Science*, 3514:204–211, Jan 2005.
8. S. Oliveira and S. C. Seok. A multilevel approach for identifying functional modules in protein-protein interaction networks. *Proceedings of IWBRA 2006, Lecture Notes in Computer Science*, 3992, 2006. to appear.

9. E. Ramadan, C. Osgood, and A. Pothen. The architecture of a proteomic network in the yeast. *Proceedings of CompLife2005, Lecture Notes in Bioinformatics*, 3695:265–276, 2005.
10. S.B. Seidman. Network structure and minimum degree. *Social Networks*, 5:269–287, 1983.
11. S. Skiena. *The Algorithm Design Manual*. New York:Springer-Verlag, 1998.
12. V. Spirin and L. A. Mirny. Protein complexes and functional modules in molecular networks. *Proc Natl Acad Sci U S A*, 100(21):12123–12128, October 2003.
13. H. Xiong, X. He, C. Ding, Y. Zhang, V. Kumar, and S. Holbrook. Identification of functional modules in protein complexes via hyperclique pattern discovery. In *Pacific Symposium on Biocomputing (PSB 2005)*, volume 10, pages 221–232, 2005. Available via `http://psb.stanford.edu/psb-online/`

BioPortal: A Portal for Deployment of Bioinformatics Applications on Cluster and Grid Environments

Kuan-Ching Li[1], Chiou-Nan Chen[2], Tsung-Ying Wu[3],
Chia-Hsien Wen[4], and Chuan Yi Tang[2]

[1] Parallel and Distributed Processing Center
Department of Computer Science and Information Engineering
Providence University Shalu, Taichung 43301 Taiwan
kuancli@pu.edu.tw
[2] Department of Computer Science
National Tsing Hua University Hsinchu 30013 Taiwan
{cnchen, cytang}@cs.nthu.edu.tw
[3] Grid Operation Center
National Center for High-Performance Computing
Taichung 40767 Taiwan
alex@nchc.org.tw
[4] Department of Computer Science and Information Management
Providence University Shalu, Taichung 43301 Taiwan
chwen@pu.edu.tw

Abstract. Over last few years, interest on biotechnology has increased dramatically. With the completion of sequencing of the human genome, such interest is likely to expand even more rapidly. The size of genetic information database doubles every 14 months, overwhelming explosion of information in related bioscience disciplines and consequently, overtaxing any existing computational tool for data analysis. There is a persistent and continuous search for new alternatives or new technologies, all with the common goal of improving overall computational performance. Grid infrastructures are characterized by interconnecting a number of heterogeneous hosts through the internet, by enabling large-scale aggregation and sharing of computational, data and other resources across institutional boundaries. In this research paper, we present BioPortal, a user friendly and web-based GUI that eases the deployment of well-known bioinformatics applications on large-scale cluster and grid computing environments. The major motivation of this research is to enable biologists and geneticists, as also biology students and investigators, to access to high performance computing without specific technical knowledge of the means in which are handled by these computing environments and no less important, without introducing any additional drawback, in order to accelerate their experimental and sequence data analysis. As result, we could demonstrate the viability of such design and implementation, involving solely freely available softwares.

1 Introduction

The merging of two rapidly advancing technologies, molecular biology and computer science, has resulted in a new informatics science, namely bioinformatics.

M. Daydé et al. (Eds.): VECPAR 2006, LNCS 4395, pp. 566–578, 2007.
© Springer-Verlag Berlin Heidelberg 2007

Bioinformatics includes methodologies on processing molecular biological information, in order to speedup researches in molecular biology. Modern molecular biology is characterized by huge volume of biological data. Take the classic molecular biology data type, the DNA sequence, for instance, major bioinformatics database centers including GeneBank, the NIH (National Institute of Health) genetic sequence database and its collaborating databases, the European Molecular Biology Laboratory and the DNA Data Bank of Japan, these data have reached a milestone of 100 billion bases from over 165,000 organisms [3]. Common operations on biological data include sequences analysis, protein structures predication, genome sequences comparison, sequence alignment, phylogeny tree construction, pathway research, visualization of sequence alignment results and placement of sequence databases. The most basic and important bioinformatics task is to find the set of homologies for a given sequence, since sequences are often related in functions if they are similar.

Genome research centers, such as the National Center for Biotechnology Information (NCBI) and the European Molecular Biology Laboratory (EMBL), they host enormous volume of biological information in their bioinformatics database. They also provide a number of bioinformatics tools for database search and data acquisition. With the explosion of sequence information available to researchers, computational biologists face the challenge to aid biomedical researches, that is, to invent efficient toolkits to enlarge the use of available computational cycles. Sequence comparison, multiple sequence alignment and phylogeny tree construction are the most fundamental works in biomedical research. There have been many abundant examples of bioinformatics applications that are able to provide solutions for these problems in biomedical research. Some of most extensively utilized applications for these research activities include BLAST [4][5], ClustalW [6][7] and Phylip [8].

However, bioinformatics applications typically are distributed in different individual projects and they require high performance computational environments. Biomedical researchers need to combine many works to conclude their investigation. For instance, in the south of an Asian area, once farms with many dead chickens are reported, biologist may need to identify whether it was infected by H5N1 influenza virus urgently. After obtained the chicken's testimony and RNA sequence, biologist may use BLAST tool to search and acquire other influenza virus sequences from the public database. ClustalW tool is required to compare and investigate their similarity, so then construct the phylogenic tree using Phylip tool. In the above situation, biomedical researchers need these bioinformatics applications. They may download a local version to their own computer or use them in individual server, but either one is complicated and inefficient way, due to a number of drawbacks that any similar solution may bring. Therefore, an efficient and integrated bioinformatics portal is necessary, in order to facilitate biomedical researches.

Grid computing has irresistible potential to apply supercomputing power to address a vast range of bioinformatics problems. A computational grid is a collection of distributed and heterogeneous computing nodes that has emerged as an important platform for computation intensive applications [9][10][15]. They enable large-scale aggregation and sharing of computational, data and other resources across institutional boundaries. It offers an economic and flexible model for solving massive computational problems using large numbers of computers, arranged as clusters embedded in a distributed infrastructure [11][12][13][14].

In this research paper, we integrate several important bioinformatics applications into a novel user-friendly and biologist-oriented web-based GUI portal on top of PCGrid grid computing environment [16]. The major goal in developing such GUI is to assist biologists and geneticists to access to high performance computing, without introducing additional computing drawbacks to this attempt, as to accelerate their experimental and sequence data analysis.

The remainder of this paper is organized as follows. In Section 2, a number of bioinformatics application tools available are introduced, while in Section 3 is introduced the experimental grid computing platform PCGrid, a computing environment built by interconnecting a number of computational resources located inside Providence University Campus. In Section 4, it is discussed the BioPortal bioinformatics portal workflow and implementation. Finally, in Section 5, conclusions and future works are presented.

2   Bioinformatics Applications Overview

Molecular biologists measure and utilize huge amounts of data, of various types. The intention is to use these data to:

1. reconstruct the past (e.g., infer the evolution of species),
2. predict the future (e.g., predict how some genes affect a certain disease),
3. guide bio-technology engineering (such as improving the efficiency of drug design).

Some of the concrete tasks are so complex that intermediate steps are already regarded as problem in their own and constructed an application for it. For instance, while the consensus motif of a sequence in principle determines its evolution function, one of the grand challenges in bioinformatics is to align multiple sequences among to conclude their consensus pattern and predict its function. Sequence comparison, multiple sequence alignment and phylogeny tree construction are fundamental works in biomedical research and bioinformatics. The most extensively applications for these works include BLAST, ClustalW and Phylip. BLAST is a sequence comparison and search tool, ClustalW is a progressive multiple sequence alignment tool, and Phylip is a program for inferring phylogenic tree.

The BLAST (Basic Local Alignment Search Tool) application is a widely used tool for searching DNA and protein databases for sequence similarity to identify homologs to a query sequence [20]. While often referred to as just "BLAST", this can really be thought of as a set of five sub-applications: blastp, blastn, blastx, tblastn, and tblastx.

Five sub-applications of BLAST perform the following tasks:

1. blastp: compare an amino acid query sequence against a protein sequence database,
2. blastn: compare a nucleotide query sequence against a nucleotide sequence database,
3. blastx: compares the six-frame conceptual translation products of a nucleotide query sequence (both strands) against a protein sequence database,
4. tblastn: compares a protein query sequence against a nucleotide sequence database dynamically translated in all six reading frames (both strands),

5. tblastx: compares the six-frame translations of a nucleotide query sequence against the six-frame translations of a nucleotide sequence database.

BLAST tool plays an extremely important role in NCBI GenBank database. It not only provides sequence database search, but also include many toolkits for sequence comparison. BLAST is based on Smith-Waterman local alignment algorithm [17][18], which basically identifies the best local alignment between two sequences by using dynamic programming and tracing back metrology through the sequence matrix. The mpiBLAST is a parallelized version of BLAST, developed by Los Alamos National Laboratory (LANL) [19]. The mpiBLAST segments the BLAST database and distributes it across cluster computing nodes, permitting BLAST queries to be processed on a number of computing nodes simultaneously. The mpiBLAST-g2 is an enhanced version of LANL's mpiBLAST application [21]. This enhanced application allows the parallel execution of BLAST on a grid computing environment.

ClustalW is a general purpose multiple sequence alignment program for DNA or proteins, and it produces biologically meaningful multiple sequence alignments of divergent sequences. It calculates the best match for the selected sequences, and lines them up so that the identities, similarities and differences can clearly be seen. ClustalW is one of the most popular sequences alignment packages, and it is not only a multiple sequence alignment package, but also a phylogenetic tree construction tool. The progressive alignment algorithm of ClustalW is based on three steps:

1. Calculating sequence pairwise similarity,
2. Construction of guide tree,
3. Progressive alignment of sequence.

In the first step, all pairs of sequences are aligned separately, in order to calculate a distance matrix giving the divergence of each pair of sequences. As next step, the trees are used to guide the final multiple alignment processes that are calculated from the distance matrix of step 1 using the Neighbor-Joining method [22]. In the final step, the sequences are progressively aligned according to the branching order in the guided tree. ClustalW-MPI [24] is a parallel implementation of ClustalW. All three steps have been parallelized in order to reduce the global execution time, and it runs on distributed workstation clusters as well as on traditional parallel computers [23]. The only requirement is that all computing nodes involved in ClustalW-MPI computations should have installed MPI.

Phylip is an application for inferring phylogenies tree. The tree construction algorithm is quite straightforward, and it adds species one by one to the best place in the tree and makes some rearrangement to improve the result.

3 The PCGrid Computing Infrastructure

The PCGrid grid-computing platform, standing for The Providence University Campus Grid platform, consists basically of five cluster platforms located in different floors and laboratories inside the College of Computing and Informatics (CCI) of this university. The project of constructing such grid infrastructure is aimed to increase Providence University's computational power and share the resources among investigators and researchers in fields such as bioinformatics, biochemistry, medical informatics, economy, parallel compilers, parallel software, data distribution, multicast, network security, performance analysis and visualization toolkit, computing node selection, thread migration, scheduling in cluster and grid environments, among others.

The PCGrid computing infrastructure is formed by interconnecting the cluster computing platforms via Gigabit Ethernet (1Gb/s), as illustrated in Figure 1.

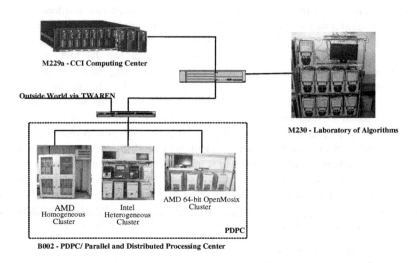

Fig. 1. The PCGrid grid computing infrastructure

The first platform is AMD Homogeneous Cluster, consisting of 17 computing nodes, where each node contains one AMD Athlon 2400+ CPU, 1GB DDR memory, 80GB HD, FedoreCore4 OS, interconnected via Gigabit Ethernet. The second cluster is Intel Heterogeneous Cluster, built up using 9 computing nodes with different CPU speed and memory size, FedoraCore2 OS, interconnected via Fast Ethernet. The third cluster platform consists of 4 computing nodes, where each computing node has one AMD 64-bit Sempron 2800+ CPU, 1GB DDR memory, 120GB HD, FedoreCore4 OS, interconnected via Gigabit Ethernet. The fourth cluster platform is IBMCluster, consisting of 9 computing nodes, where each node contains one Intel P4 3.2GHz CPU, 1 GB DDR memory, FedoraCore3 OS, 120GB HD, interconnected via Gigabit Ethernet. The fifth computing system is IBMBlade, consisting of 6 computing blades, where each blade has two PowerPC 970 1.6 GHz CPUs, 2GB DDR memory and 120GB HD, SUSE Linux OS, interconnected via Gigabit Ethernet. At moment, the total storage contains more than 6TB of storage space.

3.1 Selecting Computing Nodes to Run Parallel Applications

There are two ways to select computing nodes in PCGrid grid computing platform, either manual or automatic. In the manual process, the developer chooses the computing nodes based on CPU activities, depending on the status (busy or idle), as shown in figures 2A and 2B. If the developer persists in selecting a computing node showing RUNNING (that is, CPU in use), this job will be queued, and its execution will only be started when all selected computing nodes are idle. The alternative way to select computing nodes is automatic. All computing nodes in PCGrid platform are sorted and ranked, so that the developer selects a given condition, if he would like to select a number of computing nodes according to their speed (and idle) or he would like to select a number of computing nodes with higher network bandwidth.

All jobs submitted by any user are ranked according to user credentials, his level of priority inside the queue. The higher a user's credentials; highest is the priority to execute this user's applications in our computing platform. The queue is re-ranked every time a job is submitted to our grid platform.

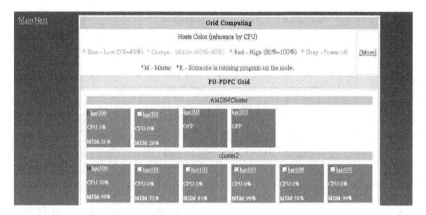

Fig. 2A. Computing Node manual selection simple mode

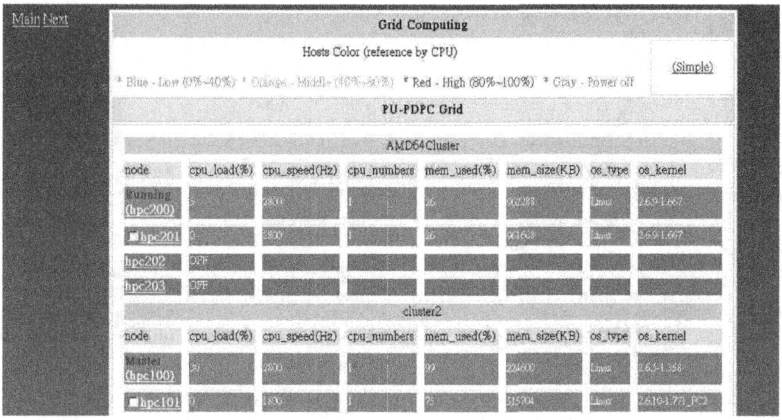

Fig. 2B. Real-time display of all computing nodes status in complete mode

3.2 Performance Visualization

We have developed a performance visualization toolkit, to display application execution performance data charts [1][2]. Performance data of sequential or parallel applications executed in PCGrid computing platform are captured and saved, and later displayed the CPU and memory utilization of that given application, as in figure 3A.

During different stages of the development of an application, the developer may want to compare the performance of different implementations of this application. For such usage on PCGrid platform, we have developed a toolkit able to perform such comparisons, as shown in figure 3B. The corresponding charts of CPU and memory utilization of each computing node involved in the computation are overlapped, to facilitate the visualization of such performance comparisons.

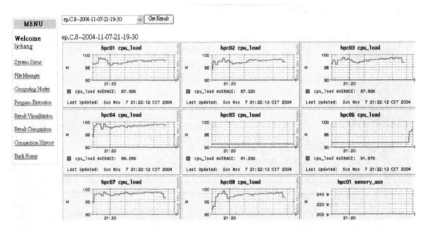

Fig. 3A. Performance data of each computing node involved in computation of PCGrid grid platform

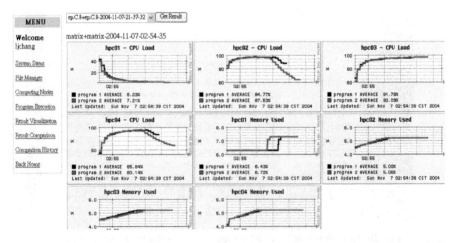

Fig. 3B. Performance comparison of two application execution results, computing node by computing node, CPU load and memory usage

4   BioPortal: A Portal for Bioinformatics Applications in Grid

We have integrated most fundamental computing applications in biomedical research and bioinformatics inside BioPortal: sequence comparison, pairwise or multiple sequence alignment and phylogeny tree construction, all in a complete workflow. We also provide an additional feature to biologists to choose automatically computing nodes to execute their parallel applications, by setting the number of computing nodes. The BioPortal will take care of selecting best computing nodes that fits users' requested computation, as described in subsection 3.1.

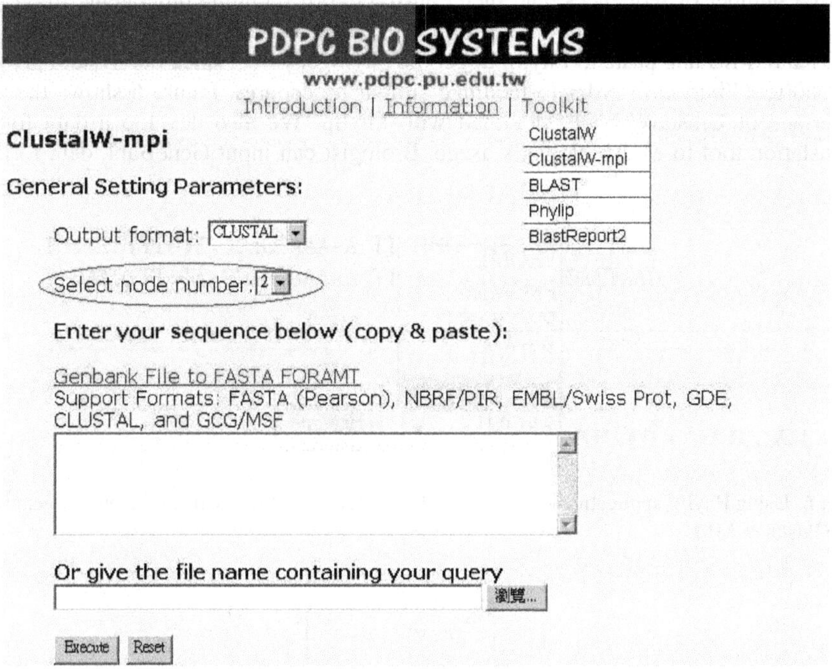

Fig. 4. BioPortal web-based GUI screenshot

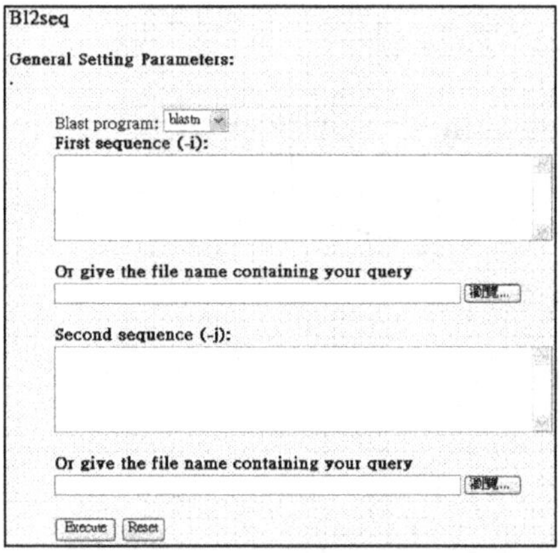

Fig. 5A. bl2seq interface 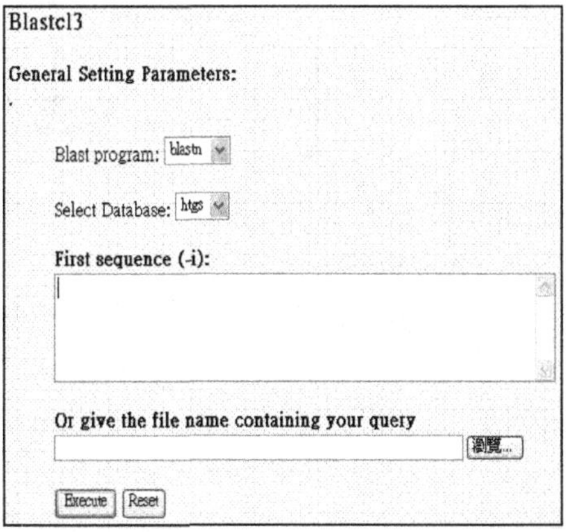   **Fig. 5B.** Blastcl3 interface

Figure 4 shows the bioinformatics portal homepage. The biologist can use bl2seq (a BLAST toolkit for two sequence comparison) to compare their own sequence with other sequences that was acquired from a bioinformatics database by blastcl3 (a NCBI BLAST client). Figure 5A and 5B show the web interface screenshot of Bl2seq and Blastcl3 respectively.

Biologists make use of ClustalW-MPI to perform multiple sequence alignment with a number of sequences, and then construct corresponding phylogenic tree using Phylip directly. Biologists do need not to copy the alignment result from the ClustalW-MPI and paste to Phylip to get the phylogeny tree, since our system provide a "shortcut" button in order to facilitate similar procedures. Figure 6 shows the web interface of ClustalW-MPI integrated with Phylip. We also develop a data format translation tool to ease biologist's usage. Biologist can input GeneBank data format,

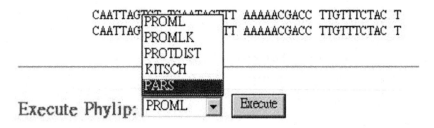

Fig. 6. Using Phylip application to construct phylogenic tree, directly from the output generated by ClustalW-MPI

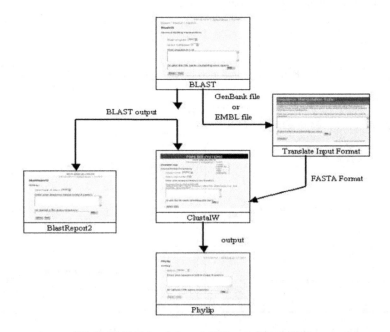

Fig. 7. BioPortal web-based GUI complete workflow

and our translation toolkit can transform it to legal FASTA format for ClustalW-MPI, as in figure 8. Detailed description of all bioinformatics services available in our BioPortal is listed in table 1, while Figure 7 shows the complete workflow of the BioPortal.

Table 1. List of bioinformatics applications provided by BioPortal

Application Tools	Description
mpiBLAST-g2	An enhanced parallel application that permits parallel execution of BLAST on Grid environments, based on GLOBUS and MPICH
Bl2seq	This application performs comparison between two sequences, using either blastn or blastp algorithms
Blastall	This application may be used to perform BLAST comparisons
BLASTcl3	A BLAST software client running on local computers that connects to BLAST servers located at NCBI, in order to perform searches and queries of NCBI sequence databases
Formatdb	It is used to format protein or nucleotide source database before these can be utilized by Blastall, Blastpgp or MEGABlast
BlastReport2	A Perl script that reads the output of Blastcl3, reformats it to ease its use and eliminates useless information
ClustalW-MPI	Parallel version of a general purpose multiple sequence alignment application for DNA or proteins, by producing meaningful multiple sequence alignment of divergent sequences
Phylip	Set of applications that performs phylogenic analyses

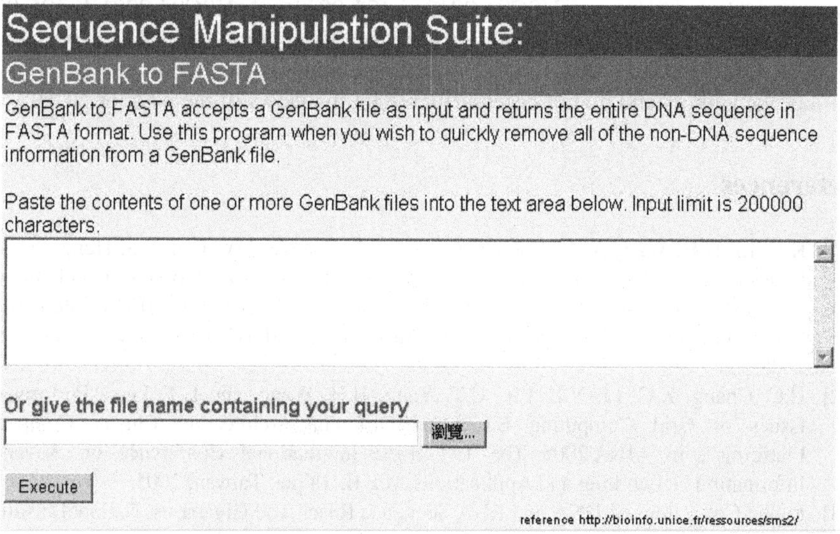

Fig. 8. Sequence data transformation toolkit

5 Conclusions and Future Work

We have constructed a campus scale computing grid platform, as also implemented a portal providing a number of well-known bioinformatics application toolkits. Not only to provide easy access of bioinformatics application toolkits to biologists and geneticists, but also large amount of computational cycles in an easy way. This portal contributes three fundamental molecular biology activities: sequence comparison, multiple sequence alignment and phylogenic tree construction, all integrated in a friendly and easy-to-use web-based GUI portal. We have solved many data inconsistency problems and finally integrated a number of different tools that are able to cooperate all together. This BioPortal not only facilitate biomedical researcher investigations and computational biology courses in graduate-level, as also it demonstrates a well-succeeded combination of high performance computing with the use of grid technology and bioinformatics.

As future work, several directions of this research are ongoing. One of goals is to develop a one-stop-shop bioinformatics portal, to provide efficient and economic computational power and cycles to biomedical researchers. At the present moment, we are in the process of integrating other well-known bioinformatics applications into this BioPortal, for instance, applications for protein structure predication and protein visualization. We expect to continuously develop on top of grid technology, so that in near future, researchers will not only be able to seamlessly utilize PCGrid computational resources, but also expand on demand to larger scale grid computing platforms, such as regional or national grid platforms.

Acknowledgements

This paper is based upon work supported in part by National Science Council (NSC), Taiwan, under grants no. NSC94-2213-E-126-005 and NSC95-2221-E-126-006-MY3, and National Center for High-Performance Computing (NCHC), Taiwan. Any opinions, findings, and conclusions or recommendations expressed in this material are those of the authors and do not necessarily reflect the views of the NSC or NCHC.

References

[1] K.C. Li, H.H. Wang, C.N. Chen, C.C. Liu, C.F. Chang, C.W. Hsu, S.S. Hung, "Design Issues of a Novel Toolkit for Parallel Application Performance Monitoring and Analysis in Cluster and Grid Environments", in I-SPAN'2005 The 8th IEEE International Symposium on Parallel Architectures, Algorithms, and Networks, Las Vegas, USA, 2005.

[2] H.C. Chang, K.C. Li, Y.L. Lin, C.T. Yang, H.H. Wang, and L.T. Lee, "Performance Issues of Grid Computing Based on Different Architecture Cluster Computing Platforms", in AINA'2005 The 19th IEEE International Conference on Advanced Information Networking and Applications, vol. II, Taipei, Taiwan, 2005.

[3] Public Collections of DNA and RNA Sequence Reach 100 Gigabases, National Institutes of Health, August 22, 2005. (http:// www.nlm.nih.gov/news/press_releases/dna_rna_100_gig.html).

[4] S.F. Altschul, W. Gish, W. Miller, E.G. Myers, and D.J. Lipman, "Basic Local Alignment Search Tool", J. Mol. Biol. 215,403-410, 1990.

[5] S. F. Altschul, T. L. Madden, A. A. Schaeffer, J. Zhang, Z. Zhang, W. Miller and D.J. Lipman, "Gapped BLAST and PSI-BLAST: A new generation of protein database search programs", Nucleic Acids Research, 25, pp. 3389-3402, 1997.

[6] D.G. Higgins, P.M. Sharp, "CLUSTAL: a package for performing multiple sequence alignment on a microcomputer", Gene, Dec 15;73(1):237-44, 1988.

[7] J.D. Thompson, D.G. Higgins, T.J. Gibson, "CLUSTAL W: improving the sensitivity of progressive multiple sequence alignment through sequence weighting, position-specific gap penalties and weight matrix choice", Nucleic Acids Res. Nov 11;22(22):4673-80, 1994.

[8] Joe Felsenstein, "PHYLIP (Phylogeny Inference Package)", version 3.5c. (http://evolution.genetics.washington.edu/phylip.html), 1993.

[9] B. Allcock, J. Bester, J. Bresnahan, A. L. Chervenak, I. Foster, C. Kesselman, S. Meder, V. Nefedova, D. Quesnal, and S. Tuecke. "Data Management and Transfer in High Performance Computational Grid Environments", Parallel Computing, vol. 28 (5), pp. 749-771, 2002.

[10] B. Allcock, J. Bester, J. Bresnahan, I. Foster, J. Gawor, J. A. Insley, J. M. Link, and M. E. Papka. "GridMapper: A Tool for Visualizing the Behavior of Large-Scale Distributed Systems", in Proceedings of 11th IEEE International Symposium on High Performance Distributed Computing (HPDC-11), 2002.

[11] M. Baker, R. Buyaa, D. Laforenza, "Grid and Grid Technologies for Wide-Area Distributed Computing", available at http://www.csse.monash.edu.au/~rajkumar/papers/ gridtech.pdf .

[12] F. Berman, A. Chien, K. Cooper, J. Dongarra, I. Foster, D. Gannon, L. Johnson, K. Kennedy, C. Kesselman, J. Mellor-Crummey, D. Reed, L. Torczon, and R. Wolski, "The GrADS Project: Software Support for High-Level Grid Application Development", International Journal of High-Performance Computing Applications, 15(4), 2002.

[13] M. Chetty, R. Buyya, "Weaving computational Grids: How analogous are they with electrical Grids?" Journal of Computing in Science and Engineering (CiSE), 2001.

[14] K. Czajkowski, I. Foster, and C. Kesselman. "Resource Co-Allocation in Computational Grids", in Proceedings of the Eighth IEEE International Symposium on High Performance Distributed Computing (HPDC-8), pp. 219-228, 1999.

[15] K. Czajkowski, S. Fitzgerald, I. Foster, and C. Kesselman. "Grid Information Services for Distributed Resource Sharing", in Proceedings of the Tenth IEEE International Symposium on High-Performance Distributed Computing (HPDC-10), 2001.

[16] K.C. Li, C.N. Chen, C.W. Hsu, S.S. Hung, C.F. Chang, C.C. Liu, C.Y. Lai, "PCGrid: Integration of College's Research Computing Infrastructures Using Grid Technology", in NCS'2005 National Computer Symposium, Taiwan, 2005.

[17] S. F. Altschul, W. Gish, W. Miller, E. W. Myers, and D. J. Lipman, "Basic local alignment search tool", Journal of Molecular Biology, vol. 215, pp. 403–410, 1990.

[18] T.F. Smith, M.S. Waterman, "Identification Of Common Molecular Subsequences", Journal of Molecular Biology, vol. 147, pp. 195-197, 1981.

[19] Los Alamos National Laboratory (http://mpiblast.lanl.gov).

[20] Heshan Lin, Xiaosong Ma, Praveen Chandramohan, Al Geist and Nagiza Samatova, "Efficient Data Access for Parallel BLAST", IEEE International Parallel & Distributed Symposium, 2005.

[21] mpiBLAST-g2, Bioinformatics Technology and Service (BITS) team, Academia Sinica Computing Centre (ASCC), Taiwan. (http://bits.sinica.edu.tw/mpiBlast/mpiBlast-g2/README.mpiBLAST-g2.html)

[22] N. Saitou, M. Nei, "The Neighbor-Joining Method: A New Method for Reconstructing Phylogenetic Trees", Molecular Biology and Evolution, 4(4), pp. 406-25, 1987.

[23] J.D. Thompson, D.G. Higgins, and T.J. Gibson, "CLUSTAL W: Improving the Sensitivity of Progressive Multiple Sequence Alignment through Sequence Weighting, Positions-Specific Gap Penalties and Weight Matrix Choice", Nucleic Acids Research, 22, pp. 4673-4680, 1994.

[24] K.B. Li, "ClustalW-MPI: ClustalW Analysis Using Distributed and Parallel Computing", Bioinformatics, 19(12), pp.1585-6, 2003.

Adaptive Distributed Metamodeling

Dirk Gorissen, Karel Crombecq, Wouter Hendrickx, and Tom Dhaene

Antwerp University, Middelheimlaan 1, 2020 Antwerp, Belgium
{dirk.gorissen,wouter.hendrickx,tom.dhaene}@ua.ac.be,
karel.crombecq@student.ua.ac.be

Abstract. Simulating and optimizing complex physical systems is known to be
a task of considerable time and computational complexity. As a result, metamodeling techniques for the efficient exploration of the design space have become
standard practice since they reduce the number of simulations needed. However, conventionally such metamodels are constructed sequentially in a one-shot
manner, without exploiting inherent parallelism. To tackle this inefficient use of
resources we present an adaptive framework where modeler and simulator interact through a distributed environment, thus decreasing model generation and
simulation turnaround time. This paper provides evidence that such a distributed
approach for adaptive sampling and modeling is worthwhile investigating. Research in this new field can lead to even more innovative automated modeling
tools for complex simulation systems.

1 Introduction

Computer based simulation has become an integral part of the engineering design
process. Rather than building real world prototypes and performing experiments, application scientists can build a computational model and simulate the physical processes at
a fraction of the original cost. However, despite the steady growth of computing power,
the computational cost to perform these complex, high-fidelity simulations maintains
pace. For example, to quote [1]:

> "...it is reported that it takes Ford Motor Company about 36-160 hrs to run one
> crash simulation [2]. For a two-variable optimization problem, assuming on
> average 50 iterations are needed by optimization and assuming each iteration
> needs one crash simulation, the total computation time would be 75 days to 11
> months, which is unacceptable in practice."

Luckily, most of these simulations can be reduced to parallel parameter sweep type applications. These consist of several instances of the simulator that are run independently
for different input parameters or datasets. Due to the inherent parallelism this can be
done in a distributed fashion thus significantly reducing "wall-clock" execution time.

For most realistic problems the high computational cost of simulator codes and
the high dimensionality of the design space simply prohibit this direct approach, thus
making these codes unusable in engineering design and multidisciplinary design optimization (MDO). Consequently, scientists have turned towards upfront approximation
methods to reduce simulation times. The basic approach is to construct a simplified

M. Daydé et al. (Eds.): VECPAR 2006, LNCS 4395, pp. 579–588, 2007.
© Springer-Verlag Berlin Heidelberg 2007

approximation of the computationally expensive simulator (e.g.: aerodynamic drag generated by a particular airfoil shape [3]), which is then used in place of the original code to facilitate MDO, design space exploration, reliability analysis, etc. [4] Since the approximation model acts as surrogate for the original code, it is often referred to as a *surrogate model* or *metamodel*. Examples of such metamodels include Kriging models, Artificial Neural Networks, Support Vector Machines, radial basis function models, polynomial and rational models.

The remainder of this paper is structured as follows: In section 2 we discuss the motivation for constructing parametrized metamodels while section 3 gives an overview of similar research efforts and related projects. Section 4 describes the design and prototype implementation of our framework and section 5 some preliminary performance results. We conclude with a critical evaluation and pointers to future work.

2   Motivation

The reasons for constructing metamodels are twofold: on the one hand metamodels are often used for efficient design space exploration, on the other hand they are used as a cheap surrogate to replace the original simulator. When performing an optimum search, the metamodel guides the search to potentially interesting regions (local minima) [3,5]. Once an adequate solution has been found, the model is discarded. When building a global, scalable model, the metamodel itself forms the object of interest. Here the goal is to construct a parametrized metamodel that can entirely replace the original objective function in a subset of the design space of interest. This is useful since the metamodel is much cheaper to evaluate. Once constructed the metamodel is retained and the objective function discarded. In this paper we are primarily concerned with the latter.

However, constructing an accurate metamodel is no trivial undertaking. In some cases it remains questionable if a usable metamodel can be constructed at all. Even if an accurate metamodel is feasible, the process of building it still requires evaluations of the original objective function. Therefore, if the process of constructing a metamodel requires, say, 80 function evaluations and each evaluation takes 10 hours, the rate at which the design space can be explored is still relatively low. Nevertheless, the authors argue that this is justifiable since it is a one time, up front investment.

To help tackle this bottleneck we propose a framework that integrates the automated building of metamodels and the adaptive selection of new simulation points (sequential design) with the distributed evaluation of the cost function. This framework will build upon previous work in modeling [6,7,8] and distributed systems [9,10].

3   Related Work

Research efforts and tools that integrate modeling and design space exploration techniques with grid computing techniques can be divided into two categories: those focussed on design optimization and those geared towards the building of standalone scalable metamodels. The first category is by far the most populous. First we have, usually commercial, integrated systems that model and optimize application specific problems. Examples are modeFRONTIER [11] for ship hulls and FlightLab [12] for aircraft.

On the other hand there are many general optimization frameworks which can be applied to different problem domains. The most notable again being Nimrod/O [13]. Nimrod/O is based on the Nimrod/G broker [14] and tackles its biggest disadvantage. This is that Nimrod/G will try to explore the complete design space on a dense grid. This is usually intractable for realistic problems. Nimrod/O performs a guided search through the design space trying to find that combination of parameters that will minimize (maximize) the model output. To this end Nimrod/O employs a number of search algorithms (e.g.: P-BFGS, Simplex, Simulated Annealing). A similar project is DAKOTA [15] from Sandia Laboratories. It is a C++ toolkit that allows a user to choose different optimization algorithms to guide the search for optimal parameter settings. Other projects include GEODISE [16], The Surrogate Management Framework (SMF), SciRun and its precursor Uintah, NetSolve, NEOS and the work by Y. S. Ong et al [17,18].

While all projects mentioned above are tailored towards optimization, they are not concerned with creating a parameterized that can be used on its own. Research efforts that do build replacement metamodels exist [19,20,5,6], notably the Multistage Bayesian Surrogate Methodology (MBSM) proposed by C.H. Amon et al [21], but fail to include concepts of distributed computing. Thus the repetitive process of evaluating the objective function while constructing the metamodel is done in a sequential fashion, and this can be extremely time consuming. We were unable to find evidence of other real projects that tackle this. Perhaps the project that comes closest to what we wish to achieve is described in [22], though it too is biased towards optimization and lacks adaptivity.

4   The Design

In this section we outline the architecture of the framework we have designed. Note that treating each individual component in depth is out of scope for this paper. Instead we will concentrate on how they fit together and interact on a more general level. More detailed analyses will follow in future publications.

A high level design diagram is shown in figure 1. The workflow is as follows: Given an application specific simulator (i.e. the objective function) and an XML configuration file containing a number of model/algorithmic tuning parameters, the modeler will build a metamodel with the user-required accuracy level. In order to do so it interacts with the simulator through the *SampleEvaluator* (SE) which executes the simulator on the grid through an existing grid middleware or broker.

4.1   The Modeler

The first component of the framework is the modeler. This component interacts with the simulator in order to construct a meta representation of the original objective function. Our modeler of choice is the Matlab M^3 Toolbox [8] and its schematic flowchart is shown in figure 2. The box on the right represents the simulation backend, the component responsible for evaluating the samples. This is the part that will be distributed. The center box depicts the core loop of the toolbox, it drives the whole modeling process and the selection of new samples. The leftmost box shows the modeler itself. It is responsible for building the polynomial/rational/Neural Network/... metamodels.

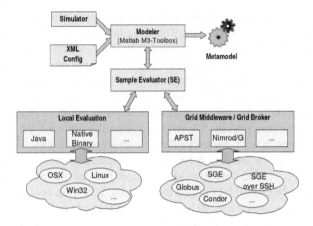

Fig. 1. High level components

The main modeling loop goes as follows: First, an initial sample distribution in the input space is chosen, and simulations are run for all points in this initial sample set. Then, several models of different complexity are built based on the locations of the sample points and the corresponding outputs of the simulator. The models are then compared over a number of measures, and ranked according to their estimated accuracy. The best models are kept, and a new set of sample locations is adaptively chosen (sequential design). The new sample set is passed on to the simulation backend, which will call the simulator for each of the new sample points. After that, the whole loop repeats itself and will continue until the toolbox has reached the desired accuracy.

For a more detailed treatment of how the modeler works (sample selection, termination criteria, etc) please refer to [7].

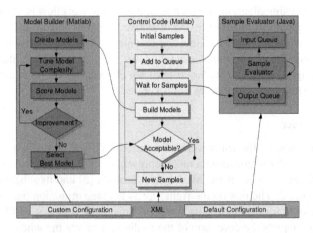

Fig. 2. The Multivariate MetaModeling toolbox (M^3)

4.2   The Grid Middleware

In order to distribute the simulation backend across heterogeneous resources an extra software layer is needed that takes care of the distribution details. In this case we have used APST [23,24] (though other middlewares such as ProActive, SGE, CoBRA, Nimrod/G and Gridbus will be supported in the future). APST can use remote machines accessed through either a Globus GRAM or ssh, remote storage accessed through a Globus GASS server, scp, ftp, sftp, or an SRB server, and queuing systems controlled by Condor, DQS, LoadLeveler, LSF, PBS, or SGE. Since APST grew from AppleS (APPlication LEvel Scheduling) [25] it also tries to schedule jobs optimally based on the characteristics of the resources used. To do this it makes use of established grid information services such as Ganglia, MDS and NWS.

APST consists of an APST client (`apst`) and an APST daemon (`apstd`), both which may be running on separate machines. The client may connect to the server and submit jobs or query the server for job status information. To submit jobs or add resources to the resource pool the client generates an XML file which is then sent to the server.

4.3   Sample Evaluator

The Sample Evaluator (SE) can be seen as a kind of Application Aware Scheduler (AAS) that forms the glue between the modeler and the middleware (of course local evaluation is supported as well). It is responsible for translating modeler requests (i.e. evaluations of datapoints) into middleware specific jobs (in this case APST `<task>` tags), polling for results, and returning them to the modeler. The SE is implemented as a set of Java interfaces and base classes that are sub classed for each of the supported middlewares. This is illustrated for the case of APST in figure 3. As can be seen from the figure there is a separate delegate for each step in the sample evaluation process: job submission, job status monitoring and processing the finished results.

As research continues, the intelligence of the SE will be constantly improved. Instead of the simple bridge it is now, we will include application specific and resource specific knowledge into the scheduling decision. Rather than requesting a batch of samples to be evaluated with equal priority the modeler assigns scores to each data sample

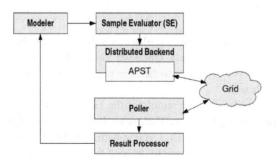

Fig. 3. APST Sample Evaluator Backend

(i.e., data samples corresponding to interesting features of the objective function, such as minima and maxima, will receive higher scores). The AAS can then take these priorities into account when making scheduling decisions. Likewise, the AAS should make use of resource information in order to achieve an optimal $task - host$ mapping (i.e., complex tasks should be scheduled on the fastest nodes).

5   Performance Comparison

5.1   Experimental Setup

In this section we illustrate the application of our M3 framework to an example from Electromagnetics (EM). We will model the the problem twice, once sequentially and once in parallel, and compare the performance. The simulator, for which we shall build a parametrized, scalable metamodel, computes the scattering parameters for a step discontinuity in a rectangular waveguide. The 3D inputs consists of input frequency, the gap height and the gap length. The (complex) outputs are the scattering parameters of this 2-port system. Figure 4 shows an approximate plot of the input-output relation at three different discrete frequencies.

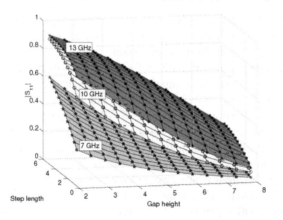

Fig. 4. A Plot of $|S_{11}|$, the Modulus of the First Scattering Parameter

While a real world example, this simulator is still relatively cheap to evaluate. One evaluation takes about 8-13 seconds. Once we are satisfied with the plumbing of the underlying framework we will turn to heavier problems with evaluation times in the order of minutes to hours.

Due to the characteristics of the simulator the exact hardware characteristics of the testbed are of little importance. Nevertheless we list them for completeness. The standalone case was run on a Pentium IV 1.9GHz with 768MB main memory. For the distributed case we used 6 hosts accessible through the local LAN. These included: four Pentium IV 2.4 GHz, one AMD Opteron 1.7 GHz, and one Pentium M 1.6GHz, each with 512MB RAM. While we regret not having been able to use 'real' grid resources we note that this is not really an issue since (1) we are currently only interested in a proof-of-concept (2) we expect the speedup (distributed vs. sequential) to increase linearly with the amount of hosts, thus adding more hosts will simply increase the speedup factor.

The M^3 toolbox and apstd ran on the same host and the APST scheduling algorithm was the default simple work-queue strategy. No grid information service was configured.

For each case the M^3 toolbox was used to build a metamodel for the objective function described above. We recorded the total time needed to build the model, the time needed to evaluate each batch of samples that is requested as the modeling progresses, and the total number of function evaluations. The results were averaged over three runs.

5.2 Test Results

Table 1 summaries the different results for each of the runs. If we first look at the average time to process one sample batch we find it is about 56 seconds in the sequential vs 14 in the distributed case. Thus we have an effective speedup factor of about 4. The size of each batch varies between about 3 to 6 samples.

We notice something similar if we look at the total execution times for each run in figure 5. The total time in the distributed case is about 4 times smaller than in the sequential case for a comparable number of function evaluations.

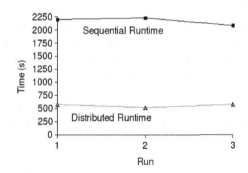

Fig. 5. Comparison Sequential and Distributed

Figure 5 seems unlogical since 6 hosts were used in the distribution. One would expect a speed up factor of 6 (or in general N, if N hosts were used). The reason is that the M^3 toolbox is not yet resource aware (*see* section 4.3) which results in an underutilization of the available compute nodes. With this improvement, together with the move to proper distributed setups involving Globus and SGE administered resources, we expect to improve the speedup significantly in the very near future. Nevertheless these figures still stand since their purpose was to illustrate the integration of distributed computing, adaptive metamodeling and sequential design.

Table 1. Test Results: Sequential (top) and Distributed (bottom)

Run	# Samples	Avg Time per Sample Batch (s)	Total Runtime (s)
1	217	56.39	2199.21
2	221	55.75	2229.86
3	206	56.29	2082.66
Avg	**214.67**	**56.14**	**2170.58**

Sequential

Run	# Samples	Avg Time per Sample Batch (s)	Total Runtime (s)
1	212	14.91	583.55
2	206	13.56	517.12
3	227	13.82	582.45
Avg	**215**	**14.1**	**561.05**

Distributed

6   Evaluation and Future Work

In this paper we have made the case for the use of distributed computing techniques while building scalable metamodels. We have presented a framework based on the M^3 toolbox and APST and contrasted its performance with the traditional approach of analyzing datapoints sequentially. The proof of principle results look very promising and warrant further extension of our framework in the future.

Future work will include:

- Move to real distributed setups involving Globus and SGE administered clusters.
- Creation of a 'real', pluggable framework where the user will be able to easily choose the modeling algorithm and the grid middleware to suit his or her application. In addition, if a required capability is not available the user should be able to plug in his own extension.
- Apply AI techniques such as genetic algorithms and machine learning algorithms to enhance the modeling, decrease the reliance on simple heuristics, and allow for more automated tuning of the modeling process.

References

1. Wang, G. G., S.S.: Review of metamodeling techniques in support of engineering design optimization. ASME Transactions, Journal of Mechanical Design (2006) in press
2. Gu, L.: A comparison of polynomial based regression models in vehicle safety analysis. In Diaz, A., ed.: 2001 ASME Design Engineering Technical Conferences - Design Automation Conference, ASME, Pittsburgh, PA. (2001)
3. Marsden, A.L., Wang, M., Dennis, J.J.E., Moin, P.: Optimal aeroacoustic shape design using the surrogate management framework: Surrogate optimization. Optimization and Engineering **Volume 5** (2004) pp. 235–262(28)
4. Simpson, T.W., Booker, A.J., Ghosh, D., Giunta, A.A., Koch, P.N., Yang, R.J.: Approximation methods in multidisciplinary analysis and optimization: A panel discussion. Structural and Multidisciplinary Optimization **27** (2004) 302–313

5. Martin, J.D., Simpson, T.W.: Use of adaptive metamodeling for design optimization. In: In Proc. of 9th AIAA/ISSMO symposium on multidisciplinary analysis and optimization, 4-6 September, Atlanta, Georgia. (2002)

6. Hendrickx, W., Dhaene, T.: Multivariate modelling of complex simulation-based systems. Proceedings of the IEEE NDS 2005 conference (2005) 212–216

7. Hendrickx, W., Dhaene, T.: Sequential design and rational metamodelling. In Kuhl, M., M., S.N., Armstrong, F.B., Joines, J.A., eds.: Proceedings of the 2005 Winter Simulation Conference. (2005) 290–298

8. Hendrickx, W., Dhaene, T.: M^3-toolbox (2005) Available on www.coms.ua.ac.be in the Software section.

9. Gorissen, D., Stuer, G., Vanmechelen, K., Broeckhove, J.: H2O Metacomputing - Jini Lookup and Discovery. In: Proceedings of the International Conference on Computational Science (ICCS), Atlanta, USA. (2005) 1072–1079

10. Hellinckx, P., Vanmechelen, K., Stuer, G., Arickx, F., J., B.: User experiences with nuclear physics calculations on H2O and on the BEgrid. In: in Proceedings of the International Conference on Computational Science (ICCS), Atlanta, USA. (2005) 1081–1088

11. modeFRONTIER: (http://www.esteco.it/products/)

12. FlightLab: (http://www.flightlab.com/)

13. Abramson, D., Lewis, A., Peachey, T., Fletcher, C.: An automatic design optimization tool and its application to computational fluid dynamics. In: Proceedings of the 2001 ACM/IEEE conference on Supercomputing (CDROM). (2001) 25–25

14. Abramson, D., Giddy, J., Kotler, L.: High performance parametric modeling with Nimrod/G: Killer application for the global grid? In: Proceedings of the International Parallel and Distributed Processing Symposium (IPDPS), Cancun, Mexico. (2000) 520– 528

15. Giunta, A., Eldred, M.: Implementation of a trust region model management strategy in the DAKOTA optimization toolkit. In: Proceedings of the 8th AIAA/USAF/NASA/ISSMO Symposium on Multidisciplinary Analysis and Optimization, Long Beach, CA. (2000)

16. Eres, M.H., Pound, G.E., Jiao, Z., Wason, J.L., Xu, F., Keane, A.J., Cox, S.J.: Implementation and utilisation of a grid-enabled problem solving environment in matlab. Future Generation Comp. Syst. **21** (2005) 920–929

17. Ng, H.K., Lim, D., Ong, Y.S., Lee, B.S., Freund, L., Parvez, S., Sendhoff, B.: A multi-cluster grid enabled evolution framework for aerodynamic airfoil design optimization. In: ICNC (2). (2005) 1112–1121

18. Ng, H.K., Ong, Y.S., Hung, T., Lee, B.S.: Grid enabled optimization. In: EGC. (2005) 296–304

19. Lehmensiek, R., Meyer, P.: Creating accurate multivariate rational interpolation models for microwave circuits by using efficient adaptive sampling to minimize the number of computational electromagnetic analyses. IEEE Trans. Microwave Theory Tech. **49** (2001) 1419–

20. De Geest, J., Dhaene, T., Faché, N., De Zutter, D.: Adaptive CAD-model building algorithm for general planar microwave structures. IEEE Transactions on Microwave Theory and Techniques **47** (1999) 1801–1809

21. Weiss L.E., Amon C.H., F.S.M.E.R.D.V.I.W.L., P.G., C.: Bayesian computer-aided experimental design of heterogeneous scaffolds for tissue engineering. Computer Aided Design **37** (2005) 1127–1139

22. Parmee, I., Abraham, J., Shackelford, M., Rana, O.F., Shaikhali, A.: Towards autonomous evolutionary design systems via grid-based technologies. In: Proceedings of ASCE Computing in Civil Engineering, Cancun, Mexico. (2005)

23. Casanova, H., Obertelli, G., Berman, F., Wolski, R.: The AppLeS parameter sweep template: User-level middleware for the grid. In: Proceedings of Supercomputing (SC 2000). (2000)

24. Casanova, H., Legrand, A., Zagorodnov, D., Berman, F.: Heuristics for scheduling parameter sweep applications in grid environments. In: Proc. 9th Heterogeneous Computing Workshop (HCW), Cancun, Mexico (2000) 349–363
25. Berman, F., Wolski, R., Casanova, H., Cirne, W., Dail, H., Faerman, M., Figueira, S., Hayes, J., Obertelli, G., Schopf, J., Shao, G., Smallen, S., Spring, N., Su, A., Zagorodnov, D.: Adaptive computing on the grid using AppLeS. IEEE Transactions on Parallel and Distributed Systems (TPDS) **14** (2003) 369–382

Distributed General Logging Architecture for Grid Environments

Carlos de Alfonso, Miguel Caballer, José V. Carrión, and Vicente Hernández

Departamento de Sistemas Informáticos y Computación,
Universidad Politécnica de Valencia,
46022 Valencia, Spain
{calfonso,micafer,jocarrion,vhernand}@dsic.upv.es

Abstract. The registry of information about the activity of applications is an important issue in Grid environments. There are different projects which have developed tools for supporting the track of the resources. Nevertheless, most of them are mainly focused in measuring CPU usage, memory, disk, etc. because they are oriented to the classical use of the Grid to share computational power and storage capacity. This paper proposes a distributed architecture which provides logging facilities in service oriented Grid environments (DiLoS). This architecture is structured so that can fit to hierarchical, flat, etc. grid deployments. The information generated during the activity in the services are scattered among the different levels of a Grid deployment, providing features such as backup of the information, delegation of the storage, etc. In order to create the hierarchy of log services, the architecture is based on discovery facilities that can be implemented in different ways, which the user may configure according to the specific deployment. A case of use is described, where the DiLoS architecture has been applied to the gCitizen project.

1 Introduction

Almost any operation which occurrs in a shared environment such as the Grid is susceptible to be registered for its late analysis. Some examples of events which are likely to be registered are the access to the services (who and when acceded), the kind of resources which are mainly used by someone who is authorised to work in the Grid, how many time has been used a resource, the changes in the state of a service, etc.

The registered information can be used later for extracting statistics about the usage of the resources in the Grid with upgrading purposes, obtaining information about the proper (or not) usage of the resources when a problem arises, debugging a distributed application, or obtaining how many resources have been used in a project for accounting purposes, among others.

In a computing environment, the most common technique for registering the activity of an application (service, daemon, etc.) in a traditional system is to create a *log file* which would contain text lines describing every action which has been performed by this application.

M. Daydé et al. (Eds.): VECPAR 2006, LNCS 4395, pp. 589–600, 2007.
© Springer-Verlag Berlin Heidelberg 2007

This model translated to a Grid environment would mean to have a lot of files pertaining to any of the resources which have been deployed. Nevertheless it would need a mechanism for mergeing these files, and thus obtaining the *log* information about the whole Grid system.

One of the main objectives of Grid technology is to provide with a great number of resources which would be frequently used by a huge number of people. In such case, the activity in the Grid, considered in a medium-long period of time, would generate a lot of registries which would be hard to store, manage or backup. Also, these data would be hard to query for obtaining useful information when analyzing the activity of the system. So, in a Grid environment, it is needed a more appropiate method for the registration of this activity.

This paper proposes a Distributed Log System (DiLoS) which defines an architecture for the registration and maintenance of the information about the activity in the system. The DiLoS system scatters the *log* files through a Virtual Organization, for obtaining features such as backup, redundancy, ease of access to logs, or decentralization, among others.

The paper is organized as follows: Section 2 discuss about the systems which are already used in Grid environment for accounting, auditory, debugging or activity registration purposes. Next, the section 3 describes the DiLoS systems, its components and the protocols used. Also it is described how the grid components would be integrated in the system. Later, the section 4 exposes a case of study about the usage of the DiLoS system in a Grid deployment oriented to eGovernment. Finally, section 5 summarizes and outlines the work which will follow to the proposal of the described architecture.

2   State of Art and Motivation

The registry of information is a feature which has been traditionally assumed in Grid environments. This characteristic would support features such as accounting or auditing. Nevertheless, up to now, most of the developments which address these issues only implement partial solutions for specific subjects of each project about Grid computing.

Most of these deployments are oriented to obtain accounting information about specific resource usage by the users (memory, disk, load, executed jobs, etc). The data is mainly used for being published for the rest of members of the Virtual Oganization with the aim of monitoring the state of the system, and scheduling tasks according that resource usage.

The next summary exposes a simple classification about the tools and architectures which are currently being deployed [1].

- **NASA-IPG [2]:** The system follows a common monitoring architecture for Grid environments, sensors to measure some characteristics of the resources, actuators to perform some process control actions and event services that provides a mechanism for forwarding sensor-collected information to other processes that are interested in that information. The event services represent the monitored data using eXtensible Markup Language (XML).

- **Heartbeart Monitor [3]:** The Globus Heatbeat Monitor (HBM) was designed to provide a mechanism for monitoring the state of processes and notifying the failure of them. It allows simultaneous monitoring both Globus system processes and application processes.
- **Netlogger [4]:** NetLogger proposes a distributed system which contains a central daemon called *netlogd* which receives information about the usage of specific resources (network, memory, cpu, etc.), from the applications in the system. On the other side, the applications are instrumented, using the NetLogger API, to generate log lines in the Universal format for Logger Messages (ULM), which contain the values about the monitorization of the usage of the resources during the execution of a set of commands. This project is mainly oriented to the analysis of high performance applications.
- **GMS [5]:** GMS supports resource monitoring, visualizing, and accounting. GMS was developed on top of existing Globus CORE. The system has been successfully deployed across the Nanyang Campus Grid. The system is able to capture Grid jobs information, organize and store the information into a relational database, and support analyzing and visualization of the data through the Web.
- **DGAS [6]:** It was originally developed within the EDG project and is now being maintained and re-engineered within the EGEE project. The Purpose of DGAS is to implement Resource Usage Metering, Accounting and Account Balancing (through resource pricing) in a fully distributed Grid environment. It is conceived to be distributed, secure and extensible.
- **APEL [7]:** APEL (Accounting Processor for Event Logs) parses batch, system and gatekeeper logs generated by a site and builds accounting records, which provide a summary of the resources consumed based on attributes such as CPU time, Wall Clock Time, Memory and grid user DN. The collection of accounting usage records is done through R-GMA. Each site publishes its own accounting data using an R-GMA primary producer using its locally assigned R-GMA server. To collect the data from all participating sites, data is streamed to a centralised database via a secondary producer. It is developed in LCG Project and used in the EGEE project.

Most of the projects commented above are mainly oriented to traditional computational Grids in which CPU usage, memory, disk capacity, load of the system or jobs in execution are the parameters that worth measuring. Nevertheless in a Grid environment also happens other events that should be tracked, such as *who accedes a services, when does it*, etc. The tools which are currently being deployed are not useful for these purpose, as they are oriented to monitor the system and not to track the services.

Moreover, current Grid trends are oriented to architectures based on the provision of general services. This means that shared resources are more heterogeneous, as they are others than CPU cycles or storage capacity. The current Grid environments are built by deploying general services, which need general *logging* facilities for enabling the track of the whole system.

Currently, there is a lack of support for these general services. The traditional systems uses *log* files to register any information which is generated by the running applications or the operations in a service from the Operative System. In this sense, Syslog [8] is the most commonly used system logging utility for UNIX systems. The applications register significant information in *log* files, which are classified by the configuration of the *log daemon*. The saved data is useful for analyzing the state of system, extracting statistics of usage or guess who has misused a resource, for instance.

3   DiLoS Architecture

The DiLoS architecture is composed by two kinds of elements. On one hand, the services which provide the log data and need to be integrated into the DiLoS architecture, and on the other hand a specific service called "Log" which will coordinate the distributed information.

The services which provide the *logs* are likely to be organized according to any distribution (such as hierarchical, cyclical, plane, etc.). Regarding the specific organization, the DiLoS Log Services (DLS) are distributed through system, so that they are properly acceded by the other services. Figure 1 outlines the architecture proposed by DiLoS, and some of the functional use cases which may happen in the system.

Each DLS is in charge of gathering the information about a set of services, which are under its scope. The scope of each DLS is defined according to the specific deployment. So, DLS do not need to be installed on each node, as each of them may gather the log information from many services. The services which provide the logs have to be configured in such way that they are able to accede the DLS that is in charge of it.

Each service saves log data in a local repository using the LOG operation. Also the DLS may have their local log repository (a DLS behaves as any other services in the system in which it is deployed).

Periodically, the services send their local registries to the DLS which is in charge of gathering its information (using the PUSH operation). Then, the DLS stores the registries of the service into a General Log repository, in order to integrate them into the Distributed Log of the Grid system.

Notwithstanding the services are the only responsible to send the information to the DLS, the data integration operation may also be initiated from the upper level, in order to *flush* the local Log repositories. The DLS would call its services for their log registries (PULL operation). This operation is part of the protocol which suggest the services to start their PUSH operation. Nevertheless PULL and PUSH operations are asynchronous. Moreover, it is not compulsory for the services to PUSH data to a DLS as a response of a PULL operation (i.e. it might not have new log entries).

On the other side, in order to provide a redundance of the information, the General Repositories managed by the DLS are also sent to other DLS which are into another scope.

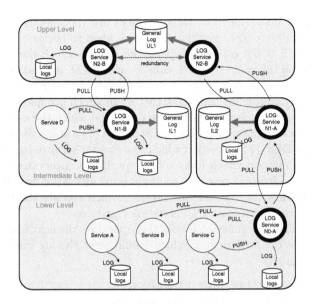

Fig. 1. DiLoS functional architecture

The architecture considers the possibility of having more than one instance of DLS for each scope, in order to enhance performance, backup capacity, load balancing and redundancy. When a service is ready to send their log registries to the DLS, the service itself is the only responsible of deciding the instance to which is going to PUSH them.

In order to integrate the services into the DiLoS architecture, the interfaces would implement the next operations, which are also provided by the API:

- PULL: It is a public method included in general services, but also in DLSs. It is exclusively invoked by a DiLoS Log service, in order to suggest the service which implements it to send the log registries to the corresponding DLS service, deployed in DiLoS architecture.
- PUSH: It is a public method implemented only by DLSs. It is invoked by any service when they need to transmit its registries to the log service. The registries PUSHed are not changed by the DLS, as the main purpose is to store them into the general repository. The connection between different DLSs is also carried out using this operation.
- LOG: It is a private method which is implemented by general services. It is called by the service itself to locally save a log. Later it would be transmitted to DiLoS log services using PUSH operations.
- QUERY: This public method returns a set of logs which accomplish with a pattern (date, user, name service).

Although the protocol for relaying the logs to a DLS is created by the pair of calls PULL/PUSH, it is important to remark that these operations are asynchronous: the DiLoS log services would require the services for log registries by using the PULL operation, but these services are the effective responsible of deciding whether to send or not the data, and when to perform the operation.

Furthermore, it is not compulsory that the DLS to which the services PUSH the logs would be the same to which calls for the PULL operation. The services are the responsible of deciding the DLS to which send their registries, by using a specific discovery system for the infrastructure in which is deployed. The discovery of the DLS services is modelled by an operation which the user may provide for its particular organization. Some examples are (1) a static file which contains the URIs of the DLS, (2) a function which searches for services in a Globus Index Service or (3) a method which links with a specific discovery architecture.

The DiLoS architecture provides a simple implementation for DLS discovery, which is based on static files for any service, which contain the list of possible DLSs to which the service can PUSH its log registries.

In order to clarify the functionality of this operation the next fragment exposes the pseudo code of the function, which would send the log entries, using the discovery system.

```
Procedure Relay (log_block)
        LLS = Discover_log_services
        If ( is empty (LLS) ) Then
                Abort_operation
        Else
                LLS(i).PUSH(log_block)
                Save_reference_log_service(LLS(i))
        End if
End Procedure
```

The usage of the specific discovery system is the key for connecting the DiLoS architecture to any particular Grid infrastructure. When the services are deployed, they have to be configured so that they create the proper organization which may be reflected in the system. As an example, a local department may use a DLS while another department should use its specific DLS; so, the services under each scope should be configured in such way that they discover its corresponding DLS when they try to PUSH their block of log registries.

3.1   Logging Policy

According to the structure proposed by DiLoS architecture, it may seem that it tries to centralise the information into a General Log repository. Nevertheless, it is only an extreme which would be possible, by applying the facilities provided by the architecture.

The effective owners of the log registries are the services themselves, as they are in a Grid environment. Nevertheless, the usage of DLS provides mechanisms for these services to delegate the storage of such information to the DiLoS Log Services, which would be part of the Grid infrastructure.

Furthermore, each service is the effective responsible of sending the information to the DLSs, but also deciding which kind of registries are going to be sent to these Log Services. In fact, the DLS are introduced for modelling situations such as backup of information or providing more storage capacity.

In some cases, the laws also enforce the electronic transactions to be stored by some entities (such as the LSSI in Spain [9]). The DLS would also be useful for implementing such policies, as the Log services may be associated to the authorities which may store the information.

Nevertheless, it is possible to isolate a set of data (preventing its relay to other scopes) from other levels, where it may be useless.

3.2 Data Saved in a Log

Each service deployed in a Grid environment is the responsible of deciding which kind of information may be registered, and thus saved into the *log file*. In this sense, in order to provide support for general services, the DiLoS architecture does not force to store a strict kind of information. This decission is supported by the fact of the deployers of the services are also the responsible of providing applications to interpret the information that is registered.

So, the DiLoS architecture allows registering any type of log information. Nevertheless, the DiLoS architecture defines three basic fields which are needed to be stored for the integration and querying for data, but also an extended field in which the services can save their data in their specific format

Every registry stored by the DiLoS log system is composed by the next fields:

1. User identity: it is the identity of the effective user which has called the service, and thus is the responsible of the operation.
2. Time stamp: it defines the time and date when the service calls the LOG function.
3. Service identification: it can be a particular identification for the service which saves the log line. It is important that this field is interpreted according to the specific system deployment. An example of deployment would use WS-naming [10] as a method for services identification.
4. LOG part: it is the extended field which encapsulates the data to be register by the service.

3.3 Use Cases

The DiLoS architecture defines the interface, the architecture and the protocol which would be used for implementing a Distributed Logging System. Nevertheless, there may be a large variety of particular deployments of the DiLoS architecture, which would implement the particularities of each system. The figure 2 explains a general example of use case.

1. At first, a general service performs an operation, and it needs to register this occurrence. So, the service uses LOG interface to save a log line in its local repository (a log file).
2. A log service called N1-B, which is placed in a upper level, requires gathering the log registries of service D. So, the log service N1-B invokes the

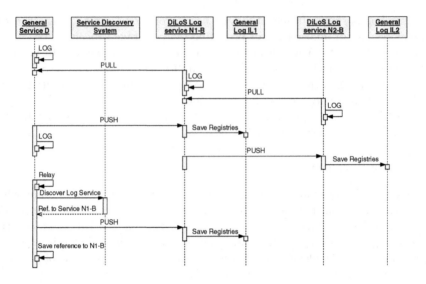

Fig. 2. DiLoS sequence diagram. Interaction among services in DiLoS Architecture.

PULL operation over the service D. As any other general service it can call its internal LOG function to store some information about its operations performed. After some time, the service D calls the PUSH interface of log service N1-B. In this way, a set of registries are sent and stored in a general log repository by the log service N1-B.

3. Moreover, the service N2-B located at level 2 needs collect data from another log services placed at lower levels. The service N2-B can search for a DLS through the discovery system. The procedure continues, as the second point calling PULL operation over the service N1-B.

4. The local repository of the general service D is full, and it is needed to perform a backup of the log entries. The service D initializes the RELAY procedure, which will discover the DLS to send the block of registries, and will perform the PUSH operation.

These would be the description of the steps which would be carried on in a general use case. It is possible to summarize the use cases in the next categories:

1. Only store and not relaying (single mode): the service store log data in a local repository (LOG), but it does not use the DLS to scatter its registries. The service works as an independent entity.

2. Relay but not storing (diskless mode): the services relays every log line generated (PUSH). The information is delegated to one or several log services.

3. Store and relay all the data (backup mode): in this mode, the DLS services work as backup systems. As in the first case all the log data is stored locally, but the information is also sent to a DLS (LOG and PUSH) in order to have a backup copy for its later recovery.

4. Store a set and relay the rest (selected mode): when the services decide that a particular information do not has to be sent to any other entity, it is stored locally this data. In other case, the information is PUSHed into another DLS (LOG and PUSH selected logs).

3.4   Application Models

The main purpose of the DiLoS architecture is to provide a General Log System for general Grid services. Any application (service, daemon, activity system, etc.) generates a set of useful information which has to be available for several reasons. DiLoS cover the requirements of most of the log models in current Grid environments, performing the specific configurations.

Some of the purposes for which this architecture would be applicable are:

– Audit procedures: It is important to collect information about a period of time with the aim of identifying the global state of the system. The DiLoS architecture provides mechanisms for storing the identity of the users which use the services, and thus identifying what happened at each moment.
– Accounting: in a Grid environment can exist specific services which provide resources to the user. These resources may be meassured, and annotated who has used them for later stablishing a price of usage and thus creating a model of exploitation. The services can use the DiLoS log services to registry log lines including information about the executed jobs, time of cpu spent for each one, disk quota, permitted or deny access, etc. and any log line would be associated to the identity of the user who has called the service.
– Debugging: An application can employ DiLoS to analyze logs, and deciding whether the behaviour of the application is what was expected or not, and thus correcting it.

4   A Particular Implementation: The gCitizen Project

As commented in previous sections, the DiLoS system can be adapted to most of the service oriented Grid deployments. The key issue is to provide the specific implementation of the functions which decide where common services have to send their log entries. As an example it is described the DiLoS customization for the middleware developed in the gCitizen project [11].

The aim of the gCitizen project is the creation of a Grid middleware for the transparent management of the information about citizens in the public administration, and the administrative procedures in which the citizens are involved, independently of the point of entry to the Administration (understood as a global entity). The vision of the project is outlined in the Figure 3.

The gCitizen project uses the Globus Toolkit 4 implementation of the WSRF standard, adding some new components which complete the architecture in order to use it in an eGovernment framework.

One of the main components of gCitizen is a General Addressing Convention (GAC), which is used to identify the services in the system. GAC follows a

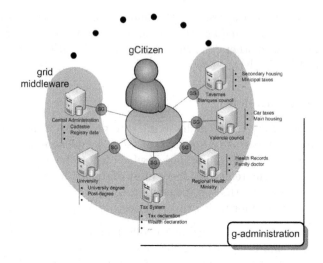

Fig. 3. gCitizen Project

similar approach to the LDAP Distinguished Names [12,13], using the standard DN fields as C, ST, O, OU, etc. and some other additional fields added to complete the hierarchy in the public administration. It uses the DN to identify the services, but also provides some semantic information about its functionality and the hierarchical location in the gCitizen system. An example of the DNs used in the gCitizen project is:

```
/C=ES/A=Valencia/ST=Valencia/O=Diputacion de Valencia/
                OU=Gestion/CN=Padron
```

In order to find the services using this addressing convention, the gCitizen project provides the Distributed Discovery Architecture (DiDA), which enables to obtain the physical location of a service, using its DN.

The DiLoS architecture has been used in the gCitizen project, applying the DiDA discovery system and the GAC convention in order to guess the proper DLS to which each service may send its log entries.

The most suitable distribution of DiLoS services in gCitizen is to set up one DLS at each administrative level, and another one for each department. Nevertheless, some of these sections would not be able to deploy the corresponding DLS, and thus would delegate the responsibility to the DLS at upper levels.

The DiLoS architecture makes easy the customization, by detaching the discovery of the corresponding DLS as a function which may be provided by the user. In this sense, a function has been specifically created for gCitizen, making use of the GAC and the DiDA architecture.

According to the distribution of the DLS, the name of the one to which a service must send its logs is obtained by changing the CN from its DN, and using the "CN=Log" instead. For the service shown previously the name of the DiLoS service of the same level it would be:

```
/C=ES/A=Valencia/ST=Valencia/O=Diputacion de Valencia/
              OU=Gestion/CN=Log
```

If this service does not exist or is unavailable, the service must go up in the levels of the hierarchy provided by its DN, searching a DLS at an upper organizational level. This scheme must be followed until it finds an available DLS or it reaches the top of the hierarchy:

```
/C=ES/A=Valencia/ST=Valencia/CN=Log
```

When a DLS needs to use the PULL operation, it must remove the CN from its DN, and contact every service which matches the remaining DN. As an example, the previous DLS would try to use the next pattern:

```
/C=ES/A=Valencia/ST=Valencia/*
```

The services in gCitizen log the external calls, the user who has performed the operation, a timestamp which indicates when it was carried out, and some specific information for each operation.

5   Conclusions and Further Work

Most of the current Grid monitoring developments are oriented to the registration of the information regarding to computing services among the Grid. Nevertheless, there is a lack of support for registering the activity in Grid deployments composed by general services.

In this sense, the DiLoS architecture has been developed. It provides a protocol and an architecture for scattering the logging information among distinct scopes in the Grid. In this sense, the DiLoS system provides elements for backing up the information, delegating its storage, etc. for the services which are deployed in the Grid.

Notwithstanding these facilities, the services are the effective owners of the information, and thus they are the responsible of deciding whether to send or not to the DiLoS Log services which are part of the logging infrastructure.

This is a complete architecture with a well defined protocol and responsibilities. It also provides an implementation which would cover common infrastructures. In order to ease the process, it is based on a discovery mechanism, which is the key to addapt DiLoS to almost any Grid organization.

Nevertheless, there are some issues which should be enhanced in order to complete the system. As an example, the recovery of the information needs a revision, in order to establish a protocol for gathering all the information which is scattered in the system. Currently, the services need to recover it from upper services, querying them about the registries generated by itself.

Another issue which would be interesting to be studied is the usage of standard formats for creating the log entries. As an example, it would be interesting to use XML further than the current *plain text* format.

Acknowledgments

The authors wish to thank the financial support received from the "Ministerio de Industria, Turismo y Comercio" to develop the project with reference FIT-350101-2004-54.

References

1. Zoltán Balaton, Peter Kacsuk, Norbert Podhorszki, and Ferenc Vajda. Comparison of representative grid monitoring tools. http://www.lpds.sztaki.hu/publications/reports/lpds-2-2000.pdf, 2000.
2. A. Waheed, W. Smith, J. George, and J. Yan. An infrastructure for monitoring and management in computational grids. In S. Dwarkadas, editor, *Proceedings of the 5th International Workshop on Languages, Compilers, and Run-Time Systems for Scalable Computers (LCR 2000)*, page 235, Rochester, NY, USA, May 2000.
3. The Globus Alliance. The globus heartbeat monitor specification. http://www-fp.globus.org/hbm/heartbeat_spec.html.
4. Brian Tierney, William E. Johnston, Brian Crowley, Gary Hoo, Chris Brooks, and Dan Gunter. The netlogger methodology for high performance distributed systems performance analysis. In *HPDC*, pages 260–267, 1998.
5. Hee-Khiang Ng, Quoc-Thuan Ho, Bu-Sung Lee, Dudy Lim, Yew-Soon Ong, and Wentong Cai. Nanyang campus inter-organization grid monitoring system. http://ntu-cg.ntu.edu.sg/pub/GMS.pdf, 2005.
6. Cosimo Anglano, Stefano Barale, Luciano Gaido, Andrea Guarise, Giuseppe Patania, Rosario M. Piro, and Albert Werbrouck. The distributed grid accounting system (dgas). http://www.to.infn.it/grid/accounting/main.html, 2004.
7. Rob Byrom, Roney Cordenonsib, Linda Cornwall, Martin Craig, Abdeslem Djaoui, Alastair Duncan, Steve Fisher, John Gordon, Steve Hicks, Dave Kant, Jason Leakec, Robin Middleton, Matt Thorpe, and Antony Wilson. John Walk. Apel: An implementation of grid accounting using r-gma. http://www.gridpp.ac.uk/abstracts/allhands2005/ahm05_rgma.pdf, 2005.
8. C. Lonvick. The BSD Syslog protocol. RFC 3164, Internet Engineering Task Force (IETF), 2001.
9. Ministerio de Industria Turismo y Comercio. Ley de Servicios de la Sociedad de la Información y de comercio electrónico. http://www.lssi.es, 2002.
10. Andrew Grimshaw and Manuel Pereira. OGSA naming working group. https://forge.gridforum.org/projects/ogsa-naming-wg, 2005.
11. Carlos de Alfonso, Miguel Caballer, and Vicente Hernández. gCitizen, Grid Technology for eGovernment Systems Integration. In *Proceedings of IADIS International Conference e-Commerce 2005*, pages 321–324, 2005.
12. W. Yeong, T. Howes, and S. Kille. Lightweight directory access protocol. RFC 1777, Internet Engineering Task Force (IETF), 1995.
13. S. Kille. A string representation of distinguished names. RFC 1779, Internet Engineering Task Force (IETF), 1995.

Interoperability Between UNICORE and ITBL

Yoshio Suzuki[1], Takahiro Minami[1], Masayuki Tani[1], Norihiro Nakajima[1],
Rainer Keller[2], and Thomas Beisel[2]

[1] Center for Computational Science and E-systems,
Japan Atomic Energy Agency
6-9-3 Higashi-Ueno, Taito-ku, Tokyo 110-0015, Japan
{suzuki.yoshio,minami.takahiro,tani.masayuki,
nakajima.norihiro}@jaea.go.jp
[2] High Performance Computing Center Stuttgart,
Universität Stuttgart
70550 Stuttgart, Germany
{keller, beisel}@hlrs.de

Abstract. The interoperability among different science grid systems is indispensable to worldwide use of a large-scale experimental facility as well as a large-scale supercomputer. One of the simplest ways to achieve the interoperability is to convert message among different science grid systems without modifying themselves. Under such consideration, the interoperability between UNICORE and ITBL (IT-Based Laboratory) has been achieved without modifying these grid systems by adopting a connection server which works as a mediator. Until international standardization is established, the method of message conversion among different science grid systems is promising as a way to establish the interoperability.

1 Introduction

Recently, there are some scientific global projects such as the International Thermonuclear Experimental Reactor (ITER) as well as the Large Hadron Collider (LHC). As the increase of global projects, the worldwide science grid environment which enables worldwide use of such a large-scale experimental facility becomes more necessary. To construct such an environment, the interoperability among different science grid systems is indispensable. A worldwide science grid environment is also expected to worldwide use of a large-scale supercomputer. In Japan, the national project (hereafter peta-scale supercomputer project) to develop and utilize the leading edge and multipurpose supercomputer begins in April 2006, where a 10 PFlops supercomputer is planned to be available around year 2010. The development and improvement of grid middleware is also situated in this project.

So far, as the most promising way to achieve international interoperability among different science grid systems, research and development of international standardization of grid systems have been promoted. The establishment of standardization makes it easier to interoperate different science grid systems. It also contributes to the enhancement of a grid system because each component of different grid systems can cooperate more flexibly with each other. However, since different types of grid system have already constructed by adopting various architectures in all the world, the

M. Daydé et al. (Eds.): VECPAR 2006, LNCS 4395, pp. 601–609, 2007.
© Springer-Verlag Berlin Heidelberg 2007

standardization of all these systems demands their modification, which prevents users from continuing the development of their applications on these grid systems.

It is critical to keep operating existing science grid environments in which the users continuously develop and execute their applications. One of the simplest ways to achieve the interoperability with keeping their operations is to convert message among different grid systems without modifying themselves. Under such consideration, the interoperability between UNICORE and IT-Based Laboratory (ITBL) [1] has been tried.

ITBL project is a national project placed as one of the e-Japan Priority Policy Program to realize the e-Japan Strategy which sets goals to make Japan the world's most advanced IT nation. ITBL project was launched at April 2001 by six institutes: the National Institute for Materials Science (NIMS), the National Research Institute for Earth Science and Disaster Prevention (NIED), Japan Aerospace Exploration Agency (JAXA), the Institute of Physical and Chemical Research (known as RIKEN), Japan Science and Technology Agency (JST), and Japan Atomic Energy Agency (JAEA) and has been carried out to March 2006 as 5 years' plan. The objective of ITBL project is to establish virtual laboratories in which researchers in various disciplines can collaboratively develop highly sophisticated simulation systems by fully utilizing computer resources located in high-speed network. To achieve this, ITBL project has been classified roughly into 4 activities; development of applications, development of system infrastructure software, maintenance of shared facility including a supercomputer, and promotion activity.

Center for Computational Science and E-systems of Japan Atomic Energy Agency (CCSE/JAEA) has developed ITBL middleware as ITBL system infrastructure software and has operated it since April 2003. Tools of ITBL middleware enable secure communication among plural supercomputers via internet (STARPC: Seamless Thinking Aide Remote Procedure Call), parallel computation between different types of supercomputer (STAMPI: Seamless Thinking Aide Message Passing Interface), job control on distributed computational environment (TME: Task Mapping Editor), parallel and distributed visualization (AVS/ITBL) and so on [1]. Researchers can use these tools on the Web browser.

Figure 1 shows the overview of the ITBL environment. At the end of the project (March 2006), 680 researchers from 89 organizations participate in ITBL project and ITBL middleware integrates about 1/5 (45 TFlops) of all the computer resources owned by the institutes and the universities for scientific research in Japan. These resources consist of 26 computers from 13 sites (10 organizations and 3 areas of JAEA). CCSE/JAEA keeps operating ITBL middleware to maintain the ITBL environment and contributes to the establishment of a backbone grid environment in the peta-scale supercomputer project launched at April 2006.

Uniform Interface to Computing Resources (UNICORE) [2] is one of the representative grid middleware. It has been developed in UNICORE project and UNICORE Plus project [2] of Germany and improved in EUROGRID project [3] and GRID project [4] of EU and now further improved in UniGrids project [5]. And UNICORE is used in High Performance Computing Center in Stuttgart (HLRS), Germany with which CCSE/JAEA has collaborates since 1999. Thus we have selected UNICORE as a first step towards international interoperability with ITBL.

Also in Japan, UNICORE has been used in National Research Grid Initiative (NAREGI) project [6]. This project has been carried out from April 2003 to March 2006, aimed at developing infrastructure software of sufficient quality to become an international standard. This infrastructure software is also expected to contribute to the development and improvement of grid middleware in the peta-scale supercomputer project. Therefore, the interoperability between UNICORE and ITBL is also meaningful to enhance the Japanese science grid environment.

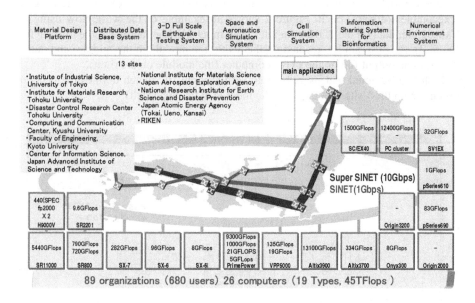

Fig. 1. Overview of the ITBL environment. 680 users from 89 organizations share 26 computers containing 19 types ones from 13 sites (11 organizations and 3 areas of JAEA). Main applications developed in ITBL project are also shown.

2 Main Function

As a main function of the interoperability between UNICORE and ITBL, we have achieved a job submission with each other. Namely, ITBL users can submit a job to any computer under UNICORE (Vsite) by using Task Mapping Editor (TME) [7, 8] which is a work flow tool on ITBL middleware. And UNICORE users can submit a job to any computer under ITBL (ITBL computer) by using a client graphic user interface (GUI) of UNICORE.

TME has a function to define, execute and monitor a scenario of job submissions, program executions and input/output file transfers over ITBL computers. Most of the applications developed on ITBL use the function of TME. Therefore, most of ITBL users can use any Vsite in the same manner. We have installed the following functions.

– Conversion of request from TME into Abstract Job Object (AJO)
– Submission of converted AJO into UNICORE

– Mutual transfer of input/output file
– Monitoring a program execution

Here, AJO is the Java object created when the job is submitted from the client GUI of UNICORE.

TME has the GUI to define programs and input/output files as an individual module and to connect them to be one scenario. Using the module which defines the program (program execution module), users can specify the computer to execute the program. In the same manner, users can specify the Vsite and execute the program (executable software) on it. Figure 2 shows an example of the work flow using the Vsite. Here, data on the ITBL computer is an input file (the module at the lower left in figure 2), the program is executed on the Vsite (the module in the middle) and data is output as an output file on the ITBL computer (the module on the right).

It should be noted that users have to set the input file which defines the program execution on the Vsite as a module (the module at the upper left).

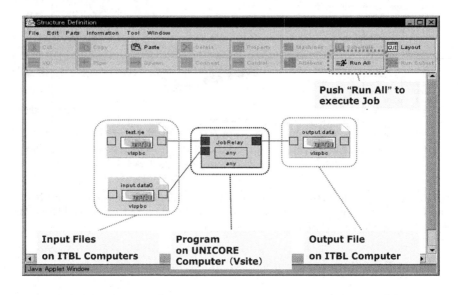

Fig. 2. The GUI of TME. A program (executable software) executed on UNICORE computer (Vsite) is defined as a module of TME.

Inversely, UNICORE users can use any ITBL computer. We have installed the following functions.

– Recognition of ITBL site as an UNICORE site (Usite)
– Conversion of AJO into ITBL's request
– Mutual transfer of input/output file
– Monitoring a program execution

Figure 3 shows an example of the work flow using the ITBL computer from the client GUI of UNICORE. Here, data on the Vsite is an input file (the module at the upper in figure 3), the program is executed on the ITBL computer (the module in the middle) and data is output as an output file on the Vsite (the module at the bottom).

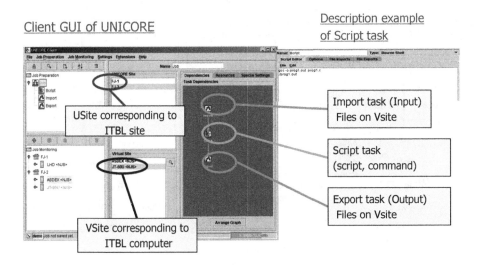

Fig. 3. The GUI of UNICORE client and the description example of script task

3 Architecture

To achieve the above function without modifying architectures of UNICORE middleware and ITBL middleware, we have adopted a connection server which works as a mediator between these two systems. Both in UNICORE and ITBL, each site has one gateway server respectively. Thus, it is appropriate to prepare the virtual server to enable to change the way of accessing the computers.

The architecture to achieve the interoperability is shown in Figure 4. To control the job on any Vsite from the client GUI of ITBL (TME), we installed in the connection server "UNICORE collaboration tool" which consists of "Job Relay Servlet", "UNICORE Collaboration Interface" and "UNICORE Collaboration Module". Here is the brief process to submit the job from the client GUI of ITBL to the Vsite.

1. The request from the client GUI of ITBL to an Usite is received by the Job Relay Servlet.
2. The Job Relay Servlet calls the UNICORE Collaboration interface on the connection server by way of RMI (Remote Method Invocation) according to the request.
3. The ITBL certificate is connected with the UNICORE certificate.
4. The UNICORE Collaboration Module generates AJO according to the request information, the authentication using this UNICORE certificate is performend on the UNICORE server, and then the AJO is submitted to the Usite. After processing at the Usite, a result of the AJO request is returned to UNICORE Collaboration Module.
5. The result is relayed to Job Relay Servlet, and is returned to the client GUI.

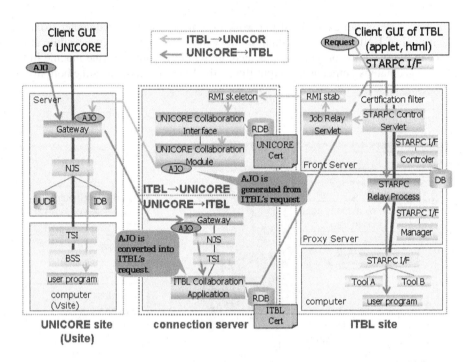

Fig. 4. Architecture of interoperability between UNICORE middleware and ITBLmiddleware

Inversely, in order to control the job on ITBL computers from the client GUI of UNICORE, we installed in the connection server the similar system to the UNICORE and "ITBL collaboration application" that manages job submission, job cancellation and status acquisition on ITBL computers based on the information from TSI (Target System Interface).

As the process to execute the program on the ITBL computer, the following procedures are executed in the connection server:

1. Gateway checks the certification attached to AJO and transfers the request to NJS (Network Job Supervisor) which is located inside the firewall.
2. NJS receives AJO certified by Gateway, checks with UUDB (UNICORE User Data Base) whether it is possible to login to the computer resource, converts (embodies) AJO into the job executable on the computer and forwards the request to TSI.
3. ITBL collaboration application converts information from TSI in order to submit the job to the ITBL computer.
4. The UNICORE certificate is connected with the ITBL certificate, and then the authentication using this ITBL certificate is performed on the ITBL server.
5. The job is submitted to the ITBL computer.
6. The result is returned to the client GUI of UNICORE.

Here, procedures from 1 to 3 are the same as those when a program is executed in UNICORE.

Finally in this section, we mention the certification method. Both UNICORE and ITBL use X.509 certificate. As the realization of this interoperability, users can use two grid environments with single sign-on, if they are allowed to use both systems (namely they have to have both certificates). Shown in the previous procedures, when ITBL users access the Vsite, ITBL certificate is converted to UNICORE certificate by UNICORE Collaboration Module. On the other hand, when UNICORE users access the ITBL computer, UNICORE certificate is converted to ITBL certificate by ITBL Collaboration Application.

ITBL adopts HTTPS protocol when users access the ITBL (Front) Server located in DMZ (DeMilitarized Zone). The connection server is also located in the same DMZ. SSL protocol is adopted for the communication between UNICORE server and communication server. Consequently, the secure communication based on SSL encryption is guaranteed.

4   Examination

We evaluate the time for job submission between two systems. To evaluate that, we use the 'date' command described in a shell script as the executing job. In addition, we use the same machine for Vsite and ITBL computer. Table 1 shows the time from beginning the operation for job submission to getting the message of job completion for four cases. In case 1 (using only UNICORE), it takes 9 seconds totally. Since the time of job execution is less than 1 second (shown in case 3), it takes about 7 seconds to get the message of job completion. In case 2 (UNICORE to ITBL), it takes 56 seconds from job submission to beginning of job execution (procedure 1, 2, 3 and 4 mentioned in the previous section). This is mainly for the authentication to access ITBL (procedure 4 mentioned in the previous section). It is 72 seconds from the beginning of job execution to getting message of job completion. The reason why it takes longer is that the message of job completion got in ITBL is converted to AJO.

Table 1. Evaluation of the time for job submission between UNICORE and ITBL

	Job submission	Beginning of job execution	Getting message of job completion
1. UNICORE only	00:01	00:02	00:09
2. UNICORE to ITBL	00:01	00:57	02:09
3. ITBL only	00:03	-	00:03
4. ITBL to UNICORE	00:10	-	01:08

It takes longer time for job submission from ITBL to UNICORE based on the similar reason. The reason why the case2 is longer than the case4 is that it takes longer time for ITBL to authenticate the certificate. Since these times are enough shorter than that of the usual job execution as well as the data transfer, these delays do not become the bottleneck. The huge data transfer caused by large-scale simulations via Internet is serious problem regardless of interoperability. This is beyond the scope of this paper and requires further study.

To further evaluate the constructed functions we have applied them into UNICORE of HLRS, Germany under the international cooperation between HLRS and CCSE/JAEA. As a result, the science grid environment enabling to interoperate SX-8 of HLRS and the 26 computers of ITBL has been constructed. Now we have been installing the assembled-structure analysis program (we call it ITBLFEM) in 4 computers: Altix3700Bx2 in Tokai Research and Development Center of JAEA, SX-6 and pSeries690 in CCSE of JAEA, and SX-8 in HLRS. ITBLFEM has been developed on ITBL environment to contribute to the safety of the nuclear power plant in case of extra large-scale earthquake. The results will be described in our future paper.

5 Related Works

There have been the GRIP and the UniGrids as related works. The GRIP has been carried out for 2 years from January 2002 to December 2003. The UniGrids, which is placed as one of the 6th framework program Specific Targeted Research Project (STREP), has been carried out for 2 years from July 2004. These projects aim at the realization of the interoperability of UNICORE and Globus [9]. The research and development for standardization have also been addressed. Both projects enable UNICORE users to utilize computers managed by Globus. Here, the standardization of these systems (especially for UNICORE in the GRIP) demands their modification.

6 Summary

We have achieved the interoperability between UNICORE and ITBL by adopting a connection server which works as a mediator between these different grid systems. The advantage of this method is that both systems need not be modified. By realizing the interoperability, ITBL users can control a job on any UNICORE computer from the GUI of TME, which is the work flow tool of ITBL, and UNICORE users can control a job on any ITBL computer from the client GUI of UNICORE.

Applying this function to UNICORE in HLRS of Germany, we have constructed the science grid environment interoperating the SX-8 in HLRS and SX-6, pSeries690, and Altix3900 in JAEA. The method of message conversion among different science grid systems is promising as a way to establish the interoperability until international standardization is established.

Acknowledgement

The authors thank Dr. Resch (HLRS), Dr. Yagawa (CCSE/JAEA), Dr. Hirayama (CCSE/JAEA), and Dr. Aoyagi (CCSE/JAEA) for their insightful advice. Part of this research carried out under the international cooperation between Center for Computational Science and E-systems in JAEA (CCSE/JAEA) and High Performance Computing Center in Stuttgart (HLRS).

References

1. Kenji Higuchi, Toshiyuki Imamura, Yoshio Suzuki, Futoshi Shimizu, Masahiko. Machida, Takayuki Otani, Yasuhiro Hasegawa, Norihiro Yamagishi, Kazuyuki Kimura, Tetsuo Aoyagi, Norihiro Nakajima, Masahiro Fukuda and Genki Yagawa: Grid Computing Supporting System on ITBL Project. High Performance Computing, Veidenbaum et al. (Eds.) 5th International Symposium ISHPC2003 (20-22 October 2003, Tokyo-Odaiba, Japan, Proceedings), LNCS2858 (2003) 245-257
2. UNICORE FORUM http://www.unicore.org/
3. EUROGRID PROJECT http://www.eurogrid.org/
4. GRID INTEROPERABILITY PROJECT http://www.grid-interoperability.org/
5. UNIGRID PROJECT http://www.unigrids.org/
6. NAREGI PROJECT http://www.naregi.org/index_e.html
7. Toshiyuki Imamura Yukihiro Hasegawa, Nobuhiro Yamagishi and Hiroshi Takemiya: TME: A Distributed resource handling tool. Recent Advances in Computational Science & Engineering, International Conference on Scientific & Engineering Computation (IC-SEC) (3-5 December 2002, Raffles City Convention Centre, Singapore) (2002) 789-792
8. Y. Suzuki, N. Matsumoto, N. Yamagishi, K. Higuchi, T. Otani, H. Nagai, H. Terada, A. Furuno, M. Chino and T. Kobayashi: Development of Multiple Job Execution and Visualization System on ITBL System Infrastructure Software and Its Utilization for Parametric Studies in Environmental Modeling. Computational Science - ICCS 2003 Sloot et al. (Eds.) International Conference (2-4 June 2003, Melbourne, Australia and St. Petersburg, Russia, Proceedings, Part III), LNCS2659 (2003) 120-129
9. The Globus Alliance http://www.globus.org/

Using Failure Injection Mechanisms to Experiment and Evaluate a Grid Failure Detector

Sébastien Monnet[1] and Marin Bertier[2]

[1] IRISA/University of Rennes I
Sebastien.Monnet@irisa.fr
[2] IRISA/INSA
Marin.Bertier@irisa.fr

Abstract. Computing grids are large-scale, highly-distributed, often hierarchical, platforms. At such scales, failures are no longer exceptions, but part of the normal behavior. When designing software for grids, developers have to take failures into account. It is crucial to make experiments at a large scale, with various volatility conditions, in order to measure the impact of failures on the whole system. This paper presents an experimental tool allowing the user to inject failures during a practical evaluation of fault-tolerant systems. We illustrate the usefulness of our tool through an evaluation of a hierarchical grid failure detector.

Keywords: Failure injection, failure detection, performance evaluation, fault tolerance, grid computing.

1 Introduction

A current trend in high-performance computing is the use of large-scale computing grids. These platforms consist of geographically distributed cluster federations gathering thousands of nodes. At this scale, node and network failures are no more exceptions, but belong to the normal system behavior. Thus grid applications must tolerate failures and their evaluation should take reaction to failures into account.

To be able to evaluate a fault-tolerant application, it is essential to test how the application reacts to failures. But such applications are often non deterministic and failures are not predictable. However, an extensive experimental evaluation requires execution reproducibility.

In this paper, we introduce a failure injection tool able to express and reproduce various failure injection scenarios. This provides the ability to extensively evaluate fault-tolerance mechanisms used by distributed applications. As an illustration, we use this tool to evaluate a failure detector service adapted to the grid architecture [7]. A failure detector is a well-known basic building block for fault-tolerant distributed systems, since most fault-tolerance mechanisms require to be notified about failures. These experiments are run over the Grid'5000 National French grid platform [1].

The remainder of this paper is composed as follows: Section 2 motivates the need of failure injection mechanisms. Section 3 describes our failure injection tool and explains how to use it. Section 4 presents the failure detector we evaluate. Section 5 illustrates the usage of our failure injection tool for a practical evaluation of the failure detector. Finally, Section 6 concludes the paper.

M. Daydé et al. (Eds.): VECPAR 2006, LNCS 4395, pp. 610–621, 2007.
© Springer-Verlag Berlin Heidelberg 2007

2   Experimenting with Various Volatility Conditions

2.1   System Model

In this paper, we suppose that the grid architecture of the system implies a hierarchical organization. It consists of clusters of nodes with high-connectivity links, typically System Area Networks (SAN) or Local Area Networks (LAN), interconnected by a global network, such as Internet. In this context, we call *local group* each pool of processes running within the same SAN or LAN, and *global network* the network which connects the local groups.

Each local group is a finite set of processes that are spread throughout a local network. The distributed system is composed of a finite set of local groups. Every process communicates only by sending and receiving messages. All processes are assumed to have a preliminary knowledge of the system's organization.

We rely on the model of partial synchrony proposed by Chandra and Toueg in [12]. This assumption fit the behavior of a typical computing grid: nodes crashes are possible and messages may be delayed or dropped by routers during network congestion.

Note that the two levels of the grid hierarchy exhibit different properties for communications (latency, bandwidth, message loss rate).

2.2   Benefits of Experimentation

A theoretical evaluation of a system can be carried out using a formal proof of a system model, which can validate the system design. However, it relies on a formal model, which is generally a simplification of the reality (taking into account only *significant* parameters). A second type of evaluation uses extensive simulations [9,15,11], which as formal proof, generally run models of the design and not the implementation itself. Finally, experimentations on real testbeds can serve as a proof of concept. Such a practical evaluation can capture aspects related, for instance, to the node specifications or to specifics of the underlying physical network. In this paper we focus on experimental evaluations.

Experimenting large-scale distributed software is difficult. The tests have to be deployed and launched on thousands of geographically distributed nodes, then the results have to be collected and analyzed. Besides, the tests have to be *reproducible*. Achieving these tasks for a large-scale environment is not a trivial task.

2.3   Controlling Volatility

In the context of large-scale, fault-tolerant distributed systems, one important aspect which needs to be controlled is *node volatility*. This section introduces a tool that provides the ability to inject failures according to pre-defined scenarii during experiments, in order to evaluate the quality of fault-tolerance mechanisms. More specifically, we illustrate how such a tool can be used in order to test a failure-detection service.

Failure Injection Requirements.   The use of failure injection mechanisms provides the ability to test fault-tolerant mechanisms with different volatility conditions. This may validate that the service provided by the software is still available when particular types of failures occur. It also provides the ability to measure the overhead introduced by the fault-tolerant mechanism to support different kinds of failures.

The experimentations are run on a testbed that is assumed to be stable. As we want to experiment with controlled failures, we assume that there are no other unexpected failures during the test (in case of a real failure, the developer will have to re-launch his test). The test tool can help the developer to introduce controlled failures during the experimentation. In order to emulate some specific scenarios and to be scalable, the test tool has to provide a simple and efficient way to describe failures distributed across thousands of nodes.

The failure injection mechanisms should be able to take only *statistical parameters* and then compute failure schedules accordingly. This allows the tester to generate a failure scenario across thousands of nodes by giving only a few parameters. The failure injection mechanisms also need to be *highly customizable* allowing the user to specify groups of nodes that should fail simultaneously. More generally, they need to provide the availability to express *failure dependencies between nodes*. This should allow the tester to emulate correlated failures. Furthermore, an important feature of a failure injector (volatility controller) is *reproducibility*. Even if failures are computed using statistical parameters, one may want to replay an execution with the same set of failures, while varying other parameters (e.g. in order to tune the fault-tolerance algorithms). While experimenting various parameters of a fault tolerance feature or testing different fault tolerance mechanisms one may want to compare different solutions within the same context.

Scenarios

Simple failure scheme. One simple way to describe a failure scheme is to assume that all the nodes have the same probability of failure and that they are independent (i.e the failure of a particular node does not depend on the failure of other ones). For instance, one may assume that the MTBF (Mean Time Between Failures) of a particular set of nodes may be one hour. The MTBF of a specific architecture can be easily observed. The developer may wish to run experiments with smaller MTBF values in order to stress the fault-tolerant mechanisms.

Correlated failures. As the underlying physical network system may be very complex, with hubs and switches, some failures may induce new ones. The crashes of some nodes may lead to the crashes of other nodes. By instance, while running a software on a cluster federation, a whole cluster may crash (Figure 1). This may be due to a power failure in one cluster room, for instance. While designing a fault-tolerant system for such an architecture, it is important to experiment its behavior while multiple failures occurs concurrently as it may happen in real executions (without failure injection mechanisms).

Accurate control. As the roles played by the different nodes in the system may not be strictly equivalent (some are more critical than others), the developer should be able to test some particular cases. For instance, one may want to experiment the simultaneous crash of two particular nodes, or the crash of a node when it is in a particular state. Typically, as illustrated by Figure 1, experimenting the failure of a node having manager capabilities may be really interesting as it may involve particular cases of the fault-tolerant algorithms.

2.4 Related Work

Researchers working in the fault-tolerance area need to inject failures during their experiments. Most often this is done in an ad-hoc manner, by manually killing nodes or by introducing a few code statements into the tested system's source code, to make failures occur. The overhead for the tester is non negligible and usually it is neither scalable nor reproducible. The goal of our failure injection mechanism is precisely to automate this task, making it easy for the testers to inject failures at *large scale* and to *control* volatility conditions.

Many research efforts focus on failure injection. However, most of them are very theoretical or focus on the problem of failure prediction [18,17,6]. In this paper we do not address the issue of *when* a failure should be injected or *what* it will induce, but we provide a practical solution to *how* to inject it. The tester may use the results of these previous research works to feed our failure injectors.

Failure injection has also been studied for simulation and emulation. For instance, [3] provides a solution to test protocols under failure injection, but it relies on a fully *centralized* approach. Our work is intended to be used for tests running on real *distributed* architectures, with the full application code.

FAIL [14] (FAult Injection Language) defines a smart way to define failures scenarios. It relies on a compiler to trap application communications to emulate failures. Our work is integrated in the test environment, not at application level, thus it allows to inject failures even when the source code is unavailable.

In contrast to previous work, we use failure mechanisms within a test tool, providing a simple way to deploy, run and fetch results of a test under various controlled volatility conditions.

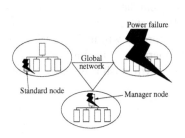

Fig. 1. Different kinds of failures

```
(00) <network analyze-class="test.Analyze">
(01)   <profile name="manager" replicas="1">
(02)        <!-- peer information -->
(03)     <peer base-name="peerA"/>
...
(11)     <bootstrap class="test.MyClass1"/>
(12)        <!-- argument -->
(13)     <arg value="x"/>
(14)   </profile>
(15)   <profile name="non-manager" replicas="20">
(16)     <peer base-name="peerB"/>
...
(23)     <bootstrap class="test.MyClass2"/>
(24)   </profile>
(25) </network>
```

Fig. 2. *JDF*'s description language

3   Our Proposal: A Flexible Failure Injection Tool

3.1   JXTA Distributed Framework (*JDF*)

We are currently developing our failure injection mechanism within the JXTA Distributed Framework (*JDF* [19,4]) tool. The *JDF* project has been initiated by Sun Microsystems, and is currently being actively developed within the PARIS Research Group [2]. JDF is a tool designed to automate the tests of JXTA-based systems. In [4] we have specified that this kind of tool should provide the ability to control the simulation of nodes' volatility. In the remaining of this section we show how to provide this ability inside *JDF*. A detailed description of *JDF* can be found in [19].

JDF allows the user to describe his test through 3 configuration files. 1) a node file containing the list of nodes on which the test is run, 2) a file storing the names and paths of the files to deploy on each node, and 3) a XML file describing the node profiles, in particular, the Java classes associated and the parameters given to these classes.

JDF's XML description file allows the tester to describe his whole system through *profiles*. Figure 2 defines two profiles, one from line 01 to 14 and one from line 15 to 24. Then multiple nodes can share a same profile. The profile named *non-manager* on Figure 2 is replicated on 20 different nodes (thanks to the *replicas* attribute). The first experimentation phase consists of the creation of these files. This phase is called *basic configuration* thereafter.

3.2   *JDF* Description Language Extension

The first requirement to fulfill in order to use failure injection is to incorporate failure information is into the *JDF* test description language.

To provide the ability to express failures dependencies (to represent correlated failures) we add a new XML tag: *failure*. This tag may have 2 attributes: 1) *grp* to indicate that all nodes having this profile are part of a same *failure group*; 2) *dep* to indicate that nodes having this profile depend, from a failure point of view, on nodes of another profile. The *grp* attribute allows to specify groups of nodes that should fail together (i.e. if one of them crashes, then all the set crashes). This can help the tester to simulate the failure of clusters, for instance. The *dep* attribute can be used to indicate that a node should crash if another one crashes (by instance to emulate the fact that the second node may serve as a gateway for the first one). For instance, in Figure 2, adding the line "*(17) <failure grp="1"/>*" in the *non-manager* profile will make all the *non-manager nodes* crash as soon as on of them crashes. Furthermore, if the line "*(18) <failure dep="manager"/>*" is added, all *non-manager nodes* will crash if the node having *manager* profile crashes.

3.3   Computing the Failure Schedule

We have developed a tool that generates a configuration file with volatility-related parameters (e.g. the global MTBF) which are given as an input to *JDF*. To do this, we introduce a new configuration file. In order to make the failure conditions reproducible, this file contains the uptimes for all nodes (i.e. the failure schedule). It is generated using the XML description file, which is necessary in order to take into account failure dependencies. This phase is called *failure schedule generation* thereafter.

The tool works as follows: it computes the first date using a given MTBF and the number of nodes (obtained from the XML description file), then it randomly chooses a node to which it assigns this first failure date. This operation is repeated until a failure date is assign to each node. Next, dependency trees are built using the information contained in the XML description file. The dependency trees are used to ensure that 1) in each failure group, nodes are assigned the smallest failure date of the group; 2) if the failure date of a node is greater than the failure date of a node on which it depends (the *dep* attribute), then the smallest date is assigned. This way, all dependencies expressed in the XML description file are satisfied.

Computing the failure schedule statically before launching the test allows the tester to easily reproduce failure conditions. The tool can be launched once to compute a failure schedule, and the same computed schedule can be used by multiple experiments.

3.4   Running Experiments with Failure Injection

Assuming that the standard *JDF* configuration files exist (i.e. the *basic configuration* phase has been done), the complexity overhead induced by the failure injection mechanisms to launch tests is very low.

To run a test by providing a MTBF value (*Simple failure scheme*) the tester has to launch a *JDF* script that will compute the failure dates before running his test (boxes A, C and E in Figure 3).

The *failure schedule generation* phase consist in executing a script with the desired MTBF value and the *JDF* standard configuration files.

As a further step, to use correlated failures, the tester needs to use our *JDF* description language extension. In this case, one *Failure configuration* phase is required to add *failure* tags in the XML configuration file (Box B in Figure 3).

Finally, the tester may need an accurate control of the failures (i.e inject a failure on a specific node at a specific time). To do this, the failure schedule has to be explicitly edited (Box D in Figure 3).

Once the schedule is computed no extra step is needed to re-execute an experiment with the same failure conditions.

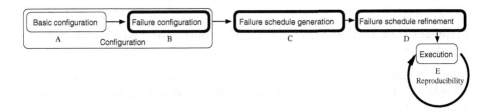

Fig. 3. Failure injection usage

3.5   Run Time Failure Injection

At deployment time, a configuration file containing the failure schedule is sent to each node. At launch time, a *killer* thread is started. This thread reads the failure date (which is actually an uptime), then waits accordingly. If the application is still running at the failure date, this thread kills it, thereby emulating a failure. Note that all the application threads are killed, but the physical node itself remains up and running. If an application uses the TCP protocol, the node will answer *immediately* that no process is currently listening to this port. If the node were really down, the application would have to wait for a TCP time-out. In the case of the UDP protocol (as for the experiments presented in this paper), this side-effect does not exist. For applications using the TCP protocol, the thread *killer* should either trap messages or really shutdown the network interface.

4   A Scalable Failure Detection Service

We used the failure injection mechanisms previously described to evaluate a scalable failure detector adapted to hierarchical grids.

4.1   Unreliable Failure Detectors

Concepts. Since their introduction by Chandra and Toueg in [12], failure detectors are becoming a basic building block for fault-tolerant systems. A failure detector is one solution to circumvent the impossibility [13] of solving deterministically the consensus in asynchronous systems in presence of failure. The aim of failure detectors is to provide information about the liveness of other processes. Each process has access to a local failure detector which maintains a list of processes that it currently suspects of having crashed. Since a failure detector is unreliable, it may erroneously add to its list a process which is still running. But if the detector later realizes that suspecting this process is a mistake, it then removes the process from its list. Failure detectors are characterized by two properties: completeness and accuracy. Completeness characterizes the failure detector capability of suspecting incorrect process permanently. Accuracy characterizes the failure detector capability of not suspecting correct processes.

We focus on the $\Diamond P$ detector, named *Eventually Perfect*, it is one of failure detector classes, which enable to solve the consensus problem (i.e. it is not the weakest). This detector requires the following characteristics:

Strong completeness: there is a time after which every process that crashes is permanently suspected by every correct process.

Eventual strong accuracy: there is a time after which correct processes are not suspected by any correct process.

Utility. A failure detector $\Diamond P$ provides the ability to solve the consensus, but it does not contradict the impossibility of Fischer, Lynch and Paterson, then it is impossible to implement it in asynchronous systems.

A failure detector has several advantages from a theoretical and a practical point of view. The first one is to abstract synchronism matter: algorithms that use a failure detector depends on failure only. The hypotheses, in terms of completeness and accuracy, describe how a failure detector detects other processes failures. These hypotheses are more natural than temporal ones but also useful.

In a practical way, the need to detect failures is a common denominator among the majority of distributed reliable applications. In fact an application must know if one of these processes has crashed: to be able to replace it in case of replication or more generally to avoid waiting infinitely its result. From this perspective, a failure detector is a specific service which provides the ability to guarantee the application vivacity. This service can be shared by several applications and then its cost is amortized.

4.2   GFD (GRID Failure Detector)

Properties. The aim of our failure detector is to propose a shared and moreover scalable detection service among several applications. In this implementation we dissociate two aspects: a basic layer which computes an estimation of the expected arrival date to provide a short detection time and an adaptation layer specific for each application. This adaptation layer guarantees the adequacy between the detection quality of service and the application needs. This architecture provides the ability to generate only one flow of messages to provide adapted detection information for all applications.

The second specificity is the hierarchical organization of the detection service in order to decrease the number of messages and the processor load [8]. It comprises two levels: a local and a global one, mapped upon the network topology. The system is composed of local groups, mapped upon SANs or LANs, bound together by a global group. Each group is a detection space: every group member watches all the other members of its group. Every local group designates at least one mandatory which will participate to the global group.

This organization implies two different failure detector types. This distinction is important since a failure does not have the same interpretation in the local context as in the global one. A local failure corresponds to the crash of a host, whereas in the global context a failure represents the crash of an entire local group. In this situation, the ability to provide different qualities of service to the local and the global detectors is a major asset of our implementation. Therefore a local group mandatory has two different failure detectors, one for the local group and one for global group.

In a local group, the failure detector uses IP-Multicast for sending periodicals *"I am alive"* messages. In SANs and LANs, IP-Multicast can be used with the broadcast property. Therefore a host only sends one message to communicate with all the other hosts. Failure detectors in a global group use UDP in order to be more compatible with the general network security policy.

5   Experimentations

We use our tool to inject failures and measure the time it takes to detect them with our failure detection service. Reproducibility is used to perform multiple experiments with the same set of failure while tuning the failure detector. The correlated failure feature is used to experiment the global level of the failure detector's hierarchy.

5.1   Experimental Setup

For all the experiments, we used the Grid'5000 platform [1], which gathers 9 clusters geographically distributed in France. These clusters are connected together through the Renater Education and Research National Network (1 Gb/s). For our preliminary experiments, we used 64 nodes distributed in 4 of these sites (Rennes, Lyon, Grenoble and Sophia). In these 4 sites, nodes are connected through a gigabit Ethernet network (1 Gb/s). This platform is hierarchical in terms of latency: a few milliseconds among the clusters, around 0.05 within each cluster.

As our failure detector is hierarchical, with a *local* and a *global* level, the 64 nodes are partitioned into 4 *local groups*, one in each cluster. Within each local group, a special node (mandatory) is responsible for the failure detection at global level (i.e cluster failures).

Even if our algorithms do not require a global clock assumption, for measurements purposes, we assume a global clock. Each node runs a *ntp* (*network time protocol*) client to synchronize its local clock and we assume that the local clock drifts are negligible (as the test period is short, of the order of a few tens of minutes). This assumption stands only for measurements purposes.

Performance Metrics.   The most important parameter of the failure detector is the **delay between heartbeats**. It defines the time between two successive emissions of an "*I am alive*" message. The failure injection is essentially characterized by the **MTBF** (*Mean Time Between Failure*) and possibly by the correlation between failures described in the test files. The experimental results are essentially expressed in terms of **detection latency**. It corresponds to the elapsed time between a node crash and the moment when the other nodes start suspecting it permanently.

5.2   Preliminary Tests

We started by evaluating the failure injection mechanisms alone. The goal is to assess its ability to inject failures according to a given MTBF following an exponential distribution. To do this, we launch 20 times a test with a MTBF value set to one minute, with no failure dependencies. Before each test, the failure schedule is recomputed in order to obtain mean failure dates. Figure 4 shows that the average number of alive nodes decrease as the time elapses. The experimental results are close to the theoretical ones obtained using an exponential distribution.

In a second experiment, we evaluated the ability of our failure injector to correctly generate *correlated failures*. There again we assume the failures follow the same exponential distribution. Besides, we add a *failure dependency*: all members of *local group 1* (located in Rennes) depend on their mandatory (i.e. they must fail if the mandatory fails). This results in a gap on Figure 5 when *Rennes' mandatory* is killed, as all nodes in Rennes fail concurrently. After this correlated failure happens, the slope of the curve is smaller. This is due to the fact that the dependency made some failures happen sooner.

We can conclude that the failure injector is able to inject failures according to a given MTBF and may also take into account correlated failures.

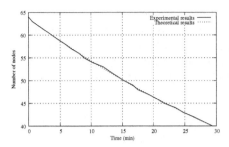

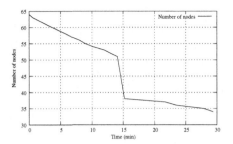

Fig. 4. Failure injection according to MTBF   **Fig. 5.** Correlated failures

5.3 Experimenting with the Failure Detector

Tradeoff: Detection Time Versus Network Stress. The failure detector is hierarchical: it provides failure detection in local groups (i.e clusters) and between these groups. We first evaluate the detection time at local level (within local groups) according to the *delay between heartbeats* of the failure detector. To do this evaluation, we set a MTBF of 30 seconds with no failure dependency, and no mandatory failures. During each run of 10 minutes, 18 nodes are killed. Figure 6 shows for each delay between heartbeats the average failure detection time in local groups. The results are very close to what we expected: theoretically, the average detection time is $(delay\_between\_heartbeats/2) + latency$ (and the maximum detection time is almost $(delay\_between\_heartbeats) + latency$). On the other hand, as the delay between heartbeats decreases, the number of messages increases, as shown by figure 7. For a fixed accuracy of the failure detection, there is a tradeoff between detection time and network stress. This is why, through adapters, our failure detector allows multiple applications to share the same heartbeat flow to minimize the network load.

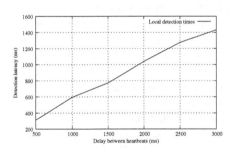

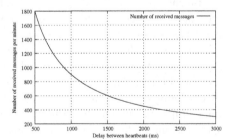

Fig. 6. Local detection times   **Fig. 7.** Network stress

Correlated Failures. The aim of this second experiment is to evaluate the detection time at the global level. At this level, the failure detection is done through the local group mandatories. When the failure of a mandatory is detected in a group, a new one is designated to replace it with an average measured nomination delay of *156ms*.

Thus, to experiment failure detection at global level, we need to use correlated failures in order to induce the crash of whole local groups. We emulate the failure of sites by introducing a failure dependency between the members of a group (i.e nodes in one site) and their mandatory. By instance, for the Rennes cluster, we add: *<failure dep="RennesInitialMandatory"/>* in the profiles of Rennes' non-initially mandatory nodes.

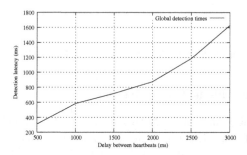

Fig. 8. Global detection times

During a 10 minutes run, 3 out of 4 mandatories are killed. Figure 8 shows the average failure detection times according to the delay between heartbeats. The results are similar to the ones obtained in local groups. The irregularity comes from the fact that less failures occur than for the previous tests. It is important to note that the correlated failures feature is mandatory to perform these measurements as a whole site should fail.

6   Conclusion and Future Work

In grid environments, building and evaluating fault-tolerant softwares is a hard task. In this paper, we present a test environment providing failure injection features, allowing the developer to control volatility without altering the application code. To illustrate this tool, we evaluate a hierarchical failure detection service. First, our experiments have show that our failure injection tool is able to provide accurate volatility control in a reproducible manner. This allowed us to evaluate a hierarchical failure detection service by emulating independent and correlated failures. In each of these two cases, we have run multiple experiments for different configurations of the failure detector. The results show that the faults are efficiently detected. To the best of our knowledge, no failure detectors have been experimented in the past using automated failure injection on grid platforms.

We plan to further enhance our test environment by adding support for message loss injection. This can be done through network emulation tools like Dummynet [16] or NIST Net [10]. The failure description language will be extended accordingly, in order to incorporate message loss description. Furthermore we will use this test environment to evaluate the fault tolerance mechanisms of higher-level grid service (e.g. the JUXMEM [5] data sharing service).

References

1. Grid'5000 project. http://www.grid5000.org.
2. The PARIS research group. http://www.irisa.fr/paris.
3. Guillermo A. Alvarez and Flaviu Cristian. Centralized failure injection for distributed, fault-tolerant protocol testing. In *International Conference on Distributed Computing Systems*, pages 0–10, 1997.
4. Gabriel Antoniu, Luc Bougé, Mathieu Jan, and Sébastien Monnet. Going large-scale in P2P experiments using the JXTA distributed framework. In *Euro-Par 2004: Parallel Processing*, number 3149 in Lect. Notes in Comp. Science, pages 1038–1047, Pisa, Italy, August 2004. Springer-Verlag.
5. Gabriel Antoniu, Jean-François Deverge, and Sébastien Monnet. How to bring together fault tolerance and data consistency to enable grid data sharing. *Concurrency and Computation: Practice and Experience*, (17), September 2006. To appear. Available as RR-5467.
6. Jean Arlat, Alain Costes, Yves Crouzet, Jean-Claude Laprie, and David Powell. Fault injection and dependability evaluation of fault-tolerant systems. *IEEE Transactions on Computers*, 42(8):913–923, 1993.
7. Marin Bertier, Olivier Marin, and Pierre Sens. Implementation and performance evaluation of an adaptable failure detector. In *Proceedings of the International Conference on Dependable Systems and Networks*, pages 354–363, Washington, DC, June 2002.
8. Marin Bertier, Olivier Marin, and Pierre Sens. Performance analysis of a hierarchical failure detector. In *Proceedings of the International Conference on Dependable Systems and Networks*, San Francisco, CA, USA, june 2003.
9. A Collaboration between researchers at UC Berkeley, LBL, USC/ISI, and Xerox PARC. The ns manual (formerly ns notes and documentation). http://www.isi.edu/nsnam/ns/doc/ns_doc.pdf, 2003.
10. Mark Carson and Darrin Santay. NIST Net - a Linux-based network emulation tool. 2004. To appear in special issue of Computer Communication Review.
11. Henry Casanova. Simgrid: A toolkit for the simulation of application scheduling. In *First IEEE/ACM International Symposium on Cluster Computing and the Grid*, pages 430–441, Brisbane, Australia, 2001.
12. Tushar Deepak Chandra and Sam Toueg. Unreliable failure detectors for reliable distributed systems. *Journal of the ACM*, 1996.
13. Michael J. Fischer, Nancy A. Lynch, and Michael S. Paterson. Impossibility of distributed consensus with one faulty process. *Journal of the ACM*, 32(2):374–382, apr 1985.
14. William Hoarau and Sébastien Tixeuil. Easy fault injection and stress testing with fail-fci, January 2006.
15. Marc Little and Daniel McCue. Construction and use of a simulation package in c++. Technical Report 437, University of Newcastle upon Tyne, June 1993.
16. Luigi Rizzo. Dummynet and forward error correction. In *1998 USENIX Annual Technical Conference*, New Orleans, LA, 1998. FREENIX track.
17. Jeffrey Voas, Frank Charron, Gary McGraw, Keith Miller, and Michael Friedman. Predicting how badly "good" software can behave. *IEEE Software*, 14(4):73–83, 1997.
18. Jeffrey Voas, Gary McGraw, Lora Kassab, and Larry Voas. A 'crystal ball' for software liability. *Computer*, 30(6):29–36, 1997.
19. JXTA Distributed Framework. http://jdf.jxta.org/, 2003.

Semantic-Based Service Trading: Application to Linear Algebra*

Michel Daydé, Aurélie Hurault, and Marc Pantel

IRIT - ENSEEIHT, 2 rue Camichel, B.P. 7122, F-31071 TOULOUSE CEDEX 7
{Michel.Dayde,Aurelie.Hurault,Marc.Pantel}@enseeiht.fr

Abstract. One of the great benefit of computational grids is to provide access to a wide range of scientific software and computers with different architectures. It is then possible to use a variety of tools for solving the same problem and even to combine these tools in order to obtain the best solution technique.

Grid service trading (searching for the best combination of software and execution platform according to the user requirements) is thus a crucial issue. Trading relies both on the description of available services and computers, on the current state of the grid, and on the user requirements. Given the large amount of services available on the Grid, this description cannot be reduced to a simple service name.

We present in this paper a more sophisticated service description similar to algebraic data type. We then illustrate how it can be used to determine the combinations of services that answer a user request. As a side effect, users do not make direct explicit calls to grid-services but talk to a more applicative-domain specific service trader.

We illustrate this approach and its possible limitations within the framework of dense linear algebra. More precisely we focus on Level 3 BLAS ([DDDH90a, DDDH90b]) and LAPACK ([ABB+99]) type of basic operations.

1 Introduction

Given all the services deployed on a grid, finding the most appropriate service or composition of services which are able to fulfill a user request is quite challenging and requires more than the knowledge of the service's signatures.

We introduce here an approach that consists in adding additional semantic information to the services in order to reduce ambiguity in their description and allow to find the services or combination of services that provide good answers to a user request using equational unification to identify all the possible choices. As a benefit, users do not need to make explicit call to specific services over the grid (such as some GridRPC call for example). The user does not need to

---

* This work has been partially supported by the French Ministery of Research throught the GRID-TLSE Project from ACI « Globalisation des Ressources Informatiques et des Données » and by the ANR (Agence Nationale de la Recherche) through the LEGO Project referenced ANR-05-CIGC-11.

M. Daydé et al. (Eds.): VECPAR 2006, LNCS 4395, pp. 622–633, 2007.
© Springer-Verlag Berlin Heidelberg 2007

know the exact name of the service he is looking for, he just has to describe the mathematical operation he wants to compute in a given applicative domain. Our service trader finds the appropriate service or combination of services (eventually it can provide the user a list of possible choices and ask him to choose the best one given the mathematical operation and not the library name). The interaction with the middleware can then be hidden behind a domain specific interface.

We take examples from dense linear algebra for the sake of simplicity, but this approach can be extended to other areas since the algorithm is generic and parameterized by the description of the application domain.

2   Problem Description

A key issue in advanced trading of services is the choice of a description formalism for the available services. The comparison between the available services and the user's requests depends on the formalism chosen for this description.

2.1   Different Approaches

The simplest description used in most SOA (Service Oriented Architecture) such as RPC, CORBA, COM, DCOM, RMI makes only use of the service signatures (input and output types of parameters). This information has the advantage to be easily available. But it is not sufficient for sophisticated trading, even if we use type isomorphisms to remove the problems of parameter position. Indeed, with such an approach there is no way to distinguish addition from multiplication as both share the same signature.

We can add keywords or meta-data to the service signature (this is currently the case in the GRID-TLSE project [PPA05, Pan04]). This formalism allows an easy comparison of the services and the request. But this description requires a preliminary agreement to define the keywords and their meaning, with all the ambiguities implied by the natural language. Another disadvantage is the difficulty to describe a complex service. How to describe without ambiguity and with keywords some Level 3 BLAS procedures such as $SGEMM$ expressed by the following formula: $\alpha * \mathbf{A} * \mathbf{B} + \beta * \mathbf{C}$?

Another approach which extends keywords and metadata is based on ontologies such as OWL. The advantage of ontologies is the possibility to have a formal description. It also provides the logic associated to reason about the descriptions. The disadvantage is that we do not control this logic which can be undecidable. The ontologies also need a preliminary agreement to define the keywords and their meaning. In the case of ontologies, this preliminary agreement is formally described thanks to relation between the different keywords, this is a main advantage over the previous approach. Moreover the definition of an ontology is not trivial and hard to achieve for a non specialist.

In the Monet[1] and HELM[2] projects the description of the computational services is based on MathML[3] and OpenMath[4] which provide an accurate description. But the comparison of services is based on RDF and ontologies which did not allow easily to adapt and combine services during the trading.

We follow the same approach as the NASA Amphion project [SWL+94] and more particularly the theorem prover SNARK (independence of the application domain, reasoning based on starting from a description of the domain). But, this project relies on «*term rewriting and the paramodulation rule for reasoning about equality*». This supposes that «*a recursive path ordering is supplied when the application domain theory is formulated*». The last constraint require that the user is familiar with complex rewriting technics. One of our main requirements is that the user should not need to know anything about the underlying technologies.

We are looking for a simpler description, with the least possible ambiguities, that can be specified by a specialist of a given domain without the help of a specialist on ontologies or the use of complex knowledge in rewriting techniques.

For all theses reasons, we have opted for a description similar to algebraic data types. The advantages of this description is the possibility of describing without ambiguity both the services and the knowledge of the main properties of the domain that are required for composing services to fulfill the user requests.

We describe in more details our semantic-based description of services in the next section. The trading algorithm is described in Section 3. Examples and possible limitations of this approach when looking for the best combination of services are reported in Section 4. We finally conclude in Section 5.

2.2   An Algebraic Data Type Based Description for Advanced Trading

As said before, the semantic used is similar to algebraic data type description [GH78]. Indeed the required information are:

- the types (or sorts) used;
- the main operators of the specific domain and their signatures (we allow overloading);
- the operators properties (such as commutativity and associativity) and the equations that link operators.

When considering dense linear algebra and basic operations such as BLAS and LAPACK, we define:

- Types: $Int, Real, Char, Matrix, \ldots$
- Operators and their signatures:

---

[1] http://monet.nag.co.uk/cocoon/monet/index.html
[2] http://helm.cs.unibo.it/
[3] http://www.w3.org/Math/
[4] http://www.openmath.org/cocoon/openmath/index.html

- Addition of matrices: $+ : Matrix \times Matrix \rightarrow Matrix$
- Multiplication of a matrix by a scalar: $* : Real \times Matrix \rightarrow Matrix$
- Matrix multiplication: $* : Matrix \times Matrix \rightarrow Matrix$
- Transpose of a matrix: $T : Matrix \rightarrow Matrix$
- Identity: $I :\rightarrow Matrix$
- Null matrix: $O :\rightarrow Matrix$
- ...

- Properties:
  - Addition $+$: commutative and associative (can be expressed directly by the corresponding equations)
  - Multiplication $*$: associative (can be expressed directly by the corresponding equations)
  - Neutral element I: $a : Matrix \quad I * a = a$
  - Absorbant element O: $a : Matrix \quad O * a = O$
  - Distributivity $*/+$:
    $a : Matrix \; b : Matrix \; c : Matrix \quad a * (b + c) = (a * b) + (a * c)$
  - Distributivity $*/+$:
    $a : Real \; b : Matrix \; c : Matrix \quad a * (b + c) = (a * b) + (a * c)$
  - ...

The last two equations can be factorized by:

$$a : \; b : Matrix \; c : Matrix \quad a * (b + c) = (a * b) + (a * c).$$

That means that the equation is valid for all the types of a for which $a*(b+c)$ and $(a * b) + (a * c)$ are well typed.

With this description, we can describe some of the Level 3 BLAS procedures in a formalism very similar to the official BLAS specification [DDDH90a].

$SGEMM$ performs one of the matrix-matrix operations:

$$\mathbf{C} = \alpha * \mathrm{op}(\mathbf{A}) * \mathrm{op}(\mathbf{B}) + \beta * \mathbf{C}$$

where α and β are scalars, op($\mathbf{A}$) and op($\mathbf{B}$) are rectangular matrices of dimensions m×k and k×n, respectively, $\mathbf{C}$ is a m × n matrix, and op($\mathbf{A}$) is $\mathbf{A}$ or $\mathbf{A}^T$.

In the trader, $SGEMM$ will be described by an XML document whose meaning is:

```
SGEMM(TRANSA:Char, TRANSB:Char, M:Int, N:Int, K:Int, ALPHA:Real,
    A:Matrix, LDA:Int, B:Matrix, LDB:Int, BETA:Real, C:Matrix, LDC:Int)
C <- ALPHA * op(TRANSA,A) * op(TRANSB,B) + BETA * C
```

Among the equations of the domain, will be: $op('n', a) = a$ and $op('t', a) = a^T$.

However, this description is not rich enough for sophisticated trading involving service combination. Some numerical properties of the matrix are very important to select a suitable Level 3 BLAS procedure. For example when considering matrix-matrix multiplication, symmetry of one of the matrices involved in the operation may lead to select $SSYMM$ rather than $SGEMM$ and similarly when dealing with a triangular matrix that is supported by $STRMM$. To take into account these properties, subtypes have been introduced in the description. Some restrictions are required about the definitions of subtypes. The relation on types must be a partial order relation (antisymmetric, transitive and reflexive), and must verify some constraints expressed in [CGL92].

To the previous description, we add:

- Types:
    - Invertible matrices: $InvMatrix < Matrix$
    - Symmetric matrices: $SymetricMatrix < Matrix$
    - Triangular matrices: $TriangularMatrix < Matrix$
    - Invertible triangular matrices:
      $InvTriangularMatrix < TriangularMatrix$,
      $InvTriangularMatrix < InvMatrix$
    - ...
- Operators and their signatures (we can specify the conservation of a property by an operator):
    - Multiplication of a symmetric matrix by a scalar:
      $* : Real \times SymetricMatrix \rightarrow SymetricMatrix$
      (the symmetric property is conserved)
    - Multiplication of a triangular matrix by a scalar:
      $* : Real \times TriangularMatrix \rightarrow TriangularMatrix$
    - Multiplication of an invertible triangular matrix by a non-zero scalar:
      $* : NzReal \times InvTriangularMatrix \rightarrow InvTriangularMatrix$
    - Transpose of a triangular matrix:
      $T : TriangularMatrix \rightarrow TriangularMatrix$
    - ...
- ...

In the examples, we give high level properties, but we can enrich the description to specify more precisely the matrix. The user which defines the application domain chosses the level of granularity of the description. It is important to notice that the impact of a more precise description is in relation with the new equations that the new properties may imply. Adding types is not very costly but it generally leads to introduce new equations which is more expensive.

We are now able to define all the services.

$SSYMM$ performs one of the matrix-matrix operations:

$$\mathbf{C} = \alpha * \mathbf{A} * \mathbf{B} + \beta * \mathbf{C}, \text{ or } \mathbf{C} = \alpha * \mathbf{B} * \mathbf{A} + \beta * \mathbf{C}$$

where α and β are scalars, $\mathbf{A}$ is an m $\times$ m symmetric matrix (only the upper or lower triangular part is used), $\mathbf{B}$ and $\mathbf{C}$ are m $\times$ n matrices.

In the trader $SSYMM$ will be described by an XML document whose meaning is:

```
SSYMM(SIDE:Char, UPLO:Char, M:Int, N:Int, ALPHA:Real, A:SymetricMatrix,
      LDA:Int, B:Matrix, LDB:Int, BETA:Real, C:Matrix, LDC:Int)
IF SIDE='l' THEN C <- ALPHA * A * B + BETA * C
IF SIDE='r' THEN C <- ALPHA * B * A + BETA * C
```

In practice the description is not exactly this one to take into account the $UPLO$ parameter. This point will be discussed later.

$STRSM$ solves one of the matrix equations:

$$\mathbf{A}*\mathbf{X}=\alpha*\mathbf{B},\ \mathbf{A^T}*\mathbf{X}=\alpha*\mathbf{B},\ \mathbf{X}*\mathbf{A}=\alpha*\mathbf{B},\ \text{or}\ \mathbf{X}*\mathbf{A^T}=\alpha*\mathbf{B}$$

where α is a scalar, $\mathbf{X}$ and $\mathbf{B}$ are m × n matrices and $\mathbf{A}$ is a unit, or non-unit, upper or lower triangular matrix. $\mathbf{B}$ is overwritten by $\mathbf{X}$.

In the trader $STRSM$ will be described by an XML document whose meaning is:

```
STRSM(SIDE:Char, UPLO:Char, TRANSA:Char, DIAG:Char, M:Int, N:Int,
      ALPHA:Real, A:InvTriangularMatrix, LDA:Int, B:Matrix, LDB:Int)
IF SIDE='l' THEN B <- ALPHA * op(TRANS,A^{-1}) * B
IF SIDE='r' THEN B <- ALPHA * B * op(TRANS,A^{-1})
```

In practice the description is not exactly this one to take into account the $UPLO$ and $DIAG$ parameters.

The matrix $\mathbf{A}$ is not necessary a triangular matrix, but can be considered as a triangular matrix ($UPLO$ indicates if it is a lower or upper triangular matrix and $DIAG$ if it is a unit matrix). This is the case when this matrix is used to store two different triangular matrices (like after a LU factorization). This problem needs more work to reach an acceptable treatment.

Currently, $STRSM$ is defined with $\mathbf{A}$ not necessary a triangular matrix, and with operation done on the upper or lower part, but it is not a good solution because the real problem is not $STRSM$ but the object which represents several objects. We must design a general solution for this problem instead of the ad-hoc approach currently in use which have an impact on all the services.

These descriptions illustrate that we can manage parameters which are both input and output. We can also specify the service in function of a given parameter.

We can now describe the services and the user's request. Our aim is to find the services or the combination of services that satisfies the client's request. For doing so, we first compute all the available services and combination of available services which answer the user request. Then, in a second step, we will chose the «best» one, according to the user's criteria. We may combine these two step for a better effectiveness.

3   Computing the Combination of Services Corresponding to an User's Request

To identify all the services and combinations of services that answer the user's problem, we compare the description of the user's problem with the description of all the services, taking into account the properties of the domain (here dense linear algebra).

3.1   The Trading Algorithm

Our comparison of two descriptions is based on equational unification [BS01] and in particular on the set of transformations of Gallier and Snyder which has been proved to be sound and complete [GS89]. This system has been adapted to add types and subtypes and also to improve the performance. The problems introduced by the overloaded functions with subtyping are treated as in [CGL92].

To control the algorithm we use two parameters: the depth of combination allowed and the number of equations applied. This second number is really critical because our algorithm has an exponential complexity for this parameter. Further improvements to our algorithm are required in the future to limit the complexity of computing the combinations of services corresponding to a request. It may be interesting to use ad-hoc treatments for properties such as commutativity, associativity, distributivity, zero element, identity element, ... The general principle of the algorithm is explained in details in [HP06].

3.2   Examples

We consider examples arising in dense linear algebra with a complete description of this domain.

For all the following examples, the results given, are some among all the results computed by the trader. The number of equations allowed and the depth of combination given are the minimum ones. If more equations are allowed to be applied and a bigger depth of combination is allowed, the number of results will grow.

Example 1. The available services are the ones from the Level 3 BLAS. The request of the user is $A : Matrix,\ B : Matrix,\ C : Matrix\quad C = A * B * C$.

One combination of services computed by the trader is:

```
Matrix p2=Any x1;
SGEMM('n','n',m?,n?,k?,1.,B,lda?,C,ldb?,0.,p2,ldc?); \\p2<-B*C
Matrix p1=Any x1;
SGEMM('n','n',m?,n?,k?,1.,A,lda?,p2,ldb?,0.,p1,ldc); \\p1<-A*p2
p1;
```

where *Any* $x1$ can be any matrix and the parameters following by a "?" are the ones we cannot determine, they will be determined later on.
To find this solution, the trader must be run with more than 5 equations allowed to be applied and a depth of combination allowed of at least 1.

Example 2. Now, the available services are the ones from the Level 3 BLAS and some from LAPACK [ABB+99] (row interchanges $SLASWP$, the Cholesky factorization $SPOTRF$ and the LU factorization $SGETRF$). The user wants to solve the linear system with multiple right-hand side members $Ax = B$ (where no property is known about A). One answer computed by the trader is:

```
InvMatrix p2=A;
Vector p6=ipiv?;
SGETRF(m?,n?,p2,lda?,p6,info?); \\p2<-fatorization LU of A (A= P*L*U)
Matrix p5=B;
SLASWP(n?,p5,lda?,k1?,k2?,p6,incx?); \\p5<-row interchanges of B
Matrix p3=p5;
STRSM('l','l','n',u?,m?,n?,1.,p2,lda?,p3,ldb?); \\solve L*x=p5; p3<-x;
Matrix p1=p3;
STRSM('l','u','n',u?,m?,n?,1.,p2,lda?,p1,ldb?); \\solve U*x=p3; p1<-x;
p1;
```

To find this solution, the trader must be run with more than 7 equations allowed
to be applied and a depth of combination allowed of at least 3.

Example 3. The example in similar conditions as the previous one but now **A** is
a symmetric positive definite matrix.

The trader computes the following compositions of services:

```
SymDefPosMatrix p2=A:SymDefPosMatrix ;
Vector p6=ipiv?;
SGETRF(m?,n?,p2,lda?,p6,info?); \\ p2<-fatorization LU of A (A= P*L*U)
Matrix p5=B;
SLASWP(n?,p5,lda?,k1?,k2?,p6,incx?); \\ p5<-row interchanges of B
Matrix p3=p5;
STRSM('l','l','n',diag?,m?,n?,1.,p2,lda?,p3,ldb?); \\solve L*x=p5; p3<-x
Matrix p1=p3;
STRSM('l','u','n',diag?,m?,n?,1.,p2,lda?,p1,ldb?); \\solve U*x=p3; p1<-x
p1;
```

To find this solution, the trader must be run with more than 7 equations allowed
to be applied and a depth of combination allowed of at least 3.

and

```
SymDefPosMatrix p2=A;
SPOTRF('u',p2,info); \\ p2<- Cholesky factorization of A (A=U{^T}*U)
Matrix p3=B;
STRSM('l','u','t',diag?,m?,n?,1.,p2,lda?,p3,ldb?); \\solve U{^T}*x=B; p3
Matrix p1=p3;
STRSM('l','u','n',diag?,m?,n?,1.,p2,lda?,p1,ldb?); \\solve U*x=p3; p1<-x
p1;
```

To find this solution, the trader must be run with more than 6 equations allowed
to be applied and a depth of combination allowed of at least 3.

and

```
SymDefPosMatrix p2=A;
SPOTRF('l',p2,info); \\p2<- Cholesky factorization of A (A=L*L{^T})
Matrix p3=B;
STRSM('l','l','n',diag?,m?,n?,1.,p2,lda?,p3,ldb?); \\solve L*x=B; p3<-x
Matrix p1=p3;
STRSM('l','l','t',diag?,m?,n?,1.,p2,lda?,p1,ldb?); \\solve U{^T}*x=p3;
p1;
```

To find this solution, the trader must be run with more than 7 equations allowed
to be applied and a depth of combination allowed of at least 3.

The first solution is the same as in the general case for **A** (i.e. **A** general
square). The other uses the fact that **A** is positive definite and replaces the LU
factorization by a Cholesky factorization which is a better solution.

Example 4. The example in similar conditions as the previous ones but now **A**
is an invertible upper triangular matrix.
    The following solution is found:

```
Matrix p1=B;
STRSM('l','u','n',diag?,m?,n?,1.,A,lda?,p1,ldb?); \\solve A*x=B; p1<-x;
p1;
```

To find this solution, the trader must be run with more than one equation allowed
to be applied and any depth of combination (since no combination is needed).

These examples illustrate the fact that the trader look for several solutions
taking into account the properties of the domain and of the parameters. All the
solutions do not have the same quality, a choice must be made among these
solutions.

4   Choosing the Solution to Be Run

The trading algorithm finds all the suitable solutions within given depth and
number of equations applied. We still have to select the one that will be executed.
Among the set of solutions produced, only the most relevant ones are kept. When
this first choice is made, we will interact with a grid middleware to finally select
the one to execute.

4.1   Discarding Solutions Without Interest

When looking for a solution, we compare the request with all the services. For a
given service, we may find a solution that is a combination of services involving
subproblems to be solved. In this case, we run again the algorithm on the sub-
problems. To avoid computation of uninteresting solutions, we do not run again
the algorithm if there is a subproblem which is the same as the initial problem.

Example: We want to compute $a + b$ and we have the service $x * y$. Then, $\{x \leftarrow a + b, \ y \leftarrow I\}$ is a solution requiring a combination, but we discard it.

We also simplify the request before running again the algorithm. This is necessary, to avoid running the algorithm on requests such that $a + O$, $a * I$, ...

4.2 Selecting the Most Relevant Solutions

Piloting Research. To improve the search of relevant results, we can explore first the most interesting services.

To decide whether a service is interesting, we consider its complexity (static information) and its availability (dynamic information). By exploring these services at the beginning of the trading process, the initial solutions found will be the most relevant ones, since they will be the least complex and they will be available.

Static information are not sufficient since we are on a Grid whose QoS can change dramatically, and we must take into account the network load, the data migration, ... Indeed, we prefer to satisfy a request with a service located on a server which has a strong availability rather than with a service located on a busy server assuming that both servers have the same performance. Improving the computation of services using these dynamic informations that can be provided by a middleware such as Diet (see section 4.3), used in the GRID-TLSE project, may be crucial for performance and will require further improvements and experiments in the trading algorithm.

Another way to find first the most relevant solutions is to change the way we traverse the research tree. Currently, we do a breadth first traversal. It may be interesting to use a more complex traversal based on a weighting of the branches. This weight will be calculated in function of the complexity of the subproblem.

Sorting Results. The obtained results must be sorted. Currently, this sort is done by considering the complexity of the services. Services which have the same complexity, are sorted in function of their parameters.

Assume that $f(x, y, x, O)$, $f(x, y, O, O)$ and $f(x, y, Any, O)$ solve the problem. The most interesting result is the last one $(f(x, y, Any, O))$ because it is the most general. $f(x, y, O, O)$ is more interesting than $f(x, y, x, O)$, because the null matrix is, in general, less complex than the user matrix. In the general case, services with same complexity will be sorted according to the increasing numbers of Any, constants and parameters given by the user within their parameters.

4.3 Interaction with a Middleware

The trader can then choose to transmit the most relevant result to the middleware which will schedule the chosen composite service. It can also choose to transmit several relevant results. The choice among the different results will be done by the middleware. Several environments provide the features needed: NetSolve [AAB+01], NINF [TNS+03], DIET [DIE], NEOS [NEO], or RCS [AGM97]. DIET is the middleware used in the GRID-TLSE project, where our work takes place.

In the case of simple service (without combination), the state of the machine where the service is located, its capacity, its availability, ... will be considered. In the case of combination of services, in addition to these information, the data dependencies must also be taken into account to evaluate the costs in term of communication between the computers running the different services. Indeed, the local execution (even on a less powerful server) might be quicker than the remote execution because of the extra overhead due to data movements.

If none of the services is satisfying to the middleware, it can ask for more results until it obtains satisfaction. More complex searches may then be started. As soon as the middleware obtains a valid solution, it executes the request (or the sequence of requests).

5   Conclusion

We have described an approach for advanced trading of services based on an algebraic data type like description of applicative domain and services. Our trading algorithm allows to compose existing services in order to satisfy the user request.

The trading algorithm first computes all the possible solutions within a given depth and a given number of equations examined. The main difficulty in that process is to limit the exponential complexity of the search for solutions by discarding the less relevant ones. Some issues are currently explored consisting in using a different strategy for searching in the solution tree: aiming at decreasing the number of branches explored, use of a cache mechanism for avoiding recomputing solutions, ... Finally, within the set of solutions computed, a selection is made by considering the complexity of the operations and their parameters.

The current trading algorithm provides the appropriate results but it is still very preliminary and further improvements on time and memory performances are required. Our goal would be to incorporate such a trading mechanisms within interactive scientific computing environments such as MATLAB or SciLAB to allow users to take advantage of grid services - when adequate - in a transparent way (without explicit calls) and to interact with a middleware to benefit of their scheduling capacity.

References

[AAB+01]   D. Arnold, S. Agrawal, S. Blackford, J. Dongarra, M. Miller, K. Sagi, Z. Shi, and S. Vadhiyar. Users' Guide to NetSolve V1.4. Computer Science Dept. Technical Report CS-01-467, University of Tennessee, Knoxville, TN, July 2001.

[ABB+99]   E. Anderson, Z. Bai, C. Bischof, L. S. Blackford, J. Demmel, Jack J. Dongarra, J. Du Croz, S. Hammarling, A. Greenbaum, A. McKenney, and D. Sorensen. *LAPACK Users' guide (third ed.)*. Society for Industrial and Applied Mathematics, Philadelphia, PA, USA, 1999.

[AGM97]   P. Arbenz, W. Gander, and J. Mori. The Remote Computational System. *Parallel Computing*, 23(10):1421–1428, 1997.

[BS01]   F. Baader and W. Snyder. Unification theory. In A. Robinson and
            A. Voronkov, editors, *Handbook of Automated Reasoning*, volume I, chap-
            ter 8, pages 445–532. Elsevier Science, 2001.

[CGL92]   Giuseppe Castagna, Giorgio Ghelli, and Giuseppe Longo. A calculus
            for overloaded functions with subtyping. In *Proceedings of the ACM
            Conference on Lisp and Functional Programming*, volume 5, pages 182–
            192, 1992.

[DDDH90a]   J. J. Dongarra, J. Du Croz, I. S. Duff, and S. Hammarling. Algorithm 679.
            a set of Level 3 Basic Linear Algebra S ubprograms. *ACM Transactions
            on Mathematical Software*, 16:1–17, 1990.

[DDDH90b]   J. J. Dongarra, J. Du Croz, I. S. Duff, and S. Hammarling. Algorithm 679.
            a set of level 3 basic linear algebra subprograms: model implementation
            and test programs. *ACM Transactions on Mathematical Software*, 16:18–
            28, 1990.

[DIE]   DIET. http://graal.ens-lyon.fr/DIET.

[GH78]   John V. Guttag and James J. Horning. The algebraic specification of
            abstract data types. *Acta Inf.*, 10:27–52, 1978.

[GS89]   J. H. Gallier and W. Snyder. Complete Sets of Transformations for
            General E-Unification. *Theor. Comput. Sci.*, 67(2-3):203–260, 1989.

[HP06]   Aurélie Hurault and Marc Pantel. Mathematical service trading based
            on equational matching. In *Proceedings of the 12th Symposium on the In-
            tegration of Symbolic Computation and Mechanized Reasoning (Calcule-
            mus 2005)*, volume 151, pages 161–177. Electronic Notes in Theoretical
            Computer Science, 21 March 2006.

[NEO]   NEOS - Server for Optimization. http://www-neos.mcs.anl.gov/neos/.

[Pan04]   M. Pantel. Test of Large Systems of Equations on the Grid: Meta-Data
            for Matrices, Computers, and Solvers. In *PMAA'04*, 2004.

[PPA05]   Marc Pantel, Chiara Puglisi, and Patrick Amestoy. Grid, Components
            and Scientific computing. In *Submission to Euro-Par 2005*, 2005.

[SWL+94]   Mark E. Stickel, Richard J. Waldinger, Michael R. Lowry, Thomas Press-
            burger, and Ian Underwood. Deductive composition of astronomical soft-
            ware from subroutine libraries. In *CADE*, pages 341–355, 1994.

[TNS+03]   Yoshio Tanaka, Hidemoto Nakada, Satoshi Sekiguchi, Toyotaro Suzu-
            mura, and Satoshi Matsuoka. Ninf-G: A Reference Implementation of
            RPC-based Programming Middleware for Grid Computing. *Journal of
            Grid Computing*, 1(1):41–51, 2003.

Management of Services Based on a Semantic Description Within the GRID-TLSE Project

Patrick Amestoy, Michel Daydé, Christophe Hamerling, Marc Pantel,
and Chiara Puglisi

TLSE Project*,
IRIT-ENSEEIHT, 2 rue Camichel, 31071 Toulouse CEDEX, France
surname.name@enseeiht.fr
http://www.irit.enseeiht.fr/tlse

Abstract. The goal of the GRID-TLSE Project is to design an expert
site that provides an easy access to a number of tools allowing comparative analysis of sparse matrix packages on a user-submitted problem, as
well as on particular matrices from the matrix collection also available
on the site.

When making available a large amount of software over a computational Grid, facilitating its deployment and its exploitation become
crucial. Within the GRID-TLSE Project, we use a software component
approach based on a high level semantic description of the scientific computing services. In this paper, we focus on one aspect of this description
of the computational services: the use of meta-data called *abstract parameters*. Our approach allows the automatic discovery and the exploitation
of new services throught the concept of *scenario*.

1 Introduction

The main goal of the GRID-TLSE Project is to design an expert site that
provides an easy access to a number of direct solvers for solving sparse linear systems, allowing their comparative analysis on user-submitted problems,
as well as on matrices from collections also available on the site. The site provides user assistance in choosing the right solver for its problems and appropriate values for the control parameters of the selected solver. It is also intended
to be a testbed for experts in sparse linear algebra. A computational Grid is
used to deal with all the runs arising from user requests. For more details see
http://www.irit.enseeiht.fr/tlse.

The expert site asks the user through a WEB interface (called WebSolve) to
describe his problem as well as, optionally, the characteristics of the computers
and the software that he plans to use. The expertise kernel (called Weaver) takes
into account the user requirements, the internal expertise scenarios and the Grid
state to build experience plans which are run using the DIET middleware [3]

* Funded by the French Ministery of Research throught ACI «Globalisation des
  Ressources Informatiques et des Données» and by the ANR (Agence Nationale de
  la Recherche) through the LEGO Project referenced ANR-05-CIGC-11.

(http://graal.ens-lyon.fr/~diet/). The results and metrics are used to produce synthetic graphics which help the user in choosing the best tools – and the corresponding value of control parameters – for his problem (according to some metric e.g. minimizing execution time).

In sparse linear algebra, similarly to other areas of scientific computing, there exists a lot of different algorithmic approaches for solving the same problem with different features and performance (e.g. several algorithmic variants for factorizing a sparse matrix).

As a consequence, the description of the computational services provided by each component is much more complex than usually advocated in software engineering (typically restricted to service name, type of input / output parameters). The computing services have functional parameters and results – as usual – but also make use of parameters and results for algorithmic control and execution metrics that depend on the numerical algorithms used. Controls (usually parameters) allow to adapt the algorithm to the user performance requirements. Metrics (usually results) provide the users insights on the results quality and on the way the computer was used.

We describe in the next sections the approach used within the GRID-TLSE Project. It has been initially designed for allowing experts in sparse linear algebra, that are not always grid computing specialists, to deploy easily software over the grid and to use it within the expertise process they describe using scenarios. This approach is generic and may be used in other areas.

2   Sparse Direct Solvers for Linear Systems

2.1   Sparse Direct Solvers

The main service used in the GRID-TLSE Project aims at solving $Ax = b$ where A is sparse using direct solvers.

The direct approach for solving $Ax = b$ consists in factorizing the matrix A into a product of simpler matrices (so called factors) and then computing the solution x. There exists different factorizations of A: $A = LU$, $A = QR$, $A = LL^{\mathsf{T}}$, $A = LDL^{\mathsf{T}}$, ...

Several algorithms can be used for solving the same linear system. They all use the same functional input parameters A and b and produce the same functional result x. However, they do not always have the same set of input / output parameters for algorithm control. They also provide execution metrics (execution time, amount of memory used, number of flops, ...) that may not be similar.

The performance of the sparse solvers depends on the exploitation of the structural and numerical properties of the matrix A and on the target computing platform characteristics. For the sake of simplicity, we focus on the LU factorization in the following sections.

2.2   Algorithm Controls and Execution Metrics

A computational service may possess a lot of input / output parameters for algorithm controls and execution metrics that may vary with its implementation.

In the general case, A is factorized into $PQ_R D_R A D_C Q_C P^\top$ where :

- D_R and D_C are diagonal scaling matrices for respectively rows and columns of A ;
- Q_R and Q_C are unsymmetric permutations for respectively rows and columns. Solvers often use only one.
- P is a symmetric permutation whose purpose is to reduce the size of the factors during the factorization of A.

The problem to be solved is then $\hat{A}\hat{x} = \hat{b}$ where $\hat{A} = PQ_R D_R A D_C Q_C P^\top$, $\hat{x} = PQ_C^\top D_C^{-1} x$ and $\hat{b} = PQ_R D_R b$. These transformations are usually computed in the first phase of the algorithm referred to as symbolic analysis. The permutations and scalings are also performed during this step. Algorithmic control parameters are tuned according to the properties of the matrix for improving execution.

Depending on the software, the permutations are either symmetric (P), unsymmetric (either Q_R or Q_C), left (PQ_R) or right ($Q_C P^\top$). Many algorithms - called orderings - are available for computing permutations, for example AMD (Approximate Minimum Degree [1]), Metis (graph partitioning [11]), MMD (Multiple Minimum Degree [12], Matrix bandwidth reduction [4]). Some packages provide several orderings and a control parameter is used to select one.

The LU factorization of $\hat{A}$ is performed next. During this factorization phase, the static symmetric ordering P can be completed by a dynamic ordering P_N (referred to as the numerical permutation) monitored using a pivoting threshold. The linear system is then $P_N \hat{A}\hat{x} = P_N \hat{b}$. The pivoting threshold is not always available as an algorithm control.

The last step ("solve") computes $\hat{x}$ using the factors L and U.

Most of the direct algorithms for solving a sparse linear problem are using these three steps (symbolic analysis, factorization and solve) in sequence. It is therefore possible to share the symbolic analysis between several factorizations (with different values for the pivoting threshold) and to share a factorization between several solves (with different values of b). One of the main benefit is to be able to use the ordering available within one sparse solver as an input for the factorization of another solver. This implies that a functional description of the package must be available to be able to call separately ordering, factorization and solve and to recover the corresponding outputs.

3   The GRID-TLSE Reflexive Approach

We use a component approach with a dynamic discovery of component characteristics. This approach relies on meta-data – called *abstract parameters* - describing all the possible features for all available service implementations. This approach is usually referred to as reflexive as it relies on services managing services. Note that one package may be deployed in several places and several versions, i.e. there may be several services implementing the same sofware.

There are two kinds of services within the GRID-TLSE Project:

- Computational services that correspond to sparse softwares or tools for processing sparse matrices (visualization, ...)
- Scenarios that are a high level level description of the expertise process. The interpretation of scenarios by the Weaver software layer generates the workflows executed over the Grid. Scenarios are specified by sparse linear algebra experts.

4   Use of Abstract Parameters for Describing Services

From the Web interface to define the objective and parameters of the user request up to the construction of scenarios, we use the same description of services based on common meta-data.

To describe a computational service, we specify:

- its functionalities: assembled/elemental entries, type of factorisations (LU, LDL^T,QR), multiprocessor, multiple Right-Hand-Side Members, ... ;
- and its algorithmic properties: unsymmetric/symmetric solver, multifrontal, left/right looking, pivoting strategy, ...

To describe a scenario, in addition to service input / output parameters (as usual), we specify:

- its execution metrics sent back by the solver executions: memory, numerical precision, execution time, ...
- its control: type of graphic visualization for post-processing, level of user (expert, non-expert, intermediate user).

More expert is the user, more control he may have on the parameters of the expertise process.

4.1   Expressing Dependencies Between Abstract Parameters

The abstract parameters are used to express constraints and/or relations that forms the basis of the expertise scenario. We can thus express qualitative and quantitative dependencies between values of metrics and control parameters within scenarios. This feature allows to limit the combinatorial explosion inherents to the expertise process. Here are some examples:

- If A symmetric and user is non-expert, then select only symmetric solver.
- Indicate that time and memory mostly depend on method and permutations but also on scaling and pivoting.
- Indicate that numerical accuracy mostly depends on pivoting but also on scaling and permutations.
- Advise orderings for QR based on $A^T A$.

- Indicate that multiple Right Hand Side option, although not available, can still be performed (simulated within computational service).
- Threshold for partial pivoting $\in [0, 1]$.

The first item illustrates how it is possible to limit the number of experiments performed over the grid: when the user is non-expert and when the targert matrix is symmetric, only symmetric solvers are experimented (while an expert user may want to run an unsymmetric solver on a symmetric matrix).

4.2   Example: Description of the MUMPS Software

We illustrate this by considering the MUMPS software ([2]). The abstract parameters describing this software (this is not an exhaustive list) looks like:

- Functional decomposition: Symbolic analysis, Factorization, Solve (the three steps are available and can be called independently)
- Control parameters: Symmetric Permutation, Unsymmetric Permutation, Pivoting Threshold
- Possible values : Symmetric Permutations available are { AMD, Metis, ... }. Unsymmetric Permutations are ...
- Metrics: estimated flops from the symbolic analysis, effective time for the whole solution, numerical precision after the solve, ...
- Dependency: numerical precision depends on the pivoting threshold values

4.3   Structuring Abstract Parameters: Illustration with Symmetric Permutations

An ordering is a heuristic to permute the graph of the initial matrix with the aim to limit the cost of the numerical factorization; the ordering has a strong impact on both the number of operations and the memory used by a solver. Orderings involves symmetric or unsymmetric permutations. We focus on the abstract parameter associated to symmetric permutations.

The abstract parameter **SymPerm** that corresponds to symmetric permutations is implemented as an enumeration of large size. One of the symmetric ordering often used is the Approximate Minimum Degree (AMD [1]) available in MUMPS and other sparse solvers. Each software may have its own implementation of the AMD ordering. One representative of the set of AMD implementations over all the sparse solvers available might be enough in most cases but they may perform differently. How to define/select a representative implementation of AMD since it may change from time to time is a quite complex issue.

Furthermore when studying the impact of using various symmetric orderings, one may not want to test all possible values of the symmetric permutation. On some matrices a subclass of orderings may be known to be superior. A non-expert user only wants to capture major differences between orderings, thus using a "good" representative of a subclass may be enough. This is a crucial issues for limiting the combinatorial complexity of this expertise process (i.e. avoiding to explore / execute all possibilities).

The "permutation" abstract parameter can be represented as a tree where:

– we define a default representative at each level of the tree,
– and a default realization for each leaf of the tree.

When managing expertise scenarios, it helps in designing more dynamic server pages by adapting the web pages to the level of the user (normal, expert, debugger), and in limiting cost of scenarios.

Figure 1 illustrates the structuration of the abstract parameter corresponding to permutations (only the symmetric permutation subtree is detailled).

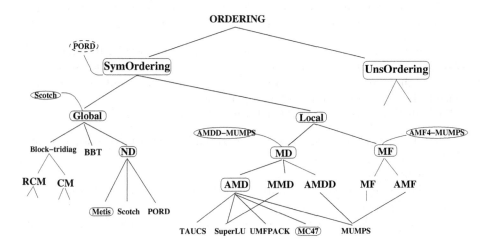

Fig. 1. Structuring the Permutation Abstract Parameter

5 Using Abstract Parameters Within the GRID-TLSE Project

The TLSE Weaver expertise kernel relies on two levels of services : the expertise scenarios exploited by users and the solvers used by scenarios and experts.

Extensibility is a key point in TLSE : new scenarios and new solvers will be integrated in the expertise site regularly. New scenarios should be able to use old solvers and new solvers should be used by old scenarios without modification. Modifications should only be required if scenarios want to use new specific features from solvers.

All services do not provide the whole set of controls and metrics. Input / output parameters should then be optional with either default values, values computed by other services or values explicitly provided. Tools may use or produce values in a slightly different manner for the same control or metric. It is therefore necessary to add a wrapper around each tool in order to adapt its real interface to the common one. New solvers may provide additional controls and metrics. Their interfaces should therefore be extensible.

The solution chosen in the TLSE Project relies on the definition of an easy to extend set of features for each service which will be wrapped around each tool. Scenarios are then using these features.

The meta-data framework used within TLSE can be summarized as follows:

- Solvers are described using meta-data and wrappers translate meta-data values to/from solver's parameters and results.
- Scenarios require solvers to provide specific meta-data and process experiments which are sets of meta-data.
- The middleware exchanges sets of meta-data with the wrappers of solvers.
- The Web interface is dynamically built from scenarios and their corresponding meta-data and solver meta-data and their values.

The service profile is composed of an abstract parameter set. It qualifies the following aspects of the service : the name of the tool; the service semantics; the functional parameters and results; the parameters and results for algorithm control; the parameters and results for execution metrics.

Each abstract parameter is defined using:

- its values (type, possible values, variation (linear, logarithmic, normal, Gaussian, ...));
- its mode : input or output;
- its constraint : mandatory, optional, with default value, with value computable by another service;
- the expertise level of the users (novice, standard, advanced, expert, manager);
- some documentation related to its purpose (several levels may be defined according to the user level);
- dependencies with other features for expressing incompatibilities, dependence upon a parameter and other constraints.

It is quite similar to an interface in the component world but extended in order to enable an easy integration of the tools that provide the same service with quite different algorithms (therefore different controls and metrics).

6   Use of Abstract Parameters Within Expertise Scenarios

The expertise scenarios are used by the expertise kernel to build experience plans according to the user request. These experience plans are worflows executed over the Grid. The results of one experience plans may be used to biuld the next experience plan and thus the workflows executed are dynamic since they may depend of results of a previous executions.

The scenarios are structured hierarchically in a dataflow like approach. Scenario inputs and outputs are connected to the sub-scenario inputs and outputs. It can also contain internal links between sub-scenario inputs and outputs. A scenario may also use internal operators for creation, modification, execution and filtering of experience plans. A given scenario may then build several internal experience plans, executes these plans, and finally produces new plans depending on the results from the previous ones. Scenarios are therefore fully dynamic and may depend on the availability of services and the results of experiences in order to generate new experiences. In order to ensure that a scenario will stop, there must be no internal cyclic links between sub-scenarios. Experience plan creation and execution operators use service description in order to assign a value to experience abstract parameters. Some instances may not qualify if some of their abstract parameters have values that are different from the ones required in the experiences.

The "Ordering sensitivity" scenario consists into studying the effect of using the available orderings on the solution of the linear systems in terms of the metrics selected by the user (execution time, memory, number of flops, ...). We only generates runs for default solvers (defined by experts), which is some kind of leaf cleaning and limits the combinatorial complexity of the expertise. The first box called "AllOrdering" corresponds the search of all available symmetric orderings. The second box, called "Exec", requires the executions of all the permutations sent by first box. The final results, in term of the metrics asked by the user, are then produced in a graphical way. The scenario is described by expert using a graphical interface called "GEOS" that is interpreted by Weaver to build experience plans. It is reported in Figure 2.

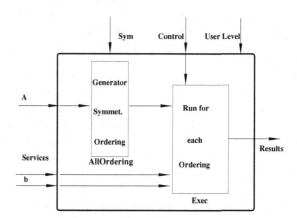

Fig. 2. Ordering Sensitivity Scenario

In the "Minimum time" scenario displayed in Figure 3, we try to identify which combination of ordering and factorization achieves the best execution time. Some branch cleaning is effected by selecting only one possibility at each level of tree of available permutations. This is expressed by the sequence of the two boxes: "AllOrdering" (used in the previous example) and "Select". We then only execute the default solvers (defined by the sparse linear algebra experts) which corresponds to some leaf cleaning.

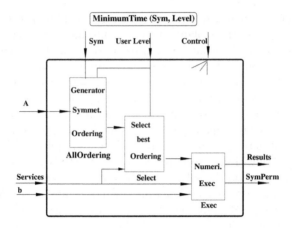

Fig. 3. Minimum Time Scenario

7   Conclusion

We have described the main aspects of the component framework used in the GRID-TLSE Project. This high level description of scientific software is used within the scenarios for generating the dynamic workflows that perform expertise. The main benefit is that adding / removing solvers does not require to update scenarios (they will be automatically discovered). New scenarios make use of all the deployed softwares.

This type of reflexive approach is commonly used for the dynamic discovery of services (for example, in the Java language or the Corba middleware). Similar approaches have been described for the use of object-oriented technologies for scientific computations in order to combine several algorithmic solutions: for example centralized and distributed matrix structures, see F. Guidec [8], E. Noulard and N. Emad [6]), the SANS (Self Adapting Numerical Software) Project (see [5]), or the Salsa Project (see [7]).

Our component framework combines two approaches : a static approach for accessing the functional parameters and the results for a given service; a dynamic approach for accessing the controls and metrics of a service. The set of meta-data used within the TLSE Project can be easily extended which is not always the case in the approaches mentioned above.

The GRID-TLSE Project focus on sparse solvers. The corresponding abstract parameters are defined using a graphical interface called PRUNE. Adding abstract parameters or specifying an entire set of new parameters is easy. As a consequence, the approach described in this paper can be extended to other areas providing that an adequate set of abstract parameters has been derived.

An important requirement in our approach is to be able to give an accurate description of the computation done by a given service according to its functional parameters and results. The service semantics could also be described using algebraic specification technologies. This semantics could then be used for service trading. This point is currently under investigation (see [10]). The use of accurate semantics allows to combine basic services in order to provide more sophisticated ones. This trading approach can also be combined with a scheduler for finding the best service combination (see [9]).

References

1. P. R. Amestoy, T. A. Davis, and I. S. Duff. An approximate minimum degree ordering algorithm. *SIAM Journal on Matrix Analysis and Applications*, 17:886–905, 1996.
2. P. R. Amestoy, I. S. Duff, and J.-Y. L'Excellent. Multifrontal parallel distributed symmetric and unsymmetric solvers. *Comput. Methods Appl. Mech. Eng.*, 184:501–520, 2000.
3. E. Caron, F. Desprez, E. Fleury, F. Lombard, J.-M. Nicod, M. Quinson, and F. Suter. Une approche hiérarchique des serveurs de calcul. *Calculateurs parallèles*, 2001.
4. E. Cuthill. Several strategies for reducing the bandwidth of matrices. In D. J. Rose and R. A. Willoughby, editors, *Sparse Matrices and Their Applications*, New York, 1972. Plenum Press.
5. J. Dongarra and V. Eijkhout. Self-adapting numerical software and automatic tuning of heuristics. In *Proceedings of the International Conference on Computational Science, June 2–4 2003, St. Petersburg (Russia) and Melbourne (Australia)*, 2003.
6. N. Emad E. Noulard. A key for reusable parallel linear algebra software. *Parallel Computing*, 27(10):1299–1319, 2001.
7. Victor Eijkhout and Erika Fuentes. A proposed standard for numerical metadata. Technical Report ICL-UT-03-02, Innovative Computing Laboratory, University of Tennessee, 2003.
8. F. Guidec. Object-Oriented Parallel Software Components for Supercomputing. In Peters D'Hollander, Joubert and Trystram, editors, *Parallel Computing: State of the Art and Perspectives. Proceedings of PARCO'95 (Parallel Computing)*, Gent, Belgium, Advances in Parallel Computing. North-Holland, 1995.
9. A. Hurault, M. Pantel, and F. Desprez. Recherche de services en algèbre linéaire sur une grille. 5-8 Avril 2005. Rencontres Francophones en Parallélisme, Architecture, Système et Composant (RenPar'16), Croisic, (France).
10. Aurélie Hurault and Marc Pantel. Mathematical service trading based on equational matching. In *Calculemus 2005, 12th Symposium on the Integration of Symbolic Computation and Mechanized Reasoning, Newcastle, United Kingdom*, July 18-19 2005.
11. G. Karypis and V. Kumar. MeTiS – *A Software Package for Partitioning Unstructured Graphs, Partitioning Meshes, and Computing Fill-Reducing Orderings of Sparse Matrices – Version 4.0*. University of Minnesota, September 1998.
12. J. W. H. Liu. Modification of the minimum degree algorithm by multiple elimination. *ACM Transactions on Mathematical Software*, 11(2):141–153, 1985.

Extending the Services and Sites of Production Grids by the Support of Advanced Portals*

Péter Kacsuk

MTA SZTAKI
Computer and Automation Research Institute of the
Hungarian Academy of Sciences
H-1518 Budapest, P.O. Box 63., Hungary
kacsuk@sztaki.hu

Abstract. Meanwhile production Grids are robust and reliable Grid systems they are not able to progress as fast as Grid research would enable it and they do not grow as fast as they were originally expected. The remedy for the first problem could be the introduction of volunteer services that can extend production Grids with new services based on the latest Grid research results. Solution for the second problem could be the extension of production Grids with volunteer Grid sites deploying either the same or different Grid middleware used by the production Grid. Advance Grid portal service is a good example for the volunteer Grid services. Combining it with other volunteer services like legacy code service, brokering, monitoring it can even help in solving the second problem by enabling the adoption of volunteer Grid sites without compromising the robustness of the core production Grid. The acceptance of the volunteer service and site concept can contribute to the long term sustainability of production Grids.

1   Introduction

In the beginning of the Grid Computing era it was not clear if the volunteer or production Grid model was more viable. The volunteer Grid model means that anyone can offer resources for a Grid system and anyone can claim resources dynamically, according to the actual needs, in order to solve a computationally intensive task. Though the model seems to be ideal, in practice, it has failed due to the significant manpower, expertise and skill that are required to deploy and maintain the necessary Grid middleware on the Grid sites. As a result volunteer Grid sites were not reliable enough causing frustration and disappointment among Grid users.

This failure of the volunteer Grid concept led to the rise of the production Grid model where building the Grid requires much more discipline and commitment from the Grid site managers. In production Grids not anybody can offer Grid resources rather only a relatively small number of institutes that are ready to sign a service level agreement (SLA) in which they accept the conditions of providing a 24/7 service. All the Grid sites should deploy the same Grid middleware that is already thoroughly tested and in this

* This research work is carried out under the FP6 Network of Excellence CoreGRID funded by the European Commission (Contract IST-2002-004265).

M. Daydé et al. (Eds.): VECPAR 2006, LNCS 4395, pp. 644–655, 2007.
© Springer-Verlag Berlin Heidelberg 2007

way they try to guarantee the robustness of the production Grid system. Such production Grids are, for example, EGEE and the UK NGS. The main advantage of the production Grids is that they are much more reliable and robust than the volunteer Grids.

However, production Grids should pay significant price for their robustness:

The number of potential Grid sites is much less than in the volunteer Grids. (For example, in the UK NGS there are only 10 sites [1].)

They grow much slower than it was expected in the volunteer model. (Production Grids apply a very conservative policy in accepting new sites and new services.)

They are very slow in adapting new Grid technologies and research results so their overall progress is much slower than it is desired. (UK NGS, for example, is still a 100% GT2-based Grid system although service-oriented Grid technologies have been available for several years. EGEE tried to adapt the service-oriented Grid technology but only in a very restricted and limited way in their new gLite middleware [2].)

What could the remedy be for all these problems? The solution could be a kind of compromise between the production and volunteer Grid model taking the advantage of both concepts. As a starting point the production Grid model should be kept since it guaranties reliability and robustness. It means that we need a Grid system where the core sites are organized as a production Grid based on strict SLA rules. However, these core sites should be extended with volunteer sites and volunteer services. Volunteer sites enable the fast growing of these Grids meanwhile volunteer services enable the fast adoption of new Grid technologies and research results. The paper will describe such a volunteer site and service adoption concept using as example advance Grid portal technology, legacy code services, brokering and Grid monitoring.

Section 2 introduces the basic concept of volunteer services using Grid portals and particularly P-GRADE portal as an example for such services. Section 3 defines the concept of volunteer Grid services and shows three different examples (legacy code services, brokering and Grid monitoring) to demonstrate the advantages of the concept. Section 4 deals with the problem of extending production Grids with volunteer Grid sites either based on the same or different middleware used by the production Grid. Finally, the section on Conclusions summarizes the main advantages of the volunteer service and site concept in the context of providing long term sustainability of production Grids.

2   Grid Portals as Volunteer Services

A volunteer Grid service is a service that can be added to a particular production Grid without any changes of the Grid middleware applied in the production Grid. A volunteer service is a simple add-on to the existing Grid middleware. A very natural candidate for such a volunteer service is a Grid portal that is built on top of the existing services, in fact, providing typically graphical user interface for accessing the middleware functionalities of the Grid.

The advantage of using portals within a production Grid is three-fold:

The user should not install anything on his/her client machine in order to use the Grid services.

The portal can be accessible from anywhere, i.e. even a mobile user can easily and conveniently access the Grid services.

The user should not learn the low level command line interface of the Grid middleware applied in the production Grid

Multi-Grid portals [3] can have even more advantages if they are used not only for a single Grid but rather for several different kinds of production Grids. In this case the main advantage is that the portal provides a unified access mechanism and interface for the connected Grids and hence the user does not have to learn the command line interface of the different connected Grids. Even more when an application should be migrated from one Grid to another one, the application porting effort could be minimized if the same portal can serve for both Grids. In this case it is the portal that should be ported between the two different Grids and not the individual applications. Advance multi-Grid portals can be connected to several Grids at the same time and can simultaneously serve several Grids by distributing the jobs of a single user among several of the connected Grids (provided that the user has certificates to several Grids). In this way a user can access several Grids at the same time and hence the power of using Grids can be tremendously enhanced.

One of the hot topics of today's Grids is the solution of interoperability among the various production Grids. Most of the research is devoted to solve this problem at the middleware level. This is a very hard issue particularly in case of the 2nd generation, resource-oriented Grid systems. Unfortunately, the overwhelming majority of the current production Grids is based on 2nd generation Grid technology. Multi-Grid portals provide a natural solution for the Grid interoperability problem from the users' point of view. The user wants to run his job

in any of the Grids that can serve it

or in the Grid that can provide the shortest response time if there are several Grids capable to run his job.

Multi-Grid portals can be connected to the brokers of different Grids and hence can contact these brokers on behalf of their users and can select the right Grid and right resources among the connected Grids. As a result the portal provides Grid transparency and Grid interoperability among the connected Grids from the user's point of view.

The usual view on Grid portals is that they simply provide a graphical user interface to submit jobs and to enable the view of Grid resources and job execution results. The other typical approach is that they are tailored to the needs of a certain end-user community (e.g. [4]) or to the technology of a particular Grid (e.g. Genius [5] for EGEE).

However, this simplified view is far not enough for the benefit of the whole Grid user community and hence more sophisticated, more advance and more generic concepts are needed to build Grid portals. A good starting point is GridSphere [6] that provides a Grid portal framework including basic functionalities necessary for any kind of Grid portal and at the same time enables the easy extension of the basic functions with new portlets by which the portal can be tailored to specific user needs and Grid requirements.

Building on this concept P-GRADE portal has further developed this idea providing a workflow-oriented generic Grid portal framework where users can develop and run DAG-like workflow [7] applications as a core functionality of the portal. Nodes of the workflows are jobs to be executed in the Grid and arcs of the DAG graph represent the necessary dependencies and file transfers among the jobs of the workflow. The portal provides all the supporting functions

to graphically edit workflows,
to manage jobs and the necessary file transfers during workflow execution,
to monitor and visualize workflow execution,
to archive workflows,
to support porting workflows between different portals,
to manage certificate proxies during workflow execution
to support mapping jobs of the workflow to various Grid resources or to Grid brokers
to manage Grids and Grid sites that can be accessed during workflow execution
to provide information system view on Grids and Grid sites

With all these built-in functionalities P-GRADE portal provides a convenient Grid environment where users can easily develop and run workflow-oriented Grid applications. Of course, single jobs are a special case of workflows and hence P-GRADE portal covers all the functionalities of single job based portals, too.

Advance Grid portal technology also means that a portal can be connected to several different Grids and provides transparent Grid access mechanism among the different Grids as it was described above. P-GRADE portal also satisfies this criterion. It is a multi-Grid portal where nodes of the workflows can simultaneously be executed in several connected Grids. As a result, in case of P-GRADE portal Grid interoperability can be realized not only at job level but also at the workflow level. Due to this feature of P-GRADE portal it was selected as the GIN VO Resource Test portal [8]. GIN VO was built by the GIN-CG [9] of OGF in order to investigate the Grid interoperability issues by connecting TeraGrid, OSG, EGEE, NorduGrid and UK NGS sites in the same VO. P-GRADE portal is the only portal currently that can provide transparent access to all these different Grid resources solving at the Grid interoperability problem at the job and workflow level [10].

In summary the advance features of P-GRADE portal provide a volunteer Grid service that significantly extends the services of existing production Grids. It is connected as volunteer service to all the major production Grids: EGEE, UK NGS, OSG, TeraGrid, NorduGrid. More than that if such a portal is connected to several of these Grids it can solve the Grid interoperability among the connected Grids.

3   Extending Production Gridswith Volunteer Services

As it was written in Section 2 a volunteer Grid service is a service that can be added to a particular production Grid without any changes of the Grid middleware applied in the production Grid. The question is if there are Grid middleware services that can be added as volunteer services to production Grids. In order to provide an answer for this question we have to give a precise definition of volunteer Grid services.

A volunteer Grid service is a service that can be added to a particular production Grid

1. without any changes of the Grid middleware
2. without installing any additional software on the core sites of the production Grid

It means that any volunteer services should be placed on additional host machines providing the volunteer service functionality. If this service (host) does not work, the production Grid works as before. If the volunteer service (host) works, users of the production Grid can access to extra Grid functionalities (services) without compromising the original services. In the rest of this section we describe three such volunteer services:

1. GEMLCA
2. GMT
3. GTBroker

3.1   GEMLCA as Volunteer Grid Service

GEMLCA (Grid Execution Management for Legacy Code Architecture) was developed for service-oriented Grid systems to enable the nearly automatic turning of legacy applications into Grid services without modifying either the source or binary of the legacy code. The GEMLCA architecture [11] in its original form assumed the availability of the legacy codes on the Grid sites and required the installation of a service-oriented Grid middleware layer (GT4 for the time being) and the GEMLCA Resource layer on every Grid site. Although this architecture concept perfectly suits to the service-oriented Grid approach it became an obstacle to provide GEMLCA as volunteer service for production Grids. There was a problem with all the three layers of GEMLCA in the current production Grids. The GEMLCA Resource layer was contradicting to the definition of volunteer services, site managers of core sites are not ready to install additional legacy codes on their resources and finally, the necessity of a service-oriented Grid layer was a problem in the current production Grid systems that do not employ service-oriented Grid layers.

Recognizing these problems GEMLCA was re-designed towards a volunteer service. The GEMLCA resource layer was removed from the Grid sites and placed into a separate GEMLCA service provider hosting environment. This hosting environment can be a single host or a set of distributed and interconnected host machines. The requirement for a service-oriented Grid layer on the Grid sites and the assumption on the availability of legacy codes were also eliminated by introducing the GEMLCA legacy code repository concept. In the new GEMLCA architecture every Grid site is represented by a GEMLCA resource located in the GEMLCA service provider hosting environment. Every GEMLCA resource is associated with a legacy code repository. If the Grid site that is represented by the GEMLCA resource is a service-oriented Grid site (e.g. GT4 site) and the legacy code is available on the Grid site, then the legacy code repository contains the description (call interface) of the legacy code service. Otherwise, the legacy code repository contains the executable legacy code and once this service is invoked the executable is transferred to the 2nd generation Grid site as a traditional job submission (meanwhile the user interface looks like a real service invocation).

With these tricks GEMLCA has been successfully adapted as a volunteer service to UK NGS, TeraGrid, OSG and EGEE. The user interface of GEMLCA was integrated with P-GRADE portal [11] and as result the workflow nodes of P-GRADE portal can be not only sequential and MPI jobs but also Grid service invocations. There are some additional advantages of integrating P-GRADE portal and GEMLCA as volunteer services for production Grids. Before such integration P-GRADE portal could solve Grid interoperability only between 2nd generation Grids. By extending the portal with GEMLCA, interoperability with even 3rd generation Grids are solved [10] at the portal, job and workflow level. Section 4.2 shows that the GEMLCA/P-GRADE portal also enables the extension of 2nd generation production Grids with 3rd generation volunteer Grid sites.

3.2   GMT as Volunteer Grid Service

In order to offer GEMLCA legacy code as volunteer services for production Grid systems, automatic testing of these services is inevitable. The GEMLCA Monitoring Toolkit (GMT) was developed to provide monitoring information based on probes concerning the status of GEMLCA resources. Using the GMT, system administrators are automatically alarmed when a test fails and can also request the execution of any test on-demand. The GMT also assists P-GRADE portal users when mapping the execution of workflow components to resources by offering only verified Grid resources when creating a new workflow or when rescuing a failed one.

The implementation of GMT is based on MDS4 (Monitoring and Discovery System) [12] that is part of the Globus distribution. MDS4 is capable to collect, store and index information about resources, respond to queries concerning the stored information using the XPath language, and control the execution of testing and information retrieval tools built as part of the GEMLCA Monitoring Toolkit. It can be extended and tailored to obtain specific information by means of polling resources, subscription to obtain notifications regarding changes to the state of specific resources, and execution of test and information collection scripts (probes).

As part of the GMT, several probes were implemented that collect information concerning the state of basic Globus services, local job manager functionality, and GEMLCA services. The probes can immediately be used as standalone tools executed automatically from the MDS by means of an XML configuration file, or manually from a command line interface, and they are also integrated into the P-GRADE portal assisting both system administrators and end-users.

Site administrators can configure the MDS4 service to run the various probes at pre-defined intervals. The results are collected by a portlet that is integrated into the P-GRADE portal. Administrators can also select a specific probe from a pull-down list displayed by a portlet and run it to verify the state of a specific service at a specific site on demand. GMT probes can also be integrated into the workflow editor of the portal to assist end-users when mapping a new workflow execution onto available Grid resources, or when rescuing and re-mapping a failed workflow. In the latest P-GRADE portal release, mapping of workflow components to underlying resources happens either manually by the end-user, or in case of LCG type Grids, by the LCG broker. The GMT aims to support manual mapping (when no LCG type broker is available) by dynamically querying the MDS4 during workflow creation time, and offering only those GEMLCA resources for mapping where the latest GMT test results were positive. Although, this does not guarantee that the resource will actually work when executing the workflow, but the probability of a successful execution will significantly be increased.

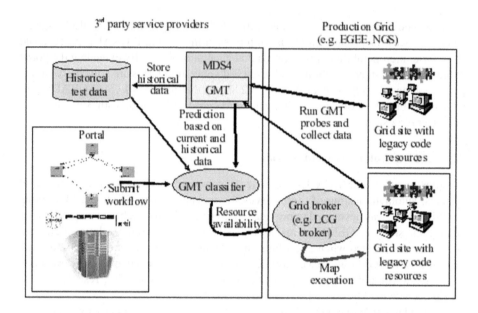

Fig. 1. GMT based resource availability prediction as volunteer service

Work is also undergoing to connect GMT to the LCG resource broker, as illustrated in Figure 1. GMT, as shown on the figure, runs regular probes on the production Grid resources and, besides updating the MDS indexing service, also creates a historical database. When the portal submits a workflow, a classifier component runs data mining algorithms on this historical data and determines which resources are "very likely to be alive". This information can be passed to the production Grid broker, for example in case of an LCG broker within the JDL (Job Description Language) file. The broker then maps the execution to the appropriate resources taking now the GMT provided information into consideration too. Notice that in this solution both MDS-4, GMT, GMT Classifier and the Historical test data database work as volunteer services for the connected production Grid that can be EGEE, UK NGS, OSG, TeraGrid, etc.

3.3   GTBroker as Volunteer Service

Most of the Globus based productions Grids (e.g. UK NGS, OSG, TeraGrid) are used without a broker that makes the life of Grid users quite hard since it is the user's task in these Grids to select the necessary Grid resource. It is not a big problem for a single task but becomes more and more difficult as the user would like to use large workflows or parameter study applications. Recognising this problem a broker called as GTBroker [13] was designed and implemented in MTA SZTAKI as volunteer Grid service for 2nd generation Globus based production Grids.

For determining the available hosts in the grid GTBroker quiries the MDS with LDAP requests. The job submission to resources is done through GRAM, and a GASS server is used to put the files needed for the job to the remote host and to get back the result files if there are any. Job requirements have to be specified in a simple RSL file. GTbroker acts in the following way: First it reads the given RSL file to get the user requirements and job properties. Then it turns to the MDS information service of the production Grid in order to get information about the resources. After getting the available nodes from the MDS the broker orders them by a predefined criterion. In the criteria one can use the following metrics: CPU speed, number of CPUs, free CPUs on the node, disk size and whether a node is a cluster. With these metrics the hosts can be ordered in a way that the ones having the best resources get higher priority than the others. Should a job fail or be pending for too much time on a node, the broker cancels and resubmits it to another host. The actual state of the jobs is tracked by the broker, that's why it is able to cancel and resubmit jobs.

Originally, the main reason for developing GTBroker was providing a way to submit up to hundreds of small jobs as fast as possible, without any failures in a 2nd generation Globus based Grid. This requirement of brokering hundreds of small jobs is extremely important for parameter study applications. After the first successful experiments more ambitious goals were defined, i.e., GTBroker should be able to

handle not only small jobs
support LCG-2 based Grids, too
work as a built-in service of P-GRADE portal

All these requirements were successfully met with the latest version of GTBroker. It can handle large number of larger jobs even in LCG-2 Grids and comparison measurements show that GTBroker significantly outperforms the official LCG-2 broker. All this is achieved without modifying any middleware software in LCG-2 production Grids or without requiring to install any additional software on the LCG-2 resources. This means that GTBroker satisfies the volunteer service criteria and hence can be used for UK NGS, OSG, TeraGrid and LCG-2 Grids.

The same way as GEMLCA and GMT can be used as independent volunteer services but can also be integrated with P-GRADE portal, GTBroker was integrated with P-GRADE portal. In this way the portal can provide high level broker services for its workflows and parameter study applications even if such broker service is not available in the connected production Grid or the available broker is a different type.

4 Extending Production Grids with Volunteer Grid Sites

As it was written in the Introduction another problem of existing production Grids is that they do not grow as fast as volunteer Grids can do due to the strict SLA requirements. Many potential sites are reluctant to sign such strict SLAs because of the lack of the necessary manpower or they want to install slightly or radically different Grid middleware on their own resources. However, production Grid management does not want to compromise the production Grid stability with enabling the join of volunteer sites based on less strict SLAs. As a result production Grids are not big enough to attract the critical mass of users that could justify their existence. Notice that a Grid is not attractive for a user if he can access a large cluster and in the Grid he cannot get access at least an order of magnitude more resources than on his own cluster. So, volunteer sites would play an important role to increase the size and attractiveness of Grid systems and this would be really important.

Again a solution for this problem could be a volunteer advanced portal service that hides the unreliable nature of volunteer Grid sites.

4.1   Volunteer Site with Less Strict SLA

The assumption is that quite a number of volunteer sites would be ready to join existing production Grids if they were not be forced to accept the strict SLAs. This, of course, would mean that many of these sites would regularly down causing frustration to the Grid users.

However, using and advanced Grid portal like P-GRADE integrated with a broker like the GTBroker we could keep the advantages of having large number of volunteer sites without causing any frustration for the users due to the unreliability of these volunteer sites. The portal users would submit their jobs and workflows through GTBroker that has the outstanding feature of tracking the status of jobs. If a job is failed on the selected resource, GTBroker is able to automatically select a new Grid resource and re-submit the job to the newly selected resource. All this is transparent to the user. He does not have to care that in the background GTBroker tried several different resources until his job successfully completed. Obviously, such resubmission increases the response time of the Grid. If a user needs a very fast response, he can give higher priority to the selection of the core Grid sites that provide more reliable service and hence shorter response time. This prioritization can easily be done by the RSL description of the job requirements.

Overall those who are concerned with speed and reliability can use the original core sites of the production Grid, those who are not so much concerned with speed and reliability can exploit the larger number of available volunteer Grid sites. All these are managed in a transparent way by a volunteer portal and broker service that are originally not available in the production Grid as a core service. So, introducing volunteer services into a production Grid can help in introducing volunteer sites as well.

4.2   Volunteer Site with Different Middleware

Many potential sites do not want to join a production Grid because they insist on their own Grid middleware installation. It is particularly true if their Grid middleware installation is more advanced than the one used in the production Grid. For example, UK NGS uses GT2 middleware and there are many universities in the UK where already GT4 middleware is installed and used. Obviously a production Grid cannot be directly extended with volunteer Grid sites if they rely on a different Grid middleware.

Again this problem could be solved by an advanced multi-Grid portal like the integrated GEMLCA/P-GRADE portal. This portal enables the access to any Grid resource that uses one of the following Grid middlewares: GT2, GT4, LCG-2, glite, NorduGrid. This feature means that if a GT4 site would like to join for example the UK NGS as a volunteer site, it is feasible and useful since any GEMLCA/P-GRADE portal user could immediately exploit the new GT4 site through the GEMLCA service of the portal. (Notice that without the volunteer portal service the original GT2 site users could not access the GT4 sites and hence GT4 sites could not support the NGS.) Moreover, the user can create a workflow where some of the nodes of the workflow are sent as jobs to the core NGS sites and some of the nodes are invoked as Grid services on the volunteer GT4 sites. In this way both the users and NGS can benefit from the availability of the new volunteer GT4 sites as shown in Figure 2.

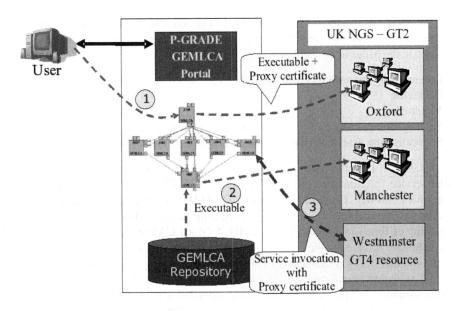

Fig. 2. Extending GT2 production Grids with GT4 resources

5   Conclusion

Volunteer services and volunteer Grid sites can play important role in future production Grid systems. There are many services needed by the production Grids (both users and Grid providers recognize their importance) but there is no manpower to develop these services by members of the production Grid. In this case, these services can be provided as volunteer services without compromising the robustness of existing services. This means that a large external research community can contribute to the development and maintenance of the new required services on a voluntary basis. If a volunteer service is widely accepted by the user community it can become officially part of the production Grid middleware stack after going through a thorough re-engineering and testing procedure like the one provided by OMII-UK.

This model of using volunteer Grid services has the following benefits:

Extends the services of production Grids without compromising their robustness.
Volunteer services should compete with each other and with the utility Grid services.
The best will be selected in a natural way to be used later as a core service.
Reduces the development cost of production Grids.

Attracts researchers to develop new Grid services.

Increases competition among Grid researchers but at the same time provides better conditions for them to test and evaluate their research results in real Grid environment by real Grid users.

Supports the collaboration of Grid researchers and Grid service/infrastructure providers.

Volunteer advanced Grid portal services integrated with other volunteer services can significantly extend the capabilities of production Grid systems. Several examples were shown in the paper in the field of

Providing legacy code services
Brokering
Monitoring
Solving Grid interoperability at the job and workflow level
Supporting the extension of production Grids with volunteer sites

P-GRADE portal is used as volunteer service for the GILDA [15] and as an official service for HunGrid [16], SEE-GRID [17], VOCE [18], EGRID [19] and many national Grids (e.g. CroGrid [20], Turkish Grid [21], etc.). Recently a volunteer service portal [23] was deployed at Worcester Polytechnic Institute (USA) to solve the interoperability of TeraGrid, OSG and SEE-GRID. A GEMLCA/P-GRADE portal [14] is used as volunteer service for the UK NGS where it extends the original GT2 sites of NGS with volunteer GT4 resources and GT4 Grids. The NGS portal is also extended with GMT providing monitoring service for the volunteer GEMLCA and GT4 resources of NGS. Another GEMLCA/P-GRADE portal extended with the GMT monitoring system is accepted as the official resource testing portal of GIN VO [22].

In summary we can say that taking into consideration all these advantages of the volunteer service and volunteer site concept this approach represents a very promising way of solving the sustainability of production Grids in a long term.

References

1. The UK NGS: http://www.grid-support.ac.uk
2. S. Burke et al.: glite 3.0 User Guide,
   https://edms.cern.ch/file/722398//gLite-3-UserGuide.pdf
3. P. Kacsuk and G. Sipos: Multi-Grid, Multi-User Workflows in the P-GRADE Grid Portal, Journal of Grid Computing, Vol. 3, No. 3-4, pp. 221-238, 2005
4. C. Blanchet and V. Lefort: GPS@: Bioinformatics grid portal for protein sequence analysis on EGEE grid, EGEE User Forum, Geneva, 2006
5. R. Barbera, A. Falzone and A. Rodolico: The GENIUS Grid Portal, Computing in High Energy and Nuclear Physics, La Jolla, California, 2003
6. http://www.gridsphere.org/
7. J. Frey: Condor DAGMan: Handling Inter-Job Dependencies,
   http://www.cs.wisc.edu/condor/dagman/, 2002
8. GIN VO Resource Test portal: https://gin-portal.cpc.wmin.ac.uk:8080/
   gridsphere/gridsphere
9. C. Catlett: Grid Interoperation Now DRAFT Charter, OGF documents, March, 2006

10. P. Kacsuk, T. Kiss and G. Sipos: Solving the Grid Interoperability Problem by P-GRADE Portal at Workflow Level, Proc. of the Grid-Enabling Legacy Applications and Supporting End User Workshop, in conjunction with HPDC'06, Paris, pp. 3-7, 2005
11. T. Delaitre, et al.: GEMLCA: Running Legacy Code Applications as Grid Services, Journal of Grid Computing, Vol. 3, No. 1-2, pp. 75-90, 2005
12. Globus Team, Globus Toolkit 4.0 Release Manuals:
    http://www.globus.org/toolkit/docs/4.0/
13. A. Kertesz: Brokering on Globus, 5th Int. Conf. of PhD Students, University of Miskolc, pp. 73-78, 2005
14. UK NGS portal:
    https://gngs-portal.cpc.wmin.ac.uk:8080/gridsphere/gridsphere
15. GILDA portal: http://portal.p-grade.hu/gilda/
16. HunGrid portal: http://portal.p-grade.hu/hungrid/
17. SEE-GRID portal: http://portal.p-grade.hu/seegrid/
18. VOCE portal: http://portal.p-grade.hu/voce/
19. EGRID portal: http://portale.egrid.it:8080/gridsphere/gridsphere
20. CroGrid portal:
    http://cro-grid-portal.irb.hr:8080/gridsphere/gridsphere
21. Turkish Grid portal:
    http://portal.grid.org.tr:8080/gridsphere/gridsphere
22. GIN VO portal: https://gin-portal.cpc.wmin.ac.uk:8080/gridsphere/gridsphere
23. WPI portal: http://pgrade.wpi.edu

PSO-Grid Data Replication Service

Víctor Méndez Muñoz[1] and Felix García Carballeira[2]

[1] Universidad de Zaragoza, CPS, Edificio Ada Byron, María de Luna, 1. 50018
Zaragoza, Spain
vmendez@unizar.es, eureka@nodo50.org
[2] Universidad Carlos III de Madrid, EPS, Edificio Sabatini, Av. de la Universidad,
30, 28911 Leganés. Madrid. Spain
fgcarbal@inf.uc3m.es

Abstract. Data grid replication is critical for improving the performance of data intensive applications. Most of the used techniques for
data replication use Replica Location Services (RLS) to resolve the logical name of files to its physical locations. An example of such service
is Giggle, which can be found in the OGSA/Globus architecture. Classical algorithms also need some catalog and optimization services. For
example, the EGEE DataGrid project, based in Globus open source components, implements for this purpose the Replica Optimization Service
(ROS) and the Replica Metadata Catalog (RMC). In this paper we propose a new approach for improving the performance of Data grid replication. With this aim, we apply Emergent Artificial Intelligence (EAI)
techniques to data replication. The paper describes a new algorithm for
replica selection in grid environments based on a PSO-LRU (Particle
Swarm Optimization) approach. For evaluating this technique we have
implemented a grid simulator called SiCoGrid. The simulation results
presented in the paper demonstrate that the new technique improve the
performance compared with traditional solutions.

1 Introduction

Grid replication of remote data is critical for data intensive enterprise and scientific applications, mostly implemented over Globus middleware[1]. Virtual Organisations are usually geographical and user affinity communities around a big
data producer, in the scale of Tera Bytes a day, with the aim of extract information from this read-only remote data, by running jobs on the Grid. On this
context replication is used for fault tolerance as well as to provide load balancing
by distributed replicas of data.

The OGSA[2] and therefore the Globus Toolkit 4.0 assumes the Giggle[3]
as a framework for constructing scalable Replica Location Services(RLS) that
allows the registration and discovery of replicas. Given a logical identifier of a
file(LFN), the RLS must to provide the physical locations of the replicas for
the file(PFN). The RLS consists of two components. Local Replica Catalogs
(LRCs) manages consistent information about logical to physical mappings on
each site or node. Replica Location Indices (RLIs) hold the information about

M. Daydé et al. (Eds.): VECPAR 2006, LNCS 4395, pp. 656–669, 2007.
© Springer-Verlag Berlin Heidelberg 2007

the mappings contained in one or more LRC. Strong consistency is not required on the RLIs, a soft protocol send LRC state information to connected RLIs, which then incorporate this information into their indices and delete time outs entries. The basic Giggle architecture on figure 1 shows two layers, but the architecture is usually configured on N layers of hierarchical RLI.

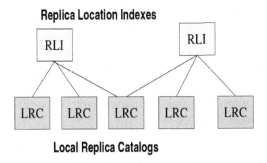

Fig. 1. Basic RLS Architecture

Many research groups have developed algorithms for replica selection and location functionalities, based on the Giggle[3] RLS architecture, using hierarchical RLS topologies, that are characterized with six parameters shown on the Table 1 of the contribs section, in wich we have added some new values that will be explained.

Other important OGSA/Globus data Grid service components are: GridFTP a not Web Service(WS) component for files transfer, Reliable File Transfer (RFT) for GridFTP monitoring, Data Replication Service (DRS) is the WS component that encapsulate the non-WS RLS and RFT for GT4, OGSA-DAI it is a WS GT4 component for relational data base and XML objects replication. Furthermore, usually it is need some aditional funtionalities, thus the EGEE DataGrid has the ROS and RMC components for the data Grid service framework.

Next section of this paper describes the related work on some aspects of data Grid service:

- Replica state of the art algorithms.
- We analyze the research branches to get some theoretical conclusions.
- We describe the features of the Grid simulators used for experimental test of this algorithms.

After related work section we explain the main contributions of our approach: a framework review for an enhanced Giggle, a better performance algorithm for replica selection based on PSO and LRU.

On the fourth section we explain the evaluation methodology and we present SiCoGrid, and on fifth we present experimental results of our improved approach to the data Grid. Finally we summarise some conclusions.

2   Related Work

Chervenak et al.[3] present some initial performance results for five canonical implementation approaches based on following Giggle configurations:

- RLS1: Single RLI for all LRCs.
- RLS2: LFN Partitioning, Redundancy, Bloom Filters.
- RLS3: Compression, Partitioning based on Collections.
- RLS4: Replica Site Partitioning, Redundancy, Bloom Filters.
- RLS5: A Hierarchical Index.

They use prototype implementations that show good scalability but does not include network simulation, the prototype is focused on disks throughput, but both disks and network could be system lack depending on study issue class.

There are some approaches that propose an economical algorithm for replica selection where the costs of file transfers are evaluated as generic equation 1:

$$cost(f, i, j) = f(bandwidht_{i,j}, size_f) \qquad (1)$$

Lamehamedi and Deelman approach[4] uses bouth hierarchical and flat propagation graphs spanning the overall set of replicas to overlay replicas on the data grid and minimizing inter-replica communications cost. Beginning on the hierarchical Giggle topology they introduce a flat-tree structure with redundant interconnections for its nodes; closer the node is to the root, more interconnections it has. The flat-tree was originally introduced by Leisersons[5] to improve the performance of interconnection networks in parallel computing systems. Lamehamedi et al. identifies on this approach that flat-tree on a ring topology suits best than hierarchical with multiple servers or peer replica applications. For simulation framework they use a network simulation[6], without consider the disks throughput, so results are limited by the premise that the system lack is on the network. Anyway they obtain rough network resource consumption evaluation comparing with the pure hierarchical RLS.

Another economic approach[7][8] understand the Grid as a market where data files represent the goods. They are purchased by Computing Elements for jobs and by Storage Element in order to make an investment that will improve their revenues in the future. The files are sold by Storage Elements to Compute Elements and to other Storage Elements. Compute Elements try to minimise the file purchase cost and the Storage Elements have the goal of maximising profits.

When a replication decision is taken, the file transfer cost is the price for the good, like the function 1 show above. The Replica Optimiser may replicate or not based on whether the replication(with associated file transfer and file deletion) will result in to reduce the expected future access cost for the local Computing

Elements. Replica Optimiser keeps track of the file requests it receives and uses an evaluation function: $E(f, r, n)$, defined in [9] that returns the predicted number of times a file f, will be request in the next n, based on the past r request history base line. The prediction function E is calculated for a new file request received on Replica Optimiser for file f. E is also calculated for every file in the storage node. If there is no file with less value than the value of new file request f, then no replication occurs. Otherwise least value file is selected for deletion an new replica is created for f.

The research group that propose this approach also present OptorSim [10] [11], the first Grid simulator that holds network and in some way disk costs. The first version was time driven but second version is event driven and it also has others scheduling improvements[12]. Results [7] present some specific realistic cases where the economic model shows marked performance improvements over traditional methods.

A Peer-to-Peer replica location service based on a distributed hash table[13] is fill on Giggle with Peer-to-Peer-RLI(P-RLI). P-RLI uses the Chord algorithm to self-organise P-RLI and it exploits the Chord overlay network to replicate P-RLI mappings. The Chord algorithm also route adaptively the P-RLI logical names with LRC sites. The replication of mappings provides a high level of reliability in the P-RLI, the consistency is stronger than in simple RLI nodes. The P-RLS performance is tested on a 16-node cluster scale with the network size. It is also tested with a simulation for larger network of P-RLI nodes, evaluating the distribution of mappings in the P-RLS network. The simulation for this test section is not a complete simulation of the P-RLS system, but rather, it focuses on how keys are mapped to the P-RLI nodes and how queries for mappings are resolved in the network.

Nowadays there are many approaches with similar methods and similar performances as state of the art above. Other descentralized adaptive replication mechanism[14] organise nodes into overlay network and distribute location information, but do not route requests. Each node that participates in the distribution network build, in time, a view of the whole system and can answer queries locally without forwarding request. Unfortunatelly this is not common on large scale scientific datasets, that suppose the most of the operative Grid infrastructures.

3   Contributions

3.1   Proposed Data Grid Service Framework

The enhanced Giggle shown on Table 1 avoid the restrictions for the flat approaches. Now it is not necessary to store the LFN mapping out of the local node. It is not necessary to implement any RLI layer on the architecture. Therefore the RLS is completely consistent. On the function used to partitioning the LFN name space, we add a entry for flat architecture with no partitioning by LFN. Every LRC manage the name space locally independent. First introduced value is for G = 0 pointing out a flat RLS composed only by LRCs and no RLI layer. There are a no partitioning actions for LFN names space ($P_L = flat$), and no partitioning the RLI name space ($P_R = flat$). For the degree of redundancy in the index space we add a new case R = 0 for the LFN mapping only on the LRC. We also have include economic and flat heuristic for possible S values (the function used to determine what LRC information to send to other catalog entities and when).

Table 1. The six parameters enhanced Giggle RLS structures and values

G		The number of RLIs
	G = 0	A flat partitioned index, only LRC on a flat layer
	G = 1	A centralised, non-redundant or partitioned index
	$G > 1$	An index that includes partitioning and/or redundancy
	$G \geq N$	A highly descentralized index
P_L		The function used to partitioning the LFN name space
	$P_L = O$	No partitioning by LFN. The RLIs must have storage to record information about all LFNs, a large number
	$P_L = hash$	Random partitioning. +load balance, -locality
	$P_L = coll$	Partitioning on collection name. -load balance, +locality
	$P_L = flat$	No partitioning by LFN. Every LRC is locally manage
P_R		Function used to partition replica site name space
	$P_R = 0$	No partitioning by site name. Indices have entries for every replica of every LFN they are responsible for.
	$P_R = IP$	Partitioning by domain name or similar.
	$P_R = flat$	There are no index for partitioning site name space.
R		The degree of redundancy in the index space
	R = 0	The LFN mapping is only on the LRC
	R = 1	No redundancy: each replica is indexed by only one RLI
	$R = G > 1$	Full index of all replicas at each RLI. Implies no partitioning, much redundancy/space overhead.
	$1 < R < G$	A highly descentralized index.
C		The function used to compress LRC information
	C = O	No compression: RLIs receives full LFN/site information
	C = bloom	RLIs receive bloom filters summaries
	C = coll	RLIs receive summaries based on collection distribution
S		Function to set what LRC information to send where
	S = full	Periodically send entire state to relevant RLIs
	S = partial	In addition, send periodic summaries of updates
	S = economic	Every economic decision send entire state to RLIs
	S = flat	Only statistical information is send for flat heuristic

At the end of the day we will have a stand alone LRC for each node, with a local location service and the need of an implicit global location interface as show figure 2, with a distributed RLS service on each node.

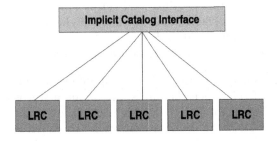

Fig. 2. Flat RLS architecture over enhaced Giggle

The data Grid service framework proposed, will complementary need a modified ROS, with flat heuristic features on two maners:

- Those algorithms like PSO that need to seend some statistical information, bind pear to pear conexion between ROS servers in each node.
- Other Emergent Artificial Intelligence (EAI) algorithms are stand alone ROS servers, and does not need any control information transfer.

There also is a distributed ROS service.

The typical RMC service si not necesary for our goal. We do not use a GUID, because each local catalog make mapping betwing LFNs and PFN in a oneness way for the local node. But the LRC will need two aditional entries for the metadata information and the original producer node of the file. So we use an enhaced RLS with some soft catalog funtionalities.

This new theoretical approach requires an heuristic that realizes enough performances with only statistical information about LRC, and a request routing scheme self described. This is our goal on the next subsection proposing PSO file location and selection scheme and LRU deletion mechanism as an alternative to traditional approaches. Our data Grid service framework is also valid for any new approach that may walk on flat heuristic way.

3.2   The Algorithm: PSO-Grid

PSO is an Emergent Artificial Intelligence technique. EAI is an Artificial Intelligence branch that uses the natural social behaviour as ant colons or PSO[15] inspired on bees swarm or birds flocks searching food. PSO has been proved as a valid approach for many different real solutions[16][17].

On Grid environments we introduce some tactic modifications, based on the strategy "follow the closer bird from the food chunk" as social PSO flavour.

- A bird flock is in a random search for food in an area.
- For each bird there is only one valid kind of food.
- The bird does not known where is the food chunk, but its known how long is from the different areas and it know how many birds are finding they food

chunk on this areas, this is called food chirp. This is the social component of our approach, thus the distance to the food chunk is calculated for each bird flock, not for individual birds.
- The strategy is to follow the closer bird flock with best success food search.

Translating this analogy to the Grid, we suppose that a file location request is a bird searching food. When the bird stand on an area it is on a Grid node, when the bird fly looking for food to another node is moving through the Remote Network. The bird takes the decision from where to search based on the flock food chirp, that is the best performance external hit ratio of different nodes. On the other hand, the food chirp will decreased across distance. If the bird is over-flying a node and find food then it will change direction to get it, if the bird arrive to destination and is no food then start again from this point. Thus the performance function for file f to node j from node i looks as following. The PSO-Grid uses a performance metric for a file replication between two nodes i, j, defined in equation 2. We use b as the identifier of the node with the best performance metric asociated to i, from the evaluated j nodes. Initially b is the producer node of the replica, that will be return by LRC soft catalog metadata information described above. We use e for the external hit ratio and c for the network cost defined below on equation 3.

$$p_{i,j} = (e_j * c_{i,j}) + ((1 - e_b) * c_{i,b}) \qquad (2)$$

The external hit ration is calculated based on N lasts external success request ratio on node j. The external ration events are the information that is sent from one ROS in each node to another. Considering network access cost we propose the following:

$$c(i,j) = lt_{i,j} * c1 + (MAX_{BW} - bw_{i,j}) * c2 \qquad (3)$$

Latency is a constant but do not mean neutral on transfers[18], the latencies are growing from one network to an other, the bandwidth on a network connection is the minimal bandwidth assigned from one network to another.

For our case $c1 = 1$ and $c2 = 0.2$. On the equation 3 $c1$ and $c2$ are coefficients that balance the relative relevance between latency and bandwidth, they also should fit with the bandwidth and latency values of the specific Grid infrastructure, and also fit with their measure relationship (ms. and MB/s.). At the end of the day latency is more important than bandwidth, because latency is always constant, and bandwidth has a variable behaviour depending on sockets allocations and number of network request in a specific moment. MAX_{BW} is the highest bandwidth of all the Grid infrastructure.

The performance function 2 is balancing the probability of find a replica in a node j with the probability of not finding on j, where we have to reply from the node with best metric b, initially the producer.

On the PSO-Grid algorithm the file request is a bird or more formally a particle, and the particles in a Site are a swarm. When the file object is found the particle died and the file reply is done with traditional routing methods. The basic algroithm is exposed on the next pseudo-code.

```
Loop
    For each particle not finding file on node i
        Initialize best node as requested file producer site,
            and best metric as performance from i to producer.
        For each node j, not i, from the Grid
            If actual performance(i,j) is less than best one.
                Get new performance(i,j) as best metric,
                    and j as best node.
            End if
        End for each j.
        Launch replica request to best node.
    End for each particle.
Until End Condition.
```

The deletion decision is taken in each node only to serve local request, using the LRU or LFU algorithm for select the file target. When a file deleted is on process to remote node reply, the node trigger a new PSO reply in the name of the in-reply remote node for the rest of the file transfer.

4   Evaluation Methodology

As we have seen on related work section, the best reliable Grid simulation shape should consider disks throughput and network traffic. For a reliable evaluation methodology we have design a Grid simulator with both of them. The complete SiCoGrid (*Simulador Completo Grid*) toolkit includes:

- Main Grid event driven simulator develop in Parsec[19].
- Workload generator, developed on C.
- E/S Subsistem: integrating DiskSim[20] on the general Grid simulator with a parsec interface.

4.1   Data Grid Simulation Design

Figure 3 presents the SiCoGrid UML design, with all the modelled components and some of the main attributes.

The toolkit includes a workload generating program, represented on the figure 3 as the *Access Pattern*. The *log* is writed on file for the given input arguments: access pattern, random seed, number of Grid clients by node, number of jobs by Grid client.

The access pattern are full file, sequential block access, random, unitary random walk, gaussian random walk, same as OptorSim simulator[10][7]. The random seed is for statistical experiment repetitions. Number of Grid Clients in a node is a component of the simultaneous request on a node. The number of jobs by grid client is a temporal lenght component of the simulation.

The *Client Data Grid*, *Storage Element* and *Computer Element* are conected across the *Local Network*. The *Grid node* is directly conected to the RB, on

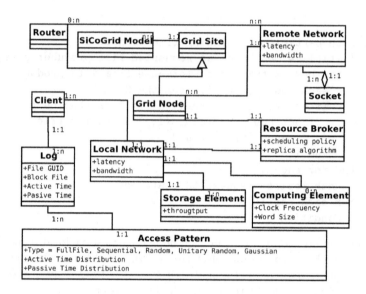

Fig. 3. SiCoGrid UML design

the same machine. Between nodes there is a remote network simulation with an infrastructure described on network configuration file.

On this environment each job will request many file blocks. The workload application return for each file request an *Active Time* and a *Passive Time*. Those times are empirical model of Web document arrivals at access link[21]. After a job get a file block response, it spend an *Active Time* for process the block part of the job, this time is calculated based on Computer Elements featured specifications on network configuration file. The last job requested block also uses an extra *Active Time* slice for process the final job computing events, and it is shaped with Weibull distribution with $a=1.46$ y $b=0.382$, and infinite mean and variance, that is characteristic of Web Services [22]. *Passive Time* is the time that the user hold between one job and another. For this parameter we use a Pareto distribution with $k=1$ and $alpha = 0.9$ with infinite mean and variance, that is a characteristic Web Service users distribution[22].

The *Grid Site* may inherit in a *Router*, and *Grid Node* with assigned Computing and Storage Elements. All Grid site components are connected to a *Local Network*, usually token media access with a hight bandwidth if we compare with *Remote Networks*.

Each *Remote Network* has aggregate various *Socket* instantiations that implements partial bandwidth of the total assigned for the Remote Network. SiCoGrid response as real systems: when the available bandwith is close to the top, more bandwith is asigned for a network transaction, when is close zero, then less bandwidth is asigned.

Pure data Grid uses computing on the client, thus all the active and passive time is consumed on the *Grid client*, as show figure 4

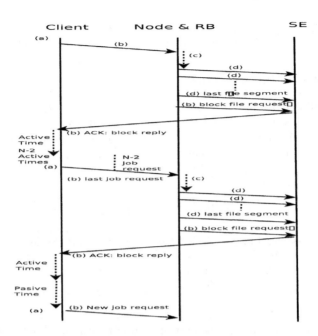

Fig. 4. Pure Data Grid Simulated Protocol

(a): The *client* read the request from the log file. (b): The *client* launch a request on the site or node, through the *Local Network* to the *Resource Broker*, that will manage the request in order to return to the clien t the appropriate data and/or computing results. (c): If the requested file is not on the site, then the RB pass the request to the *node* that depending on the replication algorithms, it route to the appropriate *Remote Network* instantiation. If the requested file is on the site, then go below to (b) RB pass the request to SE. (d): Asynchronous data replies from remote sites are received on the *node*, that send it to the corresponding SE. (b) RB pass the block request to the SE. (b)-ACK: Requested block file is send from SE to Client data Grid.

The drawback of the presented SiCoGrid is that suppose unlimited buffering resources for scheduling and replica algorithms. This is a small lack comparing with the much more significant assumptions of others Grid simulators.

One of the SiCoGrid meaningful features is that it launch a thread for any entity instantiation, taking advance over other simulators toolkits, that uses one thread for each task(task as a part of a job) in a way that avoid uses for large scale simulations.

SiCoGrid file deletion mechanism could be Least Frecuent Used(LFU) and Least Recent Used(LRU), but LRU performs better and the used on this data Grid simulation design. SiCoGrid implements the spcefic market economic model deletion scheme, that also uses LRU for secondary deletion decisions. For replica selection, location and optimisation we use three algorithms: Unconditional replication, this is the canonical that always take the replica source from the Grid

site file producer; Market Economic Model, the OptorSim scheme[10][7], Particle
Swarm Optimisation approach, as we explain on a previous sections.

4.2 Simulation Infrastructure

We have configured our SiCoGrid for a common Grid stage[23] shown on the
Table 2. These is the typical CERN datagrid specification for node tier class of
a Virtual Organisation. The storage capacity, file size, and network bandwidth
is scaled in the magnitude of twenty, for time simulation reasons. Therefore the
obtained time results will be on the same magnitude.

Table 2. Scaled Grid Stage

Tier Class	Real MB/s	Scaled Mb/s / 20	Real TB	Scaled TB / 20
1	2048	102.4	220	11
2	320	16	100	5
3	10	0.5	20	1

On the figure 5 we can see the network infrastructure used in our experiment.
The graph disposes a nomenclature where the nodes has a first number that is
the tier class, and after the point another identification number. Below there is
the storage size of the node in TB. The networks have assigned two numbers,
the first one is the latency in ms and the other is the bandwidth in MB/s.

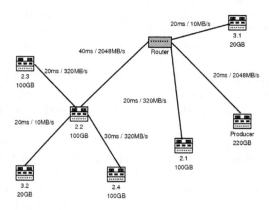

Fig. 5. Simulated Grid Stage

5 Simulation Results

Following figure 6 presents Grid simulation results based on the stage described
above, with three statistical repitions for each experiment simulation. We use

Gaussian random walk, that is the best performance for the state of the art economic OptorSim approach[10]. The job response time is scaled in the magnitude of 20 to usual jobs duration from hours to some days. There are shown different Grid sizes, as combination of the number of Grid clients by node and the number of jobs submitted by a Grid client: 4X4, 4X5, 4X6, 4X7, 5X4, 5X5, 5X6, 5X7, 6X4, 6X5, 6X6, 6X7, 7X4, 7X5, 7X6 and 7X7, so the serie increases the number of jobs for each clients by node configuration. We present results for the best performance LRU over LFU deletion scheme.

The figure 6 shows the unconditional-LRU performances with black line, the market economic modell on dark gray, and the PSO-Grid with LRU deletion on light gray. As it was expected the unconditional used for base compare, has the worst results. PSO-Grid is the best performances for all simulation series shown on the graph, and also for all the repetitions. PSO-Grid response rate is around 30% faster than canonical algorithm, and 15% faster than the market economic modell.

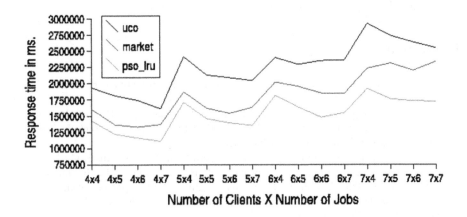

Fig. 6. Results in Simulated Grid Stage

Furthermore we compare canonical line with others, and we can see that PSO-Grid fits well with the unconditional pattern but with lower response time results. Thus PSO-Grid has good scalability features in our experiment. Market economic modell starts following this canonical pattern, but with heavy workload experiments, on the right part of the chart, it is quickly prone to join with canonical results, showing bad scalability features.

PSO-Grid performance is better due to its features: less control trafic, distributed optimization, localization and selection services, autonomous management of each node wich will fit best on user and geografical afinities, colaborative strategie against competitive strategie of the economic, that usually performs better on the long term.

6   Conclusions and Future Work

We have described two relevant contributions to the Data Grid corpus. The enhanced Giggle framework that consider flat RLS structures, opening the door to the EI and other EAI approaches for the OGSA data Grid replication architecture. Specific PSO-Grid algorithm has been proved as the better performance job response time and much better scalability features than traditional approaches, using a full network and disk subsystem simulation, with SiCoGrid toolkit.

We have open research lines for the following targets: Cyclical graph grid infrastructure simulations, other emergent EAI algorithms like Ant Colony Optimization and a depth variable correlations studies.

Acknowledgments

This work has been supported by the Spanish Ministry of Education and Science under the TIN2004-02156 contract.

References

1. Foster, I., Kesselman, C.: Globus: A metacomputing infrastructure toolkit. IJSA **11**(2) (1997) 115–128
2. Foster, I., Kesselman, C., M.Nick, J., Tuecke, S.: The physiology of the grid an open grid services architecture for distributed system integration. Technical report, Globus Proyect Draft Overwiev Paper (2002)
3. Chervenak, A.L., Deelman, E., Foster, I., Iamnitchi, A., Kesselman, C., Hoschek, W., Kunszt, P., Ripeanu, M., Schwartzkopf, B., Stockinger, H., Stockinger, K., Tierney, B.: Giggle: A framework for constructing scalable replica location services. In: Proc. of the IEEE Supercomputing Conference (SC 2002), IEEE Computer Society Press (November 2002)
4. Lamehamedi, H., Szymanski, B., shentu, Z., Deelman, E.: Data replication strategies in grid environments. In: Proceedings Of the Fifth International Conference on Algorithms and Architectures for Parallel Processing, ICA3PP02) (2002)
5. Leiserson, C.H.: Flat-trees: Universal network for hardware-efficient supercomputing. IEEE Transactions on Computers **C-34**(10) (1985) 892–901
6. http://www mash.cs.berkeley.edu/ns: Ns network simulator (1989)
7. Bell, W.H., Cameron, D.G., Capozza, L., Millar, A.P., Stockinger, K., Zini, F.: Simulation of dynamic grid replication strategies in optorsim. In: Proc. of the ACM/IEEE Workshop on Grid Computing (Grid 2002), Springer-Verlag (November 2002)
8. Cameron, D.G., Carvajal-Schiaffino, R., Millar, A.P., Nicholson, C., Stockinger, K., Zini, F.: Evaluating scheduling and replica optimisation strategies in optorsim. In: International Workshop on Grid Computing (Grid 2003), IEEE Computer Societe Press (November 2003)
9. Capozza, L., Stockinger, K., , Zini., F.: Preliminary evaluation of revenue prediction functions for economically-effective file replication. Technical report, DataGrid-02-TED-020724, Geneva, Switzerland, July 2002 (July 2002)

10. Bell, W.H., Cameron, D.G., Capozza, L., Millar, A.P., Stockinger, K., Zini, F.: Optorsim - a grid simulator for studying dynamic data replication strategies. International Journal of High Performance Computing Applications **17**(4) (2003)

11. Cameron, D.G., Carvajal-Schiaffino, R., Millar, A.P., Nicholson, C., Stockinger, K., Zini, F.: Analysis of scheduling and replica optimisation strategies for data grids using optorsim. International Journal of Grid Computing **2**(1) (2004) 57–69

12. Cameron, D.G., Carvajal-Schiaffino, R., Millar, A.P., Nicholson, C., Stockinger, K., Zini, F.: Optorsim: A simulation tool for scheduling and replica optimisation in data grids. In: International Conference for Computing in High Energy and Nuclear Physics (CHEP 2004), Interlaken (September 2004)

13. Min Cai, Ann Chervenak, M.F.: A peer-to-peer replica location service based on a distributed hash table. In: Proceedings of the High Performance Computing, Networking and Storage Conference, SCGlobal (2004)

14. Ripeanu, M., Foster, I.: A decentralized, adaptive replica location mechanism. In: 11th IEEE International Symposium on High Performance Distributed Computing (HPDC-11). (2002)

15. Shi, Y. ;Eberhart, R.: A modified particle swarm optimizer. In: Proceedings of the IEEE International Conference on Evolutionary Computation, IEEE Press. Piscataway, NY (1998) 69–73

16. Cockshott, Hartman: Improving the fermentation medium for echinocandin b production. part ii: Particle swarm optimization. Process Biochemistry **36** (2001) 661–669

17. Yoshida, Kawata, Fukuyama: A particle swarm optimization for reactive power and voltage control considering voltage security assessment. IEEE Trans. on Power Systems **15** (2001) 1232–1239

18. Cheshire, S.: It's the latency, stupid. Technical report, Stanford University (1996)

19. Leijen, D.: Parsec, a fast combinator parser. Technical report, Computer Science Department, University of Utrecht (2002)

20. R.Granger, G., L.Worthington, B., N.Patt, Y., eds.: The DiskSim Simulation Environment. Version 2.0 Reference Manual. University of Michigan (1999)

21. Deng, S.: Empirical model of www document arrivals at access link. In: Proceedings of the 1996 IEEE International Conference on Communication, IEEE-P (1996)

22. Barford, P., Crovella, M.: Generating representative web workloads. In: Network and Server Performance Evaluation In Proceedings of the 1998 ACM SIGMETRICS International Conference on Measurement and Modeling of Computer Systems, ACM SIGMETRICS (1998) 151–160

23. Ranganathan, K., Foster, I.: Identifying dynamic replication strategies for a high-performance data grid. Technical report, Departament of Computer Science, The University of Chicago (2000)

Execution Management of Scientific Models on Computational Grids

Alexandre Vassallo[1], Cristiane Oliveira[1], Carla Osthoff[1],
Halisson Brito[2], Julia Strauch[3], and Jano Souza[2,4]

[1] LNCC – Brazilian Scientific Computing Laboratory
Av. Getúlio Vargas, 333, Quitandinha, ZIP Code: 25651-075, Petrópolis, RJ, Brazil
{alex, cris, osthoff}@lncc.br
[2] COPPE/UFRJ – Systems Engineering and Computer Science Program
Federal University of Rio de Janeiro – PO Box 68511, ZIP Code: 21945-970, Rio de
Janeiro, RJ, Brazil
{hmbrito, jano}@cos.ufrj.br
[3] ENCE/IBGE – Brazilian School of Statistical Sciences, R. André Cavalcanti, 106, s. 401
ZIP Code: 20231-050, Rio de Janeiro, RJ, Brazil
juliast@ibge.gov.br
[4] IM/UFRJ – Institute of Mathematics/Federal University of Rio de Janeiro
PO Box 68511, ZIP Code: 21945-970, Rio de Janeiro, RJ, Brazil

Abstract. This paper presents ModRunner, a scientific model execution manager running on a Grid platform. ModRunner is part of the MODENA environment. Besides model execution, MODENA also deals with knowledge management in scientific models. It also works as a model library allowing for cataloguing, searching, reusing and generating scientific models. ModRunner is a simple and effective Grid Computing access system that facilitates management of independent task execution on the Grid. In this paper, we present ModRunner running over the Grid Computing middleware OurGrid. As a case study we have been using ModRunner to schedule, submit and manage tasks for the execution of Population Dynamics models.

1 Introduction

Model management has been the subject of several scientific works, ranging from model creation and execution to result analysis and model feedback, as stated by [1].

Models can generally be described as simplified representations of reality, whose goal is to abstract the reality portion which matters for the solution of a problem. Besides, models contain relevant information on phenomena or processes with the advantage of hiding irrelevant details of real problems.

In scientific work, phenomena or processes are usually complex and unknown. So, models may be used to represent them, being essential parts of any scientific experiment. An experiment usually tries to verify (either positively or negatively) some hypothesis stated by a scientist and it may have an underlying model, or even a combination of models on the phenomenon it intends to prove. So models play an important role both in research and practical applications in many fields of knowledge.

M. Daydé et al. (Eds.): VECPAR 2006, LNCS 4395, pp. 670–678, 2007.
© Springer-Verlag Berlin Heidelberg 2007

In this work we consider model execution as the steps of running "model instances", like programs or workflow definitions, in order to perform the simulation process. According to [2], the simulation process made with scientific experiments usually does the transformation of input data to produce data with added scientific value.

In order to support model execution, we present ModRunner, a simple and effective web tool to perform the execution on Grid platforms. It can be used to encapsulate tasks submission to the Grid, providing a management layer over that submission. It provides easy to use interfaces to perform management issues like capturing model parameters, obtaining remote input data, scheduling execution submissions, submitting a model to execution, keeping a history of each execution instance and storing result data.

ModRunner is part of MODENA [3], an environment for scientific model management on a Computational Grid platform. This environment has been developed to support researchers of the Geoma Project (Thematic Network for Research in Environmental Modeling of the Amazon) [4], which aims at the development of models to evaluate and foresee sustainability scenarios under different kinds of human activities and public policies for the Amazon.

MODENA is aimed at providing an infrastructure that allows geographically distributed research institutions to share data, metadata, models, knowledge, and workflow definitions, as well as to share model execution in high performance environments, through an uniform Grid Computing platform. The MODENA environment has to provide, at the same time, client and data server features to 1) reduce data, information and knowledge acquisition costs, 2) avoid data duplication, 3) reduce data processing and selection time, 4) reduce environmental data analysis and execution time, and 5) generate simulation models and environmental scenarios.

ModRunner has been developed to run either on Grid Workflow platforms, like Globus [5] or on Bag of Tasks (BoT) platforms, like OurGrid [6]. This paper presents ModRunner running on OurGrid, which is the part that is in its most advanced development stage.

We have been using population dynamics models to validate our proposals. Those models try to investigate the control of the mosquito population in order to reduce the number of malaria cases in the Amazon region.

This work is organized as follows: the second section presents the case study applied to population dynamics; the third discusses some related works in the technologies employed here; the fourth describes the Grid Computing middleware architecture used in this work; the fifth section does a brief review of the MODENA environment and presents the ModRunner task execution management system; and finally the sixth section presents final considerations and indications of future works.

2   The Case Study on Population Dynamics

Malaria is a serious public health problem around the world, affecting 40% of the population of more than 100 countries [7]. According to the World Health Organization, about 300 to 500 million new cases and 1 million deaths happen each year. In Brazil, 99% of all cases occur in the Amazon region where about 500 thousand cases are reported every year.

In the last few years, scientists have been working to create genetically modified mosquitoes in order to encapsulate the malaria plasmodium. These mosquitoes should couple to wild mosquitoes, and introduce refractory genes into wild mosquito populations.

The substitution of wild mosquito populations for genetically modified ones aims at the reduction or elimination of disease transmission, since a vaccine for malaria has not been discovered yet. This attempt should however begin only after a rigorous study on the feasibility of the control strategy and on the side effects has been made [8].

As part of the GEOMA Project, a recent work [8] aims to analyze a mathematical model consisting of a system of ordinary differential equations, which represents the main characteristics of the population dynamics for *Anopheles darlingi* in areas of the Brazilian Amazon. The model also takes into account the seasonal variation of the density of the mosquito population due to water level fluctuations.

This is an example of an application that may be shared and managed by geographically distributed researchers from distinct fields of research. So, researchers could benefit from the management issues of MODENA and the execution management issues of ModRunner.

3   Related Works

Some related works are described in the literature. Allcock et al [9] argue that service requirements involved in data transport over Grids to high-performance, distributed data-intensive applications are: i) secure, reliable and have efficient data transfer; and ii) able to register, locate, and manage multiple copies of datasets. The authors also presented the design and implementation of the GridFTP protocol which implements extensions to FTP that provide GSI security and parallel, striped, partial, and third-party transfers in a Globus environment.

Karnik and Ribbens [10] presented an approach based on a data-centric framework that offers a high-level architecture for Grid Computing Environments based on layers with clear interfaces defined in three entities, as follows:

- Model: A model, in the context of the work of [10], is a directed graph of specific executable pieces defining the control-flow and data-flow in a computation.
- Model Instance: A model instance is a model with all parameters specified.
- Simulation: A simulation is a model instance assigned to and run on a particular computational resource.

That architecture consists of three tools: job submission, parameter sweep and simulation lookup. We highlight the parameter sweep that is composed of three subsystems: i) an XML Generator that produces an XML representation of a typical input file, identifying the various parameters in it; ii) a Parameter Sweep Definition tool that allows the user to interactively indicate parameters and ranges that define an experiment, and use the XML file to produce a parameterized input file; and iii) a Sweep-Engine. This is interesting because it distinguishes the model from its representation, although this approach accepts some parameters that may not be specified until runtime.

Zang, Wu and Wang [11] presented grid workflow based on dynamic scheduling and performance evaluation implemented over standards as GCC and WFMC, which consists of user portal, resource management component, grid services management, performance management and grid workflow engine featured by dynamic scheduling.

W. Cirne et al. [12] present an OurGrid Molecular Dynamic Simulation applications platform. The platform provides an out-of-box solution to Grid users. However, the system is not integrated to any scientific model management database system.

The MODENA proposal, besides doing model management also does knowledge management, bringing a novelty and contributing with the dissemination of the knowledge about scientific models. Furthermore, it offers transparency to the user-researcher in the access to high performance environments based on the OurGrid platform. So, one of the main advantages of ModRunner is that the end-user does not need to know low level details of grid configuration or submission to have one's tasks executed in a grid environment. By using a simple Web user interface, a scientist with no deep knowledge of grid details may submit his/her models to execution, track the execution evolution and get the results, obtaining access to the benefits of using a grid environment.

4   Grid Computing Middleware Architecture

OurGrid Middleware [6] is a production-quality solution for users who want to execute Bag-of-Task (BoT) applications on a computational grid today. BoT applications are parallel applications whose tasks are independent. OurGrid provides an important platform for users that do not want to pay the cost of the installation and deployment of a heavy grid software such as Globus Middleware [5].

OurGrid design goals are to be a simple, complete and encompassing Grid Computing Platform. In other words, it means that the system has to be as close as possible to an out-of-box solution. The idea is that when a user wants to run his/her application, the last things he/she wants to be concerned with are grid details. Complete means that the system must cover the whole production cycle, from development to execution, passing by deployment and manipulation of input and output. Finally, the system is encompassing in the sense that all machines the user has access to can be used to run his/her BoT applications.

The OurGrid middleware assumes that the user has a machine, which is called the home machine, which coordinates the execution of BoT applications in OurGrid. The user submits tasks that form the application in the home machine, which is responsible for performing the tasks in the user's grid. The home machine schedules tasks to run on grid machines.

OurGrid provides simple abstraction through which the user can easily deal with the grid, hiding away the nonessential details. It schedules the application over whatever resources the user has access to, whether this access is some grid infrastructure, such as Globus or via simple remote login (such as ssh).

5 The System for Execution Management of Scientific Models

In our work, model execution management is supported by a set of tools that perform other model management tasks, within the MODENA environment. MODENA (Scientific Models Management Environment) has four layers (Fig. 1).

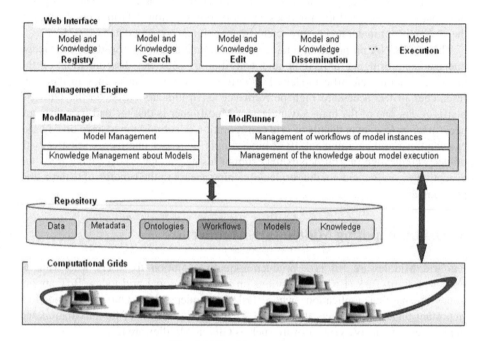

Fig. 1. MODENA Architecture

The first layer is the Web interface, where the user can have access to management functionalities. The second layer is the management one, which provides several features for knowledge management about models and execution management of model instances; the third comprises the MODENA repository layer, which contains data, metadata, workflows, models, knowledge and ontologies; and the fourth is the grid access layer, where the model instances are really executed.

The second layer of MODENA consists of two systems. The first one, named ModManager [13], comprises a system for knowledge management on scientific models, responsible for activities like capturing, retrieving, generating and exchanging data, metadata and knowledge. Fig. 2 shows the ModManager screen responsible for model metadata registration. The screen portion presented corresponds to the registration of model parameters. Equations, workflow definitions, algorithms, programs, default data, among many other model features, can also be registered.

It also enables model composition, which is the base for the workflow composition that originates chained model execution.

The second system, which is the subject of this paper, is called ModRunner, a developing tool to perform computer simulations through the execution of instances

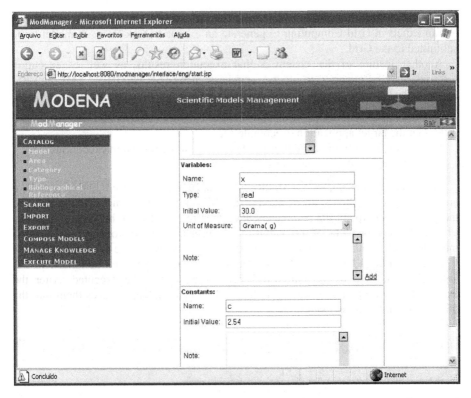

Fig. 2. Model metadata registration

of the models stored in the database. These instances may be formed by workflow definitions that represent the steps of model execution. A workflow editor permits the transformation of model instances into steps to be executed by a services processor.

ModRunner provides an easy to use interface to perform tasks like capturing model parameters (e.g. files, numeric values, string data and command line parameters), obtaining remote input data, scheduling execution submissions, submitting a model to execution, keeping a history of each execution instance and storing result data. Output data is stored at the knowledge base, where model parameters and information on file submission are also kept. Furthermore, it lets the user register qualitative data on the execution results.

The system works integrated to ModManager, meaning that the latter registers and manages model metadata while the first captures model parameters, creates a model instance, submits it to execution and stores the results. Both systems access the same database, as shown in Fig. 1.

As mentioned in section 1, ModRunner has been developed to run on different grid platforms, like Globus and OurGrid. However, this should be transparent to the user, as he/she only wants to submit his/her tasks to the grid and obtain the results, with no knowledge of the grid infrastructure. The difference is greater for the system manager, who has to decide which grid platform he/she wants to provide access to.

A good point of this system is that it provides an interface that helps an user who is not an expert in Grid Computing to generate an execution task for each model to be submitted to the Grid.

Another feature of the system is that execution results, as well as execution histories, may be exported to other researchers, in formats like CSV, XML and KO (Knowledge Objects), just as ModManager does with any model metadata [12].

In order to submit a task in ModRunner, the user has to:

1) Get the target application executable file.
2) Save the EXE file in MODENA's database system, for future executions.
3) Fill the application parameter fields.

Fig. 3 shows an example of a ModRunner task submission screen. In this case there are three parameter fields, all of them of the 'file' kind. The name of the model and the quantity and type of the parameters have been retrieved from the system database. The last field is the name of the output file the user may choose to save the results. If the field is left blank, the results will automatically be stored in the database.

After the model has been submitted, an OurGrid task execution command line is assembled and sent to the OurGrid Scheduler System to be executed. After the execution, OurGrid either sends back the results to the system or saves them into the file the user has chosen.

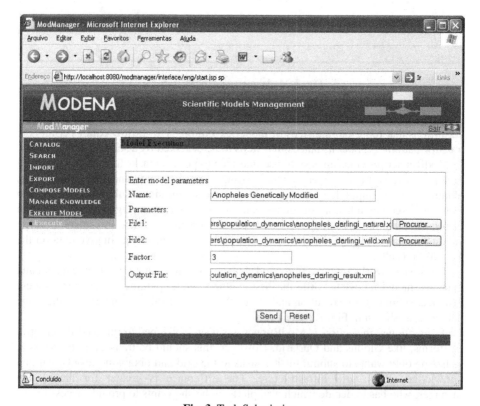

Fig. 3. Task Submission

Task submission in ModRunner lets the user either use different input data sets with the same model or use the same data set to different population dynamics models. Each simulation corresponds to different model instances running in the Grid Computing environment. So, ModRunner offers flexibility, usability, and extensibility to model execution management, allowing the user change the parameters and submit tasks to the grid environment without any knowledge about it.

The models used in the case study had to execute numerical simulations in order to satisfy the following constraints:

- Assume that the genetic manipulation does not affect the environment fitness of the mosquitoes;
- Consider transgenic heterozygous lines, so that the propagation of the malaria-refractory gene is stabilized at 56%, according to the crossing rule;
- The population density of genetically modified mosquitoes maintains the same seasonal pattern as the population density of wild-type mosquitoes;
- Numerical simulation equations are adjusted to the county of Novo Airão (Amazon state), the geographical area under study.

In order to find the closest parameters that represent the above constraints, the researcher had to test a large amount of parameters. Each task submission parameter, as well as each instance execution model, was stored in the MODENA database for future historical analysis.

6 Concluding Remarks

The contribution of this work is to present a simple and effective system that encapsulates model execution submissions for the user, apart from providing a number of facilities for model execution management on grid platforms. It also works with ModManager, another MODENA module, making a management cycle that covers from knowledge and metadata management to execution and data management.

The test with the population dynamics model was effective as the scientists were able to conduct several simulations, changing the parameters and the kind of models until they found the best model that fit their requirements, through an usable interface.

The tests also showed that the substitution of wild-type mosquitoes for genetically modified ones may take some years, depending, among other factors, on the amount of genetically modified mosquitoes introduced in the environment. Field observations should however be carried out for a sufficient large period of time to allow the detection of new variables or environmental modifications that initially were not taken into account in the mathematical model. With these new parameters, the model could be improved, tested and validated. Therefore, ModRunner has been considered an useful tool to help population dynamics researchers manage and share their results with the scientific community.

As future works we aim the integration of the system with existing model libraries. We also aim the progress of the development of the workflow management features, besides the implementation of grid services management and performance management.

Acknowledgement. The authors would like to acknowledge the assistance of CNPq (Brazilian National Council for Research) and of the GEOMA Network, for their funding, and NACAD/UFRJ (High Performance Computing Center / Federal University of Rio de Janeiro), LNCC (Brazilian Scientific Computing Laboratory) and the IST/LNCC (High Tech Institute).

References

1. Krishnan, R., Chari, K.: Model management: survey, future research directions and a bibliography. Interactive Transactions of OR/MS, (2000) 3 (1).
2. Cavalcanti, M. C., Mattoso, M., Campos, M. L., Llirbat, F., Simon, E.: Sharing Scientific Models in Environmental Applications. Proceedings of the 2002 ACM symposium on Applied computing, Madrid, Spain (2002) 453-457.
3. Brito, H., J. Strauch, Souza, J., Osthoff, C.: Scientific Models Management in Computational Grids. 17th International Scientific and Statistical Database Management Conference (SSDBM). Santa Barbara, California (2005).
4. Geoma Network. http://www.geoma.lncc.br. (2006).
5. Globus Project http://www.globus.org. (2006).
6. OurGrid Project. http://www.ourgrid.org. (2006).
7. World Health Organization. http://www.who.int/en/. (2006).
8. Wyse, A. P., Bevilacqua, L., Rafikov, M.: Population Dynamics of *An. darlingi* in the Presence of Genetically Modified Mosquitoes with Refractoriness to Malaria. (2005).
9. Allcock, B., Bester, J., Bresnahan, J., Chervenak, A. L., Foster, I., Kesselman, C., Meder, S., Nefedova, V., Quesnel, D., Tueck, S.: Data Management and Transfer in High-Performance Computational Grid Environments. Mathematics and Computer Science Division. Argonne National Laboratory. http://www.globus.org/alliance/publications/papers/dataMgmt.pdf. (2004).
10. Karnik, A., Ribbens, C. J.: Data and Activity Representation for Grid Computing. Department of Computer Science, Blacksburg, VA. http://eprints.cs.vt.edu/archive/00000598/01/hpdc.pdf. (2002).
11. Zang, S., Wu, Y., Wang, W.: Grid Workflow based on Performance Evaluation. Department of Computing and Information Technology, Fudan University, Shanghai, China. http://166.111.202.9/chinagrid/download/GCC2003/pdf/347.pdf. (2003)
12. Cirne, W., Brasileiro, F., Paranhos, D., Costa, L., Santos-Neto, E., Osthoff, C.: Building a User-Level Grid for Bag-of-Tasks Applications. Book Chapter, High Performance Computing Paradigm and Infrastructure. Wiley Series on Parallel and Distributed Computing, Albert Y.Zomaya, Series Editor. (2005).
13. Brito, H., Strauch, J., Souza, J. ModManager: a Web-based system for Knowledge Management about Scientific Models (in Portuguese). IV Brazilian Congress of Knowledge Management (KMBrasil). São Paulo (2005).

Replica Refresh Strategies in a Database Cluster*

Cécile Le Pape and Stéphane Gançarski

Laboratoire d'Informatique de Paris 6, Paris, France
`Firstname.Lastname@lip6.fr`

Abstract. Relaxing replica freshness has been exploited in database clusters to optimize load balancing. However, in most approaches, refreshment is typically coupled with other functions such as routing or scheduling, which makes it hard to analyze the impact of the refresh strategy itself on performance. In this paper, we propose to support routing-independent refresh strategies in a database cluster with mono-master lazy replication. First, we propose a model for capturing existing refresh strategies. Second, we describe the support of this model in Refresco, a middleware prototype for freshness-aware routing in database clusters. Third, we describe an experimental validation to test some typical strategies against different workloads. The results show that the choice of the best strategy depends not only on the workload, but also on the conflict rate between transactions and queries and on the level of freshness required by queries. Although there is no strategy that is best in all cases, we found that one strategy is usually very good and could be used as default strategy.

Keywords: replication, database cluster, load balancing, refresh strategy.

1 Introduction

Database clusters provide a cost-effective alternative to parallel database systems, *i.e.* database systems on tightly-coupled multiprocessors. A database cluster [10, 21, 22] is a cluster of PC servers, each running an off-the-shelf DBMS. With a large number of servers, it can reach high performances, and thus can be used as a basic block for building Grid environments, by grouping several database clusters distributed in a large scale network such as the Internet. The typical solution to obtain good load balancing in a database cluster is to replicate data at different nodes so that users can be served by any of the nodes. If the workload consists of (read-only) queries, then load balancing is relatively easy. However, if the workload includes (update) transactions in addition to queries, as it is the case in Grid environments, load balancing gets more difficult since replica consistency must be enforced. With lazy replication, a transaction updates only one replica and the other replicas are updated (refreshed) later on by separate refresh transactions [19]. By relaxing consistency, lazy replication can provide flexible transaction load balancing, in addition to query load balancing.

---

* This work was partially financed by the French ANR-ARA Respire project.

Relaxing consistency using lazy replication has gained much attention [1, 2, 18,25,22,14], even quite recently [11]. The main reason is that applications often tolerate to read data that is not perfectly consistent, and this can be exploited to improve performance. However, replica divergence must be controlled since refreshing replicas becomes more difficult as divergence increases. In [15], we addressed this problem in the context of a shared-nothing database cluster. We chose mono-master lazy replication because it is both simple and sufficient in many applications where most of the conflicts occur between transactions and queries. Transactions are simply sent to a single master node while queries may be sent to any node. Because refresh transactions at slave nodes can be scheduled in the same order as the transactions at master nodes, queries always read consistent states, though maybe stale. Thus, with mono-master replication, the problem reduces to maintaining replica freshness. A replica at a slave node is totally fresh if it has the same value as that at the master node, *i.e.* all the corresponding refresh transactions have been applied. Otherwise, the freshness level reflects the distance between the state of the replica at the slave node and that at the master node. By controlling freshness at a fine granularity level (relation or attribute), based on application requirements, we gained more flexibility for routing queries to slave nodes, thus improving load balancing.

In most approaches to load balancing, refreshment is tightly-coupled with other issues such as scheduling and routing. This makes it difficult to analyze the impact of the refresh strategy itself. For example, refreshment in [22] is interleaved with query scheduling: it is activated by the scheduler, for instance if a node is too stale to fullfill the freshness requirement of any query in the scheduler input queue. Furthermore, they do not use routing-dependent refresh: when no node is fresh enough for a query, the query execution is delayed, without guarantee on the query liveness. Many refresh strategies have been proposed in the context of distributed databases, data warehouse and database clusters. A popular strategy is to propagate updates from the source to the copies as soon as possible (ASAP), as in [3, 4, 6]. Another simple strategy is to refresh replicas periodically [5, 16] as in data warehouses [7]. Another strategy is to maintain the freshness level of replicas, by propagating updates only when a replica is too stale [24]. There are also mixed strategies. In [18], data sources push updates to cache nodes when their freshness is too low. However, cache nodes can also force refreshment if needed. In [14], an asynchronous Web cache maintains materialized views with an ASAP strategy while regular views are regenerated on demand. In all these approaches, refresh strategies are not chosen to be optimal with respect to the workload. In particular, refreshment cost is not taken into account in the routing strategy. There has been very few studies of refresh strategies and they are incomplete. For instance, they do not take into account the starting time of update propagation [23, 13] or only consider variations of ASAP [20].

This paper has three main contributions. First, we propose a model which allows describing and analyzing existing refresh strategies, independent of other load balancing issues. Second, we describe the support of this model in our Refresco prototype. Third, we describe an experimental validation based on a workload generator, to test some typical strategies against different workloads. The results show that the choice of the best strategy depends not only on the workload itself, but also on the conflict rate between transactions and queries and on the level of freshness required by queries. Although there is no strategy that is best in all cases, we found that one strategy, As Soon As Underloaded or ASAUL(0), is usually very good and could be used as default strategy. Our prototype allows to select the best strategy according to the workload type generated by the application. It is thus compliant with the OGSA-DAI [17] definition of a Data Resource Manager providing flexible and transparent access for Grid applications.

The paper is organized as follows. Section 2 describes our database cluster architecture, with emphasis on load balancing and refreshment. Section 3 defines our model to describe refresh strategies. Section 4 defines a workload model which helps defining typical workloads for experimentations. Section 5 presents our experimental validation which compares the relative performance of typical refresh policies. Section 6 concludes.

2   Database Cluster Architecture

Figure 1 gives an overview of our database cluster architecture. It preserves the autonomy of both applications and databases which can remain unchanged, which is important for Grid applications which require sites autonomy. It receives requests from the applications through a standard JDBC interface. All additional information necessary for routing and refreshing is stored and managed separately of the requests.

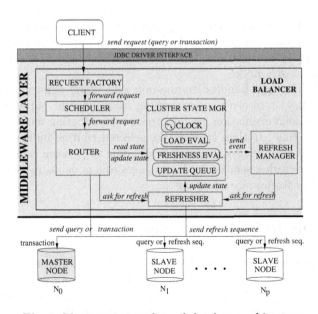

Fig. 1. Mono-master replicated database architecture

We assume that the database is fully replicated: node N_0 is the *master node* which is used to perform transactions while nodes $N_1, N_2, \ldots, N_p$ are *slaves nodes* used for queries. The master node is not necessarily a single cluster node which could be a single point of failure and a bottleneck. It is an abstraction and can be composed of several cluster nodes coordinated by any eager replication protocol such as [12]. Slave nodes are only updated through *refresh transactions* which are sent sequentially, through *refresh sequences*, according to the serialization (commit) order on the master node. This guarantees the same serialization order on slave nodes. Access to the database is through stored procedures. Each updating (resp. read-only) procedure defines a *transaction class* (resp. *query class*). A query class *potentially conflicts* with a transaction class if an instance of the transaction class may write data that an instance of the query class may read. We formally defined potential conflicts using conflict classes in [15].

The request factory enriches requests wih metadata such as parameters for stored procedures and required freshness for a query. Then it sends the requests to a FIFO scheduler. Dynamic information such as transaction commit time on the master node, data freshness on slave nodes, estimated nodes load, is maintained by the cluster state manager. The information related to each transaction is maintained until every node has executed the corresponding refresh transaction, after which it is removed.

The router implements an enhanced version of SELF (Shortest Execution Length First). Depending on application needs, the router can be switched to perform *routing-dependent (on-demand) refreshment*. To this end, it asks the freshness evaluation module to compute, for every node, the corresponding minimum refresh sequence to make the slave node fresh enough for Q, and includes the cost of the possible execution of this sequence into the cost function. After the eventual on-demand refresh is performed by the refresher on the selected node, the router sends the query to this node and updates the cluster state. Since queries are only sent to slave nodes, they do not interfere with the transaction stream on the master node.

The refresh manager handles *routing-independent refreshment*. According to the refresh policy, it receives events coming from different parts of the cluster state manager: load evaluation module, freshness evaluation module or external events such as time. It then triggers the selected routing-independent refresh policy which eventually asks the refresher module to perform refresh sequences. Whenever the refresher sends refresh sequences to a node, it updates the cluster state for further freshness evaluations.

3   Modeling Refresh Strategies

Freshness requirements are specified for *access atoms*, which represent portions of the database. Depending on the desired granularity, an access atom can be as large as the entire database or as small as a tuple value in a table. A *freshness atom* associated with an access atom a is a condition on a which bounds the staleness of a under a certain threshold t for a given *freshness measure* μ, *i.e.* such as $\mu(a) \leq t$. If we note a_i the copy of access atom a on the slave node N_i, the staleness of a_i is computed by $\mu(a_i)$ and represents the divergence between the value of a_i (on the slave node) and the value of a_0 (on the master node). The *freshness level* of a set of access atoms $\{a^1, a^2, ..., a^n\}$ is defined as the logical conjunction of freshness atoms on a^i. In [15] we introduced several freshness measures. For simplicity in this paper, we consider only measure $Age : Age(a_N)$ denotes the maximum time since at least one transaction updating a has committed on the master node and has not yet been propagated on slave node N. The *freshness level of a query Q* is a freshness level on the set of access atoms read by Q. Users determine the access atoms of the query at the granularity they desire, and define a freshness atom for each access atom. A node N is fresh enough to satisfy Q if the freshness level of Q is satisfied on N. The *freshness level of a node N* is simply the freshness level on the entire database on N.

A refresh strategy is described by the *triggering events* which raise its activation, the nodes where the refresh transactions are propagated and the number of transactions which are part of the refresh sequence. A refresh strategy may handle one or more triggering events, among:

- *Routing(N, Q)*: a query Q is routed to node N.
- *Underloaded(N, limit)*: the load of node N gets a value less than or equal to the *limit* value.
- *Stale(N, μ, limit)*: the freshness of node N for measure μ decreases below the *limit* value. In other words, the freshness level of node N for measure μ and threshold *limit* is no more satisfied. In this paper, since we only consider the *Age* measure, this parameter becomes implicit and the event can be simplified as *Stale(N, limit)* which stands for *Stale(N, Age, limit)*
- *Update_sent(T)*: a transaction T is sent to the master node.
- *Period(t)*: triggers every t seconds.

As soon as an event handled by the refresh manager is raised, the refresher computes a sequence of refresh transactions to propagate. Depending on the nature of the event, the refresh sequence is sent to a single slave node or broadcast to all slave nodes. For instance, *Routing(N, Q)* activates a refreshment only on slave node N while *Period(t)* activates a refreshment on all the slave nodes. Finally, the refresh quantity of a strategy indicates how many refresh transactions are part of the refresh sequence. This value can be minimum, *i.e.* the minimum refresh sequence which brings a node to a certain freshness. The maximum value denotes a refresh sequence containing every transaction not yet propagated to the destination. Of course, the quantity may also be arbitrary (for instance, a fixed size).

We apply our refresh model to the following strategies, which we implemented and compared, since they are the most popular in the literature.

- **On-Demand (OD).** On-Demand strategy is triggered by event *Routing(N)*. It sends a minimal refresh to node N to make it fresh enough for Q.
- **As Soon As Possible (ASAP).** ASAP strategy is triggered by a *Update_sent(T)* event. It sends a maximal refresh sequence to all the slave nodes. As ASAP strategy maintains slave nodes perfectly fresh, the refresh sequence is reduced to the transaction T which raised the event.

- **Periodic(t).** The Periodic(t) strategy is triggered by a *period(t)* event. It sends a maximum refresh sequence to all the slave nodes.
- **As Soon As Underloaded (ASAUL(limit)).** The ASAUL strategy is triggered by a *Underloaded(N,limit)* event. It sends a maximum refresh sequence to N.
- **As Soon As Too Stale (ASATS(limit)).** ASATS strategy is triggered by event *Stale(N, limit)*. It sends a maximum refresh sequence to N.

Hybrid Strategies. Refresh strategies can be combined to improve performance. Though a lot a combinations are possible, we focus here on the interaction between routing-dependent (On-Demand) and routing-independent strategies (all other strategies). Thus, for each routing-independent strategy, we derive an hybrid version which combines it with On-demand. We ran several experiments (not shown here for space limitations) to compare each basic strategy with its hybrid version. They showed that hybrid strategies always outperform basic strategies because they never trigger unnecessary refreshments. Therefore in the following, we study only hybrid strategies. In order to simplify the presentation, we use the same name as the basic strategy, since there si no ambiguity.

4   Experimental Validation

In this section, we compare the performance of hybrid refresh strategies under different workloads. After describing our experimental setup and workloads, we study the impact of conflict rate and of tolerated freshness on performance.

4.1   Experimental Setup and Workload

Our experimental validation is based on the enhanced version of the Refresco prototype, which is developed in Java 1.4.2. In order to get results independent of the underlying DBMS's behaviour, we simulated the execution of a request on a node, with 128 slave nodes, using Simjava, a process-based discrete event simulation package in Java (see http://www.dcs.ed.ac.uk/home/hase/simjava/). We chose simulation because it makes it easier to vary the various parameters and compare strategies. We also calibrated our simulator for database access using an implementation of our Refresco prototype on the 64-node cluster system of the Paris team at INRIA (http://www.irisa.fr/paris/General/cluster.htm) with PostgreSQL as underlying DBMS. In this case, for typical transactions and queries, the value of a Time Unit (TU) is approximately 10 ms.

   Our main objective is to provide a relative comparison of the refresh strategies. Therefore, we strive to keep the workload model simple, with a definition of the main parameters that impact refreshment. Note that our objective is not to capture all possible workloads which would require a much more complex workload model and is beyond the scope of this paper.

   A workload is composed of several clients. Each client is either of type *transaction* or of type *query*, *i.e.* it only sends transactions or only queries. The number of transaction clients is fixed to 16, while the number of query clients is fixed to 256. Each workload has a total duration of 10000 TU. Each request is considered as a fixed-duration job: 100 TU for queries and 5 TU for transactions. We consider that a transaction load (tl) is low (respt. high) when the transaction clients are active 1/4 (respt. 2/3) of the time. A query load (ql) is low (respt. high) when queries clients wait 300 TU (respt. 0 TU) between two queries. All the workloads are parameterized with the conflict rate (cr) and a tolerated staleness for queries (ts). We define the *conflict rate* of a workload as the proportion of potential conflicts between transactions and queries. Let $\{TC_1, TC_2, \ldots, TC_n\}$ be the application set of transaction classes and $\{QC_1, QC_2, \ldots, QC_m\}$ the application set of query classes. The conflict rate (cr) of a workload is defined by the following formula :

$$cr = \frac{\sum_{i=1}^{n} \sum_{j=1}^{m} \alpha_j \times \text{conflict}(TC_i, QC_j)}{\sum_{j=1}^{m} \alpha_j}$$

where $\text{conflict}(TC_i, QC_j)$ is equal to 1 if the transaction class TC_i potentially conflicts (see Section ??) with the query class QC_j, otherwise it is equal to 0 and α_j is the number of instances of the query class QC_j in the workload. In order to simplify, all the queries in a workload have the same *tolerated staleness*, which is the threshold of every query's freshness level. It is the maximal staleness a data on a node can have for the query to be executed on it. For instance, a workload where queries require to read perfectly fresh data has a tolerated staleness equal to 0. Thus, a workload is described as a tuple (tl, ql, cr, ts). Not all the parameters do impact on all the strategies. For instance, the ASAP strategy, which propagates immediately any transaction sent to the master node, is not sensitive to the cr and ts parameters.

4.2   Impact of Conflict Rate on Performance

Figure 2 shows the query mean response time (QMRT, average of the observed response times of queries during the experiment) of the various refresh strategies versus the conflict rate. As we focus on the conflict rate, there is no tolerated staleness (ts is fixed to 0), which is the worst case for performances. We omit workloads of type (high,low,cr,ts) and (low,high,cr,ts), but they yield similar conclusions.

Light Workloads. Figure 2(a) shows that, except for very small conflict rates, the best performance for light workloads is obtained with strategies that refresh frequently, *i.e.* maintain nodes (almost) always fresh. These strategies are ASAP (obviously) and ASAUL since nodes are idle very often. They trigger refreshment often but do not interfere much with queries because the refresh sequences are executed mostly during idle periods. In this context, ASAUL(0) is better than ASAP since it refreshes exactly during idle periods while ASAP may trigger refreshments during non-idle periods, even if such periods are rare. On the contrary, On-Demand performs rather poorly as soon as the conflict rate exceeds 0.4. Indeed, since queries are rare, it is triggered rarely. Thus, each time a query is routed, the refresher must propagate many updates (since the last refresh) before executing the query. This increases response time significantly.

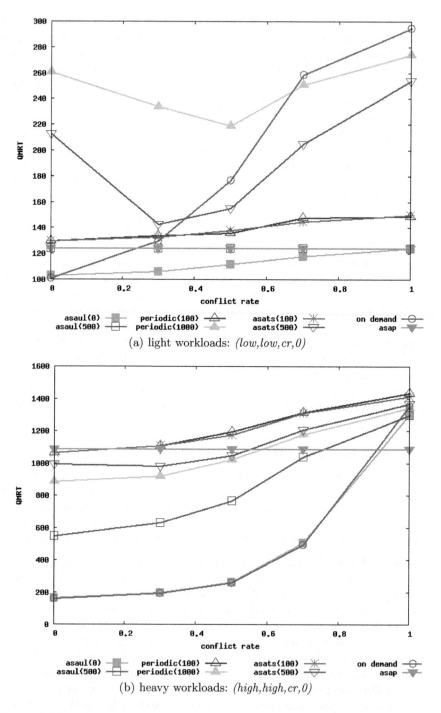

(a) light workloads: *(low,low,cr,0)*

(b) heavy workloads: *(high,high,cr,0)*

Fig. 2. Performance comparisons with varying conflict rate (tolerated staleness=0)

Heavy Workloads. In Figure 2(b) , the behavior of the strategies is quite different from that in Figure 2(a). On-Demand yields the best performance in most cases (except when the conflit rate exceeds 0.9) because refreshment is done often (the query frequency is high), but only when needed. In this context, ASAP is better only for very high conflict rates because it always refreshes. This is useless for smaller conflict rates where refresh is not frequently required. Similarly, Periodic and ASATS do not perform well. As they do not take into account the nodes load and perform maximum refresh sequences, they raise useless overhead when refreshing. We also observe that ASAUL(0) performs as On-Demand because nodes are never idle.

4.3   Impact of Tolerated Staleness on Performance

Figure 3 shows the performance (QMRT) of the various refresh strategies versus the tolerated staleness. As we focus here on the tolerated staleness, the conflict rate is fixed to 1, which is the worst case w.r.t to performances. High-low and low-high workloads give results similar to high-high and thus are omitted. A general observation is that, for all strategies except ASAP, the results are better when the tolerated staleness is higher. Obviously, when queries do not require high freshness, there is a higher probability that a node is fresh enough for any query. Thus on-demand refresh is less necessary, which speeds up query execution. This is not the case for ASAP, since it does not require on-demand refresh. When the tolerated staleness is beyond a given value, performance does not change for most strategies. This is due to the fact that all the nodes are always fresh enough for queries and thus on-demand is no more triggered. Thus, refreshing nodes is useless for queries. This is obviously the case for Periodic, but also for ASATS. In fact, ASATS also behaves periodically in this context. This is due to the fact that transactions are performed periodically on the master node, thus the freshness on slave nodes always decreases at the same speed. For light workloads, ASAUL has also a periodic behavior: when a node is idle or lightly loaded, ASAUL refreshes it and the node becomes busy. Thus, it is no longer refreshed during a given duration and gets idle. Then ASAUL refreshes it, and so on. For heavy workloads, nodes are always busy and thus, as already mentioned, ASAUL is similar to On-Demand. In particular, as nodes are never idle, ASAUL(0) performs quite the same as On-Demand. On-Demand is always sensitive to the tolerated staleness. As nodes are refreshed only when necessary, performance increases as tolerated staleness increases.

Light Workloads. Figure 3(a) shows that On-Demand is outperformed by strategies which frequently refresh nodes and thus take advantage of nodes being frequently idle. Among them, ASAUL(0) is the best since it naturally adapts to idle node events.

Heavy Workloads. Figure 3(b) shows that when the tolerated staleness is below 100 TU, ASAP is the best strategy, since frequent refreshments are necessary. From 100 up to 500, PERIODIC(100) and ASATS(100), which behave equally,

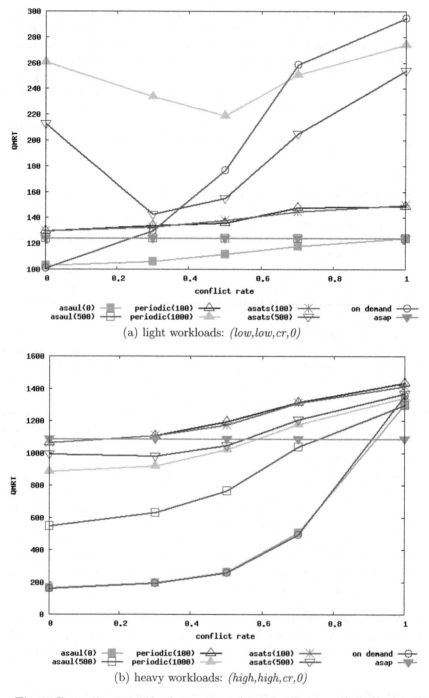

(a) light workloads: *(low,low,cr,0)*

(b) heavy workloads: *(high,high,cr,0)*

Fig. 3. Comparing strategies for varying tolerated staleness and conflict rate 1

are the best strategies. When the tolerated staleness is over 500, the overhead due to frequent refreshment is higher since nodes are never idle. Furthermore, it is useless since queries do not require high freshness. In this case, ASAUL(0) is the best strategy since it naturally adapts to idle node events. For the sake of clarity, values for a tolerated staleness over 1000 are not represesented. They remain constant for all the strategies, except for ASAUL(500) which still decreases down to 550 for a tolerated staleness of 5000, and for ASAUL(0) and On-Demand which have almost equal performance (since nodes are never idle, ASAUL(0) only triggers on-demand) and decrease down to 300, thus being the best strategies.

5   Conclusion

Relaxing replica freshness can be well exploited in database clusters to optimize load balancing. However, the refresh strategy requires special attention as the way refreshment is performed has strong impact on response time. In particular, it should be independent of other load balancing issues such as routing.

In this paper, we proposed a refresh model that allows capturing (among others) state-of-the-art refresh strategies in a database cluster with mono-master lazy replication. We distinguished between the routing-dependent (or on-demand) strategy, which is triggered by the router, and routing-independent strategies, which are triggered by other events, based on time-outs or on nodes state. We also proposed hybrid strategies, by mixing the basic strategies with the On-demand strategy. We described the support of this model by extending the Refresco middleware prototype with a refresh manager which implements the refresh strategies described in the paper. The refresh manager is independent of other load balancing functions such as routing and scheduling. In our architecture, supporting hybrid strategies is straightforward, since they are simple conjunctions of basic strategies already implemented in the refresh manager (or in the router for On-Demand).

In order to test the different strategies against different application types, we proposed a workload model which captures the major parameters which impact performance: transaction and query loads, conflict rate between transactions and queries, and level of freshness required by queries on slave nodes.

We described an experimental validation to test some typical strategies against different workloads. An important observation of our experiments is that the hybrid strategies always outperform their basic counterpart. The experimental results show that the choice of the best strategy depends not only on the workload, but also on the conflict rate between transactions and queries and on the level of freshness required by queries. Although there is no strategy that is best in all cases, we found that one strategy (ASAUL(0)) is usually very good and could be used as default strategy for the workload types we defined. As a future work, we plan to continue testing strategies against other workload types, using a richer workload model. For instance, we can assign different freshness levels for different queries in the same workload, or we can vary the ratio query/transaction in a workload, and so on. We also plan to integrate the refresh strategies into the multi-master approach presented in [9], as we suggested in [8]. The work presented in this paper can be seen as a first step toward a self-adaptable refresh strategy, which would combine different strategies by analysing on-line the incoming workload. According to the real-life applications dynamicity, our middleware should automatically adapt the refresh strategy to the current workload, using for instance machine-learning techniques.

Our approach currently works on a database cluster. As mentionned in the introduction, database clusters are good candidates to build large scale Grid environments. However, this implies that we must adress some new issues to cope with Grid application requirements. The first issue is the heterogeneity of the source. Our approach handles any relational data sources, through the use of SQL procedure and a standard JDBC driver. We must adapt it to non-relational data sources, for instance XML documents, for instance using a mediator/wrapper approach. The second issue is fault-tolerance : we must distribute our middleware over several nodes, using for instance a shared memory layer, to prevent it from being a single point of failure. Finally, we must also adapt our system to large scale distribution, by modifying the cost function used for load balancing, in order to take into account the different latencies between different sites.

References

1. R. Alonso, D. Barbará, and H. Garcia-Molina. Data caching issues in an information retrieval system. *ACM Trans. on Database Systems*, 15(3):359–384, 1990.
2. D. Barbará and H. Garcia-Molina. The demarcation protocol: A technique for maintaining constraints in distributed database systems. *VLDB Journal*, 3(3):325–353, 1994.
3. H. Berenson, P. Bernstein, J. Gray, J. Melton, E. J. O'Neil, and P. E. O'Neil. A critique of ansi isolation levels. In *ACM SIGMOD Int. Conf.*, 1995.
4. Y. Breitbart, R. Komondoor, R. Rastogi, S. Seshadri, and A. Silberschatz. Update propagation protocols for replicated databates. In *ACM SIGMOD Int. Conf.*, pages 97–108, 1999.
5. D. Carney, S. Lee, and S. Zdonik. Scalable application aware data freshening. In *IEEE Int. Conf. on Data Engineering*, 2002.
6. P. Chundi, D. J. Rosenkrantz, and S. S. Ravi. Deferred updates and data placement in distributed databases. In *IEEE Int. Conf. on Data Engineering*, pages 469–476, 1996.
7. L. S. Colby, T. Griffin, L. Libkin, I. S. Mumick, and H. Trickey. Algorithms for deferred view maintenance. In *ACM SIGMOD Int. Conf.*, pages 469–480, 1996.
8. S. Gançarski, C. Le Pape, and H. Naacke. Fine-grained refresh strategies for managing replication in database clusters. In *VLDB Wshp. on Design, Implementation and Deployment of Database Replication*, pages 47–54, 2005.
9. S. Gançarski, H. Naacke, E. Pacitti, and P. Valduriez. The leganet system: Freshness-aware transaction routing in a database cluster. *Information Systems*, To appear.
10. S. Gançarski, H. Naacke, E. Pacitti, and P. Valduriez. Parallel processing with autonomous databases in a cluster system. In *Int. Conf. On Cooperative Information Systems (CoopIS)*, 2002.

11. H. Guo, P.-A. Larson, R. Ramakrishnan, and J. Goldstein. Relaxed currency and consistency: How to say "good enough" in sql. In *ACM SIGMOD Int. Conf.*, 2004.

12. B. Kemme and G. Alonso. A new approach to developing and implementing eager database replication protocols. *ACM Trans. on Database Systems*, 25(3):333–379, 2000.

13. S. Krishnamurthy, W. H. Sanders, and M. Cukier. An adaptive framework for tunable consistency and timeliness using replication. In *Int. Conf. on Dependable Systems and Networks*, pages 17–26, 2002.

14. A. Labrinidis and N. Roussopoulos. Balancing performance and data freshness in web database servers. In *Int. Conf. on VLDB*, pages 393–404, 2003.

15. C. Le Pape, S. Gançarski, and P. Valduriez. Refresco: Improving query performance through freshness control in a database cluster. In *Int. Conf. On Cooperative Information Systems (CoopIS)*, pages 174–193, 2004.

16. H. Liu, W.-K. Ng, and E.-P. Lim. Scheduling queries to improve the freshness of a website. *World Wide Web*, 8(1):61–90, 2005.

17. S. Malaika, A. Eisenberg, and J. Melton. Standards for databases on the grid. *SIGMOD Rec.*, 32(3):92–100, 2003.

18. C. Olston and J. Widom. Offering a precision-performance tradeoff for aggregation queries over replicated data. In *Int. Conf. on VLDB*, 2000.

19. E. Pacitti, P. Minet, and E. Simon. Fast algorithms for maintaining replica consistency in lazy master replicated databases. In *Int. Conf. on VLDB*, 1999.

20. E. Pacitti and E. Simon. Update propagation strategies to improve freshness in lazy master replicated databases. *VLDB Journal*, 8(3–4):305–318, 2000.

21. U. Röhm, K. Böhm, and H.-J. Schek. Cache-aware query routing in a cluster of databases. In *IEEE Int. Conf. on Data Engineering*, 2001.

22. U. Röhm, K. Böhm, H.-J. Schek, and H. Schuldt. Fas - a freshness-sensitive coordination middleware for a cluster of olap components. In *Int. Conf. on VLDB*, 2002.

23. Y. Saito and H. M. Levy. Optimistic replication for internet data services. In *Int. Symp. on Distributed Computing*, pages 297–314, 2000.

24. S. Shah, K. Ramamritham, and P. Shenoy. Maintaining coherency of dynamic data in cooperative repositories. In *Int. Conf. on VLDB*, 1995.

25. H. Yu and A. Vahdat. Efficient numerical error bounding for replicated network services. In *Int. Conf. on VLDB*, 2000.

A Practical Evaluation of a Data Consistency Protocol for Efficient Visualization in Grid Applications⋆,⋆⋆

Gabriel Antoniu[1], Loïc Cudennec[2], and Sébastien Monnet[3]

[1] IRISA/INRIA
Gabriel.Antoniu@irisa.fr
[2] IRISA/INRIA
Loic.Cudennec@irisa.fr
[3] IRISA/University of Rennes I
Sebastien.Monnet@irisa.fr

Abstract. Data visualization is important in the context of grid applications, especially when successive refinements are iteratively realized based on intermediate results. We mainly focus on code coupling grid applications, structured as a set of distributed, autonomous, weakly-coupled codes. We consider the case where the codes are able to interact using the abstraction of a shared data space. In previous work, we have proposed an efficient visualization scheme by introducing a new operation called *relaxed read*, as an extension to the *entry consistency* model. This operation can efficiently take place without locking, in parallel with *write* operations. On the other hand, the user has to relax the consistency constraints, and accept slightly older versions of the data, whose "freshness" can however still be controlled. In this paper, we discuss and extensively evaluate the proposed consistency protocol, whose efficiency is clearly demonstrated by our experimental results.

Keywords: Data consistency, code-coupling applications, grid, visualization.

1 Introduction

With the growing demand of computing power, grid computing [11] has emerged as an appealing approach, allowing to federate and share computing and storage resources among multiple, geographically distributed sites (universities, companies, etc.). Thanks to this aggregated computing power, grids are typically useful to solve computationally intensive, parallel and/or distributed applications. In most cases, grids consist of a hierarchical federation of clusters. This hierarchy is defined in terms of hierarchical distribution, with a direct impact on the communication latency. Low-latency System-Area Networks (SANs), such as Giga Ethernet or Myrinet are often used to connect nodes within a given cluster. The various clusters may be interconnected through a higher-latency network, which can be a dedicated Wide-Area Network (WAN) whose bandwidth may reach 1 Gb/s or more.

⋆ This work was supported by the GDS project (ACI MD - French Ministry of Research, INRIA, CNRS), by the RESPIRE project of the French National Research Agency (ARA MDMSA), by the Regional Council of Brittany and by Sun Microsystems.

⋆⋆ Candidate to the Best Student Paper Award.

M. Daydé et al. (Eds.): VECPAR 2006, LNCS 4395, pp. 692–706, 2007.
© Springer-Verlag Berlin Heidelberg 2007

A particular class of applications running on grids relies on the *code-coupling* paradigm: such an application is designed as a set of (usually) parallel codes, each of which runs on a different cluster. The computation is distributed in such a way that transfers between clusters are minimized. However, some data and synchronization messages still have to be exchanged among the clusters.

Code-coupling is used in high-performance computing. Computations can be very long, and it is generally impractical to wait for the end of the application to see if the results are correct. In order to monitor the progress of the application, it is often useful to have the ability to perform an efficient visualization of the running process, without degrading the overall performance of the computation. To allow the state of the computation to be monitored, pieces of data shared by different codes need to be accessed.

In grid environments, as in other distributed systems, data sharing is a crucial issue. Currently, the most widely-used approach relies on the *explicit data access model*, where clients have to move data to computing servers. A typical example is the use of the GridFTP protocol [3]. Though this protocol provides authentication, parallel transfers, checkpoint/restart mechanisms, etc., it is still a transfer protocol which requires *explicit* data localization by the programmer. Such a low-level approach makes data management on grids rather complex. On the other hand, the concept of *transparent data access* in distributed systems through the illusion of a shared memory has intensively been studied in the context of distributed shared memory systems (DSM) since the late eighties ([12,10,4,9]). Nevertheless, DSM systems have been designed to address small scale physical architectures, usually made of tens (up to a hundred) of nodes and have usually been used on clusters. Furthermore, most of the data consistency models and protocols assume that the infrastructure is *static, without failures*. For instance, they often implicitly assume stable entities. These hypotheses are not longer valid within the grid context, where failures are part of the systems' properties. Therefore, *fault tolerance* and *volatility* increase the difficulty of designing a system providing transparent data access. The predominance of grid systems based on *explicit* transfers (GridFTP [3], IBP [8], etc.) demonstrates that transparent data sharing upon large scale architectures is still a real challenge.

In order to overcome these limitations and make a step forward towards a real virtualization of the management of large-scale distributed data, the concept of *grid datasharing service* has been proposed [5]. The idea is to provide *transparent access* to distributed grid data: in this approach, the user accesses data via global identifiers. The service which implements this model handles data localization and transfer without any help from the programmer. It transparently manages data persistence in a dynamic, large-scale, distributed environment. The data sharing service concept is based on a hybrid approach inspired by Distributed Shared Memory (DSM) systems (for transparent access to data and consistency management) and peer-to-peer (P2P) systems (for their scalability and volatility-tolerance). The JuxMem (Juxtaposed Memory) platform [5] (described in more detail in Section 2) illustrates the grid data-sharing concept. JuxMem relies on JXTA [1], a generic P2P software platform initiated by Sun Microsystems. JuxMem also serves as an experimental framework for fault-tolerance strategies and data consistency protocols.

We focus on the problem of efficient data visualization within code-coupling applications designed for grid architectures. The goal is to modify the data consistency protocol behavior in order to efficiently support the presence of a visualization process (that we call *observer*). To this purpose we have proposed an *extension* of the *entry consistency* model and a corresponding protocol that allows efficient reads, *possibly concurrent* with writes to a given data. As a counterpart, the observer has to relax the consistency constraints, and accept slightly older versions of the data, whose "freshness" can however still be controlled. The approach underlying this work has first been introduced in [6]. In this paper, we discuss and evaluate the extension of the data consistency protocol. An implementation of this strategy has been integrated within the JuxMem platform and experimented on the Grid'5000 testbed [2]. Preliminary experimental results of this work show that this solution improves the performance of the visualization observer without degrading the performance of the application that keeps reading and writing the observed data.

The next Section introduces the JuxMem grid data sharing service. Section 3 briefly describes the consistency model and explains the proposed protocol extensions. An experimental evaluation is presented in Section 4. Finally, Section 5 discusses the contribution and the future work.

2   JuxMem : A Decoupled Architecture Combining Data Consistency and Fault-Tolerance

2.1   JuxMem Overview

To experiment our approach, we have used the JuxMem software experimental platform for grid data sharing, described in [5]. From the user's perspective, JuxMem is a service providing transparent access to persistent, mutable shared data.

JuxMem has a *hierarchical* software architecture, which mirrors a hardware architecture consisting of a federation of distributed clusters. Figure 1 shows the hierarchy of the entities defined in JuxMem, consisting of a network of peer groups (`cluster` groups A, B and C on the figure), which usually correspond to clusters at the physical level. All the groups belong to a wider group, which includes all the peers which run the service (the `juxmem` group).

Each `cluster` group includes several kinds of nodes. Those which provide memory for data storage are called *providers*. Within each `cluster` group, the available providers are managed by a node called *cluster manager*. Finally, a node which simply uses the service to allocate and/or access data blocks is called *client*. A node may at the same time act as a cluster manager, as a client, and as a provider. However, for the sake of clarity, each node only plays a single role on the figure.

When allocating memory, the client has to specify on how many clusters the data should be replicated, and on how many nodes in each cluster. This results into the instantiation of a set of data replicas, associated to a group of peers called `data` group. The allocation primitive returns a global data ID, which can be used by the other nodes to identify existing data. To obtain read and/or write access to a data block, the clients only need to use this ID.

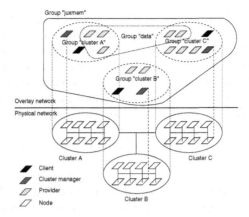

Fig. 1. Hierarchy of the entities in the network overlay defined by JuxMem

The data group is also hierarchically organized, as illustrated on Figure 2: the *Global Data Group (GDG)* gathers all provider nodes holding a replica of the same piece of data. These nodes may be distributed in different clusters, thereby increasing the data availability if faults occur. The GDG group is divided into *local data groups (LDG)*, which correspond to data copies located in the same cluster.

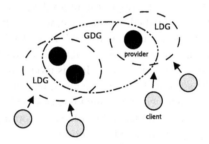

Fig. 2. JuxMem : a hierarchical architecture

In order to access a piece of data, a client has to be attached to a specific LDG. Then, when the client performs the read/write and synchronization operations, the consistency protocol layer manages data synchronization and data transmission between clients, LDGs and GDG, with the strict respect of the consistency model.

2.2   Starting Point: A Hierarchical, Fault-Tolerant Consistency Protocol

The Entry Consistency Model. To guarantee data consistency, JuxMem provides a consistency protocol that implements the entry consistency model. This model was first introduced in the Midway system [9]. As opposed to other relaxed models, it requires an explicit association of data to synchronization objects. This allows the model to leverage the relationship between a synchronization object that protects a critical section, and the data accessed within that section. A node's view of some data becomes up-to-date only when the node enters the associated critical section. This eliminates unnecessary data traffic, since only nodes that declare their intention to access data will get updated, and only the data which will be accessed will be updated. Such a concern for efficiency makes this model a good candidate in the context of scientific grid computing.

When using the entry consistency model, exclusive accesses to shared data have to be explicitly distinguished from non-exclusive accesses by using two different primitives: `acquire`, which grants mutual exclusion; `acquireRead`, which allows non-exclusive accesses on multiple nodes to be performed in parallel.

Adapting the Entry Consistency to the Grid. Existing protocols that implement the entry consistency model can not be applied directly to the grid. First, they have been designed for flat, small-scale architectures and do not cope with the hierarchical architecture of the grid (which implies a hierarchy in terms of communication latency, as previously explained). JuxMem addresses this aspect by implementing a hierarchical, consistency protocol that minimizes data traffic on long-distance, inter-cluster links. Second, traditional protocols have been designed for clusters and parallel machines, and often implicitly assume stable entities (e.g. a home node). However, failures and disconnections are part of grid's specifications. JuxMem implements a hierarchical, home-based protocol for entry consistency, where, to enhance fault tolerance, the critical role of the home is played by the a group (the LDG) at cluster level and by another group (the GDG) at global level. This protocol is described in detail in [7]. When using this protocol, if a client asks for a data access, its request may go through each level of the data group hierarchy, in order to be satisfied. For instance, when a client needs to acquire the read-lock, it sends a request to its associated LDG. If the LDG does not already have the read-lock, the LDG sends a request to the GDG. Then the lock is sent back from the GDG to the LDG and finally to the client. In this model, if a client owns a lock, its associated LDG owns the same lock. When the client modifies the data, the modifications are transmitted to the LDG when the client releases the lock, and they can further be transmitted to the GDG either immediately or lated, according to the desired level of fault-tolerance. These aspects are detailed in Section 3.

Finally, the consistency protocol gives priority to writers: a writer only has to wait that previous requests are satisfied, whereas a reader has to wait that no writer is asking for the lock. In its basic version, this strategy can cause readers starvation if two or more writers get alternatively the lock, postponing data access to readers. In order to guarantee that readers eventually access the date, a simple solution consists in setting a limit on the number of times writers actually use this priority.

3   Efficient Visualization Through Concurrent Reads and Writes

3.1   Proposed Enhancement: Relaxed Reads

We consider a scenario where an observer node reads some shared data for visualization purpose. The reads performed by this node should be efficient and low intrusive. The first idea is to favor access locality by taking advantage of the data copies located on the client node (if any), else fetch a data copy on its associated LDG, in the same cluster (if available). The second idea is to perform the read operation without acquiring a lock. This particular read operation provides the ability to have concurrent reads and writes as it does not lock the data.

The entry consistency model guarantees that the data is up-to-date only if the associated lock has been acquired. If the associated lock has not been acquired, no guarantees are provided. The approach highlighted in this paper proposes to enable relaxed reads (i.e. without acquiring a lock) for which the user application is able to keep control on the data "*freshness*". This implies that the consistency protocol implementing this extended model respects some bounds on the difference between the version of the data returned by the *rlxread* primitive and the latest version of the data (i.e. the one read after acquiring a lock). Note that this is an extenssion to the entry consistency model: the guarantees of the original model are preserved under the same conditions (i.e. when using the regular synchronization primitives); besides, new guarantees are provided in some cases where the original model does not guarantee anything. This is detailed in Section 3.4.

Therefore, for each relaxed read operation, the application specifies (as a parameter of the *rlxread* primitive) an upper bound on the difference between the latest version and the one returned by the *rlxread* primitive call.

3.2   Controlling Data Freshness

Specifying the difference between the latest version and the one returned by the *rlxread* primitive is not a trivial problem. The hierarchical aspect of the data consistency protocol does not provide the ability to retrieve the latest version in one step. For some given data, different LDGs may store different versions indeed. The LDG that owns the lock associated to the data hosts the latest version of this data while the other ones may host an older version (as LDGs do not necessarily propagate every data update to the GDG). Furthermore, even client nodes attached to the same LDG may host different versions of a given data according to the last time they access this data: the data stored by a client node is only updated when it accesses the data (using the consistency protocol primitives).

To express the difference between the latest version and the version returned by the *rlxread* primitive, we introduce two parameters that take into account the two layers of the hierarchical consistency protocol.

- The D parameter is a constant attached to each piece of data.
- The w parameter (also called *reading window*) is specified for each call to the *rlxread* primitive.

The D constant corresponds to the number of times a LDG can give the exclusive lock to its locally-attached client nodes without sending updates to the GDG. The D parameter is set when the data is allocated by the service. Setting D to a small value forces the LDG to spread updates frequently, offering the possibility to get fresher data from the other LDGs. However, this solution adds an overhead due to frequent GDG updates (releasing the lock, sending update messages, etc.). Alternatively, using a larger value lets the writers perform writes within the same cluster (associated to a given LDG), without wasting time in frequent GDG updates. The counterpart is that the data versions returned by the relaxed read in other LDGs may be a bit older. For instance, if $D = 0$ LDGs have to spread their modifications to the GDG after each release of the exclusive lock by a client. In this case, all LDGs have the same version of the data (the latest). The D parameter has been inspired by the hierarchical synchronization protocol described in [5].

The w parameter is the *reading window*. It is specified for each call of the *rlxread* primitive. It defines an upper bound on the distance between the latest version of the data and the version returned by the relaxed read. Therefore, w must be greater than or equal to D. Considering the smallest value for w (i.e. $w = D$) implies that the relaxed read returns the LDG's version. This solution offers fresher data but it also implies more network traffic when data updates occur frequently (and therefore less efficient relaxed reads). Relaxing the read (i.e. using a greater value for w), enhances the observer access speed by reducing the network traffic but the relaxed read primitive may return older versions of the data.

Note that distances D and w are positive or null and w must be greater than or equal to D. The difference $w - D$ indicates the upper bound between the version of the data stored on the client's LDG and the one returned by the relaxed read primitive on the client's node. For instance, if $D = 3$ then all the LDG can successively give the lock up to 3 times without updating the GDG. If $w = 4$ then the version of the data read by the client is either the LDG's version of the data or the previous version.

For a given data, if a client stores version V_C of the data and if V_{LDG} is the version stored on its LDG, the client can use its own version V_C as long as the following condition is satisfied (α):

$$V_c \geq V_{LDG} - (w - D)$$

This condition is checked by the LDG each time a client node performs a relaxed read.

Efficient visualization relies on the correct tuning of both D and w parameters. Therefore, a smart combination of D and w parameters has to be used depending on the type of application that is monitored and the visualization accuracy that is required.

3.3 Example

Figure 3 illustrates the roles played by w and D within the hierarchical architecture of the protocol. The d data is available in 3 different versions stored on client nodes or LDGs (V_a in one cluster, V_b and V_c in a second cluster). Several clients acquire the lock, write the data, release the lock and send updates to LDG A, increasing the V_a version (1). Every D_d lock releases within LDG A, data updates are sent to the other LDGs (i.e. to the GDG) (2). At the same time, in the second cluster, Client C performs relaxed reads, using a window w as a parameter of each access. A relaxed read request is sent from Client C to LDG B. This request contains 2 pieces of information: 1) the w parameter and 2) V_c: the version of the data owned by client C (3). Depending on the evaluation of the α condition, the LDG B sends back either its V_b version of the data or a message that allows the client to use its own version (4).

3.4 Discussion

The relaxed read proposes an *extension* of the consistency model. Entry consistency is still preserved and guarantees that clients read an up-to-date version of the data, provided they acquire the associated lock. Besides, the entry consistency model is extended by a new feature: some controls are now available when processing a read without acquiring the lock.

Note that setting $D = 0$ and $w = 0$ is not equivalent to the classic sequence of performing a read after getting a read-lock. First, during the relaxed read, the lock can be acquired by another client which can modify the data. This is not allowed in the original entry consistency model. Second, between the moment when the LDG sends the data to the client and the moment when the data is returned by the *rlxread* primitive, new versions can be produced (as the protocol allows writes to continue). Therefore, the user has to know that this approach does not offer *strict* guarantees on data freshness. Providing more guarantees would require that the LDG wait for a client acknowledgment before accepting new updates. Such an approach would however be less efficient. Furthermore, these guarantees are not necessarily needed for the problem of efficient visualization within code-coupling applications.

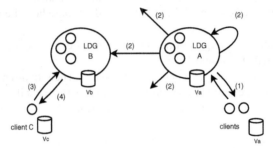

Fig. 3. A relaxed read overview

4   Evaluation

To perform an experimental evaluation of the proposed protocol, we used the Grid'5000 platform [2], which gathers 9 clusters geographically distributed in several cities of France. These clusters are connected together through the Renater Education and Research National Network (1 Gb/s). For these experiments, we used from 9 to 25 nodes in 3 of these cities (Orsay, Rennes and Toulouse). In each of these 3 clusters, nodes are locally interconnected through a Giga-Ethernet network (1 Gb/s).

Note that we do not use more nodes, as these experiments aim at evaluating the cost of the observation of a single piece of data. Even if a grid application may involve hundreds or thousands of nodes, a single piece of data is rarely accessed by more than a few tens of nodes.

4.1   A Visualization Scenario

We consider a synthetic code-coupling application running across 2 clusters located in Rennes (Cluster $C1$) and Toulouse (cluster $C2$). As illustrated by Figure 4, Cluster $C1$ runs processes that iteratively write the shared piece of data. We call these processes *writers* thereafter. On Cluster $C2$, some processes (called *readers*) perform read operations. Finally, a third cluster, located in Orsay (Cluster $C3$) is used to run a visualization process, called *observer*.

The experiments are configured as follows: each *writer* performs 50 writes, and each *reader* performs 50 reads concurrently on the same piece of data. At the same time, the *observer* on Figure 4 performs 50 observations of the piece of data. Note that the data is replicated: there is one copy in each cluster. In this example, the size of the LDGs is reduced to 1 (i.e. there is only one copy of the data in each cluster). The 3 LDGs compose the GDG for this data. The main reason that motivated this choice is that fault-tolerance is not the main goal of these experiments. Furthermore, a high replication degree would not be really relevant here, as it has a low impact on read and relaxed read operations.

The goal of these experiments is to evaluate the impact of the consistency model extension upon the visualization process. Therefore, each test is performed twice, by relying on two mechanisms for visualization: 1) using the *acquireRead/release* primitive (called *acquireRead-based visualization* thereafter); 2) using the *rlxread* primitive described in this paper, with no lock synchronization. Finally, we vary the visualization constraints by tuning w and D parameters, and measure how the visualization cost evolves.

In order to evaluate the impact of the data size in our experiments, we use 4 different sizes: 1 KB, 512 KB, 1 MB and 10 MB.

Initially, we use a single *writer* and a single *reader*. Then, in order to vary the communication patterns the number of *writers* and *readers* is gradually increased (up to 9 readers performing $9 * 50$ reads and 9 writers performing $9 * 50$ writes).

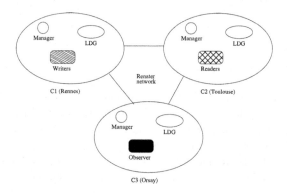

Fig. 4. Experiments configuration

4.2 Results Analysis

Benefits of the Extension. The goal of these first set of experiments is to evaluate the impact of the protocol extension even when parameters D and w are set to 0. As explain in section 3.4, this is not equivalent to reading the data through the *acquireRead* primitive, as no lock is acquired. However, this corresponds to the maximal freshness degree that is allowed by the *rlxread* primitive.

Figure 5 illustrates the impact on the visualization process. The improvement by approximately 80% is mainly explained by the fact that the visualization does not need

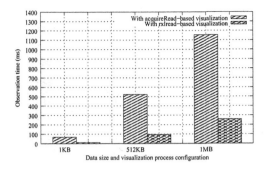

Fig. 5. Improving the observation cost

to wait for a lock. The benefit is growing with the data size: the larger the data, the longer the time to update the data and release the lock. The benefit even reaches 94% for a 10MB piece of data (not displayed on the figure for the sake of readability).

The visualization process is not the only one to take advantage of the *rlxread* primitive. The application itself shows a small improvement as it no longer has to wait for the visualization process to release its lock.

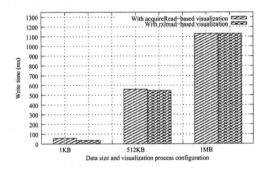

Fig. 6. Impact on the writing cost

Figures 6 and 7 respectively illustrate the gain for the writer and the reader. However, the improvement is small: in the case of the *acquireRead-based* visualization, the impact on the application is already low as the read lock is shared between the application reader and the visualization process.

Consequently, the main improvement concerns the visualization process, as shown on Figure 8, which summarizes the benefits for the reader, the writer and the observer.

Influence of D and w. In order to evaluate the impact of the D and w parameters upon the visualization and the application, we have run a second set of experiments, setting $D = 2$ and $w = 3$.

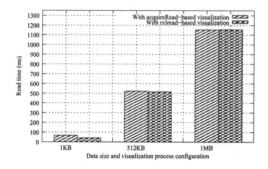

Fig. 7. Impact on the reading cost

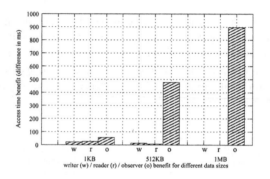

Fig. 8. Overall benefit

According to these values:

- the LDG located in Cluster $C1$ propagates updates at least every 3 writes. Therefore, the degree of fault tolerance is lower here: the latest version of the piece of data may be lost if a failure occurs in Cluster $C1$ between two update propagations. The D parameter provides the ability to tune the tradeoff between fault tolerance and data access performance.
- the LDG in Cluster $C2$ sends back the data to the observer only if the difference between its version and the observer's version is larger than 1 ($w - D$). That allows the observer not to transfer the data each time a new version is available on its LDG. Therefore, it increases performance and decreases network load while providing a sightly less accurate observation.

Figure 9 shows that relaxing the constraints on the data freshness results in an improvement for the visualization (33% for a data size of 1MB). Setting $w = 3$ reduces the probability for the observer to transfer the data. Therefore the improvement increases with the data size. On the other hand, the data returned by the *rlxread* primitive is a little bit less up-to-date.

The impact on the application is really low (almost null), as shown by figures 10 and 11.

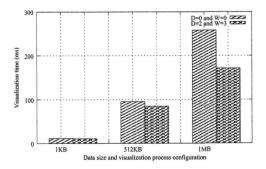

Fig. 9. Improving the observation cost (D=2 W=3)

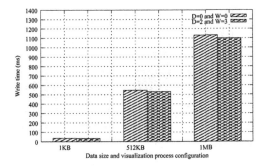

Fig. 10. Impact on the writing cost (D=2 W=3)

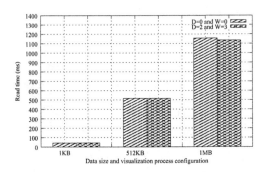

Fig. 11. Impact on the reading cost (D=2 W=3)

Varying Communication Patterns. Finally, the number of writers in Cluster $C1$ and the number of readers in Cluster $C2$ is increased in order to evaluate the impact of the number of readers and writers. Each test is run with both the *acquireRead-based* visualization (using the *acquireRead* primitive) and with the *rlxread-based* visualization. The size of the data is 1 KB. The results presented in Figure 12 show that the latency of the *rlxread* primitive is constant (and lower than in the case of *acquireRead-based*

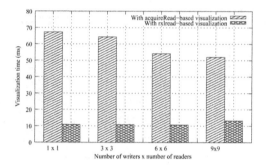

Fig. 12. Impact on the observation cost

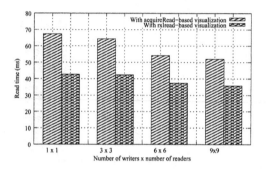

Fig. 13. Impact on the reading cost

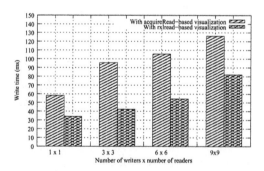

Fig. 14. Impact on the writing cost

visualization): it does not depend on the number of writers and readers. The *rlxread* primitive only induces communications between the visualization process and its LDG. The latency of the *acquireRead-based* visualization decreases while the number of readers increases: a high number of readers increases the probability that a read lock as already been given in the system. In this case, there is no need to wait for a release, the read lock can be shared by the numerous readers, providing a lower read latency.

However, as the number of writers and readers increases, the average write time grows. As the write lock is exclusive, the probability to wait for a release increases with the number of processes accessing the data using lock synchronization (i.e. except the ones using the *rlxread* primitive). However, Figures 13 and 14 show that using the *rlxread* primitive provides a significant improvement even increasing the number of writers and readers.

As for the *acquireRead-based* visualization, the latency of the read operation decreases while the number of readers increases. There again, the improvement offered by the *rlxread* primitive is significant.

5 Conclusion

Visualization is an useful feature in the context of code-coupling applications, as it may help tuning the application dynamically, while also allowing to get preliminary results, to perform demos, etc. This paper presents and evaluates an extension to the entry consistency model. We introduce the concept of *relaxed read*, that can be performed concurrently to the data accesses performed by the application. This provides the ability to achieve an efficient, and still rather accurate visualization.

Preliminary results obtained on the Grid'5000 testbed show that using the new operation (*rlxread*) is a lot more efficient and slightly less intrusive than using lock-based synchronization (e.g. through the *acquireRead* operation provided by the entry consistency model). The data version returned by the *rlxread* operation is not necessarily the most recent, however its "freshness" can be controlled and should be sufficient for visualization purposes.

We plan to further refine the approach proposed in this paper. A step forward towards transparency and self-adaptivity would consist in considering the w parameter as a hint (e.g. *not accurate*, *accurate* or *very accurate*), according to the needs of the visualization process. JuxMem may then automatically decide what exactly the w parameter should be (which expresses the "freshness degree"), by taking into account parameters like the network load or the data update rate.

References

1. The JXTA project. `http://www.jxta.org`
2. Projet Grid'5000. `http://www.grid5000.org`
3. Bill Allcock, Joe Bester, John Bresnahan, Ann L. Chervenak, Ian Foster, Carl Kesselman, Sam Meder, Veronika Nefedova, Darcy Quesnel, and Steven Tuecke. Data management and transfer in high-performance computational grid environments. *Parallel Comput.*, 28(5):749–771, 2002.
4. Cristiana Amza, Alan L. Cox, Sandhya Dwarkadas, Pete Keleher, Honghui Lu, Ramakrishnan Rajamony, Weimin Yu, and Willy Zwaenepoel. TreadMarks: Shared memory computing on networks of workstations. *IEEE Computer*, 29(2):18–28, February 1996.
5. Gabriel Antoniu, Luc Bougé, and Mathieu Jan. JuxMem: An adaptive supportive platform for data sharing on the grid. *Scalable Computing: Practice and Experience*, 6(3):45–55, November 2005. Extended version to appear in Kluwer Journal of Supercomputing.

6. Gabriel Antoniu, Loïc Cudennec, and Sébastien Monnet. Extending the entry consistency model to enable efficient visualization for code-coupling grid applications. In *Proceedings of the 6th IEEE International Symposium on Cluster Computing and the Grid (CCGrid'2006)*, May 2006. To appear.

7. Gabriel Antoniu, Jean-François Deverge, and Sébastien Monnet. How to bring together fault tolerance and data consistency to enable grid data sharing. *Concurrency and Computation: Practice and Experience*, 2006. To appear.

8. Alessandro Bassi, Micah Beck, Graham Fagg, Terry Moore, James S. Plank, Martin Swany, and Rich Wolski. The internet backplane protocol: A study in resource sharing. In *CCGRID '02: Proceedings of the 2nd IEEE/ACM International Symposium on Cluster Computing and the Grid*, page 194, Washington, DC, USA, 2002. IEEE Computer Society.

9. Brian N. Bershad, Mattew J. Zekauskas, and Wayne A. Sawdon. The Midway distributed shared memory system. In *Proceedings of the 38th IEEE International Computer Conference (COMPCON Spring '93)*, pages 528–537, Los Alamitos, CA, February 1993.

10. John B. Carter, John K. Bennett, and Willy Zwaenepoel. Implementation and performance of Munin. In *13th ACM Symposium on Operating Systems Principles (SOSP)*, pages 152–164, Pacific Grove, CA, October 1991.

11. Ian Foster, Carl Kesselman, and Steven Tuecke. The anatomy of the grid: Enabling scalable virtual organizations. *Supercomputer Applications*, 15(3):200–222, March 2001.

12. Kai Li and Paul Hudak. Memory coherence in shared virtual memory systems. *ACM Transactions on Computer Systems*, 7(4):321–359, November 1989.

Experiencing Data Grids

Nicolaas Ruberg, Nelson Kotowski, Amanda Mattos, Luciana Matos,
Melissa Machado, Daniel Oliveira, Rafael Monclar, Cláudio Ferraz,
Talitta Sanchotene, and Vanessa Braganholo

COPPE/Federal University of Rio de Janeiro, Brazil
{nicolaas, kotowski, amandasm, lrmatos, msm, danielc,
rmonclar,cferraz, talittas, vanessa}@cos.ufrj.br

Abstract. Many scientific experiments deal with data-intensive applications and the orchestration of computational workflow activities. These can benefit from data parallelism exploited in parallel systems to minimize execution time. Due to its complexity, robustness and efficiency to exploit data parallelism, grid infrastructures are widely used in some e-Science areas like bioinformatics. Workflow techniques are very important to *in-silico* bioinformatics experiments, allowing the e-scientist to describe and enact experimental process in a structured, repeatable and verifiable way. The main purpose of this paper is to describe our experience with Tavena Workbench and PeDRo, which are part of myGrid project. Taverna is provided with a workflow toolset and enactor, allowing the specification of processing units, data transfer and execution constraints. As a data entry tool, PeDRo provides a model, a controlled vocabulary and field validations for Web Services descriptions, leveraging the knowledge associated to the workflows. The main contribution of this work is a summary of some considerations drawn by our experience with the use of these tools, emphasizing its advantages and negative aspects, together with proposals for some future improvements.

1 Introduction

The development of computational infra-structures and the mass use of tools to manipulate the bioinformatics data produced by e-scientists have increased the necessity to execute *in-silico* experiments. Such experiments are usually captured by a workflow, and can be enacted using workflow engines. One of such computational strategies is myGrid [19], which exploits Grid technology to efficiently support bioinformatics applications and experiments.

However, the construction of formal data models to represent these experiments and their associate data is characterized by the use of free-text representations or semi structured data. As an example, the experiments are annotated with free-text describing the main aspects of the adopted experimental technique. These annotations are essential for a more complete analysis of the experiment, and also for future experiments.

Traditionally, several formats and formalisms have been used to construct annotation databases. Free-text is still the most common formalism. The main advantage of this approach is its expressiveness. However, the use of free-text limits search

M. Daydé et al. (Eds.): VECPAR 2006, LNCS 4395, pp. 707–718, 2007.
© Springer-Verlag Berlin Heidelberg 2007

capabilities and automated comparisons. A simple alternative would be to use a controlled vocabulary. Nevertheless, this approach would reduce the expressiveness. The most adequate option would then be to use ontologies together with a tool that allows the construction of data models and their association with the ontologies.

Different areas consider different definitions for ontologies [7,9,10]. In bioinformatics, ontology is a concise and non-ambiguous description of the relevant entities of the application domain, and of the relationship of such entities [16]. Entities may be objects, processes, functions, predicates and other application-dependant types. An ontology eliminates the uncertainty and misinterpretations of the semantics of data, programs and their relationships. Consequently, it makes it easier to create application systems in the bioinformatics domain.

Towards an engine that could not only provide an effective means of creating and enacting bioinformatics (scientific) workflows, but also deal with ontologies and the benefits that they may provide, in this paper, we describe two tools of the [my]Grid environment:

- Taverna [20], a workbench for the development and execution of workflows. Taverna allows the integration of Web Services in scientific workflows, which makes it easier to create workflows, and discover ready-to-use Web Services; and
- PeDRo [14], a tool that allows creation, manipulation and maintenance of biological ontologies.

By experiencing these tools, we provide our first contribution: a report on how good Taverna and PeDRo did concerning the aspects just highlighted. With the considerations and aspects shown in this report, we then draw our second and main contribution, which is a set of proposals for future improvements in these tools.

This paper is organized as follows. We briefly describe Taverna and PeDRo in Sections 2 and 3. Section 4 presents a report on our experience in using both of these tools. Finally, Section 5 closes this work with some final remarks and research perspectives.

2 Taverna

An initiative from the collaboration among several institutions (the European Bioinformatics Institute (EBI), IT Innovation, the School of Computer Science, University of Newcastle, Newcastle Centre for Life, School of Computer Science at the University of Manchester and the Nottingham University Mixed Reality Lab), research projects (the Biomoby project [2], Seqhound [17], Biomart [1]) and various individuals in general, Taverna [20] plays the role of a workbench for the development and execution of workflows concerning bioinformatics in the [my]Grid project.

When we use the term *"workflow"* in the [my]Grid environment, we are referring to the composition of local and remote (Web) services to achieve a biological experiment. This kind of composition is provided by defining the workflow steps using the SCUFL language. We consider a Web Service a software component that is available on the Internet and that uses a standardized XML messaging system. There should be some mechanisms so that the interested parts can easily locate services and their public interfaces.

Especially in bioinformatics, most of these *in-silico* experiments are related to the use of computational tools and databases. Almost all of these computational tools are being made available as Web services. Because of that, those who make use of such tools feel the need to orchestrate these web services in workflows as part of their *in-silico* experiments. Once the workflow is defined in the SCUFL language, each step within a workflow represents one atomic task (a Web service, for example).

One important issue that needs clarification is the main difference between business workflows and scientific workflows, since Taverna is strictly concerned with scientific workflows. According to Santos [15], scientific workflows share many characteristics of business workflows, but present some important items not found in business workflows:

1. Scientific Workflows are normally designed by scientists: Taverna's main target audience (biologists and bioinformaticians) may neither pursue a wide computational background, nor the necessary computing infrastructure or specialized staff to develop or support such workflows. Usually, Taverna users lack the knowledge of scripting or programming languages. In order to allow the ease of workflow development and usage, Taverna Workbench provides a window-based, user-friendly interface. The workflow components are added through the rovided examples, and also through the standardized data structures available, which are close to a general workflow creation language. The workflows developed in Taverna are written in the Simplified Conceptual Workflow Language (SCUFL) and enacted using the Freefluo workflow enactment engine [12].

2. Scientific Workflows are designed to prove a Hypothesis or a Theory. This way, the definition of the workflow is always a dynamic process that it is influenced by the obtained results, generating constant changes in the execution flow to achieve a desired result. The workflow will probably be re-executed many times in a day, week or month. Because of that, a mechanism that allows the scientist to save the developed workflow is needed and Taverna provides this kind of mechanism.

3. Scientific Workflows will probably be reused by other scientists: workflows already executed can be reused to reproduce an earlier experiment. The workflow tool must provide a way to recover previous workflows that can be reused or modified as needed.

4. Provenance data must be collected in order to assure high data quality: provenance data like "Responsible for the workflow execution", "Date and Time of the execution", "Annotation Data" are very important to other scientists who will re-execute the workflows in order to compare the results achieved;

5. Controlled Execution of the workflow (Partial execution): we can define a workflow as "a learning process". Because of that, scientists will only be able to decide to continue workflow execution after they have evaluated the partial results already achieved. If the results achieved are unsatisfactory, they can stop the execution and start it again with new parameters or input data. This kind of mechanism is very useful because some services included in the workflow can take a long time to execute. Because of this, the scientist must be able to stop the execution of any workflow and start it again from the point he/she has stopped it before. This way, it should be provided some "savepoints" in the workflow to mark the points in which execution can be re-started.

6. Fault Tolerance: when an error occurs, there must be a contingency plan. It is important to say that these errors are related to execution problems, like unavailable services. This way, the user must be able to define alternate services that will be executed as needed.

By default, Taverna provides some "standard" Web services that are available to the users after installation (Biomart Data Services, Soaplab Analysis Services at EBI, SOAP Services, and so on) and new Web services can be added as needed.

The Taverna Workbench is composed of four main modules: the *Advanced Model Explorer (Scufl Model Explorer)*, in which the workflow is developed following the above considerations, the *Workflow Diagram (Scufl Diagram)*, a module that presents the workflow graphically to its users, the *Available Services*, where the user is able to select or simply point to which local or remote service to use inside a workflow, and the *Enactor Launch Panel*, that presents the status of the workflow steps execution and its final result to the user [13].

Besides workflow development and execution, Taverna holds the ability to support highly complex data analysis, not only from private or local databases, but from any Web service at hand. For example, one of the workflow examples provided in Taverna Workbench allows its users to track down a gene ontology graphically. With a simple data input (an alpha-numerical code that represents the gene identification), the user submits the workflow execution, which then accesses a Web service and retrieves the gene ontology for the input provided. While the workflow is being processed, GraphViz starts to draw the result tree and associates specific roles within the ontology with colors.

3 PeDRo

The Taverna workflow environment is provided with a tool for data entry of biological data models. This tool, PeDRo [14], allows biologists to enter descriptions and ontological annotations on data sources and biological services. The data input is validated against an XML schema, and data fields are verified against a controlled vocabulary. The idea of an XML schema validation is to provide an intrinsic support to a domain metadata; and the goal of a controlled vocabulary field association is to enable an easy way to support ontologies.

The XML schema provided with Taverna/PeDRo is conceived to describe services and workflows for the purpose of discovery. The actual standards for service descriptions, UDDI, OWL-S and WSDL are not semantically rich enough to provided queries over the Taverna ontologies. Thus, such standards are extended to incorporate the concepts needed for Taverna to search and discover services in the Grid. More details on these issues are described in [14].

In order to build and to support the use of ontologies, PeDRo plays two roles: i) a data entry tool from a predefined XML schema; and ii) a quick modeling tool. As a data entry tool, PeDRo is embedded within Taverna environment as part of the Java application interface. When activated, it opens a window with a navigation tree and an edition form. On the navigation tree, elements are structured and presented accordingly to the XML schema. On the edition form, data is inserted and ontologies associated to each field. As a modeling tool, PeDRo is available as a standalone application.

It provides the same interface as in the bundle application but with more flexibility with respect to the construction of XML schema and ontologies -- both data are stored in plain text files. Therefore, to custom the XML schema, it suffices to change the XML schema text file (Figure 1), as well as to custom the ontologies, editing the respective file will incorporate the desirable property (Figure 2).

```
<xs:element name="serviceDescription">
 <xs:complexType>
  <xs:sequence>
   <xs:element name="serviceName" type="xs:string" minOccurs="0"/>
   <xs:element ref="organisation" minOccurs="0"/>
   <xs:element name="serviceType" minOccurs="0">
   <xs:simpleType>
    <xs:restriction base="xs:string">
     <xs:enumeration value="Soaplab service"/>
     <xs:enumeration value="WSDL service"/>
     ...
```

Fig. 1. XML Schema sample for Taverna ontology

```
bioinformatics_application
 Basic_Local_Alignment_Search_Tool
  tblastn
  tblastx
  blastn
  blastp
  blastx
 EMBOSS
 primer3
...
```

Fig. 2. Taverna field to Ontologies association file

4   Taverna and PeDRo Getting Together

In this section we describe our experience using the Taverna Workbench and PeDRo tools by means of a practical example. However, before proceeding, we present some difficulties found and positive aspects of Taverna.

4.1   Experiencing Taverna

We have installed and used Taverna Workbench version 1.2 in both Linux and Windows platforms and we found it very useful and easy to use. Despite the simplicity of the installation process and the effectiveness of the installation guide, some difficulties were encountered in this phase.

In order to fully experiment the Workbench we installed [my]Grid, and for that installation and experience we point out three minor faults observed: i) the configuration process; ii) security issues; and iii) lack of tools to integrate/create virtual organizations.

The first aspect that called our attention was the amount of configuration files required to run ^myGrid services, which is not an easy task, susceptible to errors. To uncomment the wrong line or miss a comment in the configuration file is enough to make some tools not to work at all. Moreover, some guidelines found within those files do not match the available information in the user's guide. Thus, for unskilled users such as e-scientists, it is difficult to install the required tools.

While configuring the XML files and properties, we noticed that security aspects are not fully observed. For example, logins and passwords are stored in plain text in XML files, which are edited by the users themselves. Since most partial results in these experiments are confidential, security can be an issue.

The third aspect observed regards the definition, construction and use of a virtual organization in a grid environment. In that sense, there is no documentation in how to setup a custom virtual organization in ^myGrid. It is not clear how to aggregate services, since Taverna does not easily provide features to publish these services. Finally, there is no authentication in the grid. Therefore there is no restriction in the use of the services provided.

Despite of these difficulties, Taverna gave us a positive impression. The tool has shown us an expressive importance to bioinformatics researchers as it offers a simple and efficient environment. This workbench is intuitive, useful and loads bioinformatics web services in its initialization. The user can verify the workflow status in real time, based on the services selected to compose the workflow activities. Also, services can be added to Taverna from specific sites which contain their definition code.

The workflow definition language (SCUFL) is simple and easy to learn. Although one can find some difficulties to use this language, it is possible to create a workflow connecting operations and filling some properties in a friendly user interface that is provided in Taverna. The user may also define them directly on a XML file. A drawback is that the SCUFL language do not complies with the *defacto* standard for Web Services Workflow BPEL4WS [4].

The real time workflow visualization is an interesting aspect observed since it prevents rework. The graphical representation of the workflow can be saved in various image formats. Many kinds of visualization are offered, from the simplest to the more complex ones, in which workflow information is exposed in the graph.

4.2   Using Taverna and PeDRo in Practice

Our strategy to test the Taverna workbench and PeDRo attributes is to cover a complete cycle of a biologist interaction with the platform. Our experiment involves: i) constructing and deploying a Web Service in the workbench; ii) describing the Web Service via the PeDRo tool; and iii) constructing and running a workflow with this Web Service.

For didactical purposes, we tested a workflow with the implementation of a simple Web Service. Its WSDL specification is presented in Fig. 3. This service receives a string as input, and echoes that string back as an output. We called it *EchoService*. The Web Service is constructed using the AXIS framework, which is the *defacto* standard for Java Web Services implementation.

```
...
<wsdl:message name="serviceMethodResponse">
  <wsdl:part name="serviceMethodReturn" type="xsd:string"/>
</wsdl:message>
<wsdl:message name="serviceMethodRequest">
  <wsdl:part name="inputArg" type="xsd:string" />
</wsdl:message>
<wsdl:portType name="EchoService">
  <wsdl:operation name="serviceMethod" parameterOrder="arg">
    <wsdl:input message="impl:serviceMethodRequest" name="serviceMethodRequest" />
    <wsdl:output message="impl:serviceMethodResponse" name="serviceMethodResponse" />
  </wsdl:operation>
</wsdl:portType>
...
```

Fig. 3. EchoService WSDL extract

Constructing and deploying the Web Service. When starting, the Taverna Workbench displays three windows: the Model Explorer, the Workflow Diagram, and the Available Services. In the first interaction, the biologist constructs a scientific workflow by picking up services on the available services window; connecting them on the model explorer window. The graphical visualization of the experiment is shown in the Workflow Diagram window.

In order to make the *EchoService* available to the workbench, we need to include it on the available services window. All the available services are displayed in a tree structure on the interface. To add a new service, we right-click the root of all services, and we select the appropriate service category on the displayed menu, which, in our case, is the "WSDL scavenger". After providing the WSDL description file address or the WSDL description URI, the service is included on the tree of services and is ready for use. In our particular case, the available services window with the *EchoService* is shown in Fig. 4.

In this first moment we observed that is necessary to re-include the service every time the workbench is restarted. This can be cumbersome if we have several customized Web Services.

Describing the Web Service via the PeDRo tool. In order to enable a semantic search over the services registered in the Grid, the Feta Engine is provided. This tool relies on an agreed ontology for the services description and an entry tool to input the data required by the ontology. With the perspective of an e-scientist, we used PeDRo to provide the semantic description of the *EchoService*. As mentioned before, PeDRo allows an annotation according to a predefined XML schema, and restricts some fields to a controlled vocabulary.

In order to provide this description, we access the PeDRo tool interface from Taverna's main menu. A form is presented so that one can provide the service description. Most of the input fields are required information to describe the service itself, such as the Web Service WSDL. However, some extra information is also necessary to better describe the service. To illustrate, in Fig. 5 we present the service description for the *EchoService*. We observe that the fields Web Service type, author, description text, and organization do not belong to the WSDL specification shown in Fig. 3, though they were included in order to increase the service semantic description.

Fig. 4. New service in Taverna Workbench

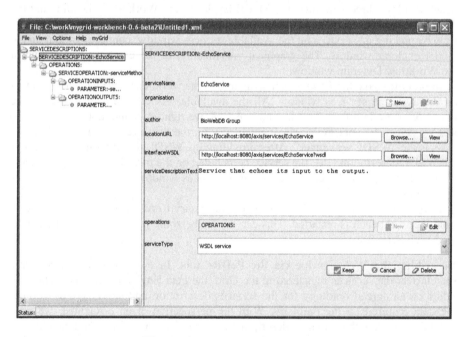

Fig. 5. Describing a Service in PeDRo

Two important aspects were observed in our experiment with PeDRo: i) although the service is already on the workbench, no description information is retrieved automatically by PeDRo; ii) in the interface, the built semantic description is ready to be published in the consortium registry site, but there is no option for a local publishing, at least not in an out of the box manner.

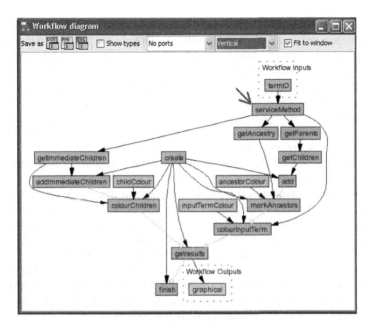

Fig. 6. Workflow with the built Web Service

Constructing and running a workflow with the EchoService. Our third step on the experiment was to include the EchoService in a workflow. We picked up one of the example workflows provided with the workbench, the ShowGeneOntology. The idea of this workflow is to retrieve the Gene ontology tree and display its graphical view for a given Gene Ontology ID. In order to have a running example with our service, we attached the input of our EchoService to the workflow input field, and attached the service output to the corresponding input services in the workflow. The resulting workflow is shown in Fig. 6.

5 Suggested Improvements and Final Remarks

In this paper, we have experienced Taverna and PeDRo. Both of these tools are focused on the bioinformatics area. Our University is involved in the BiowebDB Consortium that aims at supporting genomic workflows to provide interoperability among different analyses tools and more sensitive algorithms for distant homology detection [3]. This evaluation has motivated us to integrate some of Taverna/Pedro tools with current BiowebDB services architecture. With this idea in mind, we would like to point some problems out, and make some improvement suggestions. This is the main contribution of this paper.

Inside the Taverna Workbench we found that one of the most important topics for the development of scientific workflows is the possibility to do a controlled execution. In this tool we can steer the workflow execution by using the breakpoints feature or simply by manually pausing it. This is very useful, as we can partially or completely execute a given workflow, edit its intermediate values and even simulate a step-by-step execution by placing breakpoints at each activity to be interrupted. Although these assets provide some advantages, they do not have the necessary flexibility to enact scientific workflows completely, as we can not change the activities course at runtime according to the intermediate results.

In certain experiments, such as those concerning bioinformatics, it is almost impossible to execute the workflow in its totality, as the processing time of each web service may be enormous [12]. A workflow executed step-by-step could help to visualize errors that may have happened during the execution of a web service that is part of the workflow. Moreover, it makes it possible to cancel the workflow execution, avoiding the execution of all other processes with errors generated by previous web services, saving CPU time and reducing the cost of experiments.

However, in the Taverna development environment, the e-scientist can not find a way to dynamically choose other services to be executed on the next workflow steps depending on the results. Also, it is not possible to re-execute the workflow from a specific previous step, editing the intermediate values. It is only allowed to continue the execution from the paused step or to re-execute the entire workflow.

Another improvement opportunity is related to enabling visual workflow design through the workflow diagram. Currently, the workflow composition task is only available in Workflow Explorer module. The Workflow Diagram provides just the visualization of the created workflow, but not its edition.

We suggest the possibility to create/exclude workflow objects from the workflow project and, moreover, to edit its properties or metadata, working directly on the graph. We consider that with these improvements inside the workflow diagram module, similarly to what is provided in the workflow explorer, the workflow composition would be simplified and faster, especially for complex workflows.

After having the opportunity to analyze the Taverna Scufl Workbench environment, we would like to go further and use it in a more standard grid environment such as Globus [8]. It is not clear for us if the Taverna Team has plans to develop a Globus/^{my}Grid integration module. In our opinion this would broad the Workbench execution possibilities by taking advantage of the Globus Toolkit components, which involves failure management and wider use of grid services. However, the issues here go beyond that, since the use of Web Services in Grids still poses some problems. As defined by Foster (1998) a grid must provide security, unique identification service and quality of service [6]. Nevertheless, the current implementation of the Web Services specification does not provide these features. This is because the HTTP connection between the server and the client uses no cryptography, which means that SOAP messages are exchanged with no security. Besides, there is no user identity guarantee in service calls. The only identity is the IP address, which can be easily forged (through a HTTP proxy, for instance). Another aspect is that the HTTP and SOAP protocols have no mechanism to guarantee the provided service. This way, it is still not possible to apply the current Web Services architecture in Grids, but there are standardization proposals in Web Services involving these features. It is called Web Services Security (WS-Security) [11].

WS-Security proposes an extension to the SOAP protocol by adding message deliver guarantee, confidentiality and a unique authentication mechanism. However, WS-Security is not a standard yet [5]. This way, we can foresee a common path between Grid Services and Web Services, despite of the deficiencies to achieve the

Grid requirements in the current Web Services specification. WS-Security can be the way to get there. Globus GT4 seems to be going in this direction.

Our remarks concerning the PeDRo tool salient its characteristics as a data entry tool as well as a design tool. Pedro allowed Taverna/ [my]Grid to incorporate ontologies to the description of Web Services and workflows. It favors a simple input of Web Services description and validation against a predefined XML Schema. In addition, the decision to integrate the tool to the workbench was due to some design aspects of PeDRo. It is built in Java and made available as a package with interfaces for other Java applications. Besides defining the data model control, data validation routines can be associated to a data entry. To summarize, the benefits of PeDRo are [14]:

- it can be used for rapid data modeling and for data entry;
- it lets the creation of complete, well-formed data files;
- it supports context-sensitive help that describes the model;
- it supports controlled vocabulary (ontology) service;
- it is free and a supported open-source tool serving a user-base of scientists;
- it is simple to use and has an intuitive interface.

In other words, PeDRo allowed the Taverna development team to easily provide a model, a controlled vocabulary and field validations for Web Services descriptions on the workbench. These characteristics guarantee that the elements described with the tool will respect the requirements of the myGrid ontologies.

The main drawback to PeDRo is the lack of tools to support the data modeling. The PeDRo tool needs several configuration files, as for example, one with the validation schema, another with the contextual help, and other with the controlled vocabulary for an input field. Those configuration files are particular to the tool, which restrains changes in the ontology model; as well as the designer has to rely on other tools, e.g. XML editors, to build the XML configuration and vocabulary text files. Just the verification and integrity validation of these files are done through PeDRo's interface. Those several configuration files bring up another issue that increases the difficulty in modeling with PeDRo, those configuration files are spread in several directories. For example, the XML schema is stored in a different directory from the ontologies. As improvement, the generation of these configuration files should be done automatically through PeDRo's interface or a provided tool.

In this paper, we provided a summary on our experience using Taverna and PeDRo, both part of [my]Grid project, considering their importance to the e-science scenario. Based on these experiences, we proposed some improvements to these tools. Such suggestions aim at making easier the tasks of scientific workflow design and enaction.

References

1. BioMart Project. 2006. Available at http://www.biomart.org/.
2. BioMOBY. Available at http://biomoby.open-bio.org/.
3. BiowebDB. Available at http://www.biowebdb.org/index.html/.
4. Business Process Execution Language for Web Service version 1.1. In http://www-128.ibm.com/developerworks/library/specification/ws-bpel/, Feb 2005.

5. Foster, I. A Globus Primer. Available at http://www.globus.org/toolkit/docs/4.0/key/. 2005.

6. Foster, I.; Kesselman, C. The Grid: Blueprint for a new computing infrastructure. Morgan Kaufmann. 1998.

7. Van Heijst, G.; Schreiber, A.; Wielinga, B. Using explicit ontologies in KBS development. International Journal of Human-Computer Studies. 46. pp. 183-292. 1996.

8. Globus Toolkit. Available at http://www.globus.org/toolkit/.

9. Gruber, T. A translation approach to portable ontologies. Knowledge Acquisition, 5(2), pp. 199-220. 1993.

10. Guarino, N. Formal Ontology and Information Systems. In: International Conference on Formal Ontologies in Information Systems (FOIS). Trento, Italy, June 1998. pp. 3-15.

11. Kaler, C. et. al. Web Services Security (WS-Security). Available at http://www-128.ibm.com/developerworks/webservices/library/ws-secure/. 2002.

12. Oinn, Tom; Greenwood, Mark; Addis, Matthew; Alpdemir, M. Nedim; Ferris, Justin; Glover, Kevin; Goble, Carole; Goderis, Antoon; Hull, Duncan; Marvin, Darren; Li, Peter; Lord, Phillip; Pocock, Matthew; Senger, Martin; Stevens, Robert; Wipat, Anil; Wroe, Chris. Taverna: Lessons in creating a workflow environment for the life sciences. In: Concurrency and Computation: Practice and Experience, pp.2. 2002.

13. Oinn, Tom; Addis, Matthew; Ferris, Justin; Marvin, Darren; Senger, Martin; Greenwood, Mark; Carver, Tim; Glover, Kevin; Pocock, Matthew R.; Wipat, Anil; Li, Peter. Taverna: a tool for the composition and enactment of bioinformatics workflows. Bioinformatics Journal, 20(17), pp. 3045-3054. 2004.

14. PeDRo, dynamic form generation, XML Schema, data validation, controlled vocabulary services...; Manchester University; 2004; Available at http://pedrodownload.man.ac.uk/main.html.

15. Santos, R. T – "O Ambiente 10+C para a definição e execução de workflows in silico através de serviços web" – Master Thesis, COPPE/UFRJ, 2004. In Portuguese.

16. Schulze-Kremer, S. Ontologies for Molecular Biology. In: Pacific Symposium on Biocomputing. 1998. pp. 693-704.

17. SeqHound. Available at http://www.blueprint.org/seqhound/.

18. Silva, F.; Cavalcanti, M. Intermediate Data Management for In-Silico Workflows using Web Services. In: Workshop de Teses e Dissertações em Banco de Dados, 2005. Uberlândia, MG, Brazil.

19. Stevens, R.; Robinson, A.; Goble, C. myGrid: Personalized bioinformatics on the information grid. Bioinformatics, 19(1), pp. 302-304. 2003.

20. Taverna Project Website. 2006. Available at http://taverna.sourceforge.net/.

21. Wroe, C.; Lord, P.; Miles, S., Papay, J., Moreau, L.; Goble, C. Recycling Services and Workflows through Discovery and Reuse. Proc UK e-Science All Hands Meeting 2004, pp. 622-629. 2004.

Author Index

Lecture Notes in Computer Science

For information about Vols. 1–4317

please contact your bookseller or Springer

Vol. 4367: K. De Bosschere, D. Kaeli, P. Stenström, D. Whalley, T. Ungerer (Eds.), High Performance Embedded Architectures and Compilers. XI, 307 pages. 2007.

Vol. 4366: K. Tuyls, R. Westra, Y. Saeys, A. Nowé (Eds.), Knowledge Discovery and Emergent Complexity in Bioinformatics. IX, 183 pages. 2007. (Sublibrary LNBI).

Vol. 4364: T. Kühne (Ed.), Models in Software Engineering. XI, 332 pages. 2007.

Vol. 4362: J. van Leeuwen, G.F. Italiano, W. van der Hoek, C. Meinel, H. Sack, F. Plášil (Eds.), SOFSEM 2007: Theory and Practice of Computer Science. XXI, 937 pages. 2007.

Vol. 4361: H.J. Hoogeboom, G. Păun, G. Rozenberg, A. Salomaa (Eds.), Membrane Computing. IX, 555 pages. 2006.

Vol. 4360: W. Dubitzky, A. Schuster, P.M.A. Sloot, M. Schroeder, M. Romberg (Eds.), Distributed, High-Performance and Grid Computing in Computational Biology. X, 192 pages. 2007. (Sublibrary LNBI).

Vol. 4358: R. Vidal, A. Heyden, Y. Ma (Eds.), Dynamical Vision. IX, 329 pages. 2007.

Vol. 4357: L. Buttyán, V. Gligor, D. Westhoff (Eds.), Security and Privacy in Ad-Hoc and Sensor Networks. X, 193 pages. 2006.

Vol. 4355: J. Julliand, O. Kouchnarenko (Eds.), B 2007: Formal Specification and Development in B. XIII, 293 pages. 2006.

Vol. 4354: M. Hanus (Ed.), Practical Aspects of Declarative Languages. X, 335 pages. 2006.

Vol. 4353: T. Schwentick, D. Suciu (Eds.), Database Theory – ICDT 2007. XI, 419 pages. 2006.

Vol. 4352: T.-J. Cham, J. Cai, C. Dorai, D. Rajan, T.-S. Chua, L.-T. Chia (Eds.), Advances in Multimedia Modeling, Part II. XVIII, 743 pages. 2006.

Vol. 4351: T.-J. Cham, J. Cai, C. Dorai, D. Rajan, T.-S. Chua, L.-T. Chia (Eds.), Advances in Multimedia Modeling, Part I. XIX, 797 pages. 2006.

Vol. 4349: B. Cook, A. Podelski (Eds.), Verification, Model Checking, and Abstract Interpretation. XI, 395 pages. 2007.

Vol. 4348: S.T. Taft, R.A. Duff, R.L. Brukardt, E. Ploedereder, P. Leroy (Eds.), Ada 2005 Reference Manual. XXII, 765 pages. 2006.

Vol. 4347: J. Lopez (Ed.), Critical Information Infrastructures Security. X, 286 pages. 2006.

Vol. 4346: L. Brim, B. Haverkort, M. Leucker, J. van de Pol (Eds.), Formal Methods: Applications and Technology. X, 363 pages. 2007.

Vol. 4345: N. Maglaveras, I. Chouvarda, V. Koutkias, R. Brause (Eds.), Biological and Medical Data Analysis. XIII, 496 pages. 2006. (Sublibrary LNBI).

Vol. 4344: V. Gruhn, F. Oquendo (Eds.), Software Architecture. X, 245 pages. 2006.

Vol. 4342: H. de Swart, E. Orłowska, G. Schmidt, M. Roubens (Eds.), Theory and Applications of Relational Structures as Knowledge Instruments II. X, 373 pages. 2006. (Sublibrary LNAI).

Vol. 4341: P.Q. Nguyen (Ed.), Progress in Cryptology - VIETCRYPT 2006. XI, 385 pages. 2006.

Vol. 4340: R. Prodan, T. Fahringer, Grid Computing. XXIII, 317 pages. 2007.

Vol. 4339: E. Ayguadé, G. Baumgartner, J. Ramanujam, P. Sadayappan (Eds.), Languages and Compilers for Parallel Computing. XI, 476 pages. 2006.

Vol. 4338: P. Kalra, S. Peleg (Eds.), Computer Vision, Graphics and Image Processing. XV, 965 pages. 2006.

Vol. 4337: S. Arun-Kumar, N. Garg (Eds.), FSTTCS 2006: Foundations of Software Technology and Theoretical Computer Science. XIII, 430 pages. 2006.

Vol. 4336: V.R. Basili, H.D. Rombach, K. Schneider, B. Kitchenham, D. Pfahl, R.W. Selby, Empirical Software Engineering Issues. XVII, 194 pages. 2007.

Vol. 4335: S.A. Brueckner, S. Hassas, M. Jelasity, D. Yamins (Eds.), Engineering Self-Organising Systems. XII, 212 pages. 2007. (Sublibrary LNAI).

Vol. 4334: B. Beckert, R. Hähnle, P.H. Schmitt (Eds.), Verification of Object-Oriented Software. XXIX, 658 pages. 2007. (Sublibrary LNAI).

Vol. 4333: U. Reimer, D. Karagiannis (Eds.), Practical Aspects of Knowledge Management. XII, 338 pages. 2006. (Sublibrary LNAI).

Vol. 4332: A. Bagchi, V. Atluri (Eds.), Information Systems Security. XV, 382 pages. 2006.

Vol. 4331: G. Min, B. Di Martino, L.T. Yang, M. Guo, G. Ruenger (Eds.), Frontiers of High Performance Computing and Networking – ISPA 2006 Workshops. XXXVII, 1141 pages. 2006.

Vol. 4330: M. Guo, L.T. Yang, B. Di Martino, H.P. Zima, J. Dongarra, F. Tang (Eds.), Parallel and Distributed Processing and Applications. XVIII, 953 pages. 2006.

Vol. 4329: R. Barua, T. Lange (Eds.), Progress in Cryptology - INDOCRYPT 2006. X, 454 pages. 2006.

Vol. 4328: D. Penkler, M. Reitenspiess, F. Tam (Eds.), Service Availability. X, 289 pages. 2006.

Vol. 4327: M. Baldoni, U. Endriss (Eds.), Declarative Agent Languages and Technologies IV. VIII, 257 pages. 2006. (Sublibrary LNAI).

Vol. 4326: S. Göbel, R. Malkewitz, I. Iurgel (Eds.), Technologies for Interactive Digital Storytelling and Entertainment. X, 384 pages. 2006.

Vol. 4325: J. Cao, I. Stojmenovic, X. Jia, S.K. Das (Eds.), Mobile Ad-hoc and Sensor Networks. XIX, 887 pages. 2006.

Vol. 4323: G. Doherty, A. Blandford (Eds.), Interactive Systems. XI, 269 pages. 2007.

Vol. 4322: F. Kordon, J. Sztipanovits (Eds.), Reliable Systems on Unreliable Networked Platforms. XIV, 317 pages. 2007.

Vol. 4320: R. Gotzhein, R. Reed (Eds.), System Analysis and Modeling: Language Profiles. X, 229 pages. 2006.

Vol. 4319: L.-W. Chang, W.-N. Lie (Eds.), Advances in Image and Video Technology. XXVI, 1347 pages. 2006.

Vol. 4318: H. Lipmaa, M. Yung, D. Lin (Eds.), Information Security and Cryptology. XI, 305 pages. 2006.